Advances in Knee Ligament and Knee Preservation Surgery

Norimasa Nakamura
Robert G. Marx • Volker Musahl
Alan Getgood
Seth L. Sherman • Peter Verdonk
Editors

Advances in Knee Ligament and Knee Preservation Surgery

ISAKOS
International Society of Arthroscopy,
Knee Surgery and Orthopaedic Sports Medicine

Springer

Editors
Norimasa Nakamura
Institute for Medical Science in Sports
Osaka Health Science University
Osaka
Japan

Robert G. Marx
Orthopedic Surgery, Sports Medicine
Hospital for Special Surgery
New York
USA

Volker Musahl
MSRC, Division of Sports Medicine
University of Pittsburgh
Pittsburgh, PA
USA

Alan Getgood
Fowler Kennedy Sport Medicine Clinic
London, ON
Canada

Seth L. Sherman
Orthopaedic Surgery
Stanford University
Pao Alto, CA
USA

Peter Verdonk
Department of Orthopedic Surgery
Antwerp Orthopaedic Center
Antwerpen
Belgium

ISBN 978-3-030-84747-0 ISBN 978-3-030-84748-7 (eBook)
https://doi.org/10.1007/978-3-030-84748-7

This Springer imprint is published by the registered company Springer Nature Switzerland AG
The registered company address is: Gewerbestrasse 11, 6330 Cham, Switzerland

Foreword

Firstly I want to thank all the authors who have given up so much of their time to write this comprehensive book on knee ligament and conservation surgery. There is no financial gain to any of them, but to be successful surgeons we are judged by more than the amount of money we make. Teaching young surgeons is very fulfilling because we can all remember when we were taught and how much it meant to us. We have many ways to teach or to learn but what is great about a book is that it puts everything in order and explains where it all fits in. For me the "Why" in surgery is more important than the "How," and this book will give you an insight into the latest hot topics and controversies in knee ligament and conservation surgery. Make sure you always know "why" you are doing the surgery and for this you need basic science and anatomy and then somebody (expert) who can explain how it all comes together.

There are multiple perspectives on this variety of topics from around the globe, making for a complete guide that will help knee surgeons take care of their patients.

Willem M. van der Merwe
President ISAKOS, Sport Science Orthopaedic Centre
Cape Town
South Africa

Preface

Thanks to an incredible collection of international surgeons, this textbook attempts to tackle all of the current hot topics and controversial areas in knee ligament and knee preservation surgery. Regarding the latest in ACL surgery, we cover indications, technique, lateral augmentation, return to play, and ACL revision surgery. We also review both multi-ligament surgery and patellofemoral surgery. Complex meniscal issues are tackled along with cartilage repair, osteotomy, and biologic treatments to prevent osteoarthritis.

We include multiple perspectives on these topics from around the globe, making for a complete guide that will help every knee surgeon take better care of their patients.

Osaka, Japan	Norimasa Nakamura
New York, NY, USA	Robert G. Marx
Pittsburgh, PA, USA	Volker Musahl
London, ON, Canada	Alan Getgood
Pao Alto, CA, USA	Seth L. Sherman
Antwerpen, Belgium	Peter Verdonk

Contents

1 **Who Needs ACL Surgery?** 1
Kenneth M. Lin, Evan W. James, and Robert G. Marx

2 **Patient-Specific Graft Choice in Primary ACL Reconstruction** 11
Martin Lind and Ole Gade Sørensen

3 **Assessment of Risk Factors for Failure of ACLR: When to Address Concomitant Pathology** 21
Gian Luigi Canata, Valentina Casale, and Antonio Pastrone

4 **Technique Corner: ACLR Optimal Tunnel Placement: How to Get There?** 35
Konsei Shino, Ryohei Uchida, Hiroyuki Yokoi, Tomoki Ohori, and Tatsuo Mae

5 **Evidenced-Based Approach for Anterolateral Surgery for ACL Reconstruction** 43
Fares Uddin, Gilbert Moatshe, and Alan Getgood

6 **Why Does LET Work?** 57
João V. Novaretti and Moisés Cohen

7 **The Evidence Regarding ACL Repair** 61
Andreas Persson, Gilbert Moatshe, and Lars Engebretsen

8 **Laxity Objective Measurement Within MRI of ACL Lesions** 71
Rogério Pereira, Renato Andrade, Sofia Florim, José Alberto Duarte, and João Espregueira-Mendes

9 **Return to Sport After Anterior Cruciate Ligament Reconstruction: Criteria-Based Rehabilitation and Return to Sport Testing** 83
Kate E. Webster and Timothy E. Hewett

10 **Revision ACL Reconstruction** 95
Jonathan D. Hughes and Bryson P. Lesniak

11 **Complications of ACL Reconstruction** 107
Iftach Hetsroni, Niv Marom, and Noam Reshef

12 **Osteotomy: Slope Change Tibial Osteotomy to Address ACL Deficiency** 119
Stefano Muzzi, Camilo Muniagurria, Jordan Gruskay, and David Dejour

13 **Biologics: Post-traumatic Osteoarthritis Following Anterior Cruciate Ligament Reconstruction** 133
Sami Chergui, Antoine Denis, James Meterissian, Lee Benaroch, and Thierry Pauyo

14 **Assessment of the Multiligament Knee** 151
Marcel Betsch and Daniel B. Whelan

15 **When Do You Need to Reconstruct the Posterior Cruciate Ligament?** 167
Vishal Pai and Andy Williams

16 **Technique Corner: Posterior Cruciate Ligament Injuries** 179
Jonathan D. Hughes, Christopher M. Gibbs, Neel K. Patel, Jan-Dierk Clausen, and Volker Musahl

17 **Technique Corner: Posterolateral Corner Reconstruction** 193
Evan W. James, Kenneth M. Lin, Bruce A. Levy, and Robert G. Marx

18 **Technique Corner: MCL** 201
Robert S. Dean, Brady T. Williams, Jill K. Monson, Robert F. LaPrade, and Jorge Chahla

19 **Fracture Dislocations About the Knee** 215
Luc Rubinger, Aaron Gazendam, Seper Ekhtiari, Jeffrey Kay, Herman Johal, and Darren de SA

20 **Advances in Treating Arthrofibrosis** 243
João V. Novaretti

21 **A View of Predisposing Factors by Novel 3D Imaging Techniques for the PF Joint** 249
Yukiyoshi Toritsuka and Yuzo Yamada

22 **MPFL Reconstruction and Patellofemoral Chondral Status** 265
Keisuke Kita, Shuji Horibe, Norimasa Nakamura, and Konsei Shino

23 **Osteotomy: Coronal and Axial Plane Deformity** 273
Humza Shaikh, Rajiv Reddy, Christopher M. Gibbs, Ryan Murray, and Volker Musahl

24 **Patient-Specific Instrumentation and 3-D Osteotomy** 289
Wouter Van Genechten, Annemieke van Haver, and Peter Verdonk

25 **Save the Meniscus: Advances in Meniscal Repair Techniques** 303
Johannes Zellner and Peter Angele

26 **Meniscus Root Tears** 313
Jin Goo Kim, Dhong Won Lee, and Kyu Sung Chung

27 **Meniscus Substitution** 333
Francesca de Caro, Jonas Grammens, Wouter Van Genechten, Rene Verdonk, and Peter Verdonk

28 **Update on Indications, Techniques, and Outcomes of Meniscal Allograft Transplantation (MAT)** 341
Trevor R. Gulbrandsen, Alan G. Shamrock, and Seth L. Sherman

29 **Technique Corner: Cell-Based Cartilage Repair** 355
Joshua Wright-Chisem and Andreas H. Gomoll

30 **Technique Corner: Marrow Stimulation and Augmentation** 363
Eric D. Haunschild, Ron Gilat, Theodore Wolfson, Stephanie Wong, Nolan B. Condron, Joshua T. Kaiser, and Brian J. Cole

31 **Technique Corner: Particulate Cartilage** 375
Theresa Diermeier and Ben Rothrauff

32 **Osteochondral Allograft Transplantation** 379
C. W. Nuelle, C. M. LaPrade, and Seth L. Sherman

33 **Technique Corner: Osteochondral Autograft** 395
Alexander Hundeshagen, Benedikt Brozat, and Daniel Guenther

34 **Technical Corner: Lateral Extra-Articular Tenodesis** 405
Frederique Vanermen, Koen C. Lagae, Geert Declercq, and Peter Verdonk

35 **Clinical Application of Scaffold-Free Tissue-Engineered Construct Derived from Synovial Stem Cells** 415
Kazunori Shimomura, David A. Hart, Wataru Ando, and Norimasa Nakamura

36 **Osteotomies Around the Knee for Older Active Patients** 425
Ryohei Takeuchi, Eiji Kondo, Takenori Akiyama, Akihiko Yonekura, Ryuichi Nakamura, and Hiroshi Nakayama

37 **Current Introduction of the Biological Agent Derived from Adipose Tissue to the Treatment of Knee Osteoarthritis** 437
Wataru Ando, Isabel Wolfe, Kazunori Shimomura, Stephen Lyman, Naomasa Yokota, and Norimasa Nakamura

1 Who Needs ACL Surgery?

Kenneth M. Lin, Evan W. James,
and Robert G. Marx

1.1 Sequelae of Nonoperative Management of Acute ACL Injury

In order to understand who should get ACL surgery, we must first understand the benefits of ACL surgery. Surgical treatment of acute ACL rupture generally consists of reconstruction using a variety of graft options, including allograft and/or autograft using hamstring tendons, quadriceps tendon, or patella bone-tendon-bone graft. Historically, primary repair of the ACL was considered a viable treatment option following rupture; however, outcomes studies showed up to a 94% rate of instability at 5-year follow-up [1]. The advent of advanced suture constructs improved failure rates somewhat, but reconstruction, and not repair, is generally preferred due to more predictable outcomes in young active patients [2–4]. Therefore, in this chapter, evidence will be largely drawn from the reconstruction literature. To answer the question of which patients need ACL surgery, it is important to first understand what nonoperative management entails and how its outcomes differ from reconstruction.

K. M. Lin · E. W. James · R. G. Marx (✉)
Division of Sports Medicine and Shoulder Surgery, Hospital for Special Surgery, New York, NY, USA
e-mail: linke@hss.edu;
jamese@hss.edu; marx@hss.edu

1.1.1 Nonoperative Management Techniques

Nonoperative management for ACL ruptures is generally reserved for older patients, or those who wish to return to noncutting, straight-plane activities and do not have persistent functional instability [5]. It has been shown that nonoperative treatment has higher failure rate with younger age and higher activity level [6]. Nonoperative management largely consists of lifestyle and activity modification, as well as neuromuscular rehabilitation programs and movement pattern optimization strategies. Numerous protocols have been described, but components of programs that have been shown to lead to good physical performance and muscle strength are goal-oriented and progressive in nature: early phases focus on motion, neuromuscular control, and balance, while later phases focus on muscle strength, endurance of stabilizers, and functional performance [7–10]. Recently, the Knee Anterior Cruciate Ligament, Nonsurgical versus Surgical Treatment (KANON) trial showed that at 2- and 5-year follow-up, patients prospectively randomized to exercise program alone achieved similar rates of limb symmetry index>90% as those who underwent ACL reconstruction plus exercise [7].

N. Nakamura et al. (eds.), *Advances in Knee Ligament and Knee Preservation Surgery*,
https://doi.org/10.1007/978-3-030-84748-7_1

1.1.2 Biological Perspective

From a biological perspective, there are several important theoretical advantages to surgical treatment of acute ACL rupture. Native healing of intra-articular ligaments is limited, as the synovial healing response leads to stump retraction and lack of tissue bridging [2]. When bridging does occur, tension is often decreased due to the altered resting position of the femur and tibia, and newly formed tissue consists of fibrovascular scar rather than regeneration of native ligament and enthesis tissue [11]. For this reason, the tissue that forms through the native healing response is biomechanically, histologically, and morphologically inferior to the original ligament. Surgical reconstruction using quadrupled hamstring autograft yields a construct that is significantly stronger in tensile load and stiffness than the native ACL [12, 13]. Furthermore, the use of bone-tendon-bone grafts allows retention of a native tendon-bone insertion and relies on bone-bone healing which is more predictable [14]. During surgical reconstruction, tension of the graft construct can be directly manipulated and set. Finally, management of concurrent associated pathology, such as a meniscus tear or chondral injury, may also occur at the time of ACL reconstruction.

1.1.3 Clinical Perspective

Nonoperative management of acute ACL rupture has generally been considered inferior to surgical management for young and active patients [5, 6, 15, 16]. Clinical outcomes of importance in the setting of acute ACL injury include knee stability, prevention of subsequent repairable and irreparable meniscus tears or chondral injury, clinical outcome scores, return to work or sports, and patient satisfaction. Compared to nonoperative management, previous studies have shown that ACL reconstruction leads to improved stability and functional outcomes at 10-year [17] and even 20-year follow-up [18, 19]. It should be noted, though, that short-term outcomes have been shown to be similar [7, 20]. Nonoperative management is known to lead to persistent laxity and incomplete tissue healing on MRI. For example, van Meer et al. [21] showed in a prospective multicenter study of over 150 patients that at 2 years post-injury, only 32% of patients had improved Lachman exams (improved but not normal exam), with only 2% showing improvement on KT-1000. In this population, 60% showed improvement in fiber continuity on MRI and 44% showed resolution of empty intercondylar notch. However, all other MRI-based parameters of ACL structure and tissue quality remained abnormal, likely reflecting the presence of fibrovascular scar tissue, rather than regeneration of native ligamentous tissue, from the native healing response. It should be noted that several studies have highlighted advantages to nonoperative management. The KANON trial showed that at 5-year follow-up, patients who underwent exercise therapy alone (nonoperative treatment) had fewer knee symptoms compared to those who underwent early reconstruction followed by exercise program [8]. The authors suggest that this is because ACL reconstruction involves iatrogenic damage to the knee, such as surgical incision, graft harvest site morbidity, and bone tunnel drilling.

ACL reconstruction for acute ruptures leads to decreased rates of secondary injury, namely of the meniscus [22, 23], and thus in theory decreases the likelihood of downstream arthritis, although the long-term data to date is somewhat mixed regarding late arthritis. ACL reconstruction is known to decrease reoperation rates as it is thought to be protective against meniscal and cartilage injury [24]. In a retrospective cohort study by Sanders et al. [18] of nearly 1000 patients at mean 13.7 years follow-up, patients treated with nonoperative management for ACL rupture had a 5.4-fold increased risk of secondary meniscus tear. In this study, the nonoperatively treated cohort had a 6.0-fold increased risk of being diagnosed with arthritis. A case-control study by the same authors compared nonoperatively managed acute ACL ruptures to age- and sex-matched controls without ACL tears [25]. There was a significantly increased risk of secondary meniscal injury, osteoarthritis, and need for total knee arthroplasty in the nonoperative ACL rupture group compared to the healthy matched controls. Studies by other authors in different populations have shown similar results regarding decreased arthritis in surgically treated patients with acute ACL rupture compared to nonoperative management [26]. However, several large studies have also reported similar rates of arthritis in operative versus nonoperative ACL injury patients [17, 23, 27]. Nonetheless, despite similar rates of arthritis, these studies report fewer subsequent knee injuries and improved overall knee function in the surgical groups [17, 23, 27]. Because the operative groups tend to achieve higher function and activity, perhaps the comparison of arthritis rates is confounded, as increased activity-level or high-level sports participation may also be an independent risk factor for arthritis [28, 29]. A study comparing long-term outcomes in ACL-reconstructed knees to the contralateral "healthy" knee in the same individual showed no significant difference in radiographic arthritis (on X-ray and MRI) at 10-year follow-up [30]. However, longer-term follow-up is still needed since many patients do not convert to total knee replacement until up to 20 or 30 years after ACL reconstruction. Taken together, the results in the literature to date largely suggest that the natural history of acute ACL rupture, which is thought to lead to worsened knee function and eventually degenerative disease of the knee, is altered with surgical management, although the data surrounding risk of downstream arthritis are not definitive.

1.2 Return to Sport Following ACL Injury

Given the high prevalence of acute ACL injuries in the athletic population, return to sport is another important outcome to consider in the ACL rupture population. Following ACL reconstruction, return to similar level of sport is extremely high in elite or professional athletes. Reports from National Basketball Association (NBA) athletes suggest return to play rates of up to 88%; however, performance upon return to sport declined based on statistical performance [31–33]. Similar studies from other professional sports show high return to same level of play rates: 77% in Major League Soccer (MLS) [34], 92% (quarterbacks) and 74% (defensive players) in the National Football League (NFL) [35], and 97% in the National Hockey League (NHL) [35]. Post-return performance was similar to pre-injury level in MLS, NHL, and NFL quarterbacks, but significantly reduced in NFL defensive players. In a nonprofessional athletic population, general return to sport rates are high, but return to same level of play is less predictable. Overall return to some form of sport has been reported up to 90%, with return to pre-injury level up to 72% [36–40]. In the pediatric population, return to play after ACL reconstruction has been reported to be as high as 91%, but with high rate of second ACL injury, many of which were to the contralateral knee [41]. While the goal is to return all patients to their pre-injury level of competition, it is important to council patients that a subset of ACL reconstruction patients will struggle to attain these levels of activity.

It should be noted that while the vast majority of the literature on return to sport following ACL injury focuses on ACL reconstruction, there have been reports of return to elite sport following nonoperative treatment [42].

1.3 Patient Stratification

In developing a framework for patient selection for ACL surgery following acute injury, the general patient population should be stratified by age and activity level. With respect to age, patients can be split into pediatric and adolescent, younger adult (20–50), and older adult populations (>50). With respect to activity level, patients can be divided into high-level athletes, recreational athletes, or sedentary individuals. In addition to age and activity level, there are several other factors that play into the decision-making for ACL surgery indications. First, medical comorbidities must be considered and patients physiologically unfit for surgery should be contraindicated. Psychosocial factors, such as access to rehabilitation resources or ability to comply with postoperative restrictions, are also important to consider. Injury factors, such as chronicity, degree of laxity, and functional limitation, are important factors that can shape decision making, as they can influence the type of surgery that is performed. Similarly, the presence of other intra-articular pathology, such as repairable meniscus tears or arthritis, may be indications or contraindications for surgery, respectively, and are extremely important for predicting long-term outcomes. While balancing this constellation of factors is integral for surgical decision-making, indicating a patient for ACL surgery is based principally on the patient's age, desired activity level, and functional goals [6].

Some authors have suggested that activity level is the most important predictor for necessity to perform an ACL reconstruction [43], and that chronologic age in isolation may not be a reliable predictor [44, 45]. Several studies in the literature have assessed outcomes of ACL surgery in differing age groups, with 40 years of age as a commonly-used cutoff [44, 46–48]. Results have shown no significant difference in outcomes, although interpretations are limited by study design and heterogeneous populations with regard to operative technique, graft choice, rehabilitation, and other factors. In some populations, older (>40 years of age) patients have been shown to have greater satisfaction than younger patients [46]. Beyond being a proxy for activity level and demand on a patient's knee, age is also thought of as a proxy for the amount of degenerative change in the knee. As older patients become increasingly active, there is growing interest to expand indications for ACL reconstruction, particularly in adult patients with mild to moderate knee osteoarthritis, which will be discussed later in this chapter. A list of factors to consider when indicating patients for ACL surgery is presented in Table 1.1. To answer the question of who needs ACL surgery, in the sections below, patients will be stratified by age group, as age is universal and does not rely on various scoring systems (as activity level does). Within each age group, decision-making for ACL surgery will be discussed.

Table 1.1 Important preoperative factors to consider in ACL surgery

Patient factors
Activity level
Age
Sports participation (volume, type, level, position)
Preexisting arthritis and prior injuries
Medical comorbidities
Ability to comply with rehabilitation protocol
Injury factors
Chronicity
Concomitant injuries (meniscus, cartilage, collaterals, fractures)
Partial vs. complete rupture
Degree of instability on physical examination (Lachman, pivot shift)

1.4 Pediatric and Adolescent

In the pediatric population, the incidence of acute ACL injury and ACL reconstruction is increasing [49–51], likely due to increased participation in organized sports, improved diagnostic capabilities, and a greater awareness among doctors and families. Historically, nonoperative or delayed operative management was recommended in the pediatric and adolescent populations to avoid the rare but potentially devastating complication of physeal injury, growth arrest, and subsequent limb deformity [52]. However, studies of nonoperative management reported poor outcomes [53–55], including poor return to sports participation, high rates of subsequent knee injury and surgery, and early degenerative change. Recently, comparative studies of operative versus nonoperative management of pediatric injuries strongly favor operative management [15, 16]. Numerous techniques for surgical treatment of acute ACL injury in the pediatric patient have been described and studied in the literature [56, 57]. Outcomes following ACL reconstruction in the pediatric population are favorable and predictable, with improved stability, functional outcomes, high rate of return to sport, and low rate of physeal arrest [58, 59]. It should be noted, however, that in a subset of pediatric patients, specifically those age <14 years, with partial ruptures of <50% and a grade B pivot shift exam, have been shown to have good outcomes with nonoperative treatment [60]. Taken together, the current literature supports surgical treatment for the vast majority of acute ACL ruptures in the pediatric population to restore stability, maximize sports participation, and prevent subsequent meniscal tears and chondral injuries.

1.5 Young Adult (<40 Years)

The majority of patients with acute ACL rupture fall into the young adult category. In these patients, treatment is based on activity level and functional demands. For patients with medical contraindications to surgery in general, have sedentary occupation, and do not wish to return to jumping, cutting, or pivoting sports, nonoperative management can yield successful results [61, 62]. Outside of these groups, the vast majority of patients in this cohort should be indicated for ACL surgery in the setting of complete rupture. The extensive body of research surrounding outcomes following ACL reconstruction discussed previously strongly favors operative management in the young, active population, especially in athletic individuals [5, 17–19, 23, 25].

1.6 Older Adult (>40 Years)

Similar to other age groups, consideration for ACL surgery is largely based on activity demand and degree of instability or functional limitation. In general, as older individuals may be less active or have baseline degenerative disease, nonoperative management is a reasonable option; however, in the setting of continued symptomatic instability or further knee injury (such as meniscus tears), late ACL reconstruction may become necessary. For patients who elect to proceed with surgery, recent literature on ACL reconstruction in older patients has been shown to have good results even in those over 50 years of age [63] and 60 years of age [64].

It must be noted that there is a sub-population of middle age adults that has been reported to benefit from nonsurgical treatment of acute injury. Specifically, recreational middle age alpine skiers (mean age 42, range 30–68), with MRI evidence of complete ACL tears but who have grade pivot shift and Lachman exams 6–12 weeks after injury, can have good outcomes with conservative management at 2-year minimum follow-up [65]. Postoperative activity scores were equivalent to preoperative scores, and knee laxity had returned to normal with mean side-to-side difference in KT-1000 under 1 mm, and 10 of 11 patients had a Lachman grade 0–1+. Based on these results, the authors suggest that middle-aged skiers presenting with an acute ACL tear may be re-evaluated at 6–12 weeks following injury. If the knee is stable to Lachman and pivot shift testing, nonoperative management should be considered and the patients can expect to return to their activities.

An important consideration in the middle age adult population is a preexisting degenerative disease. With increasing activity level and recreational sports participation in an aging population, there is a growing proportion of individuals who have early arthritis but are still active. These patients are difficult to classify into traditional diagnostic groups, as their activity level would place them in the ACL reconstruction category (as opposed to nonoperative), and despite their early arthritis, may be too active or not advanced enough for total knee arthroplasty. In these patients, joint preservation techniques, such as high tibial osteotomy, are frequently performed to offload the affected compartment. Recently, there is a growing body of evidence investigating ACL reconstruction in knees with early degenerative changes. Several cohort studies have assessed outcomes following simultaneous high tibial osteotomy (HTO) and ACLR [66–69]. The results show satisfactory to good outcomes following combined HTO and ACLR, with improved alignment, stability, and outcome scores compared to preoperatively. Rates of arthritis progression are variable, with some studies reporting minimal progression [67], and others reporting higher rates of discernible progression [66]. Major demographic groups in the treatment of ACL injury are listed in Table 1.2, with their suggested treatments.

Table 1.2 Treatment of ACL injury by age

Treatment of ACL rupture by Demographic Group	
Demographic group	Treatment
Pediatric	
Complete rupture	ACL surgery
Partial rupture	Consider conservative management, especially if pre-pubescent
Young adult (age 20–40 years)	
Complete rupture with intact cartilage	ACL surgery
Alpine skier >30 years of age, stable Lachman and pivot shift at 6 weeks	Consider conservative management
Medical contraindications to surgery	Consider conservative management
Older adult (age >40 years)	
Sedentary	Consider conservative management
Medical contraindications to surgery	Consider conservative management
High activity level or participation in recreational/competitive sports	ACL surgery if no significant arthritis
Tricompartmental arthritis	Consider conservative management

1.7 Summary

As ACL rupture becomes increasingly common across all age groups with earlier sports participation and increasing activity level in the aging population, it is imperative to understand who needs ACL surgery after acute injury. There are multiple biological and clinical advantages to operative management of acute ACL injuries compared to nonoperative management. It is known that ACL reconstruction restores stability and function to the knee, and predictably leads to high rates of return to sport in elite athletes, and good but slightly lower rates in recreational athletes. Subsequent meniscal tears and chondral injuries occur less frequently and reoperation rates are reduced following ACL reconstruction. Many studies have reported decreased degenerative changes in patients who undergo surgery, although this remains controversial, as the increased activity level achieved by ACL-reconstructed patients is a likely positive confounder for the development of arthritis.

In determining who needs ACL surgery following acute injury, patient factors and associated injuries must be taken into account. The most important factor to consider is activity demand, as active individuals and athletes will require a functional ACL. ACL reconstruction is recommended for active adults, and can be performed in patients over 50 and 60 years of age, if necessary, for symptoms of recurrent instability. Age is a proxy for activity and function, but also for the remaining "lifespan" of the joint and amount of degenerative change. In younger patients it is paramount to restore functional stability to prevent further injury and arthritis. For pediatric patients and young adults, ACL reconstruction is nearly universally recommended in the setting of complete rupture. There are two notable situations in which conservative management has been shown to produce good outcomes: partial ruptures in pediatric patients and acute ACL tears in middle-age alpine skiers. As our knowledge of ACL anatomy and the biology of ligament healing continues to evolve, further high-quality comparative studies are needed to further refine ACL indications across the general population.

References

1. Raines BT, Naclerio E, Sherman SL. Management of anterior cruciate ligament injury: what's in and what's out? Indian J Orthopaed. 2017;51(5):563–75. https://doi.org/10.4103/ortho.IJOrtho_245_17.
2. Grawe B, Sugiguchi F, Bedi A, Rodeo S. Biology of anterior cruciate ligament graft healing. In: Noyes FR, Barber-Westin SD, editors. Noyes' knee disorders: surgery, rehabilitation, clinical outcomes. 2nd ed. Amsterdam: Elsevier Health Science; 2017: Chap 5.
3. Nwachukwu BU, Patel BH, Lu Y, Allen AA, Williams RJ III. Anterior cruciate ligament repair outcomes: an updated systematic review of recent literature. Arthroscopy. 2019;35(7):2233–47. https://doi.org/10.1016/j.arthro.2019.04.005.
4. Osti M, El Attal R, Doskar W, Höck P, Smekal V. High complication rate following dynamic intraligamentary stabilization for primary repair of the anterior cruciate ligament. Knee Surg Sports Traumatol Arthrosc. 2019;27(1):29–36. https://doi.org/10.1007/s00167-018-5067-3.
5. Diermeier T, Rothrauff BB, Engebretsen L, et al. Treatment after anterior cruciate ligament injury: panther symposium ACL treatment consensus

group. Knee Surg Sports Traumatol Arthrosc. 2020;28(8):2390–402. https://doi.org/10.1007/s00167-020-06012-6.
6. van der List JP, Hagemans FJA, Hofstee DJ, Jonkers FJ. The role of patient characteristics in the success of nonoperative treatment of anterior cruciate ligament injuries. Am J Sports Med. 2020;48(7):1657–64. https://doi.org/10.1177/0363546520917386.
7. Ericsson YB, Roos EM, Frobell RB. Lower extremity performance following ACL rehabilitation in the KANON-trial: impact of reconstruction and predictive value at 2 and 5 years. Br J Sports Med. 2013;47(15):980–5. https://doi.org/10.1136/bjsports-2013-092642.
8. Filbay SR, Roos EM, Frobell RB, Roemer F, Ranstam J, Lohmander LS. Delaying ACL reconstruction and treating with exercise therapy alone may alter prognostic factors for 5-year outcome: an exploratory analysis of the KANON trial. Br J Sports Med. 2017;51(22):1622–9. https://doi.org/10.1136/bjsports-2016-097124.
9. Tagesson S, Oberg B, Good L, Kvist J. A comprehensive rehabilitation program with quadriceps strengthening in closed versus open kinetic chain exercise in patients with anterior cruciate ligament deficiency: a randomized clinical trial evaluating dynamic tibial translation and muscle function. Am J Sports Med. 2008;36(2):298–307. https://doi.org/10.1177/0363546507307867.
10. Paterno MV. Non-operative care of the patient with an ACL-deficient knee. Curr Rev Musculoskelet Med. 2017;10(3):322–7. https://doi.org/10.1007/s12178-017-9431-6.
11. Galatz LM, Gerstenfeld L, Heber-Katz E, Rodeo SA. Tendon regeneration and scar formation: the concept of scarless healing. J Orthop Res. 2015;33(6):823–31. https://doi.org/10.1002/jor.22853.
12. Mehran N, Moutzouros VB, Bedi A. A review of current graft options for anterior cruciate ligament reconstruction. JBJS Rev. 2015;3(11):e2. https://doi.org/10.2106/jbjs.rvw.o.00009.
13. West RV, Harner CD. Graft selection in anterior cruciate ligament reconstruction. J Am Acad Orthop Surg. 2005;13(3):197–207.
14. Tomita F, Yasuda K, Mikami S, Sakai T, Yamazaki S, Tohyama H. Comparisons of intraosseous graft healing between the doubled flexor tendon graft and the bone-patellar tendon-bone graft in anterior cruciate ligament reconstruction. Arthroscopy. 2001;17(5):461–76. https://doi.org/10.1053/jars.2001.24059.
15. Ramski DE, Kanj WW, Franklin CC, Baldwin KD, Ganley TJ. Anterior cruciate ligament tears in children and adolescents: a meta-analysis of nonoperative versus operative treatment. Am J Sports Med. 2014;42(11):2769–76. https://doi.org/10.1177/0363546513510889.
16. Dunn KL, Lam KC, Valovich McLeod TC. Early operative versus delayed or nonoperative treatment of anterior cruciate ligament injuries in pediatric patients. J Athl Train. 2016;51(5):425–7. https://doi.org/10.4085/1062-6050.51.5.11.
17. Tsoukas D, Fotopoulos V, Basdekis G, Makridis KG. No difference in osteoarthritis after surgical and non-surgical treatment of ACL-injured knees after 10 years. Knee Surg Sports Traumatol Arthrosc. 2016;24(9):2953–9. https://doi.org/10.1007/s00167-015-3593-9.
18. Sanders TL, Kremers HM, Bryan AJ, et al. Is anterior cruciate ligament reconstruction effective in preventing secondary meniscal tears and osteoarthritis? Am J Sports Med. 2016;44(7):1699–707. https://doi.org/10.1177/0363546516634325.
19. van Yperen DT, Reijman M, van Es EM, Bierma-Zeinstra SMA, Meuffels DE. Twenty-year follow-up study comparing operative versus nonoperative treatment of anterior cruciate ligament ruptures in high-level athletes. Am J Sports Med. 2018;46(5):1129–36. https://doi.org/10.1177/0363546517751683.
20. Frobell RB, Roos EM, Roos HP, Ranstam J, Lohmander LS. A randomized trial of treatment for acute anterior cruciate ligament tears. N Engl J Med. 2010;363(4):331–42. https://doi.org/10.1056/NEJMoa0907797.
21. van Meer BL, Oei EH, Bierma-Zeinstra SM, et al. Are magnetic resonance imaging recovery and laxity improvement possible after anterior cruciate ligament rupture in nonoperative treatment? Arthroscopy. 2014;30(9):1092–9. https://doi.org/10.1016/j.arthro.2014.04.098.
22. Hagmeijer MH, Hevesi M, Desai VS, et al. Secondary meniscal tears in patients with anterior cruciate ligament injury: relationship among operative management, osteoarthritis, and arthroplasty at 18-year mean follow-up. Am J Sports Med. 2019;47(7):1583–90. https://doi.org/10.1177/0363546519844481.
23. Chalmers PN, Mall NA, Moric M, et al. Does ACL reconstruction alter natural history?: a systematic literature review of long-term outcomes. J Bone Joint Surg Am. 2014;96(4):292–300. https://doi.org/10.2106/jbjs.l.01713.
24. Dunn WR, Lyman S, Lincoln AE, Amoroso PJ, Wickiewicz T, Marx RG. The effect of anterior cruciate ligament reconstruction on the risk of knee reinjury. Am J Sports Med. 2004;32(8):1906–14. https://doi.org/10.1177/0363546504265006.
25. Sanders TL, Pareek A, Kremers HM, et al. Long-term follow-up of isolated ACL tears treated without ligament reconstruction. Knee Surg Sports Traumatol Arthrosc. 2017;25(2):493–500. https://doi.org/10.1007/s00167-016-4172-4.
26. Ajuied A, Wong F, Smith C, et al. Anterior cruciate ligament injury and radiologic progression of knee osteoarthritis: a systematic review and meta-analysis. Am J Sports Med. 2014;42(9):2242–52. https://doi.org/10.1177/0363546513508376.
27. Wellsandt E, Failla MJ, Axe MJ, Snyder-Mackler L. Does anterior cruciate ligament reconstruction improve functional and radiographic outcomes over nonoperative management 5 years after injury?

Am J Sports Med. 2018;46(9):2103–12. https://doi.org/10.1177/0363546518782698.
28. Tveit M, Rosengren BE, Nilsson J, Karlsson MK. Former male elite athletes have a higher prevalence of osteoarthritis and arthroplasty in the hip and knee than expected. Am J Sports Med. 2012;40(3):527–33. https://doi.org/10.1177/0363546511429278.
29. Driban JB, Hootman JM, Sitler MR, Harris KP, Cattano NM. Is participation in certain sports associated with knee osteoarthritis? a systematic review. J Athl Train. 2017;52(6):497–506. https://doi.org/10.4085/1062-6050-50.2.08.
30. Hoffelner T, Resch H, Moroder P, et al. No increased occurrence of osteoarthritis after anterior cruciate ligament reconstruction after isolated anterior cruciate ligament injury in athletes. Arthroscopy. 2012;28(4):517–25. https://doi.org/10.1016/j.arthro.2011.09.014.
31. Harris JD, Erickson BJ, Bach BR Jr, et al. Return-to-sport and performance after anterior cruciate ligament reconstruction in National Basketball Association Players. Sports Health. 2013;5(6):562–8. https://doi.org/10.1177/1941738113495788.
32. Nwachukwu BU, Anthony SG, Lin KM, Wang T, Altchek DW, Allen AA. Return to play and performance after anterior cruciate ligament reconstruction in the National Basketball Association: surgeon case series and literature review. Phys Sportsmed. 2017;45(3):303–8. https://doi.org/10.1080/00913847.2017.1325313.
33. Kester BS, Behery OA, Minhas SV, Hsu WK. Athletic performance and career longevity following anterior cruciate ligament reconstruction in the National Basketball Association. Knee Surg Sports Traumatol Arthrosc. 2017;25(10):3031–7. https://doi.org/10.1007/s00167-016-4060-y.
34. Erickson BJ, Harris JD, Cvetanovich GL, et al. Performance and return to sport after anterior cruciate ligament reconstruction in male Major League Soccer players. Orthop J Sports Med. 2013;1(2):2325967113497189. https://doi.org/10.1177/2325967113497189.
35. Erickson BJ, Harris JD, Cole BJ, et al. Performance and return to sport after anterior cruciate ligament reconstruction in National Hockey League Players. Orthop J Sports Med. 2014;2(9):2325967114548831. https://doi.org/10.1177/2325967114548831.
36. King E, Richter C, Jackson M, et al. Factors influencing return to play and second anterior cruciate ligament injury rates in level 1 athletes after primary anterior cruciate ligament reconstruction: 2-year follow-up on 1432 reconstructions at a single center. Am J Sports Med. 2020;48(4):812–24. https://doi.org/10.1177/0363546519900170.
37. Ardern CL, Taylor NF, Feller JA, Whitehead TS, Webster KE. Sports participation 2 years after anterior cruciate ligament reconstruction in athletes who had not returned to sport at 1 year: a prospective follow-up of physical function and psychological factors in 122 athletes. Am J Sports Med. 2015;43(4):848–56. https://doi.org/10.1177/0363546514563282.
38. Ardern CL, Webster KE, Taylor NF, Feller JA. Return to sport following anterior cruciate ligament reconstruction surgery: a systematic review and meta-analysis of the state of play. Br J Sports Med. 2011;45(7):596–606. https://doi.org/10.1136/bjsm.2010.076364.
39. McCullough KA, Phelps KD, Spindler KP, et al. Return to high school- and college-level football after anterior cruciate ligament reconstruction: a multicenter orthopaedic outcomes network (MOON) cohort study. Am J Sports Med. 2012;40(11):2523–9. https://doi.org/10.1177/0363546512456836.
40. Brophy RH, Schmitz L, Wright RW, et al. Return to play and future ACL injury risk after ACL reconstruction in soccer athletes from the multicenter orthopaedic outcomes network (MOON) group. Am J Sports Med. 2012;40(11):2517–22. https://doi.org/10.1177/0363546512459476.
41. Dekker TJ, Godin JA, Dale KM, Garrett WE, Taylor DC, Riboh JC. Return to sport after pediatric anterior cruciate ligament reconstruction and its effect on subsequent anterior cruciate ligament injury. J Bone Joint Surg Am. 2017;99(11):897–904. https://doi.org/10.2106/jbjs.16.00758.
42. Weiler R, Monte-Colombo M, Mitchell A, Haddad F. Non-operative management of a complete anterior cruciate ligament injury in an English Premier League football player with return to play in less than 8 weeks: applying common sense in the absence of evidence. BMJ Case Rep. 2015;2015:bcr2014208012. https://doi.org/10.1136/bcr-2014-208012.
43. Meuffels DE, Poldervaart MT, Diercks RL, et al. Guideline on anterior cruciate ligament injury. Acta Orthop. 2012;83(4):379–86. https://doi.org/10.3109/17453674.2012.704563.
44. Barber FA, Elrod BF, McGuire DA, Paulos LE. Is an anterior cruciate ligament reconstruction outcome age dependent? Arthroscopy. 1996;12(6):720–5. https://doi.org/10.1016/s0749-8063(96)90177-2.
45. Sloane PA, Brazier H, Murphy AW, Collins T. Evidence based medicine in clinical practice: how to advise patients on the influence of age on the outcome of surgical anterior cruciate ligament reconstruction: a review of the literature. Br J Sports Med. 2002;36(3):200–4. https://doi.org/10.1136/bjsm.36.3.200.
46. Brandsson S, Kartus J, Larsson J, Eriksson BI, Karlsson J. A comparison of results in middle-aged and young patients after anterior cruciate ligament reconstruction. Arthroscopy. 2000;16(2):178–82. https://doi.org/10.1016/S0749-8063(00)90033-1.
47. Deakon R, Zarnett M. ACL reconstruction in patients over 40 years of age using autogenous bone-patellar tendon-bone: <em>R. Timothy Deakon and M. E. Zarnett.</em> Oakville and Toronto, Ontario, Canada. Arthroscopy. 1996;12(3):388. https://doi.org/10.1016/S0749-8063(96)90153-X.
48. Viola R, Vianello R. Intra-articular ACL reconstruction in the over-40-year-old patient. Knee Surg Sports Traumatol Arthrosc. 1999;7(1):25–8. https://doi.org/10.1007/s001670050116.

49. Beck NA, Lawrence JTR, Nordin JD, DeFor TA, Tompkins M. ACL tears in school-aged children and adolescents over 20 years. Pediatrics. 2017;139(3):e20161877. https://doi.org/10.1542/peds.2016-1877.
50. Tepolt FA, Feldman L, Kocher MS. Trends in pediatric ACL reconstruction from the PHIS database. J Pediatr Orthop. 2018;38(9):e490–4. https://doi.org/10.1097/bpo.0000000000001222.
51. Dodwell ER, Lamont LE, Green DW, Pan TJ, Marx RG, Lyman S. 20 years of pediatric anterior cruciate ligament reconstruction in New York state. Am J Sports Med. 2014;42(3):675–80. https://doi.org/10.1177/0363546513518412.
52. Woods GW, O'Connor DP. Delayed anterior cruciate ligament reconstruction in adolescents with open physes. Am J Sports Med. 2004;32(1):201–10. https://doi.org/10.1177/0363546503258868.
53. Mizuta H, Kubota K, Shiraishi M, Otsuka Y, Nagamoto N, Takagi K. The conservative treatment of complete tears of the anterior cruciate ligament in skeletally immature patients. J Bone Joint Surg. 1995;77(6):890–4.
54. Graf BK, Lange RH, Fujisaki CK, Landry GL, Saluja RK. Anterior cruciate ligament tears in skeletally immature patients: meniscal pathology at presentation and after attempted conservative treatment. Arthroscopy. 1992;8(2):229–33. https://doi.org/10.1016/0749-8063(92)90041-9.
55. Moksnes H, Engebretsen L, Risberg MA. Prevalence and incidence of new meniscus and cartilage injuries after a nonoperative treatment algorithm for ACL tears in skeletally immature children: a prospective MRI study. Am J Sports Med. 2013;41(8):1771–9. https://doi.org/10.1177/0363546513491092.
56. Fabricant PD, Jones KJ, Delos D, et al. Reconstruction of the anterior cruciate ligament in the skeletally immature athlete: a review of current concepts: AAOS exhibit selection. J Bone Joint Surg Am. 2013;95(5):e28. https://doi.org/10.2106/jbjs.l.00772.
57. Al-Hadithy N, Dodds AL, Akhtar KS, Gupte CM. Current concepts of the management of anterior cruciate ligament injuries in children. Bone Joint J. 2013;95-b(11):1562–9. https://doi.org/10.1302/0301-620x.95b11.31778.
58. Cordasco FA, Mayer SW, Green DW. All-inside, all-epiphyseal anterior cruciate ligament reconstruction in skeletally immature athletes: return to sport, incidence of second surgery, and 2-year clinical outcomes. Am J Sports Med. 2017;45(4):856–63. https://doi.org/10.1177/0363546516677723.
59. Kay J, Memon M, Marx RG, Peterson D, Simunovic N, Ayeni OR. Over 90% of children and adolescents return to sport after anterior cruciate ligament reconstruction: a systematic review and meta-analysis. Knee Surg Sports Traumatol Arthrosc. 2018;26(4):1019–36. https://doi.org/10.1007/s00167-018-4830-9.
60. Kocher MS, Micheli LJ, Zurakowski D, Luke A. Partial tears of the anterior cruciate ligament in children and adolescents. Am J Sports Med. 2002;30(5):697–703. https://doi.org/10.1177/03635465020300051201.
61. Bogunovic L, Matava MJ. Operative and nonoperative treatment options for ACL tears in the adult patient: a conceptual review. Phys Sportsmed. 2013;41(4):33–40. https://doi.org/10.3810/psm.2013.11.2034.
62. Buss DD, Min R, Skyhar M, Galinat B, Warren RF, Wickiewicz TL. Nonoperative treatment of acute anterior cruciate ligament injuries in a selected group of patients. Am J Sports Med. 1995;23(2):160–5. https://doi.org/10.1177/036354659502300206.
63. Weng CJ, Yeh WL, Hsu KY, et al. Clinical and functional outcomes of anterior cruciate ligament reconstruction with autologous hamstring tendon in patients aged 50 years or older. Arthroscopy. 2020;36(2):558–62. https://doi.org/10.1016/j.arthro.2019.08.047.
64. Toanen C, Demey G, Ntagiopoulos PG, Ferrua P, Dejour D. Is there any benefit in anterior cruciate ligament reconstruction in patients older than 60 years? Am J Sports Med. 2017;45(4):832–7. https://doi.org/10.1177/0363546516678723.
65. Hetsroni I, Delos D, Fives G, Boyle BW, Lillemoe K, Marx RG. Nonoperative treatment for anterior cruciate ligament injury in recreational alpine skiers. Knee Surg Sports Traumatol Arthrosc. 2013;21(8):1910–4. https://doi.org/10.1007/s00167-012-2324-8.
66. Mehl J, Paul J, Feucht MJ, et al. ACL deficiency and varus osteoarthritis: high tibial osteotomy alone or combined with ACL reconstruction? Arch Orthop Trauma Surg. 2017;137(2):233–40. https://doi.org/10.1007/s00402-016-2604-8.
67. Jin C, Song EK, Jin QH, Lee NH, Seon JK. Outcomes of simultaneous high tibial osteotomy and anterior cruciate ligament reconstruction in anterior cruciate ligament deficient knee with osteoarthritis. BMC Musculoskelet Disord. 2018;19(1):228. https://doi.org/10.1186/s12891-018-2161-0.
68. Li Y, Zhang H, Zhang J, Li X, Song G, Feng H. Clinical outcome of simultaneous high tibial osteotomy and anterior cruciate ligament reconstruction for medial compartment osteoarthritis in young patients with anterior cruciate ligament-deficient knees: a systematic review. Arthroscopy. 2015;31(3):507–19. https://doi.org/10.1016/j.arthro.2014.07.026.
69. Stride D, Wang J, Horner NS, Alolabi B, Khanna V, Khan M. Indications and outcomes of simultaneous high tibial osteotomy and ACL reconstruction. Knee Surg Sports Traumatol Arthrosc. 2019;27(4):1320–31. https://doi.org/10.1007/s00167-019-05379-5.
70. Herzog MM, Marshall SW, Lund JL, Pate V, Mack CD, Spang JT. Trends in incidence of ACL reconstruction and concomitant procedures among commercially insured individuals in the United States, 2002–2014. Sports Health. 2018;10(6):523–31. https://doi.org/10.1177/1941738118803616.
71. Paxton ES, Stock MV, Brophy RH. Meniscal repair versus partial meniscectomy: a systematic review comparing reoperation rates and clinical outcomes. Arthroscopy. 2011;27(9):1275–88. https://doi.org/10.1016/j.arthro.2011.03.088.

Patient-Specific Graft Choice in Primary ACL Reconstruction

2

Martin Lind and Ole Gade Sørensen

2.1 Introduction

The choice of graft for anterior cruciate ligament (ACL) reconstruction has since the start of ACL surgery been a key factor for both surgical technique and the expected clinical outcomes. The three categories of grafts are autograft, allograft, and synthetic graft [1]. Autografts usually consist of either hamstrings tendon (HS), Bone patella tendon-bone (BPTB), or quadriceps tendon (QT), but also iliotibial tract and peroneus longus autograft have seen limited usage. Allografts are varied but can consist of tibialis posterior tendon, Achilles tendon, tibialis anterior tendon, BPTB, and peroneus longus tendon [2, 3].

Synthetic grafts were highly popular in the infancy of ACL reconstruction in the 1980s and 1990s. But catastrophic outcome and severe adverse effects led to these grafts being completely abandoned two decades ago [4, 5]. However, recently new synthetic grafts have been introduced both as complete grafts as the Ligament Augmentation Reconstruction System (LARS; Corin, Gloucestershire, England) or as augmentations to ACL reconstructions or repairs (Internal Brace, Arthrex, Naples, USA) [6].

The choice graft and technique to use during ACL reconstruction are based on patient's anatomy, previous surgical history, concomitant injuries as well as patient choice. Surgeon's choice is dictated by a combination of these factors including perceived functional outcome, rehabilitation speed, graft incorporation, graft availability, and donor site morbidity. The surgeon's familiarity with the graft harvest and implantation technique also influences the graft choice. Much research has been performed to identify which particular graft or technique results in the best clinical outcomes. Some of this research has been of good quality including meta-analyses, systematic reviews, and randomized controlled trials (RCT). Despite three decades of research there continues to be a wide variation in the choices made by surgeons. Limited research is available to guide surgeons to choose the best graft for ACL reconstruction when relating to patient-specific criteria such as age, sex, level of activity, and concomitant injuries. Also, long-term outcomes are not immediately available for newer techniques which makes graft choice decisions even more challenging.

The key clinical outcome parameters for presenting and comparing ACL reconstruction outcomes are the following: risk of reinjury/revision surgery, knee stability as evaluated by quantitative Lachman test, patient-reported outcomes, ability to return to sports, donor site morbidity, and functional outcomes such as muscles strength and hop tests. In the following data on these parameters will be presented.

M. Lind (✉) · O. G. Sørensen
Department of Orthopaedics, Aarhus University Hospital, Aarhus, Denmark
e-mail: martinlind@dadlnet.dk

N. Nakamura et al. (eds.), *Advances in Knee Ligament and Knee Preservation Surgery*,
https://doi.org/10.1007/978-3-030-84748-7_2

Our aim in this chapter is to present available knowledge on clinical outcome in relation to graft choice for ACL reconstruction and specifically present the knowledge of outcome impact with graft choice in relation to patient-specific factors. These data should provide a better decision platform for ACL surgeons to make optimized graft choice decisions based on current evidence.

2.2 Outcome with Bone Patella Tendon Bone Autografts

BPTB grafts for ACL reconstruction have been used since in 1969 and are still first choice in certain countries and for specific patient categories [7]. BPTB has historically been considered the gold standard for ACL reconstruction. The method of harvest includes a horizontal or longitudinal skin incision followed by resection of the mid-portion of the patella tendon with bone block at both ends with the intervening tendon as a complete unit. Thus, the graft has bone block at both ends which allows potentially superior integration of the graft into the tibial and femoral tunnels. The graft is then detached and fed through the tibial tunnel into the femur. Fixation can take place using a variety of different methods ranging from an interference fit with no fixation device to screw or suspensory fixation [8].

Revision rates for BPTB graft usage have recently been reported in high patient volume registry studies from Scandinavia and the USA demonstrating revision rates from 1.5 to 3.2% [2, 9, 10]. Knee stability evaluated as percentage of patients with normal stability defined as less than 3 mm side to side Lachman laxity is between 66 and 81%. Patient-reported outcomes subjective IKDC and KOOS scores demonstrated improvement of 15 points for KOOS with follow-up $KOOS_4$ levels of 70 and Lysholm score of 90. Donor site morbidity incidence range from 5 to 27% but was influenced by very different evaluation methods (Table 2.1).

2.3 Outcome with Hamstring Grafts

Hamstring tendons are one of the more commonly used grafts for ACL reconstruction since Lipscombe in 1982 first described the technique

Table 2.1 Literature overview of Bone Patella Tendon Bone graft for ACL reconstruction

Bone patella tendon bone graft for ACL reconstruction						
Study	Study type	Revision rate (5 years)	Knee stability	PROM	Return to sport ability	Donor site morbidity
Rahr-Wagner (2014) [10]	Registry	3.0%	81% normal (<2 mm)	KOOS4 55–70	–	–
Gifstad (2014) [9]	Registry	2.7%	–			
Maletis (2017) [2]	Registry	2.5%	–	–	–	–
Kaeding (2015) [11]	MOON group	3.2%	–	--	–	–
Mohtadi (2011) [12]	Cochrane review	2.6%	66% normal (<2 mm)	Lysholm FU 89	77%	29%
Lind (2019) [13]	Registry	1.5%	1.0 STS		–	–
Lund (2015) [14]	RCT	0%	0.8 mm STS	$KOOS_4$ 55–72 IKDC 61–70	–	40% unable to knee-walk

Outcome with BPTB graft for ACL reconstruction
STS side to side difference in instrumented sagittal knee stability

Table 2.2 Literature review of hamsting graft for ACL reconstruction

Hamsting graft for ACL reconstruction						
Study	Study type	Revision rate (5 years)	Knee stability	PROM	Return to sport ability	Donor site morbidity
Rahr-Wagner registry (2014) [10]	Registry	4.5%	84% normal (<2 mm)	$KOOS_4$ 55–71	–	–
Gifstad (2014) [9]	Registry	4.2%	–	–	–	–
Maletis (2017) [9]	Registry	2.3%	–	–	–	–
Kaeding (2015) [11]	MOON group	4.8%	–	–	–	–
Mohtadi (2011) [12]	Cochrane review	3.3%	59% normal (<2 mm)	Lysholm FU 90	81%	20%

[15]. The semitendinosus tendon with or without the gracilis tendon is harvested, typically from the ipsilateral leg. The resultant tissue is folded into a four-strand graft of 7–10 mm diameter, which is then used to reconstruct the ACL with different fixation techniques such as metal button with loops, transfixation pins, or interference screws.

Revision rates for hamstring graft usage have been reported in several valid high patient volume registry studies from both Europe and the USA. These studies describe 5-year revision rates ranging from 2.5 to 4.5% (Table 2.2). Knee stability evaluated as percentage of patients with normal stability defined as less than 3 mm side-to-side Lachman laxity, was between 59 and 84%. The patient-reported outcome scores used are numerous with Lysholm score, subjective IKDC and KOOS score being the most used instruments. For all scores patients generally experience significant improvements in score from preoperative to follow-up with Lysholm score and IKDC score 35- and 25-point improvement respectively.

Return to sport ability after hamstring ACLR for light sports was 81% as seen in a Cochrane review [12]. Donor site morbidity as evaluated by subjectively experienced anterior knee pain was seen in 20% of patients [12].

2.4 Outcome with Quadriceps Autograft

The present literature on QT autografts for ACLR has until recently been limited by small study sizes, which has prevented reporting of failure rates and outcomes from a generalized surgical population. Now national clinical registries that contain high volume data enable investigation of accurate low incidence failure rates (ACL revision). For example the Danish Knee Ligament Reconstruction Registry (DKRR), that has recently published outcome from more than 500 QT ACLRs and more than 20,000 PT and HT ACLRs with comparison of revision rates and objective clinical outcomes for these graft types alone [16]. But metanalysis and a two-level 1 RCTs have also contributed to the present outcome knowledge for QT ACLR [14, 17, 18].

QT grafts for ACL reconstruction were first described by Marshall in 1979 [19], but did not gain popularity as a graft for primary ACLR until the last 6–8 years [13]. Before this the QT graft was mainly used for ACL revision surgery and PCL reconstructions. The method of harvest includes a horizontal or longitudinal skin incision at the proximal part of the patella and distal part of the QT followed by resection of a 9–10 mm broad and 7–8 cm long tendon band from the mid-portion of the QT. The graft can involve the full tendon thickness or partial thickness by leaving the deep tendon portion that involves the vastus intermedius. The graft can be harvested with or without a 15–20 mm long bone block from the proximal aspect of the patella tendon as a complete unit. Fixation are made with metal interference screw for the bone block end and the soft tissue part fixation can take place using absorbable and non-absorbable screws or suspensory fixation.

Revision rates for QT graft usage have recently been reported in a high patient volume registry study from Denmark [13]. This study found an overall revision rate of 4.7% which was higher compared to case series and RCTs where revision rates were between 2 and 3% [18].

Knee stability as side-to-side difference quantitatively evaluated has generally been found to be very good with laxity differences from 1.1 to 2.8 mm. Normal pivot shift was seen in 75 to 85%. Patient-reported outcome score subjective IKDC and KOOS scores demonstrated improvement of 15 points for KOOS and 20 points for IKDC with follow-up $KOOS_4$ levels of 84 and IKDC levels 82 to 85. Donor site morbidity incidence ranged from 5 to 27% but was influenced by very different evaluation methods (Table 2.3).

2.5 Outcome with Allograft

The use of allograft is appealing particularly to the complete lack of donor site morbidity, reasonably good availability, and a range of graft sizes with the options of bone blocks attached to the graft. Allograft material does come with its own unique risks including risk of microbiological disease transmission and is an expensive option compared to autografts. The most commonly used allograft tendons are tibialis posterior/anterior and Achilles tendon allografts; however, patellar tendon and HT are also widely available in some countries. Other disadvantages with the use of an allograft include the immunogenic response of the host to the graft and delayed graft incorporation when compared to the autografts. A histological study assessing allografts retrieved during autopsy at 2 years after implantation demonstrated poor vascularization in the center portion of the graft, which had remained acellular [20]. Thus, unlike previous reports of good incorporation of allograft at 18 months, this study shows that graft incorporation might take 3 years or more [21]. Allografts have been widely used for primary ACLR in the USA whereas only minimally used in the rest of the world due to cost, limited availability, and legal issues.

Table 2.3 Literature review of quadriceps tendon graft for ACL reconstruction

Quadriceps tendon graft for ACL reconstruction						
Study	Study type	Revision rate (5 years)	Knee stability	PROM	Return to sport ability	Donor site morbidity
Lind (2019) ACL registry [13]	Registry	4.7%	1.8 mm STS 76% neg. Pivot shift	–	–	–
Lind (2019) [17]	RCT	0%	1.8 mm STS 84% neg. Pivot shift -	$KOOS_4$ 84 IKDC 82	–	Any complaint 27%
Lund (2015) [14]	RCT	0%	1.1 mm STS	$KOOS_4$ 65–82 IKDC 68–84	–	5% unable to knee walk
Slone (2015) [18]	Meta-analysis	0–2.9%	1.1–2.8 mm STS	85–92% good/exc Lysholm 88–93	–	5–15%

Outcome with quadriceps tendon graft ACL reconstruction
STS side-to-side difference in instrumented sagittal knee stability

Table 2.4 Allograft for ACL reconstruction

	Study type	Revision rate (5 years)	Knee stability	PROM	Return to sport ability	Donor site morbidity
Maletis [24]	Registry	3.6–13%	–	–	–	–
Keading [11]	MOON multicenter cohort	4.8%	–	–	–	–
Foster [22]	Review	8%	KT-1000: 1.6 mm	Lysholm 91		

Outcome with allograft tendon graft ACL reconstruction
STS side-to-side difference in instrumented sagittal knee stability

The literature comparing autograft and allograft have been scarce, with mainly small case series. A review of these studies concludes no differences in knee laxity and subjective outcome but higher failure rate between allografts and autografts [22, 23]. Revision rates with allograft used for primary ACLR have recently been reported in the USA in high patient volume registry studies from the MOON and MARS groups and the Kaiser Permanente ACL registry and demonstrating crude revision rates from 3.6 to 10% [11, 24]. The highest revision rates were seen for allografts that were either chemically processes or irradiated. Especially young patients under 21 years were demonstrated to have an increased risk of revision when reconstructed with an allograft with revision rates of 13% [24]. Similarly in the MOON cohort the revision rate for allograft in 20-year-old patients was found to be 10 times higher compared to BPTB autograft with revision rates of 2.5% and 25% for BPTB and allograft respectively [11, 25].

Knee stability evaluated as percentage of patients with normal stability defined as less than 3 mm side-to-side Lachman laxity was between 66 and 81%. Patient-reported outcomes, subjective IKDC and KOOS scores demonstrated improvement of 15 points for KOOS with follow-up KOOS4 levels of 70 and Lysholm score of 90 (Table 2.4).

2.6 Comparison of Graft Types Regarding Outcome

Regarding graft failure comparisons a meta-analysis found that patients undergoing primary ACL reconstruction with bone-tendon-bone autograft were less likely to experience graft rupture and/or revision ACL reconstruction than patients treated with hamstring autograft (OR, 0.83) [10, 26]. As seen in Tables 2.1 and 2.2. revision rates for patella tendon grafts ranged from 2.6 to 3.2% and hamstring from 2.5 to 4.8%, with all large volume studies finding lower revision rates for patella tendon graft compared with hamstring grafts.

Among patients who did not experience graft rupture or revision, there were no differences observed between the two graft types in graft laxity as evaluated by KT 1000 knee arthrometer, pivot shift testing, or Lachman testing. Patients who received a hamstring tendon autograft reported superior KOOS in the sport and recreation subscale (up to 7 points higher) at each follow-up compared with patients who received a patella tendon autograft [10, 16]. Patients who received a hamstring tendon autograft also had a higher Tegner Activity Scale compared with patients who received a patella tendon autograft (mean 4.9 versus mean 4.7) 1-year postoperatively [10]. Patients who received hamstring tendon autografts had increased odds of achieving functional recovery (defined as KOOS pain ≥90, symptoms ≥84, ADL ≥91, sport and recreation ≥80 and QoL ≥81) and were less likely to report subjective treatment failure (defined as a KOOS QoL <44 points) compared with patients who received patella tendon autografts [27]. A recent randomized study comparing hamstring graft, double bundle hamstring graft, and patella tendon graft found equal subjective outcome and pivot shift stability at both 2 and 5 years follow-up [28].

2.7 ACL Graft Choice and Age

The choice of graft in ACL-R in adolescents and young patient groups is important because of an increased risk of graft rupture and revision surgery. Kaeding et al. reported on young patients graft rupture risk from the MOON cohort [11]. In 2683 patients, the risk of an ipsilateral ACL graft rupture was 4.4%, highest in the young population. The odds of an ACL graft tear significantly decreased by 0.09 for every yearly increase in age. A similar correlation between younger age and risk of ACL revision surgery is found in other studies [29–31]. Persson et al. found similar results. They reported on the revision risk following ACL-R in more than 12,000 patients. The hazard ratio for ACL revision was 4.0 for the youngest age group (15–19 years of age) compared to the oldest age group (>30 years of age) [32].

Looking into study data of the effect of graft choice in the young patient group then one study found that after 5 years follow-up, the youngest age group showed a crude revision risk of 9.5% with the use of HT graft compared to 3.5% with the use of a BPTB graft for primary ACL-R [32]. Ho et al. also showed a difference in revision rate after ACL-R depending on graft choice. In 561 patients with a mean age of 15.4 years, they found that soft tissue grafts had a failure rate of 13% compared to a failure rate of 6%, when BPTB grafts were used [33]. Similar findings were reported by the MOON group. In a young, active patient group from 14–22 years of age, they reported a 2.1 increased risk of ACL revision in the HT autograft group compared to the BPTB autograft group [25].

Smaller cohort studies have shown promising outcomes scores and revision rates after the use of Q-tendon autograft in ACL-R in young patient groups [34, 35]. There are no comparative studies addressing both age and QT graft in ACL-R.

2.8 Graft Consideration in Relation to Gender

Several papers have shown an increased relative risk in females to sustain an ACL rupture compared to males, and especially young females have a very high risk of graft rupture after ACL-R if they return to contact and pivoting sports [36–38]. For these reasons there is a high interest in finding the best surgical treatment options for the female ACL patient. Most studies on the subject compare HT autografts to BPTB autografts and the results are controversy.

A review by Paterno et al. included 11 cohort studies. They reported increased A-P knee laxity in a HT autograft female group compared to a BPTB female group. Moreover, they indicated inferior knee laxity results after ACL-R in the female HT group compared with male patients undergoing the same procedure. No randomized controlled trials were included in the review [8]. Ryan et al. also reported on gender differences in outcomes after ACL-R. They included 13 studies, which all were level 2 studies or less. They reported a graft failure risk of 4.0% for males and 4.7% for females after ACL-R with use of a BPTB autograft. Hamstring graft failure risk was 6.4% for males and 9.2% for females. Meta-analysis found no difference in graft failure risk according to sex [1].

The most recent review regarding graft choice and gender was performed by Tan et al. [3]. The study reported on outcomes in female patients only, having ACL-R with either HT autograft or BPTB autografts. Fifteen studies were included in the review, three randomized controlled trials, and 12 prospective cohort studies. These studies included a total of 948 female patients with ACL-R. Almost half of the ACL-R were performed with BPTB autograft, the remaining with HT autograft. Meta-analysis found no difference in female patients between the two graft types at follow-up regarding knee laxity, pivot shift, graft rupture, or graft failure. Furthermore, no differences in objective or subjective outcome scores were found. The magnitude of anterior knee pain was the same in both groups. The study found a tendency to increased risk of anterior kneeling pain in the BPTB group compared to the HT group.

The QT has been used for ACL graft in both males and females. To our knowledge, no study address outcomes after ACL-R according to both QT graft and gender.

2.9 Graft Choice and Sports Activity

2.9.1 Pivoting Sports

The surgeon choice of graft in ACL-R differs according to the sport, the patient is planning to participate in or return to. Bradley et al. reported on the treatment trends in primary ACL-R among team physicians treating American football players. The majority (83%) would use BPTB autograft as the first choice ACL graft [39]. Similar numbers are reported by Ericksen et al. with 86% of the physicians favoring the BPTB graft [40]. The same group investigated the practice patterns among team physicians treating knee injuries in elite athletes competing in ice hockey, soccer, and alpine skiing. Seventy percent of the physicians favored the BPTB tendon as the primary graft choice for ACL-R, 14.9% would use a four-strand semitendinosus graft, whereas the quadriceps tendon autograft was chosen by 4.3% of the physicians [41].

The medical group guiding the international football association (FIFA) still advocates for the use of BPTB autograft in ACL-R in soccer players.

For several reasons, the BPTB autograft has been the graft of choice for cutting and pivoting sport. The properties of the BPTB graft might resemble the native ACL better compared to other graft types. Harvest of the patella tendon might result in a more favorable impairment of the muscle strength following ACL-R because of the sparing of the medial hamstring tendons, which are crucial in cutting and pivoting movements. Moreover, the BPTB graft might have a superior fixation potential and better ingrowth because of the bone-to-bone interface which could lead to a faster return to play.

As mentioned earlier in this chapter, a lot of studies compare outcomes after primary ACL-R after the use of different graft types. Papers reporting on outcome according to both graft type and specific sports are sparse. Gifstad et al. reported on data from the Scandinavian ACL registries. Almost 46,000 patients were undergoing primary ACL-R. They found a lower risk of ACL revision with the use of a BPTB autograft compared to HT autograft if the cause of primary ACL rupture happened at soccer, team handball, or alpine skiing [9]. The paper does not address the sport activity causing new graft rupture and subsequent ACL revision. The MOON knee group reported a cohort study of 770 high school or college athletes, aged 14–22 years, who had primary ACL-R. The patients were followed for 6 years. The majority of the patients competed in pivoting sports such as basketball, American football, and soccer prior to the primary ACL tear. The MOON knee group found a 2.1 times higher risk of ACL revision surgery if a HT autograft was used compared to a BPTB autograft. As with Scandinavian registry study, the MOON study did not report the sport activity leading to graft rupture and ACL revision [25].

The use of quadriceps tendon graft has become more and more popular in primary ACL-R and the graft has performed well in comparative studies [42]. Some studies report acceptable outcomes in patient with a high pre-operative activity score [43, 44], even in an adolescent patient group [34]. To our knowledge no studies compare outcomes after use of quadriceps tendon according to specific sports activity.

2.9.2 Recreational Sports

A wide variety of grafts is used in ACL-R among recreational athletes. There is no real evidence in the literature for stating that one graft should be superior to other graft types in primary ACL reconstruction. It seems as if the use of graft for recreational athletes is based on surgeon preferences, donor site morbidity, and patient requests.

2.10 Graft Choice and Concomitant Injuries

Concomitant injuries to ACL injury such as collateral ligament, cartilage, and meniscus injuries all influence the structure and biomechanical function of the ACL insufficient knee joint. Compromised function of these structures could therefore impact the overall knee function after ACL-R and graft choice should optimally consider and optimize knee function in relation to concomitant injuries.

Especially concomitant injury to the medial collateral ligament (MCL) which is normally treated non-surgically has raised concerns since residual valgus laxity caused by MCL deficiency increases the strain on the ACL and may jeopardize ACL graft survival [45]. Also, recent biomechanical studies have shown that the medial hamstrings are important to resist valgus forces in the MCL-deficient knee [46]. This has led to the dogma that hamstring graft should not be used in combined ACL and MCL injuries.

There is limited literature on the subject of graft choice in combined ACL and collateral ligament injuries. But one biomechanical study has demonstrated that the hamstring tendons are important valgus stabilizers in the MCL insufficient knee and therefore suggest that hamstring tendons should not be used in MCL insufficient knees [46].

The only clinical study that has investigated the issue is a registry study from Sweden that looked at revision rates in MCL injured knees when the ACL was reconstructed with either HT or BPTB grafts. The study included 622 patients with combined ACL and MCL injuries and found no difference in revision rates between HT and BPTB grafts [47].

In conclusion, there is minor clinical evidence that suggests hamstring graft usage in ACL-R in combined ACL+ MCL injured knees is safe without increased revision rates.

No studies have investigated the impact of graft choice with the presence of meniscus and cartilage injuries.

2.11 Conclusion

The graft choice for ACL-R and patient-specific factors do influence the outcome after ACL-R, but the literature in this area is not strong and mainly derived from recent registry studies that contain enough data for investigation for these factor combinations.

There are only comparative studies for hamstring and patella bone tendon bone grafts and different patient factors such as gender, age, sport types, and concomitant injuries. The key point from the literature is presented below.

Key Points

- Patella tendon graft choice results in reduced revision risk in young patients compared to hamstring graft.
- Patella tendon graft choice results in reduced revision risk in female patients compared to hamstring graft.
- Patella tendon graft choice results in reduced revision risk in athletes performing contact sport compared to hamstring graft.
- Patella tendon graft choice results in increased donor site morbidity and poorer subjective outcome compared to hamstring graft.
- Hamstring graft usage for ACL-R does not result in increased revision rates in MCL injured knees.

References

1. Ryan J, Magnussen RA, Cox CL, Hurbanek JG, Flanigan DC, Kaeding CC. ACL reconstruction: do outcomes differ by sex? A systematic review. J Bone Joint Surg Am. 2014;96:507–12.
2. Maletis GB, Chen J, Inacio MCS, Love RM, Funahashi TT. Increased risk of revision after anterior cruciate ligament reconstruction with bone-patellar tendon-bone allografts compared with autografts. Am J Sports Med. 2017;45:1333–40.
3. Tan SHS, Lau BPH, Krishna L. Outcomes of anterior cruciate ligament reconstruction in females using patellar-tendon-bone versus hamstring autografts: a systematic review and meta-analysis. J Knee Surg. 2019;32:770–87.
4. Paulos LE, Rosenberg TD, Grewe SR, Tearse DS, Beck CL. The GORE-TEX anterior cruciate ligament prosthesis. A long-term followup. Am J Sports Med. 1992;20:246–52.
5. Savarese A, Lunghi E, Budassi P, Agosti A. Remarks on the complications following ACL reconstruction using synthetic ligaments. Ital J Orthop Traumatol. 1993;19:79–86.
6. Tulloch SJ, Devitt BM, Porter T, Hartwig T, Klemm H, Hookway S, et al. Primary ACL reconstruction using the LARS device is associated with a high failure rate at minimum of 6-year follow-up. Knee Surg Sports Traumatol Arthrosc. 2019;27:3626–32.

7. Granan LP, Inacio MC, Maletis GB, Funahashi TT, Engebretsen L. Intraoperative findings and procedures in culturally and geographically different patient and surgeon populations: an anterior cruciate ligament reconstruction registry comparison between Norway and the USA. Acta Orthop. 2012;83:577–82.
8. Paterno MV, Weed AM, Hewett TE. A between sex comparison of anterior-posterior knee laxity after anterior cruciate ligament reconstruction with patellar tendon or hamstrings autograft: a systematic review. Sports Med. 2012;42:135–52.
9. Gifstad T, Foss OA, Engebretsen L, Lind M, Forssblad M, Albrektsen G, et al. Lower risk of revision with patellar tendon autografts compared with hamstring autografts: a registry study based on 45,998 primary ACL reconstructions in Scandinavia. Am J Sports Med. 2014;42:2319–28.
10. Rahr-Wagner L, Thillemann TM, Pedersen AB, Lind M. Comparison of hamstring tendon and patellar tendon grafts in anterior cruciate ligament reconstruction in a nationwide population-based cohort study: results from the danish registry of knee ligament reconstruction. Am J Sports Med. 2014;42:278–84.
11. Kaeding CC, Pedroza AD, Reinke EK, Huston LJ, Consortium M, Spindler KP. Risk factors and predictors of subsequent ACL injury in either knee after ACL reconstruction: prospective analysis of 2488 primary ACL reconstructions from the MOON cohort. Am J Sports Med. 2015;43:1583–90.
12. Mohtadi NG, Chan DS, Dainty KN, Whelan DB. Patellar tendon versus hamstring tendon autograft for anterior cruciate ligament rupture in adults. Cochrane Database Syst Rev. 2011;2011(9):CD005960. https://doi.org/10.1002/14651858.CD005960.pub2CD005960.
13. Lind M, Strauss MJ, Nielsen T, Engebretsen L. Quadriceps tendon autograft for anterior cruciate ligament reconstruction is associated with high revision rates: results from the Danish knee ligament registry. Knee Surg Sports Traumatol Arthrosc. 2019;28(7):2163–9. https://doi.org/10.1007/s00167-019-05751-5.
14. Lund B, Nielsen T, Fauno P, Christiansen SE, Lind M. Is quadriceps tendon a better graft choice than patellar tendon? A prospective randomized study. Arthroscopy. 2014;30:593–8.
15. Snook GA. A short history of the anterior cruciate ligament and the treatment of tears. Clin Orthop Relat Res. 1983;172:11–3.
16. Kvist J, Kartus J, Karlsson J, Forssblad M. Results from the Swedish national anterior cruciate ligament register. Arthroscopy. 2014;30:803–10.
17. Lind M, Nielsen TG, Soerensen OG, Mygind-Klavsen B, Fauno P. Quadriceps tendon grafts does not cause patients to have inferior subjective outcome after anterior cruciate ligament (ACL) reconstruction than do hamstring grafts: a 2-year prospective randomised controlled trial. Br J Sports Med. 2020;54:183–7.
18. Slone HS, Romine SE, Premkumar A, Xerogeanes JW. Quadriceps tendon autograft for anterior cruciate ligament reconstruction: a comprehensive review of current literature and systematic review of clinical results. Arthroscopy. 2015;31:541–54.
19. Marshall JL, Warren RF, Wickiewicz TL, Reider B. The anterior cruciate ligament: a technique of repair and reconstruction. Clin Orthop Relat Res. 1979:97–106.
20. Malinin TI, Levitt RL, Bashore C, Temple HT, Mnaymneh W. A study of retrieved allografts used to replace anterior cruciate ligaments. Arthroscopy. 2002;18:163–70.
21. Shino K, Inoue M, Horibe S, Nagano J, Ono K. Maturation of allograft tendons transplanted into the knee. An arthroscopic and histological study. J Bone Joint Surg Br. 1988;70:556–60.
22. Foster TE, Wolfe BL, Ryan S, Silvestri L, Kaye EK. Does the graft source really matter in the outcome of patients undergoing anterior cruciate ligament reconstruction? An evaluation of autograft versus allograft reconstruction results: a systematic review. Am J Sports Med. 2010;38:189–99.
23. Mehta VM, Mandala C, Foster D, Petsche TS. Comparison of revision rates in bone-patella tendon-bone autograft and allograft anterior cruciate ligament reconstruction. Orthopedics. 2010;33:12.
24. Maletis GB, Chen J, Inacio MCS, Love RM, Funahashi TT. Increased risk of revision after anterior cruciate ligament reconstruction with soft tissue allografts compared with autografts: graft processing and time make a difference. Am J Sports Med. 2017;45:1837–44.
25. Group MK, Spindler KP, Huston LJ, Zajichek A, Reinke EK, Amendola A, et al. Anterior cruciate ligament reconstruction in high school and college-aged athletes: does autograft choice influence anterior cruciate ligament revision rates? Am J Sports Med. 2020;48:298–309.
26. Chee MY, Chen Y, Pearce CJ, Murphy DP, Krishna L, Hui JH, et al. Outcome of patellar tendon versus 4-Strand hamstring tendon autografts for anterior cruciate ligament reconstruction: a systematic review and meta-analysis of prospective randomized trials. Arthroscopy. 2017;33:450–63.
27. Barenius B, Forssblad M, Engstrom B, Eriksson K. Functional recovery after anterior cruciate ligament reconstruction, a study of health-related quality of life based on the Swedish National Knee Ligament Register. Knee Surg Sports Traumatol Arthrosc. 2013;21(4):914–27. https://doi.org/10.1007/s00167-012-2162-8.
28. Mohtadi N, Chan D, Barber R, Oddone Paolucci E. A randomized clinical trial comparing patellar tendon, hamstring tendon, and double-bundle ACL reconstructions: patient-reported and clinical outcomes at a minimal 2-year follow-up. Clin J Sport Med. 2015;25:321–31.
29. Kamien PM, Hydrick JM, Replogle WH, Go LT, Barrett GR. Age, graft size, and Tegner activity level as predictors of failure in anterior cruciate ligament reconstruction with hamstring autograft. Am J Sports Med. 2013;41:1808–12.

30. Sanders TL, Pareek A, Hewett TE, Levy BA, Dahm DL, Stuart MJ, et al. Long-term rate of graft failure after ACL reconstruction: a geographic population cohort analysis. Knee Surg Sports Traumatol Arthrosc. 2017;25:222–8.
31. Webster KE, Feller JA, Leigh WB, Richmond AK. Younger patients are at increased risk for graft rupture and contralateral injury after anterior cruciate ligament reconstruction. Am J Sports Med. 2014;42:641–7.
32. Persson A, Gifstad T, Lind M, Engebretsen L, Fjeldsgaard K, Drogset JO, et al. Graft fixation influences revision risk after ACL reconstruction with hamstring tendon autografts. Acta Orthop. 2018;89:204–10.
33. Ho B, Edmonds EW, Chambers HG, Bastrom TP, Pennock AT. Risk factors for early ACL reconstruction failure in pediatric and adolescent patients: a review of 561 cases. J Pediatr Orthop. 2018;38:388–92.
34. Gagliardi AG, Carry PM, Parikh HB, Albright JC. Outcomes of quadriceps tendon with patellar bone block anterior cruciate ligament reconstruction in adolescent patients with a minimum 2-year follow-up. Am J Sports Med. 2020;48:93–8.
35. Kohl S, Stutz C, Decker S, Ziebarth K, Slongo T, Ahmad SS, et al. Mid-term results of transphyseal anterior cruciate ligament reconstruction in children and adolescents. Knee. 2014;21:80–5.
36. Prodromos CC, Han Y, Rogowski J, Joyce B, Shi K. A meta-analysis of the incidence of anterior cruciate ligament tears as a function of gender, sport, and a knee injury-reduction regimen. Arthroscopy. 2007;23:1320–1325e1326.
37. Sutton KM, Bullock JM. Anterior cruciate ligament rupture: differences between males and females. J Am Acad Orthop Surg. 2013;21:41–50.
38. Tan SHS, Kripesh A, Chan CX, Krishna L. Gender differences in intra-articular and extra-articular injuries associated with acute anterior cruciate ligament ruptures. J Knee Surg. 2019;32:616–9.
39. Bradley JP, Klimkiewicz JJ, Rytel MJ, Powell JW. Anterior cruciate ligament injuries in the National Football League: epidemiology and current treatment trends among team physicians. Arthroscopy. 2002;18:502–9.
40. Erickson BJ, Harris JD, Fillingham YA, Frank RM, Bush-Joseph CA, Bach BR Jr, et al. Anterior cruciate ligament reconstruction practice patterns by NFL and NCAA football team physicians. Arthroscopy. 2014;30:731–8.
41. Erickson BJ, Harris JD, Fillingham YA, Cvetanovich GL, Bush-Joseph C, Cole BJ, et al. Orthopedic practice patterns relating to anterior cruciate ligament reconstruction in elite athletes. Am J Orthop (Belle Mead NJ). 2015;44:E480–5.
42. Yang XG, Wang F, He X, Feng JT, Hu YC, Zhang H, Yang L, Hua K. Network meta-analysis of knee outcomes following anterior cruciate ligament reconstruction with various types of tendon grafts. Int Orthop. 2020;44(2):365–80.
43. Cavaignac E, Coulin B, Tscholl P, Nik Mohd Fatmy N, Duthon V, Menetrey J. Is quadriceps tendon autograft a better choice than hamstring autograft for anterior cruciate ligament reconstruction? A comparative study with a mean follow-up of 3.6 years. Am J Sports Med. 2017;45:1326–32.
44. Galan H, Escalante M, Della Vedova F, Slullitel D. All inside full thickness quadriceps tendon ACL reconstruction: long term follow up results. J Exp Orthop. 2020;7:13.
45. Mancini EJ, Kohen R, Esquivel AO, Cracchiolo AM, Lemos SE. Comparison of ACL strain in the MCL-deficient and MCL-reconstructed knee during simulated landing in a cadaveric model. Am J Sports Med. 2017;45:1090–4.
46. Herbort M, Michel P, Raschke MJ, Vogel N, Schulze M, Zoll A, et al. Should the ipsilateral hamstrings be used for anterior cruciate ligament reconstruction in the case of medial collateral ligament insufficiency? Biomechanical investigation regarding dynamic stabilization of the medial compartment by the hamstring muscles. Am J Sports Med. 2017;45:819–25.
47. Hamrin Senorski E, Svantesson E, Beischer S, Thomee C, Grassi A, Krupic F, et al. Concomitant injuries may not reduce the likelihood of achieving symmetrical muscle function one year after anterior cruciate ligament reconstruction: a prospective observational study based on 263 patients. Knee Surg Sports Traumatol Arthrosc. 2018;26:2966–77.

Assessment of Risk Factors for Failure of ACLR: When to Address Concomitant Pathology

3

Gian Luigi Canata, Valentina Casale, and Antonio Pastrone

3.1 Introduction

Tears of the anterior cruciate ligament (ACL) are one of the most common knee injuries in the active population. They are often associated with concomitant injuries, such as other ligamentous injuries, meniscal and chondral or osteochondral lesions. These concomitant afflictions may predispose patients to the development of functional instability, cartilage degeneration, and subsequent osteoarthritis within 10 to 15 years [1, 2]. ACL reconstruction (ACLR) is consequently one of the most common orthopedic procedures, with approximately 75,000 to 100,000 surgeries performed annually in the United States alone [3].

Surgical reconstruction techniques have evolved considerably over the years, although there is still no consensus on the optimal technique [4]. Nevertheless, very satisfying results of ACLR restoring functional knee stability have been frequently reported [5].

Several studies have evaluated and cited the most common risk factors for native ACL tears. Some examples are female sex, reduced activity level, participation in cutting sports, anatomic factors such as intercondylar notch width and tibial slope, neuromuscular control, and lower extremity biomechanics [6–12]. However, there is a relative lack of scientific data analyzing risk factors for graft failures or re-tears after an ACLR [13].

The incidence of ACLR failure has been estimated at between 2% and 6% in short term (less than 5-year) follow-up [14], while longer follow-ups (5–10 years) are associated with a higher failure risk, between 3% and 10% [15, 16]. After 10 years, the incidence stabilizes within the 5-to-10-year range [17, 18].

To minimize this rate of failure, it is fundamental to identify the major risk factors, which may include poor surgical techniques, as well as a new trauma, failure of biological incorporation, the graft type used, postoperative infections, and missed concurrent knee injuries. Further risk factors may be patient-related, such as altered neuromuscular control, age, sex, or level of activity [13].

Risk factors may also be distinguished according to time to development of recurrent postoperative instability. An early instability (<6 months after surgery) is usually caused by technical errors, failure of graft incorporation, premature return to high-demand activity, or an overly aggressive rehabilitation. A later development may be a consequence of repeated trauma to the graft, poor graft placement, generalized ligamentous laxity, and concomitant abnormalities not addressed during the ACLR [19].

G. L. Canata (✉) · V. Casale · A. Pastrone
Centre of Sports Traumatology, Koelliker Hospital, Torino, Italy
e-mail: studio@ortosport.it; canata@ortosport.it

N. Nakamura et al. (eds.), *Advances in Knee Ligament and Knee Preservation Surgery*,
https://doi.org/10.1007/978-3-030-84748-7_3

3.2 Patient Risk Factors

3.2.1 Sex

While native ACL injuries occur more frequently in women than men, some recent studies highlight that male patients have a higher risk of revision ACLR [20, 21]. However, other authors have reported different findings [22, 23], therefore an univocal consensus is still lacking. It has been widely demonstrated that female sex is a strong predictor of native ACL injury, but this has not been reported for ACL graft tears [24–27]. A possible explanation is the lower activity level of women post-surgery [5].

Another aspect to consider regarding the sex differences in ACLR results is the graft type used and the consequent presence or absence of postoperative side-to-side difference in anteroposterior (AP) knee laxity [13]. In fact, it has been largely reported that women with hamstring tendon (HS) grafts show greater AP knee laxity than those with bone-patellar tendon-bone (BPTB) grafts [25, 28–32].

In summary, evidence for considering sex as a risk factor for ACLR failure is still lacking, mostly due to confounding factors such as activity level after surgery [13]. Nevertheless, women are usually more at risk for native ACL tears than males. This discrepancy may be explained by ACL graft being stronger than native ACL, by the patient returning to a lower level of activity after surgery, or by a neuromuscular adaptation during the postoperative rehabilitation [13].

3.2.2 Age

Younger age is the most frequently and widely reported risk factor for ACLR revision [33]. Females younger than 25 years and males aged 26 to 45 years are more likely to develop a second ACL tear [34], according to current evidence on this issue [1, 5, 20, 33–35]. Increasing age, in fact, has been demonstrated to be associated with a decreased risk of ACLR revision, regardless of the postoperative level of activity [13]. According to the Grading of Recommendations and the Assessment and Development and Evaluation (GRADE) approach, age shows the highest quality evidence and correlation with ACLR failure: high relative risk increases with each incremental decrease in age [33].

3.2.3 Neuromuscular Factors

Recent studies have identified that biomechanical and neuromuscular factors are predictive of ACLR revision [35]. Some examples are hip internal rotation moment, altered postural stability, knee valgus, and asymmetric sagittal-plane knee moment during landing [12, 36]. Deficits of hip muscle external rotation torque are particularly correlated with ACLR failures: addressing the impaired hip strength may effectively reduce this risk [13].

3.2.4 Anatomical Abnormalities

Knee alignment plays a relevant role when approaching ACLR surgery. It is commonly known in fact that ACLR in knees with a varus thrust tends to fail secondarily, if the varus alignment is not addressed at the time of the reconstruction [37–39]. It has been highlighted that a varus alignment may cause gait abnormalities, such as decreased flexion moment, increased external adduction moment, increased external knee extension moment, and increased hyperextension during the stance phase [38, 40]. Furthermore, it has been reported how ACL deficiency increases the thrusting pattern in a preexisting varus alignment [38].

Recently, increased posterior tibial slope (PTS) has been reported among the anatomic risk factors for ACLR failure [41]. In particular, a slope $\geq 12°$ is associated with a higher risk of revision [42–44]. Several studies have suggested, in the case of an ACL-deficient knee, to decrease the PTS to normal values to avoid the risk of anterior tibial translation after ACLR [40, 45]. Normal PTS values may be considered as 9°–11° in the medial plateau and 6°–8° in the lateral plateau. As a further result of the PTS reduction, both the tension forces on the ACL graft decrease and the ACL re-rupture rates decrease [46].

At the same time, dynamic valgus moments are considered as risk factors for ACLR failures too [47], especially if associated with insufficiency of the posteromedial ligament complex, constituted by the medial collateral ligament (MCL) and the posterior oblique ligament (POL) [48]. This additional deficiency may result in high ACL graft forces and dynamic valgus angles. In conclusion, in the case of both knee malalignment or a TPS $\geq 12°$, a corrective osteotomy may be suggested, regardless of whether they occur during the ACLR [49–51] or the ACLR revision surgery [52].

Bony morphology, in particular the distal femur morphology, plays an important role in knee kinematics [53]. It has been recently demonstrated that an increased femoral condylar depth is associated with increased rotatory knee laxity [54], and an increased rotatory knee laxity is a well-known risk factor for ACLR revision [55].

In a recent review, the knee hyperextension has been evaluated as a risk factor for ACLR failure, as literature on this aspect is still limited [56]. The results highlight that a hyperextension $\geq 5°$ is an independent significant predictor of ACLR revision, especially in association with age younger than 26 years and the use of an allograft. Furthermore, the BPTB graft guarantees more postoperative stability than the hamstrings grafts do, especially when hyperextension and other signs of increased joint laxity are present [57, 58].

3.2.5 Body Mass Index

Some authors have evaluated the body mass index (BMI) and its relationship with graft failure. They reported that a BMI ≥ 30 kg/m^2 is associated with a lower risk of ACLR revision [26, 33, 59]. The reason could be that people with high BMI are less active, and consequently are at lower risk of revision surgery [59]. However, the literature reports contradictory results, such as no relationship between BMI and ACLR revision [4, 60], or even that obesity (BMI between 30 and 39 kg/m^2) is a risk factor in itself [61].

3.2.6 Smoking

Smoking must be included among the risk factors for ACLR failure [13]. In fact, it has been well described how the consequent release of tissue-damaging oxygen free radicals, vasoconstriction, impaired bone metabolism, and inhibition of the macrophages and fibroblasts may cause poor wound healing after surgery [62–64]. In a recent study, it has been reported that cigarette smoke is associated not only with increased complication rates after ACLR, but also worse clinical postoperative results and an increase of anterior translation [65].

3.3 Concomitant Pathologies

3.3.1 Anterolateral Ligament Complex

ACL tears are often associated with injuries to the anterolateral complex, with a consequent high risk of development of anterolateral rotatory knee instability [66, 67]. The anterolateral complex includes the superficial and deep iliotibial band (ITB), the capsule-osseous layer of the ITB, and a thickening of the lateral capsule known as the anterolateral ligament (ALL). The ALL, in particular, is an extra-articular structure of the knee and acts as a secondary restraint to anterior tibial translation and rotational instability [68, 69].

There are conflicting data about the biomechanical role of the anterolateral capsule [70].

After ACLR, a persistent instability of the knee at follow-up has been reported in 11% to 30% of cases [67]. To reduce these rates, it has been proposed to add a lateral extra-articular tenodesis (LET) to the intra-articular ACLR, especially if a high-grade preoperative pivot shift is noticed [71–73]. The LET technique was largely used to treat ACL-deficient knees, before the intra-articular and arthroscopic ACL reconstruction techniques were developed [74]. This technique has only recently been reconsidered, if added to the intra-articular ACLR, to better control rotatory laxity [73, 75–78].

To date there is no clear evidence about indications for LET; however, an ALL reconstruction or LET can be an option when a high-grade pivot shift and hyperlaxity are present and in view of resuming rotational sport activities [79].

Surgery consists of mimicking the native anatomy of ALL with an open technique using a tendon graft, usually a part of the ilio-tibial band. The Arnold-Cocker and the modified Lemaire or Andrews are three of the most widespread techniques for LET [80, 81].

The rehabilitation program consists of quick discharge and total weight bearing with crutches without immobilization [82].

LET or reconstruction of the antero-lateral ligament does not excessively constrain the lateral knee joint [83], reduces the risk of revision surgery for failure of ramp lesion, accelerate the return to sport, and improve rotational stability [71].

3.3.2 Posteromedial Ligament Complex

The ACL and the MCL together provide anterior, valgus, and anteromedial rotatory stability of the knee [84]. The MCL is known to have the ability to heal nonoperatively, and good results after nonsurgical management have been reported [85–88]. This has led many surgeons to consider not treating MCL tears even when they are associated with ACL injuries, choosing only an early or delayed ACLR [89–91]. Nevertheless, in patients with a combined injury of both ACL and MCL, leaving the MCL untreated may lead to a chronic symptomatic valgus instability [92, 93]. Furthermore, an MCL deficiency may increase the forces on the ACL graft, so restoring the properties of the MCL is suggested for better healing of both the ACL graft and the injured MCL [93–96].

Whereas the MCL is considered the primary valgus restraint in 15–90° flexion and external rotation stabilizer, the posteromedial corner (PMC), formed by the MCL and the POL, provides valgus restraint in an extended knee [97–99]. For this reason, it is clear that an injury of the PMC is significantly different and more serious than an MCL tear alone [90, 100].

Several surgical techniques have been described to treat MCL tears, including both suture repair and ligament reconstruction [84, 101–105]. There is no evidence that reconstruction is better than repair alone. To address both the MCL and the POL, other techniques have been proposed [93, 106].

3.3.3 Posterolateral Corner

Posterolateral corner (PLC) structures of the knee are basically divided into static (lateral collateral ligament, popliteofibular ligament, and popliteus tendon) and dynamic (biceps femoris and lateral head of gastrocnemius). They significantly contribute to the stability of the knee in varus and external rotation.

In case of combined ACL reconstruction and PLC lesion, in fact, the forces on graft increase, leading to higher risk of failure [107, 108]. For this reason, surgeons must diagnose and treat PLC lesions when present.

Diagnosis is set with physical examination and MRI, which is the diagnostic imaging of choice. Complete tear or avulsion of the lateral collateral ligament is the most significant predictor at MRI of posterolateral instability [109].

In the case of an acute trauma, posterolateral corner injury combined with ACL tear must be treated with surgical repair. Reconstruction with an open technique using a tendon graft is the gold standard for chronic injuries [110].

There is no consensus about the best surgical technique for these complex knee injuries, but recent literature demonstrates that a single staged, combined reconstruction is optimal with 90% of successful outcomes [111]. The Arciero technique [112] and the LaPrade technique [113] are the most used reconstruction procedures, and they do not show relevant biomechanical differences among each other [114].

Peroneal nerve lesions (25% of patients) or vascular injuries must always be assessed and treated [110].

Medium-term (6 years) clinical results have been recently reported as satisfying with both repair and reconstruction of PLC, when combined with ACL reconstruction. The only exception concerns the regained activity level, which is higher in the PLC repair group. For this reason, early diagnosis and treatment is necessary in case of posterolateral corner injuries [115].

Rehabilitation program targets are the protection of the healing process wearing a cast for 6 weeks, as well as a gradual restoration of range of motion, strength, and proprioception of the knee [116].

3.3.4 Posterior Cruciate Ligament

Simultaneous ACL and posterior cruciate ligament (PCL) rupture usually occurs in the case of knee dislocation after high impact traumas [117]. Knee dislocation is a relatively rare event (0.02%–0.2% of orthopedic injuries), and is frequently associated with neurovascular injuries or fractures that need to be treated first [118].

Once the patient has been stabilized, the surgeon must assess all the structures of the knee with MRI, which is the gold standard for diagnosis of simultaneous lesions. Clinically complete bi-cruciate lesions lead to an important AP translation instability >15 mm at all angles of knee flection [119].

Treatment consists on the reconstruction of both the cruciate ligaments and the repair of all the capsular and articular structures.

A single-stage early reconstruction with either allogenous tendon [120] or autologous graft [121] is a reproducible procedure and may be preferred for reducing the risks of infection, recovery time, and anesthesia-related complications. However, a multi-staged reconstruction decreases the risk of arthrofibrotic complications [122].

During a bi-cruciate reconstruction, particular attention should be paid to the posterolateral corner, for avoiding instability that may generate increased forces against cruciate ligament grafts and subsequent graft failure [123].

Post-operative rehabilitation is slow for this type of surgery and there is not large consensus about the best management to conform to. Generally, the knee must be protected by a posterior tibial support brace in full extension for 3 weeks, then weight bearing may be partially permitted with the assistance of crutches for 6 weeks [117].

Clinical follow-up shows significant improvement and good results, with satisfaction rate greater than 90% and return to pre-injury level activity in up to 85% of cases [117, 119–121, 123, 124].

3.3.4.1 Menisci

Clinical outcomes in ACL reconstruction mostly depend on the knee stability and the status of the menisci. Concomitant meniscus injuries at the time of ACL reconstruction have a relevant negative impact on the clinical results in terms of quality of life, pain, and recovery of sport activity at 5–10 years [125, 126].

The menisci are important secondary stabilizers of the knee and their function consists of carrying 40%–70% of the load of the joint, decreasing cartilage shear stress, and increasing joint congruency [127]. In knees without a functioning meniscus, the load on cartilage is three times higher [128–131], with an increased risk of development of long-term knee osteoarthritis [132].

Meniscal tears are frequently diagnosed in patients with ACL rupture, with a prevalence of approximately 60%, including children and adolescents [133]. The incidence of a medial meniscal tear in patients with an episode of significant instability is 10 times greater when compared to those with insignificant instability [129]. Increased medial meniscus slope [134] and greater relative anteroposterior length of the femoral condyles are associated with a higher risk of meniscal lesions in association with acute ACL rupture [135]. Other risk factors for concomitant injury are elevated BMI and surgery delay [136]. In particular, ACL reconstruction within 12 months can reduce by two times the risk of developing medial meniscal tears, but it does not change the risk for the lateral meniscus [136].

In the case of medial meniscal tear in ACL-deficient knee, meniscectomy significantly reduces knee stability and increases cartilage load [128–131]. Medial meniscal meniscus extrusion secondary to meniscal root lesion or radial lesion can raise the cartilage load by up to 126% and reduce the contact area from 17% to 64%, leading to the risk of developing knee osteoarthritis [137].

Removal of the lateral meniscus in case of tear increases shear stress 200% more than medial meniscectomy [138]. Lateral meniscus is a fundamental shock absorber due to the convexity of the lateral tibial plateau and the lateral femoral condyle, and an important restraint to anterior tibial translation during combined valgus and rotatory loads [130].

Long-term clinical outcomes of meniscal repair in ACL reconstruction are better than partial meniscectomy, because of the well-known protective function of the meniscus [139].

However, although surgeons should attempt meniscal repair when possible, literature reports 12–23% of failure after meniscal suture [140]. For this reason, the patient should always be informed about the failure risks of this procedure before surgery.

Several types of meniscal tears, such as longitudinal (including RAMP lesions), radial, and complex tears are repairable. The clinical outcome of repair depends on the size, the site of the tear, and the presence of associated injuries [141]. RAMP lesions will be discussed separately in the next paragraph.

Longitudinal vertical tears in a vascularized area (red-red zone) are the gold standard indication for repair and are usually treated with all-inside sutures [142]. Acute tears heal better especially if they are combined with ACL reconstruction [131, 143]. In the case of tears of the body and anterior horn of the meniscus, the outside-in technique can be used, while horizontal lesions can be treated by arthroscopic outside-in or open meniscal suture [142].

Root tears, especially when traumatic, can be treated by transosseous pullout reinsertion, which statistically decreases the meniscal extrusion in both unloaded and loaded positions [137].

Satisfactory clinical outcomes are described for bucket handle, and also for deep red-zone radial meniscal tear with a healing rate of 80% to 90% after repair at 1 year follow up [144–146].

Generally, repair failures in the first 6 months can be related to technical issues during surgery, while failures up to 24 months indicate a poor healing process [147].

Second-look arthroscopy is the most accurate way to verify achievement of meniscal healing; MRI has an accuracy of 62% and is not reliable in the evaluation of meniscal suture success [131]. Abnormal signal in the meniscus at 1 year after repair is frequent, however meniscal lack of extrusion or meniscal shortening are good indicators for the meniscal healing [131].

Rehabilitation after meniscal repair is slower and different than that after meniscectomy, weight bearing is usually avoided in the first 30 days and patients need to be careful especially in active flexion [148]. In addition, return to sport (86–91% of patients) is slower than in the case of meniscectomy, and should be delayed for up to 6 months [149].

3.3.4.2 Ramp Lesions

Ramp lesions are peripheral longitudinal tears of the posterior horn of the medial meniscus combined with ACL rupture. The presence of this tear in ACL rupture increases the rotational laxity of the knee [150]. Recently, ramp lesions have been in fact increasingly investigated to partially explain the frequent residual laxity in ACL reconstruction.

Approximately 26.6% to 40% of patients who undergo primary ACL reconstruction are suffering from ramp lesions [151]. In acute ACL injuries, the incidence is about 21.8%, while in chronic injuries the incidence increases to 32.8% [152]. Chronic knee instability is a potential cause of this kind of meniscal tear [152]; in revision ACL reconstruction in fact the incidence of ramp lesion is up to 39% [153].

Risk factors for developing RAMP lesions are male sex, patients aged <30 years, longer time from injury, revision ACLR, preoperative side-to-side laxity >6 mm, medial plateau edema, and concomitant lateral meniscal tears [154, 155]. The prevalence of ACL-associated ramp lesions in children and adolescents is similar to that of adult population [156].

Diagnosis may be challenging, as ramp lesions may be easily overlooked in arthroscopic examination and MRI images. The sensitivity of MRI for ramp lesions (71, 7%) is significantly lower than that for meniscal body tears (94.3%) [157]. The presence of bone marrow edema of the posteromedial tibia on MRI increases by three times the risk for RAMP lesion [155].

A systematic inspection through the intercondylar notch (Gillquist portal) or posterolateral transseptal portals is recommended during ACL reconstruction to ensure a precise diagnosis [158]. Postero-medial portal approach using 30 or 70-degree arthroscope further improves the diagnostic accuracy [152, 159].

Ramp lesions can be repaired by all-inside suture using a meniscal hook and a posterior arthroscopic portal [160]. Meniscal suture significantly increases postoperative knee stability, knee function scores, and patient satisfaction following ACL reconstruction [161]. Repair is recommended in the case of chronic lesions and those longer than 20 mm, if associated with meniscal instability, in young patients, or in ACL revision. Small stable ramp lesions treated with abrasion alone during ACL reconstruction may result in similar clinical outcomes compared with those treated with surgical repair [146]. Anterolateral tenodesis or reconstruction has a protective role on the RAMP repair [154].

3.3.5 Cartilage

Concomitant cartilage damage is commonly reported during ACL reconstruction, especially after a contact trauma or recurrent instability episodes [162]. Other risk factors are surgery delay over 12 months and high BMI [133]. Isolated cartilage lesions in ACL rupture are difficult to assess because of the frequent presence of a concomitant meniscal lesion; for this reason, there is no uniform consensus about the outcomes of ACL reconstruction in the case of chondral lesions [163].

The majority of the literature reports poorer clinical outcomes [164], and lower likelihood of achieving symmetry for extension strength at 6 months after surgery [165] in the presence of a partial or full-thickness cartilage lesion. In spite of this, a study from the Norwegian registry reports a good restoration of the knee function in both patients with a cartilage lesion (ICRS grade 3 or 4) or those without cartilage lesion 9 years after ACL reconstruction [164].

Treatment of cartilage lesions is challenging, because of its inability to regenerate or repair. Little evidence does exist of better results after surgical treatment of cartilage lesions [166]. Clinical scores 5 years after surgery do not show any differences between no treatment, debridement, or microfractures in full-thickness cartilage lesions [167]. Furthermore, small asymptomatic lesions may not necessitate surgical treatment [168]. A large full-thickness defect in young patients can be treated by attempting autologous chondrocyte implantation, scaffold-based repair, or osteochondral transplantation [169].

In the case of patellofemoral cartilage lesions in ACL reconstruction, instability and malalignment should be assessed and treated when present [170]. Currently, no consensus does exist on the post-operative rehabilitation protocol [171], which should be based on individual case evaluation.

3.4 Conclusions

ACLR is one of the most common procedures performed aiming to restore knee function. Several factors influence results and the assessment of possible associated risk factors of failure is mandatory before the operation. A careful evaluation of the patient including clinical, biomechanical, and radiological data is the best way to plan a successful treatment.

We know that an increasing percentage of athletes can resume their previous activity at the same level with an appropriate surgical technique. Associated biomechanical, ligamentous, meniscal, chondral, and osteochondral abnormalities could endanger the success of the ACLR: their proper treatment will further reduce the risk of unsatisfactory results.

References

1. Schilaty ND, Nagelli C, Bates NA, Sanders TL, Krych AJ, Stuart MJ, et al. Incidence of second anterior cruciate ligament tears and identification of associated risk factors from 2001 to 2010 using a geographic database. Orthop J Sports Med. 2017;5(8):2325967117724196.
2. Kessler MA, Behrend H, Henz S, Stutz G, Rukavina A, Kuster MS. Function, osteoarthritis and activity after ACL-rupture: 11 years follow-up results of conservative versus reconstructive treatment. Knee Surg Sports Traumatol Arthrosc. 2008;16(5):442–8.
3. Buller LT, Best MJ, Baraga MG, Kaplan LD. Trends in anterior cruciate ligament reconstruction in the United States. Orthop J Sports Med. 2015;3(1):2325967114563664.
4. Yabroudi MA, Björnsson H, Lynch AD, Muller B, Samuelsson K, Tarabichi M, et al. Predictors of revision surgery after primary anterior cruciate ligament reconstruction. Orthop J Sports Med. 2016;4(9):2325967116666039.
5. Kaeding CC, Pedroza AD, Reinke EK, Huston LJ, Spindler KP. Risk factors and predictors of subsequent ACL injury in either knee after ACL reconstruction: prospective analysis of 2488 primary ACL reconstructions from the MOON cohort. Am J Sports Med. 2015;43(7):1583–90.
6. Arendt E, Dick R. Knee injury patterns among men and women in collegiate basketball and soccer. NCAA data and review of literature. Am J Sports Med. 1995;23(6):694–701.
7. Griffin LY, Agel J, Albohm MJ, Arendt EA, Dick RW, Garrett WE, et al. Noncontact anterior cruciate ligament injuries: risk factors and prevention strategies. J Am Acad Orthop Surg. 2000;8(3):141–50.
8. Prodromos CC, Han Y, Rogowski J, Joyce B, Shi K. A meta-analysis of the incidence of anterior cruciate ligament tears as a function of gender, sport, and a knee injury-reduction regimen. Arthrosc J Arthrosc Relat Surg. 2007;23(12):1320–1325.e6.
9. Mihata LCS, Beutler AI, Boden BP. Comparing the incidence of anterior cruciate ligament injury in collegiate lacrosse, soccer, and basketball players: implications for anterior cruciate ligament mechanism and prevention. Am J Sports Med. 2006;34(6):899–904.
10. Simon RA, Everhart JS, Nagaraja HN, Chaudhari AM. A case-control study of anterior cruciate ligament volume, tibial plateau slopes and intercondylar notch dimensions in ACL-injured knees. J Biomech. 2010;43(9):1702–7.
11. LaPrade RF, Burnett QM. Femoral intercondylar notch stenosis and correlation to anterior cruciate ligament injuries. A prospective study. Am J Sports Med. 1994;22(2):198–202; discussion 203.
12. Hewett TE, Myer GD, Ford KR, Heidt RS, Colosimo AJ, McLean SG, et al. Biomechanical measures of neuromuscular control and valgus loading of the knee predict anterior cruciate ligament injury risk in female athletes: a prospective study. Am J Sports Med. 2005;33(4):492–501.
13. Duffee AR, Hewett TE, Keading CC. Patient-related risk factors for ACL graft failure. In: Revision ACL reconstruction: indications and technique. Berlin: Springer Science & Business Media; 2013. p. 1–10.
14. Wright RW, Dunn WR, Amendola A, Andrish JT, Bergfeld J, Kaeding CC, et al. Risk of tearing the intact anterior cruciate ligament in the contralateral knee and rupturing the anterior cruciate ligament graft during the first 2 years after anterior cruciate ligament reconstruction: a prospective MOON cohort study. Am J Sports Med. 2007;35(7):1131–4.
15. Drogset JO, Grøntvedt T. Anterior cruciate ligament reconstruction with and without a ligament augmentation device: results at 8-year follow-up. Am J Sports Med. 2002;30(6):851–6.
16. Roe J, Pinczewski LA, Russell VJ, Salmon LJ, Kawamata T, Chew M. A 7-year follow-up of patellar tendon and hamstring tendon grafts for arthroscopic anterior cruciate ligament reconstruction: differences and similarities. Am J Sports Med. 2005;33(9):1337–45.
17. Bourke HE, Salmon LJ, Waller A, Patterson V, Pinczewski LA. Survival of the anterior cruciate ligament graft and the contralateral ACL at a minimum of 15 years. Am J Sports Med. 2012;40(9):1985–92.
18. Shelbourne KD, Gray T. Minimum 10-year results after anterior cruciate ligament reconstruction: how the loss of normal knee motion compounds other factors related to the development of osteoarthritis after surgery. Am J Sports Med. 2009;37(3):471–80.
19. Kamath GV, Redfern JC, Greis PE, Burks RT. Revision anterior cruciate ligament reconstruction. Am J Sports Med. 2011;39(1):199–217.
20. Maletis GB, Inacio MCS, Funahashi TT. Risk factors associated with revision and contralateral anterior cruciate ligament reconstructions in the Kaiser Permanente ACLR registry. Am J Sports Med. 2015;43(3):641–7.
21. Sutherland K, Clatworthy M, Chang K, Rahardja R, Young SW. Risk factors for revision anterior cruciate ligament reconstruction and frequency with which patients change surgeons. Orthop J Sports Med. 2019;7(11):2325967119880487.
22. Ahldén M, Samuelsson K, Sernert N, Forssblad M, Karlsson J, Kartus J. The Swedish National Anterior Cruciate Ligament Register: a report on baseline variables and outcomes of surgery for almost 18,000 patients. Am J Sports Med. 2012;40(10):2230–5.
23. Desai N, Andernord D, Sundemo D, Alentorn-Geli E, Musahl V, Fu F, et al. Revision surgery in anterior cruciate ligament reconstruction: a cohort study of 17,682 patients from the Swedish National Knee Ligament Register. Knee Surg Sports Traumatol Arthrosc. 2017;25(5):1542–54.
24. Barber-Westin SD, Noyes FR, Andrews M. A rigorous comparison between the sexes of results and complications after anterior cruciate ligament reconstruction. Am J Sports Med. 1997;25(4):514–26.

25. Ferrari JD, Bach BR, Bush-Joseph CA, Wang T, Bojchuk J. Anterior cruciate ligament reconstruction in men and women: an outcome analysis comparing gender. Arthrosc J Arthrosc Relat Surg. 2001;17(6):588–96.
26. Persson A, Fjeldsgaard K, Gjertsen J-E, Kjellsen AB, Engebretsen L, Hole RM, et al. Increased risk of revision with hamstring tendon grafts compared with patellar tendon grafts after anterior cruciate ligament reconstruction: a study of 12,643 patients from the Norwegian cruciate ligament registry, 2004-2012. Am J Sports Med. 2014;42(2):285–91.
27. Salmon L, Russell V, Musgrove T, Pinczewski L, Refshauge K. Incidence and risk factors for graft rupture and contralateral rupture after anterior cruciate ligament reconstruction. Arthrosc J Arthrosc Relat Surg. 2005;21(8):948–57.
28. Gobbi A, Domzalski M, Pascual J. Comparison of anterior cruciate ligament reconstruction in male and female athletes using the patellar tendon and hamstring autografts. Knee Surg Sports Traumatol Arthrosc. 2004;12(6):534–9.
29. Bizzini M, Gorelick M, Munzinger U, Drobny T. Joint laxity and isokinetic thigh muscle strength characteristics after anterior cruciate ligament reconstruction: bone patellar tendon bone versus quadrupled hamstring autografts. Clin J Sport Med. 2006;16(1):4–9.
30. Pinczewski LA, Lyman J, Salmon LJ, Russell VJ, Roe J, Linklater J. A 10-year comparison of anterior cruciate ligament reconstructions with hamstring tendon and patellar tendon autograft: a controlled, prospective trial. Am J Sports Med. 2007;35(4):564–74.
31. Noojin FK, Barrett GR, Hartzog CW, Nash CR. Clinical comparison of intraarticular anterior cruciate ligament reconstruction using autogenous semitendinosus and gracilis tendons in men versus women. Am J Sports Med. 2000;28(6):783–9.
32. Paterno MV, Weed AM, Hewett TE. A between sex comparison of anterior-posterior knee laxity after anterior cruciate ligament reconstruction with patellar tendon or hamstrings autograft: a systematic review. Sports Med Auckl NZ. 2012;42(2):135–52.
33. Rahardja R, Zhu M, Love H, Clatworthy MG, Monk AP, Young SW. Factors associated with revision following anterior cruciate ligament reconstruction: a systematic review of registry data. Knee. 2020;27:287–99.
34. Schilaty ND, Bates NA, Sanders TL, Krych AJ, Stuart MJ, Hewett TE. Incidence of second anterior cruciate ligament tears (1990-2000) and associated factors by geographic locale. Am J Sports Med. 2017;45(7):1567–73.
35. Paterno MV, Huang B, Thomas S, Hewett TE, Schmitt LC. Clinical factors that predict a second ACL injury after ACL reconstruction and return to sport: preliminary development of a clinical decision algorithm. Orthop J Sports Med. 2017;5(12):2325967117745279.
36. Paterno MV, Schmitt LC, Ford KR, Rauh MJ, Myer GD, Huang B, et al. Biomechanical measures during landing and postural stability predict second anterior cruciate ligament injury after anterior cruciate ligament reconstruction and return to sport. Am J Sports Med. 2010;38(10):1968–78.
37. van de Pol GJ, Arnold MP, Verdonschot N, van Kampen A. Varus alignment leads to increased forces in the anterior cruciate ligament. Am J Sports Med. 2009;37(3):481–7.
38. Noyes FR, Schipplein OD, Andriacchi TP, Saddemi SR, Weise M. The anterior cruciate ligament-deficient knee with varus alignment. An analysis of gait adaptations and dynamic joint loadings. Am J Sports Med. 1992;20(6):707–16.
39. Kim S-J, Moon H-K, Chun Y-M, Chang W-H, Kim S-G. Is correctional osteotomy crucial in primary varus knees undergoing anterior cruciate ligament reconstruction? Clin Orthop. 2011;469(5):1421–6.
40. Cantivalli A, Rosso F, Bonasia DE, Rossi R. High Tibial osteotomy and anterior cruciate ligament reconstruction/revision. Clin Sports Med. 2019;38(3):417–33.
41. Cooper JD, Wang W, Prentice HA, Funahashi TT, Maletis GB. The association between Tibial slope and revision anterior cruciate ligament reconstruction in patients ≤21 years old: a matched case-control study including 317 revisions. Am J Sports Med. 2019;47(14):3330–8.
42. Lee CC, Youm YS, Cho SD, Jung SH, Bae MH, Park SJ, et al. Does posterior tibial slope affect graft rupture following anterior cruciate ligament reconstruction? Arthrosc J Arthrosc Relat Surg. 2018;34(7):2152–5.
43. Salmon LJ, Heath E, Akrawi H, Roe JP, Linklater J, Pinczewski LA. 20-year outcomes of anterior cruciate ligament reconstruction with hamstring tendon autograft: the catastrophic effect of age and posterior tibial slope. Am J Sports Med. 2018;46(3):531–43.
44. Webb JM, Salmon LJ, Leclerc E, Pinczewski LA, Roe JP. Posterior tibial slope and further anterior cruciate ligament injuries in the anterior cruciate ligament-reconstructed patient. Am J Sports Med. 2013;41(12):2800–4.
45. Herman BV, Giffin JR. High tibial osteotomy in the ACL-deficient knee with medial compartment osteoarthritis. J Orthop Traumatol. 2016;17(3):277–85.
46. Schuster P, Geßlein M, Schlumberger M, Mayer P, Richter J. The influence of tibial slope on the graft in combined high tibial osteotomy and anterior cruciate ligament reconstruction. Knee. 2018;25(4):682–91.
47. Mehl J, Otto A, Kia C, Murphy M, Obopilwe E, Imhoff FB, et al. Osseous valgus alignment and posteromedial ligament complex deficiency lead to increased ACL graft forces. Knee Surg Sports Traumatol Arthrosc. 2020;28:1119–29.
48. Mancini EJ, Kohen R, Esquivel AO, Cracchiolo AM, Lemos SE. Comparison of ACL strain in the MCL-deficient and MCL-reconstructed knee during

simulated landing in a cadaveric model. Am J Sports Med. 2017;45(5):1090–4.
49. Bonin N, Ait Si Selmi T, Donell ST, Dejour H, Neyret P. Anterior cruciate reconstruction combined with valgus upper tibial osteotomy: 12 years follow-up. Knee. 2004;11(6):431–7.
50. Noyes FR, Barber SD, Simon R. High tibial osteotomy and ligament reconstruction in varus angulated, anterior cruciate ligament-deficient knees. A two-to seven-year follow-up study. Am J Sports Med. 1993;21(1):2–12.
51. Zaffagnini S, Bonanzinga T, Grassi A, Marcheggiani Muccioli GM, Musiani C, Raggi F, et al. Combined ACL reconstruction and closing-wedge HTO for varus angulated ACL-deficient knees. Knee Surg Sports Traumatol Arthrosc. 2013;21(4):934–41.
52. Sonnery-Cottet B, Mogos S, Thaunat M, Archbold P, Fayard J-M, Freychet B, et al. Proximal tibial anterior closing wedge osteotomy in repeat revision of anterior cruciate ligament reconstruction. Am J Sports Med. 2014;42(8):1873–80.
53. Hoshino Y, Wang JH, Lorenz S, Fu FH, Tashman S. The effect of distal femur bony morphology on in vivo knee translational and rotational kinematics. Knee Surg Sports Traumatol Arthrosc. 2012;20(7):1331–8.
54. Pfeiffer TR, Burnham JM, Kanakamedala AC, Hughes JD, Zlotnicki J, Popchak A, et al. Distal femur morphology affects rotatory knee instability in patients with anterior cruciate ligament ruptures. Knee Surg Sports Traumatol Arthrosc. 2019;27(5):1514–9.
55. Magnussen RA, Reinke EK, Huston LJ, MOON Group, Hewett TE, Spindler KP. Effect of high-grade preoperative knee laxity on anterior cruciate ligament reconstruction outcomes. Am J Sports Med. 2016;44(12):3077–82.
56. Physiologic preoperative knee hyperextension is a predictor of failure in an ACL revision cohort. Am J Sports Med. 2018;46(12):2836–41.
57. Benner RW, Shelbourne KD, Gray T. The degree of knee extension does not affect postoperative stability or subsequent graft tear rate after anterior cruciate ligament reconstruction with patellar tendon autograft. Am J Sports Med. 2016;44(4):844–9.
58. Kim S-J, Moon H-K, Kim S-G, Chun Y-M, Oh K-S. Does severity or specific joint laxity influence clinical outcomes of anterior cruciate ligament reconstruction? Clin Orthop. 2010;468(4):1136–41.
59. Maletis GB, Chen J, Inacio MCS, Funahashi TT. Age-related risk factors for revision anterior cruciate ligament reconstruction: a cohort study of 21,304 patients from the Kaiser Permanente anterior cruciate ligament registry. Am J Sports Med. 2016;44(2):331–6.
60. Hettrich CM, Dunn WR, Reinke EK, MOON Group, Spindler KP. The rate of subsequent surgery and predictors after anterior cruciate ligament reconstruction: two- and 6-year follow-up results from a multicenter cohort. Am J Sports Med. 2013;41(7):1534–40.
61. Kowalchuk DA, Harner CD, Fu FH, Irrgang JJ. Prediction of patient-reported outcome after single-bundle anterior cruciate ligament reconstruction. Arthrosc J Arthrosc Relat Surg. 2009;25(5):457–63.
62. Raikin SM, Landsman JC, Alexander VA, Froimson MI, Plaxton NA. Effect of nicotine on the rate and strength of long bone fracture healing. Clin Orthop. 1998;353:231–7.
63. Porter SE, Hanley EN. The musculoskeletal effects of smoking. J Am Acad Orthop Surg. 2001;9(1):9–17.
64. Gullihorn L, Karpman R, Lippiello L. Differential effects of nicotine and smoke condensate on bone cell metabolic activity. J Orthop Trauma. 2005;19(1):17–22.
65. Novikov DA, Swensen SJ, Buza JA, Gidumal RH, Strauss EJ. The effect of smoking on ACL reconstruction: a systematic review. Phys Sportsmed. 2016;44(4):335–41.
66. Musahl V, Rahnemai-Azar AA, Costello J, Arner JW, Fu FH, Hoshino Y, et al. The influence of meniscal and anterolateral capsular injury on knee laxity in patients with anterior cruciate ligament injuries. Am J Sports Med. 2016;44(12):3126–31.
67. Ferretti A, Monaco E, Fabbri M, Maestri B, De Carli A. Prevalence and classification of injuries of anterolateral complex in acute anterior cruciate ligament tears. Arthrosc J Arthrosc Relat Surg. 2017;33(1):147–54.
68. Tavlo M, Eljaja S, Jensen JT, Siersma VD, Krogsgaard MR. The role of the anterolateral ligament in ACL insufficient and reconstructed knees on rotatory stability: a biomechanical study on human cadavers. Scand J Med Sci Sports. 2016;26(8):960–6.
69. Thein R, Boorman-Padgett J, Stone K, Wickiewicz TL, Imhauser CW, Pearle AD. Biomechanical assessment of the anterolateral ligament of the knee: a secondary restraint in simulated tests of the pivot shift and of anterior stability. J Bone Joint Surg Am. 2016;98(11):937–43.
70. Musahl V, Herbst E, Burnham JM, Fu FH. The anterolateral complex and anterolateral ligament of the knee. J Am Acad Orthop Surg. 2018;26(8):261–7.
71. Sonnery-Cottet B, Thaunat M, Freychet B, Pupim BHB, Murphy CG, Claes S. Outcome of a combined anterior cruciate ligament and anterolateral ligament reconstruction technique with a minimum 2-year follow-up. Am J Sports Med. 2015;43(7):1598–605.
72. Marcacci M, Zaffagnini S, Giordano G, Iacono F, Presti ML. Anterior cruciate ligament reconstruction associated with extra-articular tenodesis: a prospective clinical and radiographic evaluation with 10- to 13-year follow-up. Am J Sports Med. 2009;37(4):707–14.
73. Song G-Y, Hong L, Zhang H, Zhang J, Li Y, Feng H. Clinical outcomes of combined lateral extra-articular Tenodesis and intra-articular anterior cruciate ligament reconstruction in addressing high-grade pivot-shift phenomenon. Arthrosc J Arthrosc Relat Surg. 2016;32(5):898–905.

74. Kennedy JC, Stewart R, Walker DM. Anterolateral rotatory instability of the knee joint. An early analysis of the Ellison procedure. J Bone Joint Surg Am. 1978;60(8):1031–9.
75. Devitt BM, Bell SW, Ardern CL, Hartwig T, Porter TJ, Feller JA, et al. The role of lateral extra-articular Tenodesis in primary anterior cruciate ligament reconstruction: a systematic review with meta-analysis and Best-evidence synthesis. Orthop J Sports Med. 2017;5(10):2325967117731767.
76. Hewison CE, Tran MN, Kaniki N, Remtulla A, Bryant D, Getgood AM. Lateral extra-articular Tenodesis reduces rotational laxity when combined with anterior cruciate ligament reconstruction: a systematic review of the literature. Arthrosc J Arthrosc Relat Surg. 2015;31(10):2022–34.
77. Getgood AMJ, Bryant DM, Litchfield R, Heard M, McCormack RG, Rezansoff A, et al. Lateral extra-articular Tenodesis reduces failure of hamstring tendon autograft anterior cruciate ligament reconstruction: 2-year outcomes from the STABILITY study randomized clinical trial. Am J Sports Med. 2020;48(2):285–97.
78. Fonda G, Canata GL. A modified Andrews technique in antero-lateral rotatory instability of the knee. In: Surgery and arthroscopy of the knee: First European Congress of Knee Surgery and Arthroscopy, Berlin, 9–14 4 1984. Berlin: Springer Science & Business Media; 2012. p. 171.
79. Sonnery-Cottet B, Daggett M, Fayard J-M, Ferretti A, Helito CP, Lind M, et al. Anterolateral ligament expert group consensus paper on the management of internal rotation and instability of the anterior cruciate ligament—deficient knee. J Orthop Traumatol. 2017;18(2):91–106.
80. Vadalà AP, Iorio R, De Carli A, Bonifazi A, Iorio C, Gatti A, et al. An extra-articular procedure improves the clinical outcome in anterior cruciate ligament reconstruction with hamstrings in female athletes. Int Orthop. 2013;37(2):187–92.
81. Geeslin AG, Moatshe G, Chahla J, Kruckeberg BM, Muckenhirn KJ, Dornan GJ, et al. Anterolateral knee extra-articular stabilizers: a robotic study comparing anterolateral ligament reconstruction and modified Lemaire lateral extra-articular Tenodesis. Am J Sports Med. 2018;46(3):607–16.
82. Shaffer MA, Williams GN. ACL rehabilitation. In: The knee joint: surgical techniques and strategies. Berlin: Springer Science & Business Media; 2013. p. 269–90.
83. Nielsen ET, Stentz-Olesen K, de Raedt S, Jørgensen PB, Sørensen OG, Kaptein B, et al. Influence of the anterolateral ligament on knee laxity: a biomechanical cadaveric study measuring knee kinematics in 6 degrees of freedom using dynamic radiostereometric analysis. Orthop J Sports Med. 2018 Aug;6(8):2325967118789699.
84. Zhang H, Sun Y, Han X, Wang Y, Wang L, Alquhali A, et al. Simultaneous reconstruction of the anterior cruciate ligament and medial collateral ligament in patients with chronic ACL-MCL lesions: a minimum 2-year follow-up study. Am J Sports Med. 2014;42(7):1675–81.
85. Indelicato PA. Non-operative treatment of complete tears of the medial collateral ligament of the knee. J Bone Joint Surg Am. 1983;65(3):323–9.
86. Holden DL, Eggert AW, Butler JE. The nonoperative treatment of grade I and II medial collateral ligament injuries to the knee. Am J Sports Med. 1983;11(5):340–4.
87. Lundberg M, Messner K. Long-term prognosis of isolated partial medial collateral ligament ruptures. A ten-year clinical and radiographic evaluation of a prospectively observed group of patients. Am J Sports Med. 1996;24(2):160–3.
88. Reider B, Sathy MR, Talkington J, Blyznak N, Kollias S. Treatment of isolated medial collateral ligament injuries in athletes with early functional rehabilitation. A five-year follow-up study. Am J Sports Med. 1994;22(4):470–7.
89. Zaffagnini S, Bonanzinga T, Marcheggiani Muccioli GM, Giordano G, Bruni D, Bignozzi S, et al. Does chronic medial collateral ligament laxity influence the outcome of anterior cruciate ligament reconstruction?: a prospective evaluation with a minimum three-year follow-up. J Bone Joint Surg Br. 2011;93(8):1060–4.
90. Millett PJ, Pennock AT, Sterett WI, Steadman JR. Early ACL reconstruction in combined ACL-MCL injuries. J Knee Surg. 2004;17(2):94–8.
91. Grant JA, Tannenbaum E, Miller BS, Bedi A. Treatment of combined complete tears of the anterior cruciate and medial collateral ligaments. Arthrosc J Arthrosc Relat Surg. 2012;28(1):110–22.
92. Kannus P. Long-term results of conservatively treated medial collateral ligament injuries of the knee joint. Clin Orthop. 1988;226:103–12.
93. Canata GL, Chiey A, Leoni T. Surgical technique: does mini-invasive medial collateral ligament and posterior oblique ligament repair restore knee stability in combined chronic medial and ACL injuries? Clin Orthop. 2012;470(3):791–7.
94. Woo SL, Young EP, Ohland KJ, Marcin JP, Horibe S, Lin HC. The effects of transection of the anterior cruciate ligament on healing of the medial collateral ligament. A biomechanical study of the knee in dogs. J Bone Joint Surg Am. 1990;72(3):382–92.
95. Svantesson E, Hamrin Senorski E, Alentorn-Geli E, Westin O, Sundemo D, Grassi A, et al. Increased risk of ACL revision with non-surgical treatment of a concomitant medial collateral ligament injury: a study on 19,457 patients from the Swedish National Knee Ligament Registry. Knee Surg Sports Traumatol Arthrosc. 2019;27(8):2450–9.
96. Andersson C, Gillquist J. Treatment of acute isolated and combined ruptures of the anterior cruciate ligament. A long-term follow-up study. Am J Sports Med. 1992;20(1):7–12.
97. Haimes JL, Wroble RR, Grood ES, Noyes FR. Role of the medial structures in the intact and anterior cru-

ciate ligament-deficient knee. Limits of motion in the human knee. Am J Sports Med. 1994;22(3):402–9.
98. Robinson JR, Bull AMJ, Thomas RRD, Amis AA. The role of the medial collateral ligament and posteromedial capsule in controlling knee laxity. Am J Sports Med. 2006;34(11):1815–23.
99. Fischer RA, Arms SW, Johnson RJ, Pope MH. The functional relationship of the posterior oblique ligament to the medial collateral ligament of the human knee. Am J Sports Med. 1985;13(6):390–7.
100. Hillard-Sembell D, Daniel DM, Stone ML, Dobson BE, Fithian DC. Combined injuries of the anterior cruciate and medial collateral ligaments of the knee. Effect of treatment on stability and function of the joint. J Bone Joint Surg Am. 1996;78(2):169–76.
101. Borden PS, Kantaras AT, Caborn DNM. Medial collateral ligament reconstruction with allograft using a double-bundle technique. Arthrosc J Arthrosc Relat Surg. 2002;18(4):E19.
102. Gallo RA, Kozlansky G, Bonazza N, Warren RF. Combined anterior cruciate ligament and medial collateral ligament reconstruction using a single Achilles tendon allograft. Arthrosc Tech. 2017;6(5):e1821–7.
103. Dong J, Wang XF, Men X, Zhu J, Walker GN, Zheng XZ, et al. Surgical treatment of acute grade III medial collateral ligament injury combined with anterior cruciate ligament injury: anatomic ligament repair versus triangular ligament reconstruction. Arthrosc J Arthrosc Relat Surg. 2015;31(6):1108–16.
104. Kitamura N, Ogawa M, Kondo E, Kitayama S, Tohyama H, Yasuda K. A novel medial collateral ligament reconstruction procedure using semitendinosus tendon autograft in patients with multiligamentous knee injuries: clinical outcomes. Am J Sports Med. 2013;41(6):1274–81.
105. Marchant MH, Tibor LM, Sekiya JK, Hardaker WT, Garrett WE, Taylor DC. Management of medial-sided knee injuries, part 1: medial collateral ligament. Am J Sports Med. 2011;39(5):1102–13.
106. Lind M, Jakobsen BW, Lund B, Hansen MS, Abdallah O, Christiansen SE. Anatomical reconstruction of the medial collateral ligament and posteromedial corner of the knee in patients with chronic medial collateral ligament instability. Am J Sports Med. 2009;37(6):1116–22.
107. LaPrade RF, Resig S, Wentorf F, Lewis JL. The effects of grade III posterolateral knee complex injuries on anterior cruciate ligament graft force. A biomechanical analysis. Am J Sports Med. 1999;27(4):469–75.
108. Bonanzinga T, Signorelli C, Lopomo N, Grassi A, Neri MP, Filardo G, et al. Biomechanical effect of posterolateral corner sectioning after ACL injury and reconstruction. Knee Surg Sports Traumatol Arthrosc. 2015;23(10):2918–24.
109. Filli L, Rosskopf AB, Sutter R, Fucentese SF, Pfirrmann CWA. MRI predictors of posterolateral corner instability: a decision tree analysis of patients with acute anterior cruciate ligament tear. Radiology. 2018;289(1):170–80.
110. Shon O-J, Park J-W, Kim B-J. Current concepts of posterolateral corner injuries of the knee. Knee Surg Relat Res. 2017;29(4):256–68.
111. Dean RS, RF LP. ACL and posterolateral corner injuries. Curr Rev Musculoskelet Med. 2019;13:123–32.
112. Arciero RA. Anatomic posterolateral corner knee reconstruction. Arthrosc J Arthrosc Relat Surg. 2005;21(9):1147.
113. LaPrade RF, Johansen S, Wentorf FA, Engebretsen L, Esterberg JL, Tso A. An analysis of an anatomical posterolateral knee reconstruction: an in vitro biomechanical study and development of a surgical technique. Am J Sports Med. 2004;32(6):1405–14.
114. Treme GP, Salas C, Ortiz G, Gill GK, Johnson PJ, Menzer H, et al. A biomechanical comparison of the Arciero and LaPrade reconstruction for posterolateral corner knee injuries. Orthop J Sports Med. 2019;7(4):2325967119838251.
115. Westermann RW, Marx RG, Spindler KP, Huston LJ, MOON Knee Group, Amendola A, et al. No difference between posterolateral corner repair and reconstruction with concurrent ACL surgery: results from a prospective multicenter cohort. Orthop J Sports Med. 2019;7(7):2325967119861062.
116. Lunden JB, Bzdusek PJ, Monson JK, Malcomson KW, Laprade RF. Current concepts in the recognition and treatment of posterolateral corner injuries of the knee. J Orthop Sports Phys Ther. 2010;40(8):502–16.
117. Panigrahi R, Kumari Mahapatra A, Priyadarshi A, Singha Das D, Palo N, Ranjan BM. Outcome of simultaneous arthroscopic anterior cruciate ligament and posterior cruciate ligament reconstruction with hamstring tendon autograft: a multicenter prospective study. Asian J Sports Med. 2016;7(1):e29287.
118. Robertson A, Nutton RW, Keating JF. Dislocation of the knee. J Bone Joint Surg Br. 2006;88(6):706–11.
119. de Carvalho RT, Franciozi CE, Itami Y, McGarry MH, Ingham SJM, Abdalla RJ, et al. Bicruciate lesion biomechanics, part 1-diagnosis: translations over 15 mm at 90° of knee flexion are indicative of a complete tear. Knee Surg Sports Traumatol Arthrosc. 2019;27(9):2927–35.
120. Shapiro MS, Freedman EL. Allograft reconstruction of the anterior and posterior cruciate ligaments after traumatic knee dislocation. Am J Sports Med. 1995;23(5):580–7.
121. Hayashi R, Kitamura N, Kondo E, Anaguchi Y, Tohyama H, Yasuda K. Simultaneous anterior and posterior cruciate ligament reconstruction in chronic knee instabilities: surgical concepts and clinical outcome. Knee Surg Sports Traumatol Arthrosc. 2008;16(8):763–9.
122. Vicenti G, Solarino G, Carrozzo M, De Giorgi S, Moretti L, De Crescenzo A, et al. Major concern in the multiligament-injured knee treatment: a systematic review. Injury. 2019;50(Suppl 2):S89–94.

123. Fanelli GC, Giannotti BF, Edson CJ. Arthroscopically assisted combined posterior cruciate ligament/posterior lateral complex reconstruction. Arthrosc J Arthrosc Relat Surg. 1996;12(5):521–30.
124. Fanelli GC, Edson CJ. Arthroscopically assisted combined anterior and posterior cruciate ligament reconstruction in the multiple ligament injured knee: 2- to 10-year follow-up. Arthrosc J Arthrosc Relat Surg. 2002;18(7):703–14.
125. Balasingam S, Sernert N, Magnusson H, Kartus J. Patients with concomitant intra-articular lesions at index surgery deteriorate in their knee injury and osteoarthritis outcome score in the long term more than patients with isolated anterior cruciate ligament rupture: a study from the Swedish National Anterior Cruciate Ligament Register. Arthrosc J Arthrosc Relat Surg. 2018;34(5):1520–9.
126. Patterson BE, Culvenor AG, Barton CJ, Guermazi A, Stefanik JJ, Crossley KM. Patient-reported outcomes one to five years after anterior cruciate ligament reconstruction: the effect of combined injury and associations with osteoarthritis features defined on magnetic resonance imaging. Arthritis Care Res. 2020;72(3):412–22.
127. Papageorgiou CD, Gil JE, Kanamori A, Fenwick JA, Woo SL, Fu FH. The biomechanical interdependence between the anterior cruciate ligament replacement graft and the medial meniscus. Am J Sports Med. 2001;29(2):226–31.
128. Baratz ME, Fu FH, Mengato R. Meniscal tears: the effect of meniscectomy and of repair on intraarticular contact areas and stress in the human knee. A preliminary report. Am J Sports Med. 1986;14(4):270–5.
129. Joshi A, Singh N, Pradhan I, Basukala B, Banskota AK. A definition of significant instability and a scoring system for predicting meniscal tears in ACL-deficient knees. Orthop J Sports Med. 2019;7(8):2325967119866732.
130. Zheng T, Song G-Y, Feng H, Zhang H, Li Y, Li X, et al. Lateral meniscus posterior root lesion influences anterior tibial subluxation of the lateral compartment in extension after anterior cruciate ligament injury. Am J Sports Med. 2020;48(4):838–46.
131. Razi M, Mortazavi SMJ. Save the meniscus, a good strategy to preserve the knee. Arch Bone Jt Surg. 2020;8(1):1–4.
132. Curado J, Hulet C, Hardy P, Jenny J-Y, Rousseau R, Lucet A, et al. Very long-term osteoarthritis rate after anterior cruciate ligament reconstruction: 182 cases with 22-year' follow-up. Orthop Traumatol Surg Res. 2020;106(3):459–63.
133. Vavken P, Tepolt FA, Kocher MS. Concurrent meniscal and chondral injuries in pediatric and adolescent patients undergoing ACL reconstruction. J Pediatr Orthop. 2018;38(2):105–9.
134. Song G-Y, Liu X, Zhang H, Wang Q-Q, Zhang J, Li Y, et al. Increased medial meniscal slope is associated with greater risk of ramp lesion in noncontact anterior cruciate ligament injury. Am J Sports Med. 2016;44(8):2039–46.
135. Gaillard R, Magnussen R, Batailler C, Neyret P, Lustig S, Servien E. Anatomic risk factor for meniscal lesion in association with ACL rupture. J Orthop Surg. 2019;14(1):242.
136. Brambilla L, Pulici L, Carimati G, Quaglia A, Prospero E, Bait C, et al. Prevalence of associated lesions in anterior cruciate ligament reconstruction: correlation with surgical timing and with patient age, sex, and body mass index. Am J Sports Med. 2015;43(12):2966–73.
137. Seo J-H, Li G, Shetty GM, Kim J-H, Bae J-H, Jo M-L, et al. Effect of repair of radial tears at the root of the posterior horn of the medial meniscus with the pullout suture technique: a biomechanical study using porcine knees. Arthrosc J Arthrosc Relat Surg. 2009;25(11):1281–7.
138. Ahmed AM, Burke DL. In-vitro measurement of static pressure distribution in synovial joints--part I: tibial surface of the knee. J Biomech Eng. 1983;105(3):216–25.
139. Paxton ES, Stock MV, Brophy RH. Meniscal repair versus partial meniscectomy: a systematic review comparing reoperation rates and clinical outcomes. Arthrosc J Arthrosc Relat Surg. 2011;27(9):1275–88.
140. Nepple JJ, Dunn WR, Wright RW. Meniscal repair outcomes at greater than five years: a systematic literature review and meta-analysis. J Bone Joint Surg Am. 2012;94(24):2222–7.
141. Doral MN, Bilge O, Huri G, Turhan E, Verdonk R. Modern treatment of meniscal tears. EFORT Open Rev. 2018;3(5):260–8.
142. Beaufils P, Pujol N. Meniscal repair: technique. Orthop Traumatol Surg Res. 2018;104(1, Supplement):S137–45.
143. Tenuta JJ, Arciero RA. Arthroscopic evaluation of meniscal repairs. Factors that effect healing. Am J Sports Med. 1994;22(6):797–802.
144. Saltzman BM, Cotter EJ, Wang KC, Rice R, Manning BT, Yanke AB, et al. Arthroscopically repaired bucket-handle meniscus tears: patient demographics, postoperative outcomes, and a comparison of success and failure cases. Cartilage. 2020;11(1):77–87.
145. Pujol N, Beaufils P. Healing results of meniscal tears left in situ during anterior cruciate ligament reconstruction: a review of clinical studies. Knee Surg Sports Traumatol Arthrosc. 2009;17(4):396–401.
146. Duchman KR, Westermann RW, Spindler KP, Reinke EK, Huston LJ, Amendola A, et al. The fate of meniscus tears left in situ at the time of anterior cruciate ligament reconstruction: a 6-year follow-up study from the MOON cohort. Am J Sports Med. 2015;43(11):2688–95.
147. Pujol N, Lorbach O. Meniscal repair: results. In: Surgery of the meniscus. Berlin: Springer; 2016. p. 343–56.
148. Cavanaugh JT, Killian SE. Rehabilitation following meniscal repair. Curr Rev Musculoskelet Med. 2012;5(1):46–58.

149. Eberbach H, Zwingmann J, Hohloch L, Bode G, Maier D, Niemeyer P, et al. Sport-specific outcomes after isolated meniscal repair: a systematic review. Knee Surg Sports Traumatol Arthrosc. 2018;26(3):762–71.
150. Mouton C, Magosch A, Pape D, Hoffmann A, Nührenbörger C, Seil R. Ramp lesions of the medial meniscus are associated with a higher grade of dynamic rotatory laxity in ACL-injured patients in comparison to patients with an isolated injury. Knee Surg Sports Traumatol Arthrosc. 2019;28(4):1023–8.
151. Sonnery-Cottet B, Conteduca J, Thaunat M, Gunepin FX, Seil R. Hidden lesions of the posterior horn of the medial meniscus: a systematic arthroscopic exploration of the concealed portion of the knee. Am J Sports Med. 2014;42(4):921–6.
152. Kim SH, Lee SH, Kim K-I, Yang JW. Diagnostic accuracy of sequential arthroscopic approach for ramp lesions of the posterior horn of the medial meniscus in anterior cruciate ligament-deficient knee. Arthrosc J Arthrosc Relat Surg. 2018;34(5):1582–9.
153. Balazs GC, Greditzer HG, Wang D, Marom N, Potter HG, Marx RG, et al. Ramp lesions of the medial meniscus in patients undergoing primary and revision ACL reconstruction: prevalence and risk factors. Orthop J Sports Med. 2019;7(5):2325967119843509.
154. Sonnery-Cottet B, Praz C, Rosenstiel N, Blakeney WG, Ouanezar H, Kandhari V, et al. Epidemiological evaluation of meniscal ramp lesions in 3214 anterior cruciate ligament-injured knees from the SANTI study group database: a risk factor analysis and study of secondary meniscectomy rates following 769 ramp repairs. Am J Sports Med. 2018;46(13):3189–97.
155. Kumar NS, Spencer T, Cote MP, Arciero RA, Edgar C. Is edema at the posterior medial Tibial plateau indicative of a ramp lesion? An examination of 307 patients with anterior cruciate ligament reconstruction and medial meniscal tears. Orthop J Sports Med. 2018;6(6):2325967118780089.
156. Malatray M, Raux S, Peltier A, Pfirrmann C, Seil R, Chotel F. Ramp lesions in ACL deficient knees in children and adolescent population: a high prevalence confirmed in intercondylar and posteromedial exploration. Knee Surg Sports Traumatol Arthrosc. 2018;26(4):1074–9.
157. Hatayama K, Terauchi M, Saito K, Aoki J, Nonaka S, Higuchi H. Magnetic resonance imaging diagnosis of medial meniscal ramp lesions in patients with anterior cruciate ligament injuries. Arthrosc J Arthrosc Relat Surg. 2018;34(5):1631–7.
158. Buyukdogan K, Laidlaw MS, Miller MD. Meniscal ramp lesion repair by a trans-septal portal technique. Arthrosc Tech. 2017;6(4):e1379–86.
159. Gillquist J, Hagberg G. A new modification of the technique of arthroscopy of the knee joint. Acta Chir Scand. 1976;142(2):123–30.
160. Negrín R, Reyes NO, Iñiguez M, Pellegrini JJ, Wainer M, Duboy J. Meniscal ramp lesion repair using an all-inside technique. Arthrosc Tech. 2018;7(3):e265–70.
161. Gülenç B, Kemah B, Yalçın S, Sayar Ş, Korkmaz O, Erdil M. Surgical treatment of meniscal RAMP lesion. J Knee Surg. 2020;33(3):255–9.
162. Sommerfeldt M, Raheem A, Whittaker J, Hui C, Otto D. Recurrent instability episodes and meniscal or cartilage damage after anterior cruciate ligament injury: a systematic review. Orthop J Sports Med. 2018;6(7):2325967118786507.
163. de Campos GC, Nery W, Teixeira PEP, Araujo PH, de Alves W. Association between meniscal and chondral lesions and timing of anterior cruciate ligament reconstruction. Orthop J Sports Med. 2016;4(10):2325967116669309.
164. Ulstein S, Årøen A, Engebretsen L, Forssblad M, Lygre SHL, Røtterud JH. Effect of concomitant cartilage lesions on patient-reported outcomes after anterior cruciate ligament reconstruction: a nationwide cohort study from Norway and Sweden of 8470 patients with 5-year follow-up. Orthop J Sports Med. 2018 Jul;6(7):2325967118786219.
165. Everhart JS, DiBartola AC, Swank K, Pettit R, Hughes L, Lewis C, et al. Cartilage damage at the time of anterior cruciate ligament reconstruction is associated with weaker quadriceps function and lower risk of future ACL injury. Knee Surg Sports Traumatol Arthrosc. 2020;28(2):576–83.
166. Filardo G, de Caro F, Andriolo L, Kon E, Zaffagnini S, Marcacci M. Do cartilage lesions affect the clinical outcome of anterior cruciate ligament reconstruction? A systematic review. Knee Surg Sports Traumatol Arthrosc. 2017;25(10):3061–75.
167. Ulstein S, Årøen A, Engebretsen L, Forssblad M, Lygre SHL, Røtterud JH. A controlled comparison of microfracture, debridement, and no treatment of concomitant full-thickness cartilage lesions in anterior cruciate ligament-reconstructed knees: a Nationwide prospective cohort study from Norway and Sweden of 368 patients with 5-year follow-up. Orthop J Sports Med. 2018;6(8):2325967118787767.
168. Falah M, Nierenberg G, Soudry M, Hayden M, Volpin G. Treatment of articular cartilage lesions of the knee. Int Orthop. 2010;34(5):621–30.
169. Dewan AK, Gibson MA, Elisseeff JH, Trice ME. Evolution of autologous chondrocyte repair and comparison to other cartilage repair techniques. Biomed Res Int. 2014;2014:272481.
170. Moyad TF. Cartilage injuries in the adult knee. Cartilage. 2011;2(3):226–36.
171. Thrush C, Porter TJ, Devitt BM. No evidence for the most appropriate postoperative rehabilitation protocol following anterior cruciate ligament reconstruction with concomitant articular cartilage lesions: a systematic review. Knee Surg Sports Traumatol Arthrosc. 2018;26(4):1065–73.

Technique Corner: ACLR Optimal Tunnel Placement: How to Get There?

4

Konsei Shino, Ryohei Uchida, Hiroyuki Yokoi, Tomoki Ohori, and Tatsuo Mae

4.1 Introduction

Apart from tenodesis surgery, ACLR is aimed to stabilize the knee without loss of motion. According to our biomechanical studies, the more anatomically simulated the grafts are, the more efficiently they function to stabilize the knee [1, 2]. Thus it is desirable to optimally place the tunnels for the proper graft to closely mimic the native ACL by creating the tunnel apertures inside the attachment areas.

4.2 Key Issues for Anatomical ACL Reconstruction

The following issues should be addressed to achieve ACLR goal to stabilize the knee without loss of motion.

1. Graft choice and preparation

   The graft should be chosen and prepared to mimic the native ACL with oblong cross section. Thus our graft choice is a bone patellar tendon-bone (BPTB) graft with rectangular cross section for a single rectangular tunnel or multi-stranded hamstring tendon (MSHT) graft for multiple small round tunnels [3] (Fig. 4.1).
2. Bony landmark strategy

   The direct insertion areas should be exactly identified. The native ACL has direct type insertions along with fibrous cartilage layer to the femur as well as to the tibia. Hence, the areas are concave in bony surface. Thus the bony landmark strategy is advocated to strictly delineate the areas, as they are surrounded by the bony landmarks: the resident's ridge and the posterior cartilage border on the femur; the anterior ridge, the medial intercondylar ridge, the central intercondylar ridge, and the intercondylar eminence on the tibia [4–12].
3. Containing tunnel apertures inside the attachment areas

   The tunnel apertures should be created strictly inside the attachment areas. This makes it possible not only for the graft to mimic the native ACL but to reduce tunnel widening, as the bone underneath the attachment areas consists of thicker cortex [13]. Reaming of single 10-mm round tibial tunnel cause not only violate the border of the tibial attachment area but cause iatrogenic injury to the anterior horn of the lateral meniscus on the tibia [14].

K. Shino (✉) · R. Uchida · H. Yokoi · T. Ohori
Sports Orthopaedic Center, Yukioka Hospital, Osaka, Japan

T. Mae
Department of Orthopaedic Surgery, Osaka University Medical School, Osaka, Japan

N. Nakamura et al. (eds.), *Advances in Knee Ligament and Knee Preservation Surgery*,
https://doi.org/10.1007/978-3-030-84748-7_4

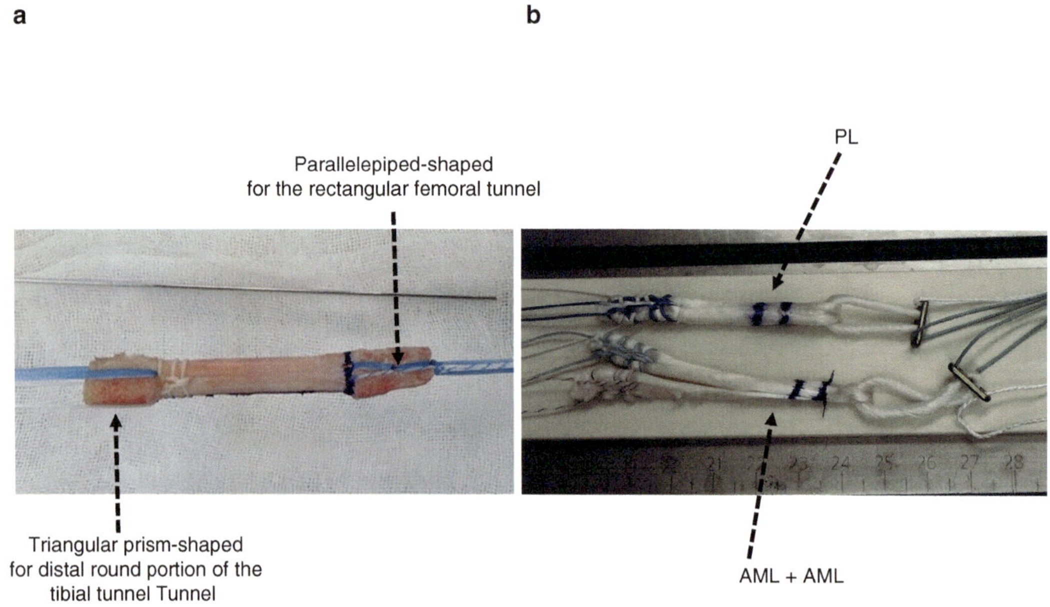

Fig. 4.1 Prepared grafts. (**a**) A bone-patellar tendon-bone (BPTB) graft for the anatomical rectangular tunnel ACLR; (**b**) Multi-stranded hamstring tendon (MSHT) grafts for the anatomical triple-bundle ACLR

4. Graft size

   The whole attachment areas may not necessarily be totally filled with grafts. The autogenous tendon grafts are greater than the native ACL in mechanical properties, and they become hypertrophic after implantation [15, 16]. Thus the grafts with smaller cross-sectional area compared to the attachment areas are sufficient.

4.3 Anatomical Bony Landmarks for the Attachment Areas

4.3.1 Femoral Attachment Area

Iwahashi et al. histologically demonstrated that the direct insertion of the ligament to the femur is located as a crescent-shaped fovea at superoposterior margin of the lateral wall of the intercondylar notch, and provided the area on 3-D CT image by reconstituting the oblique-axial CT sections [6]. Thus, the femoral insertion area of crescent shape has the following landmarks: the resident's ridge, anteriorly; posterior cartilage border, posteriorly [5–8] (Fig. 4.2).

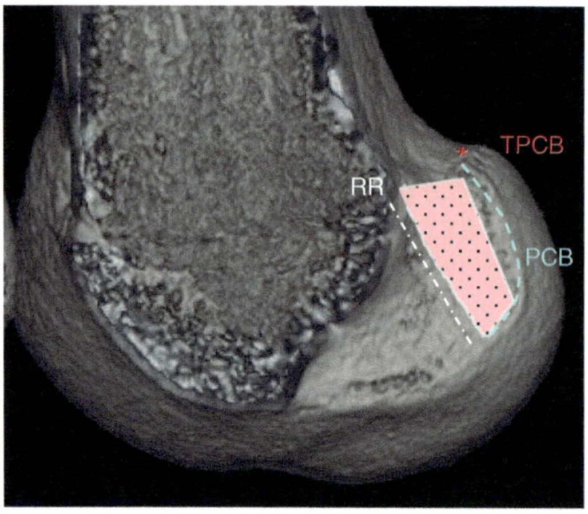

Fig. 4.2 ACL femoral attachment area of crescent/trapezoid shape on 3-D CT of the right knee. There are 2 landmarks to identify the area: *RR* the resident's ridge, *PCB* posterior cartilage border. TPCB, Top of PCB (*) is extremely important to arthroscopically identify the area

4.3.2 Tibial Attachment Area

Berg pointed out Parsons' knob (tuberculum intercondylare tertium) as the anterior border of the tibial insertion, while it is called "anterior ridge of the tibia" by Tensho [9]. Purnell and

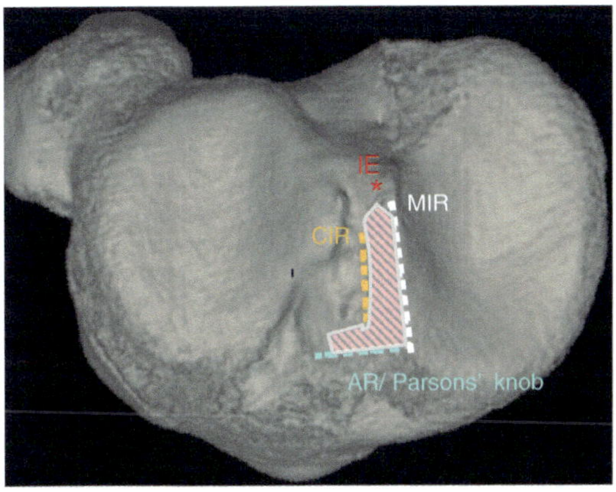

Fig. 4.3 ACL tibial attachment area of boot shape on 3-D CT of the right knee. There are 4 bony landmarks to identify the area: *MIR* medial intercondylar ridge, *AR* anterior ridge, *CIR* central intercondylar ridge, *IE* intercondylar eminence

Clancy showed that the medial intercondylar ridge could be used as the bony landmark for medial border of the tibial insertion [5]. Siebold et al. have clarified the relationship in tibial insertion between ACL and the anterior horn of the lateral meniscus [10]. Yonetani and Kusano found a distinct ridge bisecting the slope between the ACL and the anterior horn of the lateral meniscus, and named it "the central intercondylar ridge" [8, 9]. Thus, there are four landmarks for the tibial attachment area: the anterior ridge, anteriorly; the intercondylar eminence, posteriorly; the medial intercondylar ridge, medially; the central intercondylar ridge/the anterior horn of the lateral meniscus, laterally (Fig. 4.3).

4.4 Arthroscopic Delineation of the Attachment Areas Based on the Bony Landmark Strategy

4.4.1 Three Portal Technique

The anteromedial (AM) portal is an essential viewing portal to visualize the femoral or tibial attachment areas with a 45-degree oblique scope. The areas could be nearly-perpendicularly observed by rotating the scope. This makes it possible to minimize error in visualization.

The other two portals for instruments are the anterolateral (AL) portal and the far anteromedial (FAM) portal located 2–2.5 cm posterior to the anteromedial portal and just above the medial meniscus [17].

4.4.2 Femoral Attachment Area

Viewing laterally the posterior third of the lateral wall of the notch via the AM portal, the fibrous tissues including ACL stump on superior-posterior half of the lateral wall of the intercondylar notch is thoroughly removed using a radiofrequency (RF) device through the FAM portal. While mechanical shavers could be utilized, care should be taken to preserve subtle undulation of the bony surface around the attachment area. After cleaning-up, the crescent-shaped attachment area is clearly delineated by the resident's ridge, anteriorly; posterior cartilage border, posteriorly [8] (Fig. 4.4).

Even if the resident's ridge could not be clearly identified, the proximal border of the posterior cartilage margin could be used to assume the attachment area, as the long axis of the area or the ridge forms an angle of 31° to the distal femoral axis [8].

4.4.3 Tibial Attachment Area

Viewing down the ACL tibial remnant and/or attachment area via the AM portal, the stump is roughly cut to 3–5 mm in length with a mechanical shaver. Then, the fibrous tissues including ACL stump is thoroughly removed using a RF device through the AL portal while preserving the anterior horn of the lateral meniscus. After cleaning-up, the boot-shaped attachment area is clearly delineated by the anterior ridge/Parsons' knob, anteriorly; intercondylar eminence, posteriorly; the medial intercondylar ridge, medially; the central intercondylar ridge, laterally [5, 9–12]; (Fig. 4.5).

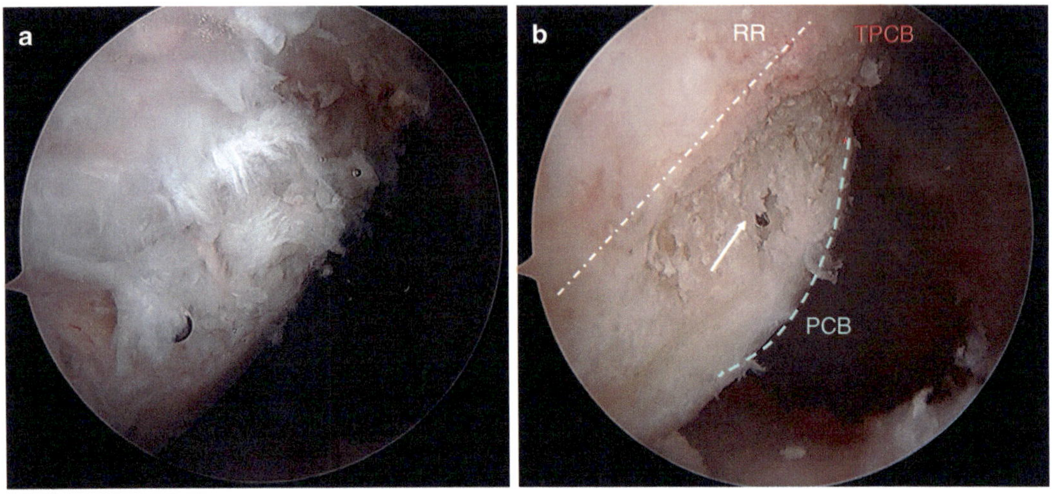

Fig. 4.4 Arthroscopic Views of the ACL femoral attachment area of the right knee through the anteromedial portal with a 45-degree oblique arthroscope directed laterally. (**a**) View before clearage of the fibrous tissue around the area. (**b**) View after clearage of the fibrous tissue around the area using RF device. There are three landmarks to identify the area: *RR* resident's ridge, *PCB* posterior cartilage border, *TPCB* Top of PCB (*). Note the tip of the guide pin at the center of the area (arrow)

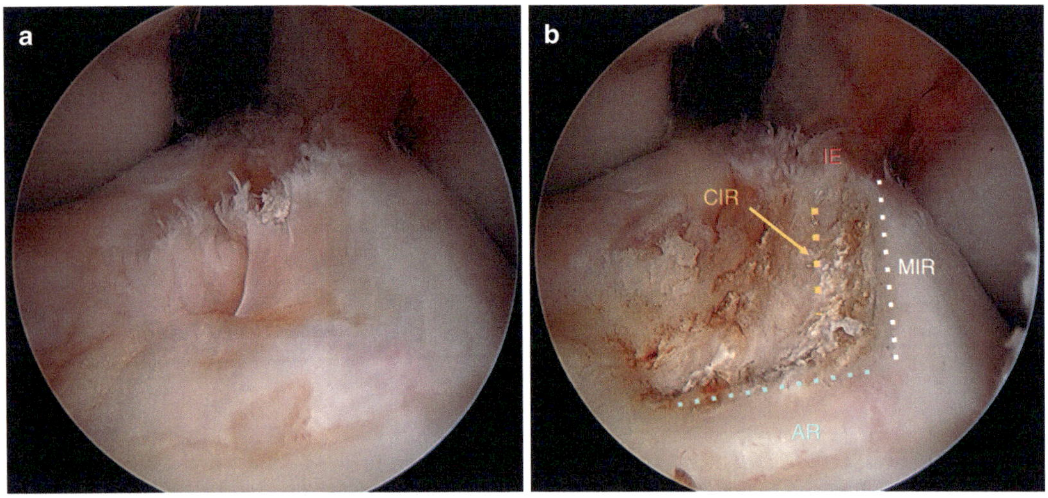

Fig. 4.5 Arthroscopic views of the ACL tibial stump or attachment area via the anteromedial portal with a 45-degree oblique arthroscope directed down. (**a**) ACL stump after roughly shaving. (**b**) View after clearage of the fibrous tissue around the area using RF device. The boot-shaped ACL tibial attachment area is clearly observed. The area has the following landmarks: *MIR* medial intercondylar ridge, *AR* anterior ridge, *CIR* central intercondylar ridge, *IE* intercondylar eminence

It is our recommendation to take plain lateral radiograph to double-check location of the tip of the guide pin (Fig. 4.6).

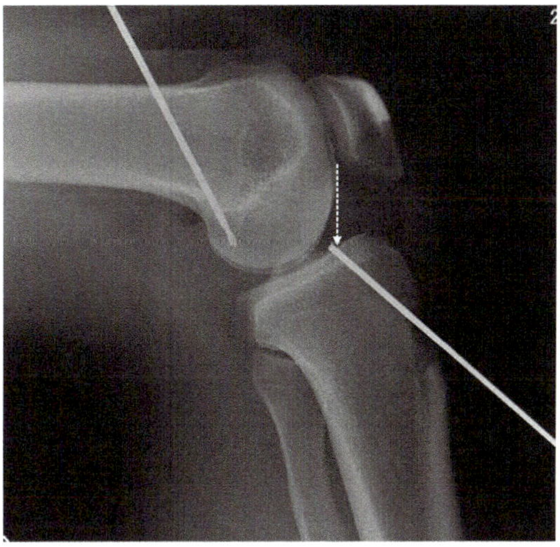

Fig. 4.6 Intra-operative lateral radiograph to double-check location of the guide pin. Note the tip of the guide pin is located in the center of the tibial attachment area (dotted arrow)

4.5 Creation of Anatomical Tunnels

To consistently create the robust femoral tunnel, outside-in approach is recommended, while small additional incisions are required [18, 19]. Once the femoral attachment area is delineated, a femoral drill guide is introduced via the AL portal to get access to the area. Two guide pins are drilled from the lateral femoral cortex to the attachment area, and then overdrilled with 4.5 to 6-mm drill bit. For the BPTB graft, the two drill holes are dilated into one rectangular tunnel (Fig. 4.7a). For MSHT graft, the two drill holes are used separately (Fig. 4.7b). While inside-out approach via FAM portal could be used if the knee could be flexed over 130°, the graft fixation on the femoral cortex may be compromised by softer cortex of the distal epiphysis.

For the tibial attachment area, the FAM portal is used for the tibial drill guide. Two to three guide pins are drilled from the medial tibial

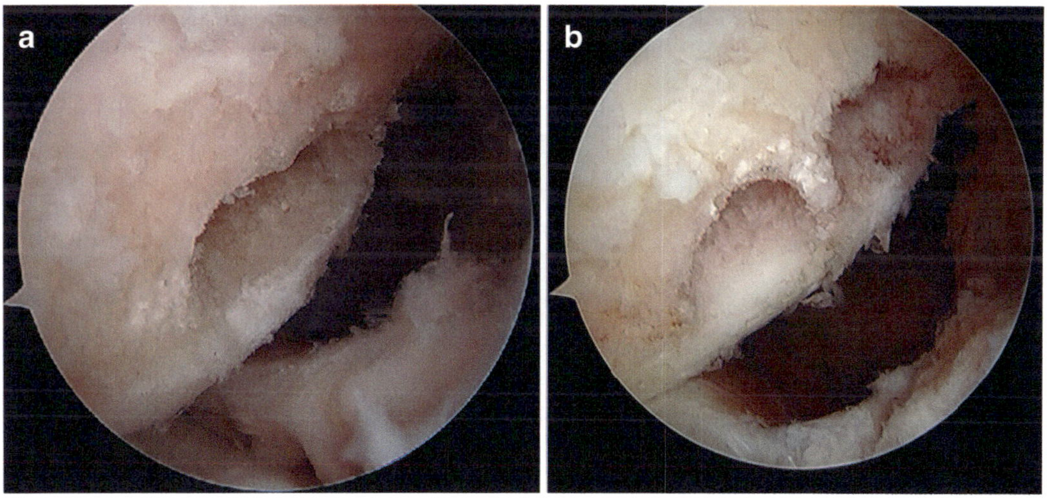

Fig. 4.7 Arthroscopic views of the anatomical femoral tunnels through the anteromedial portal. (**a**) A rectangular femoral tunnel for BPTB graft; (**b**) two round femoral tunnels for MSHT graft

metaphysis medial to the tibial tubercle to the area, and then overdrilled with 4.5- to 6-mm drill bit according to the diameter of the grafts. For the BPTB graft, the longitudinal two drill holes along the medial intercondylar ridge are dilated into one rectangular tunnel (Fig. 4.8a). For MSHT graft, the three drill holes are used separately (Fig. 4.8b).

4.6 Impingement-Free Grafting

After the tunnels are created inside the attachment areas, the proper graft: a BPTB or MSHT graft is placed according to the fiber arrangement inside the native ACL. Then the grafts do not cause impingement to the notch or the posterior cruciate ligament without notchplasty (Fig. 4.9) [3].

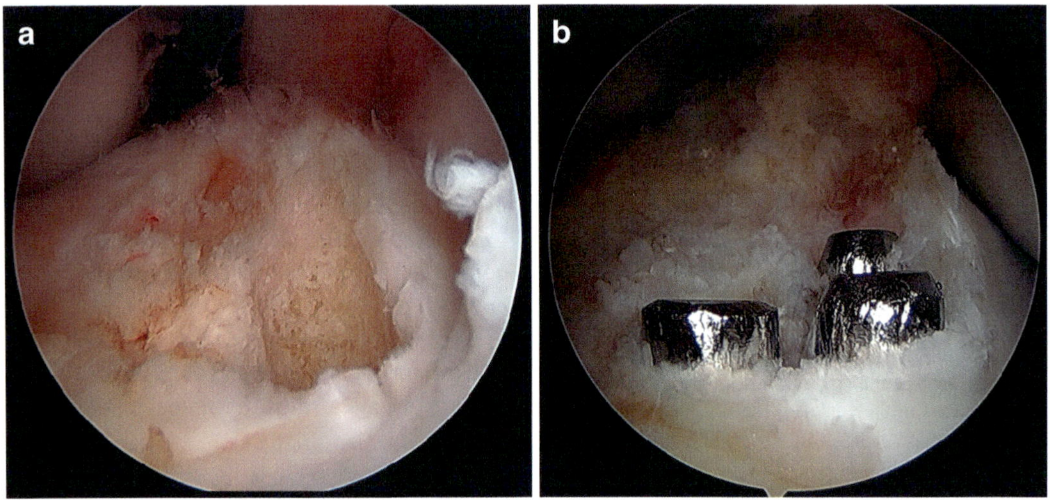

Fig. 4.8 Arthroscopic views of the anatomical tibial tunnels through the anteromedial portal. (**a**) A rectangular tibial tunnel for BPTB graft; (**b**) three round tibial tunnels for MSHT graft. A Steinmann pin is inserted into each hole

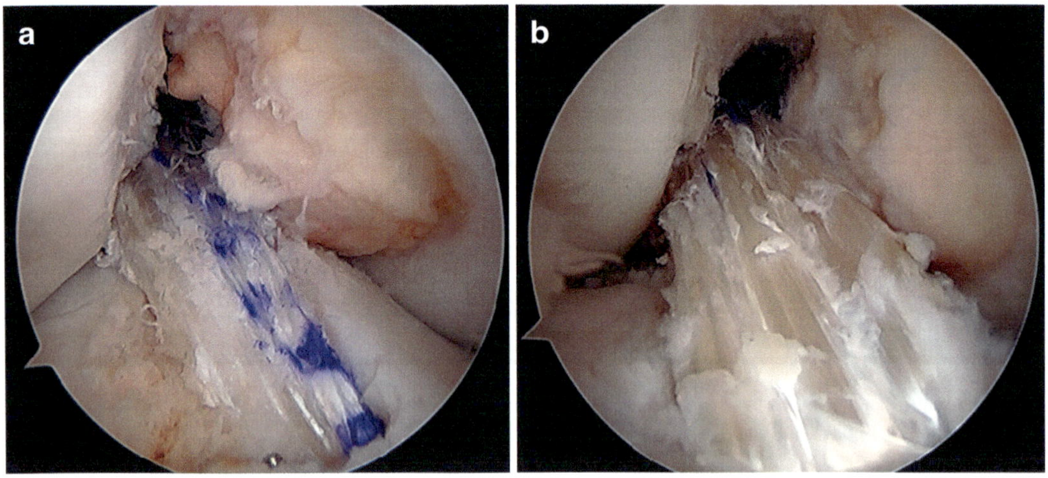

Fig. 4.9 Arthroscopic views of the anatomically-oriented grafts through the anterolateral portal: (**a**) single BPTB graft; (**b**) MSHT triple-bundle graft. Note that the grafts do not cause impingement to the notch or the posterior cruciate ligament without notchplasty

References

1. Suzuki T, Shino K, Otsubo H, Suzuki D, Mae T, Fujimiya M, Yamashita T, Fujie H. Biomechanical comparison between the rectangular -tunnel and the round-tunnel anterior cruciate ligament reconstruction procedures with a bone-patellar tendon-bone graft. Arthroscopy. 2014;30:1294–302.
2. Suzuki T, Shino K, Yamakawa S, Otsubo H, Suzuki D, Matsumura T, Fujimiya M, Fujie H, Yamashita T. A biomechanical comparison of single-, double- and triple-bundle ACL reconstructions using a hamstring tendon graft. Arthroscopy. 2019;35:896–905.
3. Shino K, Mae T, Tachibana Y. Anatomic ACL reconstruction: rectangular tunnel/bone–patellar tendon–bone or triple-bundle/semitendinosus tendon grafting. J Orthop Sci. 2015;20:457–68.
4. Berg EE. Parson's knob (tuberculum Intercondylare Tertium)—a guide to Tibial anterior cruciate ligament insertion. Clin Orthop Relat Res. 1993;292:229–31.
5. Purnell ML, Larson AI, Clancy WG. Anterior cruciate ligament insertions on the tibia and femur and their relationships to critical bony landmarks using high-resolution volume-rendering computed tomography. Am J Sports Med. 2008;36:2083–90.
6. Iwahashi T, Shino K, Nakata K, Otsubo H, Suzuki T, Amano H, Nakamura N. Direct ACL insertion to the femur assessed by histology and three-dimensional volume-rendered computed tomography. Arthroscopy. 2010;26(9 Suppl):S13–20.
7. Otsubo H, Shino K, Suzuki D, Suzuki T, Kamiya T, Watanabe K, Fujimiya M, Iwahashi T, Yamashita T. The arrangement and the attachment areas of three ACL bundles. Knee Surg Sports Traumatol Arthrosc. 2012;20:127–34.
8. Shino K, Suzuki T, Iwahashi T, Mae T, Nakata K, Nakamura N, Nakagawa S. The resident's ridge as an arthroscopic landmark for anatomical femoral tunnel drilling in ACL reconstruction. Knee Surg Sports Traumatol Arthosc. 2010;18:1164–8.
9. Tensho K, Shimodaira H, Aoki T, Narita N, Kato H, Kakegawa A, Fukushima N, Moriizumi T, FUjii M, Fujinaga Y, Saito N. Bony landmarks of the anterior cruciate ligament tibial footprint: a detailed analysis comparing 3-dimentional computed tomography images to visual and histological evaluations. Am J Sports Med. 2014;42:1433–40.
10. Siebold R, Schumacher P, Fernandez F, Smigielski R, Fink C, Brehmer A, Kirsch J. Flat midsubstance of the anterior cruciate ligament with tibial "C"-shaped insertion site. Knee Surg Sports Traumatol Arthosc. 2015;23:3136–42.
11. Kusano M, Yonetani Y, Mae T, Nakata K, Yoshikawa H, Shino K. Tibial insertions of the anterior cruciate ligament and the anterior horn of the lateral meniscus: a histological and computed tomographic study. Knee. 2017;24:782–91.
12. Yonetani Y, Kusano M, Kinugasa K, Tsujii A, Kinigasa K, Hamada M, Shino K. Tibial insertion of the anterior cruciate ligament and anterior horn of the lateral meniscus share the lateral slope of the medial intercondylar ridge: a computed tomography study in a young, healthy population. Knee. 2019;26:612–8.
13. Hutchinson MR, Ash SA. Resident's ridge: assessing the cortical thickness of the lateral wall and roof of the intercondylar notch. Arthroscopy. 2003;19:931–5.
14. LaPrade CM, Smith SD, Rasmussen MT, Wijdicks CA, Engebretsen L, Feagin JA, LaPrade RF. Consequences of tibial tunnel reaming on the meniscal roots during cruciate ligament reconstruction in a cadaveric model, part 1: the anterior cruciate ligament. Am J Sports Med. 2015;43:200–6.
15. Hamada M, Shino K, Horibe S, Mitsuoka T, Toritsuka Y, Nakamura N. Changes in cross-sectional area of hamstring anterior cruciate ligament grafts as a function of time following transplantation. Arthroscopy. 2005;21:917–22.
16. Kinugasa K, Hamada M, Yoneda K, Matuso T, Mae T, Shino K. Cross-sectional area of hamstring tendon autograft after anatomic triple bundle ACL reconstruction. Knee Surg Sports Traumatol Arthrosc. 2017;25:1219–26.
17. Shino K, Horibe S, Hamada M, Nakata K, Toritsuka Y, Mae T. Allograft anterior cruciate ligament reconstruction. Tech Knee Surg. 2002;1:78–85.
18. Hiramatsu K, Shino K, Iuchi R, Tachibana Y, Mae T. Contact area between femoral tunnel and interference screw in anatomic rectangular tunnel ACL reconstruction: a comparison of outside-in and transportal inside-out techniques. Knee Surg Sports Traumatol Arthrosc. 2018;26:519–25.
19. Shino K, Iuchi R, Tachibana Y, Ohori T, Mae T. Anatomical femoral tunnel creation: outside-in versus anteromedial portal. Ann Joint. 2017;2:34.

5 Evidenced-Based Approach for Anterolateral Surgery for ACL Reconstruction

Fares Uddin, Gilbert Moatshe, and Alan Getgood

5.1 Background

Contemporary anterior cruciate ligament (ACL) reconstruction techniques are reproducibly successful in controlling anterior tibial translation associated with ACL ruptures; however, their utility in restoring rotational laxity is variable. Several surgical advances have been made over the years in an attempt to address this rotational laxity and restore normal knee kinematics. In recent years there has been increasing trend toward anatomic single-bundle ACL reconstruction employing the placement of the femoral tunnel within the ACL footprint resulting in a lower position and more oblique orientation [1]. Theoretically this provides a biomechanical advantage for controlling tibial rotation [2], with studies demonstrating improved patient-reported outcomes as well as rotational control with this technique [3–6]. Some authors have advocated for double-bundle ACL reconstruction in order to better recreate the normal anatomy and control rotational instability. Biomechanical studies have demonstrated that double-bundle ACL reconstruction is probably superior in controlling rotational laxity [7, 8]; however, clinical studies have failed to show any significant difference when compared to single-bundle techniques [9].

Despite improved patient-reported outcomes with current ACL reconstruction techniques, an increasing amount of literature has demonstrated rising concerns over persistent instability and re-injury rates. Individuals under the age of 20 years have a re-injury rate as high as 20% (30% if contra-lateral knee injury included) [10–13]. It is not clear if rotatory laxity as measured by the pivot shift is a risk factor for repeat injury. However, a persistent positive pivot shift has been correlated with decreased patient satisfaction as well as increased functional instability [3, 14, 15]. Furthermore, a positive pivot shift may cause abnormal cartilage contact stress and increase the risk of wear [16, 17]. The literature demonstrates that current ACL reconstruction techniques are not as effective in controlling rotation as once believed [12]. Return to sport at the pre-injury level and return to competitive sports may also be as low as 63% and 44%, respectively [7]. This suggests that while ACL reconstruction with modern techniques improves functional out-

F. Uddin
Fowler Kennedy Sports Medicine Clinic, University of Western Ontario, London, ON, Canada

Bahrain Royal Guard, Bahrain Defence Force, Riffa, Bahrain

G. Moatshe
Fowler Kennedy Sports Medicine Clinic, University of Western Ontario, London, ON, Canada

OSTRC, Norwegian School of Sports Sciences, Oslo, Norway

A. Getgood (✉)
Fowler Kennedy Sports Medicine Clinic, University of Western Ontario, London, ON, Canada
e-mail: alan.getgood@uwo.ca

N. Nakamura et al. (eds.), *Advances in Knee Ligament and Knee Preservation Surgery*,
https://doi.org/10.1007/978-3-030-84748-7_5

comes in many patients, there exists a subgroup of patients in whom addressing the additional rotational laxity may be indicated in order to further improve clinical results.

This chapter will focus on the anterolateral complex (ALC) of the knee and how the use of anterolateral procedures to address persistent rotatory laxity may benefit patients undergoing both primary and revision ACL surgery.

5.2 Anatomy

Few anatomic regions in the orthopedic world have been as highly debated as the anterolateral side of the knee. In 1879, Segond was the first to describe the eponymous fracture now associated with injury to these structures [18]. At the time, he described a pearly band remaining attached to the fracture fragment. Kaplan detailed the layers and attachments of the iliotibial band (ITB) to the femur, further enhancing our understanding of this complex area [19]. He was also the first to coin the term anterolateral ligament; however, he had rather attributed it to a segment of the iliotibial band (ITB). Terry [3] subsequently divided the facia lata into its component parts and was the first to describe the deep part of the ITB as the "true anterolateral ligament of the knee" [20]. Muller, as well as Lobenhoffer, separately documented the presence of a retrograde fiber tract connecting the deep fibers of the ITB to the anterolateral tibial plateau providing static stability [21, 22].

To add to the perplexity of inherently complicated anatomy, the literature is fraught with overlapping nomenclature for various structures. Lateral capsular ligament [23], capsule-osseous layer of the ITB [20], retrograde tract fibers [22], anterior oblique band [24], and the lateral femorotibial ligament [25] are just a sample of the names used to describe a common structure. Cruels et al. are accredited with being the first to describe the anterolateral ligament (ALL) as a separate entity [25], even though confusingly, Terry et al. used the same terminology for the capsule-osseous layer of the ITB. The ALL was further refined by Vincent et al. and described as being a ligamentous structure anterior to the fibular collateral ligament (FCL); they further stated that this was the same structure previously identified in the literature as the mid third capsular ligament [26].

The rediscovery of the ALL by Claes et al. reignited much controversy and research into this topic [27]. They identified the ligament in 40 of the 41 dissected specimens and thoroughly described its structure. Interestingly, they found that it lacked connections to the overlying ITB but had a strong meniscal attachment. Dodds et al. [28] and Kennedy et al. [29] have since published the most detailed descriptions of the ALL that we know to date. The femoral attachment of the ALL has been described to be somewhat variable in position while the tibial attachment is fairly consistent. Some studies describe the femoral attachment to be proximal and posterior to the FCL origin [28, 29], others distal and anterior [27, 30], while one cadaveric study even described both variants [31]. Helito et al. in their histological study have demonstrated the ALL to have a similar well organized and dense collagenous structure to ligaments [30]. The ALL has also been described as capsular thickening present within the anterolateral complex (ALC) with collagen bundle arrangement similar to ligaments [27, 31]. The biomechanical study by Smeets et al. [32] demonstrated the ALL to have different properties to that of the adjacent capsule and similar to capsular ligaments such as the inferior glenohumeral ligament. In the pivotal study by Seebacher [33] and colleagues describing the layers of the knee, they stated that Layer 3 of the anterolateral capsule divides into superficial and deep lamina just anterior to the lateral collateral ligament (LCL) and prior to enveloping it. It would seem that many of the structures identified in previous studies may actually be synonymous with the ALL. It is now that with advancements in imaging and histology that investigators are able to more accurately characterize this structure.

The recent International ALC Consensus Group Meeting [34] attributed the various nomenclature and heterogeneity of structures in the literature to the following reasons: the lack of clear photographs and diagrams in historical papers, the use of both embalmed and fresh specimens, and differences in dissection techniques that may introduce "artifacts." The panel subsequently established six statements regarding anatomy of the anterolateral complex (ALC). They defined the ALL as being a structure within the ALC, as a capsular structure within Seebacher Layer 3, and having variable morphology between individuals, having an attachment to the lateral meniscus. They also defined the layers of the ALC as well the anatomic course and landmarks of the ALL. The layers of the ALC run from superficial to deep as follows: superficial ITB, deep ITB (with Kaplan fiber system), ALL, and capsule (Fig. 5.1a–d). The ALL originates proximal and posterior to the lateral femoral epicondyle and origin of LCL, runs superficial to LCL, and attaches to the tibia midway between Gerdy's tubercle and the head of the fibula.

5.3 Biomechanics

Cadaveric biomechanical studies have been published evaluating the kinematics of the knee following sectioning of the ALC structures. Transecting the ALL in ACL-deficient knees has been observed to significantly increase both anterior translation as well as internal rotation during the early phases of the pivot shift [35]. Using a 6-degree of freedom robot, Rasmussen et al. also demonstrated a clear increase in internal rotation following sectioning of the ALL [36]. Navigation studies evaluating dynamic pivot shift in ACL/ITB-deficient and ACL/LCL-deficient knees have shown significantly increased internal rotation laxity [37]. Similarly, Kittl et al. examined the effect of sectioning the ALL as well as the superficial and deep layers of the ITB [38]. They found that the ALL demonstrated a minor role in controlling internal rota-

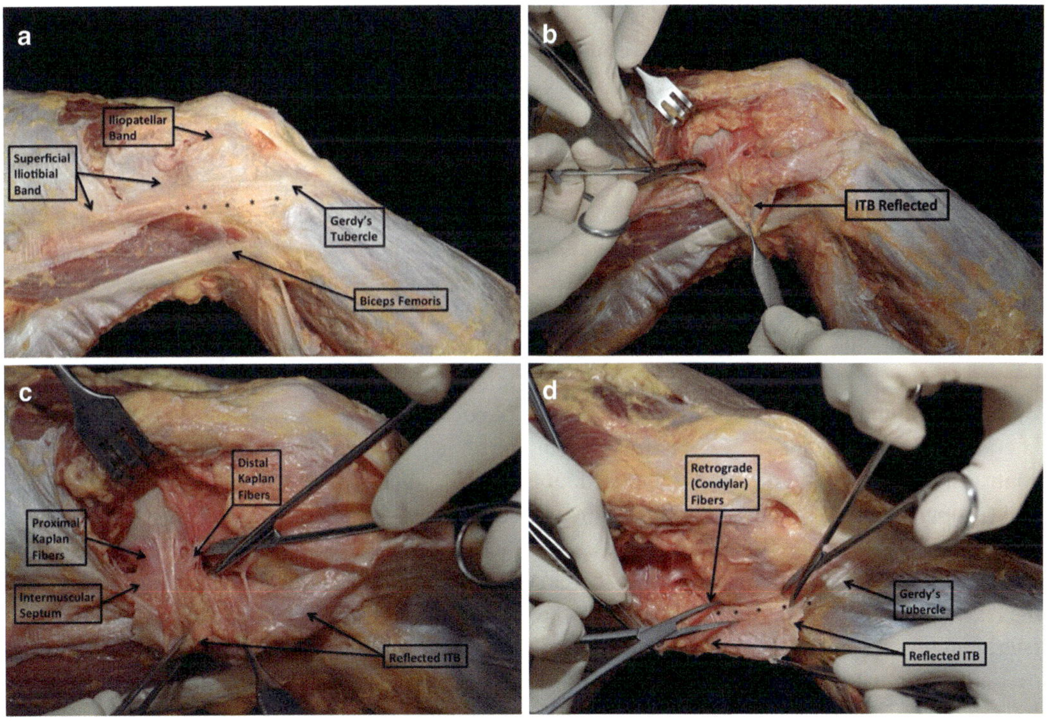

Fig. 5.1 (**a**) Lateral side of right knee; (**b**) superficial ITB incised anteriorly and reflected from anterior to posterior exposing Kaplan fiber attachments and deep ITB; (**c**) Proximal and distal Kaplan fiber attachments; (**d**) Capsulo-osseous layer of deep ITB with retrograde fiber attachment on supracondylar region of distal femur

tion in ACL-deficient knees. Conversely the ITB, specifically its deep capsule-osseous layer made a much greater contribution to internal rotation control at greater flexion angles. They also showed that the ACL contributed greatest to internal rotation control near full extension. One study evaluating anterior translation and internal rotation by the use of an optical tracking analysis and strain mapping found that the anterolateral capsule behaves more as a sheet of fibrous tissue rather than a ligamentous structure, disputing the existence of the ALL [39]. Thein et al. performed a serial sectioning study that demonstrated that the ALL only engaged in load sharing when the physiological limits of the ACL were surpassed [40]. They concluded the ALL acted as a secondary stabilizer only after the loss of the ACL. Huser et al. performed a simulated pivot shift and showed that isolated ALL sectioning in ACL intact knees resulted in no increase in tibiofemoral compartment translation. They also concluded that the ALL does not function as a primary restraint to the pivot shift [41]. This same group subsequently performed another study that demonstrated that sectioning of the ALL and ITB in ACL-deficient knees transformed 71% of the cadavers to a grade 3 pivot shift [42]. Other studies show that when combined ACL and ALL injuries are present, isolated ACL reconstruction is not sufficient to restore normal knee kinematics. In this scenario, only combined ACL and lateral extra-articular procedures were able to return kinematics back to normal [42].

Several studies have demonstrated that the lateral meniscus also plays an essential role in controlling anterolateral rotation. Increased lateral compartment anterior translation and internal rotation have been demonstrated with lateral meniscus posterior root tears [43, 44]. Whether the ALL functions as a peripheral anchor to the lateral meniscus is not yet established; however, the inframeniscal ALL fibers have been found to be significantly stiffer and stronger than their suprameniscal counterparts. The clinical implications of this are not yet fully understood [45].

The biomechanics of ALC reconstruction has been examined by numerous studies. The majority of them acknowledge the difficulty of extrapolating data from artificially created injury patterns and results to the clinical field. Spencer et al. [35] studied the effect on anterior translation and internal rotation in ACL-deficient knees with both a Lemaire type lateral extra-articular tenodesis (LET) as well as the ALL reconstruction as described by Claes et al. [27]. They found that the ALL reconstruction had little effect on controlling translation and rotation. We now know that the femoral graft position in this description was incorrect and should have been placed posterior and proximal, not anterior and distal. The LET group did show a significant reduction of internal rotation and anterior translation.

Changes in ALC reconstruction graft length based upon the attachment sites have been investigated by Kittl et al. [46]. They found that the most isometric position was a proximal and posterior attachment on the femur, with the graft passed deep to the LCL, and attached distally to Gerdy's tubercle. The authors established that a LET passed deep to the LCL would therefore be the most efficient form of reconstruction.

With regard to ALL reconstruction, a femoral insertion point posterior and proximal to the LCL origin has been demonstrated to result in minimal length change during the flexion cycle [28]. But if the anterior and distal femoral attachment site as described by Claes et al. [27] was utilized, several studies show that the ALL will tighten with flexion [46–48]. Therefore, it is clear that if an ALL reconstruction is to be of benefit to rotational control then a posterior and proximal attachment should be chosen. This would be posterior to the center of rotation of the knee and as a result the ALL graft would be tight toward full extension.

Studies have compared ALL reconstruction and LET in ACL reconstructed knees. Lateral extraarticular tenodesis graft tensioned at 20 N and passed deep to the LCL has been shown to be effective at controlling rotation with minimal over-constraint of internal rotation [49]. Both a modified Lemaire and a modified Macintosh type tenodesis, with the graft passed deep to the LCL, have been shown to restore normal knee kinematics when combined with an anatomic ACL reconstruction. In contrast, ALL reconstructions based on described anatomic techniques were only found to minimally control rotation. The same group later published a study that showed passing an LET graft deep to LCL and tensioned at different flexion angles did not result in any detrimental effects [50]. The study also reported that an ALL reconstruction technique as described by Sonnery-Cottet [51] only controlled knee laxity when the reconstruction was tensioned in full extension. Schon et al. demonstrated that a single graft tensioned at 88 N resulted in significant over-constraint of internal rotation despite the angle of fixation [52]. Ensuing studies have since suggested that 20 N to be the optimal tension for ALL reconstruction [49]. Geeslin et al. compared ALL reconstruction to a modified Lemaire LET reconstruction with varying graft tension and knee flexion parameters. They found that the Lemaire LET resulted in a significantly greater reduction in anterior translation as well as internal rotation during a simulated pivot shift [53].

However, it has been shown that in a knee with ACL and ALC injury, an anatomical reconstruction with a bone patella bone (BTB) secured in 25° of knee flexion without any ALC procedure resulted in adequate restoration of knee kinematics during a simulated pivot shift [54]. But a residual increase of internal tibial rotation of 5–7° does occur at high flexion angles when an ALC injury is concomitantly present. This cannot be controlled by an ACL reconstruction on its own.

Further to this, Herbst et al. evaluated the role of augmentation of ACL reconstruction with LET in knees with and without ALC deficiency. They found that an addition of an LET provided no further benefit in isolated ACL-deficient knees when an ACL reconstruction was performed [55]. Conversely, LET was required in the combined injuries to restore normal knee kinematics. Therefore, it can be deduced that it is important to identify ACL deficiencies in which a combined ALC injury may have occurred. Several studies have shown that in a knee exhibiting high-grade laxity, an isolated ACL injury is largely unlikely. Coexisting meniscal and ALC injuries are usually encountered and this supports the need for a combined ACL and anterolateral reconstruction [56]. Concomitant injuries in acute ACL have been reported in the literature to occur between 40 and 90% of the time [31, 57, 58].

Currently, it is difficult to determine whether one reconstruction technique is superior to another. This is largely since the methods by which the experiments are conducted, and the measurement techniques differ between studies. When utilizing an LET procedure, the recommendation is to pass the graft deep to the LCL prior to femoral fixation [49]. This provides a more optimal direction of action through the flexion range, avoids over-constraint, and allows the LCL to serve as a fulcrum. However, if utilizing an ACL and ALL reconstruction as described by Sonnery-Cottet, having the graft tensioned in extension would provide the most biomechanical advantage [46, 49].

Concerns of over constraining the lateral compartment are a potential issue with ALC procedures. Lateral compartment contact pressures following LET has been investigated and a small increase has been demonstrated [59]. This increased pressure was found to be insignificant when compared to that occurring physiologically. The clinical relevance of over-constraint is currently unknown but there is no evidence till date indicating accelerated osteoarthritis [60]. Furthermore, Shimikawa et al. also demonstrated that LET did not increase lateral compartment contact pressures even in the presence of lateral meniscectomy [61].

The ALC consensus meeting statement has recognized the potential to over constrain the normal motion of the lateral compartment. They have identified over tensioning of the graft and fixation with the tibia in external rotation to be possible reasons. As of yet there is no clinical evidence linking this procedure to accelerated osteoarthritis [62].

5.4 Clinical Evidence: An Historical Perspective

Lateral extraarticular tenodesis procedures are by no means newcomers to orthopedic surgery. In the early twentieth century it was the procedure of choice to deal with knee laxity associated with an ACL deficiency. Lemaire et al. had described a technique utilizing a 1.2 × 18 cm strip of ITB left attached distally to Gerdy's tubercle, passed deep to the LCL, through an osseous tunnel in the femur, and then anchored to a bone tunnel at Gerdy's tubercle [63] (Fig. 5.2a). Macintosh also described an ITB-based reconstruction using a 2 × 20 cm strip from the midsection, left attached distally, passed deep to the LCL, through a subperiosteal femoral tunnel just behind the LCL proximal attachment, passed through the distal intermuscular septum, then tensioned and sutured onto itself deep to the LCL [64] (Fig. 5.2b). In 1979 Arnold et al. described a modification of the Macintosh utilizing a 2 cm strip of ITB with the distal attachment intact. This was routed deep to the proximal LCL where it was sutured. The remaining graft was reflected distal over the LCL and secured to the tibia at Gerdy's tubercle with a staple at 90° flexion and in external rotation [65]. The Ellison procedure differs from the above where the 1.5 cm strip of ITB was released distally with a bone block. This was subsequently passed deep to the LCL and anchored to a bone trough in the region of the lateral patellar tendon. Combining this with a deep lateral capsular plication was determined to create a dynamic reconstruction [66] (Fig. 5.2c). Losee et al., Andrews and Sanders described further variations of isolated ITB based procedures to treat ALRI [67, 68].

The early literature describing isolated LET procedures were heterogenous due to a variety of techniques, differing rehabilitation protocols, and

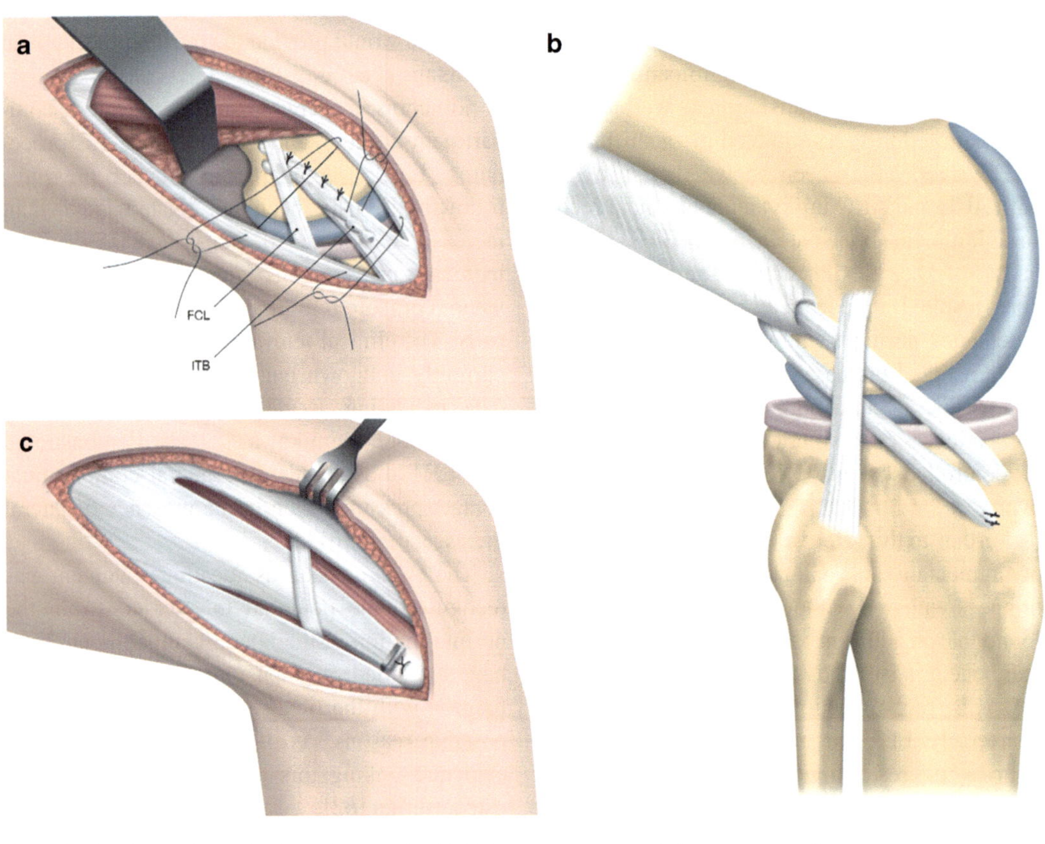

Fig. 5.2 (**a**) Original Lemaire lateral tenodesis; (**b**) Ellison procedure; (**c**) Macintosh procedure

poor outcome reporting. On the whole, results were largely poor, with reduced outcome scores and high rates of residual instability. Neyret et al. reviewed outcomes in amateur skiers under 35 and recommended against an isolated LET procedure [69]. They found poor subjective satisfaction rates with significant residual laxity. At 4.5 years, 12 of 15 patients had a positive pivot shift. Amirault et al. demonstrated 52% good to excellent outcomes with an isolated Macintosh procedure [70]. Ellison reported 44% excellent and 39% good results in their series of 18 knees; however, there was a 16.6% failure rate with 2 out of 18 knees requiring reoperation [66]. Kennedy et al. evaluated 28 patients who underwent an Ellison type reconstruction and found that only 57% of patients at 1 year had good or excellent outcomes [71]. Twenty-four patients in their series demonstrated residual laxity with a positive pivot shift assessment.

It eventually became apparent that an intraarticular procedure would indeed provide better control of knee stability. Surgeons determined that the results of their lateral reconstruction which was designed to control anterolateral rotatory stability could be extrapolated to an intraarticular reconstruction. Combined reconstructions could either be incorporated within the ACL reconstruction where a single graft is used for both intra- and extra-articular procedures or it could be augmented to an intraarticular procedure with a separate graft.

The Macintosh "over the top" reconstruction described in 1985 is an ITB-based combined intra-articular and extra-articular reconstruction [72]. Here a 4 × 25 cm strip of ITB left attached distally is passed deep to the LCL. It is then passed subperiosteally to the anterior aspect of the lateral intermuscular septum, then over the top and through the knee to reconstruct in the intra-articular ACL. Maracci et al. subsequently described a modification of this using a hamstrings autograft. The semitendinosus and gracilis tendons are harvested but left attached distally where they are sutured together [73]. The graft is then directed intraarticularly through a tibial tunnel, to the over-the-top position on the femur, and passed superficial to the LCL and secured to Gerdy's tubercle to reconstruct the extraarticular component. Lerat et al. described a "MacInJones" procedure involving the lateral third of the patellar tendon harvested in continuity with 10 cm strip of the quadriceps tendon [74]. The quadriceps component was passed deep to the LCL and secured to Gerdy's tubercle.

An expanding literature of studies examining the effect of LET procedures as an augment to an intraarticular ACL reconstruction exists. The earlier observational studies from the 1980s were largely encouraging. Bertoia reported 91% good to excellent outcomes with 97% of their cohort returning to pre-injury activity at 37 months utilizing a Macintosh over-the-top reconstruction [72]. In 1988 Dejour et al. evaluated results of an augmented BTB ACL reconstruction with a Lemaire procedure and demonstrated 83% good or excellent outcomes at a 3 year follow-up [75]. Zarin and Rowe equally demonstrated satisfactory outcomes in their review of 100 patients treated with a combined semitendinosus and ITB over-the-top reconstruction [76].

Subsequent comparative studies were however discouraging with concerns arising regarding surgical morbidity, lateral compartment over-constraint, and the possible development of osteoarthritis with LET procedures. O'Brien et al. published a retrospective series of 80 patients treated with a BTB ACL reconstruction with or without a modified Macintosh procedure followed over 4 years [77]. They showed no improvements in stability with 40% of patients complaining of chronic pain or swelling in those with combined procedures. This study had several methodological limitations; in particular it was not sufficiently powered to elicit a difference in clinical outcomes and included all age groups of patients, not those that are particularly at risk of ACL reconstruction graft failure. Strum et al. performed a retrospective study evaluating intraarticular reconstruction with combined procedures and found no additional benefit of adding an LET procedure in the 127 patients studied [78]. They, much like the previous study, did not provide any indication in patients in whom a LET procedure was performed. Anderson et al. subsequently published a randomized trial in 2001 comparing BTB, hamstring, and hamstring plus a Losee type LET procedure with a minimum of 2 years follow-up [79]. They demonstrated no additional benefit of the extra-articular procedure. In 1989 the American Orthopedic Society for Sports Medicine Conference raised concerns with the greater morbidity and higher risk of complications [80]. They issued a recommendation to abandon lateral based procedures and these as a result felt out of favor in North America. However, significant global variation existed at this time with LET augmentation still being popular in many European centers.

5.5 Clinical Evidence: The Current Concepts

An unacceptably high failure rate of ACL reconstructions in young patients has been reported recently, with reports reaching as high as 20% [10–13]. Forty-seven percent of these injuries may occur within the first year while 74% within the first 2 years following surgery [81]. Several factors may contribute to such high numbers including the surgical technique, poor neuromuscular rehabilitation, participating in high-risk pivoting sports, and early return to sports. Many of these issues are not within the full control of an orthopedic surgeon to modify, but an emphasis on improving and maintaining an optimal surgical technique is. These include appropriate graft tunnel positioning as well as addressing concomitant injuries of the meniscus, soft tissue laxity patterns, and malalignment.

Zaffagnini et al. performed a randomized controlled trial comparing the Marcacci technique as described above, a 4 strand hamstrings single-bundle ACL reconstruction, and a BTB ACL reconstruction [82]. They demonstrated improved patient outcomes with the addition of the lateral augmentation as well as a quicker return to sports. Vadala et al. explored the concept of LET as an augmentation in high-risk individuals. In their study, female patients with an ACL deficiency who had a grade 2 or 3 pivot shift preoperatively were randomized to a 4 strand hamstrings ACL reconstruction with or without an extraarticular Coker-Arnold procedure [83]. At 44.4 months follow-up 18.86% of patients in the combined ACL and LET group had residual positive pivot shift compared to 57.1% in the group with isolated ACL reconstruction.

Hewison et al. published a systematic review demonstrating that the addition of a lateral-based procedure to an ACL reconstruction improves rotational control when assessed by a pivot shift, but does not alter anterior translation nor patient-reported outcomes [84]. Rezende and colleagues published a systematic review that found ACL reconstruction augmented with LET achieved improved anteroposterior stability when measured by Lachman and KT-1000 testing [85]. Devitt et al. published a meta-analysis on the results of LET augmentation in early (<12 months) and delayed ACL reconstruction [60]. Fascinatingly, they found that LET augmentation did not improve rotational stability in the early reconstruction group but saw a statistically significant improvement in the delayed group.

Augmenting with an LET in the case of revision ACL surgery is attractive particularly if no causative factor for failure was identified. This would allow the surgeon to address residual rotational instability that may have been a cause for the failure. Furthermore, it allows the LET to shield the revision graft to excessive stress during the initial incorporation period, as has been shown by Engebretsen et al. [86, 87]. In this cadaveric study, the addition of an LET reduced graft strain by 43%. A retrospective multicenter study was performed by the French Arthroscopic society to evaluate revision ACL reconstruction with or without a LET [88]. One hundred eighty-nine patients were reviewed in which 51% had an LET performed. There were varying reconstruction techniques and the indications of augmenting with a LET were not provided. At a 2-year follow-up there was a 15% failure rate in the isolated ACL revision group and a 7% in the LET augmented group. With regard to a pivot shift evaluation, 63% in the isolated ACL revision arm had a negative test compared to 80% in those who additionally underwent a LET procedure. Ferreti et al. evaluated 30 patients who underwent a revision ACL reconstruction utilizing a quadruple strand hamstrings autograft with the Coker-Arnold modification of the Macintosh LET [89]. Their study at a 5-year follow-up showed a 10% overall failure rate with only two patients having a pivot shift greater than grade 2.

In 2015, Sonnery Cottet et al. published 2-year outcomes on augmenting a hamstring autograft ACL reconstruction with an ALL reconstruction. Out of 92 patients, only 7 had a grade 1 pivot shift and the re-rupture rate was 1% [51]. The same group subsequently published a comparative series of 502 young patients who were involved in pivoting sports, and therefore deemed to be at high risk of graft failure [90]. The combined ACL and ALL group had graft failure rates of 4%, compared to 10% and 16% in the isolated hamstrings and the isolated BTB ACL reconstruction groups, respectively.

Getgood et al. recently published the two-year results of the STABILITY Study, a prospective randomized controlled trial of 618 patients comparing ACL reconstruction with or without LET [91]. Patients 25 or younger with an ACL deficiency were included if they met 2 or more of the following criteria: grade 2 pivot shift or greater, desire to return to high-risk pivoting sports, and generalized ligamentous laxity. They found an 11% graft rupture rate in the ACL reconstruction group versus 4% in the ACL + LET group. The number needed to treat with LET to prevent 1 graft rupture was 14.3. They demonstrated a statistically significant and clinically relevant reduction in graft rupture and persistent rotatory laxity at 2 years.

To date, no clinical studies have demonstrated an increased risk of osteoarthritis with the addition of LET as historically was alleged. In their 20-year outcome report of an over-the-top hamstrings reconstruction with an LET, Zaffagnini et al. demonstrated no development of lateral compartment or patellofemoral osteoarthritis [92]. Pernin et al. in their 24.5-year follow-up found similar results and attributed development of osteoarthritis to the status of the medial meniscus and femoral chondral defects at the time of surgery [93]. A recent meta-analysis did not find any evidence of osteoarthritis in the knee at an 11-year follow-up with the addition of LET procedures [60].

5.6 Indications and Surgical Technique

Our current indications for LET include the majority of revision ACL reconstructions, young patients (<25 years old) returning to contact pivoting sports, young patients (<25 years old) having ACLR with hamstring tendon autograft, high-grade pivot shift (2 or greater), patients with knee hyperextension (>10°)/generalized ligamentous laxity (Beighton score > 4), and patients with increased posterior tibial slope (>10°).

The surgical technique the authors employ for their lateral-based procedure is an LET utilizing a modified Lemaire technique (Fig. 5.3a–f). An oblique incision is made between the lateral epicondyle and the Gerdy's tubercle approximately 5–6 cm in length (Fig. 5.3a). A 1 × 8 cm long strip of ITB is harvested, leaving the distal attachment intact (Fig. 5.3b, c). The proximal end of the graft is then whipstiched using a No. 1 Vicryl suture. The FCL is subsequently identified, a soft tissue tunnel is created deep to the FCL and superficial to the joint capsule, and the ITB graft is passed deep to the FCL (Fig. 5.3d). The graft is then secured to the supracondylar metaphyseal flare of the distal femur anterior to the intermuscular septum and proximal to the LCL origin utilizing a Richard's fixation staple (Smith & Nephew Inc) (Fig. 5.3e). Fixation is performed with the knee at 60–70^0 flexion and 0^0 rotation with minimal tension being applied to the graft. The free end of the ITB is then looped back onto itself and sutured using a No. 1 Vicryl suture. The remaining defect in the ITB is then partially closed up to but not including the transverse retinacular ligament (Fig. 5.3f).

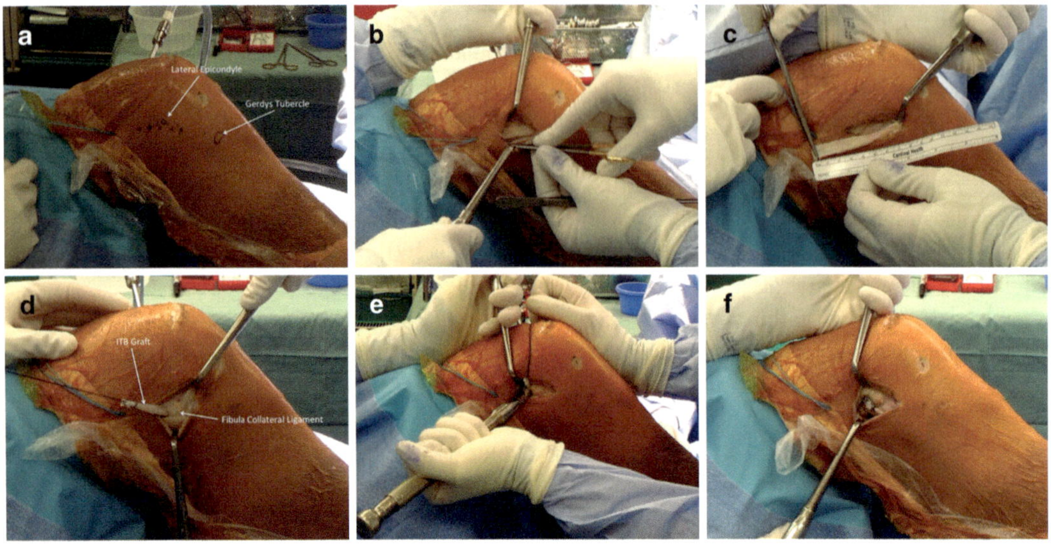

Fig. 5.3 (**a**) An intraoperative picture of a right knee flexed at about 90°. A 6 cm longitudinal incision is made approximately 1 cm posterior to the lateral femoral epicondyle. Subcutaneous tissue is divided sharply down to the level of the iliotibial band (ITB); (**b**) Iliotibial band (ITB) graft harvest—the posterior margin of the ITB is identified and an incision is made 1 cm anterior to the posterior margin, ensuring the posterior fibers of the ITB are undisturbed and an 8 cm long by 1 cm ITB strip is harvested; (**c**) An 8 cm long by 1 cm wide strip of iliotibial band (ITB) that is released proximally and freed of any deep attachments leaving it attached distally at Gerdy's tubercle—the proximal 2 cm of ITB strip are then whip stitched with #1 Vicryl suture; (**d**) The ITB strip is passed through a soft tissue tunnel under the fibular collateral ligament (FCL)—the FCL is identified by palpation, putting the knee in a figure-of-4 position may help identify the FCL because it is taut in this position. With a #15 scalpel, small incisions are made just anterior and posterior to the proximal aspect of the FCL and Metzenbaum scissors are passed deep to the FCL taking care to remain extra-capsular and prevent damage to the popliteus tendon; (**e**) The LET graft is fixed on the supracondylar flare of the lateral femoral condyle—the knee is placed in 60° of flexion with the tibia and foot in neutral rotation to avoid over-constraining the lateral joint compartment and restricting rotational freedom. The graft is held taut with minimal tension and secured to the femur with a Richards' staple (Smith and Nephew Inc., Andover, MA); (**f**) The ITB defect is closed using interrupted figure-of-eight #1 Vicryl sutures—to avoid over-constraint of the lateral patellofemoral joint we refrain from closing the ITB defect distally at the level of the transverse ligament

5.7 Conclusion

Cadaveric anatomic and biomechanical studies have demonstrated the importance of the ALC to the ACL-deficient knee and subsequent ACL reconstruction. Up until now, high-quality prospective studies evaluating the effect of ALC procedures on outcomes following ACL reconstruction have been lacking. Overall, the current literature demonstrates a pattern suggesting LET or ALL augmentation can improve rotational stability and result in decreased graft rupture rates following ACL reconstruction. It is yet to be fully determined who most benefits from an anterolateral augmentation. However, young patients returning to contact pivoting sport and undergoing a hamstring tendon autograft ACL reconstruction seem to be at high risk, and likely benefit most from LET/ALL augmentation. It is these individuals that we as clinicians need to appropriately identify and treat accordingly in order to improve results following ACL reconstruction.

References

1. Bedi A, Raphael B, Maderazo A, Pavlov H, Williams RJ. Transtibial versus anteromedial portal drilling for anterior cruciate ligament reconstruction: a cadaveric study of femoral tunnel length and obliquity. Art Ther. 2010;26(3):342–50.

2. Markolf KL, Jackson SR, McAllister DR. A comparison of 11 O'clock versus oblique femoral tunnels in the anterior cruciate ligament-reconstructed knee. Am J Sports Med. 2010;38(5):912–7.
3. Lee MC, Seong SC, Lee S, Chang CB, Park YK, Jo H, et al. Vertical femoral tunnel placement results in rotational knee laxity after anterior cruciate ligament reconstruction. Art Ther. 2007;23(7):771–8.
4. Loh JC, Fukuda Y, Tsuda E, Steadman RJ, Fu FH, Woo SL-Y. Knee stability and graft function following anterior cruciate ligament reconstruction: comparison between 11 o'clock and 10 o'clock femoral tunnel placement. Art Ther. 2003;19(3):297–304.
5. Kato Y, Maeyama A, Lertwanich P, Wang JH, Ingham SJM, Kramer S, et al. Biomechanical comparison of different graft positions for single-bundle anterior cruciate ligament reconstruction. Knee Surg Sports Traumatol Arthrosc. 2013;21(4):816–23.
6. Kondo E, Merican AM, Yasuda K, Amis AA. Biomechanical comparison of anatomic double-bundle, anatomic single-bundle, and nonanatomic single-bundle anterior cruciate ligament reconstructions. Am J Sports Med. 2011;39(2):279–88.
7. Zelle BA, Vidal AF, Brucker PU, Fu FH. Double-bundle reconstruction of the anterior cruciate ligament: anatomic and biomechanical rationale. J Am Acad Orthop Surg. 2007;15(2):87–96.
8. Yagi M, Wong EK, Kanamori A, Debski RE, Fu FH, Woo SL-Y. Biomechanical analysis of an anatomic anterior cruciate ligament reconstruction. Am J Sports Med. 2002;30(5):660–6.
9. Meredick RB, Vance KJ, Appleby D, Lubowitz JH. Outcome of single-bundle versus double-bundle reconstruction of the anterior cruciate ligament. Am J Sports Med. 2008;36(7):1414–21.
10. Shelbourne KD, Gray T, Haro M. Incidence of subsequent injury to either knee within 5 years after anterior cruciate ligament reconstruction with patellar tendon autograft. Am J Sports Med. 2009;37(2):246–51.
11. Kaeding CC, Aros B, Pedroza A, Pifel E, Amendola A, Andrish JT, et al. Allograft versus autograft anterior cruciate ligament reconstruction: predictors of failure from a MOON prospective longitudinal cohort. Sports Health. 2011;3(1):73–81.
12. Mohtadi NG, Chan DS, Dainty KN, Whelan DB. Patellar tendon versus hamstring tendon autograft for anterior cruciate ligament rupture in adults. Cochrane Database Syst Rev. 2011;(9):CD005960.
13. Magnussen RA, Lawrence JT, West RL, Toth AP, Taylor DC, Garrett WE. Graft size and patient age are predictors of early revision after anterior cruciate ligament reconstruction with hamstring autograft. Art Ther. 2012;28(4):526–31.
14. Kocher MS, Steadman JR, Briggs K, Zurakowski D, Sterett WI, Hawkins RJ. Determinants of patient satisfaction with outcome after anterior cruciate ligament reconstruction. J Bone Joint Surg Am. 2002;84(9):1560–72.
15. Ayeni OR, Chahal M, Tran MN, Sprague S. Pivot shift as an outcome measure for ACL reconstruction: a systematic review. Knee Surg Sports Traumatol Arthrosc. 2012;20(4):767–77.
16. Leiter JR, Gourlay R, McRae S, de Korompay N, MacDonald PB. Long-term follow-up of ACL reconstruction with hamstring autograft. Knee Surg Sports Traumatol Arthrosc. 2014;22(5):1061–9.
17. Jonsson H, Riklund-Ahlstrom K, Lind J. Positive pivot shift after ACL reconstruction predicts later osteoarthrosis: 63 patients followed 5-9 years after surgery. Acta Orthop Scand. 2004;75(5):594–9.
18. Segond P. Recherches cliniques et experimentales sur les epanchements sanguins du genou par entorse; 1879.
19. Kaplan E. The iliotibial tract; clinical and morphological significance. J Bone Joint Surg Am. 1958;40-A(4):817–32.
20. Terry GC, Hughston JC, Norwood LA. The anatomy of the iliopatellar band and iliotibial tract. Am J Sports Med. 1986;14(1):39–45.
21. Muller W. Functional anatomy and clinical findings of the knee joint. Helv Chir Acta. 1984;51(5):505–14.
22. Lobenhoffer P, Posel P, Witt S, Piehler J, Wirth CJ. Distal femoral fixation of the iliotibial tract. Arch Orthop Trauma Surg. 1987;106(5):285–90.
23. Hughston JC, Andrews JR, Cross MJ, Moschi A. Classification of knee ligament instabilities. Part II. The lateral compartment. J Bone Joint Surg Am. 1976;58(2):173–9.
24. Campos JC, Chung CB, Lektrakul N, Pedowitz R, Trudell D, Yu J, et al. Pathogenesis of the Segond fracture: anatomic and MR imaging evidence of an iliotibial tract or anterior oblique band avulsion. Radiology. 2001;219(2):381–6.
25. Vieira EL, Vieira EÁ, Teixeira da Silva R, dos Santos Berlfein PA, Abdalla RJ, Cohen M. An anatomic study of the iliotibial tract. Art Ther. 2007;23(3):269–74.
26. Vincent J-P, Magnussen RA, Gezmez F, Uguen A, Jacobi M, Weppe F, et al. The anterolateral ligament of the human knee: an anatomic and histologic study. Knee Surg Sports Traumatol Arthrosc. 2012;20(1):147–52.
27. Claes S, Vereecke E, Maes M, Victor J, Verdonk P, Bellemans J. Anatomy of the anterolateral ligament of the knee. J Anat. 2013;223(4):321–8.
28. Dodds AL, Halewood C, Gupte CM, Williams A, Amis AA. The anterolateral ligament. Bone Joint J. 2014;96-B(3):325–31.
29. Kennedy MI, Claes S, Fuso FAF, Williams BT, Goldsmith MT, Turnbull TL, et al. The anterolateral ligament. Am J Sports Med. 2015;43(7):1606–15.
30. Helito CP, Demange MK, Bonadio MB, Tírico LEP, Gobbi RG, Pécora JR, et al. Anatomy and histology of the knee anterolateral ligament. Orthop J Sports Med. 2013;1(7):2325967113513546.
31. Caterine S, Litchfield R, Johnson M, Chronik B, Getgood A. A cadaveric study of the anterolateral ligament: re-introducing the lateral capsular ligament. Knee Surg Sports Traumatol Arthrosc. 2015;23(11):3186–95.
32. Smeets K, Slane J, Scheys L, Forsyth R, Claes S, Bellemans J. The anterolateral ligament has similar

biomechanical and histologic properties to the inferior glenohumeral ligament. Art Ther. 2017;33(5):1028–1035.e1.
33. Seebacher JR, Inglis AE, Marshall JL, Warren RF. The structure of the posterolateral aspect of the knee. J Bone Joint Surg Am. 1982;64(4):536–41.
34. Getgood A, Brown C, Lording T, Amis A, Claes S, Geeslin A, et al. The anterolateral complex of the knee: results from the international ALC consensus group meeting. Knee Surg Sports Traumatol Arthrosc. 2019;27(1):166–76.
35. Spencer L, Burkhart TA, Tran MN, Rezansoff AJ, Deo S, Caterine S, et al. Biomechanical analysis of simulated clinical testing and reconstruction of the anterolateral ligament of the knee. Am J Sports Med. 2015;43(9):2189–97.
36. Rasmussen MT, Nitri M, Williams BT, Moulton SG, Cruz RS, Dornan GJ, et al. An in vitro robotic assessment of the anterolateral ligament, part 1. Am J Sports Med. 2016;44(3):585–92.
37. Sonnery-Cottet B, Lutz C, Daggett M, Dalmay F, Freychet B, Niglis L, et al. The involvement of the anterolateral ligament in rotational control of the knee. Am J Sports Med. 2016;44(5):1209–14.
38. Kittl C, El-Daou H, Athwal KK, Gupte CM, Weiler A, Williams A, et al. The role of the anterolateral structures and the ACL in controlling laxity of the intact and ACL-deficient knee. Am J Sports Med. 2016;44(2):345–54.
39. Guenther D, Rahnemai-Azar AA, Bell KM, Irarrázaval S, Fu FH, Musahl V, et al. The anterolateral capsule of the knee behaves like a sheet of fibrous tissue. Am J Sports Med. 2017;45(4):849–55.
40. Thein R, Boorman-Padgett J, Stone K, Wickiewicz TL, Imhauser CW, Pearle AD. Biomechanical assessment of the anterolateral ligament of the knee : a secondary restraint in simulated tests of the pivot shift and of anterior stability. J Bone Joint Surg Am. 2016;98(11):937–43.
41. Huser LE, Noyes FR, Jurgensmeier D, Levy MS. Anterolateral ligament and iliotibial band control of rotational stability in the anterior cruciate ligament–intact knee: defined by tibiofemoral compartment translations and rotations. Art Ther. 2017;33(3):595–604.
42. Noyes FR, Huser LE, Levy MS. Rotational knee instability in ACL-deficient knees: role of the anterolateral ligament and iliotibial band as defined by tibiofemoral compartment translations and rotations. J Bone Joint Surg Am. 2017;99(4):305–14.
43. Lording T, Corbo G, Bryant D, Burkhart TA, Getgood A. Rotational laxity control by the anterolateral ligament and the lateral meniscus is dependent on knee flexion angle: a cadaveric biomechanical study. Clin Orthop Relat Res. 2017;475(10):2401–8.
44. Shybut TB, Vega CE, Haddad J, Alexander JW, Gold JE, Noble PC, et al. Effect of lateral meniscal root tear on the stability of the anterior cruciate ligament–deficient knee. Am J Sports Med. 2015;43(4):905–11.
45. Corbo G, Norris M, Getgood A, Burkhart TA. The infra-meniscal fibers of the anterolateral ligament are stronger and stiffer than the supra-meniscal fibers despite similar histological characteristics. Knee Surg Sports Traumatol Arthrosc. 2017;25(4):1078–85.
46. Kittl C, Halewood C, Stephen JM, Gupte CM, Weiler A, Williams A, et al. Length change patterns in the lateral extra-articular structures of the knee and related reconstructions. Am J Sports Med. 2015;43(2):354–62.
47. Bell KM, Rahnemai-Azar AA, Irarrazaval S, Guenther D, Fu FH, Musahl V, et al. In situ force in the anterior cruciate ligament, the lateral collateral ligament, and the anterolateral capsule complex during a simulated pivot shift test. J Orthop Res. 2018;36(3):847–53.
48. Zens M, Niemeyer P, Ruhhammer J, Bernstein A, Woias P, Mayr HO, et al. Length changes of the anterolateral ligament during passive knee motion. Am J Sports Med. 2015;43(10):2545–52.
49. Inderhaug E, Stephen JM, Williams A, Amis AA. Biomechanical comparison of anterolateral procedures combined with anterior cruciate ligament reconstruction. Am J Sports Med. 2017;45(2):347–54.
50. Inderhaug E, Stephen JM, Williams A, Amis AA. Anterolateral Tenodesis or anterolateral ligament complex reconstruction: effect of flexion angle at graft fixation when combined with ACL reconstruction. Am J Sports Med. 2017;45(13):3089–97.
51. Sonnery-Cottet B, Thaunat M, Freychet B, Pupim BH, Murphy CG, Claes S. Outcome of a combined anterior cruciate ligament and anterolateral ligament reconstruction technique with a minimum 2-year follow-up. Am J Sports Med. 2015;43(7):1598–605.
52. Schon JM, Moatshe G, Brady AW, Serra Cruz R, Chahla J, Dornan GJ, et al. Anatomic anterolateral ligament reconstruction of the knee leads to overconstraint at any fixation angle. Am J Sports Med. 2016;44(10):2546–56.
53. Geeslin AG, Moatshe G, Chahla J, Kruckeberg BM, Muckenhirn KJ, Dornan GJ, et al. Anterolateral knee extra-articular stabilizers: a robotic study comparing anterolateral ligament reconstruction and modified Lemaire lateral extra-articular Tenodesis. Am J Sports Med. 2018;46(3):607–16.
54. Noyes FR, Huser LE, Jurgensmeier D, Walsh J, Levy MS. Is an anterolateral ligament reconstruction required in ACL-reconstructed knees with associated injury to the anterolateral structures? A robotic analysis of rotational knee stability. Am J Sports Med. 2017;45(5):1018–27.
55. Herbst E, Arilla FV, Guenther D, Yacuzzi C, Rahnemai-Azar AA, Fu FH, et al. Lateral extra-articular Tenodesis has no effect in knees with isolated anterior cruciate ligament injury. Art Ther. 2018;34(1):251–60.
56. Musahl V, Rahnemai-Azar AA, Costello J, Arner JW, Fu FH, Hoshino Y, et al. The influence of meniscal and anterolateral capsular injury on knee laxity in patients with anterior cruciate ligament injuries. Am J Sports Med. 2016;44(12):3126–31.
57. Ferretti A, , Monaco E, , Fabbri M, et al. Prevalence and classification of injuries of anterolateral complex

in acute anterior cruciate ligament tears. Art Ther 2017;33(1):147–154.
58. Helito CP, Helito PVP, Costa HP, Demange MK, Bordalo-Rodrigues M. Assessment of the anterolateral ligament of the knee by magnetic resonance imaging in acute injuries of the anterior cruciate ligament. Art Ther. 2017;33(1):140–6.
59. Inderhaug E, Stephen JM, El-Daou H, Williams A, Amis AA. The effects of anterolateral Tenodesis on tibiofemoral contact pressures and kinematics. Am J Sports Med. 2017;45(13):3081–8.
60. Devitt BM, Bouguennec N, Barfod KW, Porter T, Webster KE, Feller JA. Combined anterior cruciate ligament reconstruction and lateral extra-articular tenodesis does not result in an increased rate of osteoarthritis: a systematic review and best evidence synthesis. Knee Surg Sports Traumatol Arthrosc. 2017;25(4):1149–60.
61. Shimakawa T, Burkhart TA, Dunning CE, Degen RM, Getgood AM. Lateral compartment contact pressures do not increase after lateral extra-articular Tenodesis and subsequent subtotal meniscectomy. Orthop J Sports Med. 2019;7(6):2325967119854657.
62. Getgood A, Brown C, Lording T, et al. The anterolateral complex of the knee: results from the International ALC Consensus Group Meeting. Knee Surg Sports Traumatol Arthrosc. 2019;27:166–76.
63. Lemaire M. Chronic knee instability. Technics and results of ligament plasty in sports injuries. J Chir (Paris). 1975;110(4):281–94.
64. MacIntosh DL, Darby TA. Lateral substitution reconstruction. In: Proceedings of the Canadian Orthopaedic Association. J Bone Joint Surg Br. 1976;58:142.
65. Arnold JA, Coker TP, Heaton LM, Park JP, Harris WD. Natural history of anterior cruciate tears. Am J Sports Med. 1979;7(6):305–13.
66. Ellison AE. Distal iliotibial-band transfer for anterolateral rotatory instability of the knee. J Bone Joint Surg Am. 1979;61(3):330–7.
67. Losee RE, Johnson TR, Southwick WO. Anterior subluxation of the lateral tibial plateau. A diagnostic test and operative repair. J Bone Joint Surg Am. 1978;60(8):1015–30.
68. Andrews JR, Sanders R. A "mini-reconstruction" technique in treating anterolateral rotatory instability (ALRI). Clin Orthop Relat Res. 1983;172:93–6.
69. Neyret P, Palomo JR, Donell ST, Dejour H. Extra-articular tenodesis for anterior cruciate ligament rupture in amateur skiers. Br J Sports Med. 1994;28(1):31–4.
70. Amirault JD, Cameron JC, MacIntosh DL, Marks P. Chronic anterior cruciate ligament deficiency. Long-term results of MacIntosh's lateral substitution reconstruction. J Bone Joint Surg Br. 1988;70(4):622–4.
71. Kennedy JC, Stewart R, Walker DM. Anterolateral rotatory instability of the knee joint. An early analysis of the Ellison procedure. J Bone Joint Surg Am. 1978;60(8):1031–9.
72. Bertoia JT, Urovitz EP, Richards RR, Gross AE. Anterior cruciate reconstruction using the MacIntosh lateral-substitution over-the-top repair. J Bone Joint Surg Am. 1985;67(8):1183–8.
73. Marcacci M, Zaffagnini S, Iacono F, Neri MP, Loreti I, Petitto A. Arthroscopic intra- and extra-articular anterior cruciate ligament reconstruction with gracilis and semitendinosus tendons. Knee Surg Sports Traumatol Arthrosc. 1998;6(2):68–75.
74. Lerat JL, Dupre La Tour L, Herzberg G, Moyen B. Review of 100 patients operated on for chronic anterior laxity of the knee by a procedure derived from the Jones and MacIntosh methods. Value of dynamic radiography for the objective analysis of the results. Rev Chir Orthop Reparatrice Appar Mot. 1987;73(Suppl 2):201–4.
75. Dejour H, Walch G, Neyret P, Adeleine P. Results of surgically treated chronic anterior laxities. Apropos of 251 cases reviewed with a minimum follow-up of 3 years. Rev Chir Orthop Reparatrice Appar Mot. 1988;74(7):622–36.
76. Zarins B, Rowe CR. Combined anterior cruciate-ligament reconstruction using semitendinosus tendon and iliotibial tract. J Bone Joint Surg Am. 1986;68(2):160–77.
77. O'Brien SJ, Warren RF, Wickiewicz TL, Rawlins BA, Allen AA, Panariello R, et al. The iliotibial band lateral sling procedure and its effect on the results of anterior cruciate ligament reconstruction. Am J Sports Med. 1991;19(1):21–4; discussion 4–5.
78. Strum GM, Fox JM, Ferkel RD, Dorey FH, Del Pizzo W, Friedman MJ, et al. Intraarticular versus intraarticular and extraarticular reconstruction for chronic anterior cruciate ligament instability. Clin Orthop Relat Res. 1989;245:188–98.
79. Anderson AF, Snyder RB, Lipscomb AB Jr. Anterior cruciate ligament reconstruction. A prospective randomized study of three surgical methods. Am J Sports Med. 2001;29(3):272–9.
80. Pearl AJ. In: Bergfeld JA, editor. Extra-articular reconstruction in the anterior cruciate ligament deficient knee. Rosemont, IL: American Orthopaedic Society for Sports Medicine; 1992.
81. Webster KE, Feller JA. Exploring the high reinjury rate in younger patients undergoing anterior cruciate ligament reconstruction. Am J Sports Med. 2016;44(11):2827–32.
82. Zaffagnini S, Marcacci M, Lo Presti M, Giordano G, Iacono F, Neri MP. Prospective and randomized evaluation of ACL reconstruction with three techniques: a clinical and radiographic evaluation at 5 years follow-up. Knee Surg Sports Traumatol Arthrosc. 2006;14(11):1060–9.
83. Vadalà AP, Iorio R, De Carli A, Bonifazi A, Iorio C, Gatti A, et al. An extra-articular procedure improves the clinical outcome in anterior cruciate ligament reconstruction with hamstrings in female athletes. Int Orthop. 2013;37(2):187–92.
84. Hewison CE, Tran MN, Kaniki N, Remtulla A, Bryant D, Getgood AM. Lateral extra-articular Tenodesis reduces rotational laxity when combined with anterior

cruciate ligament reconstruction: a systematic review of the literature. Art Ther. 2015;31(10):2022–34.
85. Rezende FC, de Moraes VY, Martimbianco AL, Luzo MV, da Silveira Franciozi CE, Belloti JC. Does combined intra- and extraarticular ACL reconstruction improve function and stability? A meta-analysis. Clin Orthop Relat Res. 2015;473(8):2609–18.
86. Draganich LF, Reider B, Ling M, Samuelson M. An in vitro study of an intraarticular and extraarticular reconstruction in the anterior cruciate ligament deficient knee. Am J Sports Med. 1990;18(3):262–6.
87. Engebretsen L, Lew WD, Lewis JL, Hunter RE. The effect of an iliotibial tenodesis on intraarticular graft forces and knee joint motion. Am J Sports Med. 1990;18(2):169–76.
88. Trojani C, Beaufils P, Burdin G, Bussiere C, Chassaing V, Djian P, et al. Revision ACL reconstruction: influence of a lateral tenodesis. Knee Surg Sports Traumatol Arthrosc. 2012;20(8):1565–70.
89. Ferretti A, Conteduca F, Monaco E, De Carli A, D'Arrigo C. Revision anterior cruciate ligament reconstruction with doubled semitendinosus and gracilis tendons and lateral extra-articular reconstruction. Surgical technique. J Bone Joint Surg Am. 2007;89(Suppl 2 Pt 2):196–213.
90. Sonnery-Cottet B, Saithna A, Cavalier M, Kajetanek C, Temponi EF, Daggett M, et al. Anterolateral ligament reconstruction is associated with significantly reduced ACL graft rupture rates at a minimum follow-up of 2 years: a prospective comparative study of 502 patients from the SANTI study group. Am J Sports Med. 2017;45(7):1547–57.
91. Getgood AMJ, Bryant DM, Litchfield R, Heard M, McCormack RG, Rezansoff A, et al. Lateral extra-articular Tenodesis reduces failure of hamstring tendon autograft anterior cruciate ligament reconstruction: 2-year outcomes from the STABILITY study randomized clinical trial. Am J Sports Med. 2020;48(2):285–97.
92. Zaffagnini S, Marcheggiani Muccioli GM, Grassi A, Roberti di Sarsina T, Raggi F, Signorelli C, et al. Over-the-top ACL reconstruction plus extra-articular lateral Tenodesis with hamstring tendon grafts: prospective evaluation with 20-year minimum follow-up. Am J Sports Med. 2017;45(14):3233–42.
93. Pernin J, Verdonk P, Si Selmi TA, Massin P, Neyret P. Long-term follow-up of 24.5 years after intra-articular anterior cruciate ligament reconstruction with lateral extra-articular augmentation. Am J Sports Med. 2010;38(6):1094–102.

Why Does LET Work?

6

João V. Novaretti and Moisés Cohen

6.1 Introduction

Several lateral extra-articular tenodesis (LET) procedures have been described for treatment of anterior cruciate ligament (ACL) injury [1–5]. Since rotatory knee instability and ACL re-rupture are relevant concerns after ACL reconstruction, there has been an increase in the use of the LET procedure in combination with ACL reconstruction in an attempt to address these issues. In this chapter, we discuss the biomechanical and clinical rationale of why LET procedures work and may assist surgeons to achieve better outcomes after ACL reconstruction.

6.2 Biomechanics

The anterolateral complex (ALC) is a combination of structures on the lateral side of the knee that has been related to the anterolateral rotatory stability [6, 7]. Thus, most biomechanical studies use an ALC deficiency model to investigate the biomechanical functions of various LET procedures, although ALC injury may not be commonly present in patients with ACL injuries. A previous study investigated the contribution of the ALC structures and the ACL in restraining knee kinematics in response to a simulated pivot-shift test in the ACL-intact and ACL-deficient knee [8]. The iliotibial tract (ITT) resisted 31% of drawer force with the ACL cut at 30° of knee flexion and was the primary restraint of internal rotation for both the intact and ACL-deficient knee from 30° to 90° of knee flexion while the other ALC structures provided no significant contribution. Further, the ITT provided 72% of the restraint at 45° during the pivot-shift test for the ACL-deficient group. In conclusion, the ITT showed significant contributions in restraining anterior subluxation of the lateral tibial plateau and tibial internal rotation, which are related to the anterolateral rotatory instability, while the other ALC structures had a minor role. Thus, it may be appropriate to identify and reconstruct ITT injuries in case of suspected anterolateral rotatory instability.

Another study investigated the biomechanical effects of anterolateral capsule injury and modified Lemaire LET in ACL-deficient and ACL-reconstructed knees in response to physical examinations [9]. Anterior translation of the lateral knee compartment and internal tibial rotation during the pivot-shift test were highest in combined ACL-deficient and anterolateral capsule-deficient knees (12.3 mm and 16.3°, respectively). Combined ACL reconstruction and LET reduced the anterior translation of the lateral knee compartment during the pivot-shift test with the

J. V. Novaretti · M. Cohen (✉)
Orthopaedics and Traumatology Sports Center (CETE), Department of Orthopaedics and Traumatology, Paulista School of Medicine (EPM), Federal University of São Paulo, São Paulo, Brazil

N. Nakamura et al. (eds.), *Advances in Knee Ligament and Knee Preservation Surgery*,
https://doi.org/10.1007/978-3-030-84748-7_6

presence of an anterolateral capsule injury while isolated ACL reconstruction did not. However, combined ACL reconstruction and LET led to overconstraint of internal tibial rotation when the anterolateral capsule was intact. Thus, surgeons should consider the potential benefits of LET with ACL reconstruction in the presence of an anterolateral capsule injury as well as the potential deleterious effects in an isolated ACL injury reported in this cadaveric study. A separate study evaluated the Macintosh LET with 20 N and 80 N of graft tension after anterolateral soft tissue transection [10]. Although no overconstraint was observed when the LET was performed with 20 N tensioning, grafts tensioned with 80 N caused significant overconstraint of the knee and increase in lateral tibiofemoral compartment contact pressures.

One study investigated the effects of ACL reconstruction and a type of LET procedure using a gracilis-tendon autograft on joint motion in ACL-deficient knees and in combined ACL and anterolateral capsule-deficient knees [11]. At higher flexion angles, the LET procedure added rotational stability to ACL reconstruction alone in a knee with combined ACL and anterolateral capsule deficiency. Yet, anterior tibial translation was not affected by the addition of the LET procedure. Meanwhile, the additional LET procedure was unnecessary for the isolated ACL-deficient knee, as an ACL reconstruction alone was able to restore the kinematics of the knee. Of note, 2 of 7 specimens showed decreased internal tibial rotation with the combination of ACL reconstruction and LET procedure.

A recent study investigated the effects of a LET procedure with a semitendinosus graft on tibiofemoral compartment contact area and pressures, knee kinematics, and forces [12]. Intact, anterolateral capsule deficient and post-LET knee states were tested. No overconstraint of the knee and no increase in contact pressure nor decrease in contact area in the lateral tibiofemoral compartment were observed after LET with deficient anterolateral capsule. In situ force in the ACL decreased after LET while in situ force in the LET graft was higher than that of the native anterolateral capsule. The lack of knee overconstraint without significant increases in lateral compartment pressures indicates that if an LET with semitendinosus graft is not over tensioned, accelerated degenerative changes in the lateral compartment should not be expected after this procedure. Additionally, LET reduces the in situ force in the ACL in the setting of ALC injury possibly providing a protective effect to the ACL. Yet, when evaluating knee kinematics, LET did not restore all kinematics to the intact knee state. Another study tested LET with a semitendinosus graft to investigate the effects of different knee flexion angles of graft fixation and observed overconstraint of the knee at all flexion angles [13]. A separate study also evaluated LET with a semitendinosus graft and compared it with the Lemaire LET [14]. Significant overconstraint of the knee was observed with both techniques.

6.3 Clinical Studies

Earlier results with different LET procedures combined with ACL reconstruction were discouraging and raised the concern of accelerated knee osteoarthritis due to overconstraint of the knee. One clinical study in 1990 reported chronic lateral knee pain and significant associated morbidity with extra-articular augmentation of ACL reconstruction [15]. A randomized study including 72 patients with ACL injury from 2000 to 2002 had significantly worse subjective, objective, and functional results when ACL reconstruction was combined with LET (sling technique [16]) when compared to isolated ACL reconstruction [17]. In a separate long-term case series, 71% of patients at 24-year follow-up had moderate or severe degenerative changes on radiographs after combined intra- and extra-articular ACL reconstruction with an iliotibial band graft left attached in the tibial insertion [18]. Therefore, these techniques were mainly abandoned for several years with the advances of intra-articular ACL reconstruction techniques. Yet, the persistence of rotatory knee instability after isolated intra-articular ACL reconstruction with current techniques in some cases led to a recent resurgence of the LET procedures in combination with ACL reconstruction.

A recent randomized controlled trial with 618 patients investigated whether the addition of a modified Lemaire LET to a single-bundle, hamstring ACL reconstruction could reduce the risk of ACL reconstruction failure in young, active patients at 2 years of follow-up [19]. Clinical failure was considered when persistent asymmetric pivot-shift or graft failure was detected. The combined ACL reconstruction + LET group had significantly less clinical failure (25% vs. 40%) and less graft rupture (4% vs. 11%) than the ACL reconstruction group. Of note, a number of patients in the ACL reconstruction + LET group reported hardware irritation that necessitated staple removal and also had higher pain scores in the first 3 months compared with the ACL reconstruction group. The results of less graft failure with the addition of a combined extra-articular procedure are in agreement with previous nonrandomized comparative study [20]. With a cohort of 512 young patients participating in pivoting sports, authors investigated outcomes after ACL reconstruction with either patellar tendon autograft or hamstring tendon autograft, the latter with or without combined anterolateral ligament (ALL) reconstruction [20]. At a minimum follow-up of 2 years, the rate of graft failure with ACL reconstruction with hamstring graft combined with ALL reconstruction was 2.5 times less than with ACL reconstruction with patellar tendon autograft and 3.1 times less than with ACL reconstruction with hamstring graft. Yet, no differences were observed with or without the ALL reconstruction in functional knee scores.

A previous randomized controlled trial with 110 patients compared isolated ACL reconstruction and combined ACL with ALL reconstruction with semitendinosus for the ACL and gracilis tendon for the ALL [21]. Interestingly, the ACL and ALL reconstruction group showed significantly less anterior translation using an instrumented knee laxity testing (KT-1000) than the isolated ACL reconstruction group. However, no difference between groups was observed in functional knee scores as well as in clinical examination findings. In contrast, another recent study observed better patient-reported outcomes and return to multi-directional sports after combined LET using the iliotibial band with ACL reconstruction compared with isolated ACL reconstruction [22]. In this study, authors performed LET only in cases with at least one major criterion (high-grade pivot shift and revision ACL reconstruction) or ≥2 minor criteria (hyperlaxity, age <20 years, failed contralateral ACL reconstruction and elite athlete). Additionally, No significant difference in re-operation rate or type of surgery was found between the two groups.

6.4 Conclusions

The current literature shows that each LET procedure may have unique biomechanical behaviors and therefore different results. Different testing systems, loading conditions, fixation locations, type, and tensioning of the graft and fixation angles used in biomechanical studies may all play a role in the conflicting results observed across the literature. Thus, the biomechanical results of one type of LET procedure should not be extrapolated to others. Overall, different LET procedures have shown a role in adding rotational stability to ACL reconstruction when an injury to ALC structures is found combined with ACL injury. Yet, LET procedures should be used with caution for cases of isolated ACL injury since biomechanical studies have shown that it may overconstraint the knee in this scenario.

The exact mechanism of action of LET procedures in the clinical setting cannot be answered, as stated in the recent randomized controlled trial on this subject [19]. One hypothesis is that LET may reduce load on the ACL graft as shown in biomechanical studies [12] and thus may provide a protective effect to the ACL graft while it heals during the ligamentization phase. The added rotatory stability after LET to ACL reconstruction shown in biomechanical studies has yet to be objectively confirmed in clinical studies. Therefore, further clinical studies are needed to evaluate the mid- and long-term outcomes of different LET procedures for rotatory knee instability and may improve our understanding about how LET works.

References

1. Arnold JA. A lateral extra-articular tenodesis for anterior cruciate ligament deficiency of the knee. Orthop Clin North Am. 1985;16:213–22.
2. Bertoia JT, Urovitz EP, Richards RR, Gross AE. Anterior cruciate reconstruction using the MacIntosh lateral-substitution over-the-top repair. J Bone Joint Surg Am. 1985;67:1183–8.
3. Kennedy JC, Stewart R, Walker DM. Anterolateral rotatory instability of the knee joint. An early analysis of the Ellison procedure. J Bone Joint Surg Am. 1978;60:1031–9.
4. Lemaire M. Chronic knee instability. Technics and results of ligament plasty in sports injuries. J Chir (Paris). 1975;110:281–94.
5. Losee RE, Johnson TR, Southwick WO. Anterior subluxation of the lateral tibial plateau. A diagnostic test and operative repair. J Bone Joint Surg Am. 1978;60:1015–30.
6. Vieira EL, Vieira EA, da Silva RT, Berlfein PA, Abdalla RJ, Cohen M. An anatomic study of the iliotibial tract. Arthroscopy. 2007;23:269–74.
7. Musahl V, Herbst E, Burnham JM, Fu FH. The anterolateral complex and anterolateral ligament of the knee. J Am Acad Orthop Surg. 2018;26:261–7.
8. Kittl C, El-Daou H, Athwal KK, et al. The role of the anterolateral structures and the ACL in controlling laxity of the intact and ACL-deficient knee. Am J Sports Med. 2016;44:345–54.
9. Herbst E, Arilla FV, Guenther D, et al. Lateral extra-articular tenodesis has no effect in knees with isolated anterior cruciate ligament injury. Arthroscopy. 2018;34:251–60.
10. Inderhaug E, Stephen JM, El-Daou H, Williams A, Amis AA. The effects of anterolateral tenodesis on tibiofemoral contact pressures and kinematics. Am J Sports Med. 2017;45:3081–8.
11. Guenther D, Irarrazaval S, Bell KM, et al. The role of extra-articular tenodesis in combined ACL and anterolateral capsular injury. J Bone Joint Surg Am. 2017;99:1654–60.
12. Novaretti JV, Arner JW, Chan CK, et al. Does lateral extra-articular tenodesis of the knee affect ACL graft in situ forces and tibiofemoral contact pressures? Arthroscopy. 2020;36(5):1365–73.
13. Schon JM, Moatshe G, Brady AW, et al. Anatomic anterolateral ligament reconstruction of the knee leads to overconstraint at any fixation angle. Am J Sports Med. 2016;44:2546–56.
14. Geeslin AG, Moatshe G, Chahla J, et al. Anterolateral knee extra-articular stabilizers: a robotic study comparing anterolateral ligament reconstruction and modified lemaire lateral extra-articular tenodesis. Am J Sports Med. 2018;46:607–16.
15. Sgaglione NA, Warren RF, Wickiewicz TL, Gold DA, Panariello RA. Primary repair with semitendinosus tendon augmentation of acute anterior cruciate ligament injuries. Am J Sports Med. 1990;18:64–73.
16. Marcacci M, Zaffagnini S, Iacono F, Neri MP, Loreti I, Petitto A. Arthroscopic intra- and extra-articular anterior cruciate ligament reconstruction with gracilis and semitendinosus tendons. Knee Surg Sports Traumatol Arthrosc. 1998;6:68–75.
17. Zaffagnini S, Bruni D, Russo A, et al. ST/G ACL reconstruction: double strand plus extra-articular sling vs double bundle, randomized study at 3-year follow-up. Scand J Med Sci Sports. 2008;18:573–81.
18. Yamaguchi S, Sasho T, Tsuchiya A, Wada Y, Moriya H. Long term results of anterior cruciate ligament reconstruction with iliotibial tract: 6-, 13-, and 24-year longitudinal follow-up. Knee Surg Sports Traumatol Arthrosc. 2006;14:1094–100.
19. Getgood AMJ, Bryant DM, Litchfield R, et al. Lateral extra-articular tenodesis reduces failure of hamstring tendon autograft anterior cruciate ligament reconstruction: 2-year outcomes from the STABILITY study randomized clinical trial. Am J Sports Med. 2020;48:285–97.
20. Sonnery-Cottet B, Saithna A, Cavalier M, et al. Anterolateral ligament reconstruction is associated with significantly reduced ACL graft rupture rates at a minimum follow-up of 2 years: a prospective comparative study of 502 patients from the SANTI study Group. Am J Sports Med. 2017;45:1547–57.
21. Ibrahim SA, Shohdy EM, Marwan Y, et al. Anatomic reconstruction of the anterior cruciate ligament of the knee with or without reconstruction of the anterolateral ligament: a randomized clinical trial. Am J Sports Med. 2017;45:1558–66.
22. Rowan FE, Huq SS, Haddad FS. Lateral extra-articular tenodesis with ACL reconstruction demonstrates better patient-reported outcomes compared to ACL reconstruction alone at 2 years minimum follow-up. Arch Orthop Trauma Surg. 2019;139:1425–33.

The Evidence Regarding ACL Repair

7

Andreas Persson, Gilbert Moatshe, and Lars Engebretsen

7.1 ACL Repair: Last Millennium and Current Status

7.1.1 Historical Pearls

A 41-year-old miner was admitted to the General Infirmary at Leeds, 1895, complaining of instability and weakness of the right knee. He sustained the injury 9 months earlier, together with multiple rib fractures and a fracture of the left leg, falling from a height at work. Surgeon A. W. Mayo Robson performed open surgery, confirming the clinical diagnosis of bicruciate ligament injury. He found both ligaments avulsed from their proximal attachments and sutured them to their upper attachments with catgut sutures. The immediate postoperative period was without complications and 6 years after surgery the patient reported that "his leg is perfectly strong," but with little aching on the medial side after being overworked especially in cold and damp weather. Mayo Robson claimed that repair by operation "is both feasible and hopeful as to its ultimate result" [1]. This is the first described suture repair of the anterior cruciate ligament (ACL), and in the following years, at the beginning of the twentieth century, there were only a few more reports of surgical management of this injury.

In 1913, the German assistant surgeon dr H. Goetjes reported a review of in total 30 cases of cruciate ligament tears or tibial spine fractures [2]. In case of a substance tear of the ACL with possibility to adapt the ruptured end, he recommended to reconnect the stumps with a suture. If not adaptable, he recommended the stumps to be resected. This treatment philosophy was debated, and in 1916 Sir Robert Jones stated that "The operation of stitching the ligaments is absolutely futile. Natural cicatricial tissue, allowed to mature without being stretched, is the only reliable means of repair" [3]. The first augmented repair was described in 1927 by Erwin Payr of Leipzig. He, like Goetjes, acknowledged stump atrophy, but described usage of an augment with a strip of fascia tensor latae to fill the tissue defect [4].

In 1938, Ivar Palmer of Stockholm published his thesis in which he accounted for all aspects of knee injuries and their treatment [5]. He recommended an early repair for all ACL ruptures, but also acknowledged that some patients have long-standing cruciate ligament injuries without any symptoms. The general approach at the time was either to do nothing, or to excise the torn ligament.

A. Persson (✉)
Department of Orthopedic Surgery, Martina Hansens Hospital, Bærum, Norway

Oslo Sports Trauma Research Center, Oslo, Norway

G. Moatshe · L. Engebretsen
Oslo Sports Trauma Research Center, Oslo, Norway

Department of Orthopaedic Surgery, Oslo University Hospital, Oslo, Norway
e-mail: lars.engebretsen@medisin.uio.no

N. Nakamura et al. (eds.), *Advances in Knee Ligament and Knee Preservation Surgery*,
https://doi.org/10.1007/978-3-030-84748-7_7

O'Donoghue popularized the ACL suture repair in the USA. In 1950, he stated that "only by complete repair of all structures will these injuries show maximum recovery" [6]. He found best results in his patients when operated early after injury, independent of the severity of the injury and concluded that "surgery should not be reserved for those cases in which conservative treatment has failed" [7]. At that time, arthroscopy was still not widely recognized as a treatment or diagnostic tool. Only in 1976, Ejnar Eriksson recommended acute arthroscopy at clinical suspicion of and ACL tear, and suture if the ligament was found ruptured [8]. Further positive short-term results from numerous authors led to ACL suture being considered the treatment of choice in the 1970s and 1980s.

Some worrying outcomes were acknowledged, often cited Feagin and Curl presented their 5-year results from their cohort of 64 cadets in 1976. Most patients in their previous report with 2-year follow-up had good to excellent results, but after 5 years their results were discouraging; "It is doubtful to us that surgery in its present form will ever prove successful, and therefore we are justified to consider tendon transfer or appropriate synthetic substitution" [9]. As an answer to the possible problems with the ACL suture, there was an increased interest by industry to develop synthetic ligaments made out of novel materials such as gore-tex, carbon, and dacron in the 1970s. The results were often disastrous with intraarticular foreign body reactions and high failure rates [10–12]. John C. Kennedy developed together with 3M the Kennedy-Ligament Augmentation Device. It was made out of braided polypropylene fibers designed to share the load and protect the host tissue during its initial fragile healing phase. He claimed that "The evolution of synthetic materials to aid in the repair of ACL instability is in an embryonic stage of development" [13].

In 1989, Andersson et al. published the first prospective randomized study on the topic of ACL repair. In this 3-armed prospective randomized study, the results of the ACL repair were similar to those where the injured ACL was left untreated, and an augmented repair had better overall function and higher activity level [14]. When Engebretsen et al. published a randomized controlled study comparing ACL repair to a Kennedy-LAD augmented repair or bone-patellar tendon-bone (BTB) reconstruction in 1990, the results were clear—the repair group deteriorated and the augmented repair did not do as well as the BTB reconstruction at 2-year follow-up [15]. Several reports simultaneously showed consistently good or excellent results for the ACL reconstruction reported by Clancy, Shelbourne, and Eriksson among others [16–19], led to a shift in treatment and ACL repair with or without augment was no longer considered as a treatment option in the end of the 1990s.

7.1.2 Shift of Focus: Repair Re-invented?

The interest for the ACL repair has been increasing in recent years. By comparing hits on PubMed for the search strings "ACL reconstruction" and "primary ACL repair," we find the current exponential increase in literature (Fig. 7.1). The difference between those two search strings was present already in the 1980s, before the suture technique was abandoned. Thomas Kuhn introduced the concept *Paradigm shift* in 1962 as a "... fundamental change in the basic concepts and experimental practices of a scientific discipline" [20]. One can argue that the *experimental practice* was changed from the 1980s, with more research conducted on ACL reconstructions compared with ACL repair; hence, there was a paradigm shift. This is a logical consequence and development with the knowledge present at that time. However, this does not mean that all factors have been considered, and that there are no more questions to be answered regarding *who* could benefit from an ACL repair, and *how* it potentially should be performed. In the light of the development of arthroscopic surgery and todays' improved framework to conduct studies, it could be reasonable to ask;

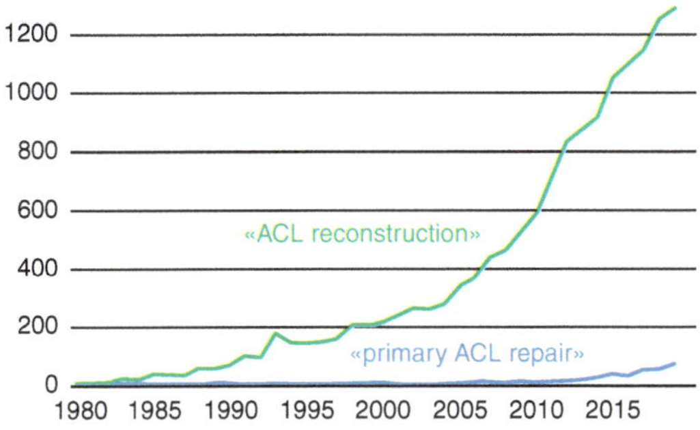

Fig. 7.1 Hits on PubMed for the strings "ACL reconstruction" and "primary ACL repair" from 1980 through 2019

1. Are there new surgical techniques or modern postoperative treatment options that have documented an acceptable long-term outcome after ACL repair?

or

2. Is there any evidence that subgroups of patients would benefit from a repair rather than a reconstruction?

And if not;

3. Are the possible benefits with an ACL repair compared with reconstruction large enough to defend its resurge in scientific studies and clinical practice?

Based on the current literature, these questions will be discussed in the following sections.

7.2 What Has Changed?

During the last decades there has been a change in the treatment of a typical ACL deficient patient. A broader availability of medical services and diagnostic tools has improved patient outcomes. Magnetic resonance imaging enhances the possibility to assess concomitant injuries and ACL tear patterns opening for a diverse treatment approach to different injury patterns.

Originally, the suture of the ACL was performed through open surgery while modern surgical technique is mainly arthroscopically assisted with the advantage being decreased surgical trauma. In addition, rehabilitation protocols have changed drastically, from a postoperative standard of 4–6 weeks of immobilization in a cast/splint [21, 22] to immediate weight-bearing and full range of motion for the repaired isolated ACL injury [23]. Postoperative patellofemoral pain, joint stiffness, and general loss of function [9, 24, 25] are symptoms that are likely to be improved by modern rehabilitation protocols. The recommended time for return-to-play for athletes was not always described in the historical literature and did not seem to have been given the same importance as in modern treatment. This could be important for allowing sufficient time for healing and rehabilitation. Strand et al. suggested that even though the failure rate at long-term follow-up is unacceptably high, the results when performed with new arthroscopic technique and modern rehabilitation should be investigated [26].

The treatment of concomitant injuries at the time of surgery will affect outcomes. O'Donoghue stated in 1960 "... the damaged or displaced meniscus must be removed.." [7], in contrast to today's slogan; "save the meniscus" [27–29]. For some patients, the biomechanical support offered by a repaired meniscus may improve outcome in ACL repairs in the same way as for ACL reconstructed patients [30].

7.2.1 Techniques: What's New?

Recent advances in knowledge of ACL healing and tissue engineering have led to an evolution of surgical techniques involving ACL repair. A group of researchers lead by Dr. M. M. Murray in Boston hypothesizes that the reason for that the ACL does not heal, in contrast to the ligament healing of the MCL, is because of the negative effects of intra-articular plasminogen degrading the formation of a bridging blood clot formation at the rupture site [31]. After substantial background research, including animal testing, they have developed an open technique using a xenograft as a provisional collagen scaffold soaked in the patient's full blood allowing for cell ingrowth and proliferation at the ACL rupture site, combined with a mechanical suture repair [32]. The results from their first-in-human study comparing the Bridge-Enhanced Anterior Cruciate Ligament Repair (BEAR) technique with hamstring reconstructions presented promising results, but with short follow-up [33]. A recent randomized controlled trial demonstrated noninferiority of subjective knee function and instrumented laxity in the BEAR group to patients treated with ACL reconstruction (ACLR) at 2 years [34]. However, it is worth noting that 14% of the patients in the BEAR group compared with 6% of the patients in the ACLR group required a conversion to ACLR/revision ACLR during the first 2 years. Even though the difference in hard failure rates did not reach statistical significance ($p = 0.32$), a 14% revision rate to ACLR in the young population already at 2 years requires further discussion. This is important especially when considering most of the earlier studies also reported good outcomes at short term follow-up that was not maintained at long term follow-up.

In Switzerland, a "dynamic intraligamentary stabilization" (DIS) device (Ligamys; Mathys Medical) has been developed and tested in an animal model [35]. Their theory is that the biologic healing capacity of the ACL is dependent on 8 weeks of immobilization, and to overcome the stiffness problems of traditional immobilization the repair is supported by braided polyethylene wire connected to a dynamic screw-spring mechanism and the patients are allowed full range of motion postoperatively. Similarly, static repair augmentation with a braided polyethylene tape (Internal Brace, Arthrex) is supposed to support the healing ligament and allow for early rehabilitation.

In a recent systematic review by Hoogeslag et al., the authors concluded that "The current overall level and quality of evidence regarding contemporary ACL suture repair are poor" and they currently find no indication for usage of a suture repair technique for the acute ACL injury. However, further they discuss whether a "bridging collagen scaffold may improve future outcomes for all ACL suture repair groups" [36].

7.3 Did We Miss Something?

7.3.1 Tear Site: What's the Evidence?

Present advocates of ACL repair hold the location of the ACL tear as an important factor which could affect outcome, and indicate their belief that this has been forgotten in the practice away from ACL repair. In frequently cited historical studies in favor of abandoning repair [9, 15], all ACL tear types were included. Other highly cited historical papers by Weaver and Genelin et al. [37, 38] on the topic have been presented to show evidence for better outcomes with proximal ACL ruptures, while others claim that tissue quality is also of importance in order to perform a successful repair [39].

7.3.2 Proximal Tears: Historical Literature (→ 1990)

In a retrospective report from Weaver et al., they investigated the results from patients who had undergone either isolated ACL repair or in combination with medial collateral ligament (MCL) repair in the skiing village of Aspen [37]. They found 304 patients treated at their clinic during the period 1976–1979, and in total 104 (31%) of those patients came for clinical examination at a mean follow-up of 42 months. In 92 of those patients, information regarding the type of ACL injury was available in the operation notes. The success rate according to cruciate disruption site were reported to be 79%, 23%, 100%, and 100% in the groups "avulsed from femur" ($n = 66$), "interstitial tear" ($n = 13$), "anterior tibial spine avulsion" ($n = 10$), and "distal avulsion" ($n = 3$), respectively. The comparison of the numbers between "avulsed from femur" and "interstitial tears" is the part of the paper that is used to highlight the success of repair of type 1 tears (avulsed from femur). In contrast, the authors' conclusion of the study was that "... patients do not do well following primary ACL repair" and "... even in the most favorable groups the success rate is not high enough to deny the patients supplementary procedures...".

Genelin et al. reported on patients operated between 1982 and 1984 [38]; at 5–7 years follow-up 42 patients (86%) of the original 49 were available for reexamination. The average age at surgery was 27 years. A "totally stable knee" was found in only 29% of the patients; however, 81% had a side-to-side difference with a KT-1000 of less than 3 mm and 43% of the patients reported "no subjective complaints whatsoever." Only 6 out of 12 patients with "totally stable knees" had no subjective complaints. The authors discussed the discrepancy between patient's satisfaction and knee stability and if this could be improved by a different post-operative treatment. Even though 29% of the patients were clinically stable, they claimed to have better results than similar studies, and hypothesized that this could be due to their selection of patients with proximal ACL tears. As there was no control group in this study, one can only speculate if similar patient satisfaction and knee stability would be obtained in patients treated non-surgically.

Sherman et al. performed an extensive retrospective review of patients operated in 1979–1984 with the Marshall multiple suture technique [40]. Of the original 106 patients, 50 were available for reexamination at a mean follow-up of just above 5 years. They reported a higher proportion of stable knees (negative Lachman's test and negative pivot-shift test) compared with Genelin et al. (46% vs. 29%). Amongst the patients with stable knees, there was a higher proportion of type 1 ACL tears (femoral avulsion rupture) compared with type IV ACL tears (midsubstance rupture) (35% vs 17%, respectively). However, this comparison was not statistically significant.

Recently, Van der List and DiFelice performed a systematic review analyzing historic literature aiming to investigate the role of tear location on the outcome of ACL repairs [41]. For the different outcomes, they analyzed the correlation between the percentage of proximal tears in the individual studies to the percentage of success of the different outcome measures of interest. Despite that only "patient satisfaction" had a significant moderate positive correlation (correlation coefficient 0.56, $p = 0.01$), the authors concluded that the "Tear location seems to have played a role on the outcomes of open primary ACL repairs." They found no statistically significant correlation when investigating stability testing (KT-1000, Lachman test, anterior drawer test, pivot shift test), Lysholm scores, Tegner scores, return to sports or failure rates. In addition, there are several points that induces a great risk of bias in this systematic review; the heterogeneity of the included studies, such as common data definitions, standardized collected outcome measures, and time-points for these, and most studies are non-comparable retrospective cohort studies or case series.

In summary, there is very weak evidence that proximal tears have better outcome compared to tears at other locations from a historical perspective.

7.3.3 Proximal Tears: Modern Literature (1991 →)

The majority of the literature available on arthroscopic ACL repair is based on studies from the last decade. To the author's knowledge, there is only one study that reports on a "classic" ACL repair with an isolated suture pull-out technique [42]. Most studies report on a heterogeneous group of proximal ruptures and mid-substance ruptures [43], and only one study compares outcome between tear patterns [44]. Current literature is focused on the previously mentioned three repair techniques previously mentioned (BEAR implant, Ligamys, and Internal Brace) in addition to non-augmented repair using suture anchors placed in the femoral ACL footprint.

7.3.3.1 Internal Brace

Jonkergouw et al. reported retrospectively on 27 consecutive patients treated with a ACL suture anchor repair augmented with InternalBrace, exclusively on proximal tears [45]. At a minimum of 2-year follow-up, 92% of the patients were available and 2 of the patients (8%) were reported to have a re-tear of their ACL, and another two patients required removal of tibial hardware. Smith et al. presented results on two patients (5 and 6 years old) with proximal ACL suture repair combined with temporary augmentation with non-absorbable braided tape (FiberTape, Arthrex, Naples, FL, USA), a static augment similar to Internal Brace [46]. The authors reported excellent outcome at 1 and 2 years. Dabis et al. also argued that the procedure is safe and has excellent short-term outcomes for the adolescent population following their patients for 2 years [47]. Even though some results for treating proximal tears with a static augment are encouraging, their relevance is still limited due to their short follow-up time and small patient populations.

7.3.3.2 Ligamys

For the DIS technique with Ligamys, two studies report outcomes after repair of proximal ruptures. Ateschrang et al. included 47 patients followed for 12 months [48]. At final follow-up, 37 patients (68%) were examined and a rather high proportion of failures (KT-1000 >5 mm and/or <50% of ACL restoration volume upon arthroscopic re-evaluation) were detected (14%). In a study by Ahmad et al. [49], 71 patients were treated for acute ACL ruptures in the period 2011–2013 with the same technique. Thirteen were lost to follow-up and four patients were excluded due to mid-substance or distal ruptures. At a median of 6.3 years, 57 patients were available for follow-up ($n = 48$; physical follow-up, $n = 13$; telephonic follow-up), an overall 70% survivorship was reported. However, 29 patients (60%) were reported to have an "intact DIS" in the objective IKDC measure, and the real failure rate is therefore somewhat confusing. Even though the follow-up is mid to long term, the failure rate is high and there is a high drop-out.

7.3.3.3 BEAR Implant

There is no study on the BEAR implant technique that includes only proximal ACL tears.

7.3.3.4 Suture-Anchor Technique

In a systematic review of clinical outcomes after proximal repairs with either a suture pull-out technique or repair by suture anchors, Houck et al. concluded that the literature is limited (six studies included) with inconsistent failure rates (0–25% at 28–79 months' follow-up [50]. The outcome measures (failure/reoperation rates) also differed slightly between the studies, limiting the possibility of comparison.

DiFelice et al. reported their results for patients with a mean follow-up of 6 years in a retrospective review [51]. Eleven patients were originally included, and the ACL stump was fixed with 2 suture anchors. At final follow-up, one patient was classified as a failure (9%) due to an early re-rupture, and one patient had slight laxity at clinical evaluation (Lachman 1+, Pivot shift test grade 1A) but with good subjective outcome.

Hoffmann et al. included 13 patients with a mean follow-up of 6.6 years in their retrospective study [52]. They used one anchor in addition to microfracture of the ACL footprint, and ten patients were available for clinical examination. Of those patients, 2 had a side-to-side difference of >3 mm at Rolimeter testing at 30° of flexion (20%), and one more patient not available for clinical examination reported 1 give-way episode/month.

In contrast to the older patients in DeFelice's and Hoffmann's cohorts (average 43 and 37 years, respectively), Bigoni et al. reported on five patients with an average age of 9 years at surgery [53]. The proximal tears were reinserted with bioabsorbable suture anchors, and at a mean follow-up of 3.6 years one patient was reported to have a grade 1 Lachman test. The mean KT-1000 side-to-side difference was 3 mm (2–4 mm) for the whole group, which may indicate that there was a significant laxity in many of those small adolescent knees.

Overall, there are only a few non-comparative studies for proximal tears performed with modern techniques and their results and design is not robust enough to guide present clinical practice. Furthermore, most of the studies included few patients, did not have a control group and had a short follow up.

7.4 Possible Advantages with ACL Suture Repair

If the outcomes would be the same, the advantages for the patients with ACL suture compared with ACL reconstruction are intriguing. No donor site morbidity and less surgical trauma would lead to a faster recovery. Furthermore, preserving the native ACL tissue may maintain proprioceptive functions [54], which is important in neuromuscular control and overall knee function. From the surgeons' perspective, a failed repair can be treated almost as a primary reconstruction with all autografts available and no bone loss.

7.5 Future Studies

Taking the possible advantages of an ACL suture repair into account, one can argue that despite the discouraging results so far, we should aim to find subgroups where these techniques are successful. While bearing this in mind, a shift back to an accepted indication for early surgery for more patients would lead to an increase in sales for industry, and perhaps overtreatment of patients who would manage with non-operative management.

Most of the current literature on ACL repair is based on few patients with heterogeneous tear patterns, different repair techniques, and short-term follow-up. Until larger studies with robust and stringent methodology and longer term follow-up are available, it is difficult to determine the role of ACL repair in clinical practice. Anterior cruciate ligament repair should be practiced in a controlled and preferably in a multicenter study setting, in order to improve the current knowledge.

7.6 Conclusions

- The interest in ACL suture repair has increased the last decade.
- Historic or contemporary literature does not clearly support that the outcome is better for suturing proximal ACL tears as opposed to other tear patterns.
- There is a risk of bias duc to industry involvement in studies performed evaluating ACL repair.
- There are theoretical advantages with suture repair compared with ACL reconstruction, and subgroups of patients may have good outcomes with repair at long follow-up.
- There is an urgent need for independent comparative randomized studies comparing suture repair techniques to ACL reconstruction, or to non-operative treatment.
- Currently, ACL suture repair should not be used in everyday clinical practice, only in the setting of a research study.

References

1. MayoRobson AW. VI. ruptured crucial ligaments and their repair by operation. Ann Surg. 1903;37(5):716–8.
2. Goetjes H. Über verletzungen der ligamenta cruciata des kniegelenks. Dtsch Z Chir. 1913;123:221–89.
3. Jones R. Disabilities of the knee joint. Br Med J. 1916;2:169–73.
4. Payr E. Der heutige Stand der Gelenkchirurgie Verhandlungen der Deutschen Gesellschaft für Chirurgie, 21st Congress. Arch klin Chir. 1927;148:404–521.
5. Palmer I. On the injuries to the ligaments of the knee joint: a clinical study. Acta Chir Scand. 1938;53(Suppl):41–56.
6. O'Donoghue DH. Surgical treatment of fresh injuries to the major ligaments of the knee. J Bone Joint Surg Am. 1950;32 A(4):721–38.
7. O'Donoghue DH. Surgical treatments of injuries to the knee. Clin Orthop Relat Res. 1960;18(11).
8. Eriksson E. Sports injuries of the knee ligaments: their diagnosis, treatment, rehabilitation, and prevention. Med Sci Sports. 1976;8(3):133–44.
9. Feagin JA Jr, Curl WW. Isolated tear of the anterior cruciate ligament: 5-year follow-up study. Am J Sports Med. 1976;4(3):95–100.
10. Indelicato PA, Pascale MS, Huegel MO. Early experience with the GORE-TEX polytetrafluoroethylene anterior cruciate ligament prosthesis. Am J Sports Med. 1989;17(1):55–2.
11. Wredmark T, Engstrom B. Five-year results of anterior cruciate ligament reconstruction with the Stryker Dacron high-strength ligament. Knee Surg Sports Traumatol Arthrosc. 1993;1(2):71–5.
12. Woods GA, Indelicato PA, Prevot TJ. The Gore-Tex anterior cruciate ligament prosthesis. Two versus three year results. Am J Sports Med. 1991;19(1):48–55.

13. Kennedy JC. Application of prosthetics to anterior cruciate ligament reconstruction and repair. Clin Orthop Relat Res. 1983;172:125–8.
14. Andersson C, et al. Surgical or non-surgical treatment of acute rupture of the anterior cruciate ligament. A randomized study with long-term follow-up. J Bone Joint Surg Am. 1989;71(7):965–74.
15. Engebretsen L, et al. A prospective, randomized study of three surgical techniques for treatment of acute ruptures of the anterior cruciate ligament. Am J Sports Med. 1990;18(6):585–90.
16. Clancy WG Jr, et al. Anterior cruciate ligament reconstruction using one-third of the patellar ligament, augmented by extra-articular tendon transfers. J Bone Joint Surg Am. 1982;64(3):352–9.
17. Franke K. Clinical experience in 130 cruciate ligament reconstructions. Orthop Clin North Am. 1976;7(1):191–3.
18. Eriksson E. Reconstruction of the anterior cruciate ligament. Orthop Clin North Am. 1976;7(1): 167–79.
19. Shelbourne KD, et al. Anterior cruciate ligament injury: evaluation of intraarticular reconstruction of acute tears without repair. Two to seven year followup of 155 athletes. Am J Sports Med. 1990;18(5):484–8. discussion 488–9.
20. Kuhn TS. The structure of scientific revolutions. University of Chicago Press; 1962.
21. Strand T, et al. Knee function following suture of fresh tear of the anterior cruciate ligament. Acta Orthop Scand. 1984;55(2):181–4.
22. Marshall JL, Warren RF, Wickiewicz TL. Primary surgical treatment of anterior cruciate ligament lesions. Am J Sports Med. 1982;10(2):103–7.
23. van Melick N, et al. Evidence-based clinical practice update: practice guidelines for anterior cruciate ligament rehabilitation based on a systematic review and multidisciplinary consensus. Br J Sports Med. 2016;50(24):1506–15.
24. Kaplan N, Wickiewicz TL, Warren RF. Primary surgical treatment of anterior cruciate ligament ruptures. A long-term follow-up study. Am J Sports Med. 1990;18(4):354–8.
25. Fetto JF, Marshall JL. The natural history and diagnosis of anterior cruciate ligament insufficiency. Clin Orthop Relat Res. 1980;147:29–38.
26. Strand T, et al. Long-term follow-up after primary repair of the anterior cruciate ligament: clinical and radiological evaluation 15-23 years postoperatively. Arch Orthop Trauma Surg. 2005;125(4):217–21.
27. Lubowitz JH, Poehling GG. Save the meniscus. Arthroscopy. 2011;27(3):301–2.
28. Seil R, Becker R. Time for a paradigm change in meniscal repair: save the meniscus! Knee Surg Sports Traumatol Arthrosc. 2016;24(5):1421–3.
29. Pujol N, Beaufils P. Save the meniscus again! Knee Surg Sports Traumatol Arthrosc. 2019;27(2):341–2.
30. Robb C, et al. Meniscal integrity predicts laxity of anterior cruciate ligament reconstruction. Knee Surg Sports Traumatol Arthrosc. 2015;23(12):3683–90.
31. Murray MM, et al. Histological changes in the human anterior cruciate ligament after rupture. J Bone Joint Surg Am. 2000;82(10):1387–97.
32. Murray MM, Fleming BC. Biology of anterior cruciate ligament injury and repair: kappa delta ann doner Vaughn award paper 2013. J Orthop Res. 2013;31(10):1501–6.
33. Murray MM, et al. Bridge-enhanced anterior cruciate ligament repair: two-year results of a first-in-human study. Orthop J Sports Med. 2019;7(3):2325967118824356.
34. Murray MM, et al. Bridge-enhanced anterior cruciate ligament repair is not inferior to autograft anterior cruciate ligament reconstruction at 2 years: results of a prospective randomized clinical trial. Am J Sports Med. 2020;48(6):1305–15.
35. Kohl S, et al. Anterior crucial ligament rupture: self-healing through dynamic intraligamentary stabilization technique. Knee Surg Sports Traumatol Arthrosc. 2013;21(3):599–605.
36. Hoogeslag RAG, et al. Efficacy of nonaugmented, static augmented, and dynamic augmented suture repair of the ruptured anterior cruciate ligament: a systematic review of the literature. Am J Sports Med. 2020;48(14):3626–37.
37. Weaver J, et al. Primary knee ligament repair–revisited. Clin Orthop Relat Res. 1985;(199):185–91.
38. Genelin F, et al. Late results following proximal reinsertion of isolated ruptured ACL ligaments. Knee Surg Sports Traumatol Arthrosc. 1993;1(1):17–9.
39. Higgins RW, Steadman JR. Anterior cruciate ligament repairs in world class skiers. Am J Sports Med. 1987;15(5):439–47.
40. Sherman MF, et al. The long-term followup of primary anterior cruciate ligament repair. Defining a rationale for augmentation. Am J Sports Med. 1991;19(3):243–55.
41. van der List JP, DiFelice GS. Role of tear location on outcomes of open primary repair of the anterior cruciate ligament: a systematic review of historical studies. Knee. 2017;24(5):898–908.
42. Mukhopadhyay R, et al. ACL femoral avulsion repair using suture pull-out technique: a case series of thirteen patients. Chin J Traumatol. 2018;21(6):352–5.
43. Nwachukwu BU, et al. Anterior cruciate ligament repair outcomes: an updated systematic review of recent literature. Arthroscopy. 2019;35(7):2233–47.
44. Ateschrang A, et al. Improved results of ACL primary repair in one-part tears with intact synovial coverage. Knee Surg Sports Traumatol Arthrosc. 2019;27(1):37–43.
45. Jonkergouw A, van der List JP, DiFelice GS. Arthroscopic primary repair of proximal anterior cruciate ligament tears: outcomes of the first 56 consecutive patients and the role of additional internal bracing. Knee Surg Sports Traumatol Arthrosc. 2019;27(1):21–8.
46. Smith JO, et al. Paediatric ACL repair reinforced with temporary internal bracing. Knee Surg Sports Traumatol Arthrosc. 2016;24(6):1845–51.

47. Dabis J, et al. Paediatric proximal ACL tears managed with direct ACL repair is safe, effective and has excellent short-term outcomes. Knee Surg Sports Traumatol Arthrosc. 2020;28(8):2551–6.
48. Ateschrang A, et al. Recovery of ACL function after dynamic intraligamentary stabilization is resultant to restoration of ACL integrity and scar tissue formation. Knee Surg Sports Traumatol Arthrosc. 2018;26(2):589–95.
49. Ahmad SS, et al. Seventy percent long-term survival of the repaired ACL after dynamic intraligamentary stabilization. Knee Surg Sports Traumatol Arthrosc. 2020;28(2):594–8.
50. Houck DA, et al. Primary arthroscopic repair of the anterior cruciate ligament: a systematic review of clin ical outcomes. Arthroscopy. 2019;35(12):3318–27.
51. DiFelice GS, van der List JP. Clinical outcomes of arthroscopic primary repair of proximal anterior cruciate ligament tears are maintained at mid-term follow-up. Arthroscopy. 2018;34(4):1085–93.
52. Hoffmann C, et al. Primary single suture anchor refixation of anterior cruciate ligament proximal avulsion tears leads to good functional mid-term results: a preliminary study in 12 patients. J Orthop Surg Res. 2017;12(1):171.
53. Bigoni M, et al. Arthroscopic anterior cruciate ligament repair for proximal anterior cruciate ligament tears in skeletally immature patients: surgical technique and preliminary results. Knee. 2017;24(1):40–8.
54. Kennedy JC, Alexander IJ, Hayes KC. Nerve supply of the human knee and its functional importance. Am J Sports Med. 1982;10(6):329–35.

8 Laxity Objective Measurement Within MRI of ACL Lesions

Rogério Pereira, Renato Andrade, Sofia Florim, José Alberto Duarte, and João Espregueira-Mendes

8.1 Laxity Versus Instability

Joint laxity is an objective and measurable parameter. Within human joints, we may have physiological laxity (normal laxity) or pathological laxity (abnormal laxity). Common language and definitions are crucial to enable a clear and constructive communication and scientific discussion. Back in 2006, during the works of the Anterior Cruciate Ligament (ACL) Study Group, John Feagin addressed the audience making a simplified but pretty clear distinction between instability and joint laxity, often used in an interchangeable manner. He stated that "instability is a symptom described by a patient, whereas laxity is an objective finding" [1]. Instability is present when the individual describes the joint as unstable when moving, walking, running, jumping, or twisting. Frequently, patients will refer that the joint "gives way". Biomechanically joint laxity is the passive response of a joint to an externally applied force or torque [2]. The presence of abnormal laxity may or may not exist along with instability. The joint laxity profile varies among individuals. Differences in joint laxity have been reported related to sex [3–5], bone morphology

R. Pereira (✉)
Clínica Espregueira—FIFA Medical Centre of Excellence, Porto, Portugal

Dom Henrique Research Centre, Porto, Portugal

Faculty of Sports, University of Porto, Porto, Portugal

Health School, University Fernando Pessoa, Porto, Portugal
e-mail: rp@espregueira.com

R. Andrade
Porto Biomechanics Laboratory (LABIOMEP), Faculty of Sports, University of Porto, Porto, Portugal
e-mail: randrade@espregueira.com

S. Florim
Centro Hospitalar Vila Nova de Gaia/Espinho, Gaia, Portugal

J. A. Duarte
Faculty of Sports, University of Porto, Porto, Portugal
e-mail: jarduarte@fade.up.pt

J. Espregueira-Mendes
Clínica Espregueira—FIFA Medical Centre of Excellence, Porto, Portugal

Dom Henrique Research Centre, Porto, Portugal

3B's Research Group–Biomaterials, Biodegradables and Biomimetics, University of Minho, Headquarters of the European Institute of Excellence on Tissue Engineering and Regenerative Medicine, Guimarães, Portugal

ICVS/3B's–PT Government Associate Laboratory, Braga/Guimarães, Portugal

School of Medicine, University of Minho, Braga, Portugal
e-mail: jem@espregueira.com

N. Nakamura et al. (eds.), *Advances in Knee Ligament and Knee Preservation Surgery*,
https://doi.org/10.1007/978-3-030-84748-7_8

and morphometrics [6–9], in presence of ligament or menisci injury [10–14], and outcomes of surgery [15–18], among others.

8.2 Measurement of Joint Laxity

When measuring any parameter, or you do it or you do not within an acceptable range. Resolution, precision, and accuracy of the measuring device are critical to enable the development of a screening system with clinical usefulness supported by its sensitivity and specificity. It should not be uncritically accepted the existence of important disparities in measurements outcomes (both under arthrokinematics and clinical views) that are obtained by different professionals, techniques, or devices [19] since they may mislead inappropriate interventions or absence of it. Both arthrokinematic and clinical outcomes may hinder the safety and efficacy that should support, by default, the clinical interventions. Precision and accuracy of joint laxity measurements should fit within physics conformity frames and not in general practices frames. Once it is considered a specific parameter, testing significance and setting, for the same person within a determined anatomofunctional status and period of human development, the outcomes expected must be reproducible and accurate. Methodological rigor is indispensable for research validity, usefulness of joint laxity measurements, and especially safety and effectiveness. We acknowledge though that majority of tests and testing principles yield value. Even facing different outcomes when assessing a same parameter (e.g., knee sagittal joint laxity as a quantity under observation) with different techniques or instruments, they may yield clinical useful information if the resolution and precision are suitable, and the accuracy is sufficiently close to its actual value. This may however be deceptive if the outcomes do not comprehend magnitudes indexed to normal and abnormal ranges. Take as an example that we acquired a thermometer which maximum scaling is 37 degrees Celsius (98.6 degrees Fahrenheit), it may yield an outstanding reliability, but there is no room for validity.

Knee joint laxity assessment has several angles that are worthy of research. We may allocate factors that interfere with knee joint multiplanar laxity envelope, to the individual intrinsic factors and the interplay of these with the choices and actions taken in case of need (e.g., treatment option in case of ACL ligament tear). Existing knee joint laxity may differ between uninjured and injured knees [20–22], either among different people in the same clinical condition [23], ontogenetic status of development [3, 24], biological circumstances [24, 25], sex [3–5], and different patterns of ACL tears [19].

There is great clinical and preventive potential to characterize and quantify the multiplanar knee joint laxity envelope. We are in need of studies that accurately assess joint laxity within different biological, pathological, or clinical conditions. This should be accomplished using laximetry—i.e., objective measurement of joint laxity—and, eventually, combining tests and/or equipment either for screening and to improve diagnosis [26]. The two main categories of laximetry are stress imaging and arthrometry. These two techniques classically aim to describe and quantify the displacement of the tibia in relation to the femur within the sagittal and transversal planes. These techniques often use cut-off values as dichotomic screening tools to elicit one of two diagnostic results: ruptured or not ruptured. The joint laxity data derive from an applied external force that aims to quantify bony displacement, either in unilateral or by side-to-side difference (SSD). Beyond the dichotomous application, laximetry can become an important diagnosis and profiling tool of different patterns of ACL tears (partial or total rupture) [19] and their interference in knee arthrokinematics, treatment decision, and surgical planning, prognostic purposes, or to quantify post-operative joint laxity.

8.3 Clinical Examination Combined with Laximetry and Imaging

While manual clinical examination is paramount for diagnosis, it is subjective both in the technique and interpretation [27]. The sensitivity and specificity of instrumented joint laxity measurements seem to increase with the combination of standard clinical examination in a two-step assessment process. Increased accuracy is found when combining clinical examination with Telos™ stress radiography [28]. The exposure to radiation and lack of ability to provide imaging evidence of soft tissue injury of stress radiography make the combined use of magnetic imaging resonance (MRI) and instrumented joint laxity testing the obvious next step to accurately measure knee joint laxity [22, 29].

The MRI alone does not provide for biomechanical competence data and joint laxity underestimation associated with some laximeters can mislead algorithm of treatment. Accurate laximetry combined with MRI will overcome barriers in anatomical and biomechanical competence assessment of apparent remaining intact ACL fibers or bundles or inserted grafts. When combining MRI with instrumented joint laxity, we sum up the visualization of anatomical structural evidence of injury and the ligament functional competence. The devices that are compatible with MRI use intrinsic anatomical landmarks [30] as references to measure the bony displacement and calculate the knee joint laxity.

Attempts also have been made, within a single-step assessment process, coupling measuring devices with manual elicited testing maneuvers, as the pivot shift, to confer objectivity and quantification [27]. Yet, the subjective variability in the testing technique persists due to disparities among professionals when eliciting the motion and determining which parameters use. These disparities are predominantly dependent on the assessor skills, training and experience, being present even in the often-used clinical tests as the pivot-shift [31].

The quest for accuracy and clinical usefulness should be the main goal of researchers and health care providers. The type of information is not so important—whereas from static or dynamic sagittal and/or rotatory testing, instrumented or not, separately or coupled with imaging assessment, under anesthesia or unanesthetized—but the validity, reliability, and accuracy should remain our focus. There are however premises learned from decades of research that pinpoint that settings and parameters are critical in joint laxity objective assessment and quantification.

8.4 Joint Laxity After Single ACL or Combined with Other Anterolateral Structures Injury

The diagnosis of different patterns of ACL tear is important for precision health care. New knowledge and evidence gathered in different domains as anatomy, biomechanics, pathomechanics, reinjury rates, and surgical techniques related to ACL injury and treatment should support customized risk management interventions and surgical planning.

Observed anterior translation and internal rotation of the tibia varies due to different ACL injury patterns. Partial or total tears, partial tears involving either the anteromedial or the posterolateral bundles and, it is also believed, that part of the abnormal rotatory joint laxity originates from additional injury to the anterolateral soft tissue structures. This is well documented in studies of biomechanical testing of cadaveric specimens. Lagae et al. [32] have recently reported different patterns of knee joint laxity after sectioning different anterolateral soft tissue structures which potentially mimic injuries subsequent to knee trauma, as the anterolateral ligament (ALL) and the deep fibers of the iliotibial band (ITB). Cutting the ACL did not significantly increase tibial internal rotation laxity significantly compared to the intact knee at any flexion angle. In the ACL-deficient knee, sectioning the ALL significantly increased the anterior laxity only at 20° to 30° of knee flexion, and only significantly increased internal rotation at 50° of knee flexion. A large increase in internal rotatory laxity is found however between 20° and 100° of flexion after sectioning the deep fibers of the ITB (including the Kaplan fibers), specifically the proximal and distal bundles [33] and the condylar strap [34]. This goes in line with the findings of Godin et al. [33] that support the role of the proximal and distal Kaplan fibrous bundles in rotational knee stability. The proximal and distal (Kaplan) bundles are 22.5 mm apart at the distal femur and revealed a mean maximum load during pull-to-failure testing of 71.3 N and 170.2 N, respectively. Later, Landreau et al. [34] identified a third and more distal bundle of deep ITB fibers attaching to the femur between the distal Kaplan fibers and the epicondyle, which they named as "condylar strap". Even lacking biomechanical analysis, the qualitative evaluation of behavior in internal rotation revealed a tenodesis effect of the ITB which may add to anterolateral knee stability.

Several legit questions and concerns of translation to the clinical practice arise when interpreting these valuable anatomic and biomechanical studies. While biomechanical evidence suggests an important role of ITB in anterolateral instability control, injury frequency of deep ITB fibers in the setting of acute ACL tear [35, 36] is low compared to that of the ALL [37]. Yet, in presence of Segond fractures [38], where the ITB seems to be attached approximately in half of the cases and even in the absence of a Segond fracture [39], an ITB injury is a good marker for ACL injury. In fact, Lagae et al. [32] have shown that an isolated ACL anatomic reconstruction restored anterior tibial translation, but the remaining and significant internal rotatory laxity was only normalized after adding an extra-articular lateral tenodesis. Inderhaug et al. [40] also showed us that isolated ACL reconstruction does not restore normal kinematics, ACL combined with ALL reconstruction resulted in abnormal rotational joint laxity and that adding a lateral extra-articular tenodesis (MacIntosh or Lemaire) restored the knee internal rotation laxity to its native values. Other studies have also highlighted the importance of deep fibers of the ITB in controlling rotational joint laxity, but with a minimal influence of the ALL [41–43]. This makes us think of a potential overlooking behavior in MRI patterns in the setting of ACL injury and of the utility to combine the assessment of the ligament structural integrity and its functional competence within the same examination.

Correlational studies involving MRI and surgical exploration of the anterolateral complex (ALC) have shown high incidence in the setting of acute ACL-injured knees. However, MRI alone has low sensitivity, specificity, and accuracy for the diagnosis of ITB injury. The ITB was considered abnormal in approximately 31% of the cases [36]. Giving the number of cases, low diagnostic values of MRI alone, and relevance of ITB injury on rotatory joint laxity [32], the PKTD can play a role in functional diagnosis workflow of these additional injuries through joint laxity profiling. Rotatory joint laxity assessment within MRI may also be of particular importance in presence of Segond fractures since different structures of the ALC can be detached along with the bone avulsed fragment. The ITB often detaches along with the fragment with frequency depending on the dimensions and volume of the fragment as distance sparing it from the center of Gerdy's tubercle [38, 44]. It is important to identify the patients with injury of the anterolateral structures, that if combined with increased rotatory joint laxity, are candidates to concomitant procedures such as lateral extra-articular tenodesis to better control the tibial internal rotation [45] and decrease the risk of graft failure [46].

8.5 Partial ACL Tears: MRI Diagnosis, Instrumented Joint Laxity Discrimination and Assessment of Biomechanical Competence

The MRI has high diagnostic accuracy for complete ACL tears [47]. Even the novel fully automated deep learning MRI techniques show high accuracy in identifying ACL tears [48]. However, when used to diagnose partial tears, the MRI is not capable to reliably detect partial tears [49] showing a high rate of false positives [47], even when using 3-Tesla MRI machines [50, 51]. Indeed, the MRI has low correlation with arthroscopic findings in cases of partial ACL tears [28, 52] and does not assess the functional competence of the intact ACL bundle. The instrumented joint laxity assessment is able to discriminate and document significant differences in mean SSD anterior tibial displacement in partial ACL tears [28]. Near one-third of patients treated arthroscopically for ACL injuries display a partial tear, being 14.1% classified as intact posterolateral bundle, 4.0% as intact anteromedial bundle, and 12.4% as posterior cruciate ligament healing. The SSD tibial displacement between ACL complete tear and all types of partial tears was significantly greater with Telos (mean 7.4 mm in total vs. 4.0 mm in partial ACL tears) than with the Rolimiter (mean 5.3 mm in total vs. 2.6 mm in partial ACL tears) [19]. The underestimation of joint laxity using the Rolimiter can hamper the desired accuracy for treatment decision and follow-up. The GeNouRoB, also comparing to Telos device, has showed a reasonably high diagnostic accuracy for ACL partial tears using a 2.5 mm cut-off (sensitivity of 84% and specificity of 81%) [53].

The ability to discriminate total from partial tears can be decisive for the surgical planning because the preservation of the ACL remnants enables anatomical landmarks for tunnel positioning [54] and provides vascular and mechanical benefits to the graft [55–57]. A selected group of patients with partial ACL tears may also respond well to conservative treatment [58–61] and in these cases it is crucial to assess the intact bundle competence. In cases of suspected partial tears, we use MRI instrumented-assessment to evaluate if there is any associated abnormal joint laxity [62].

8.6 Post-operative Knee Joint Laxity

Residual sagittal [63] and rotatory joint laxity [30, 64] as well as abnormal rotational motion [65–68] often persist after ACLR and are a common cause of poor long-term outcomes [69–71]. Residual knee joint laxity may disclose differences after ACL reconstruction procedures that might be related to the surgical technique [72–74], graft choice [75], concomitant procedures [16], graft tension or fixation angle [76, 77], and healing [78]. Residual anterior knee joint laxity 6 months following primary ACL reconstruction is associated with younger age (<30 years old), preoperative anterior laxity (SSD >5 mm), hamstring tendon graft, and resection of the medial meniscus [79]. Residual rotatory joint laxity measured by the pivot shift at 1 year after ACL reconstruction is associated with knee hyperextension and greater preoperative pivot shift under anesthesia. Age, gender, Lachman test, KT-1000 measurement, single-bundle vs. double-bundle, meniscus injury sites, and meniscus surgery were not predictors of residual rotational joint laxity [80].

Despite the evolution of surgical techniques, residual joint laxity should be a concern because it increases the ACL peak strain and has a fourfold increased risk for ACL injury for every 1.3 mm increase in SSD in anterior-posterior tibial displacement [81]. When athletes display residual joint laxity that is combined with neuromuscular deficits common in patients who tear the ACL [82]—such as weakness of hip external rotators—they will be exposed to a higher risk of reinjury during sport-specific tasks that involve pivoting or landing where the strain applied to graft is increased.

The use of accurate multiplanar laximetry techniques is important to monitor the postoperative outcomes. Restoration of knee stability is the main goal of surgical reconstruction and post-operative joint laxity evaluation should therefore always take part of a complete follow-up assessment. Despite the current literature on the importance of knee joint laxity on the treatment outcomes [83], but only 6% of studies use laxity-based assessment as a criterion for the return to sport decision [84]. In our experience, we use the MRI instrumented-assessment [85] that, in addition to the other often reported clinical and physical impairment-based objective criteria, supports our decision on when the athlete is ready to return to unrestricted sports.

8.7 MRI Instrumented-Assessment of Knee Joint Laxity

The Porto Knee Testing Device (PKTD) is an MRI-safe knee joint laxity testing device, made of polyurethane-based mixed resins, for the measurement of sagittal and rotatory knee joint laxity

(Fig. 8.1). The PKTD operates through two movable platforms that are activated by plunger mechanisms. One platform induces an antero-posterior translation and the other internal or external rotation of the leg. These two movable platforms can operate isolated or in combination, allowing to measure isolated sagittal and rotatory joint laxity, or the two simultaneously combined. The operator can control the magnitude of load transmission and adjust for different degrees of knee flexion.

We combine the PKTD assessment with MRI visualization to objectively assess the knee joint laxity. After applying postero-anterior and/or rotatory stress, we measure the tibial displacement in the medial and lateral plateaus relative to the resting baseline position (Fig. 8.2). The tibial displacement is used as an isolated measure—i.e., the total amount of displacement—and is also compared with the contralateral knee.

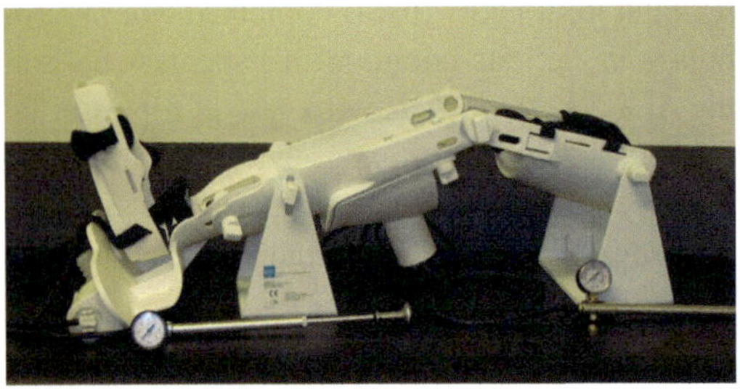

Fig. 8.1 Photograph of the Porto-Knee Testing Device (PKTD)

The PKTD is a valid tool to assess ACL complete tears. The sagittal joint laxity is correlated with the KT-1000 and the rotational joint laxity is correlated with the pivot shift results [29]. While combining the anterior tibial displacement in both the medial and lateral plateaus, we obtain the most specific measure (94%); when combining the tibial internal and external rotation in the lateral plateau, we obtain the most sensitive measure (93%) [22].

The ability to visualize soft tissues concomitantly with accurate objective joint laxity measurement [30] allows to correlate the structural integrity of the ligament with its functional competence. Eventually, we can establish multiplanar knee joint laxity cluster profiles that may be associated with specific injury patterns [37, 86–88], time between injury and surgery [89], different ACL reconstruction surgical

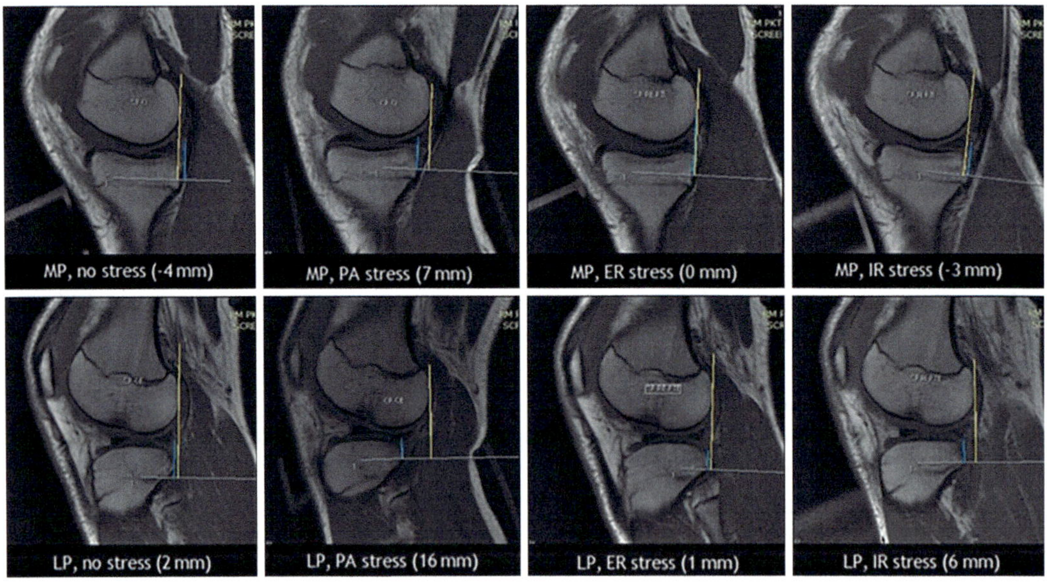

Fig. 8.2 PKTD exam of an ACL total rupture. *MP* medial plateau, *LP* lateral plateau, *PA* posteroanterior translation, *ER* external tibial rotation, *IR* internal tibial rotation. Blue line indicates tangent line to the posterior tibial plateau and orange line indicates tangent line to the posterior femoral condyle

techniques outcomes [68, 90], or anatomic features such as bone morphology or morphometrics [91–95]. The PKTD can have an important role in establishing these multiplanar knee joint laxity cluster profiles as it combines the assessment of both "anatomy" and "function" [96]. For instance, the MRI visualization might identify a partial ACL tear with an intact bundle, that after the PKTD assessment can reveal incompetent to provide stability to the knee (Fig. 8.3) [62]. We may find also injury of the anterolateral structures of the knee that, if combined with abnormal rotational joint laxity, may require the addition of a lateral extra-articular tenodesis. When examining external tibial rotation laxity at 30 degrees of flexion, it may identify cases with posterolateral corner injury that may have been undetected during the dial test (Fig. 8.4) [97]. Using the PKTD, we can identify these subclinical groups that may require differentiated or additional surgical intervention and thus refine our surgical indications and

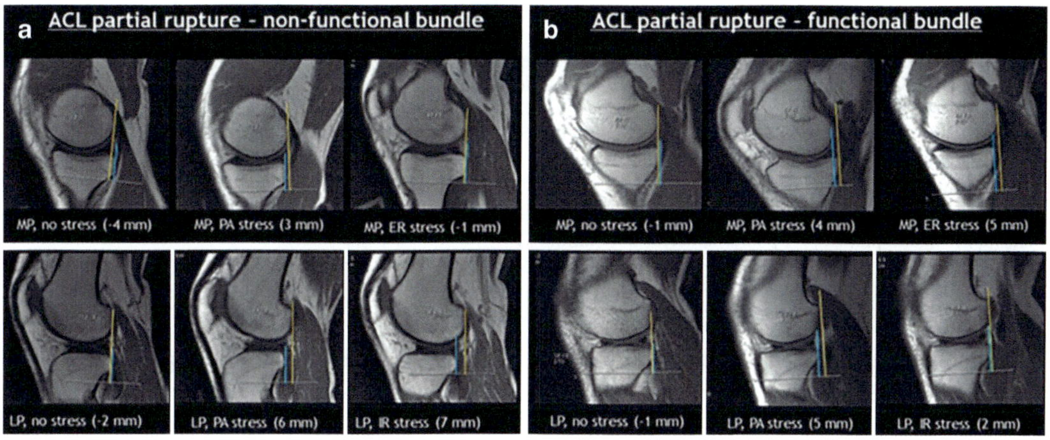

Fig. 8.3 PKTD exam of two cases of ACL partial rupture. (**a**) Partial ACL rupture with an intact, but non-functional bundle; (**b**) partial ACL rupture with an intact and functional bundle. *MP* medial plateau, *LP* lateral plateau, *PA* posteroanterior translation, *ER* external tibial rotation, *IR* internal tibial rotation. Blue line indicates tangent line to the posterior tibial plateau and orange line indicates tangent line to the posterior femoral condyle

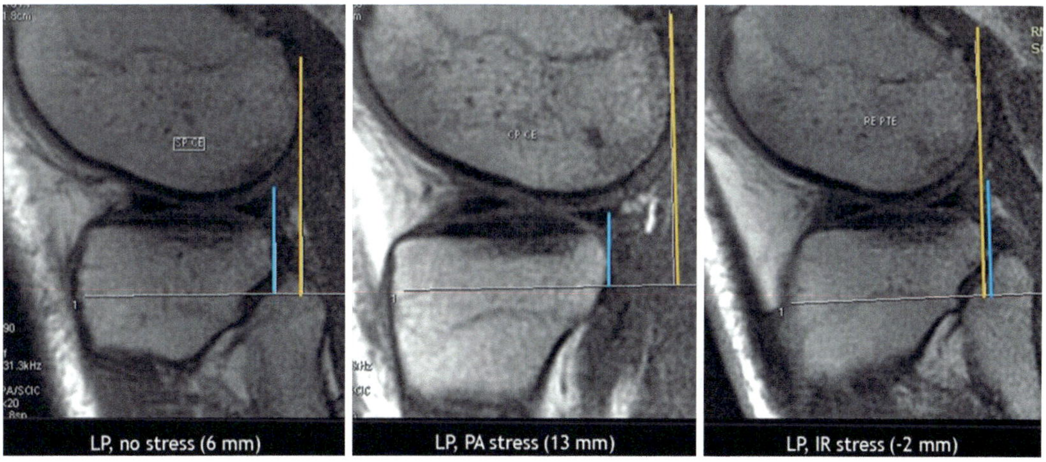

Fig. 8.4 PKTD exam showing increased external rotation that was undetected under the dial test. *LP* lateral plateau, *PA* posteroanterior translation, *ER* external tibial rotation. Blue line indicates tangent line to the posterior tibial plateau and orange line indicates tangent line to the posterior femoral condyle

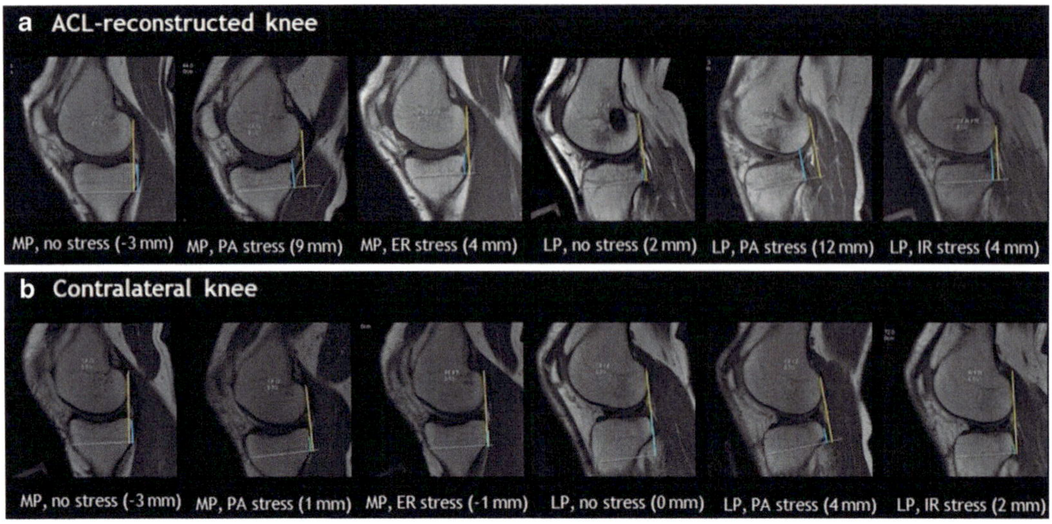

Fig. 8.5 PKTD exam comparing an ACL-reconstructed knee (**a**) with its contralateral healthy knee (**b**). From the PKTD examination, we can observe that there is still residual laxity that requires reintervention. *MP* medial plateau, *LP* lateral plateau, *PA* posteroanterior translation, *ER* external tibial rotation, *IR* internal tibial rotation. Blue line indicates tangent line to the posterior tibial plateau and orange line indicates tangent line to the posterior femoral condyle

individualize the treatment. In the follow-up of conservative or surgical approaches, the PKTD also plays an important role in the prospective monitoring of knee joint laxity and identify those with residual joint laxity (Fig. 8.5) [98]. It will provide useful information for the decision to clear the athletes to unrestricted sporting activities or those that may require further rehabilitation or surgical reintervention [85].

8.8 Conclusions

The PKTD is an MRI-safe knee joint laxity testing device which enables assessment of isolated or combined anteroposterior and rotatory joint laxity. Accurate assessment of multiplanar tibial displacement with imaging visualization can establish joint laxity cluster profiles that may correlate with specific injury patterns. Joint laxity can vary in quantity and in quality if there is an isolated ACL injury or there is additionally injury to peripheral structures, such as the ALC (especially the ALL and deep fibers of the ITB) or the menisci. Combining the MRI visual inspection of anatomical injury with the mechanical capability using the PKTD, we are able to accurately assess and characterize the knee joint multiplanar laxity and thus support treatment decisions and customized interventions while aiming for superior outcomes. The restoration of passive sagittal and transversal knee stability is the main purpose of surgical interventions addressing ACL reconstruction and pre- and post-operative measurements should therefore be systematically performed to support orthopedic precision medicine.

References

1. Feagin J. Personal communication. Hawaii: ACL Study Group; 2006.
2. Musahl V, Hoshino Y, Becker R, Karlsson J. Rotatory knee laxity and the pivot shift. Knee Surg Sports Traumatol Arthrosc. 2012;20:601–2.
3. Tagesson S, Witvrouw E, Kvist J. Differences in knee joint stabilization between children and adults and between the sexes. Am J Sports Med. 2013;41:678–83.
4. Beynnon BD, Bernstein IM, Belisle A, Brattbakk B, Devanny P, Risinger R, Durant D. The effect of estradiol and progesterone on knee and ankle joint laxity. Am J Sports Med. 2005;33:1298–304.

5. Pfeiffer TR, Kanakamedala AC, Herbst E, Nagai K, Murphy C, Burnham JM, Popchak A, Debski RE, Musahl V. Female sex is associated with greater rotatory knee laxity in collegiate athletes. Knee Surg Sports Traumatol Arthrosc. 2018;26:1319–25.
6. Andrade R, Vasta S, Sevivas N, Pereira R, Leal A, Papalia R, Pereira H, Espregueira-Mendes J. Notch morphology is a risk factor for ACL injury: a systematic review and meta-analysis. J ISAKOS. 2016;1:70–81.
7. Saita Y, Schoenhuber H, Thiébat G, et al. Knee hyperextension and a small lateral condyle are associated with greater quantified antero-lateral rotatory instability in the patients with a complete anterior cruciate ligament (ACL) rupture. Knee Surg Sports Traumatol Arthrosc. 2019;27:868–74.
8. Branch T, Stinton S, Sharma A, Lavoie F, Guier C, Neyret P. The impact of bone morphology on the outcome of the pivot shift test: a cohort study. BMC Musculoskelet Disord. 2017;18:463.
9. Rahnemai-Azar AA, Abebe ES, Johnson P, Labrum J, Fu FH, Irrgang JJ, Samuelsson K, Musahl V. Increased lateral tibial slope predicts high-grade rotatory knee laxity pre-operatively in ACL reconstruction. Knee Surg Sports Traumatol Arthrosc. 2017;25:1170–6.
10. Grassi A, Di Paolo S, Lucidi GA, Macchiarola L, Raggi F, Zaffagnini S. The contribution of partial Meniscectomy to preoperative laxity and laxity after anatomic single-bundle anterior cruciate ligament reconstruction: in vivo kinematics with navigation. Am J Sports Med. 2019;47:3203–11.
11. Dejour D, Pungitore M, Valluy J, Nover L, Saffarini M, Demey G. Preoperative laxity in ACL-deficient knees increases with posterior tibial slope and medial meniscal tears. Knee Surg Sports Traumatol Arthrosc. 2019;27:564–72.
12. Cristiani R, Rönnblad E, Engström B, Forssblad M, Stålman A. Medial meniscus resection increases and medial meniscus repair preserves anterior knee laxity: a cohort study of 4497 patients with primary anterior cruciate ligament reconstruction. Am J Sports Med. 2018;46:357–62.
13. Musahl V, Rahnemai-Azar AA, Costello J, Arner JW, Fu FH, Hoshino Y, Lopomo N, Samuelsson K, Irrgang JJ. The influence of meniscal and anterolateral capsular injury on knee laxity in patients with anterior cruciate ligament injuries. Am J Sports Med. 2016;44:3126–31.
14. van Lieshout WAM, Martijn CD, van Ginneken BTJ, van Heerwaarden RJ. Medial collateral ligament laxity in valgus knee deformity before and after medial closing wedge high tibial osteotomy measured with instrumented laxity measurements and patient reported outcome. J Exp Orthop. 2018;5:49.
15. Samuelsen BT, Webster KE, Johnson NR, Hewett TE, Krych AJ. Hamstring autograft versus patellar tendon autograft for ACL reconstruction: is there a difference in graft failure rate? A meta-analysis of 47,613 patients. Clin Orthop Relat Res. 2017;475:2459–68.
16. Grassi A, Signorelli C, Lucidi GA, Raggi F, Macchiarola L, Roberti Di Sarsina T, Marcheggiani Muccioli GM, Filardo G, Zaffagnini S. ACL reconstruction with lateral plasty reduces translational and rotatory laxity compared to anatomical single bundle and non-anatomical double bundle surgery: an in vivo kinematic evaluation with navigation system. Clin Biomech (Bristol, Avon). 2019;69:1–8. https://doi.org/10.1016/j.clinbiomech.2019.06.012.
17. Kim SJ, Choi CH, Kim SH, Lee SK, Lee W, Kim T, Jung M. Bone-patellar tendon-bone autograft could be recommended as a superior graft to hamstring autograft for ACL reconstruction in patients with generalized joint laxity: 2- and 5-year follow-up study. Knee Surg Sports Traumatol Arthrosc. 2018;26:2568–79.
18. Nyland J, Collis P, Huffstutler A, Sachdeva S, Spears JR, Greene J, Caborn DNM. Quadriceps tendon autograft ACL reconstruction has less pivot shift laxity and lower failure rates than hamstring tendon autografts. Knee Surg Sports Traumatol Arthrosc. 2020;28:509–18.
19. Panisset JC, Ntagiopoulos PG, Saggin PR, Dejour D. A comparison of Telos™ stress radiography versus Rolimeter™ in the diagnosis of different patterns of anterior cruciate ligament tears. Orthop Traumatol Surg Res. 2012;98:751–8.
20. Imbert P, Belvedere C, Leardini A. Human knee laxity in ACL-deficient and physiological contralateral joints: intra-operative measurements using a navigation system. Biomed Eng Online. 2014;13:86.
21. Mouton C, Theisen D, Meyer T, Agostinis H, Nührenbörger C, Pape D, Seil R. Noninjured knees of patients with noncontact acl injuries display higher average anterior and internal rotational knee laxity compared with healthy knees of a noninjured population. Am J Sports Med. 2015;43:1918–23.
22. Espregueira-Mendes J, Andrade R, Leal A, Pereira H, Skaf A, Rodrigues-Gomes S, Oliveira JM, Reis RL, Pereira R. Global rotation has high sensitivity in ACL lesions within stress MRI. Knee Surg Sports Traumatol Arthrosc. 2017;25:2993–3003.
23. Sernert N, Kartus JT, Ejerhed L, Karlsson J. Right and left knee laxity measurements: a prospective study of patients with anterior cruciate ligament injuries and normal control subjects. Arthroscopy. 2004;20:564–71.
24. Shultz SJ, Sander TC, Kirk SE, Perrin DH. Sex differences in knee joint laxity change across the female menstrual cycle. J Sports Med Phys Fitness. 2005;45:594–603.
25. Charlton WP, Coslett-Charlton LM, Ciccotti MG. Correlation of estradiol in pregnancy and anterior cruciate ligament laxity. Clin Orthop Relat Res. 2001;(387):165–70.
26. Hegedus EJ, Cook C, Lewis J, Wright A, Park JY. Combining orthopedic special tests to improve diagnosis of shoulder pathology. Phys Ther Sport. 2015;16:87–92.

27. Ahldén M, Hoshino Y, Samuelsson K, Araujo P, Musahl V, Karlsson J. Dynamic knee laxity measurement devices. Knee Surg Sports Traumatol Arthrosc. 2012;20:621–32.
28. Dejour D, Ntagiopoulos PG, Saggin PR, Panisset JC. The diagnostic value of clinical tests, magnetic resonance imaging, and instrumented laxity in the differentiation of complete versus partial anterior cruciate ligament tears. Arthroscopy. 2013;29:491–9.
29. Espregueira-Mendes J, Pereira H, Sevivas N, Passos C, Vasconcelos JC, Monteiro A, Oliveira JM, Reis RL. Assessment of rotatory laxity in anterior cruciate ligament-deficient knees using magnetic resonance imaging with Porto-knee testing device. Knee Surg Sports Traumatol Arthrosc. 2012;20:671–8.
30. Tashiro Y, Okazaki K, Miura H, Matsuda S, Yasunaga T, Hashizume M, Nakanishi Y, Iwamoto Y. Quantitative assessment of rotatory instability after anterior cruciate ligament reconstruction. Am J Sports Med. 2009;37:909–16.
31. Musahl V, Hoshino Y, Ahlden M, Araujo P, Irrgang JJ, Zaffagnini S, Karlsson J, Fu FH. The pivot shift: a global user guide. Knee Surg Sports Traumatol Arthrosc. 2012;20:724–31.
32. Lagae KC, Robberecht J, Athwal KK, Verdonk PCM, Amis AA. ACL reconstruction combined with lateral monoloop tenodesis can restore intact knee laxity. Knee Surg Sports Traumatol Arthrosc. 2020;28:1159–68.
33. Godin JA, Chahla J, Moatshe G, Kruckeberg BM, Muckenhirn KJ, Vap AR, Geeslin AG, LaPrade RF. A comprehensive reanalysis of the distal iliotibial band: quantitative anatomy, radiographic markers, and biomechanical properties. Am J Sports Med. 2017;45:2595–603.
34. Landreau P, Catteeuw A, Hamie F, Saithna A, Sonnery-Cottet B, Smigielski R. Anatomic study and reanalysis of the nomenclature of the anterolateral complex of the knee focusing on the distal iliotibial band: identification and description of the condylar strap. Orthop J Sports Med. 2019;7:2325967118818064.
35. Daggett M, Claes S, Helito CP, Imbert P, Monaco E, Lutz C, Sonnery-Cottet B. The role of the anterolateral structures and the acl in controlling laxity of the intact and acl-deficient knee: letter to the editor. Am J Sports Med. 2016;44:14–5.
36. Monaco E, Helito CP, Redler A, Argento G, De Carli A, Saithna A, Helito PVP, Ferretti A. Correlation between magnetic resonance imaging and surgical exploration of the anterolateral structures of the acute anterior cruciate ligament-injured knee. Am J Sports Med. 2019;47:1186–93.
37. Ferretti A, Monaco E, Redler A, Argento G, De Carli A, Saithna A, Helito PVP, Helito CP. High prevalence of anterolateral ligament abnormalities on MRI in knees with acute anterior cruciate ligament injuries: a case-control series from the SANTI study group. Orthop J Sports Med. 2019;7:2325967119852916.
38. Helito PVP, Bartholomeeusen S, Claes S, Rodrigues MB, Helito CP. Magnetic resonance imaging evaluation of the anterolateral ligament and the iliotibial band in acute anterior cruciate ligament injuries associated with segond fractures. Arthroscopy. 2020; https://doi.org/10.1016/j.arthro.2020.02.005.
39. Khanna M, Gupte C, Dodds A, Williams A, Walker M. Magnetic resonance imaging appearances of the capsulo-osseous layer of the iliotibial band and femoral attachments of the iliotibial band in the normal and pivot-shift ACL injured knee. Skelet Radiol. 2019;48:729–40.
40. Inderhaug E, Stephen JM, Williams A, Amis AA. Biomechanical comparison of anterolateral procedures combined with anterior cruciate ligament reconstruction. Am J Sports Med. 2017;45:347–54.
41. Stentz-Olesen K, Nielsen ET, de Raedt S, Jørgensen PB, Sørensen OG, Kaptein B, Søballe K, Stilling M. Reconstructing the anterolateral ligament does not decrease rotational knee laxity in ACL-reconstructed knees. Knee Surg Sports Traumatol Arthrosc. 2017;25:1125–31.
42. Jenny JY, Puliero B, Schockmel G, Harnoist S, Clavert P. Minimal influence of the anterolateral knee ligament on anterior and rotational laxity of the knee: a cadaveric study. Eur J Orthop Surg Traumatol. 2018;28:955–8.
43. Smith PA, Thomas DM, Pomajzl RJ, Bley JA, Pfeiffer FM, Cook JL. A biomechanical study of the role of the anterolateral ligament and the deep iliotibial band for control of a simulated pivot shift with comparison of minimally invasive extra-articular anterolateral tendon graft reconstruction versus modified lemaire reconstruction after anterior cruciate ligament reconstruction. Arthroscopy. 2019;35:1473–83.
44. Shaikh H, Herbst E, Rahnemai-Azar AA, Bottene Villa Albers M, Naendrup JH, Musahl V, Irrgang JJ, Fu FH. The segond fracture is an avulsion of the anterolateral complex. Am J Sports Med. 2017;45:2247–52.
45. Slette EL, Mikula JD, Schon JM, Marchetti DC, Kheir MM, Turnbull TL, LaPrade RF. Biomechanical results of lateral extra-articular tenodesis procedures of the knee: a systematic review. Arthroscopy. 2016;32:2592–611.
46. Getgood AMJ, Bryant DM, Litchfield R, et al. Lateral extra-articular tenodesis reduces failure of hamstring tendon autograft anterior cruciate ligament reconstruction: 2-year outcomes from the STABILITY study randomized clinical trial. Am J Sports Med. 2020;48:285–97.
47. Ghasem Hanafi M, Momen Gharibvand M, Jaffari Gharibvand R, Sadoni H. Diagnostic value of oblique coronal and oblique sagittal magnetic resonance imaging (MRI) in diagnosis of anterior cruciate ligament (ACL) tears. J Med Life. 2018;11:281–5.
48. Liu F, Guan B, Zhou Z, et al. Fully automated diagnosis of anterior cruciate ligament tears on knee MR images by using deep learning. Radiol Artif Intell. 2019;1:180091.
49. Lawrance JA, Ostlere SJ, Dodd CA. MRI diagnosis of partial tears of the anterior cruciate ligament. Injury. 1996;27:153–5.

50. Van Dyck P, De Smet E, Veryser J, Lambrecht V, Gielen JL, Vanhoenacker FM, Dossche L, Parizel PM. Partial tear of the anterior cruciate ligament of the knee: injury patterns on MR imaging. Knee Surg Sports Traumatol Arthrosc. 2012;20:256–61.
51. Van Dyck P, Vanhoenacker FM, Gielen JL, Dossche L, Van Gestel J, Wouters K, Parizel PM. Three tesla magnetic resonance imaging of the anterior cruciate ligament of the knee: can we differentiate complete from partial tears? Skelet Radiol. 2011;40:701–7.
52. Jog AV, Smith TJ, Pipitone PS, Toorkey BC, Morgan CD, Bartolozzi AR. Is a partial anterior cruciate ligament tear truly partial? A clinical, arthroscopic, and histologic investigation. Arthroscopy. 2020; https://doi.org/10.1016/j.arthro.2020.02.037.
53. Lefevre N, Bohu Y, Naouri JF, Klouche S, Herman S. Validity of GNRB® arthrometer compared to Telos™ in the assessment of partial anterior cruciate ligament tears. Knee Surg Sports Traumatol Arthrosc. 2014;22:285–90.
54. Shimodaira H, Tensho K, Akaoka Y, Takanashi S, Kato H, Saito N. Remnant-preserving tibial tunnel positioning using anatomic landmarks in double-bundle anterior cruciate ligament reconstruction. Arthroscopy. 2016;32:1822–30.
55. Georgoulis AD, Pappa L, Moebius U, Malamou-Mitsi V, Pappa S, Papageorgiou CO, Agnantis NJ, Soucacos PN. The presence of proprioceptive mechanoreceptors in the remnants of the ruptured ACL as a possible source of re-innervation of the ACL autograft. Knee Surg Sports Traumatol Arthrosc. 2001;9:364–8.
56. Lee BI, Min KD, Choi HS, Kim JB, Kim ST. Arthroscopic anterior cruciate ligament reconstruction with the tibial-remnant preserving technique using a hamstring graft. Arthroscopy. 2006;22:340.e1–7.
57. Gohil S, Annear PO, Breidahl W. Anterior cruciate ligament reconstruction using autologous double hamstrings: a comparison of standard versus minimal debridement techniques using MRI to assess revascularisation. A randomised prospective study with a one-year follow-up. J Bone Joint Surg Br. 2007;89:1165–71.
58. Ekås GR, Moksnes H, Grindem H, Risberg MA, Engebretsen L. Coping with anterior cruciate ligament injury from childhood to maturation: a prospective case series of 44 patients with mean 8 years' follow-up. Am J Sports Med. 2019;47:22–30.
59. Thoma LM, Grindem H, Logerstedt D, Axe M, Engebretsen L, Risberg MA, Snyder-Mackler L. Coper classification early after anterior cruciate ligament rupture changes with progressive neuromuscular and strength training and is associated with 2-year success: the delaware-Oslo ACL cohort study. Am J Sports Med. 2019;47:807–14.
60. Wellsandt E, Failla MJ, Axe MJ, Snyder-Mackler L. Does anterior cruciate ligament reconstruction improve functional and radiographic outcomes over nonoperative management 5 years after injury? Am J Sports Med. 2018;46:2103–12.
61. Colombet P, Dejour D, Panisset JC, Siebold R. Current concept of partial anterior cruciate ligament ruptures. Orthop Traumatol Surg Res. 2010;96:S109–18.
62. Ohashi B, Ward J, Araujo P, Kfuri M, Pereira H, Espregueira-Mendes J, Musahl V. Partial anterior cruciate ligament ruptures: knee laxity measurements and pivot shift. In: Doral MM, Karlsson J, editors. Sports injuries: prevention. Treatment and Rehabilitation: Diagnosis; 2015. p. 1245–58.
63. Nakanishi Y, Matsushita T, Nagai K, Araki D, Kanzaki N, Hoshino Y, Matsumoto T, Niikura T, Kuroda R. Greater knee joint laxity remains in teenagers after anatomical double-bundle anterior cruciate ligament reconstruction compared to young adults. Knee Surg Sports Traumatol Arthrosc. 2020; https://doi.org/10.1007/s00167-020-05910-z.
64. Izawa T, Okazaki K, Tashiro Y, Matsubara H, Miura H, Matsuda S, Hashizume M, Iwamoto Y. Comparison of rotatory stability after anterior cruciate ligament reconstruction between single-bundle and double-bundle techniques. Am J Sports Med. 2011;39:1470–7.
65. Tagesson S, Öberg B, Kvist J. Static and dynamic tibial translation before, 5 weeks after, and 5 years after anterior cruciate ligament reconstruction. Knee Surg Sports Traumatol Arthrosc. 2015;23:3691–7.
66. Tashman S, Collon D, Anderson K, Kolowich P, Anderst W. Abnormal rotational knee motion during running after anterior cruciate ligament reconstruction. Am J Sports Med. 2004;32:975–83.
67. Tashman S, Kolowich P, Collon D, Anderson K, Anderst W. Dynamic function of the ACL-reconstructed knee during running. Clin Orthop Relat Res. 2007;454:66–73.
68. Georgoulis AD, Ristanis S, Chouliaras V, Moraiti C, Stergiou N. Tibial rotation is not restored after ACL reconstruction with a hamstring graft. Clin Orthop Relat Res. 2007;454:89–94.
69. Kamath GV, Redfern JC, Greis PE, Burks RT. Revision anterior cruciate ligament reconstruction. Am J Sports Med. 2011;39:199–217.
70. Corsetti JR, Jackson DW. Failure of anterior cruciate ligament reconstruction: the biologic basis. Clin Orthop Relat Res. 1996:42–9.
71. Vergis A, Gillquist J. Graft failure in intra-articular anterior cruciate ligament reconstructions: a review of the literature. Arthroscopy. 1995;11:312–21.
72. Lee S, Kim H, Jang J, Seong SC, Lee MC. Comparison of anterior and rotatory laxity using navigation between single- and double-bundle ACL reconstruction: prospective randomized trial. Knee Surg Sports Traumatol Arthrosc. 2012;20:752–61.
73. Li YL, Ning GZ, Wu Q, Wu QL, Li Y, Hao Y, Feng SQ. Single-bundle or double-bundle for anterior cruciate ligament reconstruction: a meta-analysis. Knee. 2014;21:28–37.
74. Kilinc BE, Kara A, Oc Y, Celik H, Camur S, Bilgin E, Erten YT, Sahinkaya T, Eren OT. Transtibial vs anatomical single bundle technique for anterior cruciate ligament reconstruction: a Retrospective Cohort Study. Int J Surg. 2016;29:62–9.

75. Cristiani R, Sarakatsianos V, Engström B, Samuelsson K, Forssblad M, Stålman A. Increased knee laxity with hamstring tendon autograft compared to patellar tendon autograft: a cohort study of 5462 patients with primary anterior cruciate ligament reconstruction. Knee Surg Sports Traumatol Arthrosc. 2019;27:381–8.
76. Koga H, Muneta T, Yagishita K, Ju YJ, Sekiya I. The effect of graft fixation angles on anteroposterior and rotational knee laxity in double-bundle anterior cruciate ligament reconstruction: evaluation using computerized navigation. Am J Sports Med. 2012;40:615–23.
77. Dhaher YY, Salehghaffari S, Adouni M. Anterior laxity, graft-tunnel interaction and surgical design variations during anterior cruciate ligament reconstruction: a probabilistic simulation of the surgery. J Biomech. 2016;49:3009–16.
78. Hofbauer M, Soldati F, Szomolanyi P, Trattnig S, Bartolucci F, Fu F, Denti M. Hamstring tendon autografts do not show complete graft maturity 6 months postoperatively after anterior cruciate ligament reconstruction. Knee Surg Sports Traumatol Arthrosc. 2019;27:130–6.
79. Cristiani R, Forssblad M, Engström B, Edman G, Stålman A. Risk factors for abnormal anteroposterior knee laxity after primary anterior cruciate ligament reconstruction. Arthroscopy. 2018;34:2478–84.
80. Ueki H, Nakagawa Y, Ohara T, et al. Risk factors for residual pivot shift after anterior cruciate ligament reconstruction: data from the MAKS group. Knee Surg Sports Traumatol Arthrosc. 2018;26:3724–30.
81. Myer GD, Ford KR, Paterno MV, Nick TG, Hewett TE. The effects of generalized joint laxity on risk of anterior cruciate ligament injury in young female athletes. Am J Sports Med. 2008;36:1073–80.
82. Paterno MV, Schmitt LC, Ford KR, Rauh MJ, Myer GD, Huang B, Hewett TE. Biomechanical measures during landing and postural stability predict second anterior cruciate ligament injury after anterior cruciate ligament reconstruction and return to sport. Am J Sports Med. 2010;38:1968–78.
83. Rohman EM, Macalena JA. Anterior cruciate ligament assessment using arthrometry and stress imaging. Curr Rev Musculoskelet Med. 2016;9:130–8.
84. Burgi CR, Peters S, Ardern CL, Magill JR, Gomez CD, Sylvain J, Reiman MP. Which criteria are used to clear patients to return to sport after primary ACL reconstruction? A scoping review. Br J Sports Med. 2019;53:1154–61.
85. Andrade R, Pereira R, Bastos R, Duarte H, Pereira H, Rodrigues-Gomes S, Espregueira-Mendes J. MRI-based laxity measurement for return to play. In: Musahl V, Karlsson J, Krutsch W, Mandelbaum B, Espregueira-Mendes J, d'Hooghe P, editors. Return to play in football. Berlin, Heidelberg: Springer; 2018. p. 205–15.
86. Zaffagnini S, Signorelli C, Bonanzinga T, Grassi A, Galán H, Akkawi I, Bragonzoni L, Cataldi F, Marcacci M. Does meniscus removal affect ACL-deficient knee laxity? An in vivo study. Knee Surg Sports Traumatol Arthrosc. 2016;24:3599–604.
87. Andrade R, Rebelo-Marques A, Bastos R, Zaffagnini S, Seil R, Ayeni OR, Espregueira-Mendes J. Identification of normal and injured anterolateral ligaments of the knee: a systematic review of magnetic resonance imaging studies. Arthroscopy. 2019;35:1594–1613.e1.
88. Bastos R, Andrade R, Vasta S, Pereira R, Papalia R, van der Merwe W, Rodeo S, Espregueira-Mendes J. Tibiofemoral bone bruise volume is not associated with meniscal injury and knee laxity in patients with anterior cruciate ligament rupture. Knee Surg Sports Traumatol Arthrosc. 2019;27:3318–26.
89. Yanagisawa S, Kimura M, Hagiwara K, Ogoshi A, Nakagawa T, Shiozawa H, Ohsawa T. Factors affecting knee laxity following anterior cruciate ligament reconstruction using a hamstring tendon. Knee. 2017;24:1075–82.
90. Rupp S, Müller B, Seil R. Knee laxity after ACL reconstruction with a BPTB graft. Knee Surg Sports Traumatol Arthrosc. 2001;9:72–6.
91. Vasta S, Andrade R, Pereira R, Bastos R, Battaglia AG, Papalia R, Espregueira-Mendes J. Bone morphology and morphometry of the lateral femoral condyle is a risk factor for ACL injury. Knee Surg Sports Traumatol Arthrosc. 2018;26:2817–25.
92. Fernandes MS, Pereira R, Andrade R, Vasta S, Pereira H, Pinheiro JP, Espregueira-Mendes J. Is the femoral lateral condyle's bone morphology the trochlea of the ACL? Knee Surg Sports Traumatol Arthrosc. 2017;25:207–14.
93. Napier RJ, Garcia E, Devitt BM, Feller JA, Webster KE. Increased radiographic posterior tibial slope is associated with subsequent injury following revision anterior cruciate ligament reconstruction. Orthop J Sports Med. 2019;7:2325967119879373.
94. Cooper JD, Wang W, Prentice HA, Funahashi TT, Maletis GB. The association between tibial slope and revision anterior cruciate ligament reconstruction in patients ≤21 years old: a matched case-control study including 317 revisions. Am J Sports Med. 2019;47:3330–8.
95. Dejour D, Saffarini M, Demey G, Baverel L. Tibial slope correction combined with second revision ACL produces good knee stability and prevents graft rupture. Knee Surg Sports Traumatol Arthrosc. 2015;23:2846–52.
96. Pereira H, Gomes S, Vasconcelos JC, Soares L, Pereira R, Oliveira JM, Reis RL, Espregueira-Mendes J. MRI laxity assessment. In: Musahl V, Karlsson J, Kuroda R, Zaffagnini S, editors. Rotatory knee instability. Cham: Springer; 2017. p. 49–61.
97. Andrade R, Duarte H, Pereira R, Arias C, Bastos R, Rodrigues-Gomes S, Oliveira JM, Reis RL, Espregueira-Mendes J. An advanced device for multiplanar instability assessment in MRI. In: Margheritini F, Espregueira-Mendes J, Gobbi A, editors. Complex knee ligament injuries. Springer; 2019. p. 27–33.
98. Andrade R, Pereira H, Espregueira-Mendes J. ACL treatment in 2016–controversy and consensus. Asian J Arthroscopy. 2016;1:3–10.

9 Return to Sport After Anterior Cruciate Ligament Reconstruction: Criteria-Based Rehabilitation and Return to Sport Testing

Kate E. Webster and Timothy E. Hewett

9.1 Introduction

Most anterior cruciate ligament (ACL) injuries occur during sports participation. Therefore, the goal for most athletes who undergo ACL reconstruction surgery is to return their preinjury sport at a similar level of performance and in the absence of further injury. However, it has become apparent that for many athletes, these goals are not always attained. Rates of return to sport are often lower than expected and younger athletes are at considerable risk for sustaining multiple ACL injuries. This chapter will explore current knowledge regarding return to sport after ACL reconstruction, with a focus on criteria-based rehabilitation and return to sport testing.

9.2 Return to Sport Rates and Factors that Influence Return to Sport

An initial systematic review that investigated return to sport rates after ACL reconstruction surgery showed that while 82% of patients returned to some form of sport, only 63% were participating at their pre-injury level at follow-up [1]. These return rates contrasted with the finding that 90% of patients rated normal or nearly normal on impairment-based measures. This review was subsequently updated, and in the update similar rates were noted; 65% returned to the pre-injury level of sport and 55% returned to competition sport [2]. Comparable findings have also been reported for reviews in which only patients with revision reconstruction have been included [3]. Perhaps not surprisingly, elite-level athletes have the highest return rates whereby 83% return to their pre-injury sport [4], followed by younger aged athletes, in whom it has been reported that 81% return to competition sport [5]. Despite higher return rates in some select groups, the overall message is that a return to sport is not guaranteed following ACL reconstruction surgery.

Many factors influence whether an individual will return to sport after ACL reconstruction, and include demographic and social characteristics, as well as surgical and rehabilitation factors [6]. There are empirical data to show that males have higher return rates, usually in the order of approximately 10%, and that younger athletes can have up to 30% higher return rates [1, 7]. A positive psychological response and higher levels of motivation during rehabilitation have also been associated with higher rates of returning to pre-injury sport [8–11]. It is also highly relevant to consider the role of patient expectations and work in this

K. E. Webster (✉)
School of Allied Health, Human Services and Sport, La Trobe University, Melbourne, VIC, Australia
e-mail: k.webster@latrobe.edu.au

T. E. Hewett
Hewett Consultants, Rochester, MN, USA
e-mail: hewett.timothy@mayo.edu

N. Nakamura et al. (eds.), *Advances in Knee Ligament and Knee Preservation Surgery*,
https://doi.org/10.1007/978-3-030-84748-7_9

area has shown that for a first ACL injury, over 80% of patients expect to be able to return to their previous level of sport [12, 13]. As this is higher than what return rates are, many athletes will not realize their goals, and may therefore need support and advice to readjust their expectations.

9.3 Return to Sport and Second ACL Injury

When planning a return to sport after ACL reconstruction surgery patients are often concerned about the risk of sustaining the same injury again. This is logical, as return to sport puts the individual at risk of both ACL graft rupture and rupture of the contralateral ACL. Therefore, there has been much effort to determine predictors, such as patient and surgical variables, that can be used to identify at risk individuals. Numerous variables for re-injury have been considered, often with mixed findings [7]. However, most studies do report on sex and age.

Findings in relation to patient sex and re-injury are not straightforward. Sex as a risk factor for graft rupture has either shown no influence or males, particularly younger males have been shown to be at greatest risk [14–16]. In contrast, females appear at greatest risk for subsequent contralateral ACL injury [17, 18]. Why such sex difference may occur is not clear and requires further investigations that consider potential confounders such as the type of sport played as well as the amount of exposure.

There is a large amount of evidence to show that younger athletes are at significantly increased risk for both graft ruptures and contralateral ACL injuries [19]. Although the definition, or cut-off, for 'younger' has varied (usually from <16 years to <25 years), both large cohort studies and registry databases have confirmed this increased risk. The rate of second ACL injuries in the younger population is concerningly high, with cohort studies reporting that between 20% and 30% of younger athletes sustain a second ACL injury [16, 20–22]. It is therefore important to understand the reasons why younger patients are at such increased risk so that this can form part of any rehabilitation strategy or return to play decision making.

It is well accepted that it is unlikely that age itself is the risk factor, but rather a proxy for multiple factors [23]. In this regard, there are a few salient aspects where younger patients tend to differ from their 'older' counterparts. These are that younger patients are more likely to return to sport and they are also more likely to return earlier after surgery [24]. Furthermore, when they do return, the sports they play are high-risk sports for knee injury [25]. Recent studies have also shown that younger athletes resume these high-risk sports with strength and functional deficits [26, 27]. Data from a Swedish Rehabilitation Outcome Registry showed that 50% of adolescent patients (15–20 years) had resumed strenuous sports by 8 months post ACL reconstruction surgery despite only 29% having achieved satisfactory muscle function [26]. A similar study by Toole et al. [28] reported that only 14% of adolescent patients (mean age 17 years) met recommended strength and functional thresholds when cleared to return at 8 months post-surgery. As such, the rehabilitation of and timing for return to sport needs to be carefully considered in this group.

9.4 When Should Athletes Return to Sport?

Whatever surgical technique or rehabilitation program is used, perhaps the most difficult question to answer is when is it safe for the athlete to return to sport? There are two issues we are concerned about in relation to this question. The first is graft rupture/failure which was discussed in the previous section. Damage to the rest of the knee, both in the short and longer term, also should be considered.

Animal studies have shown that there are distinct phases of graft maturation with early graft necrosis and subsequent hypercellularity and revascularization being potential risk periods for reinjury [29]. While there is less information in humans, it has been suggested that there are the same phases, but that these occur over a slower

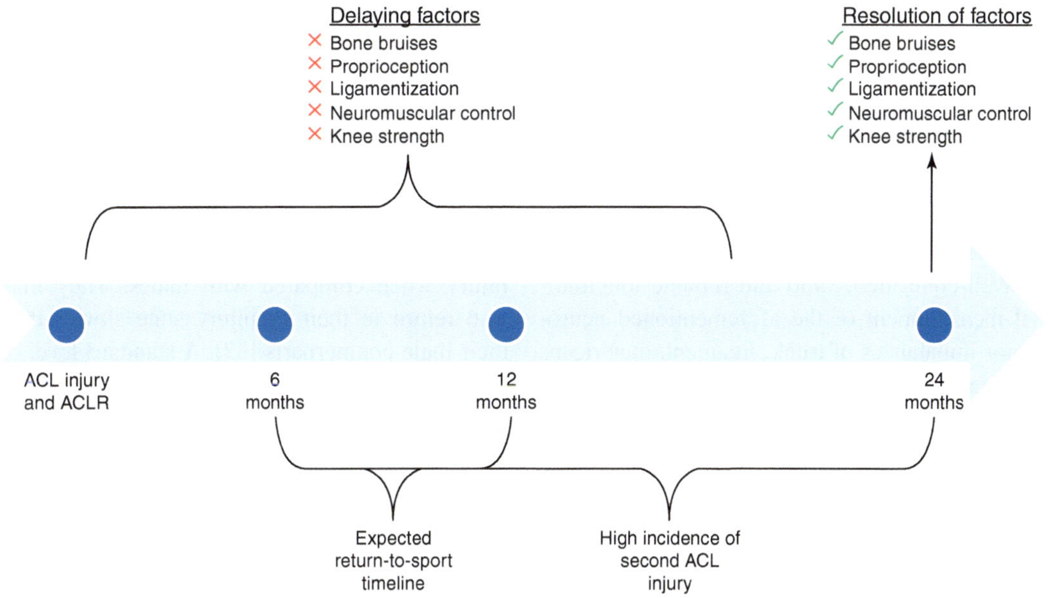

Fig. 9.1 Schematic showing that the time of recovery of baseline joint health and function is not indicated until 2 years post ACL injury or reconstruction surgery. From Nagelli and Hewett [32]

time frame than in animals [30]. For example, the period of remodelling, where the graft is most at risk, corresponds roughly to the 4- to 12-month post-surgery time point. This also corresponds to the time when many athletes are considering a return to their pre-injury sport, which for most, includes pivoting and twisting movements. Data from hamstring tendon grafts also indicate that remodelling may take as long as 12–24 months [31]. This is the same time point where registry datasets have shown that the peak of second ACL injuries occurs.

In addition to these biological considerations, it is common for athletes to demonstrate deficits in neuromuscular control and knee extension strength that do not reach baseline levels until at least 2 years after surgery. To avoid reinjury, such functional deficits should ideally be resolved before return to sport is considered. Therefore, Nagelli and Hewett [32] suggested that patients wait for 2 years before making a return to sport and Fig. 9.1 shows a schematic diagram of both biological and functional factors which are subject to change by having such a delay. While the theory behind this notion is evidence based, it is also recognized that there is yet to be evidence to confirm that such a delay would make a difference. It is also not clear whether athletes, parents or coaches would be willing to adhere to such long timeframes and these may significantly impact on career prospects for some young athletes. There is also wide individual variation in the recovery of ACL injury and reconstruction surgery and therefore an approach that is both time and criteria based may be preferable.

9.5 Criteria-Based Rehabilitation

One approach to address readiness to return to sport is to target modifiable neuromotor deficits known to be associated with both first and second ACL injury in phased rehabilitation programs and set criteria for progression from one phase to the next, which include return to sport. These modifiable deficits include trunk, ligament, quadriceps, and limb/leg dominance patterns [33].

These neuromotor deficits of trunk, ligament, quadriceps, and limb/leg dominance are fortunately readily observable and serially measurable after ACL rupture and reconstruction and should be measured longitudinally during the return to sport (RTS) process [33]. The athlete should understand that although the surgical reconstruction of the ligament may have created a mechanically stable joint, they may not have a functionally stable knee during dynamic movements and that they must be serially tested for any potential functional deficits [34, 35] and these deficits need to be addressed before return to play is advised.

Well-controlled, valid and reliable longitudinal measurement of the aforementioned neuromotor imbalances of trunk, ligament, quadriceps and leg dominance are critical for the allowable progression of post-injury and post-operative progression of the injured athlete through both the early and late stages of rehabilitation. As such, coordination between active muscular and passive ligamentous control of tibiofemoral motion in all three planes of motion must be assessed and enhanced [33]. Fortunately, such neuromotor imbalances, trunk dominance, for example, are readily observable and measurable post ACL rupture and reconstruction. A dynamically stable knee joint, which is prepared to return to demands of competitive play, must coordinate passive mechanical function of the ACL graft with appropriate neuromuscular control of the hip, knee and entire kinetic chain, and most especially of the trunk or "core," to provide muscular dampening of joint loads and reduce strain on the graft [34]. A progressive, functional testing and rehabilitation program is required to provide the athlete with an effective means of facilitation of positive adaptations to the much needed and often impaired proprioceptive function of the knee joint. Proper ACL return to play testing and training can provide the athlete with a dynamically functional joint that is prepared to respond to the extreme forces generated during athletic competition and reduce the risk of a second or even further subsequent injury and will optimally prepare even the young, highly active high-risk athlete to achieve pre-injury performance levels safely [36].

Neuromuscular control deficits, for example, quadriceps dominance, are readily observable and measurable after ACL rupture and well into the postoperative rehabilitation period [37]. Rehabilitation professionals should serially measure these deficits and focus rehabilitative interventions in these areas. Paterno et al. [38] demonstrated the observable and measurable presence of quadriceps dominance by a straight knee and a "stiffening" strategy on balance testing in athletes at high risk for a second ACL injury. Female athletes may have a greater loss of single-leg postural balance at return from a knee injury when compared with males. They may also return to their preinjury state slower than their male counterparts [37]. A standard level of proficiency in postural balance before return to sport is important to protect athletes from a second ACL injury. Athletes that do not demonstrate postural balance within two standard deviations of normal have a significantly higher risk of injury [39]. Balance and proprioceptive training past the acute postsurgical rehabilitation phase is a necessary requirement, not only for restoration of functionality, but for its prophylactic effect on ligament reinjury [37, 38, 40, 41].

The observed presence of significant leg dominance, as evidenced and measured by the absence of limb symmetry following ACL reconstruction, may be due to imbalances between muscular strength and joint kinematics between contralateral lower extremity measures in athletes at high risk for first and second ACL ruptures. For example, females may generate lower hamstrings torques on the non-dominant than in the dominant leg [42]. More specifically, adolescent female athletes have significant side-to-side differences in maximum knee valgus angle compared to males during a box drop vertical jump [43]. Side-to-side imbalances in muscular strength, flexibility and coordination are important predictors of increased injury risk [44–46]. Side-to-side balance in strength and flexibility is important for the prevention of injuries and when imbalances are present, the athlete may be more injury-prone [44]. Patients with muscle strength imbalances may exhibit a higher incidence of first and second ACL injury [45]. We [46] developed a model to predict ACL injury risk with high sensitivity and specificity. Half of the parameters in the predictive model were leg dominance indicators of side-to-side differences in lower limb kinematics and kinetics [46]. Side-to-side limb neuromotor imbalances likely increase risk for both limbs. Over-reliance on the non-ACL ruptured and reconstructed limb can put greater stress and torques on that knee, while the involved operated limb is also likely at increased risk due to an inability of the musculature on that side to effectively absorb the high forces associated with sporting activities.

A more advanced level of testing and training that should be utilized to observe, measure and target ligament, quadriceps, leg and trunk dominance patterns is the utilization of unanticipated cutting movements. Prior to teaching unanticipated landing and cutting, rehabilitation professionals, coaches and athletes should first consider the work of Paterno et al. [38]. Trunk dominance may be evidenced by the observation of a net negative hip internal rotation impulse on a drop vertical jump with 3D biomechanical testing or visually observable internal hip rotation. Ligament dominance can be observed by inward hip and knee collapse during the drop vertical jump. Quadriceps dominance by a straight knee and a "stiffening" strategy on balance testing. Leg dominance can be observed and measured by side-to side differences in relative quadriceps and hamstring activation via peak knee flexion/extension moments on a drop vertical jump during 3D biomechanical testing, peak torque on dynamometer testing or with standardized hop testing [33, 38, 47, 48].

Limb or leg dominance may not be fully corrected by simple single-faceted sagittal plane training and conditioning protocols that do not incorporate cutting manoeuvres will not provide similar levels of external varus/valgus or rotational loads that are seen during sport-specific cutting manoeuvres [49]. Testing and training programs that measure and teach safe levels of knee abduction/adduction moment dominated valgus and varus stresses may induce more muscle dominant neuromuscular adaptations [50]. Such adaptations can better prepare an athlete for more multi-directional sports movements that can improve their performance and reduce risk of second ACL injury [42, 51, 52]. High-risk athletes perform cutting techniques with decreased knee flexion and increased knee abduction/adduction valgus/varus angles [53].

Ligament dominance, as evidenced by increased "valgus" knee abduction loads, can double when performing unanticipated cutting manoeuvres similar to those utilized in sport [54]. Thus the endpoint of training designed to reduce knee abduction/adduction ACL loading via valgus and varus torques can be gained through training the athlete to use movement techniques that produce the low abduction and adduction knee joint moments [50]. Training that incorporates unanticipated movements can reduce knee joint loads [55]. In addition, training individuals to pre-activate the neuro-musculature that surrounds and controls the position of the knee joint prior to ground contact may facilitate appropriate kinematic adjustments and ACL loads may be reduced [54, 56]. Training the athlete to employ safe cutting techniques in unanticipated sport situations may instil technique adaptations that will more readily transfer onto the field of play. If achieved, the ligament dominant athlete may become muscular dominant, reducing their future risk of ACL injury [42, 52].

9.6 Return to Sport Testing

Another approach to assess readiness to return to sport that can be applied to either criteria or time-based rehabilitation is to use a set of criteria or 'test battery' to give the athlete clearance for return to sport. This is typically used at the final stages of rehabilitation and athletes who pass are cleared to return. Not surprisingly, the content of such testing is varied, and several consensus statements and clinical practice guidelines have been put forward. van Melick et al. [57] attempted to reach a consensus in regard to which criteria should be used to determine the moment of return to play. It was recommended that an extensive test battery for both quantity and quality of movement should be performed, and that the test battery itself should include a series of strength tests, hop tests and measurement of quality of movement. A limb symmetry index of greater than 90% was suggested as a pass criterion, but it was also suggested that this could be increased to 100% for patients planning a return to pivoting or contact sports. This seems a reasonable approach; however, as younger athletes are also more likely to meet criteria that are used to indicate readiness to return to sport, there may be a role for adjusting 'pass' thresholds for these criteria based on age as well.

An additional consensus statement on return to sport concluded that, for any injury, the return to sport decision should always use information gained from a battery of tests and should assess direction change and reactive agility, as well as psychological readiness [58]. In this regard, it is interesting to note that more recent studies have attempted to cover a broad range of risk factors, often suggesting that 15–20 different tests be used [59, 60]. This is likely due to uncertainty regarding what the most important factors for various outcomes, such as player performance or risk of further injury, are. However, this approach may cause an unnecessary burden and use of resources for both the patient and clinician. It has therefore been suggested that the focus should be on fewer but important factors and that five factors should be enough as any one factor would ideally account for at least 20% of the predictive variance [61].

It is also worth discussing whether return to sport tests is designed to determine whether the patient is capable of returning to play or whether they are designed to determine whether it is safe. These are often used interchangeably, and the questions of safety and capability cannot necessarily be addressed in the same way [62]. Most return to sport testing is done with the aim of assessing safety, i.e., whether the patient can make a return without suffering a further ACL injury. However, the same or similar measurements can also be used as assessments of how much of their functional capacity a patient has regained and the likelihood of making a return to sport. For example, meeting return to sport criteria at 6 months post-operatively has been associated with higher rates of returning to preinjury levels of activity up to 2 years [63]. The remainder of this chapter will focus on return to sport testing to assess the safety of returning to sport.

9.6.1 Return to Sport Testing for Making a Safe Return to Play

As it is becoming increasingly common for return to sport testing to be used as part of the decision-making process for a safe return to sport following ACL reconstruction surgery, for such testing to be of value to both the clinician and the patient, its validity should be known, or at least scientifically assessed. Despite the many discussions about return to sport testing following ACL reconstruction, the evidence surrounding it is relatively limited and to some extent contradictory. However, one aspect for which there is consistency is that the proportion of patients who pass return to sport testing is actually rather low. This was highlighted in a recent systematic review which showed a 23% pass rate from 8 studies with 876 patients who were tested before returning to sport at between 5- and 10-month post-surgery [64]. Most of these studies used >90% limb symmetry as the threshold for a pass. Of potential concern, the same pass rate was also reported for three studies (234 patients) in whom the patients had already resumed playing strenuous sports.

The same systematic review also determined whether passing RTS test batteries reduced subsequent rates of any knee injury [64]. The combined results from two studies showed a 72% reduction in risk with passing RTS criteria. However, this reduction was not statistically significant, and 95% confidence intervals were large (from 93% reduction in risk to 21% increase in risk). One of the studies also reported that the risk of subsequent knee injury markedly reduced for each month an athlete delayed returning to sport, until the 9-month mark, and this has been frequently cited [65]. Of the individual components in the RTS test battery, having a quadriceps strength deficit prior to return to sport was the most significant predictor of further knee reinjury. However, since only 18 patients (out of 106) actually passed the RTS testing criteria, caution needs to be applied when interpreting these results and overall it is difficult to draw firm conclusions.

Regarding passing RTS testing and subsequent ACL injury, two reviews showed that passing test batteries did not reduce the risk for all subsequent ACL injuries [64, 66]. However, one review found that passing an RTS test battery did significantly reduce the risk for subsequent graft rupture by 60%, although it increased the risk for a subsequent contralateral ACL injury by 235% [64]. This highlights the conflicting findings that are reported in the literature. Of the included studies in these meta-analyses, only two showed significant results [67, 68]. These also had the largest patient sample sizes, and as such made the greatest contribution to the weighting of the meta-analyses. It is therefore relevant to have a closer look at both.

Kyritsis et al. [67] recorded graft ruptures in elite male athletes and reported that those who did not meet all 7 RTS criteria had a four times greater risk of graft rupture. The hamstring-to quadriceps ratio of the involved leg alone was also highly associated with graft rupture, with a ten times greater risk for every 10% difference in strength. Unfortunately, the contralateral side was not reported on. In comparison, Sousa et al. [68] did not find a reduced risk for graft rupture in their group who passed criteria at 6 months, but they did find a significantly increased risk for contralateral injuries. As the patients who passed RTS testing were cleared for an early return, compared to those who failed and were advised to wait, the authors suggested that this increase in contralateral ACL injuries may be related to an increased activity level in the patients who had passed the criteria. This is a logical assumption; however, a close look at the Kaplan-Meier survival curves presented in Sousa et al. shows that there were few early second ACL injuries in the cohort. For graft ruptures, the earliest occurred after 36 months from surgery and for contralateral ACL injury there are no differences between the RTS pass and fail groups until approximately 20–30 months from surgery. Therefore, despite one group (those who passed RTS testing at 6 months) being cleared for an earlier return this does not seem to have had an impact on the timing of subsequent ACL injuries. As such, the impact of any potential difference in early exposure may in fact be minimal and this needs further exploration [69].

Unfortunately, relatively few studies that investigate RTS testing and further ACL injury report return to sport rates along with re-injury rates. Even when they do, the definition of return to sport varies from one study to another. For example, Grindem et al. [65] classified any reported participation, including training, as having returned. If the patients who pass RTS test are not returning to comparable levels of activity or sport as those who fail testing, it is difficult to draw conclusions about the benefit of meeting the RTS criteria. It is also highly likely that other factors come into play in the time interval between when RTS testing is conducted and when further ACL injuries occur. It may not be meaningful to relate an injury that occurs back to an RTS test that occurred many years earlier. There has been little discussion in the literature as to what may be an appropriate follow-up time period following RTS testing [64].

The frequency of conducting RTS testing has also received little attention. A recent study did nonetheless conclude that it is pointless to undertake knee strength assessments that are closer than 2 months apart [70]. As knee strength outcomes are one variable which has more frequently been associated with reinjury [65, 67] such information can serve to reduce the pressures therapists may feel to provide continual testing. However, as this was a cross-sectional study much work is needed to validate this initial data and extend it to other types of tests.

Whether clinical measures can be used to identify groups at high risk of reinjury has also been investigated [71, 72]. While some risk factors have been identified, such as increased knee laxity and performance on hop for distance tests, no matter what testing is done there are still athletes who suffer a further injury without the presence of risk factors and don't fit into a 'high-risk classification'. Therefore, although some athletes may be able to be identified who go onto have a second injury, the accuracy of the prediction is low, and many athletes who are considered to be at low risk also sustain further injuries. Therefore, despite the increasing popularity surrounding

RTS testing, there are many questions that remain unanswered. Until further research data are available, caution should be exercised when using information from RTS testing to provide advice to patients—particularly at an individual patient level—regarding their risk for subsequent injury, if or when they choose to return to sport.

9.6.2 The Psychological Aspect

While much emphasis has been placed on tests for physical function and strength, as previously noted, the importance of addressing psychological factors has been recognized. Although there has been much less empirical work in this area, what is available shows promise. Two initial pilot studies showed that fear of reinjury was associated with a second ACL injury. In the first, Tagesson and Kvist [73] conducted a battery of assessments in a cohort of 19 patients before surgery and at 5 weeks after ACL reconstruction which included; fear of reinjury, knee confidence, patient-reported function, activity levels, static and dynamic tibial translation and muscle strength. They followed the group for 5 years and found that those who went on to have a second ACL injury had greater fear of reinjury and greater static tibial translation in both knees compared to those who remained uninjured. In the second pilot study, Paterno et al. [74] assessed 40 patients who had been cleared to return to sport and tracked them for 12 months to identify second ACL injuries. Patients with a greater fear of injury were 13 times more likely to suffer a second ACL injury. These authors suggested that it may be important to measure self-reported fear of movement and incorporate this into return to sport discharge criteria to reduce the risk of further ACL injury.

In a subsequent larger study, McPherson et al. [75] had 429 athletes complete the anterior cruciate ligament return to sport after injury (ACL-RSI) scale, a measure of psychological readiness for returning to sport, both before and at 12 months after ACL reconstruction, and then followed the cohort for a minimum 2 years to determine further injury. Given that most ACL injuries occur in the context of sport, only those who had made a return to sport were included in the final analysed cohort of 329 patients. For this group, when measured at 12 months, younger injured patients ($\leq$20 years) had significantly lower psychological readiness than younger non-injured patients. A follow-up study in the same young cohort showed that those who re-injured had little improvement in ACL-RSI scores from preoperative to 12 months post-surgery, whereas those not injured showed a 20-point increase [76]. While this work needs further validation and replication, it would appear prudent to consider a psychological aspect alongside any physical return to sport assessment.

9.7 Conclusions

Many athletes do not return to their prior level of sport following ACL reconstruction surgery. Of those that do, a proportion will suffer a second ACL injury and younger patients who return to strenuous sports are at high risk of this. Criteria-based return to sport rehabilitation has been shown to be of significant benefit and should be preferred over solely time-based rehabilitation. Despite much interest in return to sport testing there is currently insufficient evidence that it can be used to provide advice to individual patients regarding their risk for further ACL injury.

ACL rupture and reconstruction should not be a career-ending injury; however, there are considerable and serious obstacles to overcome during rehabilitation and prior to return to sport. Surgical management with appropriate early physical rehabilitation can bring an athlete back to baseline functional level which allows the athlete to safely return to competitive play. However, the neuromotor imbalances of trunk, ligament, quadriceps, leg dominance may continue to exist and be observable and longitudinally measurable during dynamic sports manoeuvres out to and even beyond 2 years in competitive athletes if not appropriately assessed and treated. Therefore, additional focus on longitudinal measurement and maximization of strength along with minimization of the aforementioned neuromotor imbalances is necessary to ensure successful rehabilitative management and return to sport following ACL reconstruction of the young, highly active high-risk competitive athlete.

References

1. Ardern CL, Webster KE, Taylor NF, Feller JA. Return to sport following anterior cruciate ligament reconstruction surgery: a systematic review and meta-analysis of the state of play. Br J Sports Med. 2011;45(7):596–606.
2. Ardern CL, Taylor NF, Feller JA, Webster KE. Fifty-five per cent return to competitive sport following anterior cruciate ligament reconstruction surgery: an updated systematic review and meta-analysis incluing aspects of physcial functioning and contextual factors. Br J Sports Med. 2014;48(21):1543–52.
3. Grassi A, Zaffagnini S, Marcheggiani Muccioli GM, Neri MP, Della Villa S, Marcacci M. After revision anterior cruciate ligament reconstruction, who returns to sport? A systematic review and meta-analysis. Br J Sports Med. 2015;49(20):1295–304.
4. Lai CCH, Ardern CL, Feller JA, Webster KE. Eighty-three per cent of elite athletes return to preinjury sport after anterior cruciate ligament reconstruction: a systematic review with meta-analysis of return to sport rates, graft rupture rates and performance outcomes. Br J Sports Med. 2018;52(2):128–38.
5. Kay J, Memon M, Marx RG, Peterson D, Simunovic N, Ayeni OR. Over 90% of children and adolescents return to sport after anterior cruciate ligament reconstruction: a systematic review and meta-analysis. Knee Surg Sports Traumatol Arthrosc. 2018;26(4):1019–36.
6. Feller JA, Webster KE. Return to sport following anterior cruciate ligament reconstruction. Int Orthop. 2013;37(2):285–90.
7. Webster KE, Feller JA, Leigh W, Richmond A. Younger patients are at increased risk for graft rupture and contralateral injury after anterior cruciate ligament reconstruction. Am J Sports Med. 2014;42(3):641–7.
8. Ardern CL, Taylor NF, Feller JA, Whitehead TS, Webster KE. Psychological responses matter in returning to preinjury level of sport after anterior cruciate ligament reconstruction surgery. Am J Sports Med. 2013;41(7):1549–58.
9. Kvist J, Ek A, Sporrstedt K, Good L. Fear of re-injury: a hindrance for returning to sports after anterior cruciate ligament reconstruction. Knee Surg Sports Traumatol Arthrosc. 2005;13(5):393–7.
10. Langford JL, Webster KE, Feller JA. A prospective longitudinal study to assess psychological changes following anterior cruciate ligament reconstruction surgery. Br J Sports Med. 2009;43(5):377–81.
11. Webster KE, Feller JA, Lambros C. Development and preliminary validation of a scale to measure the psychological impact of returning to sport following anterior cruciate ligament reconstruction surgery. Phys Ther Sport. 2008;9(1):9–15.
12. Webster KE, Feller JA. Expectations for return to Preinjury sport before and after anterior cruciate ligament reconstruction. Am J Sports Med. 2019;47(3):578–83.
13. Feucht MJ, Cotic M, Saier T, Minzlaff P, Plath JE, Imhoff AB, et al. Patient expectations of primary and revision anterior cruciate ligament reconstruction. Knee Surg Sports Traumatol Arthrosc. 2016;24(1):201–7.
14. Bourke HE, Gordon DJ, Salmon LJ, Waller A, Linklater J, Pinczewski LA. The outcome at 15 years of endoscopic anterior cruciate ligament reconstruction using hamstring tendon autograft for 'isolated' anterior cruciate ligament rupture. J Bone Joint Surg. 2012;94B(5):630–7.
15. Shelbourne KD, Gray T, Haro M. Incidence of subsequent injury to either knee within 5 years after anterior cruciate ligament reconstruction with patellar tendon autograft. Am J Sports Med. 2009;37(2):246–51.
16. Webster KE, Feller JA. Exploring the high Reinjury rate in younger patients undergoing anterior cruciate ligament reconstruction. Am J Sports Med. 2016;44(11):2827–32.
17. Maletis GB, Inacio MC, Funahashi TT. Risk factors associated with revision and contralateral anterior cruciate ligament reconstructions in the Kaiser Permanente ACLR registry. Am J Sports Med. 2015;43(3):641–7.
18. Snaebjornsson T, Hamrin Senorski E, Sundemo D, Svantesson E, Westin O, Musahl V, et al. Adolescents and female patients are at increased risk for contralateral anterior cruciate ligament reconstruction: a cohort study from the Swedish National Knee Ligament Register based on 17,682 patients. Knee Surg Sports Traumatol Arthrosc. 2017;25(12):3938–44.
19. Wiggins AJ, Grandhi RK, Schneider DK, Stanfield D, Webster KE, Myer GD. Risk of secondary injury in younger athletes after anterior cruciate ligament reconstruction: a systematic review and meta-analysis. Am J Sports Med. 2016;44(7):1861–76.
20. Morgan MD, Salmon LJ, Waller A, Roe JP, Pinczewski LA. Fifteen-year survival of endoscopic anterior cruciate ligament reconstruction in patients aged 18 years and younger. Am J Sports Med. 2016;44(2):384–92.
21. Paterno MV, Rauh MJ, Schmitt LC, Ford KR, Hewett TE. Incidence of second ACL injuries 2 years after primary ACL reconstruction and return to sport. Am J Sports Med. 2014;42(7):1567–73.
22. Dekker TJ, Godin JA, Dale KM, Garrett WE, Taylor DC, Riboh JC. Return to sport after pediatric anterior cruciate ligament reconstruction and its effect on subsequent anterior cruciate ligament injury. J Bone Joint Surg Am. 2017;99(11):897–904.
23. Grindem H, Engebretsen L, Axe M, Snyder-Mackler L, Risberg MA. Activity and functional readiness, not age, are the critical factors for second anterior cruciate ligament injury - the Delaware-Oslo ACL cohort

study. Br J Sports Med. 2020; https://doi.org/10.1136/bjsports-2019-100623.
24. Webster KE, Feller JA. A research update on the state of play for return to sport after anterior cruciate ligament reconstruction. J Orthop Traumatol. 2019;20(1):10.
25. Webster KE, Feller JA, Whitehead TS, Myer GD, Merory PB. Return to sport in the younger patient with anterior cruciate ligament reconstruction. Orthop J Sports Med. 2017;5(4):2325967117703399.
26. Beischer S, Hamrin Senorski E, Thomee C, Samuelsson K, Thomee R. Young athletes return too early to knee-strenuous sport, without acceptable knee function after anterior cruciate ligament reconstruction. Knee Surg Sports Traumatol Arthrosc. 2018;26(7):1966–74.
27. Hamrin Senorski E, Svantesson E, Beischer S, Thomee C, Thomee R, Karlsson J, et al. Low 1-year return-to-sport rate after anterior cruciate ligament reconstruction regardless of patient and surgical factors: a prospective cohort study of 272 patients. Am J Sports Med. 2018;46(7):1551–8.
28. Toole AR, Ithurburn MP, Rauh MJ, Hewett TE, Paterno MV, Schmitt LC. Young athletes cleared for sports participation after anterior cruciate ligament reconstruction: how many actually meet recommended return-to-sport criterion cutoffs? J Orthop Sports Phys Ther. 2017;47(11):825–33.
29. Scheffler SU, Unterhauser FN, Weiler A. Graft remodeling and ligamentization after cruciate ligament reconstruction. Knee Surg Sports Traumatol Arthrosc. 2008;16(9):834–42.
30. Claes S, Verdonk P, Forsyth R, Bellemans J. The "ligamentization" process in anterior cruciate ligament reconstruction: what happens to the human graft? A systematic review of the literature. Am J Sports Med. 2011;39:2476–83.
31. Pauzenberger L, Syre S, Schurz M. "Ligamentization" in hamstring tendon grafts after anterior cruciate ligament reconstruction: a systematic review of the literature and a glimpse into the future. Arthroscopy. 2013;29(10):1712–21.
32. Nagelli CV, Hewett TE. Should return to sport be delayed until 2 years after anterior cruciate ligament reconstruction? Biol Funct Considerat Sports Med. 2017;47(2):221–32.
33. Hewett TE, Ford KR, Hoogenboom BJ, Myer GD. Understanding and preventing acl injuries: current biomechanical and epidemiologic considerations. 2010. N Am J Sports Phys Ther. 2010;5(4):234–51.
34. Laskowski ER, Newcomer-Aney K, Smith J. Refining rehabilitation with proprioceptive training: expediting return to play. Phys Sportsmed. 1997;25:89–102.
35. Lephart SM, Fu FH. Proprioception and neuromuscular control in joint stability. Champaign, IL: Human Kinetics; 2000. p. 439.
36. Griffin LY, Agel J, Albohm MJ, Arendt EA, Dick RW, Garrett WE, et al. Noncontact anterior cruciate ligament injuries: risk factors and prevention strategies. J Am Acad Orthop Surg. 2000;8(3):141–50.
37. Hewett TE, Paterno MV, Myer GD. Strategies for enhancing proprioception and neuromuscular control of the knee. Clin Orthop Relat Res. 2002;402:76–94.
38. Paterno MV, Schmitt LC, Ford KR, Rauh MJ, Myer GD, Huang B, et al. Biomechanical measures during landing and postural stability predict second anterior cruciate ligament injury after anterior cruciate ligament reconstruction and return to sport. Am J Sports Med. 2010;38(10):1968–78.
39. Tropp H, Odenrick P. Postural control in single-limb stance. J Orthop Res. 1988;6(6):833–9.
40. Caraffa A, Cerulli G, Projetti M, Aisa G, Rizzo A. Prevention of anterior cruciate ligament injuries in soccer. A prospective controlled study of proprioceptive training. Knee Surg Sports Traumatol Arthrosc. 1996;4(1):19–21.
41. Tropp H, Ekstrand J, Gillquist J. Factors affecting stabilometry recordings of single limb stance. Am J Sports Med. 1984;12(3):185–8.
42. Hewett TE, Stroupe AL, Nance TA, Noyes FR. Plyometric training in female athletes. Decreased impact forces and increased hamstring torques. Am J Sports Med. 1996;24(6):765–73.
43. Ford KR, Myer GD, Hewett TE. Valgus knee motion during landing in high school female and male basketball players. Med Sci Sports Exerc. 2003;35(10):1745–50.
44. Knapik JJ, Bauman CL, Jones BH, Harris JM, Vaughan L. Preseason strength and flexibility imbalances associated with athletic injuries in female collegiate athletes. Am J Sports Med. 1991;19(1):76–81.
45. Baumhauer J, Alosa D, Renstrom A, Trevino S, Beynnon B. A prospective study of ankle injury risk factors. Am J Sports Med. 1995;23(5):564–70.
46. Hewett TE, Myer GD, Ford KR, Heidt RS Jr, Colosimo AJ, McLean SG, et al. Neuromuscular control and valgus loading of the knee predict ACL injury risk in female athletes. Am J Sports Med. 2005;33(4):492–501.
47. Myer GD, Paterno MV, Ford KR, Hewett TE. Neuromuscular training techniques to target deficits before return to sport after anterior cruciate ligament reconstruction. J Strength Cond Res. 2008;22(3):987–1014.
48. Myer GD, Paterno MV, Ford KR, Quatman CE, Hewett TE. Rehabilitation after anterior cruciate ligament reconstruction: criteria-based progression through the return-to-sport phase. J Orthop Sports Phys Ther. 2006;36(6):385–402.
49. Lloyd DG, Buchanan TS. Strategies of muscular support of varus and valgus isometric loads at the human knee. J Biomech. 2001;34(10):1257–67.
50. Lloyd DG. Rationale for training programs to reduce anterior cruciate ligament injuries in Australian football. J Orthop Sports Phys Ther. 2001;31(11):645–54.
51. Cahill BR, Griffith EH. Effect of preseason conditioning on the incidence and severity of high school football knee injuries. Am J Sports Med. 1978;6(4):180–4.
52. Hewett TE, Riccobene JV, Lindenfeld TN, Noyes FR. The effect of neuromuscular training on the

incidence of knee injury in female athletes: a prospective study. Am J Sports Med. 1999;27(6):699–706.
53. Malinzak RA, Colby SM, Kirkendall DT, Yu B, Garrett WE. A comparison of knee joint motion patterns between men and women in selected athletic tasks. Clin Biomech. 2001;16(5):438–45.
54. Besier TF, Lloyd DG, Ackland TR, Cochrane JL. Anticipatory effects on knee joint loading during running and cutting maneuvers. Med Sci Sports Exerc. 2001;33(7):1176–81.
55. Myer GD, Ford KR, Palumbo JP, Hewett TE. Neuromuscular training improves performance and lower-extremity biomechanics in female athletes. J Strength Cond Res. 2005;19(1):51–60.
56. Neptune RR, Wright IC, van den Bogert AJ. Muscle coordination and function during cutting movements. Med Sci Sports Exerc. 1999;31(2):294–302.
57. van Melick N, van Cingel RE, Brooijmans F, Neeter C, van Tienen T, Hullegie W, et al. Evidence-based clinical practice update: practice guidelines for anterior cruciate ligament rehabilitation based on a systematic review and multidisciplinary consensus. Br J Sports Med. 2016;50(24):1506–15.
58. Ardern CL, Glasgow P, Schneiders A, Witvrouw E, Clarsen B, Cools A, et al. Consensus statement on return to sport from the First World Congress in Sports Physical Therapy, Bern. Br J Sports Med. 2016;50:14853–64.
59. Panariello R, Stump T, Allen A. Rehabilitation and return to play following anterior cruciate ligament reconstruction. Oper Tech Sports Med. 2017;26:181–93.
60. Ellman MB, Sherman SL, Forsythe B, LaPrade RF, Cole BJ, Bach BR Jr. Return to play following anterior cruciate ligament reconstruction. J Am Acad Orthop Surg. 2015;23(5):283–96.
61. Hewett TE, Webster KE, Hurd WJ. Systematic selection of key logistic regression variables for risk prediction analyses: a five-factor maximum model. Clin J Sport Med. 2019;29(1):78–85.
62. Feller JA, Webster KE. Where are we with return-to-sport testing following ACL reconstruction? Orthop Traumatol Surg Res. 2019;105(6):1037–8.
63. Nawasreh Z, Logerstedt D, Cummer K, Axe M, Risberg MA, Snyder-Mackler L. Functional performance 6 months after ACL reconstruction can predict return to participation in the same preinjury activity level 12 and 24 months after surgery. Br J Sports Med. 2018;52(6):375.
64. Webster KE, Hewett TE. What is the evidence for and validity of return-to-sport testing after anterior cruciate ligament reconstruction surgery? A systematic review and meta-analysis. Sports Med. 2019;49(6):917–29.
65. Grindem H, Snyder-Mackler L, Moksnes H, Engebretsen L, Risberg MA. Simple decision rules can reduce reinjury risk by 84% after ACL reconstruction: the Delaware-Oslo ACL cohort study. Br J Sports Med. 2016;50(13):804–8.
66. Losciale JM, Zdeb RM, Ledbetter L, Reiman MP, Sell TC. The association between passing return-to-sport criteria and second anterior cruciate ligament injury risk: a systematic review with meta-analysis. J Orthop Sports Phys Ther. 2019;49(2):43–54.
67. Kyritsis P, Bahr R, Landreau P, Miladi R, Witvrouw E. Likelihood of ACL graft rupture: not meeting six clinical discharge criteria before return to sport is associated with a four times greater risk of rupture. Br J Sports Med. 2016;50(15):946–51.
68. Sousa PL, Krych AJ, Cates RA, Levy BA, Stuart MJ, Dahm DL. Return to sport: does excellent 6-month strength and function following ACL reconstruction predict midterm outcomes? Knee Surg Sports Traumatol Arthrosc. 2017;25(5):1356–63.
69. Webster KE, Hewett TE. Return-to-sport testing following ACL reconstruction revisited. Br J Sports Med. 2020;54(1):2–3.
70. Bodkin SG, Rutherford MH, Diduch DR, Brockmeier SF, Hart JM. How much time is needed between serial "return to play" assessments to achieve clinically important strength gains in patients recovering from anterior cruciate ligament reconstruction? Am J Sports Med. 2020;48(1):70–7.
71. Paterno MV, Huang B, Thomas S, Hewett TE, Schmitt LC. Clinical factors that predict a second ACL injury after ACL reconstruction and return to sport: preliminary development of a clinical decision algorithm. Orthop J Sports Med. 2017;5(12):2325967117745279.
72. Webster KE, Feller JA. Clinical tests can be used to screen for second anterior cruciate ligament injury in younger patients who return to sport. Orthop J Sports Med. 2019;7(8):2325967119863003.
73. Tagesson S, Kvist J. Greater fear of re-injury and increased tibial translation in patients who later sustain an ACL graft rupture or a contralateral ACL rupture: a pilot study. J Sports Sci. 2016;34(2):125–32.
74. Paterno MV, Flynn K, Thomas S, Schmitt LC. Self-reported fear predicts functional performance and second ACL injury after ACL reconstruction and return to sport: a pilot study. Sports Health. 2018;10(3):228–33.
75. McPherson AL, Feller JA, Hewett TE, Webster KE. Psychological readiness to return to sport is associated with second anterior cruciate ligament injuries. Am J Sports Med. 2019;47(4):857–62.
76. McPherson AL, Feller JA, Hewett TE, Webster KE. Smaller s. Am J Sports Med. 2019;47(5):1209–15.

Revision ACL Reconstruction

10

Jonathan D. Hughes and Bryson P. Lesniak

10.1 Introduction

10.1.1 Background

The rate of failure after anterior cruciate ligament (ACL) reconstruction remains low, with a reported failure rate of 3–14% [1–6]. However, when failures do occur, patient outcomes are less favorable after revision ACL reconstruction. A recent prospective cohort study showed that after revision ACL reconstruction, 38% of patients met subjective failure criteria, while only 20% of patients undergoing primary ACL reconstructions met subjective failure criteria [7]. Data from the Multicenter Orthopedic Outcomes Network (MOON) and Multicenter ACL Revision Study (MARS) groups demonstrated meniscal and cartilage injury in 90% of patients at the time of revision ACL reconstruction, and these patients had a 1.7 times greater risk of Outerbridge grade 3 or 4 cartilaginous lesions than patients undergoing primary ACL reconstruction [8, 9]. Additionally, these data concluded patients undergoing multiple revision ACL reconstructions had lower activity scores, increased risk of chondral injuries in the patellofemoral and medial compartments, and had a high rate of non-traumatic, recurrent graft tears [10]. Various studies have demonstrated that patient-reported outcomes, activity levels, and return to sport rates are worse after revision ACL reconstruction compared to primary ACL reconstruction [11, 12].

J. D. Hughes (✉) · B. P. Lesniak
Department of Orthopaedic Surgery, University of Pittsburgh, UPMC Freddie Fu Sports Medicine Center, Pittsburgh, PA, USA
e-mail: Hughesjd3@upmc.edu; lesniakbp@upmc.edu

10.1.2 Factors Contributing to Failure

The most important aspect when approaching failure of ACL reconstruction is understanding the reason for failure. Although reasons for failure are multifactorial, repeat traumatic events or technical errors are the most common causes for failure. Recent data from the MARS group demonstrated the most common reason for failure after primary ACL reconstruction was a traumatic event causing graft failure. However, when looking at patients undergoing multiple revisions, technical error was reported to be the most common error [10]. A separate study from the MARS database showed that femoral tunnel malposition was a contributing cause of failure in 48% of cases and the only cause of failure in 25% of cases. They concluded that anterior and vertical femoral tunnel malposition was the most common error [13]. A retrospective review reported anterior femoral tunnel placement in 36% of revision cases [14]. Anterior femoral placement results in excessive graft tension,

N. Nakamura et al. (eds.), *Advances in Knee Ligament and Knee Preservation Surgery*,
https://doi.org/10.1007/978-3-030-84748-7_10

leading to loss of flexion and stretching of the graft. Vertical femoral tunnel placement, although adequately restoring anteroposterior stability, may lead to continued rotational instability and failure. On the tibial side, an anteriorly placed tibial tunnel may lead to loss of knee extension due to graft impingement against the intercondylar notch. A posteriorly placed tibial tunnel may lead to impingement against the posterior cruciate ligament. Medial or lateral tibial tunnel placement may also lead to graft impingement on the intercondylar notch, as well as possible injury to the cartilage on the tibial plateau.

Graft choice has also been implicated in causes for ACL reconstruction failure. Numerous studies have demonstrated allograft is associated with a higher rate of failure in young patients, with one study showing a failure rate of 23.1% and a reoperation rate of 38% [5, 15, 16]. A study from the MARS cohort reported a 4.4 times greater risk of failure with allograft compared to autograft [15], while a recent meta-analysis found allograft had a significantly higher risk of rupture than autograft, with an odds ratio of 5.03 [17]. When comparing autograft options, various studies have demonstrated equivalent graft failure among quadriceps tendon (QT), patellar tendon (BPTB), and hamstrings tendon (HT) [18–20]. However, recent literature has indicated a higher failure rate with HT compared to BPTB in younger patients [21–23].

Graft size can be a cause of ACL reconstruction failure. One retrospective review concluded that HT graft size less than 8 mm in diameter had an increased re-tear risk [24], while a recent systematic review concluded that HT graft diameter greater than 8 mm reduced the failure rate [25]. A recent study determined for HT graft diameters within 7 mm and 9 mm, there was a 0.82 times lower likelihood of revision with every 0.5 mm increase in graft diameter [26]. However, for patients with small intercondylar notches or bony morphology, smaller grafts may be beneficial and decrease rate of failure in this subset of patients [27, 28]. Therefore, each case must be individualized to the particular patient.

10.2 Preoperative Workup

10.2.1 History and Physical Exam

The most important aspect of the preoperative workup is elucidating the cause for ACL reconstruction failure. A detailed history should be taken, including mechanism of injury, activity level and sports participation, antecedent pain or feelings of instability in the knee, and surgical history on the knee. Prior clinical and operative notes should be reviewed, if available. Additional information to obtain includes prior surgical history on either knee, history of infections and bleeding disorders, risk factors for osteoporosis, personal or family history of hyperlaxity, and history of smoking. All these factors can play a role in ACL reconstruction failure, and aid in future surgical planning.

A thorough physical exam should be performed, including the contralateral, healthy knee. Range of motion, alignment, muscle strength and girth, and scars from prior surgery should be documented for bilateral extremities. The knee rotational and anteroposterior stability should be evaluated and compared to the contralateral side. The injured knee should be assessed for physical signs of latent infection. Any signs of hyperlaxity should also be noted. A pivot shift of bilateral knees should be performed and compared, as high-grade rotatory knee laxity is an inclusion in the recently published STABILITY trial discussing the addition of further stabilization procedures such as lateral extra-articular tenodesis (LET) [29].

10.2.2 Imaging Studies

Initial radiographic evaluation includes standing anterior-posterior, weight bearing 45° flexion posterior-anterior, lateral, and Merchant views. The weight bearing 45° flexion posterior-anterior views can be utilized to measure the angle of the prior femoral tunnel, while the lateral view can be used in conjunction with the 45° flexion posterior-anterior views to measure the angle of the prior tibial tunnel. The type and position of fixation devices can be identified on these views. These fixation devices, including interference screws, staples, and buttons, are important for operative planning, as these may need to be removed during revision surgery. Additionally, full-length standing radiographs should be obtained to evaluate overall mechanical alignment in the coronal and sagittal planes. The current magnetic resonance imaging (MRI) as well as the MRI from the initial injury should be reviewed in detail to look for concomitant pathology such as meniscus tears, chondral injury, anterolateral capsule injuries, and additional ligamentous injuries.

The authors prefer to obtain a thin-cut computed tomography (CT) scan with three-dimensional (3D) reconstructions in all revision settings. These images can provide crucial information on the position of the tunnels, tunnel widening, position of radiolucent fixation devices, and concomitant bony pathology. One study reported bone tunnels seen on CT were also identified on radiographs, but there can be discrepancies in the measurements of the femoral tunnels between the imaging modalities [30]. Therefore, CT scans with 3D reconstructions can add valuable and more accurate measurements for preoperative planning, as well as understanding the spatial orientations of the tunnels [31].

Table 10.1 The REVision using Imaging to guide Staging and Evaluation (REVISE) in ACL Reconstruction Classification

Type 1A	One-stage revision, tibial and femoral tunnels have acceptable size, position, and trajectory
Type 1B	One-stage revision with modifications needed to tunnel(s)
1B-F	Femoral tunnel needs modification
1B-T	Tibial tunnel needs modification
1B-FT	Femoral and tibial tunnels need modification
Type 2	Two-stage revision
2-W	Tunnel widening or overlap
2-M	Presence of malalignment
2-I	Infection

This classification system evaluates the prior tibial and femoral tunnels after ACL reconstruction and determines the need for a one-stage versus two-stage revision ACL reconstruction. *ACL* anterior cruciate ligament

10.3 Surgical Planning

After a thorough review of the available imaging studies, a surgical plan can be developed that addresses tunnel position, possible hardware removal, mechanical alignment, and any meniscal deficiency. Revision ACL reconstruction can be performed as a one-stage or two-stage procedure, which can usually be determined preoperatively through the initial workup. The location and size of the prior tunnels on the tibial and femoral sides is the main determinant of one-stage versus two-stage procedures. The location of these tunnels can be accurately assessed via CT scans with 3D reconstructions and categorized according to the The REVision using Imaging to guide Staging and Evaluation (REVISE) in ACL Reconstruction Classification (Table 10.1) [32]. A REVISE type 1A includes tibial and femoral tunnels that have acceptable size, position, and trajectory (Fig. 10.1). These cases are approached as a one-stage revision without the need to revise the prior tunnels. Type 1B also refers to a one-stage revision, but requires modifications to the femoral (F) and/or tibial (T) tunnels (Fig. 10.2). This classification type can be sub-classified based on the tunnel needing revised (1B-F for femoral tunnel, 1B-T for tibial tunnel, and 1B-FT for both tibial and femoral tunnels). Type 2 refers to a two-stage revision, due to significant tunnel widening or overlap, poor tunnel position, malalignment, or infection (Fig. 10.3). This classification type can be subdivided into W, M, and I modifiers to identify tunnel widening or overlap (2-W), presence of malalignment (2-M), and infection (2-I). Prior studies have demonstrated that ~34% of revision ACL reconstructions follow Type 1A, 50% follow Type 1B, and 15% follow Type 2 [9, 33]. Hardware should be left in place, if possible, to prevent the creation of bone defects, if it does not interfere with tunnel placement.

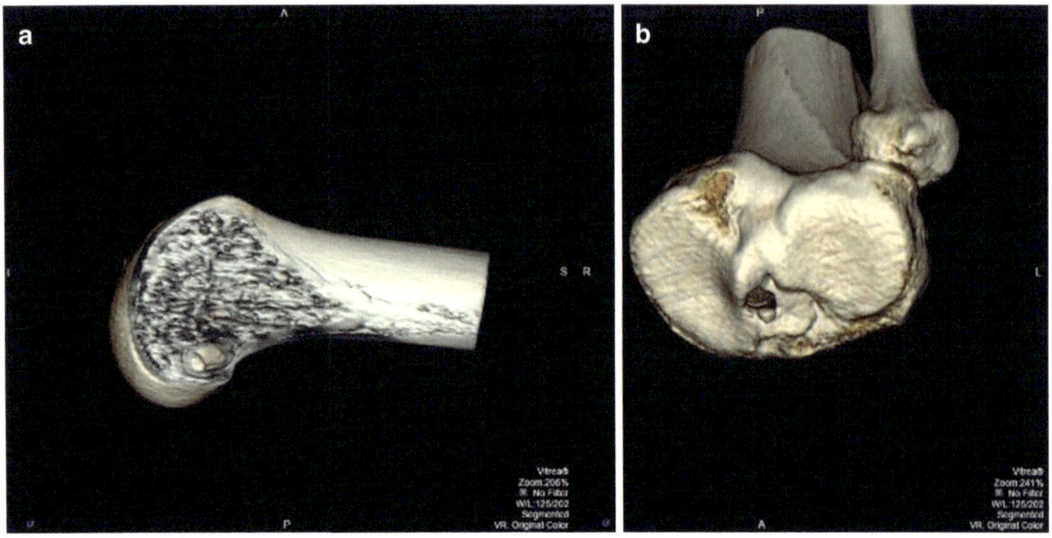

Fig. 10.1 REVision using Imaging to guide Staging and Evaluation (REVISE) Type 1A. Three-dimensional computed tomography (3-D CT) reconstruction of a prior anterior cruciate ligament reconstruction (ACL-R). The sagittal femur (**a**) and axial tibia (**b**) demonstrate acceptable location and size of the prior femoral and tibial tunnels. These tunnels can be utilized again for the revision ACL-R

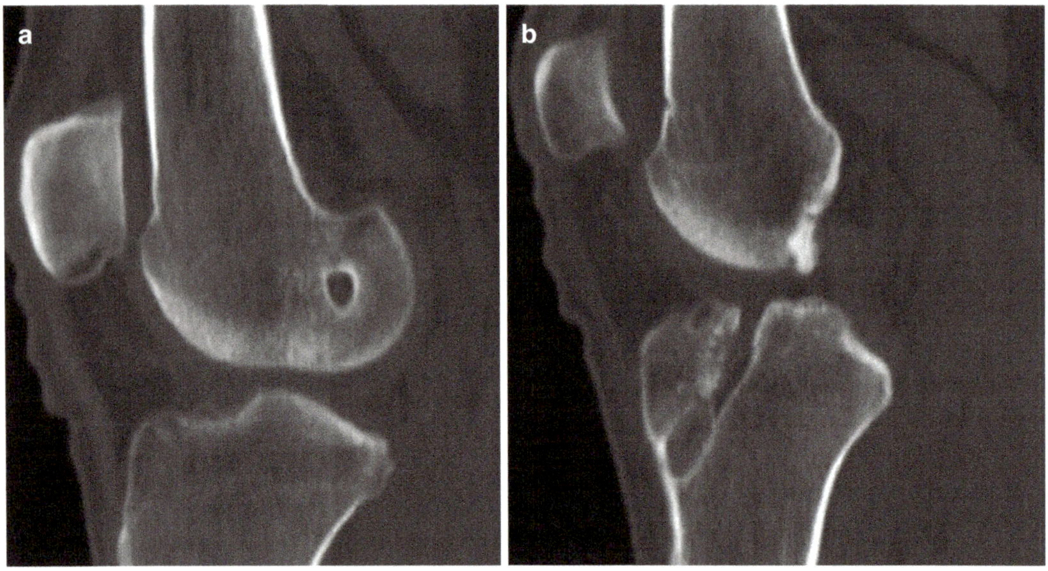

Fig. 10.2 REVision using Imaging to guide Staging and Evaluation (REVISE) Type 1B-F. A sagittal computed tomography of a prior anterior cruciate ligament reconstruction (ACL-R) demonstrates a vertical and anterior tunnel on the femur (image **a**) and an appropriately placed tibial tunnel (image **b**). A new femur tunnel will need to be drilled, while the prior tibial tunnel can be used again for the revision ACL-R

10.3.1 Type 1A

Tunnels that are appropriately placed can be used again in the revision setting, even with widening identified preoperatively. Hardware removal may be necessary if interference screws were utilized. Cortical fixation devices, staples, and cortical screws can be avoided during surgery and therefore most often do not need to be removed. While addressing screw removal, a curette or burr should

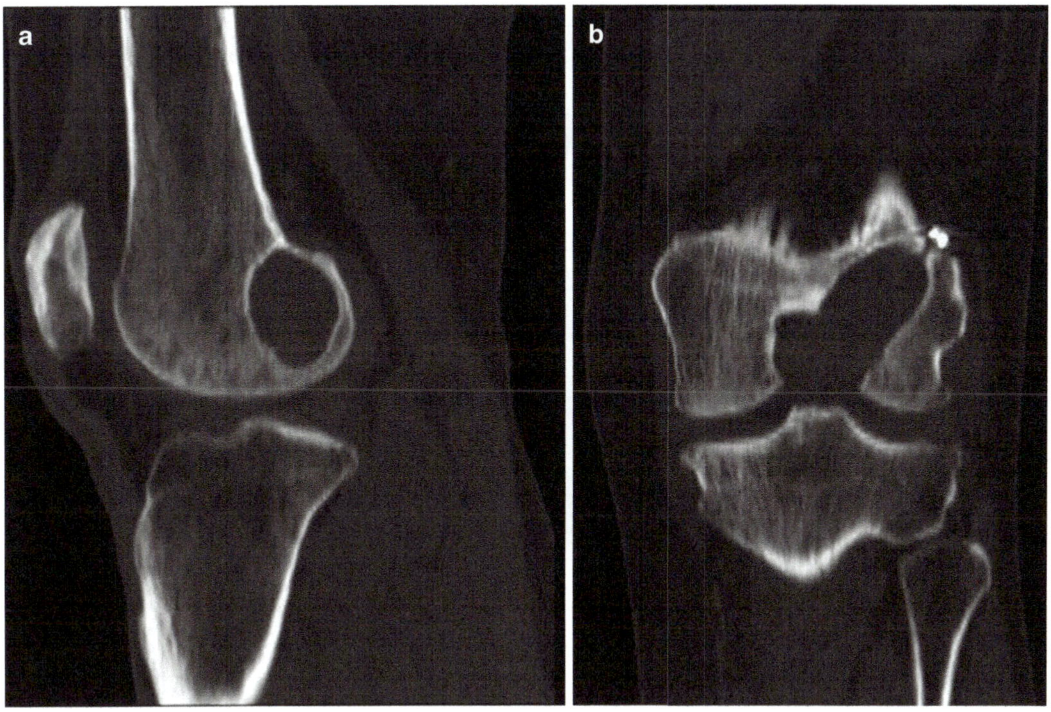

Fig. 10.3 REVision using Imaging to guide Staging and Evaluation (REVISE) Type 2-W. A sagittal computed tomography of a prior anterior cruciate ligament reconstruction (ACL-R) demonstrates tunnel widening and osteolysis on the femur tunnel on the sagittal (image **a**) and coronal (image **b**) images. This will require a two-stage procedure, with bone grafting at the index surgery

be used to remove all bone around the head of the screw. If the type of prior screw is unknown, one should have screw removal sets available. Once the screw is removed, the remaining soft tissue in the tunnel is debrided and the tunnel is dilated to the appropriate size (Fig. 10.4). The authors' preference is to debride the index tunnels by starting with a small reamer and sequentially increasing the diameter until healthy, bleeding bone is exposed in the tunnel. Placing a dry arthroscope into the tunnel between reamings can help determine when fresh bleeding bone is exposed circumferentially in the tunnel. A larger soft tissue graft can be used with adjustable or continuous loop cortical fixation devices, especially if the posterior wall is incompetent on the femoral side. Interference screws can be utilized if the surgeon prefers if there is no wall breach. On the femoral side, if the tunnel remains larger than the graft after tunnel debridement, or a defect remains from the screw removal, an interference screw or bone graft can also be placed into the tunnel (Fig. 10.4). Grafts with bone blocks can also be used, either autograft or allograft, with allograft providing the ability to obtain a larger bone block if needed to fill the prior tunnel. However, the authors preferred method is to avoid the use of allograft in primary or revision ACL reconstructions when possible given the recent studies demonstrated increased failure rates with allograft tissues [4, 15, 34].

10.3.2 Type 1B

In many instances for Type 1B-F, the prior femoral tunnel is placed sufficiently anteriorly or vertically that a new tunnel can be placed without addressing the prior tunnel (Fig. 10.5). A recent case series of the MARS cohort detailed drilling an entirely new femoral tunnel in 82% of cases [13]. If the prior tunnel is widened or enlarged, bone graft or an interference screw can be placed into the tunnel to fill the void and avoid collapse between the two tunnels.

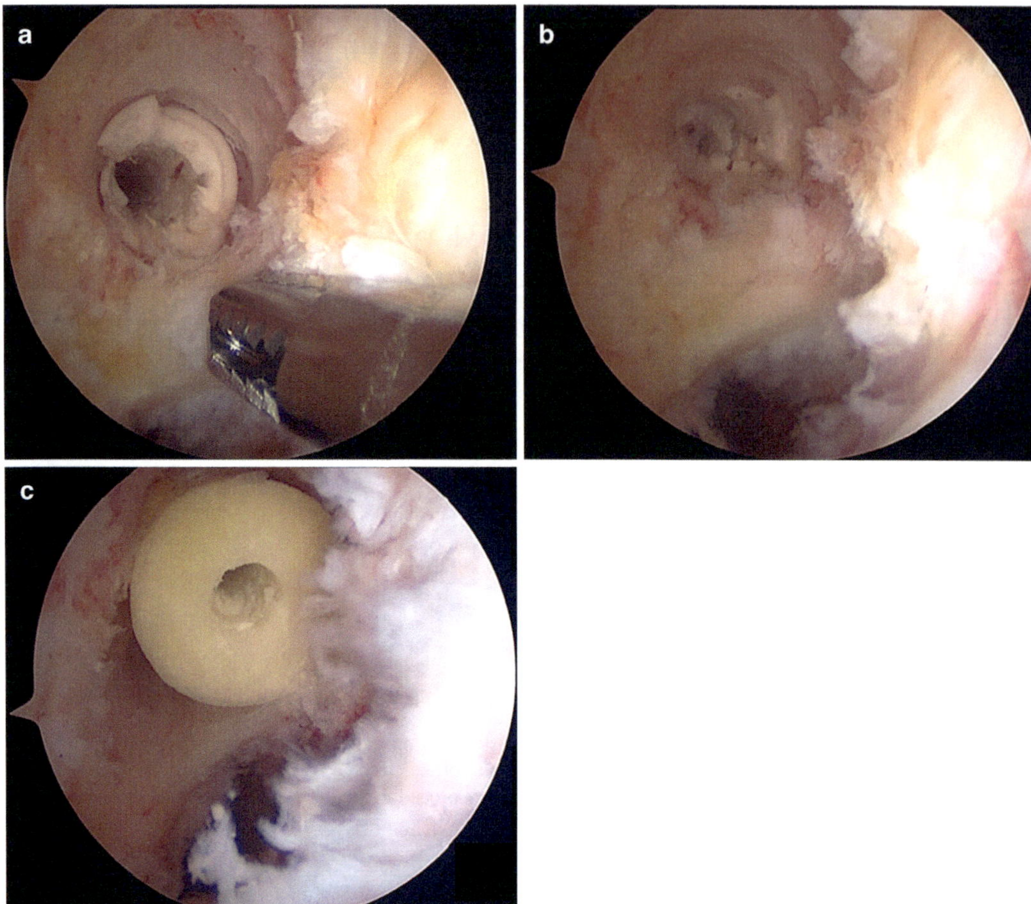

Fig. 10.4 Hardware removal and Single-stage bone grafting on the femoral side. Image (**a**) demonstrates the prior femoral tunnel screw with the prior, appropriately placed femoral tunnel distal to it (marked by the shaver). Image (**b**) shows the large femoral defect after removal of the screw and debridement of the tunnel. Image (**c**) demonstrates the bone dowel within the prior screw tunnel

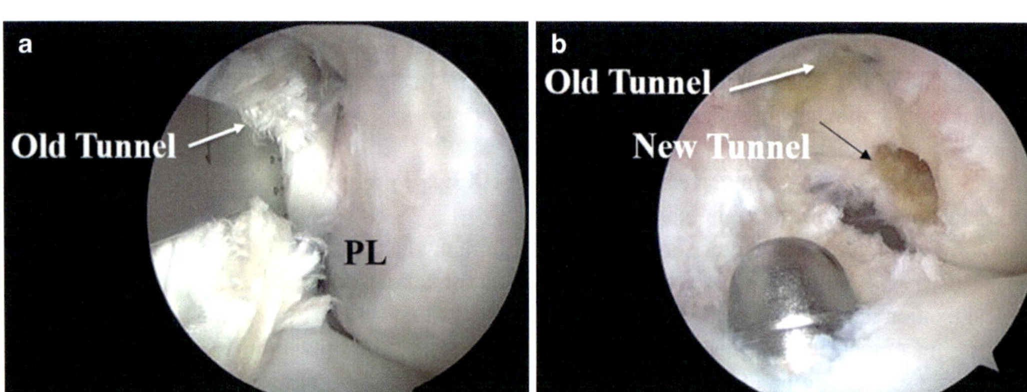

Fig. 10.5 Intra-operative images of a REVision using Imaging to guide Staging and Evaluation (REVISE) Type 1B-F. Image (**a**), as viewed from an anterolateral portal, demonstrates the prior vertically and anteriorly placed femoral tunnel. Image (**b**) demonstrates the placement of a new femoral tunnel in the anatomic footprint. Also in image (**b**), a dilator can be seen within the prior tibial tunnel, as this was placed in appropriate position and will be utilized in the revision surgery

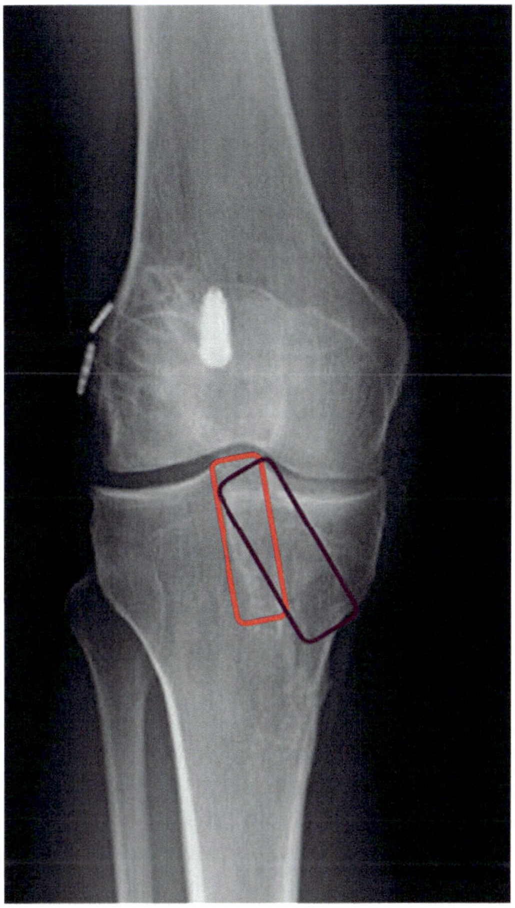

Fig. 10.6 The divergent tunnel technique. A coronal radiograph of a right knee demonstrates prior hardware from multiple anterior cruciate ligament reconstructions (ACL-R). The prior tibial tunnel (red box) aperture is appropriate but the proposed tunnel angle and direction (purple box) are new

A very common scenario in revision cases occurs when the prior tunnel is not quite anatomic, and will overlap with the new, anatomic, tunnel. In this situation, the "divergent tunnel" technique should be employed (Fig. 10.6). This involves drilling the new tunnel in a divergent angle to the previous tunnel to minimize tunnel overlap [35]. If the tunnel aperture becomes significantly widened due to the overlap, several options exist intraoperatively to address this issue. One option includes bone grafting the previous tunnel with autograft, allograft bone chips, allograft bone graft substitute, or allograft bone dowels. This option provides stability of the bone graft with the ability to drill a new tunnel through the bone graft [36, 37]. Another option includes placing a large interference screw in the widened tunnel and diverge its trajectory from the new tunnel to allow bony fixation of the screw.

If significant widening of the tunnels is present, there are two options to complete a single-stage ACL revision: the over-the-top (OTT) procedure and one-stage bone grafting. The definition of "significant tunnel widening" is up for debate. While some authors have suggested greater than 14 mm, recent literature has demonstrated inferior results with one-stage revisions with index tunnels greater than 12.5 mm diameter [38]. The OTT procedure can be an effective single-stage option in these cases and includes creating a trench in the anatomic footprint of the femur to allow graft healing. The graft is passed from the tibia through the femoral notch and posterior to the lateral femoral condyle and through the posterolateral capsule and secured to the lateral femoral condyle with a staple, bicortical screw, or other suspensory fixation devices. This technique avoids the need for a femoral tunnel and allows the graft to heal along the trench in the femur along the posterior cortex of the lateral femoral condyle (Fig. 10.7). Additionally, an Achilles allograft with a bone block can be used, allowing a large bone block to be used on the tibial side depending on the size of the prior tunnel. A recent systematic review demonstrated comparable outcomes between OTT ACL reconstruction and traditional ACL reconstruction in the primary and revision setting [39].

10.3.3 Type 2

A two-stage procedure is indicated for poorly placed tunnels that don't allow a new, anatomic tunnel to be created, tunnel widening so significant on both the tibia and femur that over-the-top procedure or one-stage bone grafting are not feasible, or significant infection. The two-stage procedure involves an initial bone-grafting procedure, or in the case of infection multiple debridements followed by bone-grafting, and then an incorporation phase allowing the bone

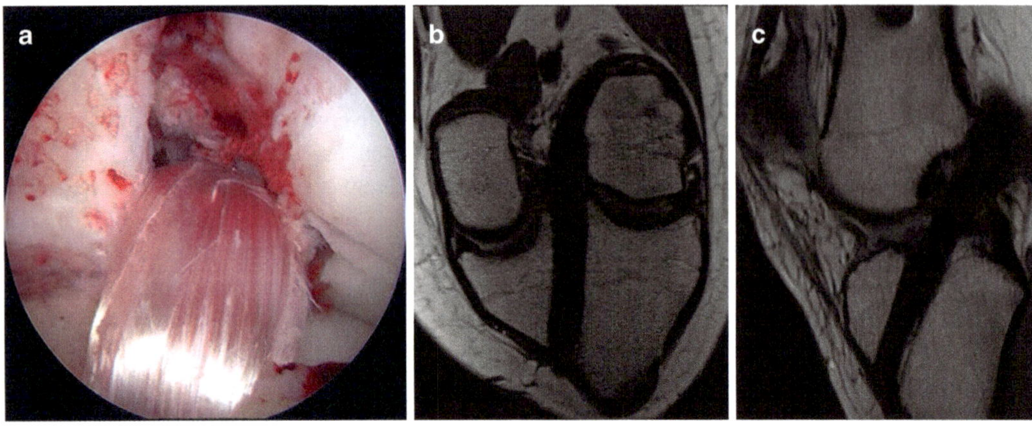

Fig. 10.7 Revision anterior cruciate ligament reconstruction (ACL-R) using the over-the-top (OTT) method. Image (**a**) is an intra-operative arthroscopic view from the anterolateral portal of the completed OTT ACL-R. Images (**b**, **c**) are postoperative coronal and sagittal computed tomography images, respectively, of a healed OTT ACL-R

graft to fully heal before the subsequent revision ACL reconstruction. At the initial surgery, hardware is removed from the tunnels and the tunnels are debrided of all soft tissue and sclerotic bone as described above. Bone graft is impacted into the tunnels. The source of bone graft can be autograft from the tibia or iliac crest, allograft, or synthetic graft. Following the primary bone grafting procedure, the patient is followed clinically for 3–4 months to allow the bone graft to fully incorporate. Repeat radiographs and CT scan are taken at the 3-month mark to ensure adequate incorporation. If the bone grafts have not fully incorporated, the patient should continue to be followed clinically with repeat imaging studies in 1–2 months to ensure adequate incorporation.

10.3.4 Graft Choice

The graft choices for revision ACL reconstruction should be individualized to each patient, with the index surgery graft choice taken into consideration. Options include ipsilateral or contralateral QT, BPTB, and HT autografts, as well as allografts. QT and BPTB autografts are advantageous, as a bone block can be harvested with either graft, allowing initial bony fixation and addressing the bony void from the index tunnel. Another advantage of QT is the flexibility of graft size. Most patient's QT can be harvested from 8 mm up to 11–12 mm diameter without difficulty or concern. However, BTB and QT harvests are limited in terms of graft length when compared to HT. Therefore, HT autografts are the authors' preferred choice for OTT procedures. As stated previously, allografts are avoided in young patients due to the increased risk of failure [4, 15, 34]. However, the Achilles allograft with a bone block is a great tool for widened tunnels or OTT procedures. The bone block can be cut to match the size of the index tibial or femoral tunnel, thus avoiding bone grafting or placement of an interference screw.

10.3.5 Additional Considerations

The cause of ACL reconstruction failure should be sought in all cases. In many cases, there is concomitant malalignment or bony morphology or unrecognized ligamentous or meniscal pathology that may be contributing to failure. Mechanical alignment should be evaluated with full-length standing radiographs. Varus malalignment may place excessive strain on the reconstructed knee, especially in the setting of a varus thrust [40]. A proximal tibial osteotomy may be warranted to address this malalignment. A more subtle malalignment is in the sagittal plane. Prior studies have demonstrated that an increased posterior tibial slope was found to predict high-grade rotatory knee laxity, while a smaller medial tibial depth and increased lateral tibial plateau slope were associated with increased risk of ACL injuries [41–43]. For cases with increased posterior tibial slope, an anterior closing wedge tibial osteotomy may be indicated. Distal femoral characteristics, such as an increased posterior femoral condylar depth, a decreased notch width, and a decreased notch width index have also been associated with persistent knee instability after ACL rupture [44–46].

Unrecognized injuries to the posterolateral or posteromedial soft tissue structures may lead to early graft failure and necessitate surgical intervention for optimal outcomes [47, 48]. Meniscus tears, especially root tears, can cause knee instability in an ACL-deficient knee [49–51]. A biomechanical study showed the medial meniscus is an important stabilizer to anterior tibial translation, while the lateral meniscus contributes to rotatory knee stability [52]. If these characteristics are identified preoperatively, they should be addressed appropriately with a staged or simultaneous meniscal transplant. Lastly, there is a subset of patients who are at high risk of ACL reconstruction failure, including those that have hyperlaxity, are returning to high impact, pivoting activities, and are young. A recent multicenter, randomized control trial compared ACL reconstruction with and without LET in these young patients. The authors found decreased rates of failure and rotatory knee laxity in the ACL reconstruction with LET group [29]. While further studies are ongoing to examine this intervention and patient population, surgeons should consider LET in the high-risk ACL patient.

10.4 Conclusion

Revision ACL reconstruction is an increasingly common procedure, but may present challenges to the operating surgeon. Recent literature has shown that outcomes after revision ACL reconstruction are worse than primary ACL reconstruction, highlighting the importance of approaching revision ACL surgery with care and expertise. The most important aspect of the preoperative workup is elucidating the cause of failure of the index ACL reconstruction, and addressing those causes in the revision setting. A thorough preoperative workup is essential to ensure optimal outcomes.

References

1. Ahlden M, Samuelsson K, Sernert N, Forssblad M, Karlsson J, Kartus J. The Swedish National Anterior Cruciate Ligament Register: a report on baseline variables and outcomes of surgery for almost 18,000 patients. Am J Sports Med. 2012;40:2230–5.
2. Annear PT, Rohr EJ, Hille DM, Gohil S, Ebert JR. No clinical difference in 10-year outcomes between standard and minimal graft debridement techniques in patients undergoing anterior cruciate ligament reconstruction using autologous hamstrings: a randomized controlled trial. Knee Surg Sports Traumatol Arthrosc. 2019;27:516–23.
3. MOON Knee Group, Spindler KP, Huston LJ, Zajichek A, Reinke EK, Amendola A, et al. Anterior cruciate ligament reconstruction in high school and college-aged athletes: does autograft choice influence anterior cruciate ligament revision rates? Am J Sports Med. 2020;48(2):298–309.
4. Kaeding CC, Pedroza AD, Reinke EK, Huston LJ, Spindler KP. Risk factors and predictors of subsequent ACL injury in either knee after ACL reconstruction: prospective analysis of 2488 primary ACL reconstructions from the MOON cohort. Am J Sports Med. 2015;43:1583–90.
5. Maletis GB, Inacio MC, Funahashi TT. Risk factors associated with revision and contralateral anterior cruciate ligament reconstructions in the Kaiser Permanente ACLR registry. Am J Sports Med. 2015;43:641–7.
6. Wright RW, Magnussen RA, Dunn WR, Spindler KP. Ipsilateral graft and contralateral ACL rupture at five years or more following ACL reconstruction a systematic review. J Bone Joint Surg (Am). 2011;93(12):1159–65.
7. Lind M, Menhert F, Pedersen AB. Incidence and outcome after revision anterior cruciate ligament reconstruction: results from the Danish registry for knee ligament reconstructions. Am J Sports Med. 2012;40:1551–7.
8. Borchers JR, Kaeding CC, Pedroza AD, Huston LJ, Spindler KP, Wright RW, et al. Intra-articular findings in primary and revision anterior cruciate ligament reconstruction surgery: a comparison of the

MOON and MARS study groups. Am J Sports Med. 2011;39:1889–93.
9. MARS Group. Descriptive epidemiology of the multicenter ACL revision study (MARS) cohort. Am J Sports Med. 2010;38:1979–86.
10. Chen JL, Allen CR, Stephens TE, Haas AK, Huston LJ, Wright RW, et al. Differences in mechanisms of failure, intraoperative findings, and surgical characteristics between single-and multiple-revision ACL reconstructions: a MARS cohort study. Am J Sports Med. 2013;41:1571–8.
11. Reinhardt KR, Hammoud S, Bowers AL, Umunna B-P, Cordasco FA. Revision ACL reconstruction in skeletally mature athletes younger than 18 years. Clin Orthop Relat Res. 2012;470:835–42.
12. Wright R, Spindler K, Huston L, Amendola A, Andrish J, Brophy R, et al. Revision ACL reconstruction outcomes: MOON cohort. J Knee Surg. 2011;24:289–94.
13. Morgan JA, Dahm D, Levy B, Stuart MJ. Femoral tunnel malposition in ACL revision reconstruction. J Knee Surg. 2012;25:361–8.
14. Trojani C, Sbihi A, Djian P, Potel J-F, Hulet C, Jouve F, et al. Causes for failure of ACL reconstruction and influence of meniscectomies after revision. Knee Surg Sports Traumatol Arthrosc. 2011;19:196–201.
15. Kaeding CC, Aros B, Pedroza A, Pifel E, Amendola A, Andrish JT, et al. Allograft versus autograft anterior cruciate ligament reconstruction: predictors of failure from a MOON prospective longitudinal cohort. Sports Health. 2011;3:73–81.
16. van Eck CF, Schkrohowsky JG, Working ZM, Irrgang JJ, Fu FH. Prospective analysis of failure rate and predictors of failure after anatomic anterior cruciate ligament reconstruction with allograft. Am J Sports Med. 2012;40:800–7.
17. Krych AJ, Jackson JD, Hoskin TL, Dahm DL. A meta-analysis of patellar tendon autograft versus patellar tendon allograft in anterior cruciate ligament reconstruction. Arthroscopy. 2008;24:292–8.
18. Gabler CM, Jacobs CA, Howard JS, Mattacola CG, Johnson DL. Comparison of graft failure rate between autografts placed via an anatomic anterior cruciate ligament reconstruction technique: a systematic review, meta-analysis, and meta-regression. Am J Sports Med. 2016;44:1069–79.
19. Pinczewski LA, Lyman J, Salmon LJ, Russell VJ, Roe J, Linklater J. A 10-year comparison of anterior cruciate ligament reconstructions with hamstring tendon and patellar tendon autograft: a controlled, prospective trial. Am J Sports Med. 2007;35:564–74.
20. Webster KE, Feller JA, Hartnett N, Leigh WB, Richmond AK. Comparison of patellar tendon and hamstring tendon anterior cruciate ligament reconstruction: a 15-year follow-up of a randomized controlled trial. Am J Sports Med. 2016;44:83–90.
21. Gifstad T, Foss OA, Engebretsen L, Lind M, Forssblad M, Albrektsen G, et al. Lower risk of revision with patellar tendon autografts compared with hamstring autografts: a registry study based on 45,998 primary ACL reconstructions in Scandinavia. Am J Sports Med. 2014;42:2319–28.
22. Persson A, Fjeldsgaard K, Gjertsen J-E, Kjellsen AB, Engebretsen L, Hole RM, et al. Increased risk of revision with hamstring tendon grafts compared with patellar tendon grafts after anterior cruciate ligament reconstruction: a study of 12,643 patients from the Norwegian Cruciate Ligament Registry, 2004–2012. Am J Sports Med. 2014;42:285–91.
23. Rahr-Wagner L, Thillemann TM, Pedersen AB, Lind M. Comparison of hamstring tendon and patellar tendon grafts in anterior cruciate ligament reconstruction in a nationwide population-based cohort study: results from the Danish registry of knee ligament reconstruction. Am J Sports Med. 2014;42:278–84.
24. Magnussen RA, Lawrence JTR, West RL, Toth AP, Taylor DC, Garrett WE. Graft size and patient age are predictors of early revision after anterior cruciate ligament reconstruction with hamstring autograft. Arthroscopy. 2012;28:526–31.
25. Conte EJ, Hyatt AE, Gatt CJ Jr, Dhawan A. Hamstring autograft size can be predicted and is a potential risk factor for anterior cruciate ligament reconstruction failure. Arthroscopy. 2014;30:882–90.
26. Spragg L, Chen J, Mirzayan R, Love R, Maletis G. The effect of autologous hamstring graft diameter on the likelihood for revision of anterior cruciate ligament reconstruction. Am J Sports Med. 2016;44:1475–81.
27. Görmeli CA, Görmeli G, Öztürk YB, Özdemir Z, Kahraman A. The effect of the intercondyler notch width index on anterior cruciate ligament injuries: a study on groups with unilateral and bilateral ACL injury. Orthop J Sports Med. 2014;2:2325967114S2325900204.
28. Van Eck CF, Martins CA, Vyas SM, Celentano U, van Dijk CN, Fu FH. Femoral intercondylar notch shape and dimensions in ACL-injured patients. Knee Surg Sports Traumatol Arthrosc. 2010;18:1257–62.
29. Getgood AM, Bryant D, Litchfield RB, McCormack RG, Heard M, MacDonald PB, et al. Lateral extra-articular tenodesis reduces failure of hamstring tendon autograft ACL Reconstruction-Two Year outcomes from the STABILITY study randomized clinical trial. Orthop J Sports Med. 2019;7:2325967119S2325900280.
30. Webster KE, Feller JA, Elliott J, Hutchison A, Payne R. A comparison of bone tunnel measurements made using computed tomography and digital plain radiography after anterior cruciate ligament reconstruction. Arthroscopy. 2004;20:946–50.
31. Kitamura G, Albers MBV, Lesniak BP, Rabuck SJ, Musahl V, Andrews CL, et al. 3-dimensional printed models may be a useful tool when planning revision anterior cruciate ligament reconstruction. Arthrosc Sports Med Rehabil. 2019;1:e41–6.
32. Crum RJ, Rabuck S, Ayeni O, Bedi A, Baraga M, Getgood A, et al. The REVision using imaging to guide staging and evaluation (REVISE) in ACL reconstruction classification. J Knee Surg. 2019;34(5):509–19.

33. Wolf BR, Ramme AJ, Wright RW, Brophy RH, McCarty EC, Vidal AR, et al. Variability in ACL tunnel placement: observational clinical study of surgeon ACL tunnel variability. Am J Sports Med. 2013;41:1265–73.
34. MARS Group. Effect of graft choice on the outcome of revision anterior cruciate ligament reconstruction in the Multicenter ACL Revision Study (MARS) Cohort. Am J Sports Med. 2014;42:2301–10.
35. Bach BR. Revision anterior cruciate ligament surgery. Arthroscopy. 2003;19:14–29.
36. Barrett GR, Brown TD. Femoral tunnel defect filled with a synthetic dowel graft for a single-staged revision anterior cruciate ligament reconstruction. Arthroscopy. 2007;23:796. e1-4.
37. Battaglia TC, Miller MD. Management of bony deficiency in revision anterior cruciate ligament reconstruction using allograft bone dowels: surgical technique. Arthroscopy. 2005;21:767.
38. Yoon KH, Kim JS, Park SY, Park SE. One-stage revision anterior cruciate ligament reconstruction: results according to preoperative bone tunnel diameter: five to fifteen-year follow-up. JBJS. 2018;100:993–1000.
39. Sarraj M, Shanmugaraj A, Musahl V, Lesniak BP. Over-the-top ACL reconstruction yields comparable outcomes to traditional ACL reconstruction in primary and revision settings: a systematic review. Knee Surg Sports Traumatol Arthrosc. 2019;27:427–44.
40. Noyes FR, Barber SD, Simon R. High tibial osteotomy and ligament reconstruction in varus angulated, anterior cruciate ligament-deficient knees a two-to seven-year follow-up study. Am J Sports Med. 1993;21:2–12.
41. Hashemi J, Chandrashekar N, Mansouri H, Gill B, Slauterbeck JR, Schutt RC, et al. Shallow medial tibial plateau and steep medial and lateral tibial slopes: new risk factors for anterior cruciate ligament injuries. Am J Sports Med. 2010;38:54–62.
42. Rahnemai-Azar AA, Abebe ES, Johnson P, Labrum J, Fu FH, Irrgang JJ, et al. Increased lateral tibial slope predicts high-grade rotatory knee laxity preoperatively in ACL reconstruction. Knee Surg Sports Traumatol Arthrosc. 2016; https://doi.org/10.1007/s00167-016-4157-31-7.
43. Rahnemai-Azar AA, Yaseen Z, van Eck CF, Irrgang JJ, Fu FH, Musahl V. Increased lateral tibial plateau slope predisposes male college football players to anterior cruciate ligament injury. J Bone Joint Surg Am. 2016;98:1001–6.
44. Ireland ML, Ballantyne BT, Little K, McClay IS. A radiographic analysis of the relationship between the size and shape of the intercondylar notch and anterior cruciate ligament injury. Knee Surg Sports Traumatol Arthrosc. 2001;9:200–5.
45. Pfeiffer TR, Burnham JM, Kanakamedala AC, Hughes JD, Zlotnicki J, Popchak A, et al. Distal femur morphology affects rotatory knee instability in patients with anterior cruciate ligament ruptures. Knee Surg Sports Traumatol Arthrosc. 2019;27:1514–9.
46. Uhorchak JM, Scoville CR, Williams GN, Arciero RA, St Pierre P, Taylor DC. Risk factors associated with noncontact injury of the anterior cruciate ligament: a prospective four-year evaluation of 859 West Point cadets. Am J Sports Med. 2003;31:831–42.
47. Chen FS, Rokito AS, Pitman MI. Acute and chronic posterolateral rotatory instability of the knee. J Am Acad Orthop Surg. 2000;8:97–110.
48. Engebretsen L, Lind M. Anteromedial rotatory laxity. Knee Surg Sports Traumatol Arthrosc. 2015;23:2797–804.
49. Hoshino Y, Miyaji N, Nishida K, Nishizawa Y, Araki D, Kanzaki N, et al. The concomitant lateral meniscus injury increased the pivot shift in the anterior cruciate ligament-injured knee. Knee Surg Sports Traumatol Arthrosc. 2019;27:646–51.
50. Shybut TB, Vega CE, Haddad J, Alexander JW, Gold JE, Noble PC, et al. Effect of lateral meniscal root tear on the stability of the anterior cruciate ligament-deficient knee. Am J Sports Med. 2015;43:905–11.
51. Song GY, Zhang H, Wang QQ, Zhang J, Li Y, Feng H. Risk factors associated with grade 3 pivot shift after acute anterior cruciate ligament injuries. Am J Sports Med. 2016;44:362–9.
52. Musahl V, Citak M, O'Loughlin PF, Choi D, Bedi A, Pearle AD. The effect of medial versus lateral meniscectomy on the stability of the anterior cruciate ligament-deficient knee. Am J Sports Med. 2010;38:1591–7.

Complications of ACL Reconstruction

11

Iftach Hetsroni, Niv Marom, and Noam Reshef

11.1 Introduction

ACL reconstruction is a reproducible procedure. Rate of return to sport is between 80 and 90% in elite athletes with graft rupture rates lower than 9% [1]. Nevertheless, complications do occur, ranging between 1 and 15% [2], and are important to recognize as early as possible. The definition of a "complication" however, may not be straightforward. This may refer to severe post-surgical complications such as joint infection or thromboembolic events, to complications which are specific to ligament surgery such as graft rupture or tunnel widening, and also to minor postoperative complications such as loss of skin sensation or anterior knee pain which may not interfere with the athlete's ability to return to sport. This chapter provides a spectrum of postoperative complications after ACL reconstruction which should be discussed with patients during preoperative informed consent, and also taken into consideration in decision-making junctions during and after surgery.

I. Hetsroni (✉) · N. Marom
Sports Medicine Injuries Service, Orthopedic Department, Meir General Hospital, Kfar Saba, Israel

Sackler Faculty of Medicine, Tel Aviv University, Tel Aviv, Israel

N. Reshef
Sports Medicine Injuries, Orthopedic Department, Ziv Medical Center, Zefat, Israel

Bar-Ilan University, Ramat Gan, Israel

11.2 General Complications

11.2.1 Infection

The incidence of infected ACL reconstruction ranges from 0.32% to 1.8% [3–7]. Common causes are bacteria, followed by tubercular [8] and fungal [9, 10] infections. The most common culprit is *Staphylococcus aureus* followed by *Staphylococcus coagulase* negative, responsible for 31% and 13% of the infection cases, respectively [11]. It is not unlikely that the skin flora of the operated knee is the main inoculation point. Skin incision for graft harvesting may be the main port of entry. Judd et al. [12] reported that, in 11 cases of intra-articular infection, 9 had concomitant extra-articular sites of infection, and 8 of these were at the hamstring graft harvesting site. In each case, both intra- and extra-articular cultures grew the same offending organism. Therefore, the inoculation event likely occurred at the time of surgery. Several risk factors for infection in ACL reconstruction were described and are presented in Table 11.1.

N. Nakamura et al. (eds.), *Advances in Knee Ligament and Knee Preservation Surgery*,
https://doi.org/10.1007/978-3-030-84748-7_11

Table 11.1 Risk factors for infection post ACL reconstruction surgery

Intrinsic factors	Extrinsic factors
Hamstrings autograft	Long-acting intraarticular steroid injection
Concurrent soft tissue procedures, particularly meniscus repair	Increased surgical time
Tibial fixation of hamstrings autograft with screw and washer	
Prior knee procedures	

Armstrong et al. [13] noted that long-acting, intra-articular steroids, along with increased surgical time, prior knee procedures, and concurrent soft-tissue procedures are all risk factors for infection. Meniscus repair was found to be the most prevalent concomitant procedure (66–71%) and therefore considered a risk factor for infection of ACL reconstruction [14, 15]. Judd et al. [12] found that previous knee surgery carries 1.9 higher relative risk for infection following ACL reconstruction and that the risk rises to 5.1 in revision cases. Among the graft types used for ACL reconstruction, hamstrings autograft is the most susceptible for infection. The infection rate of hamstrings autograft was reported 1.4%, higher than BPTB autograft (0.49%) and allograft (0.44%) [4]. Watterman et al. [14] found an infection rate of 0.32% in a cohort of 9511 patients. Others also reported that autograft hamstrings were the most common type of graft in the infection group, corresponding to 55% and followed by allograft (35%) and BPTB (10%). According to Judd et al. [12], all the infected cases were in reconstructions done with hamstrings autograft. The exact reason for higher infection rates in hamstrings autograft is not clear. An explanation could be the extensive extra dissection needed for the tendon harvest and the hematoma running from the muscle in a direct tunnel towards the knee. The superficial fixation of that graft was also raised as a reason. Accordingly, fixation of the hamstrings autograft with screw and washer was found to increase the risk of deep infection after ACL reconstruction compared to hamstrings autograft without screw and washer tibial fixation [12, 16]. The use of bioabsorbable screws was evaluated compared to metallic screws in two larges reviews [17, 18]. No difference was found as a risk factor for infection.

Regarding the optional use of antibiotics as a preventive measure of infection, Garamycin solution for keeping the graft between harvest and implantation does not seem to reduce the risk for infection after ACL reconstruction [19]. However, the usage of vancomycin solution showed a significant reduction in infection rate post ACL reconstruction and therefore should be considered for routine use [20, 21].

Although may appear at any point after surgery, septic arthritis following ACL reconstruction usually appears between 2 weeks (acute) and 2 months (subacute) after the reconstruction [22]. Arthrocentesis and synovial fluid culture remain the gold standard in the diagnosis of septic arthritis. Arthrocentesis should be performed upon the establishment of a substantial suspicion and prior to antibiotics administration. Synovial fluid evaluation should be obtained for WBC and cell count, glucose and protein levels, direct microscopy, and gram stain and bacterial cultures. Although post-operative infection can be seen with synovial fluid WBC levels as low as 25,000 cells, it is acceptable that WBC level over 100,000 cells in the synovial fluid is considered septic arthritis [15, 23, 24]. Bacteriological cultures taken prior to antibiotic administration are positive in high rates [12, 15] and are important not only to confirm the diagnosis but to conduct the accurate antibiotic treatment. Negative culture with the presence of continuous clinical and laboratory signs of infection should raise the possibility of fungal or tubercular infection and specific cultures should be obtained accordingly.

Arthrocentesis should be performed followed by blood cultures—prior to antibiotic administration. That will be followed by arthroscopic lavage and debridement which should be carried out as soon as the diagnosis of septic arthritis is made. Tissue biopsy and fibrin tissue culture should be taken. Graft removal is a controversial issue. Most of the studies are in favor of graft preservation as much as possible, and graft removal only if grossly damaged or if substantial infected tissue is found within the graft itself [3, 6, 7, 14, 25]. Graft integrity should be evaluated clinically and arthroscopically. Arthroscopically, the fibrinous film surrounding the graft should be debrided gently without harming the graft. This is a good sign for potent retainable graft [12, 15]. Extensive lavage and synovectomy should be performed during arthroscopy with debridement of the gutters, suprapatellar pouch, and the anterior compartment. Inspection of the posterior compartment should be mandatory, and even minimal signs of infected tissue should be treated with posterior compartment synovectomy using posteromedial and posterolateral portals. The number of arthroscopic lavages is not fixed, and depends on the clinical improvement of the patient. The use of continuous irrigation system, close suction system, and antibiotic beads were all described [11, 25, 26]. However, there is not enough data to support a routine use of them. Intravenous administration of broad-spectrum antibiotics, tailored to the causing bacteria in consultation with infectious disease specialist [12, 27] should be initiated after cultures were taken. Treatment period should be from 3 to 6 weeks and third generation Cephalosporin or Vancomycin is usually recommended [12]. If the graft and hardware were removed, there should be a gap of at least 3 months [28] to a year [15] from the event of infection to the surgery of revision. Inflammatory markers and laboratory tests should be normalized prior to revision surgery [28]. Although not specified as a criteria we recommend that blood culture and knee aspiration culture should be obtained and documented sterile before performing the revision ACL reconstruction surgery.

11.2.2 Arthrofibrosis

Arthrofibrosis has been recognized as an adverse outcome after ACL reconstruction [29–31] and a major factor associated with patient dissatisfaction [32]. While multiple factors are associated with this outcome, time interval from injury to ACL reconstruction has been pointed out by several investigators as a leading risk factor. Shelbourne et al. [30] originally reported in a retrospective analysis of 169 autologous BPTB ACL reconstructions in young athletes that patients who had surgery within the first week or between 8 and 21 days from the injury had significantly increased incidence of arthrofibrosis compared to patients who had their ligament reconstruction at more than 3 weeks from the injury (i.e., up to 17% vs. 0%, respectively). Of note, follow-up time was 3 months only. It should be noticed however, that in cases where ACL reconstruction was performed between 8 and 21 days from the injury, accelerated postoperative rehabilitation program resulted in substantial decrease in the incidence of arthrofibrosis. Inferior outcomes were observed in the "early" reconstruction group also in regard to strength isokinetic tests. The authors concluded that delaying ACL reconstruction at least 3 weeks from the injury will result in earlier return to strength and in significantly decreased incidence of arthrofibrosis. These outcomes were reproduced by Wasilewski et al. [31] who performed a retrospective analysis of 87 autologous hamstrings ACL reconstructions with concomitant ITB tenodesis who were divided into three groups based on timing of surgery. Follow-up was reported up to 18 months. They showed that arthrofibrosis was found in 22% of reconstructions performed within 1 month from injury compared to 0% when reconstruction was performed between 1 and 6 months or 12.5% when reconstruction was performed after 6 months from injury. They also showed inferior Quadriceps torque in the "early" reconstructions. Of note, the standard rehabilitation protocol used in their study mandated substantial motion limitations and included immobilization at 30° knee flexion for 7 to 10 days postoperatively, followed by braced motion from 20° to 60° for a few additional weeks. They also pointed out that recovery after ACL reconstruction performed within 1 month from injury was significantly slower compared to recovery when reconstruction was performed later than 1 month from injury.

Cosgarea et al. [29] performed a retrospective analysis of 191 consecutive autologous BPTB ACL reconstructions and similarly to Shelbourne et al. [30] and Wasilewski et al. [31] showed that surgery performed within the first 3 weeks of injury had a significantly higher incidence of arthrofibrosis compared to surgery performed later than 3 weeks from injury (21% vs. 9%, respectively). However, an important finding of their study was that incidence of arthrofibrosis decreased from more than 20% to less than 3% when postoperative rehabilitation protocol was changed from bracing in 45° flexion for 7 days before the initiation of passive extension to bracing in full extension immediately after surgery.

Mayr et al. [33] performed a retrospective analysis of risk factors for arthrofibrosis after ACL reconstruction in 223 patients, 75% of which had their reconstruction with autologous BPTB graft. They also demonstrated that incidence of arthrofibrosis was increased in cases where reconstruction was performed within 4 weeks from injury, but that irritated knee (swelling, effusion, hyperthermia) and lack of full ROM before surgery were more important risk factors for the development of arthrofibrosis than time interval from injury to surgery. In other words, when surgery was performed later than 4 weeks from injury but the knee was irritated, there was an increased risk for the development of arthrofibrosis compared to when surgery was performed within the first 4 weeks from the injury.

The first prospective randomized clinical trial that investigated the incidence of arthrofibrosis in "early" versus "delayed" ACL reconstruction was performed by Meighan et al. [34]. They studied a small series of athletic patients that underwent ACL reconstruction using autologous quadrupled hamstrings graft and used similar postoperative rehabilitation protocols for both groups. The "early" reconstruction group had surgery within 2 weeks from injury, and the "delayed" group had surgery between 8 and 12 weeks from injury. Although loss of knee motion was more pronounced at 2 weeks after the operation in the "early" group, at 1 year follow-up there were no differences in knee motion, nor there were differences between the groups in relation to IKDC, Lysholm, Tegner scores, and examination of Quadriceps and hamstrings muscle power and torque.

The outcomes of "early" versus "delayed" ACL reconstruction were also investigated in a systematic review and meta-analysis by Smith et al. [35]. This meta-analysis could not identify any significant differences in the incidence of arthrofibrosis or in any functional outcome score or activity level outcome scores between reconstructions performed "early" compared to those performed "late".

In summary, ACL reconstruction may preferably be performed later than 3 weeks after the injury to lower potential risk of knee arthrofibrosis. In cases that require earlier intervention, such as lack of knee extension due to displaced bucket handle meniscus tear or anteriorly flipped ACL fragment that blocks extension, attention should be applied during the immediate postoperative protocol to maintain full extension and start knee flexion as early as possible.

11.2.3 Thromboembolic Events

Thromboembolic disease is potentially a common complication in orthopedic surgery. Without receiving prophylaxis, the reported incidence is over 50% for total knee replacement [36, 37]. Symptomatic cases may range from swelling of the calf and soreness to a post-phlebitic syndrome and in severe cases, thromboembolism. While arthroscopic meniscectomy carries a low risk, with a reported incidence of 0.3% without thromboprophylaxis [38], cruciate ligament reconstruction has a four times higher risk of DVT than other non-major arthroscopic surgery [39, 40]. Recently, a large prospective series of nearly 1000 patients with ACLR [41] had an overall reported incidence of thromboembolic complications reaching 0.6%. However, for the specific complication of symptomatic pulmonary embolism after outpatient arthroscopic procedures of the knee which is extremely rare, with a reported incidence of 2.8 cases for every 10,000 arthroscopies, ACL reconstruction by itself was not a risk factor [42]. For this specific complication, significant risk factors were age over 40 years, operative time over 90 min, history of malignancy, and female gender [42]. The diagnosis of

DVT should be based on clinical suspicion and evaluation, followed by supporting additional tests, such as the D-dimer blood test and imaging tests [43, 44]. The clinical prediction tool published by Wells et al. [43] has been assessed and validated in multiple clinical studies and can accurately categorize outpatients as low, moderate, or high clinical probability. In terms of the preferred imaging modality, ultrasound imaging was utilized in the majority of the studies investigating rates of DVT after ACL Reconstruction [39] and is considered the test of choice [44]. In their systematic review reporting on rates of thromboembolic events following ACL reconstruction, Erickson et al. [39] reported that included studies screening for DVT were conducted during the period of post-operative day 3 to 28, which is the expected post-operative period when ambulation is most affected by surgery and therefore poses the higher risk for a DVT to occur. While clear recommendations for thromboprophylaxis after arthroscopic knee procedures lack evidence [45], wise approach to avoid overtreatment with possible bleeding complications and at the same time decrease the likelihood of symptomatic life-threatening thromboembolic disease after ACL reconstruction would therefore be considering the following: decreasing operative times to below 90 min, avoiding unnecessary tourniquet use, counseling to patients to cease contraceptive pills and smoking a few weeks prior to surgery, and tailoring postoperative prophylaxis to patients at risk such as those with prior thromboembolic disease and patients with medical history of malignancy.

11.2.4 Recurrent Postoperative Hemarthrosis

Recurrent postoperative hemarthrosis is rare after ACL reconstruction. Potential causes which should be considered include blood thinners administered for thromboprophylaxis, coagulation cascade anomalies (i.e., hemophilia, factor VII deficiency, factor XI deficiency, Von Willebrand's disease), iatrogenic vascular injuries, and others. Minor vessels such as geniculate branches may be injured during surgery. Tsubosaka et al. [46] described a case of pseudoaneurysm of the articular branch of the descending genicular artery presenting as medial pulsating mass 2 days after the reconstruction. This was treated by embolization after CT-angiography confirmed the diagnosis. Lamo-Espinosa et al. [47] reported a case of inferior lateral genicular artery lesion as a result of anterior horn lateral meniscectomy presenting as knee swelling and pulsatile knee bleeding 1 day after surgery. This was treated by selective embolization.

In summary, postoperative recurrent hemarthrosis is rare but should be discussed with patients as a potential complication which might delay the rehabilitation process. When recurrent knee aspirations following surgery are not effective, unusual etiologies should be thought of.

11.3 Complications Related to Autograft Harvest

The three most commonly used autografts will be reviewed: (1) Bone-Patellar tendon-bone, (2) Hamstrings tendons, and (3) Quadriceps tendon.

11.3.1 Bone-Patella Tendon-Bone Autograft-Related Complications

BPTB autograft is still considered the gold-standard graft for many surgeons when planning surgery for highly active young adults, promising the lowest graft failure and revision rates [48]. Nevertheless, there are a few potential complications which have been discussed in relation to this type of graft. These include the risk of patella fracture and the risk of "anterior knee pain" and persistent quadriceps weakness. With regard to patella fracture, the risk seems extremely low with incidence of 0.3% in a recent large prospective cohort of nearly 1000 cases [41]. Several strategies to lower the risk of fractures were previously described and include harvesting the central part of the patella and not more than 30% of the patella thickness, not using osteotomes for the initial bone cuts, and the use of minimal required force if using a mallet in this area [49]. With "anterior knee pain", despite the concerns that patellar tendon harvest is likely to result in more anterior knee pain compared to other autografts [50], Rousseau et al. [41] recently showed that after 2 years follow-up the incidence of anterior knee pain is only about 3% and similar in BPTB and hamstrings autograft. Yet, the specific complication of kneeling pain is higher with patellar tendon reconstruction and may reach 10% at a 5-year follow-up [48]. This should therefore be taken into consideration in sports which require kneeling, wrestlers for example which need to bend forward on their knees frequently during their sports. With regard to persistent quadriceps weakness, it has been suggested that there is continued weakness of up to 6% in the quadriceps on the reconstructed compared with the uninjured side when using BPTB autograft with current ACL reconstruction techniques [51]. However, the clinical significance of these findings is unclear.

11.3.2 Hamstrings Autograft-Related Complications

Hamstrings tendons are commonly used as graft source in ACL reconstruction. This might have several potential morbidities, including: (1) Sensory nerve injury; (2) Hamstrings retraction and persistent crumps; (3) Persistent hamstrings weakness; and (4) Residual medial side knee laxity. A fifth and serious morbidity is knee infection which was discussed earlier in this chapter and shown to be at higher rates when using Hamstrings autograft compared to other graft sources for ACL reconstruction.

11.3.2.1 Sensory Nerve Injury

The infrapatellar branch of the saphenous nerve passes medially to laterally, just over the pes anserinus. Incision made over that area carries the risk for sensory branch injury. Franz et al. [52] reported 14% of injury of the sensory branch of the saphenous nerve during anteromedial harvesting of the hamstring tendon. Haviv et al. reported incidence of 58% decreased sensation of the sartorius branch of the saphenous nerve in patients with autograft hamstrings tendon [53]. 25% had full recovery after 8 months. Patients nevertheless considered loss of sensation as minimal complication. Oblique anteromedial incision, single semitendinosus tendon graft technique, and popliteal (posteromedial) harvesting approach may reduce the risk of this complication.

11.3.2.2 Hamstring Retraction and Persistent Cramps

Although the regeneration of the hamstring tendons was described, it does not happen consistently and the structure of the muscle is known to be altered. Konrath et al. [54] showed that in 35% of the cohort there was a regeneration of the Semitendinosus and Gracillis tendon. The size of the Semitendinosus and Gracillis muscle was lower when the tendons did not regenerate. The overall medial thigh muscle volume was decreased compared to the contralateral thigh. This was in direct correlation to decreased knee flexion strength. Nakamae et al. [55] described two cases of unsuccessful hamstrings tendon regeneration presenting severe posterior thigh pain and muscle bulk retraction. Janssen et al. [56] described full regeneration of the Gracillies tendon and 66% regeneration of the semitendinosus tendon in their cohort 1 year after ACL reconstruction. Patient that did not regenerate the tendon showed muscle retraction and decrease in cross-sectional area. Without good regeneration, the muscles tend to be smaller and retracted which might lead to posterior thigh cramps.

11.3.2.3 Persistent Hamstrings Weakness

Konrath et al. [54] showed decreased knee flexion strength in 65% of the patients who had undergone ACL reconstruction using hamstring tendons autograft. Rogowski et al. [57] showed that at 6 months after ACL reconstruction, hamstrings tendon group had weakness in flexion compared to patellar tendon group. Bourne et al. [58] showed that, in elite Australian Football female athletes, at 1–10 years after single leg ACL reconstruction with hamstrings tendon, there was a significant decrease in eccentric knee flexion strength compared to the uninjured leg. Yet, all players returned to their previous activity level. To conclude hamstrings tendon harvesting may lead to hamstrings muscle weakness. Yet, although weakness may persist years after the reconstruction, the clinical significance of such weakness in returning to sports is unclear.

11.3.2.4 Medial Side Laxity

Hamstring tendons contribute to the dynamic stability of the knee. Sacrificing the hamstring tendon as a graft for ACL reconstruction may contribute to medial side instability. Toor et al. [59] demonstrated in a cadaveric study that the pes anserinus contributes to control movements of valgus stress, translation, and rotation. Unloading the pes anserinus may therefore alter movements of anterior translation, external rotation, and valgus stress. Yet, using a single tendon graft seems to decrease such negative effect on medial knee constraints. In the presence of concomitant medial instability due to MCL injury, surgeons should address the MCL lesion or consider a different graft rather than autograft hamstrings for ACL reconstruction.

11.3.3 Quadriceps Autograft-Related Complications

Quadriceps tendon autografts have gained interest in recent years [60–62]. It is suggested that Quadriceps tendon harvest does not compromise the extensor mechanism [60–63] and is "safe". However, three principal complications have been described: (1) *Bleeding and hematoma formation*: It has been suggested by Slone et al. [62] that lateral violation of the Quadriceps muscle, where the perforating vessels exist can lead to significant hematoma and even a compartment syndrome, if this bleeding is not recognized. Additionally, extravasation of intraarticular blood through a full-thickness quadriceps harvest can cause a hematoma anterior to the quadriceps. A centralized graft harvest within the Quadriceps tendon, partial thickness harvest, and no harvest proximal to the myotendinous junction of the Rectus Femoris (6 to 8 cm proximal to the tendon insertion on the patella) can minimize the risk; (2) *Cosmetic deformity of the distal thigh*: This uncommon complication is in most cases the result of retraction of the Rectus Femoris muscle after full or partial-thickness graft harvest. There seems to be a correlation with extensive tendon harvest and violation of the myotendinous junction of the Rectus Femoris [62]. Despite the obvious distal thigh deformity, it does not appear to have functional implications; (3) *Patella fracture*: Patella fracture is a rare complication of a combined Quadriceps tendon and patella bone plug harvest with a reported incidence of 0.03% [62, 64]. As with Bone patellar tendon bone harvest, it is recommended to avoid the use of osteotomes for the initial bone cuts, harvest the central part of the patella and avoid harvesting more than 30% of the patella thickness [49, 64]. Despite the recent resurgence of interest in Quadriceps tendon autograft, it is still the least studied and least used graft for ACL reconstruction among surgeons today [62]. Future studies may introduce additional harvest complications and may also be able to categorize complications based on Quadriceps autograft characteristics (i.e. full vs. partial thickness, length of harvest, with vs. without bone plug).

11.4 Other Complications

11.4.1 Recurrence of Anterior Laxity

Graft instability after ACL reconstruction can be the consequence of multiple factors, including inaccurate tunnel positioning, overlooked or unaddressed concomitant ligamentous instabilities, low-quality or low-volume graft tissue, unsecured graft fixation techniques, hardware failure, return to pivoting activities before graft maturation or before fulfilling return-to-sport criteria, among others. Meta-analyses, large cohorts, and recent prospective studies [41, 48, 50] have shown that graft rupture is around 5% for autologous BPTB and around 10% for hamstrings autograft. Graft failure may be higher also with the use of allografts compared to autografts, and particularly in active young adults [65].

11.4.2 Hardware Failure

Secured and stable graft fixation is of paramount importance for a successful ACL reconstruction. While the graft fixation implant is considered the weakest link in the graft fixation complex, it plays a major role in its effectiveness [66]. Any technical error during the surgical placement of the implant may lead to hardware failure and this includes using defective hardware, mal-positioning and/or mal-deployment of the hardware and failure to achieve good and secured fixation after the implanted device is utilized. It has been suggested that some fixation devices are more prone to failure than others [66] or may have worse adverse outcomes [67, 68], though it is important to note that all devices have possible risks and complications. Suspensory/cortical fixation devices, for example, are applying fixation away from the joint, which may allow graft movement within the tunnel [69], compromise the integrity of the graft fixation complex, and lead to loss of stability and graft failure [66]. Animal studies reported that ACL graft fixation in the tunnel and closer to the joint improves tendon-to-bone healing [69, 70] and can provide less graft tunnel motion and tunnel enlargement [69]. Nevertheless, using new-generation devices, utilizing sockets instead of tunnels, confirming proper device deployment and proper graft tensioning are all possible ways to minimize the risks. It is also crucial to make sure no violation of the cortex has happened while establishing the socket/tunnel and passing the cortical fixation device through the cortex, as this will lead to migration of the device and failure of fixation. On the other hand, intra-tunnel interference fixation devices may influence graft rotation/orientation while being applied, break within the joint, cause tunnel widening while inserted, and may lead to osteolysis when bio-absorbable screws are used [66]. Confirming optimal position and trajectory prior and during the interference device insertion, inserting the device while applying constant tension on the graft and under continuous visualization, assessing for adequate resistance while the device is utilized and post-surgical radiographs when metal devices are used are all possible ways to minimize the risks. With this fixation method it is crucial to make sure all tunnel walls (especially the posterior femoral wall) are preserved and were not compromised during rimming or passing the graft since this may lead to unstable construct and failure. While fixation techniques and devices are numerous and new generations of fixation devices are still introduced, it is the surgeon responsibility to be aware of the risks and possible complications of the used hardware and take them into consideration while planning the surgery, executing the surgery, and during the post-surgery period. Whenever such complication is encountered, it should be addressed immediately. Additional hardware complication is pain related to hardware material. A recent large cohort study [41] reported that this can affect up to 10% of patients and may require a second procedure for hardware removal. A supplementary tibial fixation for hamstrings autograft was the reason for pain in most cases. Choosing low-profile fixation devices and avoiding a supplementary tibial fixation in older patients or patients with disorders affecting bone mineral density have all been proposed as ways to minimize the risk.

11.4.3 Tunnel Widening

Tunnel widening after ACL reconstruction is an imaging finding that may or may not be associated with clinical symptoms. It may be related to tears and insufficiency of the reconstructed ligament and may complicate the surgical technique during revision surgery. There are several theories that suggest potential culprits. Graft type, graft fixation device, type of screw fixation or suspensory fixation, surgical technique of tunnel placement, and tunnel angle were all evaluated as potential reasons for tunnel widening. Li et al. [71] evaluated the graft bending angle (GBA- the angle between the femoral tunnel and the line of the intra-articular tract of the graft) as a cause for delayed graft maturation and tunnel widening. They found that high GBA (between 60° and 70°) was associated with higher increased graft signal and wider femoral tunnel at 12 months after surgery. No correlation with functional outcome measures was found. Regarding bone patellar tendon bone, due to the bone to bone interface in the tibial and femoral tunnels, it is widely acceptable that bone to bone healing should not result in tunnel widening. However, Struewer et al. [72] showed that 17% of the patients after this graft type had tunnel widening of up to 40% and 12% of the patients had widening up to 50%. Yet, no correlation to clinical outcome measurement was noted. Nevertheless, the phenomenon was mostly investigated in soft tissue grafts which have greater tunnel widening than bone patellar grafts [73]. In addition, soft tissue allograft is related to more pronounced tunnel widening, both femoral and tibial side, when compared to hamstrings autograft. Again, no difference was found in the clinical outcome measurements [73, 74]. Type of fixation device is also a factor studied as a cause for tunnel widening. Choi et al. [75] and Ahn et al. [76] found no difference in tunnel widening nor in clinical outcome between fixed loop and adjustable loop fixation devices. Comparing titanium and absorbable screws showed similar clinical results but bioabsorbable screws were associated with greater tunnel widening and bone cyst formation compared to titanium screws [77]. In summary, bone tunnel widening is a common phenomenon. The actual influence on the clinical outcome is not clear and suspected to be negligible. Soft tissue grafts, and particularly allografts, seem to present a bit wider tunnels compared to bone patellar tendon bone autografts. The main importance of that parameter seems to be higher complexity during ACL revision when required.

11.5 Summary

This chapter reviewed a spectrum of minor and major postoperative complications of ACL reconstruction. With awareness of surgeons about these potential complications, decision-making during pre-, intra-, and postoperative periods can be optimized in order to avoid some of these and also to properly address such challenges when encountered.

References

1. Lai CCH, Ardern CL, Feller JA, Webster KE. Eighty-three per cent of elite athletes return to preinjury sport after anterior cruciate ligament reconstruction: a systematic review with meta-analysis of return to sport rates, graft rupture rates and performance outcomes. Br J Sports Med. 2018;52:128–38.
2. Cinque ME, Chahla J, Moatshe G, DePhillipo NN, Kennedy NI, Godin JA, LaPrade RF. Outcomes and complications after primary anterior cruciate ligament reconstruction are similar in younger and older patients. Orthop J Sports Med. 2017;5:2325967117729659.
3. Barbara K, Alan I, Goran V, Saša J. Knee infection following anterior cruciate ligament reconstruction: a cohort study of one thousand, eight hundred and ninety one patients from the single-centre database. Int Orthop. 2020;44(5):869–75. https://doi.org/10.1007/s00264-020-04500-5.
4. Barker JU, Drakos MC, Maak TG, Warren RF, Williams RJ 3rd, Allen AA. Effect of graft selection on the incidence of postoperative infection in anterior cruciate ligament reconstruction. Am J Sports Med. 2010;38(2):281–6. https://doi.org/10.1177/0363546509346414.
5. Maletis GB, Inacio MC, Reynolds S, Desmond JL, Maletis MM, Funahashi TT. Incidence of postoperative anterior cruciate ligament reconstruction infections: graft choice makes a difference. Am J Sports Med. 2013;41(8):1780–5. https://doi.org/10.1177/0363546513490665.
6. Schuster P, Schulz M, Immendoerfer M, Mayer P, Schlumberger M, Richter J. Septic arthritis after arthroscopic anterior cruciate ligament reconstruction: evaluation of an arthroscopic graft-retaining treatment protocol. Am J Sports Med. 2015;43(12):3005–12. https://doi.org/10.1177/0363546515603054.
7. Torres-Claramunt R, Pelfort X, Erquicia J, Gil-González S, Gelber PE, Puig L, Monllau JC. Knee joint infection after ACL reconstruction: prevalence, management and functional outcomes. Knee Surg Sports Traumatol Arthrosc. 2013;21(12):2844–9. https://doi.org/10.1007/s00167-012-2264-3.
8. Nag HL, Neogi DS, Nataraj AR, Kumar VA, Yadav CS, Singh U. Tubercular infection after arthroscopic anterior cruciate ligament reconstruction. Arthroscopy. 2009;25(2):131–6. https://doi.org/10.1016/j.arthro.2008.09.009.
9. Muscolo DL, Carbo L, Aponte-Tinao LA, Ayerza MA, Makino A. Massive bone loss from fungal infection

after anterior cruciate ligament arthroscopic reconstruction. Clin Orthop Relat Res. 2009;467(9):2420–5. https://doi.org/10.1007/s11999-009-0714-0.
10. Sun L, Zhang L, Wang K, Wang W, Tian M. Fungal osteomyelitis after arthroscopic anterior cruciate ligament reconstruction: a case report with review of the literature. Knee. 2012;19(5):728–31. https://doi.org/10.1016/j.knee.2011.10.007.
11. Van Tongel A, Stuyck J, Bellemans J, Vandenneucker H. Septic arthritis after arthroscopic anterior cruciate ligament reconstruction: a retrospective analysis of incidence, management and outcome. Am J Sports Med. 2007;35(7):1059–63.
12. Judd D, Bottoni C, Kim D, Burke M, Hooker S. Infections following arthroscopic anterior cruciate ligament reconstruction. Arthroscopy. 2006;22(4):375–84.
13. Armstrong RW, Bolding F, Joseph R. Septic arthritis following arthroscopy: clinical syndromes and analysis of risk factors. Arthroscopy. 1992;8(2):213–23.
14. Waterman BR, Arroyo W, Cotter EJ, Zacchilli MA, Garcia EJ, Owens BD. Septic arthritis after anterior cruciate ligament reconstruction: clinical and functional outcomes based on graft retention or removal. Orthop J Sports Med. 2018;6(3):2325967118758626. https://doi.org/10.1177/2325967118758626.
15. Williams RJ 3rd, Laurencin CT, Warren RF, Speciale AC, Brause BD, O'Brien S. Septic arthritis after arthroscopic anterior cruciate ligament reconstruction. Diagnosis and management. Am J Sports Med. 1997;25(2):261–7.
16. Hurvitz AP, Prentice HA, Funahashi TT, Maletis GB. Screw and sheath tibial fixation associated with a higher likelihood of deep infection after hamstring graft anterior cruciate ligament reconstruction. Am J Sports Med. 2020;48(4):806–11. https://doi.org/10.1177/0363546520902716.
17. Debieux P, Franciozi CE, Lenza M, Tamaoki MJ, Magnussen RA, Faloppa F, Belloti JC. Bioabsorbable versus metallic interference screws for graft fixation in anterior cruciate ligament reconstruction. Cochrane Database Syst Rev. 2016;7:CD009772. https://doi.org/10.1002/14651858.CD009772.pub2.
18. Shen C, Jiang SD, Jiang LS, Dai LY. Bioabsorbable versus metallic interference screw fixation in anterior cruciate ligament reconstruction: a meta-analysis of randomized controlled trials. Arthroscopy. 2010;26(5):705–13. https://doi.org/10.1016/j.arthro.2009.12.011.
19. Yazdi H, Moradi A, Herbort M. The effect of gentamicin in irrigating solutions on articular infection prophylaxis during arthroscopic ACL reconstruction. Arch Orthop Trauma Surg. 2014;134(2):257–61. https://doi.org/10.1007/s00402-013-1910-7.
20. Bohu Y, Klouche S, Sezer HB, Herman S, Grimaud O, Gerometta A, Meyer A, Lefevre N. Vancomycin-soaked autografts during ACL reconstruction reduce the risk of post-operative infection without affecting return to sport or knee function. Knee Surg Sports Traumatol Arthrosc. 2020;28(8):2578–85. https://doi.org/10.1007/s00167-020-05879-9.
21. Schuster P, Schlumberger M, Mayer P, Eichinger M, Geßlein M, Reddemann F, Richter J. Soaking of the graft in vancomycin dramatically reduces the incidence of postoperative septic arthritis after anterior cruciate ligament reconstruction. Knee Surg Sports Traumatol Arthrosc. 2020;28(8):2587–91. https://doi.org/10.1007/s00167-020-05882-0.
22. Sonnery-Cottet B, Thaunat M, Archbold P, Issartel B, Cadet ER. Management of septic arthritis following anterior cruciate ligament reconstruction: a review of current practices and recommendations. J Am Acad Orthop Surg. 2014;22(5):271–3. https://doi.org/10.5435/JAAOS-22-05-271.
23. Margaretten ME, Kohlwes J, Moore D, Bent S. Does this adult patient have septic arthritis? JAMA. 2007;297(13):1478–88.
24. Schollin-Borg M, Michaëlsson K, Rahme H. Presentation, outcome, and cause of septic arthritis after anterior cruciate ligament reconstruction: a case control study. Arthroscopy. 2003;19(9):941–7.
25. Schulz AP, Götze S, Schmidt HG, Jürgens C, Faschingbauer M. Septic arthritis of the knee after anterior cruciate ligament surgery: a stage-adapted treatment regimen. Am J Sports Med. 2007;35(7):1064–9.
26. McAllister DR, Parker RD, Cooper AE, Recht MP, Abate J. Outcomes of postoperative septic arthritis after anterior cruciate ligament reconstruction. Am J Sports Med. 1999;27(5):562–70.
27. Indelli PF, Dillingham M, Fanton G, Schurman DJ. Septic arthritis in postoperative anterior cruciate ligament reconstruction. Clin Orthop Relat Res. 2002;398:182–8.
28. Hantes ME, Raoulis VA, Doxariotis N, Drakos A, Karachalios T, Malizos KN. Management of septic arthritis after arthroscopic anterior cruciate ligament reconstruction using a standard surgical protocol. Knee. 2017;24(3):588–93.
29. Cosgarea AJ, Sebastianelli WJ, DeHaven KE. Prevention of arthrofibrosis after anterior cruciate ligament reconstruction using the central third patellar tendon autograft. Am J Sports Med. 1995;23:87–92.
30. Shelbourne KD, Wilcken JH, Mollabashy A, DeCarlo M. Arthrofibrosis in acute anterior cruciate ligament reconstruction. The effect of timing of reconstruction and rehabilitation. Am J Sports Med. 1991;19:332–6.
31. Wasilewski SA, Covall DJ, Cohen S. Effect of surgical timing on recovery and associated injuries after anterior cruciate ligament reconstruction. Am J Sports Med. 1993;21:338–42.
32. Kocher MS, Steadman JR, Briggs K, et al. Determinants of patient satisfaction with outcome after anterior cruciate ligament reconstruction. J Bone Surg Am. 2002;84:1560–72.
33. Mayr HO, Weig TG, Plitz W. Arthrofibrosis following ACL reconstruction – reasons and outcome. Arch Orthop Trauma Surg. 2004;124:518–22.

34. Meighan AA, Keating JF, Will E. Outcome after reconstruction of the anterior cruciate ligament in athletic patients. A comparison of early versus delayed surgery. J Bone Joint Surg Br. 2003;85:521–4.
35. Smith TO, Davies L, Hing CB. Early versus delayed surgery for anterior cruciate ligament reconstruction: a systematic review and meta-analysis. Knee Surg Sports Traumatol Arthrosc. 2010;18:304–11.
36. Douketis JD, Eikelboom JW, Quinlan DJ, Willan AR, Crowther MA. Short duration prophylaxis against venous thromboembolism after total hip or knee replacement: a meta-analysis of prospective studies investigating symptomatic outcomes. Arch Int Med. 2002;162:1465–71.
37. Stringer MD, Steadman CA, Hedges AR, et al. Deep vein thrombosis after elective knee surgery: an incidence study in 312 patients. J Bone Joint Surg (Br). 1989;71-B:492–7.
38. Krych AJ, Sousa PL, Morgan JA, et al. Incidence and risk factor analysis of symptomatic venous thromboembolism after knee arthroscopy. Arthroscopy. 2015;31:2112–8.
39. Erickson BJ, Saltzman BM, Campbell KA, et al. Rates of deep venous thrombosis and pulmonary embolus after anterior cruciate ligament reconstruction: a systematic review. Sports Health. 2015;7:261–6.
40. Struijk-Mulder MC, Ettema HB, Verheyen CC, Buller HR. Deep vein thrombosis after arthroscopic anterior cruciate ligament reconstruction: a prospective cohort study of 100 patients. Arthroscopy. 2013;29:1211–6.
41. Rousseau R, Labruyere C, Kajetanek C, Deschamps O, Makridis KG, Djian P. Complications after anterior cruciate ligament reconstruction and their relation to the type of graft: a prospective study of 958 cases. Am J Sports Med. 2019;47(11):2543–9.
42. Hetsroni I, Lyman S, Do H, Mann G, Marx RG. Symptomatic pulmonary embolism after outpatient arthroscopic procedures of the knee: the incidence and risk factors in 418,323 arthroscopies. J Bone Joint Surg (Br). 2011;93:47–51.
43. Wells PS, Anderson DR, Rodger M, Forgie M, Kearon C, Dreyer J, Kovacs G, Mitchell M, Lewandowski B, Kovacs MJ. Evaluation of D-dimer in the diagnosis of suspected deep-vein thrombosis. N Engl J Med. 2003;349:1227–35.
44. Wells PS, Owen C, Doucette S, Fergusson D, Tran H. Does this patient have deep vein thrombosis? JAMA. 2006;295(2):199–207. https://doi.org/10.1001/jama.295.2.199.
45. Zheng G, Tang Q, Shang P, Pan XY, Liu HX. No effectiveness of anticoagulants for thromboprophylaxis after non-major knee arthroscopy: a systematic review and meta-analysis of randomized controlled trials. J Thrombosis Thrombolysis. 2018;45:562–70.
46. Tsubosaka M, Matsushita T, Kuroda R, Matsumoto T, Kurosaka M. Pseudoaneurysm of the articular branch of the descending genicular artery following double-bundle anterior cruciate ligament reconstruction. Knee Surg Sports Traumatol Arthrosc. 2017;25:2721–4.
47. Lamo-Espinosa JM, Llombart Blanco R, Valenti JR. Inferior lateral genicular artery injury during anterior cruciate ligament reconstruction surgery. Case Rep Surg. 2012;2012:457198.
48. Mohtadi NG, Chan DS. A randomized clinical trial comparing patellar tendon, hamstring tendon, and double-bundle ACL reconstructions. Patient-reported and clinical outcomes at 5-year follow-up. J Bone Joint Surg Am. 2019;101:949–60.
49. Lee GH, McCulloch P, Cole BJ, Bush-Joseph CA, Bach BR Jr. The incidence of acute patellar tendon harvest complications for anterior cruciate ligament reconstruction. Arthroscopy. 2008;24(2):162–6.
50. Reinhardt KR, Hetsroni I, Marx RG. Graft selection for anterior cruciate ligament reconstruction: a level I systematic review comparing failure rates and functional outcomes. Orthop Clin North Am. 2010;41(2):249–62.
51. Keays SL, Bullock-Saxton JE, Keays AC, Newcombe PA, Bullock MI. A 6-year follow-up of the effect of graft site on strength, stability, range of motion, function, and joint degeneration after anterior cruciate ligament reconstruction. Am J Sports Med. 2007;35(5):729–39.
52. Franz W, Baumann A. Minimally invasive semitendinosus tendon harvesting from the popliteal fossa versus conventional hamstring tendon harvesting for ACL reconstruction: a prospective, randomised controlled trial in 100 patients. Knee. 2016;23(1):106–10. https://doi.org/10.1016/j.knee.2015.09.001.
53. Haviv B, Bronak S, Rath E, Yassin M. Nerve injury during anterior cruciate ligament reconstruction: a comparison between patellar and hamstring tendon grafts harvest. Knee. 2017;24(3):564–9. https://doi.org/10.1016/j.knee.2017.03.009.
54. Konrath JM, Vertullo CJ, Kennedy BA, Bush HS, Barrett RS, Lloyd DG. Morphologic characteristics and strength of the hamstring muscles remain altered at 2 years after use of a hamstring tendon graft in anterior cruciate ligament reconstruction. Am J Sports Med. 2016;44(10):2589–98.
55. Nakamae A, Ochi M, Deie M, Adachi N. Unsuccessful regeneration of the semitendinosus tendon harvested for anterior cruciate ligament reconstruction: report of two cases. Orthop Traumatol Surg Res. 2012;98(8):932–5. https://doi.org/10.1016/j.otsr.2012.07.011.
56. Janssen RP, van der Velden MJ, Pasmans HL, Sala HA. Regeneration of hamstring tendons after anterior cruciate ligament reconstruction. Knee Surg Sports Traumatol Arthrosc. 2013;21(4):898–905. https://doi.org/10.1007/s00167-012-2125-0.
57. Rogowski I, Vigne G, Blache Y, Thaunat M, Fayard JM, Monnot D, Sonnery-Cottet B. Does the graft used for ACL reconstruction affect the knee muscular strength ratio at six months postoperatively? Int J Sports Phys Ther. 2019;14(4):546–53.
58. Bourne MN, Bruder AM, Mentiplay BF, Carey DL, Patterson BE, Crossley KM. Eccentric knee flexor

weakness in elite female footballers 1-10 years following anterior cruciate ligament reconstruction. Phys Ther Sport. 2019;37:144–9. https://doi.org/10.1016/j.ptsp.2019.03.010.
59. Toor AS, Limpisvasti O, Ihn HE, McGarry MH, Banffy M, Lee TQ. The significant effect of the medial hamstrings on dynamic knee stability. Knee Surg Sports Traumatol Arthrosc. 2019;27(8):2608–16. https://doi.org/10.1007/s00167-018-5283-x.
60. Hurley ET, Calvo-Gurry M, Withers D, Farrington SK, Moran R, Moran CJ. Quadriceps tendon autograft in anterior cruciate ligament reconstruction: a systematic review. Arthroscopy. 2018;34(5):1690–8.
61. Mouarbes D, Menetrey J, Marot V, Courtot L, Berard E, Cavaignac E. Anterior cruciate ligament reconstruction: a systematic review and meta-analysis of outcomes for quadriceps tendon autograft versus bone-patellar tendon-bone and hamstring-tendon autografts. Am J Sports Med. 2019;47(14):3531–40.
62. Slone HS, Romine SE, Premkumar A, Xerogeanes JW. Quadriceps tendon autograft for anterior cruciate ligament reconstruction: a comprehensive review of current literature and systematic review of clinical results. Arthroscopy. 2015;31(3):541–54.
63. Letter M, Baraga MG, Best TM, et al. Comparison of neuromuscular firing patterns of the superficial quadriceps in soft tissue quadriceps tendon versus bone-patellar tendon-bone ACL autografts. Orthop J Sports Med. 2019;7(12):2325967119887674.
64. Sheean AJ, Musahl V, Slone HS, et al. Quadriceps tendon autograft for arthroscopic knee ligament reconstruction: use it now, use it often. Br J Sports Med. 2018;52(11):698–701.
65. Kaeding CC, Aros B, Pedroza A, Pifel E, Amendola A, Andrish JT, Dunn WR, Marx RG, McCarty EC, Parker RD, Wright RW, Spindler KP. Allograft versus autograft anterior cruciate ligament reconstruction: predictors of failure from a MOON prospective longitudinal cohort. Sports Health. 2011;3(1):73–81.
66. Eysturoy NH, Nissen KA, Nielsen T, Lind M. The influence of graft fixation methods on revision rates after primary anterior cruciate ligament reconstruction. Am J Sports Med. 2018;46(3):524–30.
67. Mascarenhas R, Saltzman BM, Sayegh ET, et al. Bioabsorbable versus metallic interference screws in anterior cruciate ligament reconstruction: a systematic review of overlapping meta-analyses. Arthroscopy. 2015;31(3):561–8.
68. Pereira H, Correlo VM, Silva-Correia J, Oliveira JM, Reis RL, Espregueira-Mendes J. Migration of "bioabsorbable" screws in ACL repair. How much do we know? A systematic review. Knee Surg Sports Traumatol Arthrosc. 2013;21(4):986–94.
69. Ibrahim SA, Abdul Ghafar S, Marwan Y, et al. Intratunnel versus extratunnel autologous hamstring double-bundle graft for anterior cruciate ligament reconstruction: a comparison of 2 femoral fixation procedures. Am J Sports Med. 2015;43(1):161–8.
70. Weiler A, Hoffmann RF, Bail HJ, Rehm O, Sudkamp NP. Tendon healing in a bone tunnel. Part II: histologic analysis after biodegradable interference fit fixation in a model of anterior cruciate ligament reconstruction in sheep. Arthroscopy. 2002;18(2):124–35.
71. Li H, Liu S, Sun Y, Li H, Chen S, Chen J. Influence of graft bending angle on graft maturation, the femoral tunnel, and functional outcomes by 12 months after anterior cruciate ligament reconstruction. Orthop J Sports Med. 2019;7(11):2325967119882663. https://doi.org/10.1177/2325967119882663.
72. Struewer J, Efe T, Frangen TM, Schwarting T, Buecking B, Ruchholtz S, Schüttler KF, Ziring E. Prevalence and influence of tibial tunnel widening after isolated anterior cruciate ligament reconstruction using patella-bone-tendon-bone-graft: long-term follow-up. Orthop Rev (Pavia). 2012;4(2):e21. https://doi.org/10.4081/or.2012.e21.
73. Bhullar R, Habib A, Zhang K, de Sa D, Horner NS, Duong A, Simunovic N, Espregueira-Mendes J, Ayeni OR. Tunnel osteolysis post-ACL reconstruction: a systematic review examining select diagnostic modalities, treatment options and rehabilitation protocols. Knee Surg Sports Traumatol Arthrosc. 2019;27(2):524–33. https://doi.org/10.1007/s00167-018-5142-9.
74. Robbrecht C, Claes S, Cromheecke M, Mahieu P, Kakavelakis K, Victor J, Bellemans J, Verdonk P. Reliability of a semi-automated 3D-CT measuring method for tunnel diameters after anterior cruciate ligament reconstruction: a comparison between soft-tissue single-bundle allograft vs. autograft. Knee. 2014;21(5):926–31. https://doi.org/10.1016/j.knee.2014.05.003.
75. Choi NH, Yang BS, Victoroff BN. Clinical and radiological outcomes after hamstring anterior cruciate ligament reconstructions: comparison between fixed-loop and adjustable-loop cortical suspension devices. Am J Sports Med. 2017;45(4):826–31. https://doi.org/10.1177/0363546516674183.
76. Ahn HW, Seon JK, Song EK, Park CJ, Lim HA. Comparison of clinical and radiologic outcomes and second-look arthroscopic findings after anterior cruciate ligament reconstruction using fixed and adjustable loop cortical suspension devices. Arthroscopy. 2019;35(6):1736–42. https://doi.org/10.1016/j.arthro.2019.01.051.
77. Arama Y, Salmon LJ, Sri-Ram K, Linklater J, Roe JP, Pinczewski LA. Bioabsorbable versus titanium screws in anterior cruciate ligament reconstruction using hamstring autograft: a prospective, blinded, randomized controlled trial with 5-year follow-up. Am J Sports Med. 2015;43(8):1893–901. https://doi.org/10.1177/0363546515588926.

Osteotomy: Slope Change Tibial Osteotomy to Address ACL Deficiency

12

Stefano Muzzi, Camilo Muniagurria,
Jordan Gruskay, and David Dejour

12.1 Introduction

As the incidence of anterior cruciate ligament (ACL) reconstructions continues to increase, the rate of revision surgery continues to climb. Despite technical and rehabilitation advances in primary ACL surgery, the rate of revision and even re-revision remains higher than desired.

Primary anterior cruciate ligament reconstruction (ACLR) generally leads to good outcomes and has a revision rate of between 1.6% and 2.1% [1]. With follow-up longer than 10 years, the ACL graft rupture rate increases to 6% and clinical failure occurs in approximately 10% of ACLR cases (range, 2–26%) [2]. Revisions of previously reconstructed ACLs show even higher re-rupture rates of approximately 13.7% [3]. Furthermore, outcomes with subsequent revisions can be dismal, with one study showing less than one-third of patients returning to their prior levels of activity [4]. Additionally, based on long-term follow-up, osteoarthritis develops in 21–48% of patients after combined ACL, meniscus, and cartilage injuries but in only 0–13% of patients with isolated ACL rupture [5].

S. Muzzi (✉) · D. Dejour
Lyon-Ortho-Clinic, Clinique de la Sauvegarde, Lyon, France

C. Muniagurria
Hospital Alemán de Buenos Aires (Deutsches Hospital), Buenos Aires, Argentina

J. Gruskay
Hospital for Special Surgery, New York, NY, USA

Despite a relatively high failure rate of revision ACL reconstructions, additional reconstructive procedures are not commonly performed. This is for multiple reasons, including decreased patient activity and expectations with increasing age and concerns of undergoing a further surgical procedure [6].

Since the outcomes of revision ACL procedures remain poor compared to the results of primary reconstructions [7–10] it is crucial to identify and address the factors that may have contributed to graft failure. These factors can be divided in "extrinsic" or "intrinsic". Extrinsic factors, related to surgical technique and rehabilitation, are often determined according to the patients' intended sports activities, [11, 12] while intrinsic factors include specific anatomic features such as an increased posterior tibial slope (PTS), a narrow intercondylar notch, hyperlaxity or gender.

12.2 Biomechanics

The factors that contribute to the abnormal knee kinematics after ACL injury and reconstruction remain unclear. Bone shape has been implicated in the development of hip and knee osteoarthritis, although there is little knowledge about the effects of bone shape on knee kinematics.

Many risk factors have already been identified for ACL injuries. These include a decreased notch width, generalized joint laxity, subtalar pronation, hormonal factors, body mass index,

N. Nakamura et al. (eds.), *Advances in Knee Ligament and Knee Preservation Surgery*,
https://doi.org/10.1007/978-3-030-84748-7_12

knee recurvatum, and increased Q angle. Previously, it has been demonstrated that anterior tibial translation (ATT) increased 0.6 mm per degree of posterior tibial slope (PTS) in ACL-deficient knees [13] and, more recently, studies have shown that increased PTS is an independent risk factor for primary ACL injuries [14].

The normal PTS is within the range of 5–7°, depending on the measurement technique, and it is considered pathologic if it exceeds 12° [15].

An increase in PTS leads to an anterior shift of the tibia's resting position, which not only intensifies the stress on the ACL, but can also lead to abnormal loading of the knee, resulting in damage to menisci and articular cartilage [16].

Andrew et al. [17], in their cadaveric study, quantified the effect of changes in sagittal plane tibial slope on ACLR graft force at varying knee flexion angles. They proved with axial loading that the PTS had an independently significant, linearly increasing effect on graft force regardless of flexion angle (coefficient = 0.92, SE = 0.08, $P < 0.001$). Meanwhile, significantly higher graft force was observed at 0° of flexion as compared with all other flexion angles for the loaded (all $P < 0.001$) and unloaded (all $P < 0.001$) conditions.

In a more recent clinical study, Dejour et al. [18] have proved that static ATT increased significantly in knees with tibial slopes >7°, by approximately 0.3 mm per degree, and that dynamic ATT increased significantly in knees with tibial slope ≥12°, by approximately 0.2 mm per degree. These findings confirm the relationship between tibial slope and ATT first described by Dejour [19, 20] and confirmed more recently by Schatka et al. [21].

It is important to highlight that these studies did not find an association between the pivot shift test and tibial slope, possibly because the latter is generally measured at the medial compartment, while rotational stability may depend more on lateral tibial slope [14, 22].

Of course, meniscal status is another significant contributor to knee stability and anterior tibial translation control that has been well established in the literature.

The role of the menisci in limiting ATT can be explained by considering them as part of a 'soft tissue slope' which increases the coverage and reduces the bony slope [23].

The association of medial meniscal tears with dynamic ATT, as well as the pivot shift has been corroborated by cadaveric studies [24, 25] gait analyses and simulations, which detected a significant impact of medial meniscal tears on ATT during gait [24, 26, 27].

Samuelsen et al. [28] found a significant interaction between PTS and posterior medial meniscal root tear (PMMR), where a PMMR tear was observed to potentiate the effect that increased PTS has on ACLR graft forces at 30° of knee flexion. A PMMR tear also led to a significant increase in ACLR graft force when compared with the intact state, whereas the meniscal repair state was not significantly different from the intact state. This finding corroborates and highlighted the importance of the medial meniscus as a secondary stabilizer of the knee.

Meanwhile, it is important to notice that Dejour et al. [18] recently found that the lateral meniscus was not a major component of either static anteroposterior or rotational knee stability in ACL-deficient knees.

Taking all these biomechanical considerations into account, a slope-reducing osteotomy procedure can significantly reduce ACL force and anterior tibial translation with tibio-femoral compression (TFC) alone as well as combined with anterior force or valgus moment [29].

12.3 Measurement

Radiographic examination includes an anteroposterior view (in monopodal stance when it is possible at 20° of flexion), a Rosenberg view (postero-anterior at 35–40° of knee flexion), a true lateral view (in monopodal stance at 25–30° of knee flexion), and an axial view at 30° of knee flexion. Assessments are made using true lateral views of the knee under fluoroscopic control to ensure that the femoral condyles are superimposed.

As for the PTS measurement, there is little consensus on the ideal references to be used. Many anatomical references have been described but there is little information about the values measured with the different methods, and, for this reason, it is difficult to compare the measurements of different studies. The authors utilize the

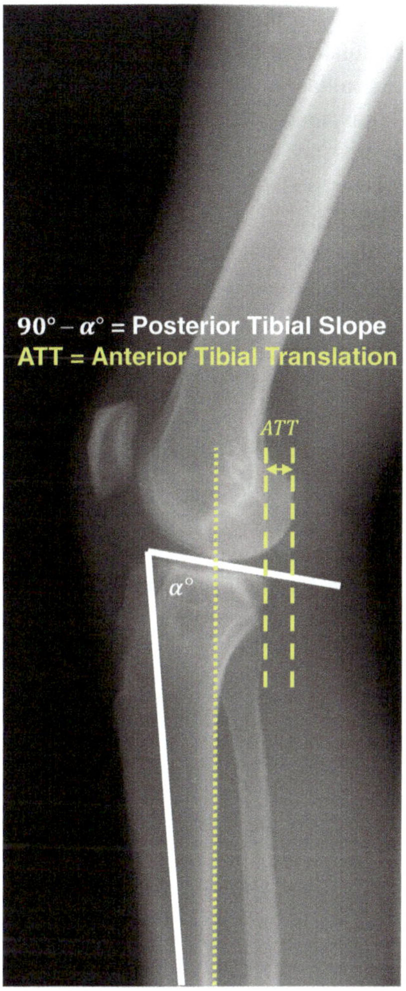

Fig. 12.1 Posterior tibial slope (PTS) and anterior tibial translation (ATT) measurements

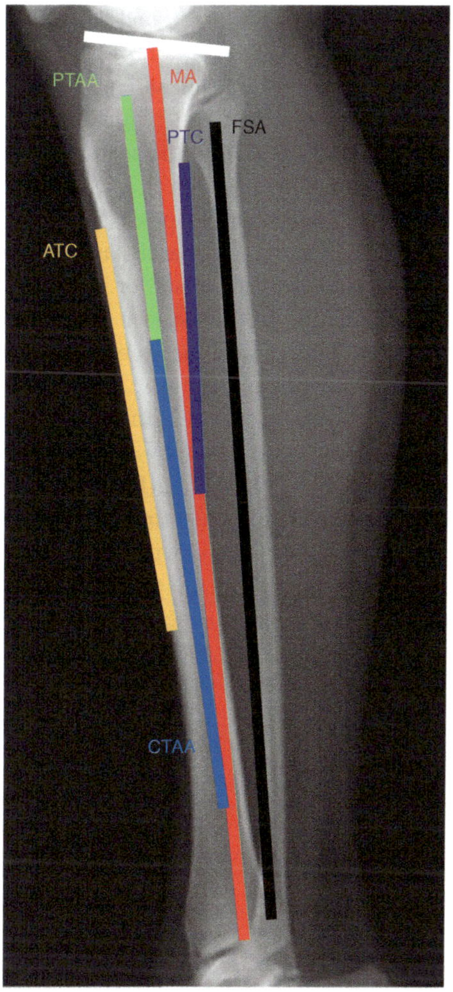

Fig. 12.2 Different anatomical references for PTS measurement: *ATC* anterior tibial cortex, *PTA* proximal tibial axis, *CTA* central tibial axis, *MA* mechanical axis, *PTC* posterior tibial cortex, *FA* fibular axis

method described by Dejour and Bonnin [19], using a goniometer with a precision of ±1° to measure the angle between the line perpendicular to the proximal tibial diaphyseal axis and the line tangent to the most superior points at the anterior and posterior edges of the medial tibial plateau (Fig. 12.1). According to this method, the physiological PTS measures ~7°. Other anatomical references, however, have been described and applied as the longitudinal axis.

Jae Ho Yoo et al. [30] have utilized the mechanical axis (MA, a line connecting the midpoints of the tibia plateau and the tibia plafond), the anterior tibial cortex (ATC, a line connecting 2 points on the anterior tibial cortex at 5 and 15 cm distal to the knee joint line), the proximal tibial anatomical axis (PTAA, a line connecting midpoints of outer cortical diameter at 5 and 15 cm distal to the knee joint line), the central tibial anatomical axis (CTAA, a line connecting midpoints of outer cortical diameter at 10 cm distal to the knee joint line and at 10 cm proximal to the ankle joint line), the posterior tibial cortex (PTC, a line connecting 2 points on the posterior tibial cortex at 5 and 15 cm distal to the knee joint line), and the fibular shaft axis (FSA, a line connecting midpoints of outer cortical diameter of proximal and distal ends of the fibular diaphysis) (Fig. 12.2). They evaluated 90 knees in 60 consecutive female patients and found that the mean PTS varied up to 5° based on the axis chosen. It was 10.6° with the mechanical axis, 13.8° with the anterior tibial cortex, 10.8° with the proximal anatomical axis, 12.9° with the central anatomical axis, 7.8° with the posterior tibial cortex, and 9.5° with the fibular shaft diaphysis.

The ATT is defined as the distance between two lines parallel to the posterior tibial cortex: the first, tangent to the posterior aspect of the medial tibial plateau, and the second, tangent to the posterior femoral condyles. Static ATT is measured on monopodal weight-bearing radiographs with the knee flexed by 20° (Fig. 12.1). The dynamic tibial translation (DTT) measure is realized by using the Telos™ stress device (Telos GmbH, Marburg, Germany) with 150 N at 20° of knee flexion and the side-to-side difference (SSD) between the injured and healthy knee is then calculated (Fig. 12.3).

The patellar height is expressed by the Caton–Deschamps index [31, 32].

The presence of radiographic signs of osteoarthritis is graded following the classification of Ahlback [15].

It is also important to consider the presence of meniscal lesions or previous menisectomy that could exacerbate the effects of a high PTS. The recent study of Lustig et al. [23] demonstrated how "soft tissue tibial slope", measured using magnetic resonance imaging (MRI), is influenced by menisci which shift the tibial slope toward the horizontal.

In this regard, it has to be mentioned that much of what is known about the geometry of the asymmetric, three-dimensional, osseous portion

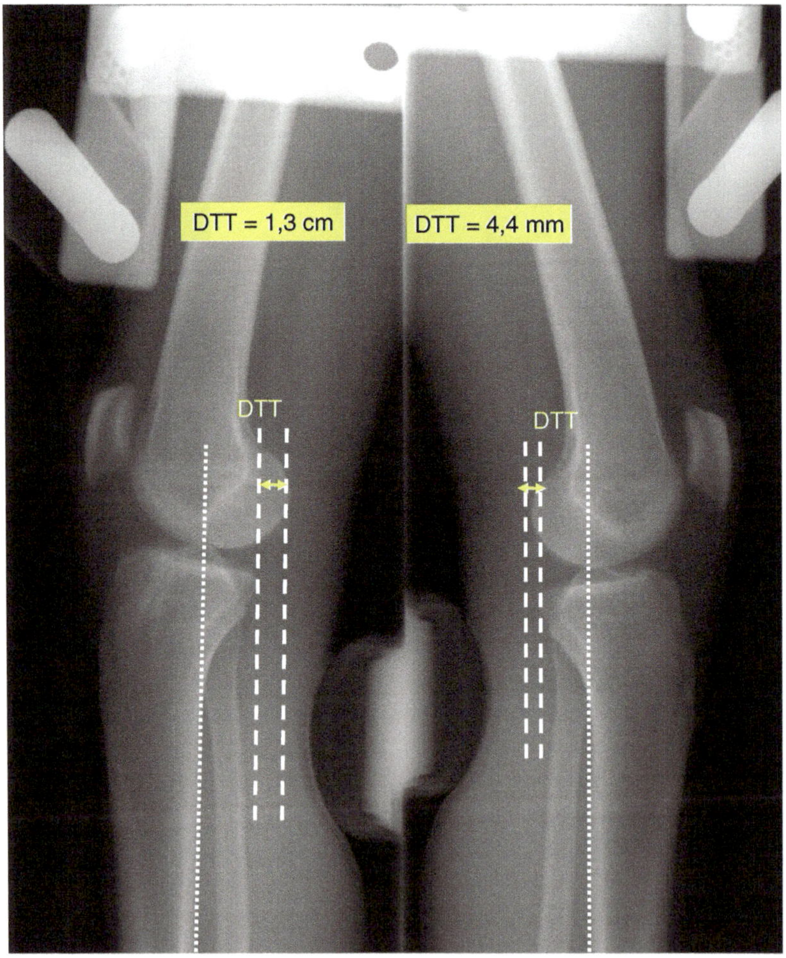

Fig. 12.3 Dynamic tibial translation (DTT)

of the tibial plateau is based on two-dimensional measurements obtained from lateral radiographs. With that approach, it is difficult to differentiate between the medial and lateral aspects of the plateau because they are superimposed.

As such, Hashemi et al. [33] performed the slope measurements with the accepted radiographic methods using MRI (Fig. 12.4a–c), which allowed them to characterize the slope of the tibial plateau at the center of the articular

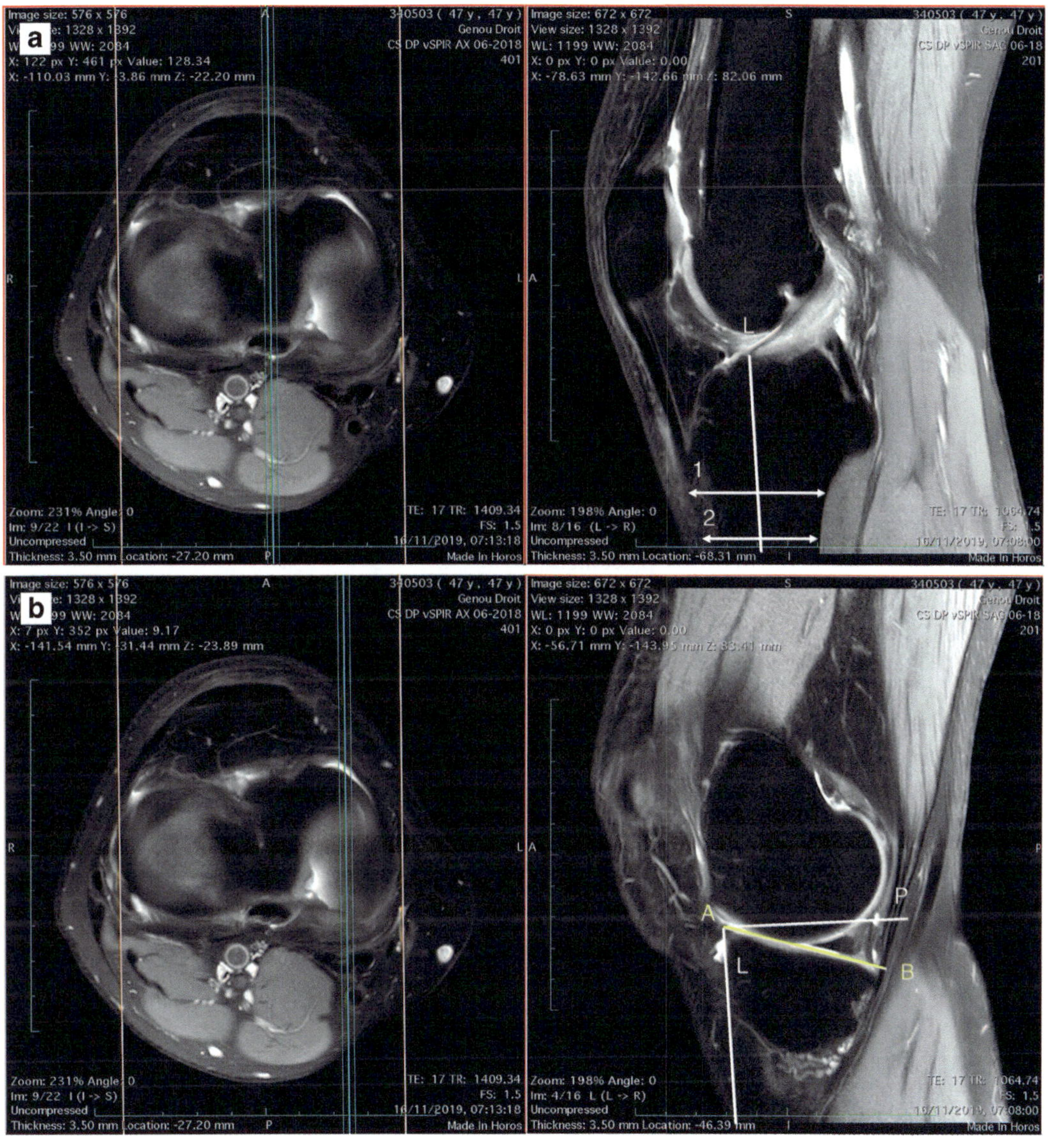

Fig. 12.4 Magnetic resonance images illustrating the method used to determine the medial and lateral tibial slopes. (**a**) The sagittal plane (represented by the green lines in the axial view at the center of the tibial spines) was used to determine the orientation of the diaphyseal axis in the sagittal plane. The axis L is the line connecting the midpoints of the lines 1 and 2 drawn at 4–5 cm from the joint line. (**b**) The axis L is copied and pasted in the sagittal plane (represented by the green lines in the axial view) that can clearly show the orientation of the tibia. The peak anterior and posterior points on the tibial plateau are identified (A and B) and a line perpendicular to the axis is drawn (P). The slope of the line extending between A and B represents the medial tibial slope, and it is measured by the angle between the line itself and the line P. (**c**) The same procedure is utilized in order to measure the lateral tibial plateau slope

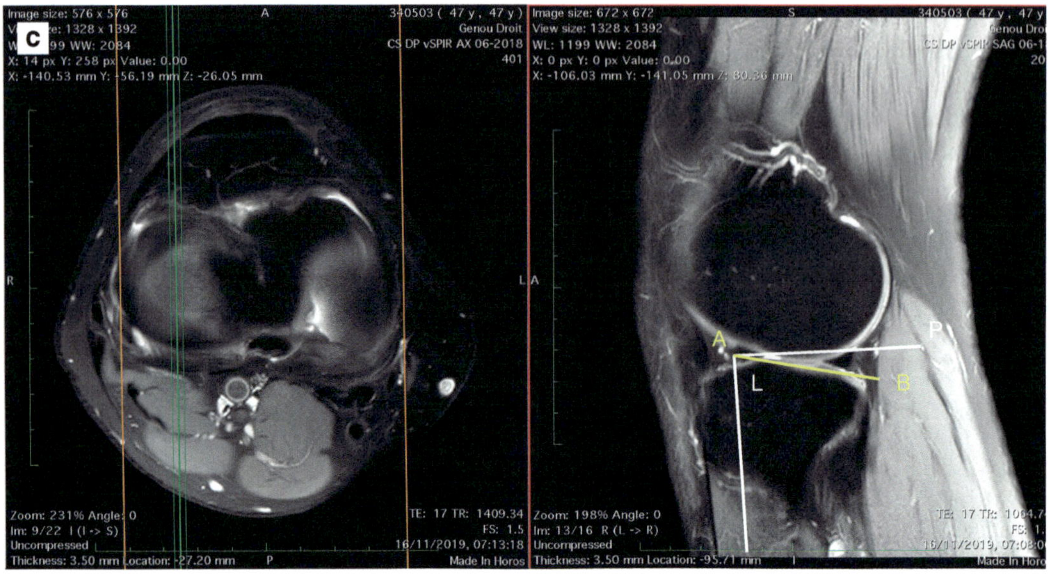

Fig. 12.4 (continued)

surfaces of both the medial and lateral compartments.

Furthermore, they advocate for measuring the depth of the concavity of the medial compartment to better characterize the complex three-dimensional geometry of the tibial plateau. A deep medial plateau will constrain the femoral condyle to a greater extent and result in increased resistance to displacement of the tibia relatively to the femur. Conversely, the combination of a high medial tibial slope and low depth of concavity may be associated with a decreased resistance to displacement of the tibia relative to the femur, placing the knee at increased risk for ligament injury.

These recent findings indicate that a surgeon will need to combine the assessment of radiographic images and MRI in order to have a more complete idea of the three-dimensional geometry of the "soft tissue tibial slope" and the knee kinematics for each patient.

12.4 Clinical Application

The importance of the adverse effect of a high PTS on the ACL biomechanics has been confirmed by many clinical studies, evaluating the correlation between PTS and the rate of native and grafted ACLs (re-)rupture.

Regarding primary ACL injuries, Sonnery-Cottet et al. [34] performed a case-control study comparing a study group of 50 patients who had sustained an isolated, complete rupture of the ACL and a control group of 50 patients who had consulted for other reasons to determine whether there was a difference in the PTS and notch width index (NWI) between them. They found a statistically significantly steeper PTS and narrower NWI in the patients with a ruptured ACL than in the uninjured group. Waiwaiole et al. [35] reviewed medical records of 221 patients who underwent MRI of the knee between January 2003 and December 2009 and separated them into two subgroups: a study group of 107 subjects who had undergone surgery for ACL injury and a control group of 114 patients diagnosed with patellofemoral syndrome. They found significantly greater values for lateral PTS (6° ± 4°; $P < 0.001$) and medial PTS (7° ± 4°; $P = 0.002$) in the study group compared with controls (5° ± 3° and 5° ± 4°, respectively). They found also a statistically significant relationships between young age and lateral PTS with ACL injury. Zeng et al. [36] performed a case-control study containing 146 patients in total (73 noncontact ACL injuries and 73 meniscus injuries) and found that the mean PTS of the ACL-injured group was significantly higher than that of the control group ($P < 0.001$).

Historically, anatomic features such as an increased PTS or narrow intercondylar notch have not been routinely addressed during ACL reconstructions, despite the biomechanical studies documenting their negative effect on the ACL graft. As such, there are many studies that investigated the correlation between them and the risk of ACL graft failures.

Webb et al. [37] enrolled 200 consecutive patients who underwent primary ACLR with hamstring autografts in a prospective longitudinal study over 15 years and found that the mean tibial slope was significantly greater in patients with ACL graft tear (50 of them) compared with patients with no further injury (9.9° vs 8.5°). The mean PTS for those with both an ACL graft and contralateral ACL rupture was 12.9°. Patients with a tibial slope of 12° or higher had an odds ratio of further ACL injuries increased by a factor of 5, to an incidence of 59%. Christensen et al. [14] compared, in a case-control study, 35 patients with early (within 2 years) failure of primary ACLR with 35 control patients who underwent ACLR with a minimum of 4 years of clinical follow-up and no evidence of graft failure. They compared their lateral tibial posterior slope (LTPS) and they found it to be significantly higher in the early ACL failure group (8.4° vs 6.5°; $p = 0.012$), with a 1.6, 2.4, and 3.8 odds ratio for graft failure considering respectively a 2°, 4°, and 6° increase in the LTPS value. The most striking correlation between an increasing LTPS and graft failure was observed in women. In this population, a 4° change in the slope increased the risk of graft failure nearly five times, and a 6° change resulted in over ten times the increased risk. Lately, Salmon et al. [38] further highlighted the negative effect that a steep PTS has on ACL grafts. They reviewed 179 patients who underwent isolated primary ACLR with hamstring autograft and they found out that ACL graft survival was significantly affected by age <18 years at the time of reconstruction (hazard ratio, 3.3; 95% CI, 1.7–6.4; $P = 0.001$) and a PTS of 12° or more (hazard ratio, 3.1; 95% CI, 1.5–5.9; $P = 0.001$), while contralateral ACL injury was significantly affected only by a PTS of 12° or more (hazard ratio, 7.3; 95% CI, 3–18; $P = 0.001$). In particular, adolescents with a tibial slope of 12° or more were 11 times more likely to rupture their ACL graft and seven times more likely to rupture their contralateral ACL than were adults with tibial slopes of 12° or less. At 20 years, the ACL survival for adolescents with a PTS of 12° or more was 22%, showing the catastrophic effect of tibial slope on further ACL injuries.

The negative correlation between PTS and ACL graft survival has also been shown in cases of combined ACL reconstruction and alignment correction osteotomies addressing chronic cases of early osteoarthritis and instability. It is well known that bony procedures treating coronal deformity can incidentally change the PTS and affect the sagittal balance. Schuster et al. [39] retrospectively evaluated 50 cases of combined high tibial osteotomy (HTO), ACL reconstruction, and chondral resurfacing (CR, abrasion plus microfracture), and analyzed the graft failure rates in relation to the tibial slope. They found that graft insufficiency was strongly dependent on tibial slope, with a failure rate of 7% in cases of postoperative tibial slope <7.5°, 24% in cases of slope between 7.5° and 12.5°, and 36% in cases of slope >12.5°.

Finally, some studies have shown a high PTS in rare cases of congenital absence of the ACL. Frikha et al. [40] presented a descriptive analysis of 8 knees with congenital agenesia of the ACL in five patients of the same family and they found that the tibial slope was increased in all knees (mean 20.6°).

All of this evidence draws attention on PTS as an important and independent risk factor for ACL (re)rupture and instability. It becomes crucial, then, to address this feature in cases of ACL-deficiency.

A high PTS can be associated with deformities in the coronal plane (frequently a double varus), and in these cases it is recommended to combine the ACL reconstruction with a HTO.

In general, the tibial slope tends to increase after opening-wedge HTO while it has an inverse tendency after closing-wedge HTO [41, 42]. Arun et al. [43] retrospectively analyzed data from 30 patients who underwent arthroscopic ACLR along with medial opening-wedge osteotomy, measuring preoperative and postoperative tibial slopes. They found that patients who had a PTS decrease of 5° or more postoperatively had the best results compared with the others in terms of ACL graft survival and functional outcomes. Thus, they suggested that placing the tricortical graft posterior to midline in the opening wedge could reduce the PTS and, consequently, the stress on the graft, leading to better functional outcomes.

However, not all patients who present with an increased PTS has a coronal plane deformity or arthritis issues. In order to properly treat these patients, it has been suggested that decreasing the PTS could potentially protect ACL grafts and reduce the risk of failed revisions. As such, several authors have described a deflexion osteotomy performed with an anterior closing-wedge osteotomy.

That is a complex and technically demanding procedure, and limited studies can be found in literature about it.

In the first article, Dejour et al. [13] reported a series of 22 knees with chronic anterior laxity and excessive PTS (average 16.5°). Four of them were isolated tibial deflexion osteotomies while the other 18 were combined with ACLR. Better clinical results were observed in the latter group. PTS was corrected to an average of 7° postoperatively. ATT in monopodal stance decreased from 12.5 mm preoperatively to 3 mm at last follow-up.

Sonnery-Cottet et al. [6] retrospectively evaluated five patients after a slope-reducing anterior closing-wedge osteotomy in combination with an ACL re-revision after a mean 32 months follow-up. All patients presented an excessive PTS as an intrinsic risk factor for graft failure. The mean PTS decreased from 13.6° preoperatively to 9.2° postoperatively and the anterior laxity measured with the KT-1000 arthrometer decreased from 10.4 mm to 2.8 mm. As regards clinical outcomes, Lysholm score and IKDC score both improved from preoperatively to the last follow-up and the mean Tegner activity score reached the same level as before the last ACL injury (7.4 and 7.2).

Finally, Dejour et al. [15] reported the outcomes, at a minimum of 2-year follow-up, of nine patients that underwent second revision ACLR combined with tibial deflexion osteotomy for a high PTS value (all patients had a PTS > 12°). The mean PTS decreased from 13.2° ± 2.6° (median 13°; range 12–18°) preoperatively to 4.4° ± 2.3° (median 4°; range 2–8°) postoperatively. With regard to the clinical outcomes, the mean Lysholm score was 73.8 ± 5.8 (median 74; range 65–82), and the IKDC-SKF was 71.6 ± 6.1 (median 72.8; range 62.2–78.5) showing satisfactory results and suggesting that tibia slope correction protects reconstructed ACLs from fatigue failure.

12.5 Indications and Contraindications

In patients with multiple ligament tears and surgeries, it is of paramount importance to carefully analyze the causes of the previous failures and to address the risk factors in order to avoid another rerupture. The authors, therefore, recommend a correction of the PTS with an anterior closing-wedge osteotomy in patients with a failed ACLR and a PTS > 12°. Contraindications include hyperextension (>10°), a significant deformity in the coronal plane, and end-stage osteoarthritis. In patients with more than 10° of hyperextension, it is not possible to perform the osteotomy because it would generate an excessive *genu recurvatum* while patients with lower degree of hyperextension (0–10°) are treated like all the other. Since there are very few studies reporting outcomes of slope reduction techniques, there is still no consensus regarding their indications and contraindications.

12.6 Preoperative Planning

All the patients must undergo a standard radiographic evaluation as explained before. The PTS must be carefully measured and the amount of slope correction is estimated in order to obtain a PTS between 3° and 5° considering 1 mm of resection equal to 1° of correction.

12.7 Surgical Technique

In the literature, there are two surgical techniques described to perform a tibial deflexion osteotomy. Originally, Dejour et al. [13] proposed an approach to the osteotomy site above the patellar tendon insertion. On the contrary, Sonnery-Cottet et al. [6] proposed to detach the tibial tubercle to reach the osteotomy site 4 to 5 cm distal to the joint line. The main advantages of the original procedure are that it maintains the patellar height while keeping both the patellar tendon and tibial tuberosity intact, though it is technically challenging to position the osteotomy at the appropriate level above the anterior tibial tuberosity. The alternative procedure allows more comfortable exposure of the osteotomy site, but it requires detachment of the patellar tendon and the tibial tuberosity, which may be associated with increased morbidity and tougher rehabilitation.

In the technique proposed by the authors, the patient is positioned in the supine position and a tourniquet is placed high on the thigh. A lateral post at the level of the tourniquet maintains the leg position in the frontal plane, and a distal support holds the knee at 90° of flexion, allowing full range of motion when desired (Fig. 12.5).

The first step is the harvesting of the autograft, depending on where the previous grafts were taken (hamstrings, patellar tendon, or quadriceps tendon).

Second, the knee is arthroscopically assessed through the anterolateral and anteromedial portals, in order to evaluate the status of the cartilage and the menisci, as well as the shape of the intercondylar notch (a notchplasty is performed if needed) and previous tunnel positions. Third, the femoral and tibial tunnels are drilled to match the graft diameter, but the graft is not yet inserted. Fourth, all meniscal interventions (either meniscal suture or removal) are performed if required. Fifth, the tibial deflexion osteotomy is performed through an anterior longitudinal incision medial to the tibial tuberosity. The deep medial collateral ligament and the iliotibial band on Gerdy's tubercle are detached up to the posterior part of the tibia, in order to expose the place of the osteotomy site, and the patellar tendon insertion on the tibia. Detachment of the tibial tubercle is not nec-

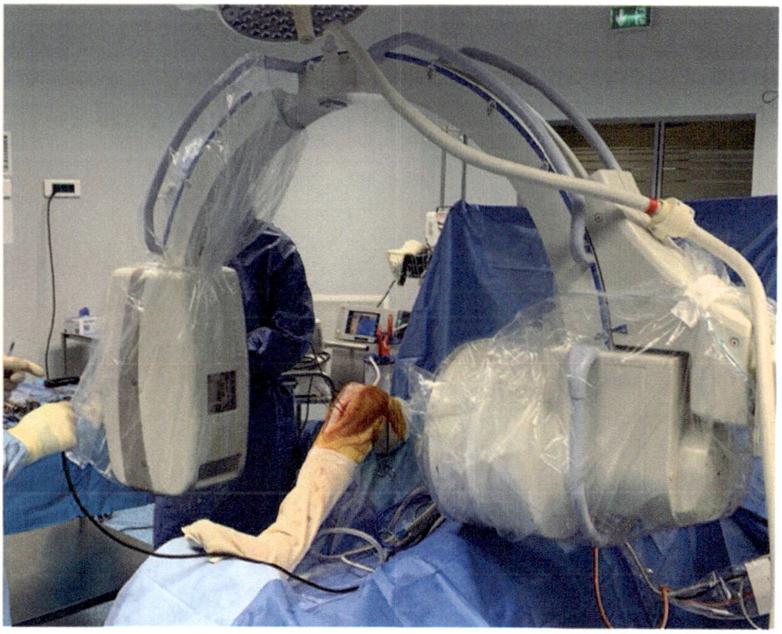

Fig. 12.5 The knee is positioned at 90° of flexion with fluoroscopy ready to be utilized during surgery

essary because the osteotomy is performed at the level of the insertion of the patellar tendon and a biplanar osteotomy will be done. Two monocortical parallel K-wires are positioned under fluoroscopy on both sides of the patellar tendon aimed towards the posterior tibial cortex, 1 cm below the joint line. The level of the osteotomy always starts at the superior margin of the patellar tendon insertion and continues distally. Other two monocortical parallel K-wires are positioned distally in accordance with the amount of correction decided at the preoperative planning. They are orientated distally to proximally aiming to the end of the first two K-wires and their position is controlled under fluoroscopy (Fig. 12.6). The K-wires are placed until their tips are 10 mm away from the tibial plateau surface just below the tibial insertion of the posterior cruciate ligament. These 4 K-wires will serve as the guide for the bone cuts for the closing wedge osteotomy (Fig. 12.7). The first bone cut will be exactly behind the tibial tubercle in the coronal plane in order to perform a biplanar osteotomy. The tibia is then osteotomized cutting under the proximal pins (to be sure not to violate the joint cartilage), keeping the posterior cortex intact as a hinge. The patellar tendon is protected from the oscillating saw using spreaders. Before the distal osteotomy is executed, the measurement of the correction is confirmed. The distal osteotomy is performed with a convergent orientation toward the posterior part of the proximal osteotomy. The proper orientation of the saw is assessed under fluoroscopic guidance. In patients with a slightly varus-valgus deformity, the osteotomy can also be 2-dimensional with an anterior and lateral-medial based wedge respectively.

After making the distal cut, the anterior wedge of bone is removed (Fig. 12.8a, b). With the intact posterior cortex acting like a hinge, the osteotomy gap can be decreased by manipulation of the limb. To decrease the gap, it is possible to push down on the proximal tibial plateau, or simply extend the leg. Extending the leg will exert pressure through the femoral condyles onto the tibial plateau. The amount of slope correction is confirmed and measured under fluoroscopy. Then the osteotomy fixation is achieved using two staples on both sides of patellar tendon (Figs. 12.9 and 12.10). After osteotomy fixation (checked at fluo-

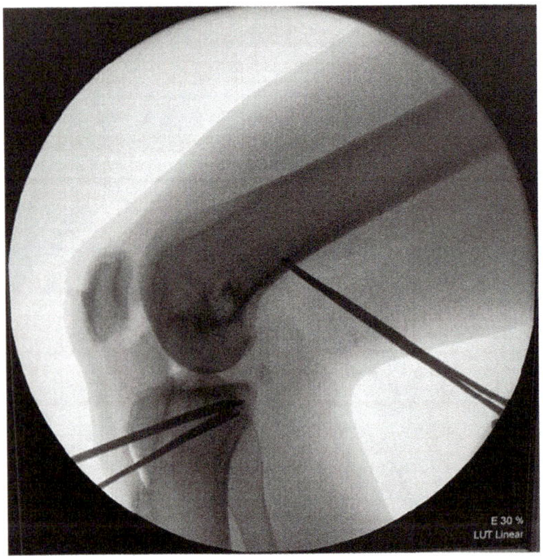

Fig. 12.6 The 4 K wires are positioned under fluoroscopy, aiming just below the tibial insertion of the posterior cruciate ligament

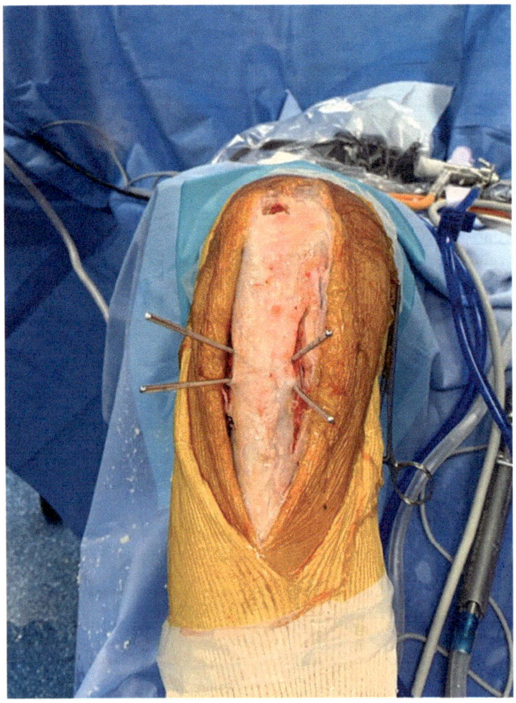

Fig. 12.7 The 4 K wires will serve as the guide for the bone cuts

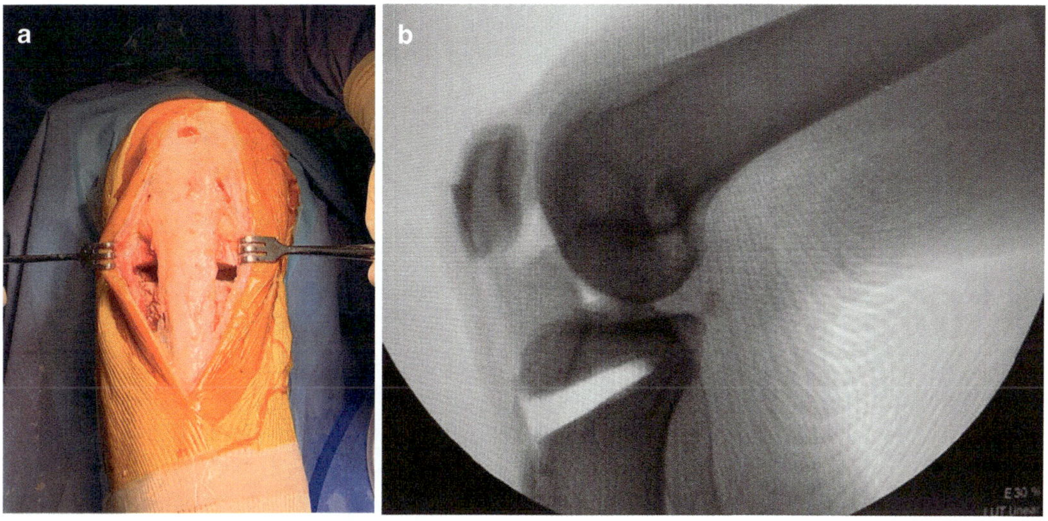

Fig. 12.8 (**a**) The anterior wedge of bone is removed. (**b**) and the osteotomy is checked under fluoroscopy

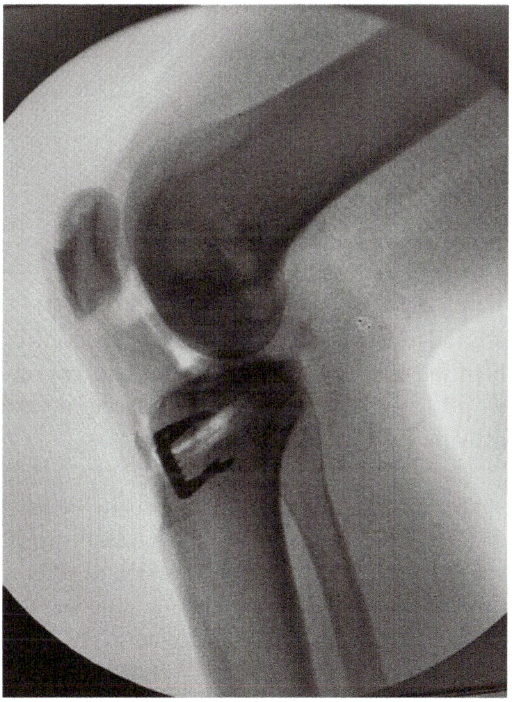

Fig. 12.9 Osteotomy fixation

roscopy), the final step is completing the ACL graft. The tibial tunnel is gently hand-redrilled to debride tibial tunnel from bone fragments. The graft is then pulled and fixed using interference screws or suspensory fixation. The fixation is done at 90° of flexion in order to give a little bit of stiffness to the knee in order to compensate for the negative effect of the "genu recurvatum" caused by the osteotomy. As a final step, a double fixation of the graft is obtained by passing the sutures under the staples and tying them together firmly.

12.8 Postoperative Rehabilitation

After surgery, patients are immobilized in an extension brace during transfers, in order to address the "genu recurvatum", and weight bearing is not allowed for the first 3 weeks. According to the patient tolerance, nonaggressive rehabilitation is immediately begun with passive and active motion exercises and full range of motion. The main objectives in the first phase are reduction in knee swelling, quadriceps control, and recovery of range of motion, while always avoiding hyperextension. After 3 weeks, weight bearing is gradually progressed in the extension brace with the goal of full weight bearing at day 45. Afterward, patients are placed on a standard phase 2 ACL protocol from days 45 to 90, with swimming and cycling activities. They are finally moved to phase 3 for 3 months (3–6 months post- operation), during which the schedule comprises a pro-

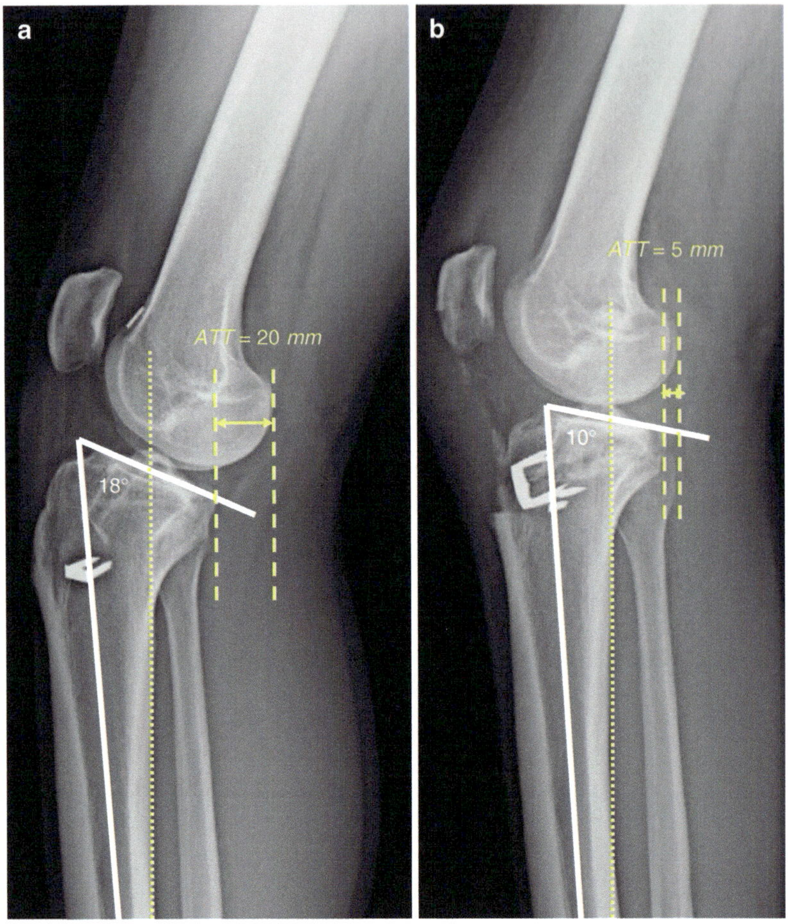

Fig. 12.10 Comparison between the PTS and the ATT from (**a**) preoperative and (**b**) postoperative in the lateral view

gressive return-to-sport program. After 6 months, isokinetic and functional tests are performed: patients can return to full sports activities if they have a good quadriceps/hamstrings ratio and muscle recovery comparable with the contralateral side.

12.9 Conclusions

Preliminary outcomes of ACLR combined with anterior closing-wedge osteotomy in patients with ACL re-ruptures and high PTS show promising results in restoring good knee stability, satisfactory functional levels, and reduced recurrent failures. The authors emphasize the importance of this procedure since it is the only way to properly address the increased forces produced by high values of PTS to which ACL grafts are subjected, even while standing and during normal activities like walking [44].

Given the mounting evidence from recent studies of the strong correlation between high PTS and ACL re-rupture rate as well as the promising results obtained by the deflexion osteotomy procedure, it is worth discussing whether this intervention may also be indicated in the high-risk patient with primary ACL ruptures (PTS > 15° and/or ATT > 10 mm in monopodal weight-bearing).

Larger series and longer follow-up are needed, however, to confirm the efficacy of this technique, considering that still few studies are published on this topic.

References

1. Björnsson H, Andernord D, Desai N, et al. No difference in revision rates between single- and double-bundle anterior cruciate ligament reconstruction: a comparative study of 16,791 patients from the Swedish National Knee Ligament Register. Arthroscopy. 2015;31:659–64.
2. Crawford SN, Waterman BR, Lubowitz JH. Long-term failure of anterior cruciate ligament reconstruction. Arthroscopy. 2013;29:1566–71.
3. Wright RW, Gill CS, Chen L, et al. Outcome of revision anterior cruciate ligament reconstruction: a systematic review. J Bone Joint Surg Am. 2012;94:531–6.
4. Griffith TB, Allen BJ, Levy BA, Stuart MJ, Dahm DL. Outcomes of repeat revision anterior cruciate ligament reconstruction. Am J Sports Med. 2013;41:1296–301.
5. Øiestad BE, Engebretsen L, Storheim K, Risberg MA. Knee osteoarthritis after anterior cruciate ligament injury: a systematic review. Am J Sports Med. 2009;37:1434–43.
6. Sonnery-Cottet B, Mogos S, Thaunat M, Archbold P, Fayard J-M, Freychet B, Chambat P. Proximal tibial anterior closing wedge osteotomy in repeat revision of anterior cruciate ligament reconstruction. Am J Sports Med. 2014;42(8):1873–80.
7. Colombet P. Knee laxity control in revision anterior cruciate ligament reconstruction versus anterior cruciate ligament reconstruction and lateral tenodesis: clinical assessment using computer-assisted navigation. Am J Sports Med. 2011;39(6):1248–54.
8. Gifstad T, Drogset JO, Viset A, Grøntvedt T, Hortemo GS. Inferior results after revision ACL reconstructions: a comparison with primary ACL reconstructions. Knee Surg Sports Traumatol Arthrosc. 2013;21(9):2011–8.
9. Lind M, Lund B, Fauno P, Said S, Miller LL, Christiansen SE. Medium to long-term follow-up after ACL revision. Knee Surg Sports Traumatol Arthrosc. 2012;20(1):166–72.
10. Wright RW, Gill CS, Chen L, Brophy RH, Matava MJ, Smith MV, Mall NA. Outcome of revision anterior cruciate ligament reconstruction: a systematic review. J Bone Joint Surg Am. 2012;94(6):531–6.
11. Bien DP. Rationale and implementation of anterior cruciate ligament injury prevention warm-up programs in female athletes. J Strength Cond Res. 2011;25(1):271–85.
12. Ebben WP, Fauth ML, Petushek EJ, Garceau LR, Hsu BE, Lutsch BN, Feldmann CR. Gender-based analysis of hamstring and quadriceps muscle activation during jump landings and cutting. J Strength Cond Res. 2010;24(2):408–15.
13. Dejour D, Kuhn A, Dejour H. Tibial deflexion osteotomy and chronic anterior laxity: a series of 22 cases. Rev Chir Orthop. 1998;84:28–9.
14. Christensen JJ, Krych AJ, Engasser WM, Vanhees MK, Collins MS, Dahm DL. Lateral tibial posterior slope is increased in patients with early graft failure after anterior cruciate ligament reconstruction. Am J Sports Med. 2015;43(10):2510–4.
15. Dejour D, Saffarini M, Demey G, Baverel L. Tibial slope correction combined with second revision ACL produces good knee stability and prevents graft rupture. Knee Surg Sports Traumatol Arthrosc. 2015;23(10):2846–52.
16. van de Pol GJ, Arnold MP, Verdonschot N, van Kampen A. Varus alignment leads to increased forces in the anterior cruciate ligament. Am J Sports Med. 2009;37:481–7.
17. Bernhardson AS, et al. Tibial slope and its effect on force in anterior cruciate ligament grafts: anterior cruciate ligament force increases linearly as posterior tibial slope increases. Am J Sports Med. 2019;47(2):296–302.
18. Dejour D, Pungitore M, Valluy J, Nover L, Saffarini M, Demey G. Preoperative laxity in ACL-deficient knees increases with posterior tibial slope and medial meniscal tears. Knee Surg Sports Traumatol Arthrosc. 2019;27(2):564–72.
19. Dejour H, Bonnin M. Tibial translation after anterior cruciate ligament rupture. Two radiological tests compared. J Bone Joint Surg Br. 1994;76(5):745–9.
20. Dejour H, Walch G, Neyret P, Adeleine P. Results of surgically treated chronic anterior laxities. Apropos of 251 cases reviewed with a minimum follow-up of 3 years. Rev Chir Orthop Reparatrice Appar Mot. 1988;74(7):622–36.
21. Schatka I, Weiler A, Jung TM, Walter TC, Gwinner C. High tibial slope correlates with increased posterior tibial translation in healthy knees. Knee Surg Sports Traumatol Arthrosc. 2018;26(9):2697–703.
22. Rahnemai-Azar AA, Abebe ES, Johnson P, Labrum J, Fu FH, Irrgang JJ, Samuelsson K, Musahl V. Increased lateral tibial slope predicts high-grade rotatory knee laxity preoperatively in ACL reconstruction. Knee Surg Sports Traumatol Arthrosc. 2017;25(4):1170–6.
23. Lustig S, Scholes CJ, Leo SP, Coolican M, Parker DA. Influence of soft tissues on the proximal bony tibial slope measured with two-dimensional MRI. Knee Surg Sports Traumatol Arthrosc. 2013;21(2):372–9.
24. Ahn JH, Bae TS, Kang KS, Kang SY, Lee SH. Longitudinal tear of the medial meniscus posterior horn in the anterior cruciate ligament-deficient knee significantly influences anterior stability. Am J Sports Med. 2011;39(10):2187–93.
25. Stephen JM, Halewood C, Kittl C, Bollen SR, Williams A, Amis AA. Posteromedial meniscocapsular lesions increase tibiofemoral joint laxity with anterior cruciate ligament deficiency, and their repair reduces laxity. Am J Sports Med. 2016;44(2):400–8.
26. Ali AA, Harris MD, Shalhoub S, Maletsky LP, Rullkoetter PJ, Shelburne KB. Combined measurement and modeling of specimen-specific knee mechanics for healthy and ACL-deficient conditions. J Biomech. 2017;57:117–24.
27. Lorbach O, Kieb M, Herbort M, Weyers I, Raschke M, Engelhardt M. The influence of the medial menis-

cus in different conditions on anterior tibial translation in the anterior cruciate deficient knee. Int Orthop. 2015;39(4):681–7.
28. Samuelsen BT, Aman ZS, Kennedy MI, Dornan GJ, Storaci HW, Brady AW, LaPrade RF. Posterior medial meniscus root tears potentiate the effect of increased tibial slope on anterior cruciate ligament graft forces. Am J Sports Med. 2019;48(2):334–40.
29. Yamaguchi KT, Cheung EC, Markolf KL, Boguszewski DV, Mathew J, Lama CJ, Petrigliano FA. Effects of anterior closing wedge tibial osteotomy on anterior cruciate ligament force and knee kinematics. Am J Sports Med. 2018;46(2):370–7.
30. Yoo JH, Chang CB, Shin KS, Seong SC, Kim TK. Anatomical references to assess the posterior tibial slope in total knee arthroplasty: a comparison of 5 anatomical axes. J Arthroplast. 2008;23(04):586–92.
31. Caton J. Method of measuring the height of the patella. Acta Orthop Belg. 1989;55:385–6.
32. Caton J, Deschamps G, Chambat P, et al. Patella infera. Apropos of 128 cases. Rev Chir Orthopédique Réparatrice Appar Mot. 1982;68:317–25.
33. Hashemi J, Chandrashekar N, Gill B, et al. The geometry of the tibial plateau and its influence on the biomechanics of the tibiofemoral joint. J Bone Joint Surg Am. 2008;90(12):2724–34. https://doi.org/10.2106/JBJS.G.01358.
34. Sonnery-Cottet B, Archbold P, Cucurulo T, et al. The influence of the tibial slope and the size of the intercondylar notch on rupture of the anterior cruciate ligament. J Bone Joint Surg Br. 2011;93(11):1475–8.
35. Waiwaiole A, Gurbani A, Motamedi K, Seeger L, Sim MS, Nwajuaku P, Hame SL. Relationship of ACL injury and posterior tibial slope with patient age, sex, and race. Orthop J Sports Med. 2016;4(11):2325967116672852.
36. Zeng C, Yang T, Wu S, Gao SG, Li H, Deng ZH, Zhang Y, Lei GH. Is posterior tibial slope associated with noncontact anterior cruciate ligament injury? Knee Surg Sports Traumatol Arthrosc. 2016;24(3):830–7.
37. Webb JM, Salmon LJ, Leclerc E, Pinczewski LA, Roe JP. Posterior tibial slope and further anterior cruciate ligament injuries in the anterior cruciate ligament–reconstructed patient. Am J Sports Med. 2013;41:2800–4.
38. Salmon LJ, Heath E, Akrawi H, Roe JP, Linklater J, Pinczewski LA. 20-year outcomes of anterior cruciate ligament reconstruction with hamstring tendon autograft: the catastrophic effect of age and posterior tibial slope. Am J Sports Med. 2018;46(3):531–43.
39. Schuster P, Geßlein M, Schlumberger M, Mayer P, Richter J. The influence of tibial slope on the graft in combined high tibial osteotomy and anterior cruciate ligament reconstruction. Knee. 2018;25(4):682–91.
40. Frikha R, Dahmene J, Ben Hamida R, Chaieb Z, Janhaoui N, Laziz Ben Ayeche M. Congenital absence of the anterior cruciate ligament: eight cases in the same family. Rev Chir Orthop Reparatrice Appar Mot. 2005;91(7):642–8.
41. El-Azab H, Halawa A, Anetzberger H, Imhoff AB, Hinterwimmer S. The effect of closed- and open-wedge high tibial osteotomy on tibial slope: a retrospective radiological review of 120 cases. J Bone Joint Surg Br. 2008;90:1193–7.
42. Hohmann E, Bryant A, Imhoff AB. The effect of closed wedge high tibial osteotomy on tibial slope: a radiographic study. Knee Surg Sports Traumatol Arthrosc. 2006;14:454–9.
43. Arun GR, Kumaraswamy V, Rajan D, et al. Long-term follow up of single-stage anterior cruciate ligament reconstruction and high tibial osteotomy and its relation with posterior tibial slope. Arch Orthop Trauma Surg. 2016;136:505–11.
44. Shelburne KB, Kim HJ, Sterett WI, Pandy MG. Effect of posterior tibial slope on knee biomechanics during functional activity. J Orthop Res. 2011;29(02):223–31.

13 Biologics: Post-traumatic Osteoarthritis Following Anterior Cruciate Ligament Reconstruction

Sami Chergui, Antoine Denis, James Meterissian, Lee Benaroch, and Thierry Pauyo

13.1 Introduction

Anterior cruciate ligament (ACL) tears are amongst the most common injuries treated in orthopedic surgery [1]. While both conservative and surgical treatments are available, ACL reconstruction (ACLR) is usually recommended for healthy patients who participate in pivoting sports [2]. The frequency of ACLR surgeries is consistently increasing as the annual incidence of this surgical treatment rose from 68.6/100000 person-years in 2010 to 74.6/100000 person-years in 2014 in the United States [1, 3].

One of the possible consequences that can arise following surgical ACL reconstruction is post-traumatic osteoarthritis (PTOA). There is some debate surrounding the specific etiologies of PTOA; however, traumatic injury and iatrogenic trauma arising from the surgical intervention are known to be major players in the development of PTOA [4]. This traumatic etiology differentiates PTOA from OA as the latter tends to be caused by chronic cellular or matrix-derived factors [5]. The patient population in PTOA tends to be younger and more active than the patients affected by OA [5, 6]. Besides the etiology and patient population, the pathophysiology and progression of both diseases is believed to be very similar.

PTOA is a common adverse event following ACLR as its prevalence among patients after reconstruction (44%) is reported to be greater than those who are ACL-deficient (37%) [7]. More recently, a meta-analysis by Chen et al. showed that the prevalence of PTOA following ACLR is even greater than previously reported with 51.6% of patients experiencing this postoperative complication within 10 years following surgery [8]. PTOA also has important individual and societal implications as patients with a history of knee ligament reconstruction tend to undergo total knee arthroplasty 9 years prior to patients who had other knee surgeries [9]. Lower-extremity PTOA alone costs the United States (US) healthcare system over 3 billion dollars annually [10].

The initial management of PTOA typically includes lifestyle modifications, weight management, unloader bracing, physical therapy, or intra-articular injections [11]. In more advanced cases and when conservative treatment fails, surgical treatment such as cartilage repair, debridement, osteotomy, or total knee replacement is

S. Chergui · A. Denis · L. Benaroch · T. Pauyo (✉)
McGill University Health Centre,
Montreal, QC, Canada

Shriners Hospital for Children—Canada,
Montreal, QC, Canada
e-mail: sami.chergui@mail.mcgill.ca;
antoine.denis@mail.mcgill.ca;
thierry.pauyo@mcgill.ca

J. Meterissian
University of Montréal, Montreal, QC, Canada

N. Nakamura et al. (eds.), *Advances in Knee Ligament and Knee Preservation Surgery*,
https://doi.org/10.1007/978-3-030-84748-7_13

undertaken [12]. Currently, most first-line treatments offered are pharmacologic in nature and are focused on improving symptoms and joint function. However, none of these medications appear to modify the progression of PTOA by tackling the underlying cause, which is cartilage injury [13]. Since PTOA can be a disabling condition that may greatly affect patients' quality of life, more emphasis is starting to be placed on the use of orthobiologics and other injectables as an adjunct to ACLR in order to prevent and potentially reverse articular degradation. Orthobiologics are a category of treatments based on substances naturally derived from the human body (platelet-rich plasma, amniotic products, or stem cells). These differ from injectables such as hyaluronic acid, corticosteroids, or monoclonal antibodies which are not treatments produced from external sources and not directly produced by the body [14].

This chapter will discuss the intra-articular injectable biologic agents that are administered as adjuncts to ACLR in order to prevent or delay the development of PTOA.

13.2 Platelet-Rich Plasma (PRP)

Introduced in the field of orthopedics in the 1990s, platelet-rich plasma (PRP) is a blood product that has a concentration of platelets that is 3–15 times higher than normal blood concentration [15, 16]. Various methods are available to produce this orthobiologic, and they all differ in terms of platelet concentration, leukocyte levels, cost, and time necessary for production. PRP is shown to be an effective adjunct to various orthopedic procedures such as total knee arthrotomy (TKA) and meniscal repair. A meta-analysis conducted by Ma et al. showed that administration of PRP during TKA significantly decreases intraoperative blood loss [13 (switch old source w new)]. A systematic review by Muchedzi et al. also observed that PRP led to statistically lower post-TKA VAS pain scores [17]. However, both authors found no significant differences in terms of function, quality of life, or length of hospital stay [16, 17]. The chemokines and cytokines within the PRP solution are reported to accelerate healing by regulating local inflammation and by increasing the deposition of proteoglycans and type II collagen. The various contents in PRP also promote tissue repair through the stimulation of medicinal signaling cells, macrophages, fibroblasts, and chondrocytes. Furthermore, PRP promotes the proliferation, differentiation, communication, and chemotaxis of these cells [18].

PRP products can be classified in two categories according to their cellular concentrations: leukocyte-rich PRP (LR-PRP) and leukocyte-poor PRP (LP-PRP). LR-PRP and LP-PRP are differentiated by containing leukocyte concentrations higher or lower, respectively, compared to human baseline levels [19]. LR-PRP tends to increase the release of pro-inflammatory mediators such as tumor necrosis factor alpha (TNF-a), interferon gamma (INF-y), interleukin-6 (IL-6), and IL-1B. In contrast, the anti-inflammatory cytokines IL-4 and IL-10 increase in the presence of LP-PRP [20–22].

13.2.1 Preparation and Administration

During the preparation of the PRP product, blood is usually drawn from the antecubital vein using preferably an 18-gauge needle to prevent damage to the platelets. About 30–60 mL of whole blood needs to be extracted in order to obtain 3–6 mL of PRP [23]. Three methods are available to produce PRP using whole blood: single centrifugation, double spin centrifugation, and plasmapheresis followed by centrifugation. They can produce a solution with a platelet concentration up to 3, 8, and 15 times, respectively, higher than normal physiologic levels. However, the higher concentration obtained with plasmapheresis also comes with a much higher financial cost. Several administration methods such as graft coating are available. But intra-articular injections are the most common method of administration [16].

13.2.2 Outcomes

PRP has multiple potential applications in the context of ACLR. Most notably, PRP is used by surgeons as an adjunct to ACL surgery to help with graft-to-bone healing, graft maturation, as well as preventing postsurgical pain and PTOA. PRP is observed to be safe as only minor complications are reported. The main adverse events observed are moderate pain, swelling, and mild effusion lasting a few days at the target joint [24].

The effects on ACL graft healing is the most studied outcome regarding PRP use during ACLR. Graft maturation following ACLR was evaluated via MRI by Seijas et al. who revealed a beneficial role of PRP in stimulating faster and more complete graft remodeling within 12 months [25]. A review on ACLR by Andriolo et al. also highlighted how PRP can improve the graft's mechanical properties such as linear stiffness and tensile load [26]. However, Figueroa et al. conducted a randomized control trial (RCT) and observed no significant advantage in the use of PRP with ACLR in terms of graft maturation after 14 months [27].

There is no specific evidence regarding the use of PRP for PTOA in the setting of ACLR, as research efforts focus currently on osteoarthritis unrelated to trauma. Sampson et al. found that three sets of PRP injections at 4-week intervals improve KOOS and VAS scores significantly for up to 12 months in knee OA patients [28]. Moreover, Kon et al. reported that PRP injections at 21-day intervals yield significant improvements of IKDC and VAS scores in knee OA at 12-month follow-up. However, it was seen that the benefits decline after 6 months while remaining significantly better compared to baseline [29]. Hence, PRP is thought to improve patient-reported outcomes and pain scores on a short-term basis. One could extrapolate that these results would positively impact patient with PTOA in the setting of ACL reconstruction. However, more research is needed to support its use in patient with PTOA.

LR-PRP and LP-PRP have been compared to hyaluronic acid (HA) in the context of knee OA through a meta-analysis conducted by Riboh et al. that compiled six RCTs and two prospective studies. The group observed that only LP-PRP yields significantly superior WOMAC scores compared to HA and placebo [30]. Furthermore, an RCT conducted by Filardo et al. demonstrated that, when compared to HA, LR-PRP did not lead to statistically significant improvements in IKDC, KOOS, EQ-VAS, or Tegner scores. However, the PRP group experiences significantly higher rates of post-injection swelling and pain [31]. These studies consequently suggest that LP-PRP is more efficacious comparatively to LR-PRP in patients suffering from knee OA. These findings are probably related to the fact that LP-PRP promotes the release of anti-inflammatory factors contrarily to LR-PRP as previously mentioned [20–22].

When looking at the effects of single vs multiple injections, Vilchez-Cavazos et al., in a meta-analysis, found no difference between a single or multiple PRP injections in terms of pain relief based on the VAS score in knee OA. However, multiple injections provided improved function with higher WOMAC and IKDC scores [32]. While more research is needed on dosing regimen, this meta-analysis suggests that the number of injections may vary according to the goals of care (pain control vs function). Even if many studies concluded that PRP is advantageous, Altamura et al. demonstrated that PRP administration without surgical intervention leads to a poor rate of return to sport for patients with knee OA. In this prospective study, patient-reported outcomes (IKDC, VAS, Tegner scores) significantly improved, but only 48.9% of patients returned to the same level of physical activity after 24 months [33]. Thus, PRP injections alone might not be an ideal treatment for patient with PTOA who wish for a successful return to sport. In this light, in patients with PTOA, combining PRP with ACL reconstruction could provide an interesting avenue to potentially improve outcomes and return to sports rates.

There currently are no guidelines for the use of PRP with ACLR or PTOA. The American Academy of Orthopedic Surgeons (AAOS) does not find the evidence strong enough to recommend its use in patients with OA of the knee [24]. A major obstacle to the implementation of PRP treatment is the heterogeneity in PRP complex preparation and administration protocol. Different PRP products are available, with variations in preparation, platelet concentration and cellular content. This renders the standardization of PRP challenging. Despite the lack of clear evidence or guidelines, PRP is still widely used by clinicians for OA patients unresponsive to first-line treatment due to its simple preparation technique, low costs, noninvasiveness, and safety.

13.3 Hyaluronic Acid (HA)

In healthy articular cartilage, hyaluronic acid (HA) is one of many glycosaminoglycans found in the synovial fluid and extracellular matrix (ECM). This compound is naturally secreted by chondrocytes, fibroblasts, and synoviocytes. HA greatly contributes to joint maintenance by acting as a lubricant during slow movements and as a shock absorber during rapid or high-impact movements [34]. Additionally, HA can reduce general inflammation through the regulation of macrophage proliferation and phagocytosis, leukocyte chemotaxis, and cellular reactionary cascades [35]. HA is consequently hypothesized to have disease-modifying effects by acting upon the inflammatory processes of OA.

13.3.1 Preparation and Administration

HA preparations are most commonly derived from bacterial sources, such as streptococcus, as it is associated with the least amount of side effects and is the most cost-effective option currently available compared to animal sources (mostly avian) [36]. Thus, it is not classified as an orthobiologic. Several preparations of injectable HA are available for clinical use. An important difference is that HA products can be naturally derived and not cross-linked or artificially cross-linked to increase the molecular weight [37]. The fundamental differences between the various preparation methods are the molecular weight and the duration of treatment (number of injections). Migliore et al. observed that high molecular weight preparations (6,000,000–7,000,000 Da) result in a better increase in fluid retention in the joint. They consequently proposed that this can lead to a stronger anti-inflammatory effect [38].

13.3.2 Outcomes

HA has been approved by the FDA in 1997 and used, similarly to PRP, to promote graft repair, control pain and prevent further degeneration of the knee. For now, HA is only classified as a symptom-modifying agent for OA [34]. However, during the earliest uses of HA, the pain relief experienced by patients in studies lasted multiple months, which is much longer than the half-life of HA. Intra-articular HA is consequently hypothesized to have disease-modifying and preventative effects in the treatment of osteoarthritis and PTOA [39]. Despite this treatment's potential, there is a lack of studies for the prevention of PTOA in patients with ACL injuries.

Experiments in animal models exploring this hypothesis revealed that HA injections trigger multiple mechanisms potentially beneficial to the osteoarthritic knee. It was found in rabbit models that underwent ACLR that HA can improve tissue healing, angiogenesis, and production of cartilage matrix components [39]. HA injections in canine models of ACLR also seemed to prevent chondrocyte apoptosis and decrease the secretion of local inflammatory cytokines. Other common benefits observed in this animal model include reduced lymphocyte motility, inhibition of chondrodegradative enzymes, and significant improvements in gross morphology [39]. Similar experiments focused on the mechanisms of HA in humans with OA and found that these injections improve chondrocyte density and metabolism while reducing inflammation and edema [39]. Thus, it is believed that HA could promote the restauration of cartilage as well as reverse the inflammatory cartilaginous destruction present in OA and PTOA. However, the progression of PTOA seems unaffected by HA in ACL-deficient rat models. This suggests that proper reconstruction and stabilization of the knee is necessary for HA's therapeutic effect [40].

Currently, no studies have explored the use of intra-articular injections of HA to prevent PTOA following ACLR in humans. Due to this lack of studies, papers on the effects of HA on ACLR outcomes not specific to PTOA will be discussed. In an RCT, Huang et al. divided 120 patients into a control group that received saline and three other groups that received intra-articular HA at 4, 8, and 12 weeks post-ACLR, respectively. After 1 year, HA groups experienced significantly better range of motion, Lysholm scores, ambulation speed, and muscle peak torque compared to the saline control group. The patients receiving HA at 8 weeks following ACLR had significantly superior Lysholm scores compared to the other two HA groups [41]. Chau et al. conducted an RCT to compare in 32 patients the outcomes of ACLR with HA injections to ACLR alone. The greatest difference was found 2 days postoperatively with significantly better KOOS scores and diminished postoperative swelling in the HA group. However, the improvements in the two groups equalized after 2 weeks [42]. Moreover, Di Martino et al. performed an RCT involving 60 patients and demonstrated that a single injection HA administered 1 day after ACLR does not provide statistically significant clinical improvements compared to saline in terms of SF-36, IKDC, VAS, or Tegner scores [43]. Wang et al. also observed in OA patients over 2 years that HA knee injections every 6 months led to significantly better cartilage preservation on MRI compared to control patients that only used NSAIDs, analgesics, or physical therapy [44]. This cartilage preservation on MRI could justify the improvements of clinical scores seen in other trials and shows that HA might be associated with a change in the progression of OA.

The efficacy of high molecular weight (HMW) and low molecular weight (LMW) variants of HA in OA have been compared through multiple studies. A meta-analysis by Hummer et al. showed that HMW HA leads to pain relief that is statistically and clinically superior to LMW HA on the VAS pain scale [45]. Atamaz et al. have also demonstrated through their randomized study that HMW HA yields significantly better scores on the VAS pain scale and WOMAC functional scales compared to LMW HA [46]. Furthermore, Bahrami et al. have observed that a single injection of HMW HA has similar effects than 3 weekly injections of LMW HA in terms of WOMAC, Lequesne, and VAS scales [47]. Thus, it seems that HMW HA is the ideal alternative for patients suffering from knee OA as it seems to yield significantly better effects while being more convenient by requiring fewer doses. However, randomized trials by Lee et al. and Gigis et al. show no significant effect of molecular weight of HA on VAS or WOMAC scores [48, 49]. Shewale et al. also found that the use of HMW or LMW HA does not affect the likelihood of OA patients requiring eventual surgery [50]. Despite these results, the evidence still seems overall to be in favor of administering HMW HA for knee OA patients. Consequently, HMW HA might be a more interesting treatment option than LMW HA for PTOA as well.

HA was compared to other treatment modalities as well for the treatment of OA. Bannuru et al. carried out a meta-analysis compiling seven RCTs on knee OA and showed that HA has significantly superior pain relief compared to corticosteroids after 8 weeks post-injection. However, they observed that corticosteroids have good short-term effects and are more effective than HA in the first 2 weeks following injection [51]. Function restauration and stiffness were similar for both treatments. Furthermore, Wu et al. compiled 10 RCTs in a meta-analysis and reported significantly better results with PRP compared to HA when recording WOMAC, VAS, and IKDC scores in patients with OA [52]. Conversely, Filardo et al. conducted an RCT and observed no significant difference in IKDC, KOOS, EQ-VAS, or Tegner scores between OA patients treated with PRP or HA. But this team did find significantly higher rates of self-limited episodes of post-injection swelling and pain in the PRP group [31]. Lamo-Espinosa also conducted an RCT on patients with knee OA and found that intra-articular stem cell injections yielded better pain relief on the VAS score compared to HA injections after 12 months [53]. Moreover, a meta-analysis of RCTs performed by Miller et al. demonstrated slight knee pain and function improvements that are statistically superior with HA injections compared to NSAIDs in OA patients within 26 weeks. But even if the difference was statistically significant, it was not great enough to be clinically significant [54]. Thus, the evidence is still mitigated on whether or not HA is the optimal intra-articular treatment for OA or PTOA patients.

Pseudosepsis is a rare but important adverse condition occurring following HA injection and is also called severe acute inflammatory reaction (SAIR) [f]. It is characterized by (1) intra-articular knee infusion and pain with an acute onset 24–72 h following the HA injection, (2) more than one injection in the past, (3) absence of calcium pyrophosphate crystals and infectious agents in the synovial fluid, (4) elevated mononuclear cells (mostly macrophages) in the synovial fluid, and (5) requires treatment (nonsteroidal anti-inflammatories, intra-articular steroid injection, arthrocentesis) [55]. Pseudosepsis is hypothesized to be an immunologic reaction [56]. Cross-linked hyaluronic products have been shown through animal studies to lead to the development of significantly more serum antibodies in primates and rabbits compared to non-crosslinked HA [55, 57]. Consequently, the former has been proposed to be more strongly associated with pseudosepsis. However, a meta-analysis conducted by Bannuru et al. shows that there are no significant differences in the rates of complications, including pseudosepsis, between various HA products even if they differed in their cross-linkage, molecular weight, or source (bacterial fermentation or avian) [58].

Due to the debatable evidence, the American College of Rheumatology has no recommendations on the use of HA in OA, and the American Academy of Orthopedic Surgeons (AAOS) discourages the use of HA to treat knee OA [24]. But despite HA not being supported by these organizations, it is still widely used as HA injections are considered to be safe for the patient since adverse events are rare. Only 2–4% of patients in clinical trials reported transient inflammation at the joint [24]. Moreover, Miller et al. observed that 19.8% of OA patients using HA reported adverse events (local pain or inflammation, gastrointestinal symptoms, headache) compared to 29.0% of OA patients using NSAIDs [54]. Further research must be conducted to fully understand the scope of HA's physiological action on the osteoarthritic knee following ACLR, and whether different formulations, dosages, and timing of administration improve the patient's outcome.

13.4 Intra-Articular Corticosteroid Injection

Since the early 1950s, intraarticular corticosteroid injections are frequently used for multiple rheumatological conditions and OA [59]. In recent years, increasing expertise in biochemical analysis of articular fluids has led to the recognition that biochemical disturbances are the earliest indicators of osteoarthritis (OA) disease, especially in post-traumatic osteoarthritis (PTOA) [60]. These disturbances and pro-inflammatory responses are also found early in knee synovial fluid after ACL injury and are known to ignite the cascade leading to cartilage damage [61]. As a result, administration of corticosteroids has been believed to prevent OA from progressing to significant cartilage damage by reducing the inflammatory cascade following trauma [62].

13.4.1 Preparation and Administration

Different corticosteroid injections are available for use. The various options can be classified in two groups: particulate and non-particulate corticosteroids. Particulate corticosteroids are not water-soluble and consequently aggregate on the joint. This causes the drug to remain in synovial fluid and be constantly released over longer periods of time. As an example, methylprednisolone acetate, one of the most commonly injected steroids, has an effect that lasts around 7 days. On the other side, non-particulate steroids are water-soluble, meaning that they are cleared quickly from the joint and have shorter lasting effects. Corticosteroids are usually administered alongside a local anesthetic. Fluoroscopy and ultrasound can improve accuracy of the injection if needed [63]. Since these compounds are not naturally found in the human body, they are not considered as orthobiologics.

13.4.2 Outcomes

Corticosteroid intra-articular injections are frequently used in the context of acute or chronic inflammatory processes in order to alleviate pain while inhibiting the release of inflammatory mediators [24]. However, intra-surgical administration of corticosteroids is rare due to the significant increase in risk of postoperative infection in the knee [64]. Despite widespread use in patients with OA, very few studies have been published on the effects of corticosteroids for the prevention of PTOA in patients with ACL injuries. Lattermann et al. studied with an RCT the effect of corticosteroid administration within days following ACL rupture. It was found that C-telopeptide of type II collagen (CTX-II) is the only 1 out of 12 biochemical markers with a concentration significantly lower in the treatment group. It is a marker associated with collagen type II breakdown. Moreover, there were no differences between patient-reported outcomes (KOOS, IKDC, VAS pain, PCS) in placebo and intervention groups after 5 weeks [62]. Bellamy et al. carried out a meta-analysis compiling 27 trials to evaluate the efficacy of corticosteroid injections for OA and demonstrated that corticosteroids correlate with only small to moderate improvements to the WOMAC score until 6 weeks following the injection. They found no evidence of improvements after 13 weeks post-injection [65]. However, Raynauld et al. noted in their RCT significant improvements of knee pain and stiffness in the corticosteroid group based on the VAS and WOMAC scores [66]. Thus, despite showing limited effects on patient-reported scales, corticosteroid injections show promise by affecting markers of chondral degeneration. Further exploration is needed to determine if these biochemical changes can translate in decreased onset or severity of PTOA in the context of ACLR.

Commonly used corticosteroid agents for intra-articular injection include methylprednisolone acetate, triamcinolone hexacetonide, triamcinolone acetonide, and betamethasone. No corticosteroid has been shown to result in significantly superior outcomes compared to others [67–69]. However, these studies were not tailored specifically to patients with PTOA following ACL repair. Corticosteroids were also studied alongside HA products by Bannuru et al. who found corticosteroids to deliver significantly better short-term pain relief to OA patients over a 2-week period following injection compared to HA [51].

Side effects have been reported after the use of corticosteroid injections. Most commonly, reactive flares may occur 6–12 hours following the injection and resolve within 1–3 days [70]. An adverse event that is important to consider while using intra-articular corticosteroids is chondrotoxicity. Dragoo et al. observed within in vitro models that triamcinolone, betamethasone sodium phosphate, and betamethasone acetate lead to a significant decrease in chondrocyte viability after a single injection dose [71]. A systematic review by Wernecke et al. noted that corticosteroid toxicity is dose- and time-dependent, where beneficial effects can start to be overshadowed by negative outcomes with high doses and prolonged duration of treatment [72]. To further this point, McAlindon et al. conducted an RCT in which a 2-year regimen of intra-articular triamcinolone leads to significant cartilage loss without significant pain improvement compared to saline injections [73]. They consequently suggested that intra-articular corticosteroids may worsen osteoarthritic disease without any worthwhile symptomatic benefits. However, Raynauld et al. performed an RCT using triamcinolone as well and observed no significant cartilage loss compared to the saline control group after 2 years [66]. Thus, corticosteroids can be an appropriate and safe solution if low dosages and low treatment durations are prioritized.

Local anesthetics have also been found to have chondrotoxic effect that is dependent on the dose, duration, and type of local anesthetic used [74]. This is crucial to keep in mind because local anesthetics and corticosteroids have been observed to have a compounded deleterious effect on cartilage when used together. Jayaram et al. have shown through a systematic review that adding corticosteroids to local anesthetics significantly aggravated chondrotoxicity compared to local anesthetics alone within in vitro and in vivo models [75]. Consequently, the simultaneous administration of corticosteroids and local anesthetics in the knee should be avoided as much as possible.

The current evidence supporting the injection of intra-articular corticosteroids is weak. Consequently, AAOS finds the evidence to be inconclusive and has not issued recommendations on the subject [24]. More high-level studies with conclusive results are necessary to assess the clinical potential of corticosteroids in PTOA.

13.5 Medicinal Signaling Cells

Cartilage degradation present in conditions such as PTOA is a challenging condition to repair due partly to the inability of chondrocytes to self-regenerate. Consequently, medicinal signaling cells (MSCs) have been one of the most studied potential solutions for osteoarthritis. These cells spark interest because they could potentially induce cartilage repair and relieve symptoms through cytokine release, cell-to-cell interactions, and chondrocyte repopulation [76]. The injected stem cells are believed to reduce inflammation by releasing anti-inflammatory factors. This orthobiologic treatment also exerts an immunoregulatory effect that blocks T-cell function at the joint by inhibiting TNF-a and INF-y [76].

13.5.1 Preparation and Administration

Medicinal signaling cells that are used for treatment of OA can either be autologous to the patient or allogeneic. Allogeneic cells are the most used due to convenience and lower costs. Autologous cells might be safer and have lower risks of adverse reactions but the increased cost and necessity to harvest the cells invasively are important aspects to consider [77]. MSCs are harvested in various tissues such as bone marrow, adipose tissue, the spleen, synovial fluid, and lung tissue. Bone marrow is an optimal collection site for stem cells. Indeed, a higher concentration of bone-marrow-derived stem cells (BMSC) can be obtained in less volume compared to other sources such as adipose tissue and peripheral blood. Also, the bone marrow is also relatively easy to access for collection and can be offered as an outpatient procedure [78]. MSCs can be administered in various ways but they are most commonly injected in the target joint or surgically implanted using an artificial scaffold [78].

13.5.2 Outcomes

The use of medicinal signaling cells for tissue regeneration is still in its beginnings and shows promise in various fields ranging from autoimmune diseases to musculoskeletal conditions. MSCs are currently used as adjuncts to many knee surgeries such as ACLR and meniscectomy as they are believed to improve graft healing and prevent complications [76].

In rabbit models, when used in conjunction with ACLR, BMSCs administered at the tendon-bone junction were found to form a fibrocartilage that closely resembles that of normal ACL with improved biochemical properties [79]. However, it was subsequently shown in human adults that noncultivated BMSCs do not appear to accelerate graft-to-bone healing in ACLR when compared to a group of patients who underwent ACLR without additional BMSCs [80].

Stem cells have been studied by various groups for OA, but the efficiency of stem cells for PTOA post-ACLR seems to be a knowledge gap that still has not been explored. Garay-Mendoza et al. tested the efficacy of BMSCs against OA while using acetaminophen (a common OA treatment) as a control. The VAS scale improved significantly at the 1-week, 1-month, and 6-month follow-ups. The WOMAC score improved significantly at 1 month and 6 months. Thus, the use of BMSCs is associated with significant improvement in terms of knee pain and quality of life of OA patients. The procedure is also associated with a small rate of adverse events. In that study, only 1 out of 30 patients in the BMSC group reported joint swelling and pain [78].

Other sources of medicinal signaling cells have also been explored. Adipose tissue is an abundant source and easy to access. Cattaneo et al. evaluated the injection of micro-fragmented adipose tissue in osteoarthritic knees while undergoing corrective meniscectomy or chondral shaving. A steady and statistically significant improvement of all KOOS scores within 1, 3, 6, and 12 months follow-up was observed, with KOOS sport and quality of life being the most improved scores. 92% of the patients clinically improved and 100% of them were satisfied with the treatment [81]. No adverse events nor relevant complications were recorded. The result of the study pointed to adipose-derived medicinal signaling cells (ADMSCs) as a safe and beneficial adjuvant in the surgical treatment of degenerative knee chondropathies. While there are no studies on the use of ADMSCs in ACLR to treat PTOA, their effects in OA show that they might have potential benefits in PTOA and that more research is warranted on the subject.

Limited comparison has been done on the differences in clinical outcomes between bone-marrow and adipose-derived stem cells. Huang et al. showed with an in vitro study that BMSCs have greater chondrogenic differentiation and produce more cartilage compared to ADMSCs in a noninflammatory environment [82]. However, Pagani et al. demonstrated recently in vitro that ADMSCs have increased chondrogenic potential compared to BMSCs in an inflammatory setting such as in OA [83]. These results suggest that ADMSCs may be more effective at regenerating chondrocyte populations and regenerating cartilage in the setting of OA. Mautner et al. compared these two types of stem cells and found no statistical differences in EQOL, VAS, or KOOS score improvements in 110 OA patients [84].

The use of medicinal signaling cells has also been explored in conjunction with PRP for adult OA. Bastos et al. compared the use of BMSCs with and without PRP. The KOOS score improved significantly within 12 months. However, there were no statistically significant differences in KOOS score improvements between both groups at the 12-month endpoint [85]. This study suggests that BMSCs on their own are an effective treatment for knee OA and that the addition of PRP might not be warranted.

Centeno et al. have observed a prevalence of serious adverse events of 1.5% in patients treated with stem cells for degenerative joint conditions [86]. Neoplasm, neurologic symptoms, and vascular events were the most common serious adverse events seen by that group. Pain was the most common side effect overall and affected 29% of patients after treatment [86]. Even though stem cells seem safe when administered properly, AAOS has not issued guidelines or clear recommendations regarding the use of stem cells in the treatment of OA or PTOA [87]. Due to their large potential and relatively low risks, stem cells are still an important subject of interest that continues to be explored.

13.6 Amniotic Suspension Allografts (ASA)

The placenta, amniotic fluid, and amniotic membranes have long been a strong subject of interest in regenerative medicine. The field of orthopedics has in the past paid a particular interest towards the amnion, which is the inner layer of the fetal membrane. It is deprived of vasculature, neurons, or lymphatic ducts and exhibits low immunogenic potential [88]. This membrane contains collagen type I, III, V, and VI, making it a durable surface against mechanical stress [88]. The cells composing the amnion have also been suggested to play an anti-inflammatory and antimicrobial effect by affecting levels of various bioactive compounds. In the setting of OA, amniotic membranes IL-1 and IL-10, two cytokines that inhibit the progression of inflammation and cartilage damage [89, 90].

13.6.1 Preparation and Administration

Various amniotic products are available and are mostly in the form of allografts or injectables. The amnion is obtained from voluntary donors through elective uncomplicated cesarian section since vaginal delivery leads to exposure to bacteria from the vaginal flora [88]. The membrane is then treated with antifungals and antibiotics (against gram-positive and gram-negative) before being divided in smaller sections and stored. The tissue is either stored as cryopreserved human amniotic membrane (CHAM) or as dry human amniotic membrane (DHAM). The main difference between both preparation methods is that DHAM can be stored at room temperature contrarily to CHAM that needs to be stored at −80 °C [91]. Dehydrated human amniotic/chorionic membrane (DHACM) is another type of preserved membrane that contains both tissue from the chorion and amnion [91]. These tissues will then be manipulated in different ways depending on the desired product. Various injectable formulations, called amniotic suspension allografts (ASA), are available with different methods of production. One of injectables is micronized DHACM (μ-DHACM). It is produced via the PURION® process that allows the donated tissue to be devitalized and dehydrated while conserving bioactive compounds [92].

13.6.2 Outcomes

Amniotic products have been used for over a century in ophthalmology and dermatology [93]. The use of amniotic products in the field of orthopedics is still not a widespread practice. However, there is a growing amount of literature showing its potential in multiple situations such as prevention of postoperative scarring, plantar fasciitis, tendon repair, or OA [88, 94].

Animal models suggest that this treatment can potentially have a therapeutic effect on OA and slow down cartilage degradation. Marino-Martinez et al. have observed that rabbit knees treated with amniotic membrane injections had significantly healthier cartilage after 6 weeks compared to untreated knees. Untreated knees showed significantly greater rates of hypertrophy, cracks, cell clusters, and structural loss [95]. Raines et al. also showed in rats that one high dose injection led to significantly improved cartilage thickness and volume as well as decreased degradation after 4 weeks compared to untreated rats [94]. Similarly, Willett et al. noted after 3 weeks decreased proteoglycan loss and significantly less advanced cartilage erosion in rats injected with μ-DHACM [92]. Consequently, ASA might be an interesting product to slow down the rate of cartilage degradation and delay the onset of PTOA following ACLR.

The literature on the use of ASA to treat human knees affected with OA is very limited. Only two studies are available on the subject. Vines et al. conducted an open-label prospective feasibility study where six patients were administered a single dose of intra-articular ASA. The baseline KOOS, IKDC, and Single Assessment Numeric Evaluation (SANE) scores were 43.35, 41.70, and 51.25, respectively. After 12 months, these scores increased to 70.23, 64.40, and 85.80, respectively. However, due to the small sample size and lack of control group, no statistical analysis was made to establish any significant conclusions [96]. Farr et al. conducted an RCT comparing single injections of ASA, HA, and saline in OA patients. The group found that ASA had the lowest failure rate at 13.20% compared to 68.80% for HA and 75.00% for saline. The ASA group also showed significantly greater improvements on VAS, KOOS-pain, and KOOS-activities daily living scores compared to both other groups after 6 months. ASA patients also experienced significantly better improvements in KOOS-symptoms scores when compared to HA at 3 months and saline at 6 months [97]. Consequently, the RCT shows that ASA can be an effective treatment against OA that it potentially yields even better outcomes than other intra-articular injectables already in use.

Although very limited, the current literature on ASA suggests that they are relatively safe to use. Both previously mentioned studies on ASA did not identify any inflammatory reactions or other side effects following treatment administration. Amniotic cells have also been proposed to not possess tumorigenic potential. An in vivo study by Miki et al. noted that injected amniotic cells were not tumorigenic in immunodeficient mice [98]. Amniotic products are also thought to not be immunogenic due to not expressing human leukocyte antigen class II [99]. This was further supported by Akle et al. who documented that no immunologic reactions were provoked when amniotic cells were injected in the forearms of healthy human volunteers [100]. There is still a gap regarding the use of amniotic products in OA, resulting in no recommendations to be made by AAOS regarding their use. However, promising initial findings warrant further research for the potential development of guidelines for ASA. Further investigations could also show this treatment's potential for PTOA following ACLR.

13.7 Monoclonal Antibodies

Chronic pain is one of the most debilitating and important symptoms of OA, yet the least well studied [101]. The inability to treat this chronic pain in individuals with OA leads to reduced functional activity and a markedly diminished quality of life [102]. In OA, articular degradation produces classical inflammatory molecules such as prostaglandins, bradykinins, cytokines, and chemokines [103]. These molecules have been shown to trigger the nociceptive pathway [104–107]. Continuous stimulation from these molecules may lead to peripheral nociception sensitization, effectively decreasing the threshold of the stimuli required for activation [103]. The inhibition of these neural pathways has been proposed as means to diminish pain associated with OA and PTOA.

Nerve growth factor (NGF), a member of the neurotrophin family, was found to play a pivotal role in the development of sympathetic and sensory neurons responsible for nociception and temperature sensation [104, 106]. NGF has been found to be expressed in the subchondral bone of patients with OA, connecting NGF to osteoarthritic pain [105, 107]. Novel monoclonal antibodies are hypothesized to be able to alleviate chronic pain by preventing NGF from binding to its receptor, tropomyosin-related kinase-A (TrkA) [108].

13.7.1 Preparation and Administration

Some of the most notable examples of human anti-NGF antibodies are tanezumab and fasinumab. Ideal dosages and durations of treatments are still being determined through various trials. Intra-articular, subcutaneous, and intravenous injections are still being considered as potential administration possibilities for both antibodies [109].

13.7.2 Outcomes

Tanezumab is by far the most studied monoclonal antibody for the treatment of OA. An RCT by Schnitzer et al. with 696 patients compared subcutaneous administration of tanezumab using three groups of patients with OA (two groups with different antibody doses and one control group). The WOMAC pain subscale, WOMAC function, and global assessment were significantly better in the tanezumab groups compared to placebo. There was no significant difference found between the two tanezumab doses. The use of acetaminophen as rescue medication was similar in all three groups as well [110]. Furthermore, Kan et al. and Chen et al. performed two meta-analyses encompassing 14 RCTs in total and observed that tanezumab has significant positive effects on pain, function, and overall assessment when compared to placebo [111, 112].

Fasinumab has also only been weakly studied. Only one randomized control trial was available on its use for OA. An RCT by Dakin et al. divided 342 patients in four groups that got different doses of fasinumab and one placebo group. All fasinumab groups experienced significantly better pain, function, and overall assessment scores. The clinical improvements did not seem to share a clear relationship with the doses of antibody [113].

Alongside the positive clinical outcomes, OA trials found that monoclonal antibodies can cause serious adverse events. Hochberg et al. reported an association between an increase in tanezumab dosing and higher rates of osteonecrosis, which was even more evident when combined with NSAID use [114]. However, Chen et al. demonstrated no significant increase of serious adverse events in the tanezumab groups but rates of discontinued treatment due to paresthesia, arthralgia, hypesthesia and peripheral edema were significantly greater [111]. Kan et al. also showed that tanezumab significantly increased the rates of peripheral neuropathy [112]. Fasinumab is believed to potentially have similar side effects to tanezumab. The FDA considered the risks severe enough to place a temporary hold on the clinical use of NGF antibodies in 2010. The FDA had instituted another temporary restriction on their use in 2012 when it was observed that these antibodies caused damage to the sympathetic nervous system of certain animal models. Their use has been, however, reinstituted by the FDA with the implementation of additional measures to protect patients [109].

Although significant clinical benefits have been observed, monoclonal antibodies are still a novel treatment that has not been deeply explored. Consequently, no AAOS recommendation had been found regarding their use in OA or PTOA [115]. Moreover, no work has been done on the use of monoclonal antibodies as an adjunct to ACLR. Further studies are required to fully determine the clinical advantages and the risk of adverse effects associated with this treatment.

13.8 Conclusion

The use of injectable modalities in the treatment of knee PTOA after ACLR seems to be a promising approach to control symptoms and improve knee function by correcting the underlying pathological processes at the molecular level. Orthobiologics (PRP, amniotic products, MSCs) and other intra-articular treatments (HA, corticosteroids, monoclonal antibodies) spark the most interest in the literature. However, research on the use of these modalities in PTOA following ACLR is still limited, and their efficacy in this setting still has not been clearly demonstrated. Furthermore, there are currently no guidelines regarding the use of biologics in PTOA. The majority of the evidence relates to their use in the context of osteoarthritis, but their potential use in post-traumatic osteoarthritis is not a far leap. Despite a lack of clear guidelines or consensus on their clinical impact, orthobiologics and injectables are commonly used as adjuncts to multiple surgical procedures or chronic conditions in orthopedics. While intra-articular treatments are promising adjuncts for surgeons treating post-traumatic knee arthritis after ACLR, more studies are required to improve our understanding of their clinical benefits and applications.

Summary Table

Treatment type	Orthobiologic?	Preparation	Outcomes	Safety
Platelet-rich plasma (PRP)	Yes	– Blood centrifugation – Plasmapheresis (more expensive) – Two types: Leukocyte rich (LR) and leukocyte poor (LP) PRP	– Significant positive effects on pain and function – LP-PRP possibly more effective than LR-PRP for OA – Possibly not best option if return to sport is the treatment goal – No recommendations by AAOS	– Swelling, pain, effusion have been reported

Treatment type	Orthobiologic?	Preparation	Outcomes	Safety
Hyaluronic acid (HA)	No	– Bacterial fermentation – Animal sources (avian mostly) – Two types: High molecular weight (HMW) and low molecular weight (LMW) HA	– Significant positive effects on pain and function – HMW HA possibly more effective than LMW HA for OA – Possibly less efficient than PRP or MSCs – No recommendations by AAOS	– Low risk of pseudosepsis – Most common: Local pain or inflammation, gastrointestinal symptoms, headache
Corticosteroids	No	– Two types: Particulate and non-particulate	– Significant positive effects on pain and function – Possible better short-term pain relief than HA – No recommendations by AAOS	– Risk of infection if used intraoperatively – Risk of chondrotoxicity if high doses or long durations – Possible reactive flares
Medicinal Signalling cell (MSC)	Yes	– Allogeneic cells most commonly used – Bone marrow most common source. Adipose tissue also important source	– Significant positive effects on pain and function – Similar efficacy of bone marrow and adipose stem cells – No recommendations by AAOS	– Pain most common – Low risk of serious adverse event (neoplasm, neurologic, or vascular mostly)
Amniotic suspension allografts (ASA)	Yes	– Cells from amniotic membranes – Used as tissue allograft or ASA injectable	– Significant positive effects on pain and function – No recommendations by AAOS	– No adverse events recorded for now – Non-teratogenic or immunogenic
Monoclonal antibodies	No	– Various antibody options are available	– Significant positive effects on pain and function – No recommendations by AAOS	– Risk of osteonecrosis, paresthesia, arthralgia, hypesthesia and peripheral edema

References

1. Sanders TL, Maradit Kremers H, Bryan AJ, et al. Incidence of anterior cruciate ligament tears and reconstruction: a 21-year population-based study. Am J Sports Med. 2016;44(6):1502–7.
2. Musahl V, Karlsson J. Anterior cruciate ligament tear. N Engl J Med. 2019;380(24):2341–8.
3. Herzog MM, Marshall SW, Lund JL, Pate V, Mack CD, Spang JT. Incidence of anterior cruciate ligament reconstruction among adolescent females in the United States, 2002 through 2014. JAMA Pediatr. 2017;171(8):808–10.
4. Thomas AC, Hubbard-turner T, Wikstrom EA, Palmieri-smith RM. Epidemiology of posttraumatic osteoarthritis. J Athl Train. 2017;52(6):491–6.
5. Martel-pelletier J, Barr AJ, Cicuttini FM, et al. Osteoarthritis. Nat Rev Dis Primers. 2016;2:16072.
6. Carbone A, Rodeo S. Review of current understanding of post-traumatic osteoarthritis resulting from sports injuries. J Orthop Res. 2017;35(3):397–405.
7. Luc B, Gribble PA, Pietrosimone BG. Osteoarthritis prevalence following anterior cruciate ligament reconstruction: a systematic review and numbers-needed-to-treat analysis. J Athl Train. 2014;49(6):806–19.
8. Chen T, Wang S, Li Y, Ai C, Jiang F, Chen S. Radiographic osteoarthritis prevalence over ten years after anterior cruciate ligament reconstruction. Int J Sports Med. 2019;40(11):683–95.
9. Brophy RH, Gray BL, Nunley RM, Barrack RL, Clohisy JC. Total knee arthroplasty after previous knee surgery: expected interval and the effect on patient age. J Bone Joint Surg Am. 2014;96(10):801–5.
10. Brown TD, Johnston RC, Saltzman CL, Marsh JL, Buckwalter JA. Posttraumatic osteoarthritis: a first

estimate of incidence, prevalence, and burden of disease. J Orthop Trauma. 2006;20(10):739–44.
11. DeRogatis M, Anis HK, Sodhi N, et al. Non-operative treatment options for knee osteoarthritis. Ann Transl Med. 2019;7(Suppl 7):S245.
12. Rönn K, Reischl N, Gautier E, Jacobi M. Current surgical treatment of knee osteoarthritis. Arthritis. 2011;2011:454873.
13. Kramer WC, Hendricks KJ, Wang J. Pathogenetic mechanisms of posttraumatic osteoarthritis: opportunities for early intervention. Int J Clin Exp Med. 2011;4(4):285–98.
14. Dhillon MS, Behera P, Patel S, Shetty V. Orthobiologics and platelet rich plasma. Indian J Orthop. 2014;48(1):1–9.
15. Figueroa D, Figueroa F, Calvo R, Vaisman A, Ahumada X, Arellano S. Platelet-rich plasma use in anterior cruciate ligament surgery: systematic review of the literature. Arthroscopy. 2015;31(5):981–8.
16. Ma J, Sun J, Guo W, Li Z, Wang B, Wang W. The effect of platelet-rich plasma on reducing blood loss after total knee arthroplasty: a systematic review and meta-analysis. Medicine. 2017;96(26):e7262.
17. Muchedzi TA, Roberts SB. A systematic review of the effects of platelet rich plasma on outcomes for patients with knee osteoarthritis and following total knee arthroplasty. Surgeon. 2018;16(4):250–8.
18. Marmotti A, Rossi R, Castoldi F, Roveda E, Michielon G, Peretti GM. PRP and articular cartilage: a clinical update. Biomed Res Int. 2015;2015:542502.
19. Le ADK, Enweze L, DeBaun MR, Dragoo JL. Current clinical recommendations for use of platelet-rich plasma. Curr Rev Musculoskelet Med. 2018;11(4):624–34.
20. Sundman EA, Cole BJ, Fortier LA. Growth factor and catabolic cytokine concentrations are influenced by the cellular composition of platelet-rich plasma. Am J Sports Med. 2011;39(10):2135–40.
21. Civinini R. Growth factors in the treatment of early osteoarthritis. Clin Cases Miner Bone Metab. 2013;10(1):26–9.
22. Carballo CB, Nakagawa Y, Sekiya I, Rodeo SA. Basic science of articular cartilage. Clin Sports Med. 2017;36(3):413–25.
23. Foster TE, Puskas BL, Mandelbaum BR, Gerhardt MB, Rodeo SA. Platelet-rich plasma: from basic science to clinical applications. Am J Sports Med. 2009;37(11):2259–72.
24. Ayhan E, Kesmezacar H, Akgun I. Intraarticular injections (corticosteroid, hyaluronic acid, platelet rich plasma) for the knee osteoarthritis. World J Orthop. 2014;5(3):351–61.
25. Seijas R, Ares O, Catala J, Alvarez-Diaz P, Cusco X, Cugat R. Magnetic resonance imaging evaluation of patellar tendon graft remodelling after anterior cruciate ligament reconstruction with or without platelet-rich plasma. J Orthop Surg (Hong Kong). 2013;21:10–4.
26. Andriolo L, Di Matteo B, Kon E, Filardo G, Venieri G, Marcacci M. PRP augmentation for ACL reconstruction. Biomed Res Int. 2015;2015:371746.
27. Figueroa D, Melean P, Calvo R, Vaisman A, Zilleruelo N, Figueroa F, Villalón I. Magnetic resonance imaging evaluation of the integration and maturation of semitendinosus-gracilis graft in anterior cruciate ligament reconstruction using autologous platelet concentrate. Arthroscopy. 2010;26:1318–25.
28. Sampson S, Reed M, Silvers H, Meng M, Mandelbaum B. Injection of platelet-rich plasma in patients with primary and secondary knee osteoarthritis: a pilot study. Am J Phys Med Rehabil. 2010;89:961–9.
29. Kon E, Buda R, Filardo G, Di Martino A, Timoncini A, Cenacchi A, Fornasari PM, Giannini S, Marcacci M. Plateletrich plasma: intra-articular knee injections produced favorable results on degenerative cartilage lesions. Knee Surg Sports Traumatol Arthrosc. 2010;18:472–9.
30. Riboh JC, Saltzman BM, Yanke AB, Fortier L, Cole BJ. Effect of leukocyte concentration on the efficacy of platelet-rich plasma in the treatment of knee osteoarthritis. Am J Sports Med. 2016;44(3):792–800.
31. Filardo G, Di Matteo B, Di Martino A, et al. Platelet-rich plasma intra-articular knee injections show no superiority versus viscosupplementation: a randomized controlled trial. Am J Sports Med. 2015;43(7):1575–82.
32. Vilchez-Cavazos F, Millán-alanís JM, Blázquez-saldaña J, et al. Comparison of the clinical effectiveness of single versus multiple injections of platelet-rich plasma in the treatment of knee osteoarthritis: a systematic review and meta-analysis. Orthop J Sports Med. 2019;7(12):2325967119887116.
33. Altamura SA, Di Martino A, Andriolo L, et al. Platelet-rich plasma for sport-active patients with knee osteoarthritis: limited return to sport. Biomed Res Int. 2020;2020:8243865.
34. Axe JM, Snyder-mackler L, Axe MJ. The role of viscosupplementation. Sports Med Arthrosc Rev. 2013;21(1):18–22.
35. Tamoto K, Tada M, Shimada S, Nochi H, Mori Y. Effects of high-molecular-weight hyaluronates on the functions of Guinea pig polymorphonuclear leukocytes. Semin Arthritis Rheum. 1993;22(6 Suppl 1):4–8.
36. Zakeri A, Rasaee MJ, Pourzardosht N. Enhanced hyluronic acid production in Streptococcus zooepidemicus by over expressing HasA and molecular weight control with Niscin and glucose. Biotechnol Rep (Amst). 2017;16:65–70.
37. Hamburger MI, Lakhanpal S, Mooar PA, Oster D. Intra-articular hyaluronans: a review of product-specific safety profiles. Semin Arthritis Rheum. 2003;32(5):296–309.
38. Migliore A, Giovannangeli F, Granata M, Laganà B. Hylan g-f 20: review of its safety and efficacy in the management of joint pain in osteoarthritis.

Clin Med Insights Arthritis Musculoskelet Disord. 2010;3:55–68.

39. Goldberg VM, Buckwalter JA. Hyaluronans in the treatment of osteoarthritis of the knee: evidence for disease-modifying activity. Osteoarthr Cartil. 2005;13(3):216–24.
40. Teeple E, Elsaid KA, Jay GD, et al. Effects of supplemental intra-articular lubricin and hyaluronic acid on the progression of posttraumatic arthritis in the anterior cruciate ligament-deficient rat knee. Am J Sports Med. 2011;39(1):164–72.
41. Huang MH, Yang RC, Chou PH. Preliminary effects of hyaluronic acid on early rehabilitation of patients with isolated anterior cruciate ligament reconstruction. Clin J Sport Med. 2007;17(4):242–50.
42. Chau JY, Chan WL, Woo SB, et al. Hyaluronic acid instillation following arthroscopic anterior cruciate ligament reconstruction: a double-blinded, randomised controlled study. J Orthop Surg (Hong Kong). 2012;20(2):162–5.
43. Di Martino A, Tentoni F, Di Matteo B, et al. Early viscosupplementation after anterior cruciate ligament reconstruction: a randomized controlled trial. Am J Sports Med. 2016;44(10):2572–8.
44. Wang Y, Hall S, Hanna F, et al. Effects of hylan G-F 20 supplementation on cartilage preservation detected by magnetic resonance imaging in osteoarthritis of the knee: a two-year single-blind clinical trial. BMC Musculoskeletal Disord. 2011;12:195–203.
45. Hummer CD, Angst F, Ngai W, et al. High molecular weight Intraarticular hyaluronic acid for the treatment of knee osteoarthritis: a network meta-analysis. BMC Musculoskelet Disord. 2020;21(1):702.
46. Atamaz F, Kirazli Y, Akkoc Y. A comparison of two different intra-articular hyaluronan drugs and physical therapy in the management of knee osteoarthritis. Rheumatol Int. 2006;26(10):873–8.
47. Bahrami MH, Raeissadat SA, Cheraghi M, Rahimi-Dehgolan S, Ebrahimpour A. Efficacy of single high-molecular-weight versus triple low-molecular-weight hyaluronic acid intra-articular injection among knee osteoarthritis patients. BMC Musculoskelet Disord. 2020;21(1):550.
48. Lee P, Kim Y, Lim Y, et al. Comparison between high and low molecular weight hyaluronates in knee osteoarthritis patients: open-label, randomized, multicentre clinical trial. J Int Med Res. 2006;34(1):77–87.
49. Gigis I, Fotiadis E, Nenopoulos A, Tsitas K, Hatzokos I. Comparison of two different molecular weight intra-articular injections of hyaluronic acid for the treatment of knee osteoarthritis. Hippokratia. 2016;20(1):26–31.
50. Shewale AR, Barnes CL, Fischbach LA, Ounpraseuth ST, Painter JT, Martin BC. Comparison of low-, moderate-, and high-molecular-weight hyaluronic acid injections in delaying time to knee surgery. J Arthroplast. 2017;32(10):2952–7.e21.
51. Bannuru RR, Natov NS, Obadan IE, Price LL, Schmid CH, Mcalindon TE. Therapeutic trajectory of hyaluronic acid versus corticosteroids in the treatment of knee osteoarthritis: a systematic review and meta-analysis. Arthritis Rheum. 2009;61(12):1704–11.
52. Wu Q, Luo X, Xiong Y, et al. Platelet-rich plasma versus hyaluronic acid in knee osteoarthritis: a meta-analysis with the consistent ratio of injection. J Orthop Surg (Hong Kong). 2020;28(1):2309499019887660.
53. Lamo-Espinosa JM, Mora G, Blanco JF, et al. Intra-articular injection of two different doses of autologous bone marrow mesenchymal stem cells versus hyaluronic acid in the treatment of knee osteoarthritis: multicenter randomized controlled clinical trial (phase I/II). J Transl Med. 2016;14(1):246.
54. Miller LE, Fredericson M, Altman RD. Hyaluronic acid injections or Oral nonsteroidal anti-inflammatory drugs for knee osteoarthritis: systematic review and meta-analysis of randomized trials. Orthop J Sports Med. 2020;8(1):2325967119897909.
55. Goldberg VM, Coutts RD. Pseudoseptic reactions to hylan viscosupplementation: diagnosis and treatment. Clin Orthop Relat Res. 2004;419:30–137.
56. Aydin M, Arikan M, Togral G, Varis O, Aydin G. Viscosupplementation of the knee: three cases of acute Pseudoseptic arthritis with painful and irritating complications and a literature review. Eur J Rheumatol. 2017;4(1):59–62.
57. Bucher W, Otto T, Hamburger MI. Differentiation of hyaluronate products by qualitative differences in their immunogenicity in rabbits: possible mechanism for product-specific severe adverse reactions? Arthritis Rheum. 2002;46:2543–4.
58. Bannuru RR, Osani M, Vaysbrot EE, McAlindon TE. Comparative safety profile of hyaluronic acid products for knee osteoarthritis: a systematic review and network meta-analysis. Osteoarthr Cartil. 2016;24(12):2022–41.
59. Benedek TG. History of the development of corticosteroid therapy. Clin Exp Rheumatol. 2011;29(5 Suppl 68):S5–S12.
60. Lohmander LS, Englund PM, Dahl LL, Roos EM. The long-term consequence of anterior cruciate ligament and meniscus injuries: osteoarthritis. Am J Sports Med. 2007;35(10):1756–69.
61. Sieker JT, Ayturk UM, Proffen BL, Weissenberger MH, Kiapour AM, Murray MM. Immediate administration of intraarticular triamcinolone acetonide after joint injury modulates molecular outcomes associated with early synovitis. Arthritis Rheumatol. 2016;68(7):1637–47.
62. Lattermann C, Jacobs CA, Proffitt Bunnell M, et al. A multicenter study of early anti-inflammatory treatment in patients with acute anterior cruciate ligament tear. Am J Sports Med. 2017;45(2):325–33.

63. Yaftali NA, Weber K. Corticosteroids and hyaluronic acid injections. Clin Sports Med. 2019;38(1):1–15.
64. Cancienne JM, Gwathmey FW, Werner BC. Intraoperative corticosteroid injection at the time of knee arthroscopy is associated with increased postoperative infection rates in a large medicare population. Arthroscopy. 2016;32(1):90–5.
65. Bellamy N, Campbell J, Robinson V, Gee T, Bourne R, Wells G. Intraarticular corticosteroid for treatment of osteoarthritis of the knee. Cochrane Database Syst Rev. 2005;2:CD005328.
66. Raynauld JP, Buckland-wright C, Ward R, et al. Safety and efficacy of long-term intraarticular steroid injections in osteoarthritis of the knee: a randomized, double-blind, placebo-controlled trial. Arthritis Rheum. 2003;48(2):370–7.
67. Bain LS, Balch HW, Wetherly JM, Yeadon A. Intraarticular triamcinolone hexacetonide: double-blind comparison with methylprednisolone. Br J Clin Pract. 1972;26(12):559–61.
68. Pyne D, Ioannou Y, Mootoo R, Bhanji A. Intra-articular steroids in knee osteoarthritis: a comparative study of triamcinolone hexacetonide and methylprednisolone acetate. Clin Rheumatol. 2004;23(2):116–20.
69. Valtonen EJ. Clinical comparison of triamcinolone-hexacetonide and betamethasone in the treatment of osteoarthrosis of the knee-joint. Scand J Rheumatol Suppl. 1981;41:1–7.
70. Gerwin N, Hops C, Lucke A. Intraarticular drug delivery in osteoarthritis. Adv Drug Deliv Rev. 2006;58(2):226–42.
71. Dragoo JL, Danial CM, Braun HJ, Pouliot MA, Kim HJ. The chondrotoxicity of single-dose corticosteroids. Knee Surg Sports Traumatol Arthrosc. 2012;20(9):1809–14.
72. Wernecke C, Braun HJ, Dragoo JL. The effect of intra-articular corticosteroids on articular cartilage: a systematic review. Orthop J Sports Med. 2015;3(5):2325967115581163.
73. McAlindon TE, Lavalley MP, Harvey WF, et al. Effect of intra-articular triamcinolone vs saline on knee cartilage volume and pain in patients with knee osteoarthritis: a randomized clinical trial. JAMA. 2017;317(19):1967–75.
74. Kreuz PC, Steinwachs M, Angele P. Single-dose local anesthetics exhibit a type-, dose-, and time-dependent chondrotoxic effect on chondrocytes and cartilage: a systematic review of the current literature. Knee Surg Sports Traumatol Arthrosc. 2018;26(3):819–30.
75. Jayaram P, Kennedy DJ, Yeh P, Dragoo J. Chondrotoxic effects of local anesthetics on human knee articular cartilage: a systematic review. J Inj Funct Rehabilit. 2019;11(4):379–400.
76. Gomez-salazar M, Gonzalez-galofre ZN, Casamitjana J, Crisan M, James AW, Péault B. Five decades later, are mesenchymal stem cells still relevant? Front Bioeng Biotechnol. 2020;8:148.
77. Mautner K, Carr D, Whitley J, Bowers R. Allogeneic versus autologous injectable mesenchymal stem cells for knee osteoarthritis: review and current status. Tech Orthop. 2019;34(4):244–56.
78. Garay-Mendoza D, Villarreal-martínez L, Garza-bedolla A, et al. The effect of intra-articular injection of autologous bone marrow stem cells on pain and knee function in patients with osteoarthritis. Int J Rheum Dis. 2018;21(1):140–7.
79. Lim JK, Hui J, Li L, Thambyah A, Goh J, Lee EH. Enhancement of tendon graft osteointegration using mesenchymal stem cells in a rabbit model of anterior cruciate ligament reconstruction. Arthroscopy. 2004;20(9):899–910.
80. Silva A, Sampaio R, Fernandes R, Pinto E. Is there a role for adult non-cultivated bone marrow stem cells in ACL reconstruction? Knee Surg Sports Traumatol Arthrosc. 2014;22(1):66–71.
81. Cattaneo G, De Caro A, Napoli F, Chiapale D, Trada P, Camera A. Micro-fragmented adipose tissue injection associated with arthroscopic procedures in patients with symptomatic knee osteoarthritis. BMC Musculoskelet Disord. 2018;19(1):176.
82. Huang JI, Kazmi N, Durbhakula MM, et al. Chondrogenic potential of progenitor cells derived from human bone marrow and adipose tissue: a patient-matched comparison. J Orthop Res. 2005;23:1383–9.
83. Pagani S, Borsari V, Veronesi F, et al. Increased chondrogenic potential of mesenchymal cells from adipose tissue versus bone marrow-derived cells in osteoarthritic in vitro models. J Cell Physiol. 2017;232:1478–88.
84. Mautner K, Bowers R, Easley K, Fausel Z, Robinson R. Functional outcomes following microfragmented adipose tissue versus bone marrow aspirate concentrate injections for symptomatic knee osteoarthritis. Stem Cells Transl Med. 2019;8(11):1149–56.
85. Bastos R, Mathias M, Andrade R, et al. Intra-articular injections of expanded mesenchymal stem cells with and without addition of platelet-rich plasma are safe and effective for knee osteoarthritis. Knee Surg Sports Traumatol Arthrosc. 2018;26(11):3342–50.
86. Centeno CJ, Al-sayegh H, Freeman MD, Smith J, Murrell WD, Bubnov R. A multi-center analysis of adverse events among two thousand, three hundred and seventy-two adult patients undergoing adult autologous stem cell therapy for orthopaedic conditions. Int Orthop. 2016;40(8):1755–65.
87. Chu CR, Rodeo S, Bhutani N, et al. Optimizing clinical use of biologics in orthopaedic surgery: consensus recommendations from the 2018 AAOS/NIH U-13 conference. J Am Acad Orthop Surg. 2019;27(2):e50–63.
88. Huddleston HP, Cohn MR, Haunschild ED, Wong SE, Farr J, Yanke AB. Amniotic product treatments: clinical and basic science evidence. Curr Rev Musculoskelet Med. 2020;13(2):148–54.

89. Hao Y, Ma DH-K, Hwang DG, Kim W-S, Zhang F. Identification of antiangiogenic and antiinflammatory proteins in human amniotic membrane. Cornea. 2000;19(3):348–52.
90. Koh JW, Shin YJ, Oh JY, et al. The expression of timps in cryo-preserved and freeze-dried amniotic membrane. Curr Eye Res. 2007;32(7–8):611–6.
91. Malhotra C, Jain AK. Human amniotic membrane transplantation: different modalities of its use in ophthalmology. WJT. 2014;4(2):111.
92. Willett NJ, Thote T, Lin AS, et al. Intra-articular injection of micronized dehydrated human amnion/chorion membrane attenuates osteoarthritis development. Arthritis Res Ther. 2014;16(1):R47.
93. Heckmann N, Auran R, Mirzayan R. Application of amniotic tissue in orthopedic surgery. Am J Orthop (Belle Mead NJ). 2016;45(7):E421–5.
94. Raines AL, Shih M-S, Chua L, Su C-W, Tseng SCG, O'Connell J. Efficacy of particulate amniotic membrane and umbilical cord tissues in attenuating cartilage destruction in an osteoarthritis model. Tissue Eng Part A. 2017;23(1–2):12–9.
95. Marino-Martinez I, Martinez-Castro A, Pena-Martinez V, et al. Human amniotic membrane intra-articular injection prevents cartilage damage in an osteoarthritis model. Exp Ther Med. 2019;17(1):11–6.
96. Vines J, Aliprantis A, Gomoll A, Farr J. Cryopreserved amniotic suspension for the treatment of knee osteoarthritis. J Knee Surg. 2015;29(06):443–50.
97. Farr J, Gomoll AH, Yanke AB, Strauss EJ, Mowry KC, ASA Study Group. A randomized controlled single-blind study demonstrating superiority of amniotic suspension allograft injection over hyaluronic acid and saline control for modification of knee osteoarthritis symptoms. J Knee Surg. 2019;32(11):1143–54.
98. Miki T. Amnion-derived stem cells: in quest of clinical applications. Stem Cell Res Ther. 2011;2(3):25.
99. Miki T, Strom SC. Amnion-derived pluripotent/multipotent stem cells. Stem Cell Rev. 2006;2(2):133–41.
100. Akle CA, Welsh KI, Adinolfi M, Leibowitz S, Mccoll I. Immunogenicity of human amniotic epithelial cells after transplantation into volunteers. Lancet. 1981;318(8254):1003–5.
101. Goldring MB. The link between structural damage and pain in a genetic model of osteoarthritis and intervertebral disc degeneration: a joint misadventure. Arthritis Rheum. 2009;60(9):2550–2.
102. Centers for Disease Control and Prevention (CDC). Prevalence of disabilities and associated health conditions among adults--United States, 1999. MMWR Morb Mortal Wkly Rep. 2001;50(7):120–5.
103. Miller RE, Tran PB, Obeidat AM, et al. The role of peripheral nociceptive neurons in the pathophysiology of osteoarthritis pain. Curr Osteoporos Rep. 2015;13(5):318–26.
104. Sokolove J, Lepus CM. Role of inflammation in the pathogenesis of osteoarthritis: latest findings and interpretations. Ther Adv Musculoskelet Dis. 2013;5(2):77–94.
105. Miller RJ, Jung H, Bhangoo SK, White FA. Cytokine and chemokine regulation of sensory neuron function. Handb Exp Pharmacol. 2009;194:417–49.
106. Levi-montalcini R. The nerve growth factor 35 years later. Science. 1987;237(4819):1154–62.
107. Walsh DA, Mcwilliams DF, Turley MJ, et al. Angiogenesis and nerve growth factor at the osteochondral junction in rheumatoid arthritis and osteoarthritis. Rheumatology (Oxford). 2010;49(10):1852–61.
108. Enomoto M, Mantyh PW, Murrell J, Innes JF, Lascelles BDX. Anti-nerve growth factor monoclonal antibodies for the control of pain in dogs and cats. Vet Rec. 2019;184(1):23.
109. Jayabalan P, Schnitzer TJ. Tanezumab in the treatment of chronic musculoskeletal conditions. Expert Opin Biol Ther. 2017;17(2):245–54.
110. Schnitzer TJ, Easton R, Pang S, et al. Effect of tanezumab on joint pain, physical function, and patient global assessment of osteoarthritis among patients with osteoarthritis of the hip or knee: a randomized clinical trial. JAMA. 2019;322(1):37–48.
111. Chen J, Li J, Li R, et al. Efficacy and safety of tanezumab on osteoarthritis knee and hip pains: a meta-analysis of randomized controlled trials. Pain Med. 2017;18(2):374–85.
112. Kan SL, Li Y, Ning GZ, et al. Tanezumab for patients with osteoarthritis of the knee: a meta-analysis. PLoS One. 2016;11(6):e0157105.
113. Dakin P, Dimartino SJ, Gao H, et al. The efficacy, tolerability, and joint safety of fasinumab in osteoarthritis pain: a phase IIb/III double-blind, placebo-controlled. Randomiz Clin Trial Arthritis Rheumatol. 2019;71(11):1824–34.
114. Hochberg MC. Serious joint-related adverse events in randomized controlled trials of anti-nerve growth factor monoclonal antibodies. Osteoarthr Cartil. 2015;23(Suppl 1):S18–21.
115. Brown GA. AAOS clinical practice guideline: treatment of osteoarthritis of the knee: evidence-based guideline, 2nd edition. J Am Acad Orthop Surg. 2013;21(9):577–9.

Assessment of the Multiligament Knee

14

Marcel Betsch and Daniel B. Whelan

14.1 Physical Examination

14.1.1 Acute Assessment of Multiligament Knee

Multiligament knee injuries (MLKI), which are defined as a tear of at least two of the four major ligament structures of the knee, can lead to significant morbidity [1]. These types of injuries require an extensive force, and they are frequently associated with a knee dislocation or subluxation of the knee. Knee dislocations often result in a MLKI; however, not all MLKI are knee dislocations. Rates of knee dislocations in the current literature range from 0.001 to 0.013% per year; however, it is believed that the actual incidence is higher due to spontaneous knee reduction and missed injuries [2].

Because of the nature of MLKI, a prompt examination of injured patients is key to guide proper treatment and to rule out associated serious neurovascular injuries, joint malposition or open dislocations. Often these patients suffer from multisystem traumatic injuries that should be triaged using the Advanced Trauma and Life Support protocol [2]. Initial treatment of patients with knee dislocations, if necessary, should include resuscitation and stabilization of their vital signs followed by the assessment of limb viability. In cases of a grossly dislocated knee, it is necessary to reduce the knee immediately with clear documentation of the neurovascular status pre- and post-reduction. After successful reduction of the dislocated knee, it should be immobilized to keep the knee reduced, maintain neurovascular function and decrease swelling. Thereafter, an examination of the knee should be conducted.

MLKI comprise of a wide range of ligamentous and intra-articular injury patterns, as well as severe injuries to the neurovascular structures of the affected limb. Because of the complexity of these and associated injuries, a standardized and methodical approach to their assessment is recommended including inspection, palpation, range of motion, neurovascular exam and clinical tests.

14.1.1.1 Inspection

Every inspection should include a visual examination of the affected leg to rule out deformity, malalignment, evidence of active bleeding, open injuries, soft tissue swelling, ecchymosis, skin mottling or blisters [3]. MLKI due to low-energy trauma, for example, during sports activities, may be overlooked since they usually result in less soft tissue injury than high-energy traumas

M. Betsch (✉)
Department of Orthopaedics and Trauma Surgery, University Hospital Mannheim, Medical Faculty University Heidelberg, Mannheim, Germany

D. B. Whelan
Division of Orthopaedic Surgery, University of Toronto, St. Michael's Hospital, Toronto, ON, Canada
e-mail: Daniel.Whelan@unityhealth.to

N. Nakamura et al. (eds.), *Advances in Knee Ligament and Knee Preservation Surgery*,
https://doi.org/10.1007/978-3-030-84748-7_14

and may, therefore, present with a relatively benign appearing limb. Even more challenging is the fact that approximately 50% of the dislocated knees spontaneously reduce prior to their presentation in the emergency room [4, 5]. In cases where the joint capsule is damaged, there may not be a joint effusion present during initial assessment, rather than synovial fluid and blood leaking into the adjacent soft tissues [3, 6]. Additionally, in patients with a high BMI, the deformity and swelling may be obscured by excess tissue around the knee [6]. In cases with a posterior knee dislocation, oftentimes bruising and hematoma can be found over the anterior tibia due to a dashboard-type injury mechanism, which is frequently associated with a posterior cruciate ligament (PCL) and posterolateral corner (PLC) injury [7]. In patients with an open knee dislocation, which occurs with an incidence of 5–30%, urgent surgical treatment with irrigation and debridement is required to prevent infection and further complications [8, 9]. Open knee dislocations may result in a worse outcome because of the secondary injuries to the soft tissue enveloping the knee, which often require open or staged reconstruction of knee structures. Approximately 4% of knee dislocations are irreducible, which was first described in 1906 by Ruppanner [10]. The majority of these cases are posterolateral and lateral dislocations with entrapment of medial knee structures [11]. The "dimple sign" named by Reckling and Peltier is one of the hallmark signs of a posterolateral knee dislocation associated with irreducibility of a knee dislocation [12]. This clinical sign is presented by the medial femoral condyle buttonholing through the medial capsule so that soft tissues and skin become trapped. Different medial knee structures can invaginate the joint causing this sign, such as the MCL, vastus medialis, adductor magnus tendon and, therefore, oftentimes open reduction is required to successfully reduce the knee joint (Fig. 14.1a, b).

14.1.1.2 Palpation

After the inspection of the injured limb, we recommend palpation of all bony landmarks and soft tissues around the knee. The examiner should palpate for any swelling, gaps and points of tenderness in order to identify injuries. Focus should be laid on the palpation of the fibular head, since tenderness here can indicate an avulsion of the lateral collateral ligament (LCL) with a respective injury to the lateral knee structures. Crepitus and tenderness over the medial side of the knee on the other hand may indicate a medial-sided

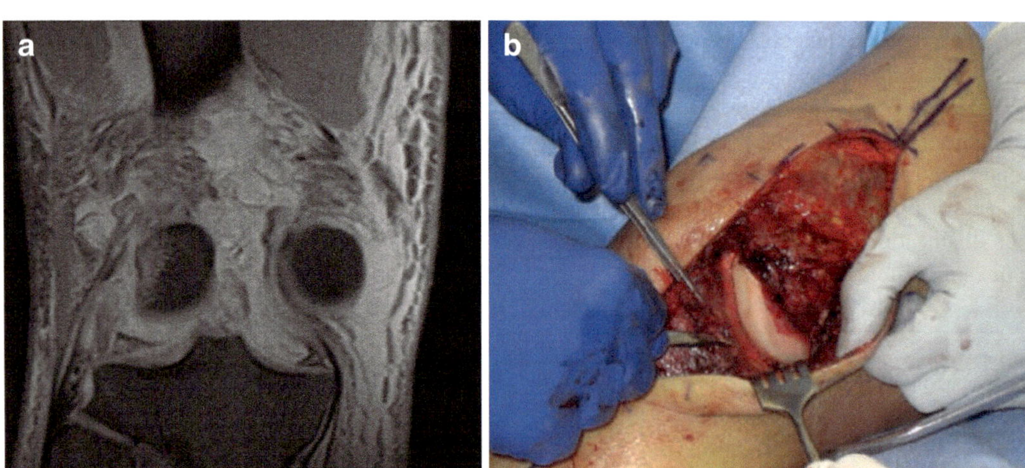

Fig. 14.1 (**a**) Coronal MRI and (**b**) intraoperative photo demonstrating "buttonholing" of the medial femoral condyle through the medial capsule with incarceration of the MCL and medial capsule in the joint preventing reduction. (From C. Fanelli (Ed) The Multiple Ligament Injured Knee, Springer, 2019. With permission)

tibial plateau fracture, which can be associated with a knee dislocation. The position of the patella after a knee dislocation should also be examined, since knee dislocations can also be associated with a patellar dislocation. An anterior knee dislocation with an associated ACL injury can be diagnosed by the prominence of the femoral condyles posteriorly. On the other hand, a "posterior sag" sign, which is described by the posterior sag of the tibia which diminishes the contour of the tibial tubercle, is associated with PCL injuries and should create awareness for neurovascular injuries. A varus recurvatum found in full knee extension after knee dislocation is a common finding with posterolateral corner injuries. Injuries to the knee extensor mechanism with a knee dislocation are rare; however, if missed they can lead to unfavourable outcome. Therefore, the assessment and palpation of the extensor mechanism should be included in the clinical examination of MLKI. It can be difficult or impossible to evaluate range of motion of the acutely injured knee due to associated pain. After a MLKI, it can take weeks of intensive physiotherapy to regain knee range of motion.

14.1.2 Vascular Assessment

Failure to diagnose a vascular injury in the setting of a MLKI can result in severe consequences for the patient, including limb amputation. Incidence rates of vascular injury in patients with a knee dislocation vary between 4.8 and 65% [13, 14]. Examiners should be concerned for possible vascular injuries especially in high-energy traumas. A recent systematic review showed that 80% of patients with a vascular injury due to a knee dislocation required surgery and 12% ended in an amputation of the affected limb [14]. Every vascular exam should start by inspecting the leg for capillary refill, warmth, skin colour, and actively bleeding wounds. The pulses of the dorsalis pedis and posterior tibial artery should be palpated, documented and compared to the contralateral limb. However, the examiner should be aware that the presence of palpable pulses does not rule out a vascular injury [15]. Especially, because there is evidence that shows that non-occlusive intimal tears of the popliteal artery can occur with a normal initial presentation, they can go on to cause an occlusive thrombus 48–72 h after the time of injury [16]. In cases of a knee dislocation with absent or abnormal pulses, we recommend to perform a closed reduction followed by a re-evaluation of the vascular status. The direction of the knee dislocation may indicate the mechanism of a possible vascular injury. An anterior dislocation of the knee joint may lead to overstretching of the popliteal artery, which is due to the anatomical tethering of the artery at the adductor hiatus and at the entrance of the gastrocnemius-soleus complex. Posterior dislocations can lead to a contusion of the artery with resulting intimal damage by the posterior tibial plateau. Previously, it was shown that popliteal artery injuries are most commonly caused by a posterior knee dislocation, occurring in up to 44% of the cases [17]. In patients that present with a "reduced" knee after a dislocation and bi-cruciate ligament injury, an arterial injury should be suspected with high incidence as in a frank knee dislocation with a tear of both cruciate ligaments [18]. The ankle-brachial index (ABI), which is performed with a Doppler probe and a blood pressure cuff, is recommended as a screening tool if ANY physical sign of vascular injury is present at ANY time during the assessment. For this test, the ratio between the systolic blood pressure of the affected limb and of the arm is calculated. Uncompromised arterial flow is defined by an ABI >0.9 [15, 19]. However, these values can be falsely inflated in patients with peripheral arterial disease [20]. A study by Mills et al. showed that the ABI can be used to identify vascular injuries that require surgical intervention after knee dislocation with 100% sensitivity and specificity [15]. However, in the setting of an inconclusive or abnormal ABI, we strongly recommend that a computed tomography (CT) angiogram should be performed followed by a vascular surgery consultation. This is of great importance, because the lack of limb perfusion for longer than 6–8 h increases the risk of amputation [17].

14.1.3 Neurologic Assessment

Nerve injuries occur in up to 25% of all knee dislocations, which was previously confirmed in a systematic review by Medina et al. [21]. Most commonly, the peroneal nerve (14–26% of knee dislocations), rather than the tibial nerve, is involved in MLKI [22]. The higher frequency of peroneal nerve injuries in knee dislocations is likely due to the anatomic location of the nerve, which passes around the proximal fibula. Of importance is also the direction of the knee dislocation, because lateral and posterolateral dislocations increase the likelihood of sustaining a common peroneal nerve injury by stretching it (45% incidence) [6, 23–25]. In the setting of MLKI, a thorough and focused neurological examination and documentation, with testing of sensibility and motor function of the whole limb is recommended. However, this might be difficult in an acute setting due to pain. Sensory assessment in all nerve distributions of the lower leg and foot should be performed. Strength of all muscles innervated by the peroneal and tibial nerves should be evaluated, which produces movements related to ankle dorsiflexion and plantar flexion, foot inversion and great toe extension. Injury to the peroneal nerve will result in loss of sensation at the dorsum and lateral aspects of the foot, including the first web space as well as drop foot and an altered gait pattern. The severity of nerve injuries can vary from neuropraxia to complete nerve disruption, and findings from the neurological exam can help prognosticate the chance of recovery. Factors that are associated with peroneal nerve function recovery are younger age and the absence of a fracture [26]. Eighty-seven percent of partial common peroneal nerve injuries fully recovered compared to 38% of complete nerve injuries [27]. We recommend repeat neurologic examination with complete documentation during every step of MLKI treatment and in particular when a closed reduction of a knee dislocation is necessary to rule out iatrogenic nerve injury.

14.1.4 Clinical Exam and Special Tests

In every patient an exam of knee stability and ligamentous structures should be performed. However, assessment of the injured knee can be challenging in the acute setting because of pain, hematoma, or associated injuries. Despite these challenges, we recommend an examination of every knee dislocation with thorough documentation of gross knee instability and laxity, which can indicate disruption of multiple ligaments. A more definitive examination of the injured knee can be performed once swelling and pain have decreased. A comprehensive evaluation of the knee should include the following structures: ACL, PCL, MCL, LCL, PLC and posteromedial corner (PMC).

14.1.5 Anterior Cruciate Ligament

The Lachman test is considered the "gold standard" for assessing anterior knee stability because of its high sensitivity [28]. This test is performed with the knee in 30° of flexion and the amount of anterior translation is graded and compared to the uninjured knee. The Lachman test can be false positive in patients with a PCL injury and an intact ACL, because of the posterior tibial drop back that occurs with a PCL injury. The examiner should also pay close attention to keep the leg in neutral rotation during this test, because excessive rotation may result in false-positive findings. Furthermore, this test may be difficult to interpret in the setting of MLKI. The anterior drawer test can also be used to assess for ACL injuries; however, this test has lower sensitivity and specificity in patients with concomitant knee injuries, in swollen knees, and in cases with protective hamstring muscle spasms. The advantage of this test is that it can be combined with internal and external rotation to also check for rotational instability of the PCL.

14.1.6 Posterior Cruciate Ligament

The integrity of the PCL can be assessed by the posterior sag and posterior drawer test. The posterior sag test is a passive test, where the hips and knees are kept at 90° of flexion, and a posterior translation or sag of the proximal tibia relative to the distal femur is noted (Fig. 14.2). For the posterior drawer test, we position the patient in 45° of hip and 80° of knee flexion with the feet being flat on the examining table. Care must be taken to keep the leg in neutral rotation to decrease the risk of false-positive findings. The posterior drawer test is performed by directing a posterior force on the tibia while quantitatively and qualitatively assessing the posterior translation of the tibia in relation to the femur. A PCL tear can be presented by excessive posterior tibial translation, a soft end feel or the combination of both findings.

14.1.7 Collateral Ligaments

Examination of the collateral ligaments is performed with controlled valgus and varus forces. The knee should be evaluated at both full extension and 30° of flexion and compared with the uninjured knee to rule out pre-existing symmetric physiologic knee laxity. A valgus force that results in an excessive medial opening of the knee at 30° of flexion indicates an isolated MCL injury [29]. If there is significant medial opening with valgus forces in the fully extended position, this can indicate MCL, ACL, PCL or PMC combined injuries. It should also be noted that the ACL is a secondary restraint to medial joint opening. Lateral joint opening with varus forces at 30° of knee flexion can be a sign of an isolated LCL injury. Lateral opening at both full extension and 30° of flexion suggests an injury not only to the LCL, but also of the lateral capsule and PCL [6].

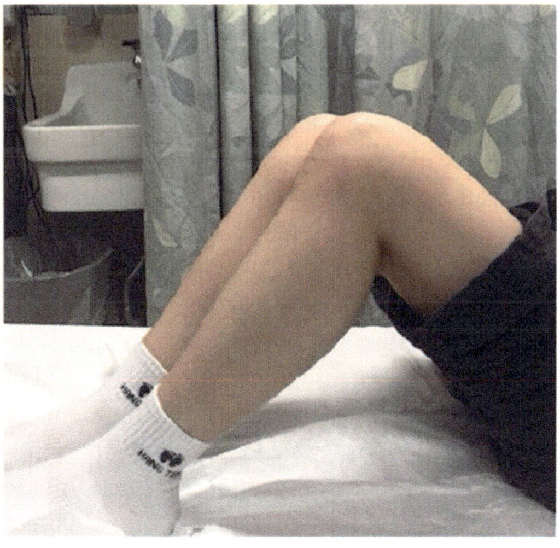

Fig. 14.2 Tibial posterior sag evident on patient's left side, demonstrating PCL insufficiency. (From C. Fanelli (Ed) The Multiple Ligament Injured Knee, Springer, 2019. With permission)

14.1.8 Posterolateral and Posteromedial Corner Injuries

Antero-medial and antero-lateral rotational instabilities can be assessed by the Slocum test, which is a modification of the anterior drawer test. The knee should be positioned at 90° of flexion with both external and internal rotational positions of the tibia to evaluate the integrity of the PLC and PMC, respectively [7, 29, 30]. A further test to check for the integrity of the PLC is the so-called external rotation recurvatum test. During this test, the examiner holds the great toe of each foot while the patient's knees move towards full passive extension. This test is considered to be positive, when the injured knee stays in varus alignment, hyperextension, and external rotation compared to the uninjured leg [31]. However, care must be taken not to re-dislocate the knee during this test. Posterolateral instability can also be assessed by the dial test. This test is performed with the patient in the prone position and the knee held at both 30° and 90° of flexion. An external rotational force is applied to the foot and the thigh-foot angle is measured. Differences greater than 10° between the legs are considered significant. An increase of external rotation at 30° of knee flexion, but not at 90° is suggestive of a PCL and PLC injury. Increased external rotation just at 30° and not at 90° of knee flexion on the other hand is consistent with an isolated PLC injury [6, 7]. It must be noted that a positive dial test at 30° and 90° may also be significant for PMC instability [32].

14.2 Chronic Presentation of Multiligament Knee

MLKI cannot be accurately identified in the acute setting especially in multi-trauma patients and in cases when the patients present with an already reduced knee dislocation in the emergency department. Although the approach to a chronic MLKI assessment is different than that to an acute injury, every evaluation should include a detailed history, clinical exam supplemented with appropriate radiographic studies to identify the injured structures. In some cases, patients with MLKI are presented to surgeons, weeks or months after their injury, either due to initial misdiagnosis, geographic proximity, or a previous trial of conservative treatment before seeing a specialist. The assessment of a chronic MLKI should include all necessary diagnostic steps and exams of an acute injury, including special tests for ACL, PCL, MCL, and LCL injuries as described above. However, the concern for an acute vascular injury has usually passed. We also recommend a thorough assessment of the peripheral nerves and their status should be documented and compared to initial reports to determine if there has been any change or recovery in neurologic function over time. The assessment of chronic MLKI should also include the identification of concomitant injuries to the knee including cartilage and meniscus pathologies. Furthermore, we do recommend an evaluation of leg alignment under static and dynamic conditions, because varus instability or thrust can lead to graft stretching or failure if misdiagnosed. In these cases, a realignment osteotomy may be necessary prior to or concomitant with the ligamentous reconstruction. In a chronic setting, we also routinely perform the external rotation recurvatum test as described above, to evaluate for possible PCL injuries (Fig. 14.3). Additionally, the dial test and Slocum test should be applied to patients presenting in a chronic setting as described above.

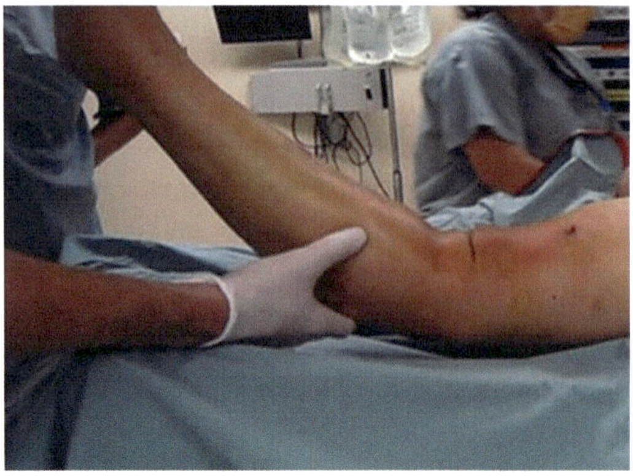

Fig. 14.3 Grossly positive external rotation recurvatum test performed under anaesthesia with significant hyperextension in a chronic injury. (From C. Fanelli (Ed) The Multiple Ligament Injured Knee, Springer, 2019. With permission)

14.3 Imaging of MLKI

14.3.1 Plain Radiographs

Diagnostic imaging should always begin with standard anterior-posterior, lateral, 30° anterior-posterior, skyline and intercondylar notch views of both knees. These views should be obtained to rule out knee dislocation or subluxation, assess bony alignment, and to evaluate for insertion site bony avulsions of the cruciate ligaments, collateral ligaments complexes, and the extensor mechanisms (e.g. Segond fractures, avulsion of fibular head, tibial spine). In MLKI with an associated knee dislocation, plain radiographs are essential to assess the direction of the dislocation. Fractures of the tibial plateau and less commonly of the femoral condyles are also associated with MLKIs. In obvious knee dislocations, as a result of a low-energy trauma, an immediate reduction without prior imaging can be carried out because of the lower likelihood of associated fractures. However, in high-energy knee dislocations, we do recommend radiographs prior to the reduction, because of the higher chance of associated fractures. Three-foot alignment views of both legs should be performed in order to obtain information regarding malalignment. Isolated ligamentous injuries usually do not affect leg alignment; however, a combined cruciate ligament and posterolateral corner injury can lead to varus malalignment over time.

14.3.2 Stress Radiographs

The assessment of knee joint laxity by physical examination alone has been reported to be subjective, inaccurate and poorly reproducible [33, 34]. In addition, the experience of the physician, patient's pain and concomitant knee injuries may bias the interpretation of the physical exam [35, 36]. Therefore, we recommend the use of stress radiographs in order to objectively measure knee joint laxity. Stress radiographs are defined as the visual measurement of a resultant joint translation captured on x-rays in the presence of a directionally applied force. In the sub-acute and chronic setting, these radiographs can be useful to objectively measure joint laxity. However, many considerations should be considered when performing stress radiographs in order to achieve reliable and reproducible measurements. Focus should be on the consistent patient positioning, correct identification of anatomical reference points, how to reproducibly apply forces, and what measurements suggest significant laxity [37]. A valgus stress applied to both knees individually and compared to an AP radiograph can be used to evaluate MCL laxity (Fig. 14.4). The difference in opening of the medial joint space with and without valgus stress is measured and compared to the uninjured knee. For the assessment of the LCL, a varus stress is applied and the difference in opening of the lateral side is also compared to the contralateral knee. Medial or lateral joint opening greater than 3 mm with valgus or respectively varus stress is considered pathologic. In cases with a joint opening of more than 5 mm, reconstruction of the injured structures should be considered [36, 38]. Stress radiographs can also be used for the assessment of PCL tears. PCL stress radiographs are performed with the patient kneeling on an edge so that the patient's weight stresses the injured PCL. PCL incompetency is demonstrated by excessive posterior translation of the tibia compared to the uninjured knee. Posterior tibial translation greater than 5 mm is considered abnormal, 10 mm or greater generally requires treatment. Bilateral skyline x-rays can also be used to quantify the posterior sag by comparing the position of the tibial plateau relative to the anterior femur [39] (Fig. 14.5).

14.3.3 Computed Tomography (CT)

There is an acute indication for vascular imaging in MLKIs where a vascular injury is suspected. Previously, angiograms were performed in the setting of knee dislocation to diagnose vascular injuries. More recently, angiography has been replaced by CT angiograms, since they are less invasive, fast, and readily available. When there is clinical suspicion for a vascular injury or the ABI is inconclusive or abnormal, a CT angiogram should be performed. CT scans are less accurate in diagnosing ligament injuries than MRI; however, they can be helpful in order to provide more detailed imaging of bony avulsions, for example, in case of a PCL tibial avulsion, or other fractures.

14.3.4 Magnetic Resonance Imaging (MRI)

MRI is considered the gold standard in the diagnosis and formulation of a treatment plan for MLKI. Moreover, MRI scans can also help to evaluate chondral defects, meniscal tears, capsular tears, bone marrow oedema, muscle injuries and occult fractures. A study by Bui et al. showed that 75% of patients after a knee dislocation had signs of bone bruising and 25% showed meniscal tears [40]. In addition, the MRI can be helpful in the planning of the surgical strategy, repair versus reconstruction, and in determining the amount of graft needed for reconstruction [41]. Previously, it was shown that MRI can diagnose ligamentous and meniscal injuries in patients with a knee dislocation in the realm of 85–100%, exceeding by far that of physical exam [42]. We do recommend to perform MRI scans in patients that require urgent surgical management with an irreducible

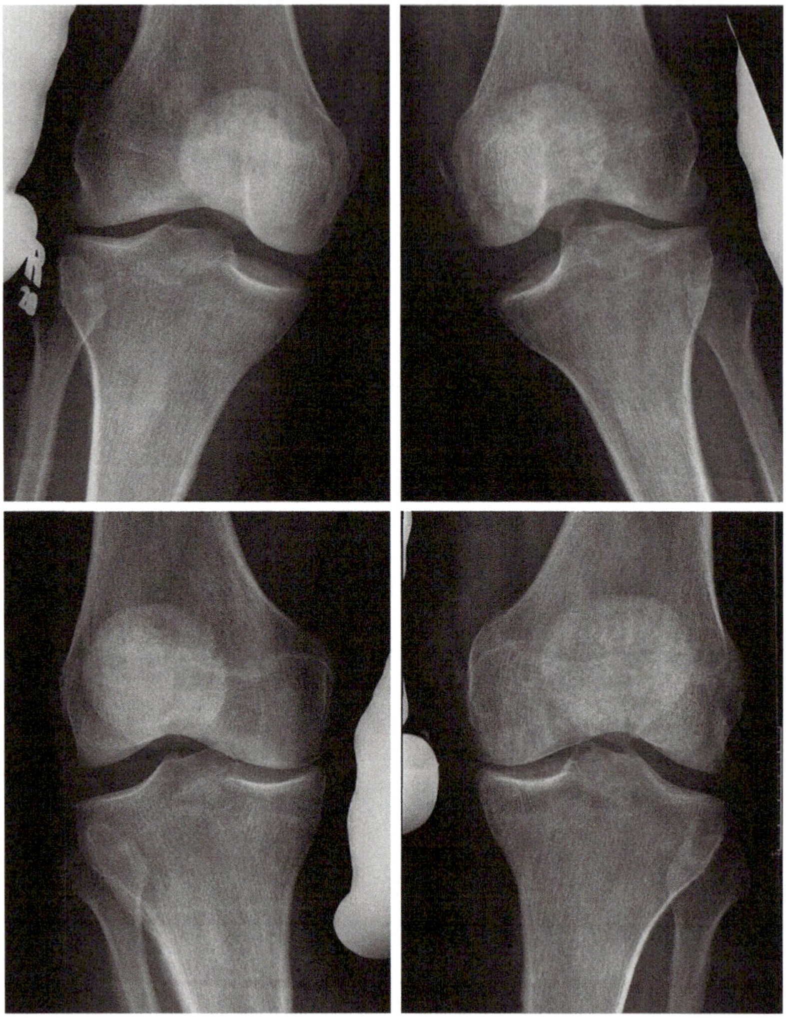

Fig. 14.4 Valgus stress x-rays with medial opening on the left knee. Varus stress x-rays with lateral opening on the right knee. (From C. Fanelli (Ed) The Multiple Ligament Injured Knee, Springer, 2019. With permission)

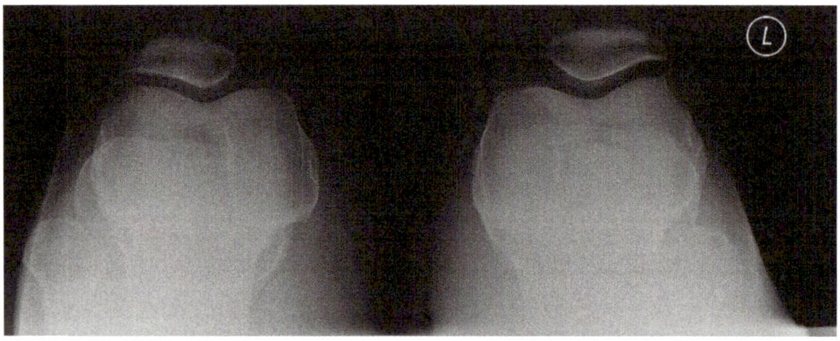

Fig. 14.5 Bilateral skyline views with tibial sag on the right knee. (From C. Fanelli (Ed) The Multiple Ligament Injured Knee, Springer, 2019. With permission)

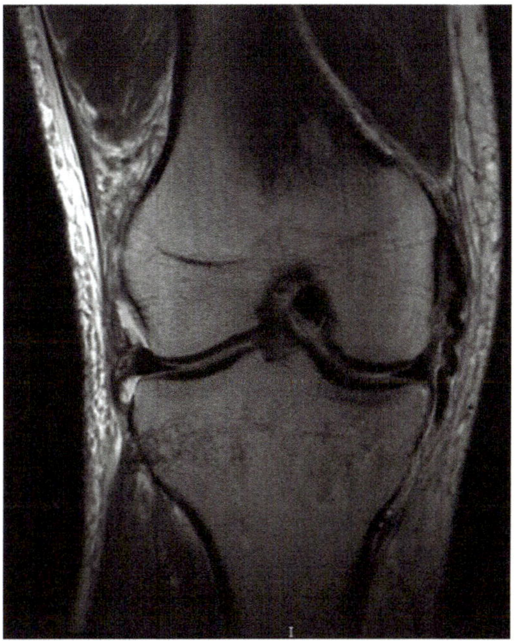

Fig. 14.6 Coronal T1-weighted MRI cut showing tibial-sided MCL avulsion. (From C. Fanelli (Ed) The Multiple Ligament Injured Knee, Springer, 2019. With permission)

or open knee dislocation or an acute vascular injury. Coronal MRI cuts of the knee are usually used to assess the integrity of the medial and lateral collateral ligaments, whereas sagittal cuts are most helpful in the evaluation of the ACL and PCL (Fig. 14.6).

14.4 Controversies in the Treatment of MLKI

14.4.1 Early Versus Late Ligament Reconstruction

In the past, MLKIs were treated with prolonged immobilization in a hinged brace or in a splint [43]. Improved functional outcomes after surgical management of MLKIs compared to nonoperative treatment have been shown in a meta-analysis by Dedmond and Almekinders in 2001 and in a systematic review by Peskun et al. in 2011 [43, 44]. Based on these findings, surgical treatment is recommended in order to restore pain-free and stable knee function. The timing of surgery is influenced by the nature of the injury, multiple-system injuries, vascular status of the injured limb, open injuries, stability after reduction, skin conditions and the surgeon's preference [45, 46]. All MLKIs that are associated with vascular injuries, open knee dislocations, compartment syndrome, and irreducible knee dislocations should be treated urgently. In patients with an acute arterial injury, urgent vascular intervention is necessary to avoid distal limb ischemia, and the unstable knee should be stabilized with a spanning external fixator [47]. Open knee dislocations also require urgent surgical treatment with debridement, irrigation and soft-tissue management to reduce the risk of infection and associated complications. Irreducible knee dislocations should be taken to the operating room urgently for an open reduction to prevent point loading of the articular cartilage and further tension and injury to neurovascular structures.

However, in all other cases of MLKIs, the timing of the operative treatment is still a topic for debates. So far, there is no clear consensus on the definition of acute and chronic MLKIs. Previously, the 3-week mark was set as the critical time to acutely treat MLKIs with improved outcomes [23, 48, 49]. Other authors have used the 6-week mark to delineate between acute and chronic [50].

Many authors advocate early surgical treatment within the first 3 weeks after the injury [23, 51]. It is believed that early operative intervention leads to better functional and clinical outcomes with lower risks of further meniscal and chondral injuries [52, 53]. Additionally, authors state the importance of returning the knee to its anatomical state and alignment before tissue necrosis and scarring occur [48, 49]. In cases where injured ligaments or the joint capsule needs to be repaired, it is preferably done before tissues retract and when anatomical structures can still be easily identified. Harner et al. reported on 19 patients with a knee dislocation that were treated <3 weeks (acute) and on 12 patients treated >3 weeks after injury (chronic) [23]. After a mean follow-up of 44 months, patients in the acute treatment groups scored significantly higher in the Knee Outcome Survey Sports Activity Score, and there was a trend towards higher scores in the Lysholm Score and Knee Outcome Survey Activities of Daily Living score. However, knee range of motion at the final follow-up was similar between the two groups, and more patients in the acute treatment group required manipulation under anaesthesia because of arthrofibrosis. This was also confirmed by Tzurbakis et al. in a study of 44 patients with knee dislocations [54]. Patients in the acute group scored higher on the International Knee Documentation Committee (IKDC) subjective and symptom subgroups. No differences were found for the overall IKDC and Lysholm scores, and the final knee range of motion was also similar between the acute and chronic groups. In a retrospective study of 22 knee dislocations, Liow et al. also reported higher Lysholm scores and Tegner activity rating scores for the acute treatment group, with no differences in knee range of motion [51]. Finally, Fanelli et al. did not find any significant differences between acutely and chronically managed knee dislocations in terms of Lysholm score, Tegner activity level score, and Hospital for Special Surgery (HSS) scores at a minimum of 24 months [55].

A systematic review by Mook et al. suggested that early surgical treatment of knee dislocations may lead to more complications compared with delayed treatment [56]. Hohmann et al. performed in 2017 a systematic review and meta-analysis comparing early versus late surgical treatment of multiligament knee injuries [57]. A total of eight studies with 260 patients were included in this review, and the pool estimate for clinical outcome showed that earlier surgical intervention results in superior clinical outcomes compared to late reconstruction; however, only a trend towards improved knee range of motion was found for the early intervention group. So far there do not exist high-quality level 1 evidence studies guiding the decision of surgical timing in MLKIs. Prospective, high-quality studies are necessary to clearly identify the benefits of early versus late ligamentous reconstruction after knee dislocation.

14.4.2 Ligament Repair Versus Reconstruction

There is growing evidence that surgical management of MLKIs leads to superior clinical and functional outcomes compared to conservative treatment. However, there is still some controversy on whether ligament repair or reconstruction should be performed. In general, it is recommended to repair ligaments around the knee only in an acute setting. Three weeks seems to be the threshold to determine between "acute" and "chronic" injuries [58, 59]. After 3 weeks of the injury, reconstruction of the injured structures is preferred, because of scarring, granulation, and retraction of the ligament stumps [23, 51].

Over the last years, there have been multiple studies that compared outcomes between repair and reconstruction of MLKIs. Most surgeons prefer reconstruction of the cruciate ligaments, while repair of the corners (PMC and PLC) seems to be mixed as to whether this is superior to reconstruction.

Medial collateral ligament injuries in the setting of a MLKI should be repaired or reconstructed if it is unstable during examination under anaesthesia. This was confirmed in a review by Kovachevich et al. who showed similar outcomes for both techniques [60]. Halinen et al. also found comparable results in patients with either conservative or surgical treatment of their MCL injuries with early ACL reconstruction [61]. Combined reconstruction of both ACL and MCL may reduce the risk of graft failure and loosening; however, it can also lead to a significant decrease in knee range of motion and arthrofibrosis.

The posteromedial corner (PMC) contains the posterior oblique ligament (POL), the oblique popliteal ligament, fibres of the semimembranosus, and the posterior horn of the medial meniscus. So far there are conflicting reports as to repair or reconstruct the PMC in MLKIs. Different techniques have been described to reconstruct or repair the PMC, with no clear evidence to which technique is superior. Within 3 weeks after the injury, the medial structures are considered robust enough to facilitate a repair [62]. However, a study by Stannard et al. showed that repair of the PMC leads to a significantly greater failure rate [63]. In a retrospective study by King et al., it was reported that reconstruction of the PMC results in significantly improved outcomes relative to repair [64].

The posterolateral corner of the knee is often injured in the setting of MLKIs. In a study by Stannard et al., 56 patients with a PLC tear were managed either by repair or reconstruction [65]. In their study, the failure rate of repair was 37% compared to 9% in the reconstruction group. This was also confirmed in a recent systematic review by Geeslin et al. that showed a 38% failure rate with acute PLC repair compared to a 9% failure rate with reconstruction [66]. There are various anatomical and non-anatomical techniques for PLC reconstruction described in the literature, with paucity of high-level evidence to recommend the best reconstructive technique. In acute MLKIs, with a bony avulsion of the lateral collateral ligament, repair of the ligament can lead to satisfactory outcome [67].

Our approach to ACL tears in MLKIs is to perform a staged reconstruction using an anatomical single bundle reconstruction with an ipsilateral bone-patellar-bone graft. Mariani et al. compared direct repair of ACL and or PCL injuries with reconstruction after knee dislocations [49]. In their study, they found improved stability and better range of motion in patients with reconstruction compared to repair after a mean follow-up of 6.9 years. In addition, Owens et al. treated patients with knee dislocations and ACL/PCL injuries in their retrospective study with primary repair [68]. All patients in this study were able to return to their previous work with little or no activity modification; however, 5 out of 25 patients required treatment due to arthrofibrosis of the knee.

PCL injuries can be treated with transtibial or tibial inlay single bundle or double bundle techniques. Recent biomechanical studies have demonstrated that double PCL reconstruction can better restore the native knee kinematics than single bundle reconstruction [69, 70]. However, there was no difference in clinical outcomes [71, 72]. For acute bony avulsions of the PCL, we recommend direct repair, since studies have shown favourable clinical outcomes with this technique [68, 73]. In a systematic review, Levy et al. found that PCL reconstruction leads to better outcomes than repair in the setting of MLKIs; however, like most controversies in MLKIs, more high-level evidence specific to these complex patients is needed [59].

14.4.3 Rehabilitation After MLKIs

Although knee dislocations are rare, their clinical significance has been well-documented. Due to the high-energy mechanism of knee dislocations, they often result in concomitant limb- and life-threatening injuries. Operative treatment is considered superior to conservative management for MLKIs, since surgically treated patients are more likely to return to work and sports [43, 74]. The great variability in surgical treatment options for MLKIs has led to a lack of studies examining the timing and composition of rehabilitation protocols. Because of this, rehabilitation after knee dislocation is poorly understood and remains a subject of intense debate. Rehabilitation after MLKIs should focus on early immobilization of the knee and a gradual return to activities. Risks of too early and aggressive rehabilitation are failure of the repaired or reconstructed knee structures. Therefore, the treating physicians must balance these two competing factors by frequent examinations. Recovery after MLKIs usually takes 9–12 months of intensive rehabilitation before patients can return to full activities. This extensive rehab period allows for proper graft healing and incorporation in order to prevent graft failure. Every rehab plan after MLKIs should emphasise on functional outcomes including regaining of motion, function and strength as well as on graft protection. In cases with PCL reconstruction, we do recommend the use of a dynamic PCL brace throughout the whole rehabilitation period (9–12 months) in order to protect the reconstructed ligament.

During the initial phase of rehabilitation, focus should be on supportive measures such as pain, inflammation, and effusion control.

Following the STaR Trial investigators, rehabilitation after MLKI surgery can be divided into three phases: (a) tissue protection, (b) restoration of motor control, and (c) optimization of knee function [75]. During the tissue protection, rehab phase patients should focus on the restoration of the knee function without over-stressing the repaired/ reconstructed knee structures, as well as on gait training with appropriate assistive devices. During the first 4–6 weeks after the surgery, knee flexion should be limited to 60° and neutral knee extension. By 12 weeks post-surgery, patients should achieve 90° of knee flexion. For patients with PCL and PLC reconstructions, initial flexion exercises should be performed passively, either prone or supine, to avoid a posterior sag, which could stretch out the reconstructed ligaments. Stationary cycling without resistance is recommended in individuals that have achieved 90° of knee flexion and neutral extension not earlier than 6–7 weeks after the surgery [50, 76]. Recommendations for weight bearing after MLKI surgery are highly variable. Weight bearing should initially be limited and performed in a knee immobilizer to prevent excessive knee motion. The knee brace and crutches should be used for at least 6 weeks, but the patients are able to bear little weight through the crutches if tolerable. Even during the tissue protection phase, patients should start with exercises to restore quadriceps activation as well as gluteal sets, ankle pumps, and four-way straight leg raises. It is recommended to gradually introduce force to the knee and not to provide excessive stress, and therefore, resisted exercises should follow a clear timeline.

During the restoration of motor-control phase, the repaired or reconstructed tissues can be loaded in a gradual fashion. The goals of this phase are the restoration of nearly symmetrical muscle strength, full knee range of motion and normal gait, with a return to activities of daily living. It is recommended that external resistance should be added when patients can perform a bilateral squat without pain and a step down from a 7-in. step [77]. Stationary cycling for aerobic conditioning can begin 10 weeks after the operation when the patients have achieved the necessary knee range of motion and control. Elliptical machines are recommended once the patient can walk independently with sufficient quadriceps strength. All training loads should be progressed slowly, by 10–20% per week.

Goal of the final phase of rehabilitation is the optimization of knee function so that patients can return to their pre-morbid knee function level. During this phase, tissue considerations are no longer relevant, and the physical therapist can advance knee function without restrictions of tissue protection. Exercises during this phase should include cycling or elliptical trainer for conditioning, and bilateral weight training with free weights. However, running should not begin before the 16th week after the surgery, and hopping, cutting and pivoting sport specific drills should not be performed before the 20th postoperative week.

In summary, we believe that rehabilitation after MLKIs strongly depends on the condition of patient and his/her comorbidities, the pattern of injury, and the quality of the reconstruction. Therefore, we advocate rehabilitation protocols that consider inflammation, knee range of motion, and muscle strength. Careful rehabilitation can maximize the clinical outcome after MLKIs.

References

1. Brautigan B, Johnson DL. The epidemiology of knee dislocations. Clin Sports Med. 2000;19(3):387–97.
2. Mills WJ, Tejwani N. Heterotopic ossification after knee dislocation: the predictive value of the injury severity score. J Orthop Trauma. 2003;17(5):338–45.
3. Skendzel JG, Sekiya JK, Wojtys EM. Diagnosis and management of the multiligament-injured knee. J Orthop Sports Phys Ther. 2012;42(3):234–42.
4. Seroyer ST, Musahl V, Harner CD. Management of the acute knee dislocation: the Pittsburgh experience. Injury. 2008;39(7):710–8.

5. Shelbourne KD, Klootwyk TE. Low-velocity knee dislocation with sports injuries. Treatment principles. Clin Sports Med. 2000;19(3):443–56.
6. Manske RC, Prohaska D. Physical examination and imaging of the acute multiple ligament knee injury. N Am J Sports Phys Ther. 2008;3(4):191–7.
7. Devitt BM, Whelan DB. Physical examination and imaging of the lateral collateral ligament and posterolateral corner of the knee. Sports Med Arthrosc Rev. 2015;23(1):10–6.
8. King JJ, et al. Surgical outcomes after traumatic open knee dislocation. Knee Surg Sports Traumatol Arthrosc. 2009;17(9):1027–32.
9. Scarcella NR, et al. Clinical and functional results of 119 patients with knee dislocations. J Orthop Trauma. 2017;31(7):380–6.
10. Kilicoglu O, et al. Muscular buttonholing: an unusual cause of irreducible knee dislocation. Arthroscopy. 2001;17(6):E22.
11. Quinlan AG, Sharrard WJ. Postero-lateral dislocation of the knee with capsular interposition. J Bone Joint Surg Br. 1958;40-B(4):660–3.
12. Reckling FW, Peltier LF. Acute knee dislocations and their complications. 1969. Clin Orthop Relat Res. 2004;(422):135–41.
13. Hoover NW. Injuries of the popliteal artery associated with fractures and dislocations. Surg Clin North Am. 1961;41:1099–112.
14. McCoy GF, et al. Vascular injury associated with low-velocity dislocations of the knee. J Bone Joint Surg Br. 1987;69(2):285–7.
15. Mills WJ, Barei DP, McNair P. The value of the ankle-brachial index for diagnosing arterial injury after knee dislocation: a prospective study. J Trauma. 2004;56(6):1261–5.
16. Wascher DC. High-velocity knee dislocation with vascular injury. Treatment principles. Clin Sports Med. 2000;19(3):457–77.
17. Green NE, Allen BL. Vascular injuries associated with dislocation of the knee. J Bone Joint Surg Am. 1977;59(2):236–9.
18. Wascher DC, Dvirnak PC, DeCoster TA. Knee dislocation: initial assessment and implications for treatment. J Orthop Trauma. 1997;11(7):525–9.
19. Klineberg EO, et al. The role of arteriography in assessing popliteal artery injury in knee dislocations. J Trauma. 2004;56(4):786–90.
20. Aboyans V, et al. The association between elevated ankle systolic pressures and peripheral occlusive arterial disease in diabetic and nondiabetic subjects. J Vasc Surg. 2008;48(5):1197–203.
21. Medina O, et al. Vascular and nerve injury after knee dislocation: a systematic review. Clin Orthop Relat Res. 2014;472(9):2621–9.
22. Frassica FJ, et al. Dislocation of the knee. Clin Orthop Relat Res. 1991;263:200–5.
23. Harner CD, et al. Surgical management of knee dislocations. J Bone Joint Surg Am. 2004;86(2):262–73.
24. Ríos A, et al. Results after treatment of traumatic knee dislocations: a report of 26 cases. J Trauma. 2003;55(3):489–94.
25. Niall DM, Nutton RW, Keating JF. Palsy of the common peroneal nerve after traumatic dislocation of the knee. J Bone Joint Surg Br. 2005;87(5):664–7.
26. Peskun CJ, et al. Risk factors for peroneal nerve injury and recovery in knee dislocation. Clin Orthop Relat Res. 2012;470(3):774–8.
27. Woodmass JM, et al. A systematic review of peroneal nerve palsy and recovery following traumatic knee dislocation. Knee Surg Sports Traumatol Arthrosc. 2015;23(10):2992–3002.
28. Benjaminse A, Gokeler A, van der Schans CP. Clinical diagnosis of an anterior cruciate ligament rupture: a meta-analysis. J Orthop Sports Phys Ther. 2006;36(5):267–88.
29. Lundquist RB, et al. Posteromedial corner of the knee: the neglected corner. Radiographics. 2015;35(4):1123–37.
30. Slocum DB, Larson RL. Rotatory instability of the knee. Its pathogenesis and a clinical test to demonstrate its presence. J Bone Joint Surg Am. 1968;50(2):211–25.
31. Hughston JC, Norwood LA. The posterolateral drawer test and external rotational recurvatum test for posterolateral rotatory instability of the knee. Clin Orthop Relat Res. 1980;147:82–7.
32. Griffith CJ, et al. Medial knee injury: part 1, static function of the individual components of the main medial knee structures. Am J Sports Med. 2009;37(9):1762–70.
33. Benvenuti JF, et al. Objective assessment of the anterior tibial translation in Lachman test position. Comparison between three types of measurement. Knee Surg Sports Traumatol Arthrosc. 1998;6(4):215–9.
34. Katz JW, Fingeroth RJ. The diagnostic accuracy of ruptures of the anterior cruciate ligament comparing the Lachman test, the anterior drawer sign, and the pivot shift test in acute and chronic knee injuries. Am J Sports Med. 1986;14(1):88–91.
35. Kastelein M, et al. Assessing medial collateral ligament knee lesions in general practice. Am J Med. 2008;121(11):982–8.e2.
36. Laprade RF, et al. Correlation of valgus stress radiographs with medial knee ligament injuries: an in vitro biomechanical study. Am J Sports Med. 2010;38(2):330–8.
37. Rohman EM, Macalena JA. Anterior cruciate ligament assessment using arthrometry and stress imaging. Curr Rev Musculoskelet Med. 2016;9(2):130–8.
38. James EW, Williams BT, LaPrade RF. Stress radiography for the diagnosis of knee ligament injuries: a systematic review. Clin Orthop Relat Res. 2014;472(9):2644–57.
39. Puddu G, et al. The axial view in evaluating tibial translation in cases of insufficiency of the posterior cruciate ligament. Arthroscopy. 2000;16(2):217–20.

40. Bui KL, et al. Knee dislocations: a magnetic resonance imaging study correlated with clinical and operative findings. Skelet Radiol. 2008;37(7):653–61.
41. LaPrade RF, et al. The magnetic resonance imaging appearance of individual structures of the posterolateral knee. A prospective study of normal knees and knees with surgically verified grade III injuries. Am J Sports Med. 2000;28(2):191–9.
42. Twaddle BC, et al. MRI in acute knee dislocation. A prospective study of clinical, MRI, and surgical findings. J Bone Joint Surg Br. 1996;78(4):573–9.
43. Dedmond BT, Almekinders LC. Operative versus nonoperative treatment of knee dislocations: a meta-analysis. Am J Knee Surg. 2001;14(1):33–8.
44. Peskun CJ, Whelan DB. Outcomes of operative and nonoperative treatment of multiligament knee injuries: an evidence-based review. Sports Med Arthrosc Rev. 2011;19(2):167–73.
45. Fanelli GC, Orcutt DR, Edson CJ. The multiple-ligament injured knee: evaluation, treatment, and results. Arthroscopy. 2005;21(4):471–86.
46. Fanelli GC, et al. Treatment of combined anterior cruciate-posterior cruciate ligament-medial-lateral side knee injuries. J Knee Surg. 2005;18(3):240–8.
47. Lachman JR, Rehman S, Pipitone PS. Traumatic knee dislocations: evaluation, management, and surgical treatment. Orthop Clin North Am. 2015;46(4):479–93.
48. Engebretsen L, et al. Outcome after knee dislocations: a 2-9 years follow-up of 85 consecutive patients. Knee Surg Sports Traumatol Arthrosc. 2009;17(9):1013–26.
49. Mariani PP, et al. Comparison of surgical treatments for knee dislocation. Am J Knee Surg. 1999;12(4):214–21.
50. Geeslin AG, LaPrade RF. Outcomes of treatment of acute grade-III isolated and combined posterolateral knee injuries: a prospective case series and surgical technique. J Bone Joint Surg Am. 2011;93(18):1672–83.
51. Liow RY, et al. Ligament repair and reconstruction in traumatic dislocation of the knee. J Bone Joint Surg Br. 2003;85(6):845–51.
52. Smith TO, Davies L, Hing CB. Early versus delayed surgery for anterior cruciate ligament reconstruction: a systematic review and meta-analysis. Knee Surg Sports Traumatol Arthrosc. 2010;18(3):304–11.
53. Ercin E, et al. Importance of restricting sportive activity and time from injury to surgery in anterior cruciate ligament reconstruction. Open Orthop J. 2015;9:427–31.
54. Tzurbakis M, et al. Surgical treatment of multiple knee ligament injuries in 44 patients: 2-8 years follow-up results. Knee Surg Sports Traumatol Arthrosc. 2006;14(8):739–49.
55. Fanelli GC, Giannotti BF, Edson CJ. Arthroscopically assisted combined posterior cruciate ligament/posterior lateral complex reconstruction. Arthroscopy. 1996;12(5):521–30.
56. Mook WR, et al. Multiple-ligament knee injuries: a systematic review of the timing of operative intervention and postoperative rehabilitation. J Bone Joint Surg Am. 2009;91(12):2946–57.
57. Hohmann E, Glatt V, Tetsworth K. Early or delayed reconstruction in multi-ligament knee injuries: a systematic review and meta-analysis. Knee. 2017;24(5):909–16.
58. Fanelli GC, Edson CJ. Combined posterior cruciate ligament-posterolateral reconstructions with Achilles tendon allograft and biceps femoris tendon tenodesis: 2- to 10-year follow-up. Arthroscopy. 2004;20(4):339–45.
59. Levy BA, et al. Decision making in the multiligament-injured knee: an evidence-based systematic review. Arthroscopy. 2009;25(4):430–8.
60. Kovachevich R, et al. Operative management of the medial collateral ligament in the multi-ligament injured knee: an evidence-based systematic review. Knee Surg Sports Traumatol Arthrosc. 2009;17(7):823–9.
61. Halinen J, et al. Operative and nonoperative treatments of medial collateral ligament rupture with early anterior cruciate ligament reconstruction: a prospective randomized study. Am J Sports Med. 2006;34(7):1134–40.
62. Tibor LM, et al. Management of medial-sided knee injuries, part 2: posteromedial corner. Am J Sports Med. 2011;39(6):1332–40.
63. Stannard JP, et al. Posteromedial corner injury in knee dislocations. J Knee Surg. 2012;25(5):429–34.
64. King AH, et al. Surgical outcomes of medial versus lateral multiligament-injured, dislocated knees. Arthroscopy. 2016;32(9):1814–9.
65. Stannard JP, et al. The posterolateral corner of the knee: repair versus reconstruction. Am J Sports Med. 2005;33(6):881–8.
66. Geeslin AG, Moulton SG, LaPrade RF. A systematic review of the outcomes of posterolateral corner knee injuries, part 1: surgical treatment of acute injuries. Am J Sports Med. 2016;44(5):1336–42.
67. Laprade RF, et al. Improving outcomes for posterolateral knee injuries. J Orthop Res. 2014;32(4):485–91.
68. Owens BD, et al. Primary repair of knee dislocations: results in 25 patients (28 knees) at a mean follow-up of four years. J Orthop Trauma. 2007;21(2):92–6.
69. Kennedy NI, et al. Kinematic analysis of the posterior cruciate ligament, part 1: the individual and collective function of the anterolateral and posteromedial bundles. Am J Sports Med. 2013;41(12):2828–38.
70. Wijdicks CA, et al. Kinematic analysis of the posterior cruciate ligament, part 2: a comparison of anatomic single- versus double-bundle reconstruction. Am J Sports Med. 2013;41(12):2839–48.
71. Li Y, et al. Comparison of single-bundle and double-bundle isolated posterior cruciate ligament reconstruction with allograft: a prospective, randomized study. Arthroscopy. 2014;30(6):695–700.
72. Yoon KH, et al. A prospective randomized study comparing arthroscopic single-bundle and double-bundle

posterior cruciate ligament reconstructions preserving remnant fibers. Am J Sports Med. 2011;39(3):474–80.
73. Frosch KH, et al. Primary ligament sutures as a treatment option of knee dislocations: a meta-analysis. Knee Surg Sports Traumatol Arthrosc. 2013;21(7):1502–9.
74. Wong CH, et al. Knee dislocations-a retrospective study comparing operative versus closed immobilization treatment outcomes. Knee Surg Sports Traumatol Arthrosc. 2004;12(6):540–4.
75. Lynch AD, et al. Current concepts and controversies in rehabilitation after surgery for multiple ligament knee injury. Curr Rev Musculoskelet Med. 2017;10(3):328–45.
76. Quelard B, et al. Isolated posterior cruciate ligament reconstruction: is non-aggressive rehabilitation the right protocol? Orthop Traumatol Surg Res. 2010;96(3):256–62.
77. Piva SR, et al. Reliability of measures of impairments associated with patellofemoral pain syndrome. BMC Musculoskelet Disord. 2006;7:33.

When Do You Need to Reconstruct the Posterior Cruciate Ligament?

15

Vishal Pai and Andy Williams

15.1 Introduction

PCL injuries are relatively common, particularly in the sporting world. The most common causation is a blow to the front of the proximal tibia, particularly when landing heavily on the front of the flexed knee. An alternative method is forced hyperextension. The latter can be associated with more significant injury including other structures. With pure hyperextension, the posterolateral corner is the most likely structure in line for overload, but with applied valgus as well as hyperextension, the posteromedial structures are also vulnerable.

The annual incidence of isolated PCL injury is estimated to be 2 per 100,000 persons [1]. Sporting injuries followed by traffic accidents (the typical 'dashboard' injury) were the most common cause of isolated PCL injury [2]. Whilst isolated PCL injuries are common, significant PCL ruptures as part of a combined ligament injury, such as with knee dislocation, are thankfully less common. It should be stated early in this chapter that the vast majority of PCL injuries are relatively minor and are correctly treated non-operatively.

V. Pai (✉)
Eastern Health, Melbourne, VIC, Australia
e-mail: mail@vishalpai.com

A. Williams
Fortius Clinic, London, UK
e-mail: andy.williams@fortiusclinic.com

15.2 Clinical and Kinematic Consequences

15.2.1 Clinical Consequences

The patient will often present with a history in keeping with a heavy blow to the front of the knee or hyperextension along with a feeling of discomfort in the popliteal fossa area. This pain will be worsened by flexion over 60–90° and, on occasion, by extension. Frequently, the patient is able to finish the activity they were undertaking at the time of injury (such as sport), and the symptoms then worsen for a period of 24–48 h. The more severe injuries, particularly when combined with other ligament ruptures, present more dramatically.

The classic physical sign on examination is a posterior sag/posterior drawer [3] on forced posterior displacement of the tibia with the knee at 80°. Typically, in a more chronic case, the quadriceps 'active test' [4] is positive when extending the knee from a flexed position, and the tibia is seen to move forwards. The best way of undertaking this test is with the patient's distal thigh supported by the examiner's knee placed under the patient's limb, allowing the tibia to sag posteriorly under the effect of gravity prior to active extension.

The clinical grading of PCL injuries is particularly important for determining not only the severity of the injury but also appropriate man-

N. Nakamura et al. (eds.), *Advances in Knee Ligament and Knee Preservation Surgery*,
https://doi.org/10.1007/978-3-030-84748-7_15

agement. Rather like medial collateral ligament injuries, the grading system on MRI scan has minimal clinical use in terms of determining treatment as there is little correlation with clinical grading. An MRI scan performed acutely will show what is injured [5] but not how severe the injury is [6]. The same is true for chronic PCL injuries. There may be an appearance of a ligament in continuity, looking virtually normal, despite even Grade III clinical laxity in chronic cases [7, 8].

MRI scans show a normal PCL as a well-defined continuous band of low signal intensity in all sequences. A normal PCL measures no more than 6 mm in diameter on sagittal imaging. Injury to the PCL causes abnormal widening [6]. Despite being torn, the PCL can still appear as a single continuous structure on images [9].

There are a number of clinical grading systems but the most practical is to compare the medial tibial 'step off' between knees. With the knee placed at 80–90° flexion and with comfortable neutral tibial rotation, the gap between the anteromedial tibia and the medial femoral condyle is assessed in the normal knee. This is usually 1–1.5 cm. The injured knee is then examined in the same way. Sometimes, quadriceps activity inadvertently applied by the patient will bring the tibia forward into a seemingly normal position, so it is essential that the patient is fully relaxed. In cases of major PCL laxity (Grade II and III), the tibia will immediately be seen to be sagging posteriorly. When placing the open palm on the front of the tibia and patella, the tibia is normally encountered prior to the patella. In cases of significant PCL laxity, the palm touches the tibia and patella simultaneously. The medial tibial step off is then assessed with palpation. Following this, a gentle posterior drawer force is placed on the proximal tibia and sustained until the maximum limit of posterior displacement is achieved. In the grading assessment preferred by the senior author, symmetric medial tibial step off equates to normality or a PCL sprain (i.e. no excess laxity). In Grade I, the posterior step off is present but reduced compared with the other side. Grade II indicates that the tibia has sagged back sufficient to abolish the posterior sag but not to reverse it, that is, the anteromedial tibia is level with the distal end of the medial femoral condyle. With Grade III injuries, the tibia displaces posterior to the distal medial femur. With easily palpable landmarks, the method just described is easily reproducible unlike clinical assessment of differences in millimetres of laxity used in other techniques.

Grade I injuries will heal well with non-operative treatment. Grade II injuries may or may not need surgery as described below. Grade III injuries indicate that there is likely to be at least another ligamentous injury (usually medial or lateral), and these injuries normally require surgery especially in the acute setting.

X-ray evaluation can be helpful. In the chronic setting, Skyline x-rays to assess the patellofemoral joint often reveal a posterior sag when the observer compares the position of the tibial tuberosity that is superimposed on the distal femur – usually just below the trochlear groove. Particularly when both knees are x-rayed on the same film, the side to side difference can be seen [10, 11]. In a similar manner, a more objective assessment can be undertaken using stress x-rays. A Telos device can be used to apply posterior directed force to the upper tibia with the knee at 30° and 90° and the healthy and injured knees compared. This is not easily tolerated in the acute setting due to discomfort and is more relevant to chronic laxity. A simpler method [12, 13] is to assess a lateral radiograph comparing the abnormal and normal knees with the patient kneeling so that the tibial tuberosity is at the edge of a support and allows the weight of the body and femur to slide anteriorly on the tibia.

Although the use of a stress device such as Telos is attractive and millimetre measurements are possible, the quality of measurement is highly dependent on the observer and the rotational alignment of the tibia. Nevertheless, with symmetrically aligned radiographs, the measurements can be useful. It is said that a side to side difference of more than 12 mm indicates that there must be damage to structures in addition to the PCL, for example, posterolateral or posteromedial [2, 12, 14].

15.2.2 Kinematic Consequences

There is the paradox that many posterior cruciate ligament knees have significant laxity and yet remain asymptomatic allowing full activity for prolonged periods compared to ACL-deficient knees that have little laxity but gross instability. The reason for this is related to the kinematic consequences of injury and the articular geometry. It is well established that ACL deficiency causes abnormality of kinematics in the lateral compartment [15]. For a long time, it was thought that the PCL-deficient knee caused similar posterior displacement of medial and lateral tibia on the femur. This led to the belief that PCL deficiency causes patella-femoral overload and subsequent osteoarthritis.

However, in fact, the kinematic abnormality is in the medial compartment alone. A study of living subjects standing in an open access MRI scanner allowed assessment of the kinematic consequence of isolated PCL rupture [16] to be in the medial tibiofemoral compartment. The different shapes of the articular surfaces in the medial and lateral compartments, particularly tibia, explains the paradox above. In an ACL-deficient knee, there is anterolateral instability, which allows the lateral femoral condyle to slide down the inferiorly directed sloping articular surface of the posterolateral tibia. This unstable situation produces joint subluxation and clinical instability. Whereas on the lateral tibia in the sagittal plane, there is a flat central portion 'drop-off' anteriorly and posteriorly, the medial tibia is fairly flat in its posterior half but has an anterior upslope. In the PCL-deficient knee, the medial femur will ride up the upslope of the anterior half of the medial tibia, which represents a self-stabilising situation, and thereby the patient does not have instability symptoms. However, the price is increased stress concentration with point loading and chondral overload. It was previously thought that posterior sag of the tibia meant increased pressure on the patellofemoral joint; whilst this may be true [17], the usual arthritic consequence of PCL deficiency is medial compartment osteoarthritis affecting the anterior/distal femur as an isolated chondral lesion progressing to anteromedial chondral damage on both sides of the joint as shown in Fig. 15.1 [18].

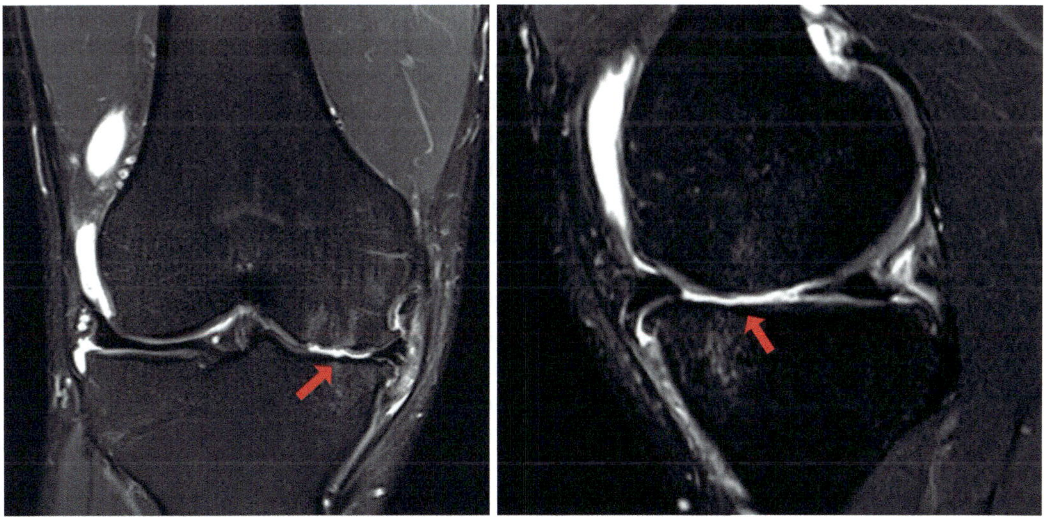

Fig. 15.1 Pattern of anteromedial osteoarthritis (red arrows) in a PCL-deficient knee

15.3 Management

The goal of treatment for ligament injury is either to provide stability and function in cases of chronic instability, or acute cases where instability is inevitable. There is also the consideration of reducing the osteoarthritis risk from the abnormal kinematics.

Only a minority of PCL injuries will come to require surgery. This must be emphasized, particularly as non-operative treatment has greatly improved with the advent of PCL-directed braces. Such braces apply an anteriorly directed force on the tibia to reduce the posterior tibial translation associated with a PCL injury [19]. PCL braces can be static or dynamic. Dynamic braces apply larger forces with an increasing flexion angle which simulates physiological loading [20].

The senior author finds that he does much less PCL surgery nowadays compared to that done a few years ago. Unlike the wholly intra-articular and intra-synovial ACL, which has little healing capacity (there is some), the PCL, being extra-synovial, can heal due to the persistence of contained haematoma in the presence of ligament fibre damage. Modern PCL braces will reduce the posterior sag and allow PCL healing with less final laxity than with previous braces or with no bracing at all.

15.3.1 Acute Isolated PCL Tears

Isolated Grade I injuries can be treated non-operatively with physiotherapy. In this group, no bracing is required, and early weight-bearing and range of motion exercises can be commenced.

A Grade II isolated PCL injury is a relative indication for surgical reconstruction in a very young active patient. However, Grade II laxity in professional and semi-professional athletes can correctly be treated non-operatively and with good outcome [21, 22]. This involves dynamic bracing for up to 12 weeks, and a rehabilitation program that would strengthen the quadriceps muscle as a priority. Avoidance of open chain hamstrings exercises, or concentric and eccentric hamstrings contraction is recommended for the first 3 months following the injury. The main challenge is convincing athletes to tolerate 12 weeks of bracing—many opt for a compromise. The senior author's current practice is to treat isolated Grade II PCL tears non-operatively, but in those patients with lesions that are still causing symptoms at 3 months, he will undertake PCL reconstruction.

Isolated Grade III PCL laxity is rare. In most cases, early PCL reconstruction/repair is required. With the advent of high-quality synthetic tapes, a combination of repair and protection of repairs by splinting them with such tapes can be an attractive and effective option [23, 24]. Repair, if possible and successful, is always preferable to reconstruction in the context of ligament surgery as only repair allows the chance of maximal if not normal proprioceptive function of the ligament. A reconstruction can improve joint proprioception by improving joint congruency allowing the motor areas of the brain to 'better recognise' the knee, but is clearly second best to successful repair. The question that is unanswered is whether or not the results of repair will actually be better than successful reconstruction.

15.3.1.1 Surgical Technique

PCL Repair

Repair is a very good option if there is an avulsion of the PCL and meniscofemoral ligaments from the medial femoral condyle. Suture tapes can be passed individually through the two bundles of the PCL and the meniscofemoral ligaments and these can be connected into the anatomic positions on the femoral footprint. The repairs can be protected by passing a synthetic tape through the tibia to the tibial insertion of the PCL, retrieved and then taken through the joint to the centre of the anterolateral bundle femoral footprint for the PCL. This construct will abolish the posterior sag on the tibia and help healing of the PCL.

PCL Reconstruction

However, most PCL injuries requiring surgery will be more suitable for a PCL reconstruction rather than a repair. The decision to perform a single or double bundle reconstruction is controversial. The senior author prefers a single bundle technique in most cases because the posteromedial bundle (PMB) of the PCL is usually functionally intact and is worth preserving. This is especially true if the meniscofemoral ligaments are intact as they lie anterior to and posterior to the PMB and may protect the PMB during the injury episode. Surgeons often completely clear the medial wall of the intercondylar notch to undertake PCL reconstruction, and this frequently means that useful posteromedial bundle tissue is lost, and the meniscofemoral ligaments can be completely defunctioned. Although more technically demanding, working above and below the PMB and anterior meniscofemoral ligament for single bundle reconstruction can be justified by preserving useful tissue, and is straightforward once practiced in the cadaver lab.

Although some studies have shown that the performance of a double bundle reconstruction more closely simulates natural kinematics of the knee with a native PCL, none of these left an intact PMB in the experimental design, and it is unclear of the integrity or otherwise of the menisco-femoral ligaments [25, 26]. Furthermore, there seems to be no significant clinical superiority of the double-bundle technique over a single bundle reconstruction [27–31]. In addition a single femoral tunnel weakens the medial femoral condyle less than two femoral tunnels [32]. Therefore, whilst respecting good results published for double-bundle PCL reconstruction [33] for the senior author, the only indication for a double bundle technique is in chronic cases or revision cases in which the PMB and menisco-femoral ligaments no longer attach to the medial femoral condyle, that is, there is a 'bare' medial wall to the notch.

The senior author respects the use of allograft, but has a strong preference for autograft since the results for allograft ACL reconstructions are inferior to PCL reconstructions [34, 35]. He tends to reconstruct the PCL using doubled bilateral gracilis and ipsilateral semitendinosus autograft tendon to make a six-strand graft. Allograft is widely used by surgeons and has the advantage of being able to be prepared away from the operation thus saving time, as well as large bulk. The role of synthetic grafts is less certain [36–38].

If surgical reconstruction is undertaken, most patients will have some degree of persistent posterior laxity after the healing phase. The graft failure rate in this setting varies from 2.3 to 30% [39]. In the senior authors experience, results have improved significantly by using a PCL-directed brace routinely. The senior authors' post-operative rehabilitation protocol involves wearing a dynamic PCL brace full time for 12 weeks post-operatively. The range of motion in this brace is restricted to 0–60° for the first 2 weeks after surgery. Thereafter, the range is increased to 0–90° in the brace from 2 to 6 weeks post-operatively. A gradual progression is made to a full range of motion from 6 to 12 weeks. Weight bearing with crutches is allowed for the first 6 weeks post-operatively. Hamstrings exercises are avoided for 12 weeks. A return to sport can be considered from 9 months after the surgery.

15.3.1.2 Results of Surgery

There is very little data on PCL repairs but with reconstruction there is reasonable data [40–42]. Hermans [41] noted an improvement in the International Knee Documentation Committee (IKDC), Lysholm and functional VAS scores with a single anterolateral bundle PCL reconstruction with a mean follow-up of 9.1 years. Of note, they found the reconstructed PCL had more laxity when compared to the uninjured side. Different graft choices were used in this study which may affect the residual graft laxity at the final follow-up.

Jackson reviewed the long-term outcome following single anterolateral bundle PCL reconstruction using a hamstrings tendon autograft, with a 10-year follow-up. They showed that PCL reconstruction resulted in improved IKDC and Lysholm scores. A significant proportion of patients increased their ability to perform moderately strenuous activities. At 10 years post-operatively, most patients had Grade I or no residual laxity. Instrumented knee testing showed a mean increase in anterior to posterior translation of 1.1 mm compared to the uninjured side.

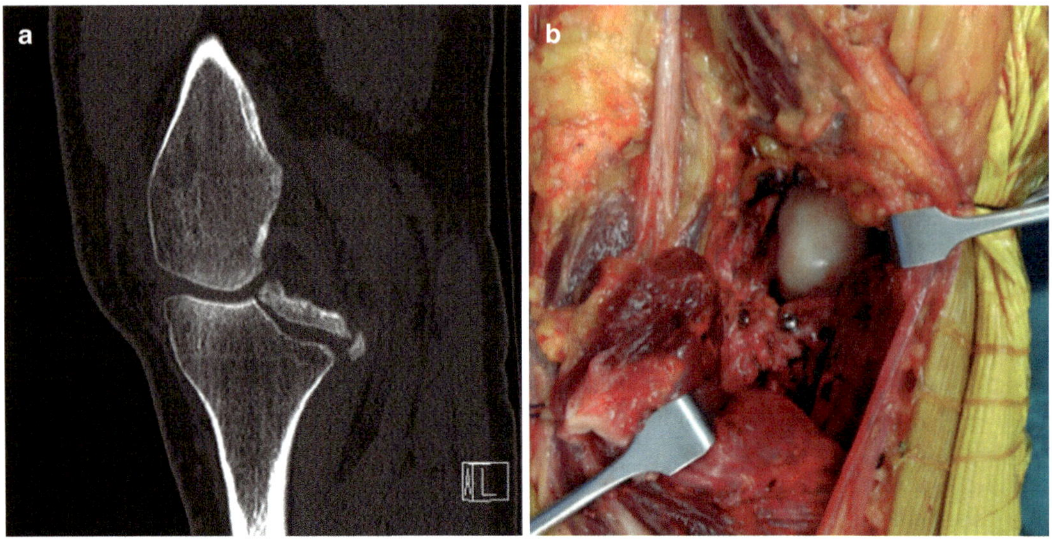

Fig. 15.2 (**a**) CT scan showing non-union of a PCL avulsion fracture. (**b**) Intra-operative photo of a posterior approach to the knee and screw fixation of the PCL avulsion fracture

Neither of the aforementioned studies used a dynamic PCL brace post-operatively. Li et al. compared the use of a standard hinged brace to a tibial support brace (TSB) for 12 weeks following a PCL reconstruction using a hamstrings graft [43]. They noted that the mean laxity was significantly lower in the group that used the TSB post-operatively. It is the senior authors' routine practice to use a PCL brace post-operatively for 12 weeks whilst the graft is healing to prevent residual laxity.

A study by the LaPrade group [25] has shown that the kinematics of a double bundle PCL reconstruction more accurately represents the native knee, particularly in flexion ranges beyond 90°. There was significantly less posterior tibial translation with a double bundle compared to a single bundle PCL reconstruction. Gwinner [44] has shown that a decreased posterior tibial slope is associated with an increase in post-operative posterior tibial translation and, therefore, increased graft laxity. Post-operative graft laxity does not seem to be influenced by the posterior tibial slope when a double bundle PCL reconstruction is performed [45]. Double bundle PCL reconstruction has been shown to have comparable subjective and functional outcomes to an isolated ACL reconstruction group in a recent cohort study [33].

15.3.2 Acute PCL Avulsion Fracture

If there is a significant bony avulsion, then fixation back to the tibia usually yields normal PCL laxity and function. The opportunity for this should not be missed. Whilst there is a place for non-operative management for absolutely undisplaced lesions, too often surgeons elect to brace these patients who then end up with permanent and excessive PCL laxity (Fig. 15.2). There is understandably fear about undertaking posterior surgery but with good dissection, exposure and surgical technique, it is straightforward, and a failure to take this opportunity of early fixation is regrettable. Open surgery was traditional with a posterior approach [46] but modern arthroscopic techniques are also possible [47].

15.3.3 Acute PCL Injuries in the Context of a Combined Ligament Injury

Most PCL surgery is undertaken for cases that have a combined ligament injury. The surgical treatment of knee dislocation and combined ligament injury is evolving. In this area, the variety of injury patterns and surgical options, and evo-

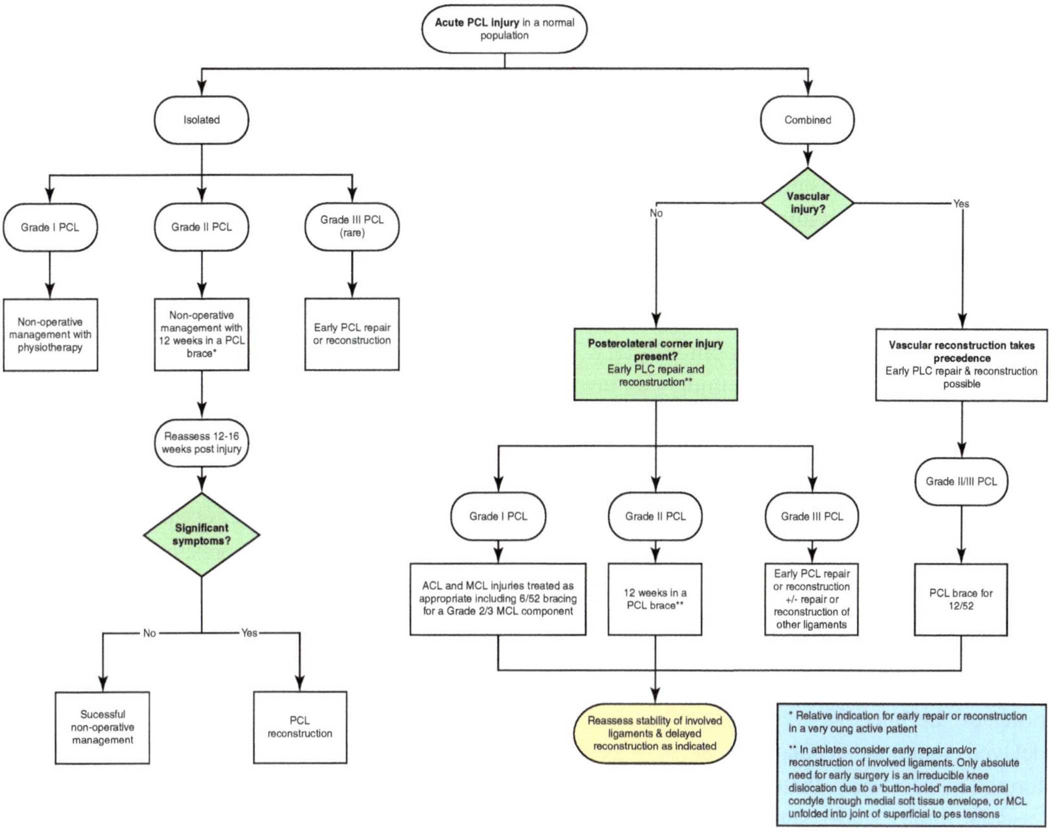

Fig. 15.3 Algorithm for management of PCL injuries

lution of these, mean that evidence base is hard to come by. Experiential-based decision making will almost certainly prevail. Acute surgery is still often necessary but less often undertaken by the senior author as he has realised that the consequence of such surgery is often stiffness. If surgery is required, a staged approach can often be beneficial to the patient or a period of bracing first [48]. A suggested approach to the management of an acute PCL injury is shown in Fig. 15.3.

15.3.3.1 Combined Ligament Injury with No Vascular Injury

Amongst the components of multiligament injury, posterolateral corner disruption requires early surgery as the results are superior to late reconstruction alone. In the acute setting, an anatomic repair of structures combined with a reconstruction to protect the repairs is more effective [49].

Most MCL injuries (even Grade III) can be treated with bracing initially [50, 51]. If persistent excess laxity persists, then a delayed surgery around 3 months later to tighten the excessively slack structures and reconstruct the MCL can be effective and rival results in the general population from early anatomic repair with reconstruction to protect the repair. It has a much lower risk of significant stiffness. With high-level athletes, the best results do come from early surgery [52, 53].

With PCL involvement, surgery is required for most Grade III lesions. Nevertheless, PCL-directed bracing can allow good healing of Grade II PCL lesions even if other parts of the injury, for example, posterolateral, need surgery. Many of these PCL lesions will heal and not need delayed surgery. This is even with a combined ACL rupture as an isolated delayed ACL reconstruction is an attractive alternative to acute multiligament reconstruction.

15.3.3.2 Combined Ligament Injury with a Vascular Injury

The incidence of vascular injury following a multiligament knee injury is 6.2–14% [54–57]. A detailed discussion of the acute management of a knee dislocation is outside the scope of this chapter. However, it goes without saying that the management of the vascular injury takes precedence over the rest of the knee injury. This may involve vascular bypass surgery to overcome the partial or complete occlusion of the popliteal artery. The acute management of the knee injury can involve placement of an external fixator or a knee brace. If an external fixator is used, it is best to ensure the knee is held just into hyperextension—communication with the vascular surgeon is important as they frequently prefer to keep the knee flexed which is certain to cause a fixed flexion deformity. The senior authors' preference is to use a PCL brace to allow motion if there is a Grade II or III PCL injury and reassesses the stability of the knee 12 weeks after the injury. A delayed ligament reconstruction will be performed as necessary. Patients with an associated vascular injury have significantly lower functional outcomes than those that do not [57]. If a vascular reconstruction has been performed and PCL reconstruction is subsequently needed, the following precautions are suggested: (1) close liaison with the vascular surgeon, (2) preoperative angiography to assess position of reconstructed vessels, and (3) tourniquet-free surgery is ideal. The tourniquet is applied but not inflated. If tourniquet is needed, it is inflated but deflated as soon as possible during the procedure.

15.3.4 Chronic PCL Injuries

Patients do still present with chronic symptoms related to PCL-insufficiency. Instability is most commonly in cases of a multiligament injury that was overlooked and was assumed to have been an isolated PCL lesion. The patients with isolated PCL lesions will present later, particularly on descending inclines if they have instability, but this is usually relatively subtle, and the patient simply feels that there is 'something wrong with the knee'. With combined ligament injury, however, instability is usually more gross, and the majority of cases will have an associated posterolateral corner insufficiency. However, posteromedial insufficiency is also not uncommon.

Another scenario is that the patient presents not with instability but rather with symptoms of a chondral lesion on the medial femoral condyle related to impingement on the upslope of the anterior tibia, as described above, or later with medial osteoarthritis. If an isolated articular cartilage lesion requires resurfacing treatment, then these techniques will fail without addressing PCL laxity that has caused the lesion. A tibial osteotomy is required if excess varus or a reversed or reduced tibial slope are present.

If a patient presents with frank medial osteoarthritis related to PCL insufficiency, then assuming the lower limb concerned is no longer than the other side and there is no valgus alignment, a medial opening wedge high tibial osteotomy is appropriate [58]. When undertaking this technique for standard medial osteoarthritis, it is notable that it is difficult to maintain the same preoperative posterior tibial slope [59, 60] as it is very easy to inadvertently increase the slope. In the context of PCL-related medial osteoarthritis, this is advantageous as increasing the posterior tibial slope tends to translate the tibia forward and allows the femur to articulate more posteriorly on healthy articular surface under weight-bearing conditions. It does, however, cause a loss of extension and, therefore, the surgeon has to be careful not to produce a fixed flexion deformity. Due to the nature of the osteotomy being an opening wedge, it is usually best to stage a PCL reconstruction if needed. Often the result is sufficiently good that delayed PCL reconstruction is no longer necessary. This can even be the case when there is associated posterolateral insufficiency when a medial opening wedge technique is used. The valgising effect of the osteotomy will tend to shift the mechanical axis into the lateral compartment, which closes down under load, and thereby counters LCL laxity. In addition, by increasing the tibial slope, the femur tends to move posteriorly. This is resisted medially by the well-fixed posterior medial meniscus, but laterally the femur slides down the inferiorly inclined posterolateral tibia with little resistance from the mobile lateral meniscus and thereby there is more posterior femoral translation laterally causing a relative external rotation of the femur/internal rotation of tibia which counters the external tibial rotation seen in posterolateral rotatory instability following lesions of the popliteus complex [61, 62].

An aspect of chronic PCL-deficient cases that is rarely considered is the problem of fixed posterior tibial translation that results from chronic capsular contracture. In the study mentioned above, using weight-bearing MRI confirming that isolated PCL insufficiency causes altered medial compartment kinematics, the study subjects had an anterior drawer force applied to their upper tibia during scanning [16]. In none could the abnormal posterior tibial subluxation be completely corrected. All of the subjects were asymptomatic with only isolated Grade 1 or 2 injuries of the PCL but despite relatively minor PCL deficits had fixed posterior subluxation. This suggests this phenomenon is more common than realized. It may be one of the reasons why isolated PCL reconstruction is less successful than ACL reconstruction. In clinically obvious cases of fixed posterior subluxation, which are thankfully rare, slope changing osteotomy will slowly correct the deformity. In chronically dislocated knees, the senior author is aware of slow relocation of the joint being possible with the use of external fixators.

15.4 Conclusion

The number of cases of PCL reconstruction in a knee surgeon's practice should be relatively small, especially given the improvements in nonoperative management with bracing. Nevertheless, nearly all Grade 3 injuries and some Grade 2 lesions will benefit from reconstruction. PCL repair is an ever-improving technique but its role is yet to be defined. When PCL surgery is needed, in competent hands, the results are pleasing but less predictable than ACL reconstruction.

References

1. Sanders TL, Pareek A, Barrett IJ, Kremers HM, Bryan AJ, Stuart MJ, et al. Incidence and long-term follow-up of isolated posterior cruciate ligament tears. Knee Surg Sports Traumatol Arthrosc. 2017;25(10):3017–23.
2. Schulz MS, Steenlage ES, Russe K, Strobel MJ. Distribution of posterior tibial displacement in knees with posterior cruciate ligament tears. JBJS. 2007;89(2):332–8.
3. Eakin CL, Cannon WD. Arthrometric evaluation of posterior cruciate ligament injuries. Am J Sports Med. 1998;26(1):96–102.
4. Daniel DM, Stone ML, Barnett P, Sachs R. Use of the quadriceps active test to diagnose posterior cruciate-ligament disruption and measure posterior laxity of the knee. J Bone Joint Surg Am. 1988;70(3):386–91.
5. Gross ML, Grover JS, Bassett LW, Seeger LL, Finerman GAM. Magnetic resonance imaging of the posterior cruciate ligament: clinical use to improve diagnostic accuracy. Am J Sports Med. 1992;20(6):732–7.
6. Rodriguez W, Vinson EN, Helms CA, Toth AP. MRI appearance of posterior cruciate ligament tears. Am J Roentgenol. 2008;191(4):W155–9.
7. Dp T, Hm F, Rd F, Dc Q, Dd B. Chronically injured posterior cruciate ligament: magnetic resonance imaging. Clin Orthop. 1997;(335):224–32.
8. Servant CTJ, Ramos JP, Thomas NP. The accuracy of magnetic resonance imaging in diagnosing chronic posterior cruciate ligament injury. Knee. 2004;11(4):265–70.
9. Akisue T, Kurosaka M, Yoshiya S, Kuroda R, Mizuno K. Evaluation of healing of the injured posterior cruciate ligament: analysis of instability and magnetic resonance imaging. Arthrosc J Arthrosc Relat Surg. 2001;17(3):264–9.
10. Thompson SM, Williams AM, Lavelle J, Church S. Assessment of posterior cruciate ligament injury using the merchant view. J Orthop Trauma. 2017;7(1):1–4.
11. Puddu G, Giannì E, Chambat P, De Paulis F. The axial view in evaluating tibial translation in cases of insufficiency of the posterior cruciate ligament. Arthrosc J Arthrosc Relat Surg. 2000;16(2):217–20.
12. Jackman T, LaPrade RF, Pontinen T, Lender PA. Intraobserver and interobserver reliability of the kneeling technique of stress radiography for the evaluation of posterior knee laxity. Am J Sports Med. 2008;36(8):1571–6.
13. Hewett TE, Noyes FR, Lee MD. Diagnosis of complete and partial posterior cruciate ligament ruptures. Stress radiography compared with KT-1000 arthrometer and posterior drawer testing. Am J Sports Med. 1997;25(5):648–55.
14. Garavaglia G, Lubbeke A, Dubois-Ferrière V, Suva D, Fritschy D, Menetrey J. Accuracy of stress radiography techniques in grading isolated and combined

posterior knee injuries: a cadaveric study. Am J Sports Med. 2007;35(12):2051–6.

15. Logan M, Dunstan E, Robinson J, Williams A, Gedroyc W, Freeman M. Tibiofemoral kinematics of the anterior cruciate ligament (ACL)-deficient weightbearing, living knee employing vertical access open "interventional" multiple resonance imaging. Am J Sports Med. 2004;32(3):720–6.
16. Logan M, Williams A, Lavelle J, Gedroyc W, Freeman M. The effect of posterior cruciate ligament deficiency on knee kinematics. Am J Sports Med. 2004;32(8):1915–22.
17. The effects of sectioning of the posterior cruciate ligament and the posterolateral complex on the articular contact pressures within the knee. Abstract. Europe PMC. [cited 2020 Mar 15]. https://europepmc.org/article/med/8501084.
18. Logan MC, Williams A, Lavelle J, Gedroyc W, Freeman M. Tibiofemoral kinematics following successful anterior cruciate ligament reconstruction using dynamic multiple resonance imaging. Am J Sports Med. 2004;32(4):984–92.
19. Heinrichs CH, Schmoelz W, Mayr R, Keiler A, Schöttle PB, Attal R. Biomechanical evaluation of a novel dynamic posterior cruciate ligament brace. Clin Biomech. 2016;33:20–5.
20. LaPrade RF, Smith SD, Wilson KJ, Wijdicks CA. Quantification of functional brace forces for posterior cruciate ligament injuries on the knee joint: an in vivo investigation. Knee Surg Sports Traumatol Arthrosc. 2015;23(10):3070–6.
21. Jacobi M, Reischl N, Wahl P, Gautier E, Jakob RP. Acute isolated injury of the posterior cruciate ligament treated by a dynamic anterior drawer brace. J Bone Joint Surg Br. 2010;92-B(10):1381–4.
22. Agolley D, Gabr A, Benjamin-Laing H, Haddad FS. Successful return to sports in athletes following non-operative management of acute isolated posterior cruciate ligament injuries. Bone Jt J. 2017;99-B(6):774–8.
23. Hopper GP, Heusdens CHW, Dossche L, Mackay GM. Posterior cruciate ligament repair with suture tape augmentation. Arthrosc Tech. 2019;8(1):e7–10.
24. Heusdens CHW, Tilborghs S, Dossche L, Van Dyck P. Primary posterior cruciate ligament repair with the novel suture tape augmentation technique. Surg Technol Int. 2019;34:469–75.
25. Wijdicks CA, Kennedy NI, Goldsmith MT, Devitt BM, Michalski MP, Årøen A, et al. Kinematic analysis of the posterior cruciate ligament, part 2: a comparison of anatomic single- versus double-bundle reconstruction. Am J Sports Med. 2013;41(12):2839–48.
26. Harner CD, Janaushek MA, Kanamori A, Yagi M, Vogrin TM, Woo SL-Y. Biomechanical analysis of a double-bundle posterior cruciate ligament reconstruction. Am J Sports Med. 2000;28(2):144–51.
27. Hatayama K, Higuchi H, Kimura M, Kobayashi Y, Asagumo H, Takagishi K. A comparison of arthroscopic single- and double-bundle posterior cruciate ligament reconstruction: review of 20 cases. Am J Orthop Belle Mead NJ. 2006;35(12):568–71.
28. Wang C-J, Weng L-H, Hsu C-C, Chan Y-S. Arthroscopic single- versus double-bundle posterior cruciate ligament reconstructions using hamstring autograft. Injury. 2004;35(12):1293–9.
29. Houe T, Jørgensen U. Arthroscopic posterior cruciate ligament reconstruction: one- vs. two-tunnel technique. Scand J Med Sci Sports. 2004;14(2):107–11.
30. Kohen RB, Sekiya JK. Single-bundle versus double-bundle posterior cruciate ligament reconstruction. Arthrosc J Arthrosc Relat Surg. 2009;25(12):1470–7.
31. Yoon KH, Bae DK, Song SJ, Cho HJ, Lee JH. A prospective randomized study comparing arthroscopic single-bundle and double-bundle posterior cruciate ligament reconstructions preserving remnant fibers. Am J Sports Med. 2011;39(3):474–80.
32. Wiley WB, Owen JR, Pearson SE, Wayne JS, Goradia VK. Medial femoral condyle strength after tunnel placement for single- and double-bundle posterior cruciate ligament reconstruction. J Knee Surg. 2007;20(3):223–7.
33. LaPrade RF, Cinque ME, Dornan GJ, DePhillipo NN, Geeslin AG, Moatshe G, Chahla J. Double-bundle posterior cruciate ligament reconstruction in 100 patients at a mean 3 years' follow-up: outcomes were comparable to anterior cruciate ligament reconstructions. 2018. [cited 2021 Jan 17]. https://journals.sagepub.com/doi/abs/10.1177/0363546517750855.
34. Maletis GB, Chen J, Inacio MCS, Love RM, Funahashi TT. Increased risk of revision after anterior cruciate ligament reconstruction with bone-patellar tendon-bone allografts compared with autografts. Am J Sports Med. 2017;45(6):1333–40.
35. Kaeding CC, Aros B, Pedroza A, Pifel E, Amendola A, Andrish JT, et al. Allograft versus autograft anterior cruciate ligament reconstruction. Sports Health. 2011;3(1):73–81.
36. Chen CP, Lin YM, Chiu YC, Wu HW, Lee CH, Tong KM, et al. Outcomes of arthroscopic double-bundle PCL reconstruction using the LARS artificial ligament. Orthopedics. 2012;35(6):e800–6.
37. Li B, Wen Y, Wu H, Qian Q, Wu Y, Lin X. Arthroscopic single-bundle posterior cruciate ligament reconstruction: retrospective review of hamstring tendon graft versus LARS artificial ligament. Int Orthop. 2009;33(4):991–6.
38. Xu X, Huang T, Liu Z, Wen H, Ye L, Hu Y, et al. Hamstring tendon autograft versus LARS artificial ligament for arthroscopic posterior cruciate ligament reconstruction in a long-term follow-up. Arch Orthop Trauma Surg. 2014;134(12):1753–9.
39. Hammoud S, Reinhardt KR, Marx RG. Outcomes of posterior cruciate ligament treatment: a review of the evidence. Sports Med Arthrosc Rev. 2010;18(4):280–91.
40. Spiridonov SI, Slinkard NJ, LaPrade RF. Isolated and combined grade-III posterior cruciate ligament tears treated with double-bundle reconstruction with use

of endoscopically placed femoral tunnels and grafts: operative technique and clinical outcomes. JBJS. 2011;93(19):1773–80.
41. Hermans S, Corten K, Bellemans J. Long-term results of isolated anterolateral bundle reconstructions of the posterior cruciate ligament: a 6- to 12-year follow-up study. Am J Sports Med. 2009;37(8):1499–507.
42. Jackson WFM, van der Tempel WM, Salmon LJ, Williams HA, Pinczewski LA. Endoscopically-assisted single-bundle posterior cruciate ligament reconstruction: results at minimum ten-year follow-up. J Bone Joint Surg Br. 2008;90(10):1328–33.
43. Li B, Shen P, Wang J-S, Wang G-B, He M, Bai L-H. Therapeutic effects of tibial support braces on posterior stability after posterior cruciate ligament reconstruction with autogenous hamstring tendon graft. Eur J Phys Rehabil Med. 2015;51(2):163–70.
44. Gwinner C, Weiler A, Roider M, Schaefer FM, Jung TM. Tibial slope strongly influences knee stability after posterior cruciate ligament reconstruction: a prospective 5- to 15-year follow-up. Am J Sports Med. 2017;45(2):355–61.
45. Bernhardson AS, DePhillipo NN, Aman ZS, Kennedy MI, Dornan GJ, LaPrade RF. Decreased posterior tibial slope does not affect postoperative posterior knee laxity after double-bundle posterior cruciate ligament reconstruction. Am J Sports Med. 2019;47(2):318–23.
46. Hooper POI, Bevan PJ, Silko C, Farrow LD. A posterior approach to open reduction and internal fixation of displaced posterior cruciate ligament tibial osseous avulsions. JBJS Essent Surg Tech. 2018;8(1):e6.
47. Arthroscopically assisted treatment of avulsion fractures of... : JBJS. [cited 2020 Mar 8]. https://journals.lww.com/jbjsjournal/Abstract/2001/05000/Arthroscopically_Assisted_Treatment_of_Avulsion.8.aspx.
48. Subbiah M, Pandey V, Rao SK, Rao S. Staged arthroscopic reconstructive surgery for multiple ligament injuries of the knee. J Orthop Surg. 2011;19(3):297–302.
49. Levy BA, Dajani KA, Morgan JA, Shah JP, Dahm DL, Stuart MJ. Repair versus reconstruction of the fibular collateral ligament and posterolateral corner in the multiligament-injured knee. 2010. [cited 2020 Mar 8]. https://journals.sagepub.com/doi/abs/10.1177/0363546509352459.
50. Mok DWH, Good C. Non-operative management of acute grade III medial collateral ligament injury of the knee: a prospective study. Injury. 1989;20(5):277–80.
51. Shelbourne KD, Porter DA. Anterior cruciate ligament-medial collateral ligament injury: nonoperative management of medial collateral ligament tears with anterior cruciate ligament reconstruction: a preliminary report. Am J Sports Med. 1992;20(3):283–6.
52. Jones M, Ball S, Williams A, Borque K. Multi-ligament knee injuries in elite athletes: return to play rates, timing, and complications. Orthop J Sports Med. 2020;8(7_Suppl 6):2325967120S00366.
53. Jones M, Mahmud T, Narvani A, Hamid I, Lewis J, Williams A. The aetiology, early and late management and outcomes of 141 multi-ligament injured knees. Orthop Proc. 2012;94-B(SUPP_XXXIX):244.
54. Ríos A, Villa A, Fahandezh H, de José C, Vaquero J. Results after treatment of traumatic knee dislocations: a report of 26 cases. J Trauma. 2003;55(3):489–94.
55. Harner CD, Waltrip RL, Bennett CH, Francis KA, Cole B, Irrgang JJ. Surgical management of knee dislocations. J Bone Joint Surg Am. 2004;86(2):262–73.
56. Boisrenoult P, Lustig S, Bonneviale P, Leray E, Versier G, Neyret P, et al. Vascular lesions associated with bicruciate and knee dislocation ligamentous injury. Orthop Traumatol Surg Res OTSR. 2009;95(8):621–6.
57. Sanders TL, Johnson NR, Levy NM, Cole PAJ, Krych AJ, Stuart M, et al. Effect of vascular injury on functional outcome in knees with multi-ligament injury: a matched-cohort analysis. JBJS. 2017;99(18):1565–71.
58. Savarese E, Bisicchia S, Romeo R, Amendola A. Role of high tibial osteotomy in chronic injuries of posterior cruciate ligament and posterolateral corner. J Orthop Traumatol. 2011;12(1):1–17.
59. Brouwer RW, Bierma-Zeinstra SMA, van Koeveringe AJ, Verhaar JAN. Patellar height and the inclination of the tibial plateau after high tibial osteotomy. The open versus the closed-wedge technique. J Bone Joint Surg Br. 2005;87(9):1227–32.
60. El-Azab H, Halawa A, Anetzberger H, Imhoff AB, Hinterwimmer S. The effect of closed- and open-wedge high tibial osteotomy on tibial slope: a retrospective radiological review of 120 cases. J Bone Joint Surg Br. 2008;90(9):1193–7.
61. Arthur A, LaPrade RF, Agel J. Proximal tibial opening wedge osteotomy as the initial treatment for chronic posterolateral corner deficiency in the Varus knee: a prospective clinical study. Am J Sports Med. 2007;35(11):1844–50.
62. LaPrade RF, Engebretsen L, Johansen S, Wentorf FA, Kurtenbach C. The effect of a proximal tibial medial opening wedge osteotomy on posterolateral knee instability: a biomechanical study. Am J Sports Med. 2008;36(5):956–60.

Technique Corner: Posterior Cruciate Ligament Injuries

16

Jonathan D. Hughes, Christopher M. Gibbs, Neel K. Patel, Jan-Dierk Clausen, and Volker Musahl

16.1 Introduction

The posterior cruciate ligament (PCL) is the largest and strongest intra-articular ligament in the knee [32, 53]. The PCL is comprised of two bundles, the anterolateral and posteromedial bundles, that function synergistically to resist posterior translation of the tibia at all degrees of knee flexion [34]. PCL injuries account for 3% of all acute knee injuries and typically occur with an acute hemarthroses in trauma patients [15]. PCL injuries typically occur after a direct injury that results in a posterior load being applied to the proximal anterior tibia, with typical mechanisms being dashboard injuries after a motor vehicle collision. However, PCL injuries rarely occur as an isolated injury, with more than 90% of tears occurring along with other knee ligament tears [75]. While these injuries are most common in trauma patients, they are frequently seen in sports such as football and skiing when there is a direct posterior load on the tibia with the foot in a plantarflexed position [15]. PCL injuries may lead to persistent instability and increase the risk of posttraumatic osteoarthritis if not addressed properly [43]. A recent retrospective review demonstrated patients with an isolated PCL tear have a significantly higher risk of developing symptomatic arthritis compared to individuals without PCL injury [62]. Despite this potential long-term disability, the treatment algorithm for complete, isolated PCL tears remains controversial. In this chapter, we will discuss the anatomy and biomechanics of the PCL, the evaluation and treatment of PCL injuries, and the outcomes following PCL injuries.

16.2 Anatomy/Biomechanics of the PCL

The PCL originates on the anterior aspect of the medial femoral condyle within the notch and inserts approximately 1 cm distal to the joint line on the posterior aspect of the tibial plateau. The PCL is comprised of two bundles, the anterolateral and posteromedial bundle. The total length is between 32 and 38 mm with an average midsubstance diameter of 11–13 mm [44]. The anterolateral bundle is the larger of the two bundles and its femoral footprint is approximately 11.0 mm from the medial arch point and 7.9 mm

J. D. Hughes (✉) · C. M. Gibbs · N. K. Patel
V. Musahl
Department of Orthopaedic Surgery, University of Pittsburgh, UPMC Freddie Fu Sports Medicine Center, Pittsburgh, PA, USA
e-mail: Hughesjd3@upmc.edu; gibbscm2@upmc.edu; musahlv@upmc.edu

J.-D. Clausen
Orthopedic Trauma Department, Hanover Medical School, Hanover, Lower Saxonyy, Germany
e-mail: clausen.jan-dierk@mh-hannover.de

N. Nakamura et al. (eds.), *Advances in Knee Ligament and Knee Preservation Surgery*,
https://doi.org/10.1007/978-3-030-84748-7_16

proximal to the distal articular cartilage [2]. The femoral footprint of the posteromedial bundle is located 11.1 mm from the medial arch point and 10.8 mm from the posterior point of the articular cartilage margin [2]. The tibial insertion of the anterolateral and posteromedial bundles are 10.7 and 4.7 mm anterior to the champagne glass drop-off of the posterior tibia, respectively [2]. The relative distance between the PCL and the popliteal fossa, mainly the popliteal artery, is important to know when reconstructing the ligament. The mean sagittal distance from the proximal PCL fovea to the popliteal artery was found to be 9.7 ± 5 mm (range 3–15 mm) with the distance increasing with increasing knee flexion [12, 48]. Establishing the landmarks for these bundles and understanding their relationships with surrounding structures is key when reconstructing the ligament as will be discussed later in this chapter.

Many studies have been performed to analyze the function of the PCL bundles and the consequences of PCL injury. The tensile strength of the PCL has reported to be between 739 and 1627 N, with the anterolateral bundle contributing significantly more than the posteromedial bundle to overall tensile strength [44]. The anterolateral bundle functions primarily to resist posterior translation of the tibia between 70° and 105° of knee flexion. The posteromedial bundle also resists posterior translation of the tibia, but it functions primarily between 0° and 15° [34]. The PCL also functions as a secondary stabilizer to rotation, especially at knee flexion angles greater than 90°. Sectioning of both bundles was required to produce greater than 10 mm of posterior tibial translation (11.7 mm) in one biomechanical study, suggesting that a complete tear is needed in order to observe a clinical grade 3 injury. Complete rupture of the PCL results in fixed anterior subluxation of the medial femoral condyle relative to the tibia and may predispose to accelerated osteoarthritis if it not properly addressed [43]. This information is important to consider when analyzing findings of the physical exam and determining the best treatment for isolated PCL injuries.

16.3 Clinical Presentation

For patients with PCL injury, a thorough clinical history most commonly identifies a description of hyperflexion of the knee with or without pretibial trauma, most frequently resulting from athletic injury or traffic accidents [15, 17]. Patients with PCL injury often present with vague symptoms such as unsteadiness, knee stiffness, swelling, or pain, and are often unable to identify the exact mechanism of injury [5, 46, 63]. In fact, many patients with PCL insufficiency present over 30 days from injury [63]. Patients with chronic PCL deficiency may present with vague complaints of increased knee pain and impaired function [8].

The physical exam should start with a careful neurovascular exam, as PCL injury may occur as a result of knee dislocation [4, 15, 35, 36]. A systematic review found an 18% rate of vascular injury and a 25% rate of nerve injury in the setting of knee dislocation [50]. Concern for vascular injury on initial exam, such as a pale or cool foot or abnormal pulses, should be further evaluated by measurement of the ankle-brachial index, and if <0.9, followed by angiography or duplex ultrasonography. Close observation of gait and weight-bearing limb alignment, as well as inspection of the injured and contralateral knees, should be performed. A mild bloody effusion, bruising in the popliteal fossa, anterior knee abrasions, and varus malalignment or thrust may be observed [17, 51, 74]. Palpation should be performed to assess for areas of tenderness or effusion, and range of motion should be evaluated, although in the acute setting this can be limited due to discomfort [5, 74]. Finally, stress testing of the cruciate and collateral ligaments to identify concomitant injuries is important as PCL injuries rarely occur in isolation [4, 74]; additional injuries to evaluate for include ligamentous or meniscal injuries and peri-articular fractures [4, 15, 36].

There are numerous special tests described to evaluate PCL deficiency. The posterior drawer test coupled with palpation of the tibia-femur step-off has been shown to be the most sensitive clinical test, with a sensitivity of 90% and speci-

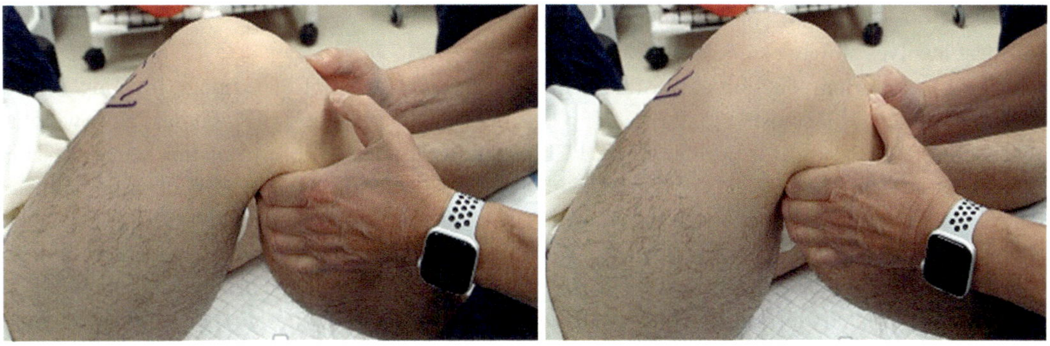

Fig. 16.1 Posterior drawer exam. It is performed by applying a posteriorly directed force to the anterior tibia and assessing the amount of posterior tibial translation. When performing this test, the examiner must ensure that the tibia is in its neutral position, as seen on the left image. On the right image, a posterior directed force on the tibia yields significant posterior tibial translation

ficity of 99% [61]. It is performed by applying a posteriorly directed force to the anterior tibia and assessing the amount of posterior tibial translation (Fig. 16.1) [74]. When performing this test, the examiner must ensure that the tibia is in its neutral position as posterior subluxation of the tibia, as would occur with a PCL injury, can lead to a false negative finding for PCL injury or a false positive finding for ACL injury with relative excess anterior translation of the tibia. The presence of such a posterior subluxation is known as the Posterior Sag Test, and is recognized by loss of contour of the tibial tubercle or posterior sagging of the tibia [74]. Classification of PCL injuries relies on the amount of posterior translation with this test. A grade 1 injury is represented by <5 mm of posterior translation or when the anterior border of the tibial plateau lies anterior to the femoral condyle; a grade 2 injury is between 5 and 10 mm of posterior translation or when the anterior border of the tibial plateau lies flush with the femoral condyles; a grade 3 injury is >10 mm of posterior translation or when the anterior border of the tibial plateau lies posterior to the femoral condyles [5, 24, 46]. Another more descriptive classification from the international knee documentation committee (IKDC) describes posterior translation of 0–2 mm, 3–5 mm, 6–10 mm, and >10 mm as "normal," "nearly normal," "abnormal," or "severely abnormal," respectively [26]. A grade 2 posterior drawer should alert the clinician to the likely presence of a PCL injury, while grade 3 indicates combined PCL and posterolateral corner (PLC) insufficiency [64].

The quadriceps active test [13] is performed with the patient in the supine position starting with the injured knee flexed to 90° and asking the patient to slide his or her foot down the table which causes the quadriceps to fire [5]. In a knee without PCL deficiency, the force of the extensor mechanism is directed slightly posterior, but with PCL deficiency, the tibia is translated posterior in relationship to the femur resulting in an anteriorly directed force with contraction of the quadriceps. For this reason, when a patient with PCL deficiency flexes the quadriceps muscle, the tibia translates anteriorly into its reduced position. Of note, this test is not accurate in patients with a history of injury or surgery on the patella or proximal tibia, such as patellectomy or tibial tubercle osteotomy, as these procedures alter the force vector which may cause inaccurate findings. The quadriceps active test has been shown to detect 97% of patients with PCL deficiency without a single false positive test [13].

A reverse pivot shift and dynamic posterior shift test have also been described [3]. The reverse pivot shift is performed with the patient supine and the knee flexed to 90°. The examiner externally rotates the femur while applying a valgus stress and extends the knee. A positive test is indicated by reduction of the tibia anteriorly around 20–40° of flexion. The dynamic posterior shift test involves slowly extending the knee and observing for anterior reduction of the tibia as the knee approaches full extension [5, 74].

The dial test is useful for evaluating for PCL and concomitant PLC injury. It is performed with the patient prone and the knee flexed to 90° while the examiner externally rotates the tibia. The test is then performed with the knee flexed to 30°. The examiner's hand (left hand while testing a left knee, and right hand testing a right knee) is placed on the dorsal thigh. The contralateral hand is placed on the ankle and foot, and an external rotation force is applied to the leg. Classically, a positive test is indicated by an increase in 10°of external rotation; a positive test at 30° of knee flexion raises concern for isolated PLC injury, while a positive test at both 30° and 90° raises concern for combined PLC and PCL injuries [5]. Grading can also be done according to the IKCD in which grade 0 is <5°, grade 1 is 6–10°, grade 2 is 11–19°, and grade 3 is >20° of rotational difference compared to the contralateral side. However, cadaveric analysis of the dial test [64] demonstrated an average of 10° of external rotation at 30° and 90° of knee flexion in intact knees. With sectioning of the PCL, external rotation increased to 15° and 16° at 30° and 90° of knee flexion, respectively. After additional sectioning of the PLC, external rotation at 30° and 90° of knee flexion increased to 22° and 28°, respectively.

The overall accuracy of clinical evaluation for diagnosis of PCL injury was assessed by blinding five sports-medicine trained orthopedic surgeons who performed a physical exam on 40 patients who had either PCL injury, ACL injury, or a normal knee [61]. Clinical examination was found to be 96% accurate with 90% sensitivity and 99% specificity for the diagnosis of PCL injury. The sensitivity of clinical examination for detection of grade 1 PCL tear was 70%, but 97% for high grade (grade 2–3) tears. Notably, KT-1000 arthrometer testing was found to be 89% accurate, but sensitivity was found to be only 33%, with 94% specificity for low-grade injuries while sensitivity was 86% with specificity of 100% for high-grade injuries.

16.4 Imaging

In the evaluation of PCL injury, imaging should begin with standard AP, lateral, 45° weight-bearing, and axial views of the knee [5]. This allows detection of overall limb alignment as well as identification of an avulsion injury or fracture. Long leg cassette views are helpful to better evaluate lower extremity alignment [51]. In the setting of chronic PCL injury, as time from injury increases, radiographic evidence of articular degeneration may be found [8]. Close scrutiny of radiographs may reveal a reverse Segond fracture of the medial tibial plateau which has been described in patients with combined PCL, medial meniscus, and MCL injuries; this finding is thought to result from valgus stress and external rotation of the flexed knee [14, 23]. Additional peri-articular fractures may also be seen [36, 51].

Multiple stress radiographic techniques have been described in the evaluation of PCL insufficiency. An AP varus stress radiograph may be useful in the evaluation of lateral ligament, PLC, and PCL injuries. A cadaveric study showed that with increase in lateral compartment gap of approximately 2.7 mm, 4.0 mm, and 7.8 mm compared to the intact state, clinicians should suspect an isolated fibular collateral ligament injury, a grade-3 PLC injury, and combined injuries to the ACL, PCL, and PLC, respectively [40]. Stress techniques for measurement of posterior laxity include use of a stress device, hamstring contraction, kneeling, and gravity to provide a posteriorly directed force on the tibia while a lateral radiograph of the knee is obtained, as well as an axial view [30]. For measurement of posterior translation on lateral radiographs, the distance between the posterior tibial plateau and femoral condyles is measured. In the Telos method, the Telos GA II stress device is used to apply a 150 N posterior load to the anterior aspect of the tibia while the patient is in the lateral decubitus position with the knee flexed to 90°, and a lateral radiograph of the knee is taken [30]. The hamstring contraction method consists of active hamstring contraction with the knee at 90° of flexion while a lateral x-ray is performed [30]. A kneeling stress radiograph involves applying the full body weight to the injured knee flexed to 90° with the tibial tubercle supported on a padded jig, and has been shown to be reliable and reproducible [28, 30]. A gravity stress view involves obtaining a lateral x-ray while the patient is supine and the hip and knee are flexed to 90° to allow gravity to apply a posterior force to the tibia [30]. In the Puddu axial view [30, 57], the patient lays supine with the knees flexed to 70°, feet in moderate plantarflexion, and the tibia in neutral rotation. With the patient holding the cassette, the x-ray beam is oriented distal to proximal parallel to the long axis of the patella. To measure posterior instability, a line is drawn tangential to the femoral condyles and the distance to the anterior tibial profile is measured and compared to the contralateral side. The use of a stress device and kneeling test has shown to be the most sensitive in detection of posterior translation; however, they were also found to be the most painful and time-intensive [30].

MRI is considered the gold standard for detection of PCL injuries, with sensitivity and specificity approaching 100% for acute injuries [16, 21, 22]. The native PCL appears as a low-signal intensity structure on T1- and T2-weighted sequences [22]. Disruption of the PCL fibers or high-signal intensity indicates PCL injury [60]. In cases of isolated injury, a mid-substance tear is most commonly seen [60]. Although MRI is accurate for acute injuries, care must be taken not to rely too heavily on MRI for diagnosis of chronic tears, as the ligament can often appear continuous with chronic PCL insufficiency [73]. However, as previously discussed, the sensitivity of detecting PCL injury is limited with low-grade injury; in these situations adjunctive tests such as an MRI may be particularly helpful [61]. As in ACL injuries, bone bruises can often be seen on MRI following PCL injury; however, the location is more variable, often occurring opposite of an MCL or PLC injury [45].

CT and Ultrasound may also be used to in the diagnosis or management of PCL injury. Although MRI has been shown to outperform CT in detecting ligamentous injuries of the knee [52], CT can be useful in the evaluation of bony injury, including avulsion injuries. A recent systematic review and meta-analysis found a 99% sensitivity and 99% specificity for the use of ultrasound in the diagnosis of PCL tears; however, there was considerable heterogeneity, as the technique is considerably affected by both patient positioning and the skill of the examiner [42].

16.5 Treatment

The treatment for isolated PCL injuries remains controversial. The treating physician must first determine whether the injury is acute or chronic, and grading needs to be performed to objectify its severity. As stated previously, if the knee has greater than 12 mm of posterior tibial translation on exam, one must consider a combined injury, which would routinely be treated surgically [28, 63]. Most authors advocate for nonoperative treatment for acute, isolated partial PCL injuries, as long-term studies have demonstrated favorable outcomes [29, 55, 66, 69]. Other studies have shown nonoperative management leads to degenerative changes in the patellofemoral and medial compartments at long term follow-up with variable outcomes [8, 19, 72]. Therefore, many surgeons recommend surgical intervention for acute, complete PCL tears or chronic, symptomatic PCL injuries.

16.5.1 Nonoperative Management

Nonoperative treatment for PCL injuries is feasible due to the inherent healing potential of the PCL after acute injury [29, 67, 68, 73]. However, many of these patients demonstrated increased posterior tibial translation after full healing. Successful healing requires minimizing posterior tibial translation and elongation of the PCL during knee flexion [29, 31]. Dynamic PCL braces apply an anterior force to the proximal posterior tibia, preventing posterior tibial translation and allowing the PCL stumps to remain approximated for improved healing. Since the PCL has variable tension with knee flexion, the brace applies increasing force with more knee flexion, replicating the anatomic forces of the PCL [29, 41]. A recent prospective study found significantly reduced mean posterior tibial translation at 24 month follow-up in all patients prescribed the dynamic PCL brace after acute injury. All patients, though, did have some residual knee laxity at final follow-up [29]. Another study demonstrated similar findings after immobilization in a cylinder cast for 6 weeks followed by treatment in a dynamic brace [31]. Although both studies detailed residual knee laxity, a recent prospective study concluded residual knee laxity did not correlate with functional outcomes [67]. Dynamic PCL braces are indicated as first-line treatment for acute, partial PCL tears. However, further long-term studies are warranted to investigate the effectiveness of dynamic PCL bracing on residual knee laxity and functional outcomes.

16.5.1.1 Nonoperative Treatment Protocol

Although specific treatment algorithms vary among surgeons, the key elements of each protocol remain similar. These key elements include preventing posterior tibial subluxation, strengthening of the quadriceps muscles, and weight bearing as tolerated. The authors' preferred protocol is as follows. After a patient is diagnosed with an acute PCL injury, they are placed in a functional PCL brace. The patient may begin weight bearing immediately, but for complete PCL tears, the treating surgeon may consider keeping the brace locked in full extension. Partial PCL tears are allowed weight bearing in an unlocked brace while avoiding hyperextensions. The brace may be unlocked for range of motion exercises and quadriceps strengthening exercises in all cases. Open chain quadriceps exercises are encouraged while avoiding hyperextension of the knee. Open chain hamstrings exercises are not permitted during the rehabilitation process. Return to sport is individualized to the athlete, but may begin as early as 2 weeks for partial tears, and 6–8 weeks in complete tears. If nonoperative management fails, then surgical intervention is warranted.

16.5.2 Operative Management

There are several described techniques for PCL reconstruction (PCL-R), including single bundle and double bundle reconstruction with autograft or allograft and transtibial tunnel and tibial-inlay techniques.

16.5.2.1 Isometric vs Anatomic PCL-R Techniques

Historically, two approaches were utilized for PCL-R, which included isometric and anatomic approaches. The isometric approach involved finding the isometric femoral attachment of the PCL that would allow minimal length changes of the graft, and thus minimize strain on the graft [20, 56]. This approach, though, was found to overconstrain the knee near extension leading to high graft tension near full knee extension and excessive graft laxity in knee flexion [59]. The anatomic approach involves placing the graft in the center of the PCL-AL bundle on the femoral side, and the anatomic center of the tibial footprint. Various studies have demonstrated that the anatomic approach, as compared to the isometric approach, leads to improved clinical outcomes and decreased posterior tibial translation [18, 49].

16.5.2.2 Transtibial Tunnel vs Tibial-Inlay PCL-R Techniques

Two techniques for tibial graft fixation have been described: transtibial tunnel and tibial-inlay techniques. The transtibial tunnel technique involves placing a bone tunnel into the anatomic PCL footprint on the tibia. This technique creates a sharp angle, or "killer turn," of the graft at the posterior aperture of the tibial tunnel, which may lead to abrasion and attritional rupture of the graft [7, 47]. The tibial-inlay technique, first introduced in 1995 [6], avoids the "killer turn" by placing a bone plug on the posterior aspect of the tibia without the need for a bone tunnel. Initially, this procedure was performed open through a posteromedial approach between the semitendinosus and medial gastrocnemius muscles. Recently, an all-arthroscopic approach has been described with good outcomes [37]. Recent studies have demonstrated improved knee stability and good clinical outcomes with both techniques [27, 37, 54].

16.5.2.3 Double-Bundle PCL-R

Double-bundle PCL-R was introduced to reconstruct the anterolateral (AL) and posteromedial (PM) bundles and restore normal anatomy and knee biomechanics [1]. Several biomechanical studies have shown that double-bundle PCL-R more closely restores the biomechanics of the normal knee than single-bundle PCL-R [25, 77]. To perform double-bundle PCL-R, either the transtibial tunnel or tibial-inlay technique can be utilized. The transtibial tunnel technique involves creating two femoral tunnels within the anatomic footprints of the AL- and PM-bundle and one tunnel on the tibial side [71]. Any graft choice can be utilized for this approach, with the bone block placed in the tibial tunnel. For the tibial-inlay technique, an Achilles allograft with bone block or quadriceps tendon autograft with bone block can be used. The soft tissue portion of the graft is split down the center to create two limbs for the double-bundle reconstruction. A trough is created on the PCL tibial footprint and the bone block is placed in this trough. The two limbs are then placed through femoral tunnels placed in the anatomic footprints of the AL- and PM-bundle. The AL-bundle is fixed at 90° of knee flexion with an anterior drawer on the tibia, and the PM-bundle is fixed in full knee extension. As stated previously, this technique can be performed open or arthroscopically depending on the preference of the treating surgeon [6, 37].

16.5.2.4 Graft Options

Graft options for PCL-R include quadriceps tendon autograft with or without bone block, bone-patellar tendon-bone (BPTB) autograft, BPTB allograft, hamstrings tendon (HT) autograft, HT allograft, and Achilles tendon allograft with bone block. A biomechanical study demonstrated a quadruple-stranded HT had significantly higher load to failure than either the BPTB or Achilles grafts, but the BPTB grafts resisted elongation significantly more than the HT [10]. A recent clinical study comparing autografts and allografts found no difference in clinical outcomes, but did report increased donor-site complications with autografts [76].

16.5.2.5 Complications

Although various complications exist for all surgical procedures, there are specific complications related to PCL-R. The most common complication after PCL-R is residual posterior tibial laxity [79]. A dreaded, albeit rare, complication involves injury to the popliteal artery. The popliteal artery is intimately associated with the posterior knee capsule and can be inadvertently damaged during tibial tunnel reaming if the surgeon breaches the far tibial cortex [11]. Additionally, during tibial tunnel reaming, the meniscal roots of the medial or lateral menisci can be damaged with errant placement of the tunnel, stressing the importance of adequate visualization of the PCL tibial footprint and use of intraoperative fluoroscopy [33].

16.5.2.6 Surgical Outcomes

A recent systematic review described outcomes after PCL-R using the transtibial tunnel technique. The authors demonstrated that 75% of patients had normal or nearly normal subjective IKDC scores. This systematic review also reported posterior knee laxity ranged between 2.0 and 5.9 mm postoperatively, which was improved from preoperative scores ranging between 8.4 and 12.3 mm. The authors reported the transtibial tunnel technique can improve posterior knee laxity by one grade; however, the procedure does not restore normal knee stability. Additionally, degenerative knee osteoarthritis was frequently encountered at final follow-up [38].

Various authors have reported significantly improved Lysholm and Tegner scores, as well as side-to-side posterior tibial translation, after the tibial-inlay technique. However, the authors reported residual laxity as compared to the native knee [65, 70].

A systematic review compared the open tibial-inlay and arthroscopic transtibial tunnel techniques. The authors found no difference in anteroposterior (AP) stability nor graft forces in biomechanical analysis between the two techniques. Clinically, the authors reported no significant difference in the Tegner and Lysholm knee scores and AP stability at short-term follow-up between the two techniques [54]. Another recent retrospective review showed no significant difference in side-to-side posterior tibial translation nor Lysholm and Tegner scores between arthroscopic transtibial tunnel and tibial-inlay techniques. The authors did report the posterior tibial translations of the transtibial tunnel and tibial-inlay techniques were 5.6 ± 2 mm and 4.7 ± 1.6 mm, respectively, which were significantly greater than a native knee [37].

Many authors have advocated for double-bundle PCL-R due to residual knee laxity after single-bundle reconstruction [37, 71, 77]. However, recent literature has shown mixed results. Various studies have shown improved postoperative subjective scores and decreased posterior tibial translation from preoperative measurements after double-bundle reconstruction. However, the posterior tibial translations ranged from 0.9 to 3.9 mm, which is still increased compared to a native knee [71, 78]. A recent systematic review and meta-analysis found significantly improved posterior tibial translation and IDKC scores in the double-bundle group compared to the single-bundle group, but no difference in Lysholm nor Tegner scores [9]. A retrospective review of single-bundle and double-bundle techniques showed improved mean side-to-side posterior tibial translation in the double-bundle group compared to the single-bundle group, but no difference in knee range of motion nor Lysholm scores [37]. Another recent systematic review of seven articles reported no difference in clinical nor functional outcomes between single-bundle and double-bundle PCL-R [58].

16.5.2.7 Author's Preferred Technique

The author's preferred technique is an arthroscopic single-bundle anatomic PCL-R using a transtibial tunnel technique. For this technique, in an isolated PCL-R, the authors will use a quadriceps tendon autograft with bone block for younger patients, while an Achilles allograft with bone block is used for older patients and multiligamentous knee injuries. On the femoral side, a 10 mm diameter tunnel is reamed in the anatomic footprint of the AL bundle (Fig. 16.2). Augmentation of residual PCL fibers is performed whenever possible, with preservation of the PM bundle. On the tibial side, the tibial footprint is fully visualized with a 70° arthroscope through a modified Gilquist and posteromedial portal. A tibial guide is then used under fluoroscopic guidance to place a guide pin in the center of the PCL tibial footprint (Fig. 16.3). A prior cadaveric study showed that aiming at the AL bundle, instead of the center of the footprint, increases the risk of injury to the medial meniscus root attachment and subsequent increased joint contact pressures comparable to a total meniscectomy [39]. A 10 mm diameter tunnel is reamed, with great care taken when reaching the far cortex, as the neurovascular bundle resides just posterior to tibial footprint. Before passing the graft, a flexible rasp is used to smooth the anterior aspect of the tibial tunnel aperture to prevent graft attritional rupture. The graft is passed into the joint through the anteromedial portal. The soft tissue portion will be placed in the tibial tunnel, while the bone block placed on the femoral side (Fig. 16.4). Fixation on the femoral side

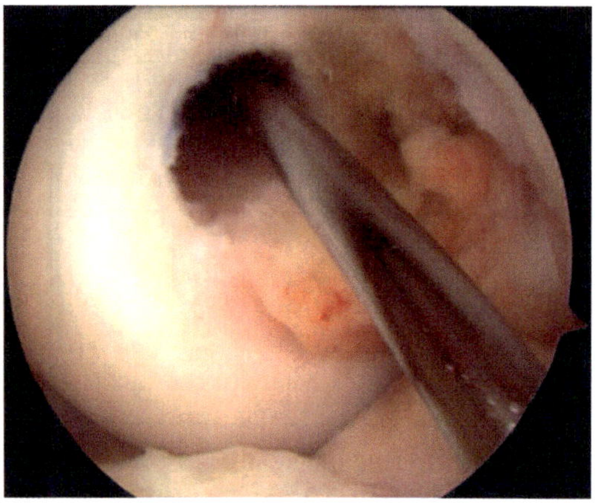

Fig. 16.2 Intraoperative femoral tunnel placement in single-bundle PCL-R on a left knee, as seen from the anterolateral portal above. The femoral tunnel has been reamed through an accessory anterolateral portal. The femoral tunnel is placed within the anatomic footprint of the anterolateral bundle on the medial femoral condyle. *PCL-R* posterior cruciate ligament reconstruction

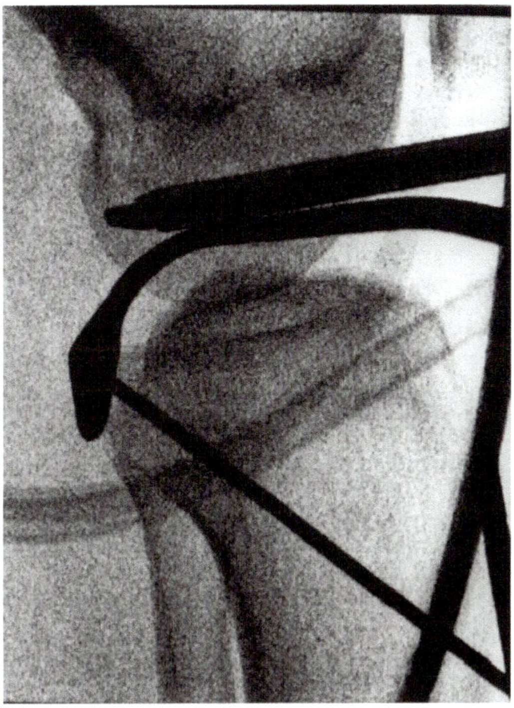

Fig. 16.3 Intraoperative fluoroscopy of a left knee. The above sagittal view demonstrates the placement of a guide pin for the tibial tunnel in a PCL-R. A tibial guide is placed through the notch onto the anatomic tibial footprint of the PCL, and a guide pin is then drilled through the tibia to the far cortex under fluoroscopy. *PCL-R* posterior cruciate ligament reconstruction

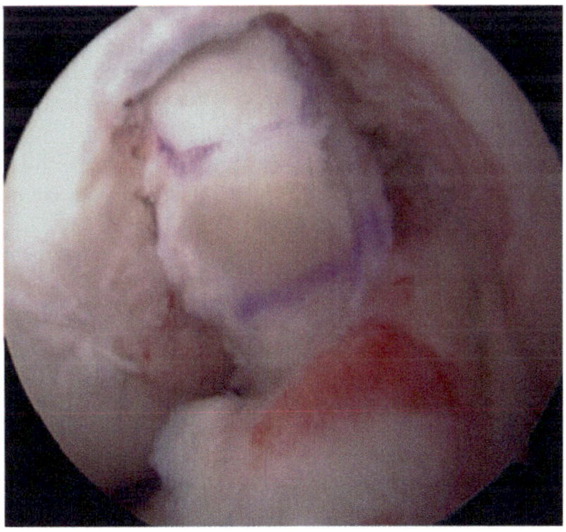

Fig. 16.4 Intraoperative arthroscopic view from the anterolateral portal of a completed PCL-R of a left knee. The quadriceps tendon autograft with bone block is seen in anatomic alignment behind the ACL, with good tension on both the PCL graft and native ACL. *PCL* posterior cruciate ligament, *PCL-R* posterior cruciate ligament reconstruction, *ACL* anterior cruciate ligament reconstruction

is with continuous loop suspensory fixation. The graft is secured on the tibial side with an interference screw on the anterior aspect of the tunnel, and tensioned with the knee at 90° of flexion and an anterior drawer placed on the tibia.

16.5.2.8 Postoperative Rehabilitation (Table 16.1)

The patient is placed in a dynamic anterior drawer brace locked in extension immediately postoperative. The patient remains non-weight-bearing for the first 2 weeks after surgery. At the 2 week mark, the patient can begin the rehabilitation program outlined in Table 16.1. The patient primarily focuses on quadriceps exercises and strengthening, as well as normalization of gait and knee range of motion. The goal of return to sport is approximately 6–9 months.

Table 16.1 Postoperative PCL rehabilitation program

2 week	• ROM from 0° to 90° • WBAT with knee locked in extension • Brace on at all times • Straight leg raises
6 week	• WBAT with brace unlocked • Closed chain quadriceps strengthening in 90° arc (leg press, wall slides) • Quadriceps isotonics with proximal pad in 90–40° arc • Closed chain stationary bike, minimal resistance for 20 min • Nordic track exercises • Isometric hamstring strengthening only
12 week	• Quadriceps isotonics – Full arc for closed chain – 90–40° arc for open chain • Isokinetic quadriceps with distal pad • Walk on treadmill (forward) and slow retrostep • Continue stationary bike • Begin straight line running at 18 weeks
24 week	• Full arc progressive resistance exercises – Concentrate on quadriceps • Agility drills • Progressive running program – Cutting and pivoting • Advanced functional exercises

At 0–2 weeks, the patient remains in brace locked in extension and is non-weight-bearing. At 2 weeks postoperative, the patient begins a formal rehabilitation program with physical therapy. The patient primarily focuses on quadriceps exercises and strengthening, as well as normalization of gait and knee range of motion. Return to sport usually occurs at the 9 month postoperative timepoint. *PCL* posterior cruciate ligament

16.6 Summary

PCL injuries are quite rare and account for around 3% of all acute traumatic injuries. The diagnosis of PCL injuries is based on a careful clinical examination and radiological analysis. It is of tremendous importance to combine clinical and radiological examinations carefully to rule out chronic PCL injuries. The treatment options include nonoperative treatment for isolated, partial PCL injuries and grade 1 and 2 PCL injuries, and operative reconstruction for high-grade acute isolated PCL injuries, chronic PCL injuries, or multiligamentous knee injuries. The surgical approach can be either open or arthroscopic, and a single or double bundle approach can be performed. Various PCL reconstruction techniques exist, including transtibial tunnel or tibial-inlay techniques. The authors prefer to perform an all-arthroscopic transtibial tunnel PCL reconstruction with quadriceps tendon autograft with a bone block.

Common complications in PCL reconstruction include persistent posterior laxity and injury to the neurovascular structures and meniscal roots during tunnel reaming. The postoperative treatment protocol should be individualized to each patient. This includes restriction of knee flexion initially with weight bearing as tolerated with the knee locked in extension, followed by free range of motion in a brace with weight bearing as tolerated. Physical therapy should focus on quadriceps muscle strengthening, with a goal to return athletes to sport in 9–12 months.

References

1. Ahmad CS, Cohen ZA, Levine WN, Gardner TR, Ateshian GA, Mow VC. Codominance of the individual posterior cruciate ligament bundles: an analysis of bundle lengths and orientation. Am J Sports Med. 2003;31:221–5.
2. Anderson CJ, Ziegler CG, Wijdicks CA, Engebretsen L, LaPrade RF. Arthroscopically pertinent anatomy of the anterolateral and posteromedial bundles of the posterior cruciate ligament. JBJS. 2012;94:1936–45.
3. Badri A, Gonzalez-Lomas G, Jazrawi L. Clinical and radiologic evaluation of the posterior cruciate ligament-injured knee, vol. 11. Humana Press Inc.; 2018. p. 515–20.
4. Becker EH, Watson JD, Dreese JC. Investigation of multiligamentous knee injury patterns with associated injuries presenting at a level I trauma center. J Orthop Trauma. 2013;27(4):226–31.
5. Bedi A, Musahl V, Cowan JB. Management of posterior cruciate ligament injuries. J Am Acad Orthop Surg. 2016;24:277–89.
6. Berg EE. Posterior cruciate ligament tibial inlay reconstruction. Arthroscopy. 1995;11:69–76.
7. Bergfeld JA, McAllister DR, Parker RD, Valdevit AD, Kambic HE. A biomechanical comparison of posterior cruciate ligament reconstruction techniques. Am J Sports Med. 2001;29:129–36.
8. Boynton MD, Tietjens BR. Long-term followup of the untreated isolated posterior cruciate ligament-deficient knee. Am J Sports Med. 1996;24:306–10.
9. Chahla J, Moatshe G, Cinque ME, Dornan GJ, Mitchell JJ, Ridley TJ, et al. Single-bundle and double-bundle posterior cruciate ligament reconstructions: a systematic review and meta-analysis of 441 patients at a minimum 2 years' follow-up. Arthroscopy. 2017;33:2066–80.
10. Chen C-H, Chou S-W, Chen W-J, Shih C-H. Fixation strength of three different graft types used in posterior cruciate ligament reconstruction. Knee Surg Sports Traumatol Arthrosc. 2004;12:371–5.
11. Cohen SB, Boyd L, Miller MD. Vascular risk associated with posterior cruciate ligament reconstruction using the arthroscopic transtibial tunnel technique. J Knee Surg. 2004;17:211–3.
12. Cosgarea AJ, Kramer DE, Bahk MS, Totty WG, Matava MJ. Proximity of the popliteal artery to the PCL during simulated knee arthroscopy–implications for establishing the posterior trans-septal portal. J Knee Surg. 2006;19:181–5.
13. Daniel DM, Stone ML, Barnett P, Sachs R. Use of the quadriceps active test to diagnose poster cruciate-ligament disruption and measure posterior laxity of the knee. J Bone Joint Surg. 1988;70-A:386–91.
14. Escobedo, Eva M., William J. Mills, and John C. Hunter. "The "reverse Segond" fracture: association with a tear of the posterior cruciate ligament and medial meniscus." American Journal of Roentgenology 178.4 (2002):979–83.
15. Fanelli GC, Edson CJ. Posterior cruciate ligament injuries in trauma patients: part II. Arthroscopy. 1995;11:526–9.
16. Fischer S, Fox J, Del Pizzo W, Friedman S, Snyder S, Ferkel R. Accuracy of diagnoses from magnetic resonance imaging of the knee. A multi-center analysis of one thousand and fourteen patients. J Bone Jt Surg. 1991;73:2–10.
17. Fowler PJ, Messieh SS. Isolated posterior cruciate ligament injuries in athletes. Am J Sports Med. 1987;15(6):553–7.
18. Galloway MT, Grood ES, Mehalik JN, Levy M, Saddler SC, Noyes FR. Posterior cruciate ligament reconstruction: an in vitro study of femoral and tibial graft placement. Am J Sports Med. 1996;24:437–45.
19. Gill TJ, DeFrate LE, Wang C, Carey CT, Zayontz S, Zarins B, et al. The effect of posterior cruciate liga-

ment reconstruction on patellofemoral contact pressures in the knee joint under simulated muscle loads. Am J Sports Med. 2004;32:109–15.
20. Grood ES, Hefzy MS, Lindenfield TN. Factors affecting the region of most isometric femoral attachments: part I: the posterior cruciate ligament. Am J Sports Med. 1989;17:197–207.
21. Gross ML, Grover JS, Bassett LW, Seeger LL, Finerman GAM. Magnetic resonance imaging of the posterior cruciate ligament. Clinical use to improve diagnostic accuracy. Am J Sports Med. 1992;20(6):732–7.
22. Grover JS, Bassett LW, Gross ML, Seeger LL, Finerman GAM. Posterior cruciate ligament: MR imaging. Radiology. 1990;174:527–30.
23. Hall FM, Hochman MG. Medial Segond-type fracture: cortical avulsion off the medial tibial plateau associated with tears of the posterior cruciate ligament and medial meniscus. Skeletal Radiol. 1997;26(9):553–5.
24. Harner CD, Höher J. Evaluation and treatment of posterior cruciate ligament injuries. Am J Sports Med. 1998;26:471–82.
25. Harner CD, Janaushek MA, Kanamori A, Yagi M, Vogrin TM, Woo SL. Biomechanical analysis of a double-bundle posterior cruciate ligament reconstruction. Am J Sports Med. 2000;28:144–51.
26. Hefti E, Müller W, Jakob RP, Stäubli HU. Evaluation of knee ligament injuries with the IKDC form. Knee Surg Sports Traumatol Arthrosc. 1993;1:226–34.
27. Hermans S, Corten K, Bellemans J. Long-term results of isolated anterolateral bundle reconstructions of the posterior cruciate ligament: a 6-to 12-year follow-up study. Am J Sports Med. 2009;37:1499–507.
28. Jackman T, LaPrade RF, Pontinen T, Lender PA. Intraobserver and interobserver reliability of the kneeling technique of stress radiography for the evaluation of posterior knee laxity. Am J Sports Med. 2008;36:1571–6.
29. Jacobi M, Reischl N, Wahl P, Gautier E, Jakob R. Acute isolated injury of the posterior cruciate ligament treated by a dynamic anterior drawer brace: a preliminary report. J Bone Joint Surg. 2010;92:1381–4.
30. Jung TM, Reinhardt C, Scheffler SU, Weiler A. Stress radiography to measure posterior cruciate ligament insufficiency: a comparison of five different techniques. Knee Surg Sports Traumatol Arthrosc. 2006;14:1116–21.
31. Jung YB, Tae SK, Lee YS, Jung HJ, Nam CH, Park SJ. Active non-operative treatment of acute isolated posterior cruciate ligament injury with cylinder cast immobilization. Knee Surg Sports Traumatol Arthrosc. 2008;16:729–33.
32. Kennedy NI, LaPrade RF, Goldsmith MT, Faucett SC, Rasmussen MT, Coatney GA, et al. Posterior cruciate ligament graft fixation angles, part 1: biomechanical evaluation for anatomic single-bundle reconstruction. Am J Sports Med. 2014;42:2338–45.
33. Kennedy NI, Michalski MP, Engebretsen L, LaPrade RF. Iatrogenic meniscus posterior root injury following reconstruction of the posterior cruciate ligament: a report of three cases. JBJS Case Connector. 2014;4:e20.
34. Kennedy NI, Wijdicks CA, Goldsmith MT, Michalski MP, Devitt BM, Årøen A, et al. Kinematic analysis of the posterior cruciate ligament, part 1: the individual and collective function of the anterolateral and posteromedial bundles. Am J Sports Med. 2013;41:2828–38.
35. Kew ME, Miller MD. Posterior cruciate ligament reconstruction in the multiple ligament injured knee. J Knee Surg. 2020;33(5):421–30.
36. Kim JG, Lim HC, Kim HJ, Hwang MH, Yoon YC, Oh JK. Delayed detection of clinically significant posterior cruciate ligament injury after peri-articular fracture around the knee of 448 patients. Arch Orthop Trauma Surg. 2012;132:1741–6.
37. Kim S-J, Kim T-E, Jo S-B, Kung Y-P. Comparison of the clinical results of three posterior cruciate ligament reconstruction techniques. JBJS. 2009;91:2543–9.
38. Kim Y-M, Lee CA, Matava MJ. Clinical results of arthroscopic single-bundle transtibial posterior cruciate ligament reconstruction: a systematic review. Am J Sports Med. 2011;39:425–34.
39. LaPrade CM, Smith SD, Rasmussen MT, Hamming MG, Wijdicks CA, Engebretsen L, et al. Consequences of tibial tunnel reaming on the meniscal roots during cruciate ligament reconstruction in a cadaveric model, part 2: the posterior cruciate ligament. Am J Sports Med. 2015;43:207–12.
40. LaPrade RF, Heikes C, Bakker AJ, Jakobsen RB. The reproducibility and repeatability of Varus stress radiographs in the assessment of isolated fibular collateral ligament and grade-III posterolateral knee injuries. An in vitro biomechanical study. J Bone Jt Surg Ser A. 2008;90:2069–76.
41. LaPrade RF, Smith SD, Wilson KJ, Wijdicks CA. Quantification of functional brace forces for posterior cruciate ligament injuries on the knee joint: an in vivo investigation. Knee Surg Sports Traumatol Arthrosc. 2015;23:3070–6.
42. Lee SH, Yun SJ. Efficiency of knee ultrasound for diagnosing anterior cruciate ligament and posterior cruciate ligament injuries: a systematic review and meta-analysis. Skelet Radiol. 2019;48: 1599–610.
43. Logan M, Williams A, Lavelle J, Gedroyc W, Freeman M. The effect of posterior cruciate ligament deficiency on knee kinematics. Am J Sports Med. 2004;32:1915–92.
44. Logterman SL, Wydra FB, Frank RM. Posterior cruciate ligament: anatomy and biomechanics. Curr Rev Musculoskelet Med. 2018;11:510–4.
45. Mair SD, Schlegel TF, Gill TJ, Hawkins RJ, Steadman JR. Incidence and location of bone bruises after acute posterior cruciate ligament injury. Am J Sports Med. 2004;32:1681–7.
46. Margheritini F, Rihn J, Musahl V, Mariani PP, Harner C. Posterior cruciate ligament injuries in the athlete an anatomical, biomechanical and clinical review. Sports Med. 2002;32(6):393–408.

47. Markolf KL, Zemanovic JR, McAllister DR. Cyclic loading of posterior cruciate ligament replacements fixed with tibial tunnel and tibial inlay methods. JBJS. 2002;84:518–24.
48. Matava MJ, Sethi NS, Totty WG. Proximity of the posterior cruciate ligament insertion to the popliteal artery as a function of the knee flexion angle: implications for posterior cruciate ligament reconstruction. Arthroscopy. 2000;16:796–804.
49. McGuire DA, Hendricks SD. Comparison of anatomic versus nonanatomic placement of femoral tunnels in Achilles double-bundle posterior cruciate ligament reconstruction. Arthroscopy. 2010;26:658–66.
50. Medina O, Arom GA, Yeranosian MG, Petrigliano FA, McAllister DR. Vascular and nerve injury after knee dislocation: a systematic review. Clin Orthop Relat Res. 2014;472(9):2621–9.
51. Montgomery SR, Johnson JS, McAllister DR, Petrigliano FA. Surgical management of PCL injuries: indications, techniques, and outcomes. Curr Rev Musculoskelet Med. 2013;6:115–23.
52. Mui LW, Engelsohn E, Umans H. Comparison of CT and MRI in patients with tibial plateau fracture: can CT findings predict ligament tear or meniscal injury? Skelet Radiol. 2007;36:145–51.
53. Pache S, Aman ZS, Kennedy M, Nakama GY, Moatshe G, Ziegler C, et al. Posterior cruciate ligament: current concepts review. Arch Bone Jt Surg. 2018;6:8–18.
54. Panchal HB, Sekiya JK. Open tibial inlay versus arthroscopic transtibial posterior cruciate ligament reconstructions. Arthroscopy. 2011;27:1289–95.
55. Patel DV, Allen AA, Warren RF, Wickiewicz TL, Simonian PT. The nonoperative treatment of acute, isolated (partial or complete) posterior cruciate ligament-deficient knees: an intermediate-term follow-up study. HSS J. 2007;3:137–46.
56. Pearsall AW, Pyevich M, Draganich LF, Larkin JJ, Reider B. In vitro study of knee stability after posterior cruciate ligament reconstruction. Clin Orthop Relat Res. 1996;327:264–71.
57. Puddu G, Giannı GE, Chambat P, De Paulis F. The axial view in evaluating tibial translation in cases of insufficiency of the posterior cruciate ligament. Arthroscopy. 2000;16(2):217–20.
58. Qi Y-S, Wang H-J, Wang S-J, Zhang Z-Z, Huang A-B, Yu J-K. A systematic review of double-bundle versus single-bundle posterior cruciate ligament reconstruction. BMC Musculoskelet Disord. 2016;17:45.
59. Race A, Amis AA. PCL reconstruction: in vitro biomechanical comparison of 'isometric' versus single and double-bundled 'anatomic' grafts. J Bone Joint Surg. 1998;80:173–9.
60. Ringler MD, Shotts EE, Collins MS, Howe BM. Intra-articular pathology associated with isolated posterior cruciate ligament injury on MRI. Skelet Radiol. 2016;45:1695–703.
61. Rubinstein RA, Donald Shelbourne K, McCarroll JR, Vanmeter CD, Rettig AC. The accuracy of the clinical examination in the setting of posterior cruciate ligament injuries. Am J Sports Med. 1994;22(4):550–7.
62. Sanders TL, Pareek A, Barrett IJ, Kremers HM, Bryan AJ, Stuart MJ, et al. Incidence and long-term follow-up of isolated posterior cruciate ligament tears. Knee Surg Sports Traumatol Arthrosc. 2017;25:3017–23.
63. Schulz MS, Russe K, Weiler A, Eichhorn HJ, Strobel HJ. Epidemiology of posterior cruciate ligament injuries. Arch Orthop Trauma Surg. 2003;123:186–91.
64. Sekiya JK, Whiddon DR, Zehms CT, Miller MD. A clinically relevant assessment of posterior cruciate ligament and posterolateral corner injuries: evaluation of isolated and combined deficiency. J Bone Jt Surg Ser A. 2008;90:1621–7.
65. Seon J-K, Song E-K. Reconstruction of isolated posterior cruciate ligament injuries: a clinical comparison of the transtibial and tibial inlay techniques. Arthroscopy. 2006;22:27–32.
66. Shelbourne KD, Clark M, Gray T. Minimum 10-year follow-up of patients after an acute, isolated posterior cruciate ligament injury treated nonoperatively. Am J Sports Med. 2013;41:1526–33.
67. Shelbourne KD, Davis TJ, Patel DV. The natural history of acute, isolated, nonoperatively treated posterior cruciate ligament injuries. Am J Sports Med. 1999;27:276–83.
68. Shelbourne KD, Jennings RW, Vahey TN. Magnetic resonance imaging of posterior cruciate ligament injuries: assessment of healing. Am J Knee Surg. 1999;12:209–13.
69. Shelbourne KD, Muthukaruppan Y. Subjective results of nonoperatively treated, acute, isolated posterior cruciate ligament injuries. Arthroscopy. 2005;21:457–61.
70. Shon OJ, Lee DC, Park CH, Kim WH, Jung KA. A comparison of arthroscopically assisted single and double bundle tibial inlay reconstruction for isolated posterior cruciate ligament injury. Clin Orthop Surg. 2010;2:76–84.
71. Spiridonov SI, Slinkard NJ, LaPrade RF. Isolated and combined grade-III posterior cruciate ligament tears treated with double-bundle reconstruction with use of endoscopically placed femoral tunnels and grafts: operative technique and clinical outcomes. JBJS. 2011;93:1773–80.
72. Strobel MJ, Weiler A, Schulz MS, Russe K, Eichhorn HJ. Arthroscopic evaluation of articular cartilage lesions in posterior cruciate ligament—deficient knees. Arthroscopy. 2003;19:262–8.
73. Tewes DP, Fritts HM, Fields RD, Quick DC, Buss DD. Chronically injured posterior cruciate ligament: magnetic resonance imaging. Clin Orthop Relat Res. 1997;335:224–32.
74. Verhulst FV, Macdonald P. Diagnosing PCL injuries: history, physical examination, imaging studies, arthroscopic evaluation. Sports Med Arthrosc Rev. 2020;28(1):2–7.
75. Vermeijden HD, Jonkergouw A, van der List JP, DiFelice GS. The multiple ligament-injured knee: when is primary repair an option? Knee. 2020;27(1):173–82.

76. Wang C-J, Chan Y-S, Weng L-H, Yuan L-J, Chen H-S. Comparison of autogenous and allogenous posterior cruciate ligament reconstructions of the knee. Injury. 2004;35:1279–85.
77. Wijdicks CA, Kennedy NI, Goldsmith MT, Devitt BM, Michalski MP, Årøen A, et al. Kinematic analysis of the posterior cruciate ligament, part 2: a comparison of anatomic single-versus double-bundle reconstruction. Am J Sports Med. 2013;41:2839–48.
78. Yoon KH, Bae DK, Song SJ, Cho HJ, Lee JH. A prospective randomized study comparing arthroscopic single-bundle and double-bundle posterior cruciate ligament reconstructions preserving remnant fibers. Am J Sports Med. 2011;39:474–80.
79. Zawodny SR, Miller MD. Complications of posterior cruciate ligament surgery. Sports Med Arthrosc Rev. 2010;18:269–74.

17 Technique Corner: Posterolateral Corner Reconstruction

Evan W. James, Kenneth M. Lin, Bruce A. Levy, and Robert G. Marx

17.1 Introduction

Posterolateral corner injuries are rare compared to other knee ligament injuries and can pose a significant challenge to the treating surgeon [1]. Diagnosis is best made using a combination of history, physical exam, and imaging. The mechanism of injury is often high energy involving either a varus directed force with knee hyperextension or excessive tibial external rotation with knee flexion [2–4]. Special physical exam maneuvers such as the varus stress test at 0° and 30° of knee flexion, prone dial test at 30° and 90° of knee flexion, posterolateral drawer test, posterolateral spin test [5], and reverse pivot shift test can be used to evaluate for posterolateral corner injuries [6]. Physical exam to evaluate for anterior cruciate ligament (ACL), posterior cruciate ligament (PCL), and medial collateral ligament (MCL) injuries is also imperative since many posterolateral corner injuries are part of a multiligament injury pattern. Plain radiographs, including knee anteroposterior (AP), lateral, tunnel, and Merchant views, and magnetic resonance imaging (MRI) are recommended to characterize the extent of intra- and extra-articular pathology [7]. In addition, stress radiographs can also be useful to help differentiate between isolated LCL tears and complete (grade III) posterolateral corner injuries [8].

The posterolateral corner consists of static and dynamic stabilizers which together confer varus and rotational stability to the knee. The three primary static stabilizers include the lateral collateral ligament (LCL), popliteofibular ligament (PFL), and popliteus tendon (PT). The LCL functions as the primary restraint to varus instability and a secondary stabilizer to internal and external rotation torques [9, 10]. Similarly, the popliteus tendon primarily provides stability to external rotation, but also functions as a secondary stabilizer against varus stress, internal rotation, and anterior translation [11, 12]. The PFL functions as a restraint to external rotation torques, with highest loads seen at 60° of flexion [13]. Other important lateral knee structures include the long head of the biceps femoris, lateral gastrocnemius tendon, iliotibial band, proximal tibiofibular ligament, lateral joint capsule, anterolateral ligament, peroneal nerve, and lateral meniscus [14, 15]. A thorough understanding of this complex anatomy is imperative when performing a posterolateral corner reconstruction.

E. W. James · K. M. Lin · R. G. Marx (✉)
Division of Sports Medicine and Shoulder Surgery, Hospital for Special Surgery, New York, NY, USA
e-mail: jamese@hss.edu; link@hss.edu; marxr@hss.edu

B. A. Levy
Department of Orthopedic Surgery and Sports Medicine, Mayo Clinic, Rochester, MN, USA
e-mail: Levy.Bruce@mayo.edu

N. Nakamura et al. (eds.), *Advances in Knee Ligament and Knee Preservation Surgery*, https://doi.org/10.1007/978-3-030-84748-7_17

17.2 Surgical Indications

Indications for posterolateral corner reconstruction include acute midsubstance tears of the LCL, PFL, and/or popliteus tendon which are generally in association with cruciate ligament injury. Rarely, isolated posterolateral corner injury can lead to functional instability and require surgery, although this is far less common. The authors recommend using a single-stage approach whenever possible since outcome studies show no difference in knee function between single and staged posterolateral corner reconstructions in the multiligament injured or dislocated knee [16–18]. Timing of posterolateral corner reconstruction is also an important variable. Emergent surgical intervention should be performed for patients with open injuries, vascular injuries, and irreducible dislocations. All other patients may undergo either acute repair or reconstruction or a combination thereof, typically defined as surgery within the first 3 weeks after injury [19–21], or delayed reconstruction, which occurs 4 weeks or later after injury. Proponents of early operative intervention argue that this approach allows for repair of some types of injuries, such as avulsion injuries, safer and easier exposure of the peroneal nerve, comparatively better clinical and functional outcomes, and decreased risk for subsequent intra-articular injuries while awaiting surgery [19, 22, 23]. Additionally, as noted above, the surgical dissection beyond 14 days after the injury is much more difficult due to scarring of the common peroneal nerve and as such, if early surgery is contemplated, the surgeon should make every effort to approach the injury within that time frame. Proponents of delayed operative intervention cite evidence that waiting allows for resolution of soft tissue swelling, better knee range of motion, and decreased risk of wound complications [24, 25]. Taken together, these factors highlight that decision-making in posterolateral corner injuries is complex and the operative plan must account for a host of variables, including repair versus reconstruction, single stage versus multistage, and acute versus delayed surgical intervention.

17.3 Surgical Technique

17.3.1 Surgical Approach

The authors' preferred technique for posterolateral corner reconstruction utilizes one Achilles tendon allograft to reconstruct the lateral collateral ligament, popliteus tendon, and popliteofibular ligament, along with a posterolateral capsular shift [17, 21, 26, 27]. Anatomic socket position and graft placement is essential. After anesthetic is administered, the patient is positioned supine on the operating table. An examination under anesthesia (EUA) is performed to assess the pattern of instability and to compare this with preoperative physical exam and imaging findings. Stress views with fluoroscopy can be helpful to confirm the diagnosis as well [28]. The extremity is then sterilely prepped and draped in the standard fashion.

Surgical exposure begins with a curvilinear laterally based incision made extending proximally along the middle of the iliotibial band and distally over Gerdy's tubercle and anterior to the fibula. The incision must be sufficiently anterior to allow for exposure of the distal femur with the knee flexed. A skin flap is raised which falls posteriorly and the biceps femoris is located. The common peroneal nerve is identified just posterior to the biceps femoris and a peroneal neurolysis is performed to allow the nerve to be protected throughout the procedure. The nerve is tagged with a vessel loop or Penrose drain to facilitate safe mobilization and protection for the duration of the procedure. The iliotibial band is incised longitudinally following the trajectory of the skin incision. Next, subperiosteal dissection is performed over the proximal lateral aspect of the fibular head using a Cobb elevator. Blunt dissection is carried posteriorly to develop the interval between the posterior border of the fibula and biceps femoris. This creates space to facilitate graft passage.

17.3.2 Graft Preparation

The graft of choice for this technique is a nonirradiated fresh-frozen Achilles tendon allograft.

The graft can be prepared at the back table either by an assistant during the surgical approach or by the surgeon at this point in the case or prior. The calcaneus allograft is trimmed to create a 9 × 20 mm bone block, and the tendinous portion is trimmed and tubularized with no. 5 suture to 7-mm in diameter for the distal half of the graft for passage through the fibular head tunnel. Passing sutures are placed to help facilitate graft passage after creation of the fibular and femoral tunnels.

17.3.3 Associated Pathology

Concomitant meniscal and/or cruciate ligament pathologies are addressed arthroscopically. This can be done prior to the open lateral surgery, or in the acute situation the open lateral surgery can be done first, allowing fluid from the subsequent arthroscopy to egress through the lateral capsular disruption which avoids fluid leak into the leg and possible compartment syndrome. If a meniscus tear is present, it can be treated either with meniscal repair, if amenable, or partial meniscectomy. If a cruciate ligament injury is present, ACL and/or PCL reconstruction tunnels are created and the graft(s) secured in the femoral tunnel(s). For PCL reconstructions, the graft may be secured in the tibia at the same time. For ACL reconstructions, fixation on the tibial side should occur after the posterolateral corner graft is secured due to a risk of creating a fixed external rotation deformity [29]. Osteochondral injuries may also be addressed at this time as needed.

17.3.4 Reconstruction Tunnels and Graft Passage

A Kirschner wire is passed from anterolateral to posteromedial in the fibular head at the insertion of the LCL to the fibular styloid at the attachment site of the PFL. Care must be taken not to over-penetrate the wire, which could cause iatrogenic injury to the popliteal artery or common peroneal nerve. The wire is over-reamed with a 7-mm reamer. Any residual soft tissue obscuring the tunnel aperture is removed with a rongeur and electrocautery. A passing suture is used to pass the graft posterior to anterior through the fibula and under the biceps femoris.

Next, attention is turned to the femur. A guidewire is placed in the anterior one-fifth of the popliteal sulcus and position is confirmed on fluoroscopy. The guidewire is over-reamed with a 9-mm reamer to create a closed socket tunnel that is 20 mm in depth. The bone plug of the Achilles tendon allograft is then advanced into the tunnel and secured with a 7 mm by 20 mm metal interference screw. Gentle traction should be applied to the graft to ensure satisfactory fixation and purchase of the interference screw. The graft is then passed down the popliteal tendon trajectory deep to the biceps towards the posterior aspect of the fibular head and then from back to front through the fibular head (Fig. 17.1).

Attention is then turned back to the femur. The femoral origin of the LCL is identified just proximal and posterior to the lateral epicondyle, approximately 18 mm from the popliteus tendon insertion, which recreates the native anatomic distance between the femoral attachments of the LCL and popliteus tendon [30]. A Beath pin is advanced at the LCL femoral origin. The graft is then draped over the Beath pin and the knee cycled through flexion and extension to evaluate for graft isometry. Once satisfactory isometry is confirmed, the pin is over-reamed with a 7-mm reamer to a depth of 50 mm, or to the far cortex. The sutures are passed through the eyelet of the Beath pin and pulled from lateral to medial in order to pass the graft into the tunnel. After definitive cruciate graft fixation, the LCL graft is tensioned while applying a valgus force with the knee in 30° flexion and gentle internal rotation and secured with a 8 × 25 mm bioabsorbable interference screw or metal soft tissue screw. This completes the ligament reconstruction portion of the procedure.

Finally, a posterolateral capsular shift is performed to further augment the reconstruction. Non-absorbable No. 2 sutures are passed through the joint capsule, advanced anteriorly, and tied to plicate the capsule. This confers additional strength to the repair construct and restores normal tension to the posterolateral joint capsule that is often torn or attenuated in posterolateral corner injuries. In the acute setting, the capsule can be repaired directly to the femur or tibia from where it has been avulsed with anchors.

The wounds are then copiously irrigated and closed in a layered fashion. The iliotibial band is closed with a nonabsorbable No. 1 suture. The peroneal nerve is again inspected to ensure that it was adequately protected for the duration of the procedure. The subcutaneous layers are closed with absorbable suture and the skin with a running subcuticular stitch. A sterile dressing is applied. A knee brace is applied with a slight valgus bend prior to the patient being awoken from anesthesia to ensure that the reconstruction remains protected. Our protocol is to provide chemoprophylaxis for deep vein thrombosis to the patient until transitioned back to full weight bearing.

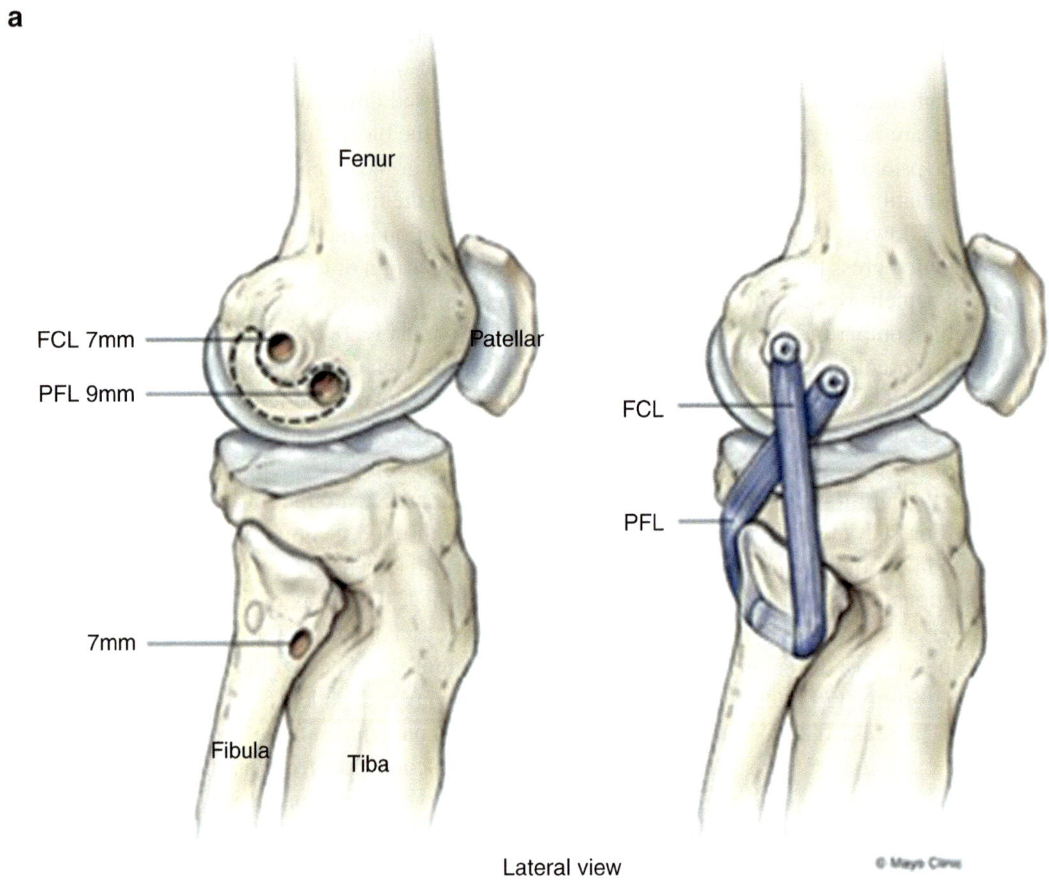

Fig. 17.1 The authors' preferred posterolateral corner reconstruction depicting (**a**) lateral view of fibular and femoral reconstruction tunnels and graft placement, and (**b**) anteroposterior view of tunnel and graft placement and the posterolateral capsular shift. (Reproduced with permission from Schechinger et al. [21])

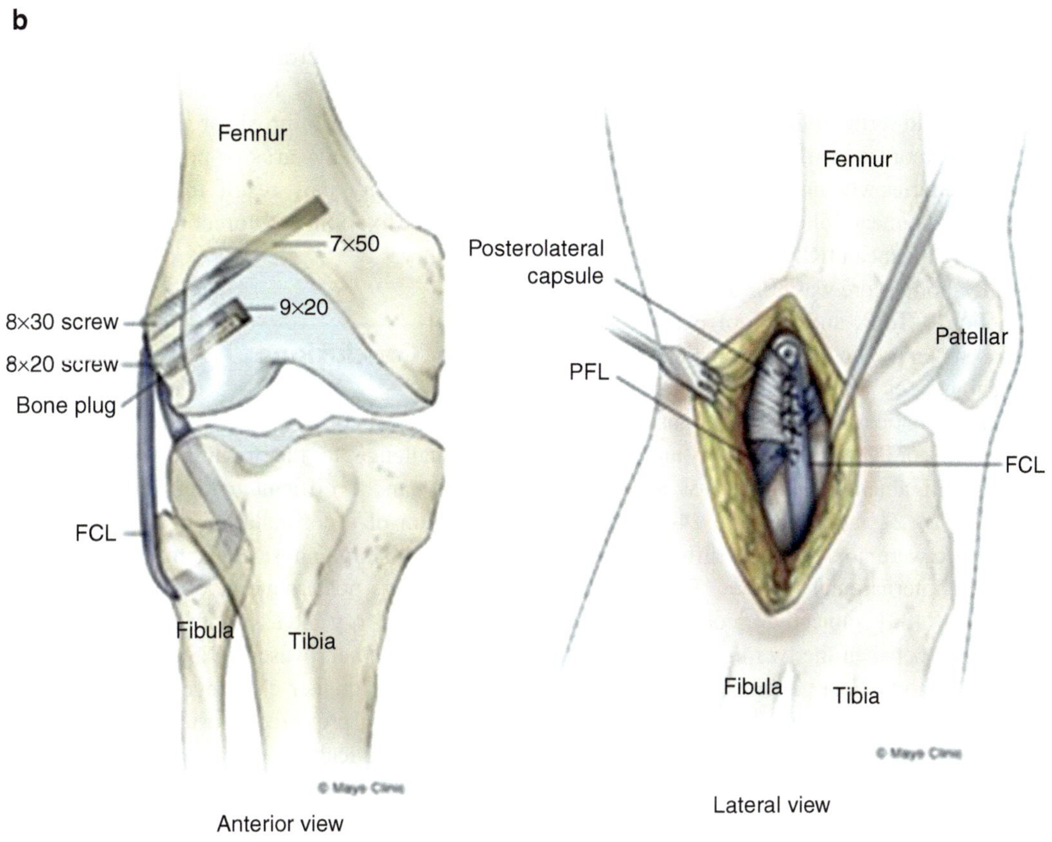

Fig. 17.1 (continued)

17.4 Rehabilitation Protocol

Postoperatively, the operative extremity is placed in either a valgus-producing unloader brace or a straight hinged knee brace with a slight valgus bend. This brace is worn continuously for the first 6 weeks before switching to a custom valgus unloader brace for the next 8–12 months [21]. Rehabilitation follows a staged approach, with sequential phases of rehabilitation focusing on range of motion, muscular endurance, muscular strength, and power [20, 31, 32]. During the first 6 weeks, knee range of motion is gradually increased to a full arc of motion. Patellar mobility can be performed along with quadriceps activation to achieve terminal extension. During weeks 2–6, weight bearing is progressively increased to full, and after that, gentle strengthening exercises are initiated. The strength program focuses first on building a muscular endurance base followed by increasing muscular strength. Once symmetric strength is achieved, the focus shifts to building power, performing functional exercises such as plyometrics, and initiation of sport-specific movements. When satisfactory gains have been achieved, the decision to return to full participation in activities and sport is based on functional testing, psychological readiness, and clearance by the treating surgeon.

17.5 Surgical Outcomes

Outcomes following this posterolateral corner reconstruction technique show excellent results at short to mid-term follow-up. Sanders et al. reported outcomes in 61 patients with multiligamentous injuries who underwent PLC reconstruction [17]. At a mean follow-up of 3.8 years (range 2–9 years), the mean IKDC score was 74.1 ± 22.3, mean Lysholm score was 80.3 ± 21.8, mean range of motion 0–126°, and 95% of patients had no increased laxity in full extension. Similarly, Schechinger et al. reported outcomes in 16 patients (7 patients with a two-ligament injury such as ACL or PCL and PLC injury and 9 patients with multiligamentous injuries) at a minimum follow-up of 24 months (range 24–75 months) [21]. There were no significant differences found in the International Knee Documentation Committee [33] subjective scores or Lysholm scores [34] between the two ligament and multiligament cohorts. Four patients demonstrated persistent 1+ varus laxity, but no appreciable functional deficits were noted. With respect to complications, one patient experienced arthrofibrosis requiring manipulation, but no patients required revision surgery. Taken together, this data supports the effectiveness of this technique for restoring posterolateral knee stability in both two ligament and multiligament injury patients. However, additional follow-up is required to further evaluate long-term outcomes.

17.6 Complications

Complications following posterolateral corner reconstruction include those associated with any open lower extremity surgery: superficial and deep infection, wound dehiscence, and deep vein thrombosis. Injury to the distal femoral and proximal fibular physis may occur in skeletally immature patients during tunnel reaming and may require non-anatomic tunnel placement [35]. Common peroneal nerve injury is rare but has been reported to occur in approximately 2% of cases [36, 37]. In addition, the tourniquet should be let down and hemostasis achieved prior to closing due to the risk of a postoperative hematoma causing compression of the common peroneal nerve at the level of the fibular neck [38]. Range of motion loss may occur due to arthrofibrosis or graft malposition. Finally, persistent varus or posterolateral rotatory instability may result from graft failure, failure of fixation, or unrecognized concomitant injury.

17.7 Conclusions

Posterolateral corner injuries represent a significant challenge to the treating surgeon. These injuries can be difficult to diagnose, requiring integration of a constellation of findings in the patient's history, physical exam, and imaging. Furthermore, posterolateral corner injuries are often high energy and associated with ACL, PCL, MCL, and/or neurovascular injuries, which can pose an even greater challenge. In this chapter, the authors' preferred posterolateral corner reconstruction technique is presented, which utilizes an Achilles tendon allograft to reconstruct the LCL, popliteus tendon, and the PFL along with a posterolateral capsular shift to provide additional stability. The technique is based on anatomic sockets and graft placement. Two- to nine-year outcome studies show excellent clinical and functional outcome scores with low risk of need for revision surgery.

References

1. LaPrade RF, Terry GC. Injuries to the posterolateral aspect of the knee. Association of anatomic injury patterns with clinical instability. Am J Sports Med. 1997;25(4):433–8.
2. Baker CL Jr, Norwood LA, Hughston JC. Acute posterolateral rotatory instability of the knee. J Bone Joint Surg Am. 1983;65(5):614–8.
3. Fleming RE Jr, Blatz DJ, McCarroll JR. Posterior problems in the knee. Posterior cruciate insufficiency and posterolateral rotatory insufficiency. Am J Sports Med. 1981;9(2):107–13.
4. Fornalski S, McGarry MH, Csintalan RP, et al. Biomechanical and anatomical assessment after knee hyperextension injury. Am J Sports Med. 2008;36(1):80–4.

5. Marx RG, Shindle MK, Warren RF. Management of posterior cruciate ligament injuries. Oper Tech Sports Med. 2009;17(3):162–6.
6. James EW, O'Brien LT, LaPrade RF, et al. Surgical approach to posterolateral chronic injury. In: Rossi R, Margheritini F, editors. Knee ligament injuries. Milan: Springer; 2014. p. 89–98.
7. Ruzbarsky JJ, Konin G, Mehta N, et al. MRI arthroscopy correlations: ligaments of the knee. Sports Med Arthrosc Rev. 2017;25(4):210–8.
8. James EW, Williams BT, LaPrade RF. Stress radiography for the diagnosis of knee ligament injuries: a systematic review. Clin Orthop Relat Res. 2014;472(9):2644–57.
9. Coobs BR, LaPrade RF, Griffith CJ, ct al. Biomechanical analysis of an isolated fibular (lateral) collateral ligament reconstruction using an autogenous semitendinosus graft. Am J Sports Med. 2007;35:1521–7.
10. Grood ES, Stowers SF, Noyes FR. Limits of movement in the human knee: effect of sectioning the posterior cruciate ligament and posterolateral structures. J Bone Joint Surg Am. 1988;70:88–97.
11. Ferrari DA, Wilson DR, Hayes WC. The effect of release of the popliteus and quadriceps force on rotation of the knee. Clin Orthop Relat Res. 2003;412:225–33.
12. LaPrade RF, Wozniczka JK, Stellmaker MP, et al. Analysis of the static function of the popliteus tendon and evaluation of an anatomic reconstruction: the "fifth ligament" of the knee. Am J Sports Med. 2010;38:543–9.
13. LaPrade RF, Tso A, Wentorf FA. Force measurements on the fibular collateral ligament, popliteofibular ligament, and popliteus tendon to applied loads. Am J Sports Med. 2004;32:1695–701.
14. Claes S, Vereecke E, Maes M, et al. Anatomy of the anterolateral ligament of the knee. J Anat. 2013;223:321–8.
15. James EW, LaPrade CM, LaPrade RF. Anatomy and biomechanics of the lateral side of the knee and surgical implications. Sports Med Arthrosc Rev. 2015;23(1):2–9.
16. Freychet B, Kennedy NI, Sanders TL, et al. No difference between single and staged posterolateral corner surgical procedures in the multiligament injured/dislocated knee. Knee Surg Sports Traumatol Arthrosc. 2020;28(7):2170–6.
17. Sanders TL, Johnson NR, Pareek, et al. Satisfactory knee function after single-stage posterolateral corner reconstruction in the multi-ligament injured/dislocated knee using the anatomic single-graft technique. Knee Surg Sports Traumatol Arthrosc. 2018;26(4):1258–65.
18. Westermann RW, Marx RG, Spindler KP, et al. No difference between posterolateral corner repair and reconstruction with concurrent ACL surgery: results from a prospective multicenter cohort. Orthop J Sports Med. 2019;7(7):2325967119861062.
19. Harner CD, Waltrip RL, Bennett CH, et al. Surgical management of knee dislocations. J Bone Joint Surg Am. 2004;86-A(2):262–73.
20. Lynch AD, Chmielewski T, Bailey L, et al. Current concepts and controversies in rehabilitation after surgery for multiple ligament knee injury. Curr Rev Musculoskelet Med. 2017;10(3):328–45.
21. Schechinger SJ, Levy BA, Dajani KA, et al. Achilles tendon allograft reconstruction of the fibular collateral ligament and posterolateral corner. Arthroscopy. 2009;25(3):232–42.
22. Levy BA, Dajani KA, Whelan DB, et al. Decision making in the multiligament-injured knee: an evidence-based systematic review. Arthroscopy. 2009;25(4):430–8.
23. Jiang W, Yao J, He Y, Sun W, et al. The timing of surgical treatment of knee dislocations: a systematic review. Knee Surg Sports Traumatol Arthrosc. 2015;23(10):3108–13.
24. Shelbourne KD, Patel DV. Timing of surgery in anterior cruciate ligament-injured knees. Knee Surg Sports Traumatol Arthrosc. 1995;3(3):148–56.
25. Smith TO, Davies L, Hing CB. Early versus delayed surgery for anterior cruciate ligament reconstruction: a systematic review and meta-analysis. Knee Surg Sports Traumatol Arthrosc. 2010;18(3):304–11.
26. Kuzma SA, Chow RM, Engasser WM, et al. Reconstruction of the posterolateral corner of the knee with Achilles tendon allograft. Arthrosc Tech. 2014;3(3):e393–8.
27. Woodmass JM, Sanders TL, Johnson NR, et al. Posterolateral corner reconstruction using the anatomical two-tailed graft technique: clinical outcomes in the multiligament injured knee. J Knee Surg. 2018;31(10):1031–6.
28. LaPrade RF, Heikes C, Bakker AJ, Jakobsen RB. The reproducibility and repeatability of varus stress radiographs in the assessment of isolated fibular collateral ligament and grade-III posterolateral knee injuries. An in vitro biomechanical study. J Bone Joint Surg Am. 2008;90(10):2069–76.
29. Wentorf FA, LaPrade RF, Lewis JL, et al. The influence of the integrity of posterolateral structures on tibiofemoral orientation when an anterior cruciate ligament graft is tensioned. Am J Sports Med. 2002;30(6):796–9.
30. LaPrade RF, Ly TV, Wentorf FA, Engebretsen L. The posterolateral attachments of the knee: a qualitative and quantitative morphologic analysis of the fibular collateral ligament, popliteus tendon, popliteofibular ligament, and lateral gastrocnemius tendon. Am J Sports Med. 2003;31:854–60.
31. Chahla J, Murray IR, Robinson J, et al. Posterolateral corner of the knee: an expert consensus statement on diagnosis, classification, treatment, and rehabilitation. Knee Surg Sports Traumatol Arthrosc. 2019;27(8):2520–9.
32. Fanelli GC, Edson CJ. Combined posterior cruciate ligament posterolateral reconstructions with Achilles

tendon allograft and biceps femoris tenodesis: 2- to 10-year follow-up. Arthroscopy. 2004;20:339–45.
33. Anderson AF, Irrgang JJ, Kocher MS, et al. The International Knee Documentation Committee Subjective Knee Evaluation Form: normative data. Am J Sports Med. 2006;34(1):128–35.
34. Tegner Y, Lysholm J. Rating systems in the evaluation of knee ligament injuries. Clin Orthop Relat Res. 1985;198:43–9.
35. Williams BT, James EW, LaPrade RF. A physeal-sparing fibular collateral ligament and proximal tibiofibular joint reconstruction in a skeletally immature athlete. Knee Surg Sports Traumatol Arthrosc. 2016;24(3):661–5.
36. LaPrade RF, Johansen S, Agel J, et al. Outcomes of an anatomic posterolateral knee reconstruction. J Bone Joint Surg Am. 2010;92(1):16–22.
37. Stannard JP, Brown SL, Farris RC, et al. The posterolateral corner of the knee: repair versus reconstruction. Am J Sports Med. 2005;33(6):881–8.
38. Girolami M, Galletti S, Montanari G, et al. Common peroneal nerve palsy due to hematoma at the fibular neck. J Knee Surg. 2013;26(Suppl 1):S132–5.

Technique Corner: MCL

18

Robert S. Dean, Brady T. Williams, Jill K. Monson, Robert F. LaPrade, and Jorge Chahla

18.1 Anatomy and Function

The PMC extends from the medial aspect of the patellar tendon to the medial border of the gastrocnemius tendon. It is comprised of five major structures including the superficial MCL (sMCL), deep MCL (dMCL), posterior oblique ligament (POL), oblique popliteal ligament (OPL), and posterior horn of the medial meniscus (PHMM) [1]. Additionally, the semimembranosus and its respective expansions provide dynamic stability to the PMC. In contrast to the lateral side of the knee, the bony anatomy of the medial femoral condyle and the medial tibial plateau articulates in a convex on concave fashion, which is inherently stable [1].

The *superficial medial collateral ligament* (sMCL) is the largest structure of the medial aspect of the knee and serves as the primary restraint to valgus forces and gapping and has a secondary role in resisting external tibial rotation, and serves as a minor restraint against anterior translation in ACL-deficient knees. The center of the femoral attachment is located 3.2 mm proximal and 4.8 mm posterior to the medial epicondyle. Distally, there are two distinct tibial attachments. The proximal tibial attachment is found directly over the anterior arm of the semimembranosus, approximately 11.2 mm distal to the joint line. The sMCL continues distally, coursing over the inferior medial genicular artery and vein and branch of the tibial nerve, before reattaching at its distal tibial insertion, 61.2 mm distal to the joint line [1]. This attachment is largely within the pes anserine bursa and constitutes a significant portion of the posterior floor of the bursa. The sMCL is approximately 9–10 cm in length (Figs. 18.1 and 18.2) [1].

The *deep medial collateral ligament* (dMCL) or mid-third medial capsular ligament is essentially a thickening of the medial joint capsule that is deep and partially adherent to the sMCL, with deep attachments to the medial meniscus [1]. The dMCL has distinct portions including a meniscofemoral portion that attaches distal and deep to the femoral attachment of the sMCL, and a meniscotibial portion that is much shorter and thicker, which attaches just distal to the edge of the articular cartilage of the medial tibial plateau (Fig. 18.3) [1].

The *posterior oblique ligament* (POL) contains three fascial attachments including the superficial, central, and capsular arms [3, 4]. Collectively, they course from the distal aspect of the semimembranosus tendon and travel longitudinally across the joint line. The POL originates on the femur 7.7 mm distal and 6.4 mm posterior to the adductor tubercle, and 1.4 mm distal and

R. S. Dean · J. K. Monson · R. F. LaPrade
Twin Cities Orthopedics, Edina, MN, USA

B. T. Williams · J. Chahla (✉)
Midwest Orthopaedics at RUSH, Chicago, IL, USA
e-mail: jorge.chahla@rushortho.com

N. Nakamura et al. (eds.), *Advances in Knee Ligament and Knee Preservation Surgery*,
https://doi.org/10.1007/978-3-030-84748-7_18

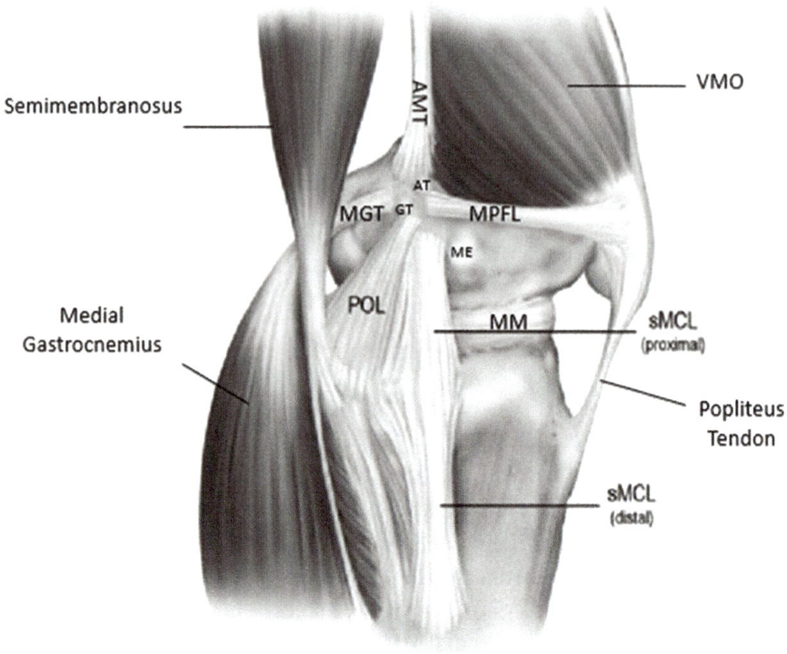

Fig. 18.1 Anatomical illustration of the major medial structures of a left knee. *AMT* adductor magnus tendon, *AT* adductor tubercle, *GT* gastrocnemius tubercle, *MPFL* medial patellofemoral ligament, *ME* medial epicondyle, *MGT* medial gastrocnemius tendon, *SM* semimembranosus, *sMCL* superficial medial collateral ligament, *POL* posterior oblique ligament, *VMO* vastus medialis obliquus muscle, *MM* medial meniscus. (Adapted with permission from Coobs et al. [2])

2.9 mm anterior to the gastrocnemius tubercle [1, 2]. Proximally, the superficial arm blends with the central arm, while distally it courses parallel and posterior to the sMCL, ultimately blending with the distal tibial expansion of the semimembranosus and its respective tibial attachment [1, 5]. The central arm of the POL is the largest and most substantial division. It originates from the distal aspect of the semimembranosus tendon, reinforcing the posteromedial capsule with attachments to the medial meniscus. Anteriorly, the central arm blends with the posterior aspect of the sMCL. Lastly, the capsular arm of the POL originates from the distal semimembranosus tendon, just posterior and lateral to the meniscofemoral capsular attachments of the central arm. The capsular arm blends with the posteromedial joint capsule and the medial aspect of the OPL [1] (Figs. 18.1 and 18.2) [1].

The *semimembranosus tendon* bifurcates into the direct arm and the anterior arm as it crosses the joint line. The direct arm is the primary attachment and inserts distally into a small groove just proximal to the tuberculum tendinis and posterior to the medial tibial crest. The anterior arm of the tendon blends with the capsular arm of the POL, which then merges with the posteromedial joint capsule with a reported attachment to the posterior inferior margin of the PHMM [1, 4, 6]. The anterior arm continues, passing deep to the superficial arm of the POL to its tibial attachment, deep to the proximal tibial attachment of the sMCL (Figs. 18.1 and 18.2).

The *oblique popliteal ligament* (OPL) is a broad fascial band that courses diagonally across the posterior aspect of the knee with an average reported length of 48 mm [1]. It originates from the capsular arm of the POL and the lateral expansion of the semimembranosus tendon and extends laterally until reaching its two attachment sites. The proximal and lateral attachment is to an osseous or cartilaginous fabella including the meniscofemoral portion of the posterolateral joint capsule and the plantaris muscle. The distal and lateral attachment is just lateral to the posterior cruciate ligament (PCL) and distal to the posterior root of the lateral meniscus.

The *posterior horn of the medial meniscus* (PHMM) is the most important weight-bearing portion of the medial meniscus and is intimately associated with the posteromedial capsule, POL, and distal expansion of the semimembranosus

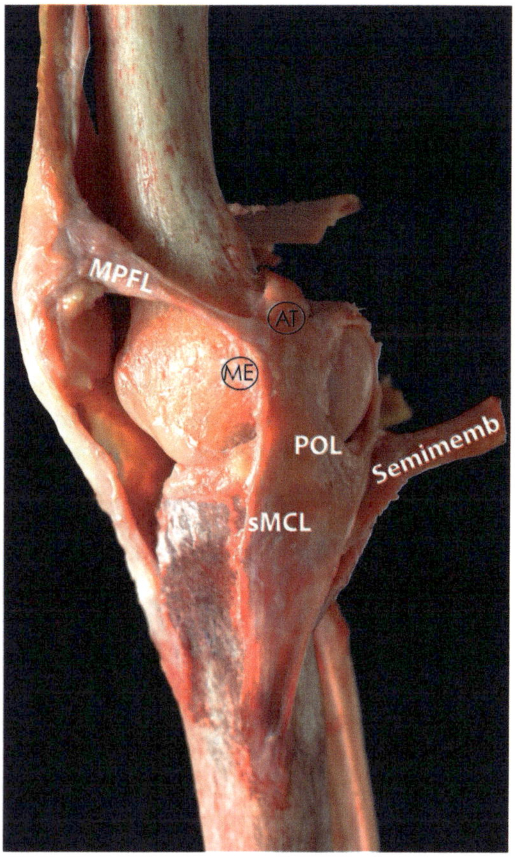

Fig. 18.2 Dissection of a right knee from a medial viewpoint. *AT* adductor tubercle, *ME* medial epicondyle, *MPFL* medial patellofemoral ligament, *POL* posterior oblique ligament, *sMCL* superficial medial collateral ligament, *Semimemb* semimembranosus

tendon [1, 7]. The posterior meniscocapsular attachment of the PHMM has an average reported length of 20.2 mm. The entire PHMM has an average total length of 21.3 mm measure along the periphery and is largely confluent with the posterior capsule [8, 9].

The *adductor magnus tendon* (AMT) attaches in a small bony depression 3.0 mm posterior and 2.7 mm proximal to the adductor tubercle. The adductor magnus tendon is rarely injured and serves as a reliable anatomical reference for medial knee reconstructions. Posteriorly, the adductor magnus tendon has a thick fascial attachment, originating distally and attaching to the proximal aspect of the medial gastrocnemius tendon and posteromedial joint capsule (Fig. 18.4) [1].

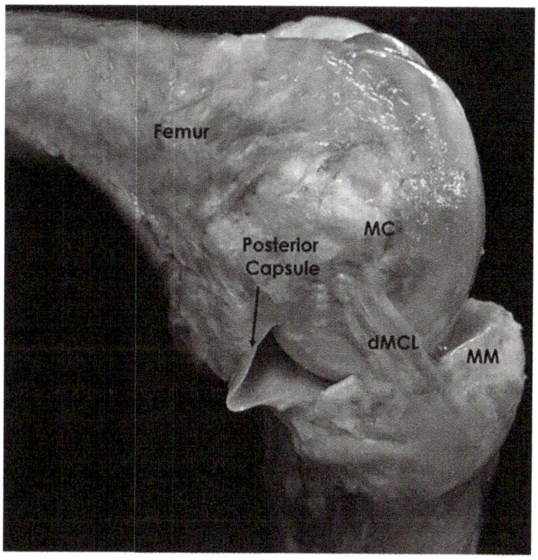

Fig. 18.3 Dissection of a left knee from a medial viewpoint. *dMCL* deep medial collateral ligament, *MC* medial femoral condyle, *MM* medial meniscus

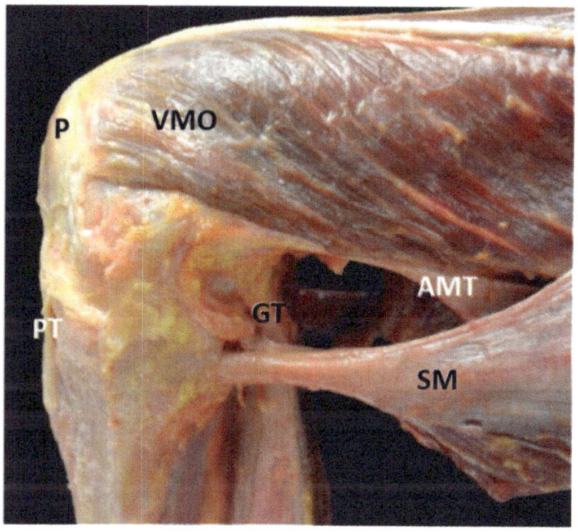

Fig. 18.4 Medial view of a right knee positioned in 90°of flexion, dissected to demonstrate the muscular and tendinous structures of the medial knee. *AMT* adductor magnus tendon, *GT* medial gastrocnemius tendon, *P* patella, *PT* patellar tendon, *SM* semimembranosus, *VMO* vastus medius obliquus muscle

18.2 Mechanism of Injury and Clinical Presentation

Patients with medial knee injuries typically present after having experienced a forced valgus stress to the knee. Severe injuries can present with side-to-side instability, valgus thrust with ambulation, and significant limitations in physical activity [10].

The joint line should be palpated, making sure to individually palpate the meniscofemoral and meniscotibial divisions. A comprehensive exam, including gait observation, should be performed to evaluate concomitant ligamentous or meniscal pathologies. Additionally, a valgus stress test should be performed with the knee in 20° of flexion and also in full extension. The medial joint line can be palpated while performing the valgus stress exam to approximate the amount of medial compartment gapping. Increased gapping with valgus stress testing at 20° is indicative of an isolated sMCL injury. Conversely, increased medial compartment gapping in full extension suggests a more severe medial-sided injury with possible cruciate ligament involvement [11]. Another medial knee specific exam maneuver is the anteromedial drawer test. Increased rotation compared to the contralateral extremity indicates an injury to the distal aspect of the sMCL, POL, and meniscotibial portion of the dMCL. The dial test should also be performed at both 30° and 90° of knee flexion [10, 12, 13].

18.3 Diagnosis and Imaging

Suspicion for medial knee injury warrants valgus stress radiographs, which are both objective and reproducible [14]. Typically, these are performed with the knee in 20° flexion with a foam bolster under the knee, and a 10 N clinician-applied valgus stress. The radiographs are assessed for side-to-side differences (SSD) in medial compartment gapping using the contralateral uninjured leg as a reference [10]. LaPrade et al. [14] reported that at 20°, an isolated grade III sMCL tear results in an average of 3.2 mm SSD in medial compartment gapping, while complete rupture of the PMC results in an average of 9.8 mm SSD in medial gapping [14]. Objective metrics such as valgus stress radiographs are particularly valuable in the chronic setting when the clinical exam findings may be equivocal (Fig. 18.5) [10, 15].

Standard anteroposterior (AP) and lateral radiographs can demonstrate avulsions near the native attachment sites of the sMCL should increase the suspicion for medial-sided injuries [16–18]. For chronic injuries, full-length weight-bearing radiographs to assess for coronal plane

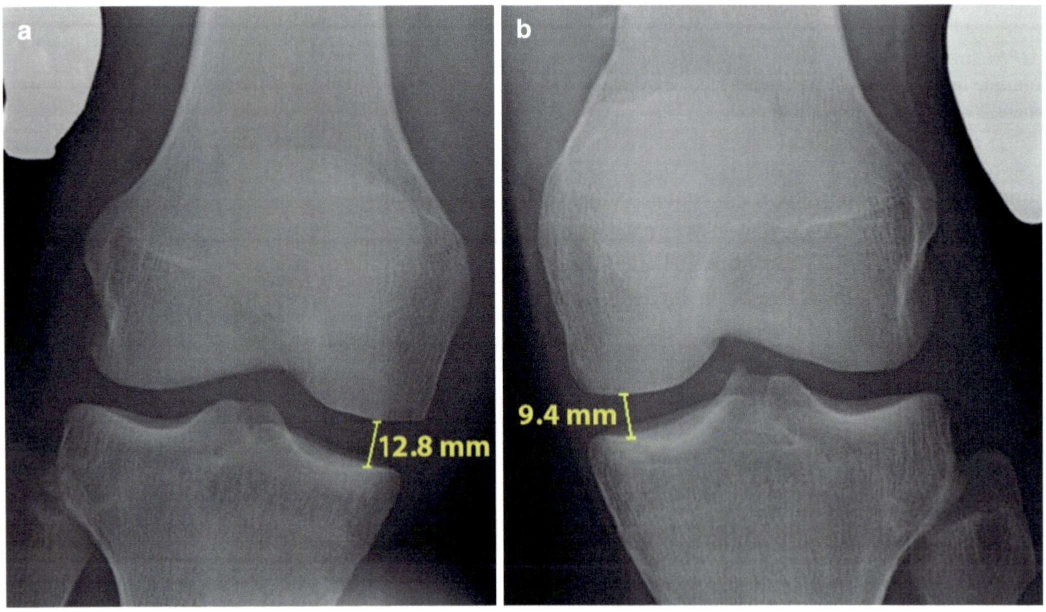

Fig. 18.5 Valgus stress radiographs obtained in 20°of flexion demonstrating medial knee laxity. (**a**) The injured right knee demonstrates 12.8 mm of medial gapping on valgus stress testing, compared to (**b**) the contralateral healthy limb (left knee) which demonstrates 9.4 mm of medial gapping. This 3.4 mm side-to-side difference is consistent with a grade III sMCL tear

alignment can be valuable given that valgus malalignment leads to an increased risk for medial instability. If valgus malalignment is identified in chronic cases, alignment correction via osteotomy can be considered prior to, or concurrent, with surgical intervention [19].

Magnetic resonance imaging (MRI) is highly sensitive with a reported sensitivity of 86% for diagnosing acute grade III sMCL injuries [20, 21]. In addition to evaluating for sMCL attenuation or increased signal intensities at the bony attachment sites, clinicians should also evaluate for lateral compartment bone bruises, which may be present in 45% of isolated sMCL tears [22].

18.3.1 Conservative Management

Grade I or II sMCL injuries can initially be treated conservatively. Early ROM promotes collagen proliferation and organization that contribute to increased tissue strength [15, 23], and helps avoid the deleterious effects of immobilization including ground substance leaching at ligament attachment sites and decreased tissue biomechanical properties [24–27].

Early goals of conservative management are to create a protective environment for healing, reduce joint inflammation, and gradually restore ROM, strong volitional quadriceps activation, and normal functional movement patterns [28, 29]. Hamstring muscle activation and strengthening should be phased in gradually, as the proximity of medial hamstring's insertion to the area of injury may contribute to pain and irritation. Postural stability and balance should also be addressed to improve limb and trunk motor control and reduce the risk of reinjury. Sport-specific training progressions may be implemented as individual muscle strength testing and athletic performance measures (Y-balance, hop testing) demonstrate first >80% limb symmetry index (LSI), and eventually >90% LSI. Objective testing should guide decision making with regard to return to sport participation [30–34]. The literature on biological augmentation is both inconclusive and controversial [35, 36].

Overall, conservative treatment of isolated sMCL tears have been reported to provide good or excellent subjective patient-reported outcomes as defined by the Hospital for Special Surgery Knee Score [37–41]. However, more severe injuries, particularly those distal to the joint line, often require surgical management.

18.3.2 Operative Treatment

Failed conservative management of sMCL tears, patients with valgus gapping in full extension, or chronic medial-sided laxity, can result in debilitating, persistent instability, ACL dysfunction and injury, weakness, and accelerated progression of osteoarthritis [42]. In these cases, surgical intervention is usually warranted. Multiple surgical techniques have been described including direct suture repair of the sMCL and POL [43], primary repair with augmentation [44], advancement of the tibial insertion site of the superficial medial collateral ligament [45], pes anserinus transfer [46], proximal advancement of the superficial medial collateral ligament [47], non-anatomic reconstruction techniques [48], and anatomic reconstruction techniques [2]. The surgical options discussed in the current chapter will focus on the authors' preferred technique, while also discussing non-anatomic reconstruction and ligament repair.

18.4 Authors' Preferred Operative Treatment

For single-stage procedures, the authors' preferred techniques are either an anatomic, double-bundle reconstruction of the PMC or an anatomic augmented repair. Biomechanically, there is no significant difference between these procedures, with both resulting in significantly improved knee stability [15, 49].

18.4.1 Anatomic Reconstruction, Double Bundle

Coobs et al. [2] developed, and biomechanically validated, an anatomic reconstruction of the PMC that recreates the sMCL, including proximal and distal segments, and the POL using two independent grafts and a total of four bone tunnels, including two femoral and two tibial tunnels [2]. Soft tissue (hamstrings or anterior tibialis) allografts or gracilis and semitendinosus autografts are the preferred graft choices for this technique. On average, the sMCL graft should be 16 cm in length and the POL graft should be 12 cm in length. The original technique describes a single anteromedial incision; however, adaptations may use three smaller incisions for this procedure. First, the distal tibial insertion of the sMCL is identified. It can be found deep to the pes anserinus bursa, approximately 6.12 cm distal to the joint line [1, 50]. Once the distal tibial footprint is identified, the hamstring tendons can be harvested if utilizing an autograft-based reconstruction. Next, a 2.4-mm eyelet guide pin is placed medial to lateral through the tibia, followed by a 7-mm tunnel reamed to a depth of 25 mm. Next, the central arm of the POL is identified. It is located along the posteromedial tibia, slightly anterior to the insertion of the semimembranosus tendon at the posterior margin of the anterior arm. With the footprint located, a guide pin is placed in an oblique direction toward Gerdy's tubercle, followed by a 7-mm tunnel reamed to a depth of 25 mm.

Next, attention is turned to the femur. In order to accurately identify the femoral landmarks, the distal attachment of the AMT is identified. The AMT is utilized to dissect down to and identify the adductor tubercle [51]. From there, the medial epicondyle can be identified 12.6 mm distal and 8.3 mm anterior to the adductor tubercle. The sMCL attachment site is 3.2 mm proximal and 4.8 posterior to the medial epicondyle. Identification of key anatomic landmarks on the femur may require fluoroscopy to assist with proper identification [52]. Once the sMCL femoral origin is identified, a guide pin is placed anterolaterally across the distal thigh. However, this tunnel should not be reamed until the femoral footprint of the POL is also identified in order to avoid tunnel convergence. The POL attachment site is 7.7 mm distal and 2.9 mm anterior to the gastrocnemius tubercle. This can be more easily identified if the posteromedial capsule is torn off the femur; however, if the capsule is still intact, a small incision can be made just posterior to the remnants of the sMCL, vertically and into the joint, to identify its femoral attachment site. Once the POL attachment site is identified, an eyelet passing pin can be placed and both femoral tunnels can be reamed with a 7-mm reamer to a depth of 25 mm. Next, any additional medial knee pathology should be addressed: This includes repair of the semimembranosus using a suture anchor, and/or repair of the dMCL using a suture anchor. From here, the grafts can be passed and tensioned using 7 × 20 bioabsorbable screw. The POL is fixed with the knee in full extension and neutral rotation, followed by the sMCL which is fixed with the knee in 20° of flexion and neutral rotation, while applying a gentle varus force (Fig. 18.6) [54].

18.4.2 Anatomic Augmented Surgical Repair with Semitendinosus Tendon Autograft

The sMCL augmented repair begins with dissection of the superficial medial knee and identification of the semitendinosus tendon at its anatomic tibial attachment for graft harvesting. The tendon is then anchored to the tibia at the sMCL distal attachment (approximately 6 cm distal to the joint line) [1] using both suture anchors and additional sutures to reattach the graft to the underlying remnant of the distal sMCL. The graft is then passed proximally, deep to the intact sartorius facia, to the femoral attachment site of the sMCL described previously. This attachment site can be found proximal and posterior to the medical epicondyle. At the anatomic footprint, a femoral tunnel is reamed with a 7-mm reamer to a depth of 35 mm. The free end of the graft is measured and 3 cm of the graft is whip-stitched. The graft

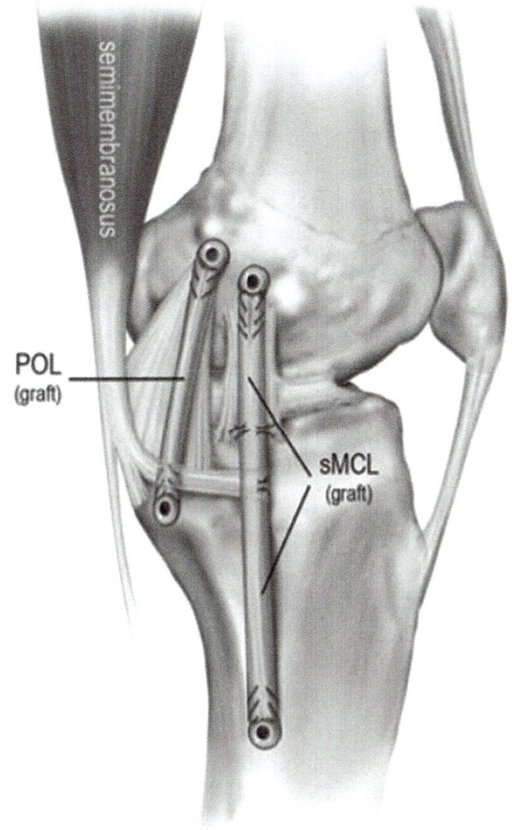

Fig. 18.6 Illustration of a left knee depicting the anatomic posteromedial corner reconstruction utilizing two grafts and four separate bone tunnels. *POL* posterior oblique ligament, *sMCL* superficial medial collateral ligament. (Reproduced with permission from Coobs et al. [53])

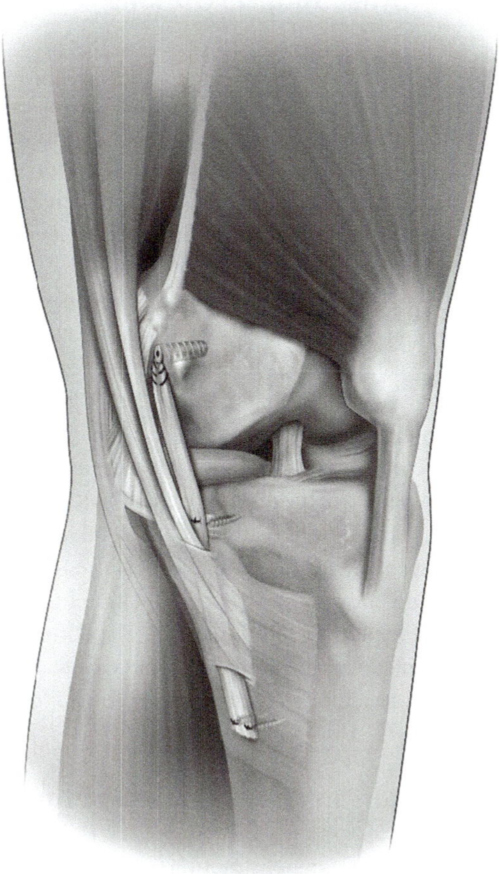

Fig. 18.7 Illustration of a left knee depicting the anatomic augmented superficial medial collateral ligament (sMCL) repair. The semitendinosus is rerouted and secured at the sMCL footprint on the medial femoral condyle. (Reproduced with permission from Wijdicks et al. [49])

is then fixed with a 7 × 25 mm bioabsorbable interference screw, while a 60 N traction force is applied with the knee in 20° of flexion, neutral rotation, and a slight varus force [55]. Finally, a suture anchor is used to anatomically restore the proximal tibial division of the sMCL, directly over the anterior arm of the semimembranosus (Fig. 18.7) [44, 45, 56].

18.4.3 Staged Surgical Management

When valgus malalignment is identified on full-length weight-bearing radiographs for chronic PMC and sMCL tears, surgeons may consider a two-stage intervention, beginning with an osteotomy to correct the alignment followed by medial-sided reconstruction as described above [57]. Osteotomy can be particularly valuable in the setting of chronic medial-sided knee injuries with valgus malalignment. One study demonstrated a 36% reduction in medial compartment gapping on valgus stress radiographs following a lateral opening-wedge distal femoral osteotomy for the correction of a valgus deformity [58].

18.5 Other Surgical Options

18.5.1 MCL Repair

Repairs have been reported to result in inferior outcomes compared to augmented repair or reconstruction techniques [59, 60]. When performed, a small incision is made over the medial epicondyle near the femoral origin of the sMCL. After careful dissection, the proximal stump and the avulsed sMCL are identified. Starting distally, the sMCL is sutured toward the proximal avulsion using a Bunnell-type pattern. Once suture passing is complete, a punch tap is used to create a threaded hole for suture anchor at the anatomic footprint. The repair sutures are passed through a suture anchor loaded with FiberTape that will function as the internal brace. The knee is flexed to 30° during fixation in order to avoid capturing the joint [61]. Repetition of this technique can be used to reapproximate other damaged structures including the POL.

18.5.2 Non-anatomic Reconstruction

Non-anatomic reconstructions have been described in the literature [48]. While there are several reported variants of this technique, one such technique places the tibial attachment more proximal to allow for the use of a shorter graft [62]. Another technique involves rerouting the semitendinosus without augmentation at the distal sMCL attachment on the tibia [63–65]. Given the subtle alterations in native anatomy associated with these procedures, it is likely that these surgeries either alter the native joint biomechanics or eventually lead to graft failure.

18.6 Rehabilitation

Postoperative rehabilitation emphasizes early range of motion as it has been shown to decrease stiffness [31]. Specific rehabilitation protocols are dependent on the concomitant surgical procedures; however, isolated PMC reconstruction rehabilitation entails non-weight bearing with crutches and the use of a stabilizing knee brace for the first 6 weeks postoperatively. During this time, patients should undergo intensive physical therapy with focus on pain and swelling reduction, progressive passive to active-assisted ROM, and quadriceps activation exercises [66–69]. Conversely, surgical reconstruction of isolated sMCL injuries allows for partial weight bearing during the first 6 weeks after operation. Compared to nonoperative management, postoperative rehabilitation is more restricted in the early recovery period due to soft tissue trauma associated with surgery. This ensures protection for adequate healing at the bone tunnel and suture graft fixation sites. Goals for therapy progression and return to activity/sport are similar to those for the nonoperative management pathway, but with slightly delayed progression. Assuming progressive functional milestones are met, including passing the Sport Performance TRAC Testing (Testing for Return to Athletic Competition), patients with isolated PMC injuries are expected to return to sport 6–9 months following PMC reconstruction. The step-by-step milestones for MCL rehabilitation are provided in Fig. 18.8.

18.7 Outcomes

Historically, repairs were the preferred surgical technique; however, clinical outcomes following surgical repair of PMC injuries in the setting of multiligament knee injuries have reported failure rates of 20%, compared to 4% following reconstruction techniques [60].

Kim et al. [70] described clinical outcomes of nonanatomic reconstructions of the PMC using a semitendinosus autograft with a preserved tibial attachment in a 24 patient case series. They reported a reduction from 7.8 mm of medial compartment gapping on valgus stress radiographs to less than 2 mm postoperatively in 91.7% (22/24) of patients. Furthermore, they reported normal or nearly normal IKDC scores in 92% of patients and mean postoperative Lysholm score of 91.9 [70]. Ibrahim et al. [71] reported 33% (5/15) of patients demonstrated 1+ residual valgus laxity at average 43 month follow-up after nonanatomic

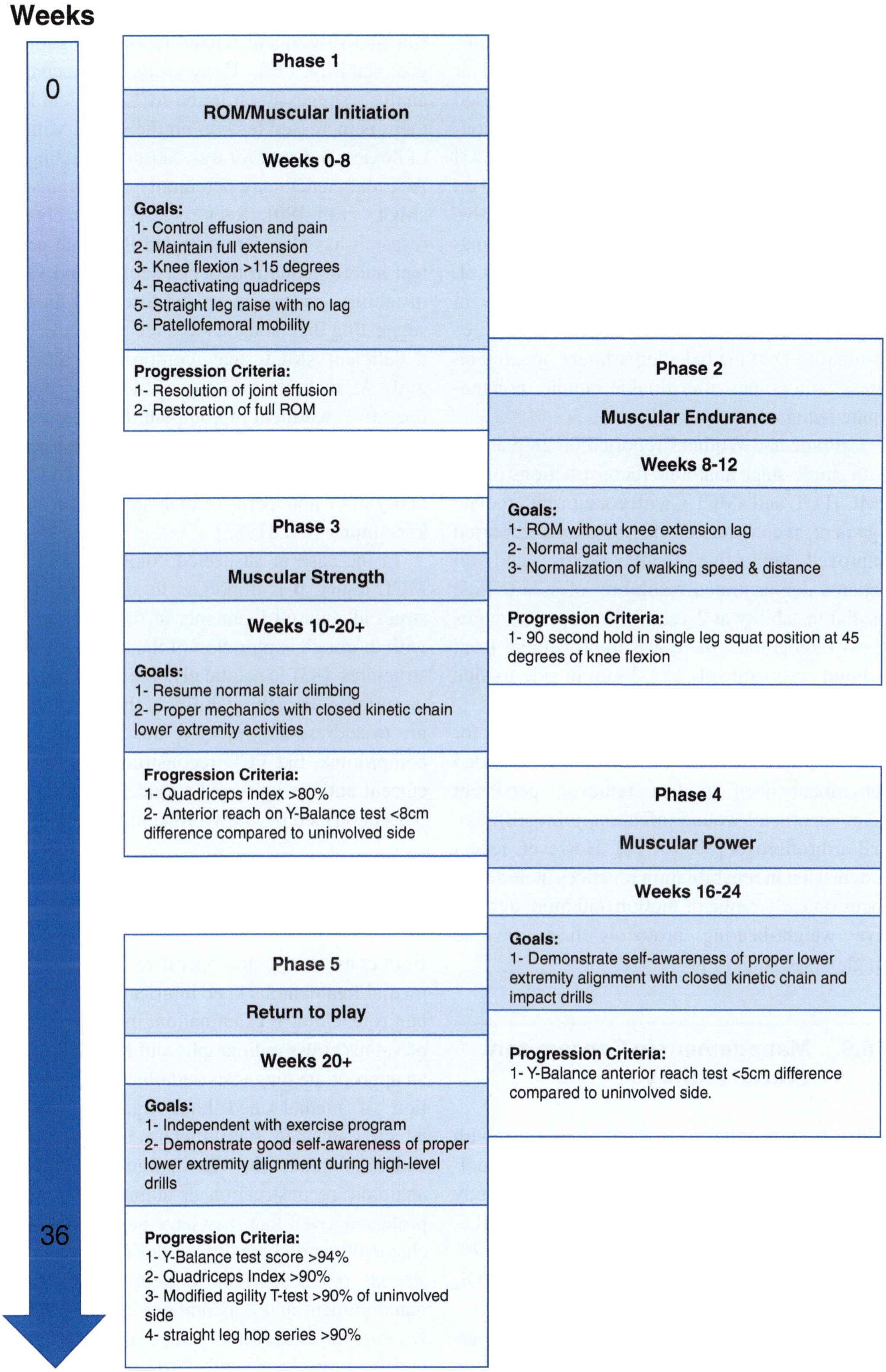

Fig. 18.8 Progressive rehabilitation protocol and timeline of milestones

sMCL reconstruction [71]. In 61 patients with grade III or IV medial instability at time of surgery, Lind et al. [72] reported 98% normal or nearly normal medial stability, and a reported 91% satisfaction rating with nonanatomic reconstruction at 2 year follow-up [72]. Liu et al. [73] reported a relative increase of 1.1 mm in SSD on stress radiographs, but excellent subjective patient reported outcomes scores after nonanatomic double-bundle reconstruction. Dong et al. [74] reported anteromedial rotary instability in 9.4% of patients, with an average of 2.9 mm of residual SSD of medial compartment opening on stress radiographs after double bundle, nonanatomic reconstruction.

LaPrade and Wijdicks reported on 28 patients with single-stage anatomic reconstructions of the PMC (POL and sMCL), with concurrent cruciate ligament reconstruction [2]. Patients reported improved subjective IKDC scores, and all patients demonstrated resolution of side-to-side medial instability at 2-year follow-up. On valgus stress radiographs, there was improvement from 6.3 mm preoperatively to 1.3 mm in side-to-side medial compartment gapping [10].

The most common reported complications for anatomic reconstruction techniques included subsequent deep implant removal, persistent pain, superficial wound infection, joint stiffness, and arthrofibrosis [50, 75, 76]. However, recent progression in rehabilitation practices including a focus on early range of motion and more aggressive weight-bearing protocols have favored single-stage surgery [77].

18.8 Management in Concomitant Cruciate Injury

It has been reported that 78% of patients with grade III MCL injuries have a concomitant cruciate injury [61, 62]. Of note, a previous study reported that in patients with combined sMCL and ACL tears confirmed operatively, 95.7% (22/23) of patients were also found to have a POL injury [78].

Previous studies have reported positive outcomes following delayed ACL reconstruction with early surgical management of sMCL injuries and subsequent rehabilitation to regain valgus stability [79]. Conversely, biomechanical analysis suggests that in the ACL-deficient knee, there is increased tension on the sMCL with 30° of flexion and anterior translation suggesting that ACL deficiency may potentially compromise the sMCL graft [80]. Reciprocally, biomechanical research has also demonstrated that both persistent anteromedial rotatory instability and valgus instability leads to increased forces on the ACL suggesting that isolated ACL reconstruction with a deficient sMCL may compromise the ACL graft. As such, the majority of literature supports operative treatment of complete PMC injuries at the time of cruciate ligament reconstruction, especially for those patients with residual valgus laxity after nonoperative management of medial knee injury [66, 81–83].

In the case of suspected combined PCL and MCL injury, it is important to surgically reconstruct all injured ligaments in the acute setting, with thorough exploration of the additional PMC structures [83]. Specifically, the POL plays an important role in the stability of the PCL and failure to address any injury to this structure could compromise the PCL reconstruction [84]. The current authors preferred surgical technique has been described by Crawford et al. [85].

18.9 Conclusion

Both conservative and operative management of medial ligamentous knee injuries have an important role. Clinical examination, including the use of valgus stress radiographs and MRI, allows for an appropriate diagnosis and objective classification of medial-sided knee ligament injuries, which can then help dictate the appropriate course of treatment. When surgery is indicated, anatomic reconstruction or augmented repair is preferred given that they have been both biomechanically and clinically validated. Furthermore, staged or concurrent correction of valgus malalignment and concomitant cruciate injuries are imperative to restore native stability and protect both cruciate and PMC reconstructions.

These anatomic reconstructions can be technically more demanding given the multiple grafts and reconstruction tunnels; therefore, relying on surgically relevant anatomical landmarks is crucial for a successful surgical intervention. Finally, postoperative therapy must utilize current rehabilitation principles focusing on range of motion, and regaining strength, while appropriately protecting the reconstruction grafts.

References

1. LaPrade RF, Engebretsen AH, Ly TV, Johansen S, Wentorf FA, Engebretsen L. The anatomy of the medial part of the knee. J Bone Jt Surg. 2007;89:2000. https://doi.org/10.2106/JBJS.F.01176.
2. Coobs BR, Wijdicks CA, Armitage BM, Spiridonov SI, Westerhaus BD, Johansen S, et al. An in vitro analysis of an anatomical medial knee reconstruction. Am J Sports Med. 2010;38:339–47. https://doi.org/10.1177/0363546509347996.
3. Hughston JC, Eilers AF. The role of the posterior oblique ligament in repairs of acute medial (collateral) ligament tears of the knee. J Bone Jt Surg Ser A. 1973;55:923–40. https://doi.org/10.2106/00004623-197355050-00002.
4. Hughston JC. The importance of the posterior oblique ligament in repairs of acute tears of the medial ligaments in knees with and without an associated rupture of the anterior cruciate ligament. Results of long-term follow-up. J Bone Jt Surg Ser A. 1994;76:1328–44. https://doi.org/10.2106/00004623-199409000-00008.
5. LaPrade RF, Morgan PM, Wentorf FA, Johansen S, Engebretsen L. The anatomy of the posterior aspect of the knee. An anatomic study. J Bone Jt Surg. 2007;89:758. https://doi.org/10.2106/JBJS.F.00120.
6. Robinson JR, Sanchez-Ballester J, Bull AMJ, Thomas R, de WM, Amis AA. The posteromedial corner revisited. An anatomical description of the passive restraining structures of the medial aspect of the human knee. J Bone Joint Surg Br. 2004;86:674–81. https://doi.org/10.1302/0301-620x.86b5.14853.
7. Dold AP, Swensen S, Strauss E, Alaia M. The posteromedial corner of the knee. J Am Acad Orthop Surg. 2017;25:752–61. https://doi.org/10.5435/JAAOS-D-16-00020.
8. DePhillipo NN, Moatshe G, Chahla J, Aman ZS, Storaci HW, Morris ER, et al. Quantitative and qualitative assessment of the posterior medial meniscus anatomy: defining meniscal ramp lesions. Am J Sports Med. 2019;47:372–8. https://doi.org/10.1177/0363546518814258.
9. Griffith CJ, LaPrade RF, Johansen S, Armitage B, Wijdicks C, Engebretsen L. Medial knee injury: part 1, static function of the individual components of the main medial knee structures. Am J Sports Med. 2009;37:1762–70. https://doi.org/10.1177/0363546509333852.
10. Laprade RF, Wijdicks CA. The management of injuries to the medial side of the knee. J Orthop Sport Phys Ther. 2012;42:221–33. https://doi.org/10.2519/jospt.2012.3624.
11. Rossi R, Dettoni F, Bruzzone M, Cottino U, D'Elicio DG, Bonasia DE. Clinical examination of the knee: know your tools for diagnosis of knee injuries. Sport Med Arthrosc Rehabil Ther Technol. 2011;3:25. https://doi.org/10.1186/1758-2555-3-25.
12. Chahla J, Moatshe G, Dean CS, LaPrade RF. Posterolateral corner of the knee: current concepts. Arch Bone Jt Surg. 2016;4:97–103.
13. Bae JH, Choi IC, Suh SW, Lim HC, Bae TS, Nha KW, et al. Evaluation of the reliability of the dial test for posterolateral rotatory instability: a cadaveric study using an isotonic rotation machine. Arthrosc J Arthrosc Relat Surg. 2008;24:593–8. https://doi.org/10.1016/j.arthro.2007.12.003.
14. LaPrade RF, Bernhardson AS, Griffith CJ, Macalena JA, Wijdicks CA. Correlation of valgus stress radiographs with medial knee ligament injuries. Am J Sports Med. 2010;38:330–8. https://doi.org/10.1177/0363546509349347.
15. Wijdicks CA, Griffith CJ, Johansen S, Engebretsen L, LaPrade RF. Injuries to the medial collateral ligament and associated medial structures of the knee. J Bone Jt Surg Ser A. 2010;92:1266–80. https://doi.org/10.2106/JBJS.I.01229.
16. Altschuler EL, Bryce TN. Pellegrini–Stieda syndrome. N Engl J Med. 2006;354:e1. https://doi.org/10.1056/NEJMicm040406.
17. Muschol M, Müller I, Petersen W, Hassenpflug J. Symptomatic calcification of the medial collateral ligament of the knee joint: a report about five cases. Knee Surg Sport Traumatol Arthrosc. 2005;13:598–602. https://doi.org/10.1007/s00167-004-0598-1.
18. Chang W-C, Huang G-S, Lee C-H, Kao H-W, Chen C-Y. Calcification of medial collateral ligament of the knee. JCR J Clin Rheumatol. 2006;12:204–5. https://doi.org/10.1097/01.rhu.0000231459.18480.a6.
19. Gelber PE, Perelli S. Treatment of the medial collateral ligament injuries. Ann Jt. 2018;3:78. https://doi.org/10.21037/aoj.2018.09.07.
20. Yao L, Dungan D, Seeger LL. MR imaging of tibial collateral ligament injury: comparison with clinical examination. Skelet Radiol. 1994;23:521–4. https://doi.org/10.1007/bf00223082.
21. Halinen J, Koivikko M, Lindahl J, Hirvensalo E. The efficacy of magnetic resonance imaging in acute multi-ligament injuries. Int Orthop. 2009;33:1733–8. https://doi.org/10.1007/s00264-008-0689-6.
22. Miller MD, Osborne JR, Gordon WT, Hinkin DT, Brinker MR. The natural history of bone bruises. Am J Sports Med. 1998;26:15–9. https://doi.org/10.1177/03635465980260011001.
23. Bonasia DE, Bruzzone M, Dettoni F, Marmotti A, Blonna D, Castoldi F, et al. Treatment of medial and posteromedial knee instability: indications, tech-

niques, and review of the results. Iowa Orthop J. 2012;32:173–83.
24. Woo SL, Gomez MA, Sites TJ, Newton PO, Orlando CA, Akeson WH. The biomechanical and morphological changes in the medial collateral ligament of the rabbit after immobilization and remobilization. J Bone Joint Surg Am. 1987;69:1200–11.
25. Newton PO, Woo SL, Kitabayashi LR, Lyon RM, Anderson DR, Akeson WH. Ultrastructural changes in knee ligaments following immobilization. Matrix. 1990;10:314–9.
26. Walsh S, Frank C, Hart D. Immobilization alters cell metabolism in an immature ligament. Clin Orthop Relat Res. 1992:277–88.
27. Padgett LR, Dahners LE. Rigid immobilization alters matrix organization in the injured rat medial collateral ligament. J Orthop Res. 1992;10:895–900. https://doi.org/10.1002/jor.1100100619.
28. Hart JM, Pietrosimone B, Hertel J, Ingersoll CD. Quadriceps activation following knee injuries: a systematic review. J Athl Train. 2010;45:87–97. https://doi.org/10.4085/1062-6050-45.1.87.
29. Kim K-M, Croy T, Hertel J, Saliba S. Effects of neuromuscular electrical stimulation after anterior cruciate ligament reconstruction on quadriceps strength, function, and patient-oriented outcomes: a systematic review. J Orthop Sport Phys Ther. 2010;40:383–91. https://doi.org/10.2519/jospt.2010.3184.
30. Johnson DL, Swenson TM, Irrgang JJ, Fu FH, Harner CD. Revision anterior cruciate ligament surgery: experience from Pittsburgh. Clin Orthop Relat Res. 1996;325:100–9. https://doi.org/10.1097/00003086-199604000-00011.
31. Arundale AJH, Bizzini M, Giordano A, Hewett TE, Logerstedt DS, Mandelbaum B, et al. Exercise-based knee and anterior cruciate ligament injury prevention. J Orthop Sports Phys Ther. 2018;48:A1–25. https://doi.org/10.2519/jospt.2018.0303.
32. Logan CA, O'Brien LT, LaPrade RF. POST operative rehabilitation of grade III medial collateral ligament injuries: evidence based rehabilitation and return to play. Int J Sports Phys Ther. 2016;11:1177–90.
33. Logerstedt DS, Scalzitti D, Risberg MA, Engebretsen L, Webster KE, Feller J, et al. Knee stability and movement coordination impairments: knee ligament sprain revision 2017. J Orthop Sports Phys Ther. 2017;47:A1–47. https://doi.org/10.2519/jospt.2017.0303.
34. Lundberg M, Messner K. Long-term prognosis of isolated partial medial collateral ligament ruptures. Am J Sports Med. 1996;24:160–3. https://doi.org/10.1177/036354659602400207.
35. Bagwell MS, Wilk KE, Colberg RE, Dugas JR. The use of serial platelet rich plasma injections with early rehabilitation to expedite grade III medial collateral ligament injury in a professional athlete: a case report. Int J Sports Phys Ther. 2018;13:520–5.
36. Eirale C, Mauri E, Hamilton B. Use of platelet rich plasma in an isolated complete medial collateral ligament lesion in a professional football (soccer) player: a case report. Asian J Sports Med. 2013;4:158–62.
37. Reider B, Sathy MR, Talkington J, Blyznak N, Kollias S. Treatment of isolated medial collateral ligament injuries in athletes with early functional rehabilitation. A five-year follow-up study. Am J Sports Med. 1994;22:470–7. https://doi.org/10.1177/036354659402200406.
38. Indelicato PA, Hermansdorfer J, Huegel M. Nonoperative management of complete tears of the medial collateral ligament of the knee in intercollegiate football players. Clin Orthop Relat Res. 1990:174–7.
39. Ballmer PM, Jakob RP. The non operative treatment of isolated complete tears of the medial collateral ligament of the knee. A prospective study. Arch Orthop Trauma Surg. 1988;107:273–6.
40. Sandberg R, Balkfors B, Nilsson B, Westlin N. Operative versus non-operative treatment of recent injuries to the ligaments of the knee. A prospective randomized study. J Bone Joint Surg Am. 1987;69:1120–6.
41. Petermann J, von Garrel T, Gotzen L. Non-operative treatment of acute medial collateral ligament lesions of the knee joint. Knee Surg Sports Traumatol Arthrosc. 1993;1:93–6.
42. Chen L, Kim PD, Ahmad CS, Levine WN. Medial collateral ligament injuries of the knee: current treatment concepts. Curr Rev Musculoskelet Med. 2008;1:108–13. https://doi.org/10.1007/s12178-007-9016-x.
43. Hughston JC, Eilers AF. The role of the posterior oblique ligament in repairs of acute medial (collateral) ligament tears of the knee. J Bone Joint Surg Am. 1973;55:923–40.
44. Gorin S, Paul DD, Wilkinson EJ. An anterior cruciate ligament and medial collateral ligament tear in a skeletally immature patient: a new technique to augment primary repair of the medial collateral ligament and an allograft reconstruction of the anterior cruciate ligament. Arthrosc J Arthrosc Relat Surg. 2003;19:e159–64. https://doi.org/10.1016/j.arthro.2003.10.031.
45. O'Donoghue DH. Reconstruction for medial instability of the knee. Late reconstr. inj. ligaments knee. Berlin: Springer; 1978. p. 66–75. https://doi.org/10.1007/978-3-642-87274-7_9.
46. Slocum DB, Larson RL. The classic: rotatory instability of the knee. Clin Orthop Relat Res. 2007;454:5–13. https://doi.org/10.1097/BLO.0b013e31802baf88.
47. Shahane SA, Bickerstaff DR. Proximal advancement of the medial collateral ligament for chronic medial instability of the knee joint. Knee. 1998;5:191–7. https://doi.org/10.1016/S0968-0160(97)10023-0.
48. Dong J, Ji G, Zhang Y, Gao S, Wang F, Chen B. Single allograft medial collateral ligament and posterior oblique ligament reconstruction: a technique to improve valgus and rotational stability. Eur J Orthop Surg Traumatol. 2014;24:1025–9. https://doi.org/10.1007/s00590-013-1265-3.
49. Wijdicks CA, Michalski MP, Rasmussen MT, Goldsmith MT, Kennedy NI, Lind M, et al. Superficial medial collateral ligament anatomic augmented repair versus anatomic reconstruction. Am J Sports Med. 2013;41:2858–66. https://doi.org/10.1177/0363546513503289.

50. LaPrade RF, Moulton SG, Nitri M, Mueller W, Engebretsen L. Clinically relevant anatomy and what anatomic reconstruction means. Knee Surg Sport Traumatol Arthrosc. 2015;23:2950–9. https://doi.org/10.1007/s00167-015-3629-1.
51. Ferrari MB, Chahla J, Mitchell JJ, Moatshe G, Mikula JD, Marchetti DC, et al. Multiligament reconstruction of the knee in the setting of knee dislocation with a medial-sided injury. Arthrosc Tech. 2017;6:e341–50. https://doi.org/10.1016/j.eats.2016.10.003.
52. Wijdicks CA, Griffith CJ, LaPrade RF, Johansen S, Sunderland A, Arendt EA, et al. Radiographic identification of the primary medial knee structures. J Bone Jt Surg Am. 2009;91:521–9. https://doi.org/10.2106/JBJS.H.00909.
53. Coobs BR, LaPrade RF, Griffith CJ, Nelson BJ. Biomechanical analysis of an isolated fibular (lateral) collateral ligament reconstruction using an autogenous semitendinosus graft. Am J Sports Med. 2007;35:1521–7. https://doi.org/10.1177/0363546507302217.
54. Laprade RF, Wijdicks CA. Development of an anatomic medial knee reconstruction. Clin Orthop Relat Res. 2012;470:806–14. https://doi.org/10.1007/s11999-011-2061-1.
55. Wijdicks CA, Brand EJ, Nuckley DJ, Johansen S, LaPrade RF, Engebretsen L. Biomechanical evaluation of a medial knee reconstruction with comparison of bioabsorbable interference screw constructs and optimization with a cortical button. Knee Surg Sport Traumatol Arthrosc. 2010;18:1532–41. https://doi.org/10.1007/s00167-010-1127-z.
56. Phisitkul P, James SL, Wolf BR, Amendola A. MCL injuries of the knee: current concepts review. Iowa Orthop J. 2006;26:77–90.
57. Encinas-Ullán CA, Rodríguez-Merchán EC. Isolated medial collateral ligament tears. EFORT Open Rev. 2018;3:398–407. https://doi.org/10.1302/2058-5241.3.170035.
58. Hetsroni I, Lyman S, Pearle AD, Marx RG. The effect of lateral opening wedge distal femoral osteotomy on medial knee opening: clinical and biomechanical factors. Knee Surg Sports Traumatol Arthrosc. 2014;22:1659–65. https://doi.org/10.1007/s00167-013-2405-3.
59. Levy BA, Dajani KA, Morgan JA, Shah JP, Dahm DL, Stuart MJ. Repair versus reconstruction of the fibular collateral ligament and posterolateral corner in the multiligament-injured knee. Am J Sports Med. 2010;38:804–9. https://doi.org/10.1177/0363546509352459.
60. Stannard J, Black B, Azbell C, Volgas D. Posteromedial corner injury in knee dislocations. J Knee Surg. 2012;25:429–34. https://doi.org/10.1055/s-0032-1322605.
61. van der List JP, DiFelice GS. Primary repair of the medial collateral ligament with internal bracing. Arthrosc Tech. 2017;6:e933–7. https://doi.org/10.1016/j.eats.2017.03.003.
62. van den Bogaerde JM, Shin E, Neu CP, Marder RA. The superficial medial collateral ligament reconstruction of the knee: effect of altering graft length on knee kinematics and stability. Knee Surg Sport Traumatol Arthrosc. 2011;19 https://doi.org/10.1007/s00167-011-1519-8.
63. Yoshiya S, Kuroda R, Mizuno K, Yamamoto T, Kurosaka M. Medial collateral ligament reconstruction using autogenous hamstring tendons: technique and results in initial cases. Am J Sports Med. 2005;33:1380–5. https://doi.org/10.1177/0363546504273487.
64. Kim SJ, Choi NH, Shin SJ. Semitendinosus tenodesis for medial instability of the knee. Arthroscopy. 2001;17:660–3. https://doi.org/10.1053/jars.2001.21846.
65. Bosworth DM. Transplantation of the semitendinosus for repair of laceration of medial collateral ligament of the knee. J Bone Joint Surg Am. 1952;34 A:196–202. https://doi.org/10.2106/00004623-195234010-00023.
66. Fanelli GC, Harris JD. Surgical treatment of acute medial collateral ligament and posteromedial corner injuries of the knee. Sports Med Arthrosc. 2006;14:78–83. https://doi.org/10.1097/01.jsa.0000212301.80496.dc.
67. Vaughn ZD, Schmidt J, Lindsey DP, Dragoo JL. Biomechanical evaluation of a 1-stage revision anterior cruciate ligament reconstruction technique using a structural bone void filler for femoral fixation. Arthrosc J Arthrosc Relat Surg. 2009;25:1011–8. https://doi.org/10.1016/j.arthro.2009.04.068.
68. Tardy N, Boisrenoult P, Teissier P, Steltzlen C, Beaufils P, Pujol N. Clinical outcomes after multiligament injured knees: medial versus lateral reconstructions. Knee Surg Sport Traumatol Arthrosc. 2017;25:524–31. https://doi.org/10.1007/s00167-016-4067-4.
69. Koga H, Muneta T, Yagishita K, Ju Y-J, Sekiya I. Surgical management of grade 3 medial knee injuries combined with cruciate ligament injuries. Knee Surg Sport Traumatol Arthrosc. 2012;20:88–94. https://doi.org/10.1007/s00167-011-1541-x.
70. Kim S-J, Lee D-H, Kim T-E, Choi N-H. Concomitant reconstruction of the medial collateral and posterior oblique ligaments for medial instability of the knee. J Bone Joint Surg Br. 2008;90-B:1323–7. https://doi.org/10.1302/0301-620X.90B10.20781.
71. Ibrahim SAR, Ahmad FHF, Salah M, Al Misfer ARK, Ghaffer SA, Khirat S. Surgical management of traumatic knee dislocation. Arthrosc J Arthrosc Relat Surg. 2008;24:178–87. https://doi.org/10.1016/j.arthro.2007.08.007.
72. Lind M, Jakobsen BW, Lund B, Hansen MS, Abdallah O, Christiansen SE. Anatomical reconstruction of the medial collateral ligament and posteromedial corner of the knee in patients with chronic medial collateral ligament instability. Am J Sports Med. 2009;37:1116–22. https://doi.org/10.1177/0363546509332498.
73. Liu X, Feng H, Zhang H, Hong L, Wang XS, Zhang J, et al. Surgical treatment of subacute and chronic valgus instability in multiligament-injured knees with superficial medial collateral ligament reconstruction using Achilles allografts. Am J Sports Med. 2013;41:1044–50. https://doi.org/10.1177/0363546513479016.

74. Dong JT, Chen BC, Men XQ, Wang F, Hao JD, Zhao JN, et al. Application of triangular vector to functionally reconstruct the medial collateral ligament with double-bundle allograft technique. Arthrosc J Arthrosc Relat Surg. 2012;28:1445–53. https://doi.org/10.1016/j.arthro.2012.03.024.
75. Stannard J, Bauer K. Current concepts in knee dislocations: PCL, ACL, and medial sided injuries. J Knee Surg. 2012;25:287–94. https://doi.org/10.1055/s-0032-1326998.
76. Noyes FR, Berrios-Torres S, Barber-Westin SD, Heckmann TP. Prevention of permanent arthrofibrosis after anterior cruciate ligament reconstruction alone or combined with associated procedures: a prospective study in 443 knees. Knee Surg Sport Traumatol Arthrosc. 2000;8:196–206. https://doi.org/10.1007/s001670000126.
77. LaPrade RF, DePhillipo NN, Cram TR, Cinque ME, Kennedy MI, Dornan GJ, et al. Partial controlled early postoperative weightbearing versus nonweightbearing after reconstruction of the fibular (lateral) collateral ligament: a randomized controlled trial and equivalence analysis. Am J Sports Med. 2018;46:2355–65. https://doi.org/10.1177/0363546518784301.
78. Halinen J, Lindahl J, Hirvensalo E, Santavirta S. Operative and nonoperative treatments of medial collateral ligament rupture with early anterior cruciate ligament reconstruction. Am J Sports Med. 2006;34:1134–40. https://doi.org/10.1177/0363546505284889.
79. Grant JA, Tannenbaum E, Miller BS, Bedi A. Treatment of combined complete tears of the anterior cruciate and medial collateral ligaments. Arthrosc J Arthrosc Relat Surg. 2012;28:110–22. https://doi.org/10.1016/j.arthro.2011.08.293.
80. Lujan TJ, Dalton MS, Thompson BM, Ellis BJ, Weiss JA. Effect of ACL deficiency on MCL strains and joint kinematics. J Biomech Eng. 2006;129:386–92. https://doi.org/10.1115/1.2720915.
81. Fanelli GC. Evaluation and treatment of medial instability of the knee. Sports Med Arthrosc. 2015;23:61–2. https://doi.org/10.1097/JSA.0000000000000070.
82. Bollier M, Smith P. Anterior cruciate ligament and medial collateral ligament injuries. J Knee Surg. 2014;27:359–68. https://doi.org/10.1055/s-0034-1381961.
83. Tandogan NR, Kayaalp A. Surgical treatment of medial knee ligament injuries: current indications and techniques. EFORT Open Rev. 2016;1:27–33. https://doi.org/10.1302/2058-5241.1.000007.
84. Weimann A, Schatka I, Herbort M, Achtnich A, Zantop T, Raschke M, et al. Reconstruction of the posterior oblique ligament and the posterior cruciate ligament in knees with posteromedial instability. Arthrosc J Arthrosc Relat Surg. 2012;28:1283–9. https://doi.org/10.1016/j.arthro.2012.02.003.
85. Crawford MD, Kennedy MI, Bernholt DL, DePhillipo NN, LaPrade RF. Combined posterior cruciate ligament and superficial medial collateral ligament knee reconstruction: avoiding tunnel convergence. Arthrosc Tech. 2019;8:e929–33. https://doi.org/10.1016/j.eats.2019.05.002.

Fracture Dislocations About the Knee

19

Luc Rubinger, Aaron Gazendam, Seper Ekhtiari, Jeffrey Kay, Herman Johal, and Darren de SA

19.1 Intra-articular Proximal Tibia and Distal Femur Fracture/Dislocations

19.1.1 Background and Mechanism of Injury

While the relationship between ligamentous injuries and knee instability is well described, there is no comprehensive classification system to describe the range of bony injuries that may occur with high-energy injuries and fracture dislocation. A high index of suspicion for ligamentous instability must be maintained when approaching periarticular fracturs around this complex joint. Periarticular fracture dislocations about the tibiofemoral joint include both distal femur fractures and proximal tibial fractures.

For tibial plateau fractures, the Hohl and Moore classification (Table 19.1) fills the void in describing true tibial fracture dislocations about the tibiofemoral joint, and draws attention to patterns with associated joint instability, which often

All images/tables are original material, created for this chapter, with the expection of Fig. 19.1 which is adapted from an open source publication, thereby not requiring copyright permission. Source of Fig. 19.1 is: Pelser, PC. (2010). Controversies in the management of tibial plateau fractures. SA Orthopaedic Journal, 9(3), 75–82.

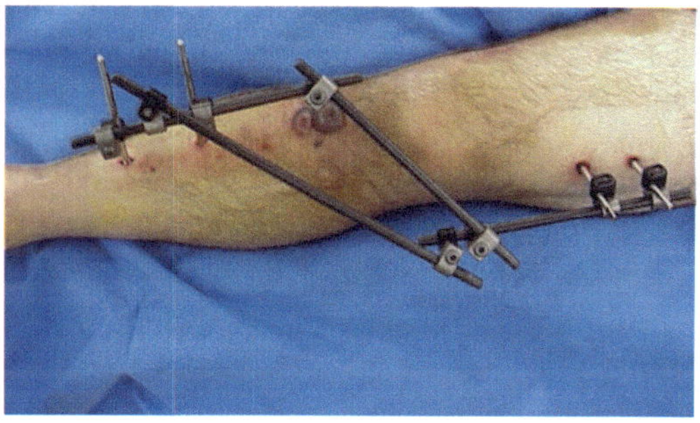

Fig. 19.1 Temporary knee-spanning external fixator. (Pelser, PC. (2010). Controversies in the management of tibial plateau fractures. SA Orthopaedic Journal, 9(3), 75–82)

Table 19.1 Hohl and Moore classification of proximal tibia fracture dislocations

Type I	Coronal split fracture
Type II	Entire condylar fracture
Type III	Rim avulsion fracture of the lateral plateau
Type IV	Rim compression fracture
Type V	Four-part fracture

L. Rubinger · A. Gazendam · S. Ekhtiari · J. Kay
H. Johal
Division of Orthopaedics, Department of Surgery, McMaster University, Hamilton, ON, Canada
e-mail: luc.rubinger@medportal.ca; aaron.gazendam@medportal.ca; seper.ekhtiari@medportal.ca; jeffery.kay@medportal.ca

D. de SA (✉)
Division of Orthopaedics, Department of Surgery, McMaster University, Hamilton, ON, Canada

Department of Orthopaedics, McMaster Children's Hospital, Hamilton, ON, Canada
e-mail: darren.desa@medportal.ca

N. Nakamura et al. (eds.), *Advances in Knee Ligament and Knee Preservation Surgery*,
https://doi.org/10.1007/978-3-030-84748-7_19

goes underrecognized when using systems such as the Schatzker classification [1, 2]. Tibial plateau fractures account for 1.7–2.0% of all fractures in adults and about 8% of fractures in the elderly [3]. These complex fractures represent a wide clinical spectrum that can be accompanied by skin and muscle compromise, neurovascular injury, compartment syndrome, ligament and meniscal tears, posterolateral corner (PLC) disruption and associated dislocation [4–8]. However, few of these fractures require separate soft tissue stabilization procedures. In a prospective cohort of 82 tibial plateau fractures, 73% had associated soft tissue injuries but only 2% required secondary soft tissue repair or reconstructive procedures [9]. Conversely, in a series of 90 consecutive multiligament knee injuries, Porrino et al. found 19 (21%) to have associated tibial plateau fracture (47% lateral plateau fractures, 37% medial plateau,16% bicondylar fractures) [4].

For distal femur fractures, there has been no definitive classification developed to describe true fracture dislocations; however, the OTA/AO classification system can be used to accurately describe the fracture pattern in terms of articular involvement and comminution. The eponymous Hoffa's fracture is a coronal plane fracture of one of the posterior femoral condyles. In the context of femoral shaft fractures, up to 30% of femoral shaft fractures have concomitant significant ligament injury [10]. In a series of 26 femoral shaft fractures, the ACL (50%) was found to be most commonly injured, followed by the MCL (31%), LCL (13%), and PCL (6%) [11]. In another series of 27 consecutive diaphyseal femur fractures who underwent MRI scans of the knee, 19% were found to have ACL injuries, 19% had grade 3 MCL injuries, 15% had grade 3 LCL injuries and 7% PCL injuries [12]. Similarly, in a series of ipsilateral femoral shaft and tibial shaft fractures (i.e. floating knees), 30% of patients had evidence of ligamentous injuries [13].

19.1.2 Clinical Presentation and Diagnostics

The first line of imaging investigations are standard orthogonal radiographs. These are typically followed by dedicated computed tomography (CT) scans to delineate fracture configuration, particularly the orientation, location and degree of displacement of depressed intra-articular fragments. Three-dimensional CT reconstruction offers a useful adjunct to the intra-articular sagittal, coronal and axial cuts to plan surgical approaches, reduction and fixation. Magnetic resonance (MR) imaging, even in the acute period, is a valuable tool to assess ligamentous, capsular, meniscal and chondral injury. MR imaging can be done following the initial injury, or, more typically, following provisional stabilization with an MRI-compatible external fixation.

19.1.3 Management Options and Evidence-Based Outcomes

Given the articular nature of these injuries, operative management is routinely necessary to restore joint stability, and limit functional impairment. In polytrauma patients, the timing of surgery is often dictated by the severity of the accompanying injuries and overall physiologic stability of the patient. General determinants such as cardiovascular, pulmonary, and neurologic function, as well as markers of response to resuscitation (i.e. lactate) play a major role in the timing and nature of acute and definitive surgical management; however, surrounding soft tissue envelope of the knee is often the definitive factor.

The high-energy nature of these injuries often precludes the use of internal fixation in the early post-injury period, given there may be surrounding soft tissue loss or rapid onset of swelling. This is illustrated as early definitive stabilization of high-energy tibial plateau fracture dislocations has been associated with a higher risk of wound breakdown and infection [14]. Alternatively, temporary external fixation, wherein a knee spanning external fixator is fixed to the femur and the tibia while the fracture zone is bridged and provisional reduction is achieved with distraction (Fig. 19.1). This technique is then followed by delayed definitive internal fixation once soft tissue swelling settles (e.g. resolution of fracture blisters, return of skin wrinkles) and has been shown to result in decreased rate of soft tissue complications [15]. This approach also facilitates the management of open wounds or vascular injuries and collection of advanced imaging (CT, MRI) with the knee in a provisionally reduced position.

For optimal external fixation, place pins on both the lateral (proximal femoral fragment) and anteriomedial (proximal tibial fragment) such that the connecting rod(s) are angled over the tibia in an oblique fashion. This provides area under the frame for swelling around the plateaus and allowing varying degrees of flexion through the knee joint should the reduction require it. Place pins at a minimum of 2 cm away from the joint line on the tibia (although farther is better in this context), and well proximal of the suprapatellar pouch on the femoral side to avoid intra-articular infection. For provisional reduction, use the half frames of the distal (tibial) and proximal (femoral) fragment as handles, and manually reduce the fracture in length with slight knee flexion (aided by a bolster), alignment and rotation.

Irrespective of the classification system used for the planning of definitive surgical treatment in tibial plateau fractures, it is important to determine the stability of the medial, lateral and posterior columns and the degree of any associated articular comminution or depression [16]. In the setting of distal femur fractures, it is equally important to determine the stability and integrity of the femoral condyles, the notch and trochlea, and whether there is a coronal plane fracture. With a 38.1% incidence of a coronal plane fracture (i.e. Hoffa fracture) in distal femur fractures with intracondylar extension, and nearly 30% of coronal plane fractures missed with plain radiographs, it is recommended to obtain CT imaging for all supracondylar-intercondylar distal femur fractures [17, 18]. This aids in determining the nature of the forces acting at the knee joint at the time of injury, and ultimately those that will need to be countered and resisted to provide a stable environment for fracture healing following fixation.

Once soft tissues allow, choice of surgical approach is paramount. Efforts must be made to utilize extensile exposures to provide adequate access to the compromised tibial column, or femoral condyle, while maximizing skin bridges and respecting the traumatized soft tissue envelope. There is insufficient evidence to broadly recommend an optimal fixation option among open reduction internal fixation (ORIF), hybrid/circular external fixation, and unilateral locked plating in proximal tibia fractures [19, 20]. For open reduction and internal fixation, anterolateral and posteromedial approaches in the supine position offer the safest and best exposure to the lateral and medial tibial columns, respectively, while prone positioning and posterior approaches may occasionally be required for select posterior column patterns. Midline anterior exposure should be avoided as a surgical approach option for proximal tibial fractures, particularly when access to more than one column is required to avoid soft tissue stripping and soft tissue complications [21]. Similarly, there is insufficient evidence to broadly recommend an optimal fixation option in the setting of intra-articular distal femur fractures between: locked plating options, dynamic condylar screws, and intramedullary fixation [22–24]. The surgical incision and approach used in the treatment of distal femur fractures will be dictated mostly by the fixation method used.

The overall goals in management are to restore articular congruity, bony alignment and stability at the knee to provide a normal mechanical axis during weight bearing in efforts to prolong lifespan of the native knee joint. Despite this, post-traumatic arthritis occurs after intra-articular fractures about the knee and causes disability in young active patients [25]. Moatshe et al. found that 42% of surgically treated knee dislocations developed OA at a minimum of 10-year follow-up [26]. Additionally, a large cohort study by Wasserstein et al. showed that 7.3% of patients treated with ORIF for a tibial plateau fracture underwent total knee arthroplasty (TKA) at 10 years [27]. This was compared with 1.8% in the matched control group. After adjustment for comorbidity in the statistical model, the risk of TKA was more than five times as likely in the tibial plateau ORIF group as in the control group, with older patients and those with bicondylar fractures having increased risk. However, the authors did not determine the role of mechanism of injury or associated knee stability, as there may have been unaddressed associated soft tissue compromise leading to advanced joint degeneration.

Although delayed, post-ORIF TKA does offer a definitive reconstructive option for those with ongoing functional compromise. It is well established that TKA for posttraumatic osteoarthritis secondary to malunion is associated with a higher rate of complications and poorer functional results than TKA for primary osteoarthritis of the knee [28]. This has led to increased interest in acute TKA for complex periarticular knee fractures, particularly as an option for elderly patients with poor bone stock and for whom prolonged non-weight-bearing status can be associated with considerable problems [29]. Interest has focused primarily on fractures of the distal femur, with some recent articles showing that TKA bypasses fracture healing issues and facilitates early mobilization and immediate weight bearing for tibial plateau fractures as well [30].

19.1.4 Case Presentation

A 59-year-old male presents to the emergency department after being struck by a car while riding his motorcycle. He had sustained an isolated right knee injury that was closed and had an intact peripheral neurovascular status. There were no clinical signs of compartment syndrome. The patient noted that he had to 'realign' his leg after the accident. The patient also had preexisting right knee pain and was scheduled to see an Orthopaedic Surgeon for 'osteoarthritis' (OA). Preoperative radiographs and CT scan images are shown in Figs. 19.2 and 19.3, respectively.

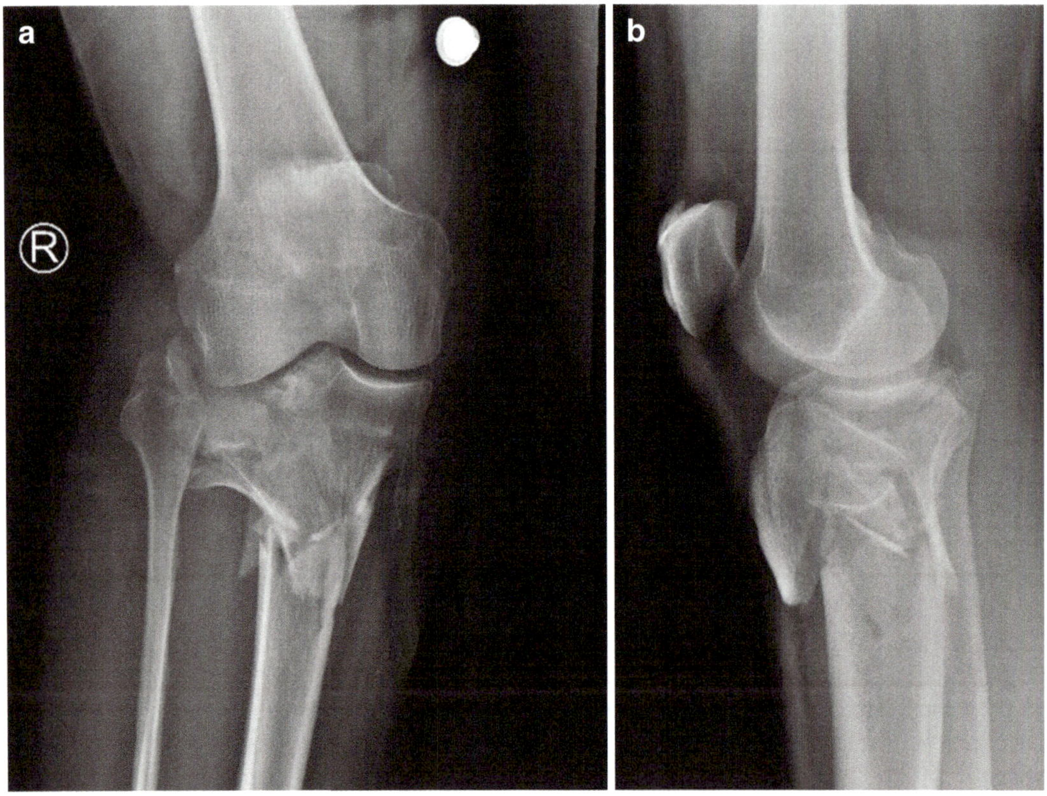

Fig. 19.2 Anteroposterior (**a**) and lateral (**b**) radiographs demonstrating preoperative right knee injury consistent with a Type V Hohl and Moore proximal tibial fracture dislocation

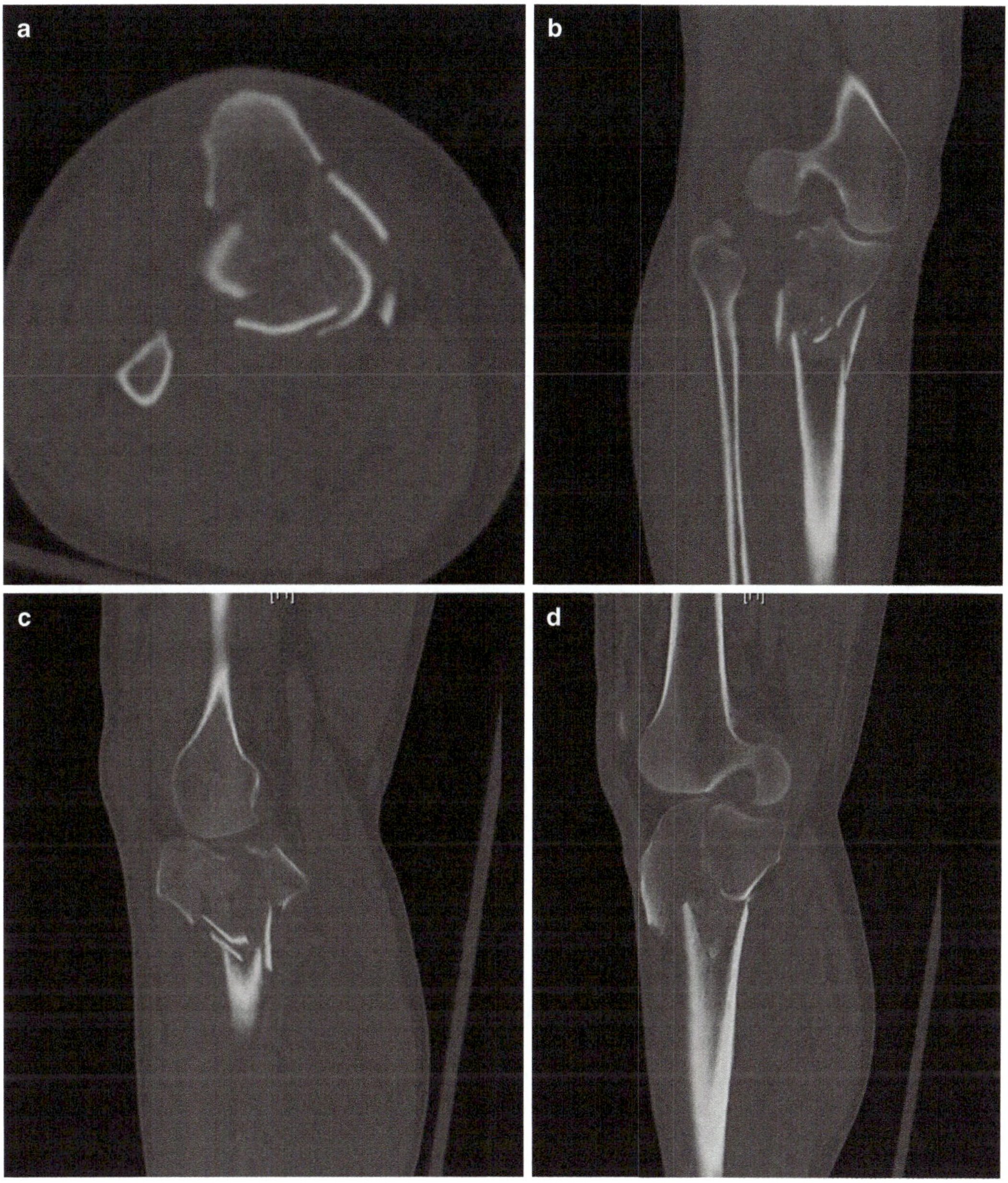

Fig. 19.3 Preoperative CT scan images of injured right knee. (**a**) Distal axial image showing comminuted diaphyseal dissociation. (**b**) Posterior-coronal image showing meta-diaphyseal dissociation and marked articular surface impaction. (**c** and **d**) Sagittal images showing marked articular surface comminution, a large tibial tubercle fragment and a large posterolateral tibial plateau fragment

The patient's history, physical examination and radiographs were consistent with a high-energy bicondylar tibial plateau fracture, with complete dissociation between the metaphyseal articular condyles and the tibial diaphysis. The tibia was shortened and in valgus alignment. In addition to condylar widening, the lateral and central joint surfaces were depressed. Further, there was significant comminution at the metaphyseal/diaphyseal junction with a large anterior tibial tubercle fragment. Given the history and the fracture pattern (Type V Hohl and Moore proximal tibial fracture-dislocation), care was taken in assessing for associated vascular injuries with serial ankle-brachial index measurements, and a full trauma team assessment was carried out to rule out non-orthopaedic injury.

19.1.4.1 Clinical Decision-Making

Given the extensive soft tissue swelling around the proximal tibia, a spanning external fixator was applied to this patient's knee within 24 h of the initial injury. After 8 days, the soft tissues had settled enough clinically, with wrinkling present on the anteromedial skin of the proximal tibia, that the patient was taken to the operating room for definitive fixation.

Bicondylar fixation was performed using a combined anterolateral and posteromedial approach. This allowed direct visualization of the fracture fragments for anatomic reduction while respecting soft tissue bridges. Further, both medial and lateral approaches were positioned at minimum 7 cm away from midline to allow for adequate skin bridges should this patient go on to need a TKA in the future. Although this patient had a history and radiographic signs consistent with mild OA of the knee, his age, bone quality and marked meta-diaphyseal comminution precluded the use of acute TKA in the treatment of this fracture. With that said, this patient will be at risk for needing a TKA in the future, and this should be incorporated into the clinical decision-making process, including the placement of incisions and management strategies to restore alignment and promote adequate bone healing.

19.1.4.2 Intraoperative Findings

Intraoperatively, the fracture was extensively comminuted, especially at the lateral tibial plateau and the lateral metaphyseal-diaphyseal junction. This called for a lateral-sided sub-meniscal arthrotomy, which revealed a lateral meniscus that was avulsed from its capsular attachments and displaced along with the depressed articular segments. The meniscus was tagged for later repair once the bony stability was restored. It should be noted that arthroscopy can be used as an adjunct to assess meniscal pathology and entrapment in the tibial plateau fracture scenario. However, it is the authors' opinion that in the fracture dislocation population, formal arthrotomy with well-visualized fracture reduction, along with open meniscal surgical management leads to more optimal fracture reduction. The fracture pattern necessitated long, bridging fixation using a lateral locking plate extending from the articular block to the diaphysis, augmented with calcium phosphate bone substitute for a large bone void that remained once the articular surface was elevated and reduced, and a medial 1/3 tubular plate to provide stability while avoiding making the construct too rigid to promote healing. Stabilization of the tibial tubercle fragment to the reconstructed columns was achieved with lag-by-technique fixation, which allowed fragment specific fixation for this challenging fracture pattern. As a final step, the lateral meniscal avulsion was repaired to the lateral capsule, and the knee was examined through a full range of motion. The ligaments were deemed stable post-fracture fixation, precluding the need for any further soft tissue procedures.

19.1.4.3 Outcome

Post-operatively, this patient was initially made non-weight bearing with no range of motion for 2 weeks. Passive range of motion exercises in a hinged knee brace began thereafter. By the 6-week mark, this patient began weight bearing with knee range of motion from 0° to 90°. Post-operative and 6-month follow-up radiographs are shown in Figs. 19.4 and 19.5, respectively.

19.1.5 Case Presentation

A 39-year-old female who presents to the emergency department after falling off her bicycle at high speeds. She had sustained a left knee injury as well as a right wrist injury. The knee injury was open and had an intact peripheral neurovascular status. On examination of the left knee, there was a large 7 cm laceration with protruding bone from both the patella and distal femur. On irrigation in the trauma bay, a complete rupture of

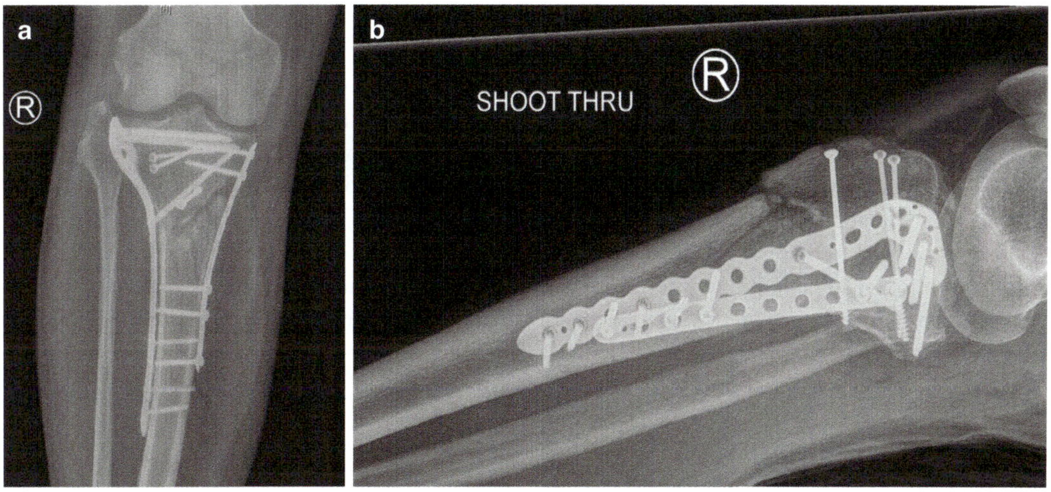

Fig. 19.4 Anteroposterior (**a**) and lateral (**b**) radiographs demonstrating immediate post-operative right knee fixation with a locked lateral plating along the diaphysis, augmented with calcium phosphate bone substitute, a medial 1/3 tubular plate and custom, lag by technique fixation of the tibial tubercle fragment

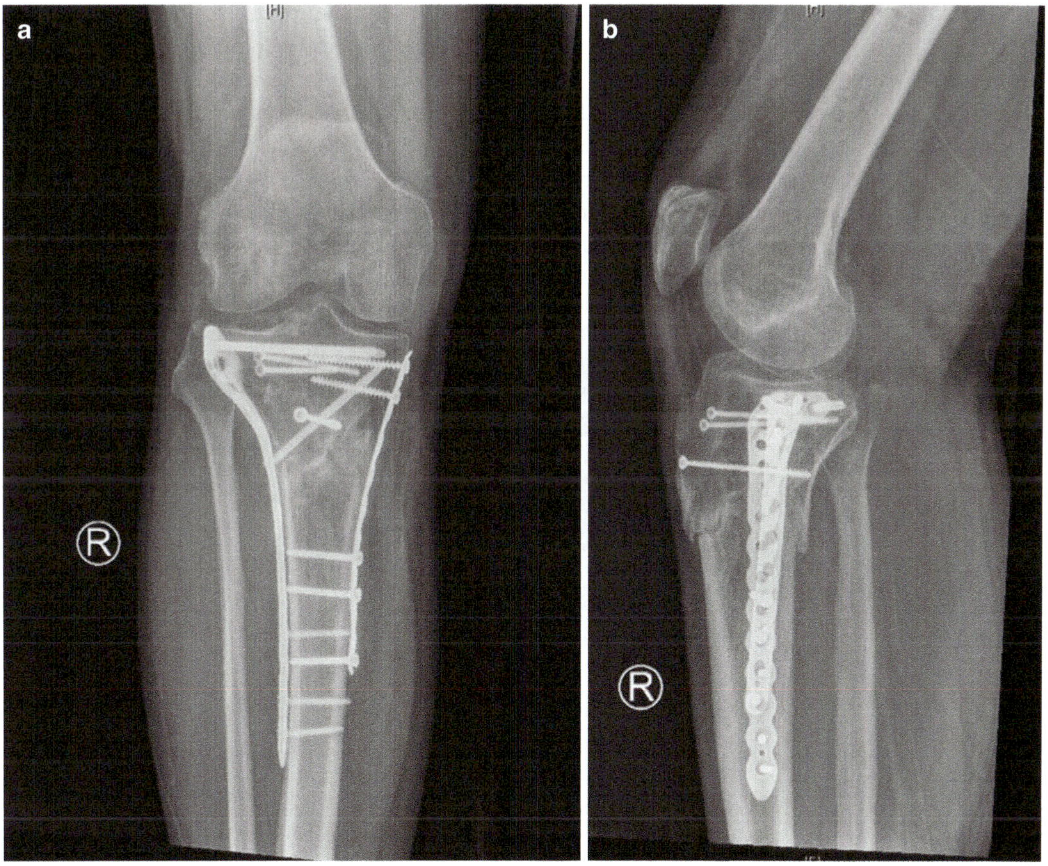

Fig. 19.5 Anteroposterior (**a**) and lateral (**b**) radiographs demonstrating immediate post-operative right knee fixation with adequate ossification and blurring of the fracture lines

the quadriceps could be palpated. Preoperative radiographs and CT scan images are shown in Figs. 19.6 and 19.7, respectively.

The patient's history, physical examination and radiographs were consistent with a high-energy open bicondylar intra-articular comminuted distal femur fracture, with associated ipsilateral patellar fracture and extensor mechanism disruption. The patient received the appropriate antibiotics, and a provisional irrigation and debridement in the trauma bay. The femur was shortened with a flexion deformity and condylar widening. Further, there was significant comminution at the metaphyseal/diaphyseal junction with bone loss. Given the history and the fracture pattern, care again was taken in assessing for associated vascular injuries with serial ankle-brachial index measurements, and a full trauma team assessment was carried out to rule out non-orthopaedic injury.

19.1.5.1 Clinical Decision-Making

Given the open nature of the fractures and the extensive soft tissue damage around the knee, the patient was brought to the operating room urgently. A thorough irrigation and debridement was carried out with normal saline and gravity flow. Fixation with a lateral locking plate was used given the intra-articular nature of the fracture with associated comminution. Fixation was achieved through an anterior approach centred over the patella that was extended laterally to allow fixation of both the distal femur and patella while incorporating the open wound for debridement.

19.1.5.2 Intraoperative Findings

After appropriate irrigation and debridement, the traumatic arthrotomy, with extension superolaterally, allowed for adequate visualization of the distal femur. The trochlear groove fragment was provisionally stabilized to the lateral condyle fragment and the lateral Hoffa fragment in an anterior to posterior plane. Anatomic reduction at these osteochondral intra-articular fracture lines was obtained and stabilized with two fully threaded cancellous screws. K-wire joysticks were used to manipulate the medial osteochondral articular block to obtain a provisional reduction relative to the lateral side. Compression across the condyles was achieved with a periarticular reduction forceps. Once stabilized the entire articular block was then provisionally fixed to the femoral shaft. Even though there was bone loss at the meta-diaphyseal junction, cortical keys were used on both the lateral and medial

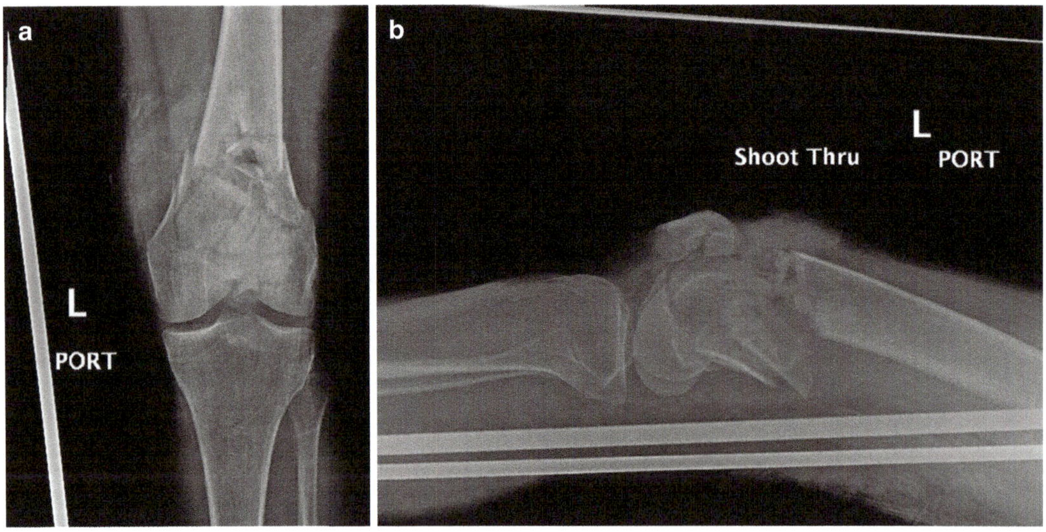

Fig. 19.6 Anteroposterior (**a**) and lateral (**b**) radiographs demonstrating preoperative left knee injury consistent with a Type 33C OTA/AO distal femoral intra-articular fracture

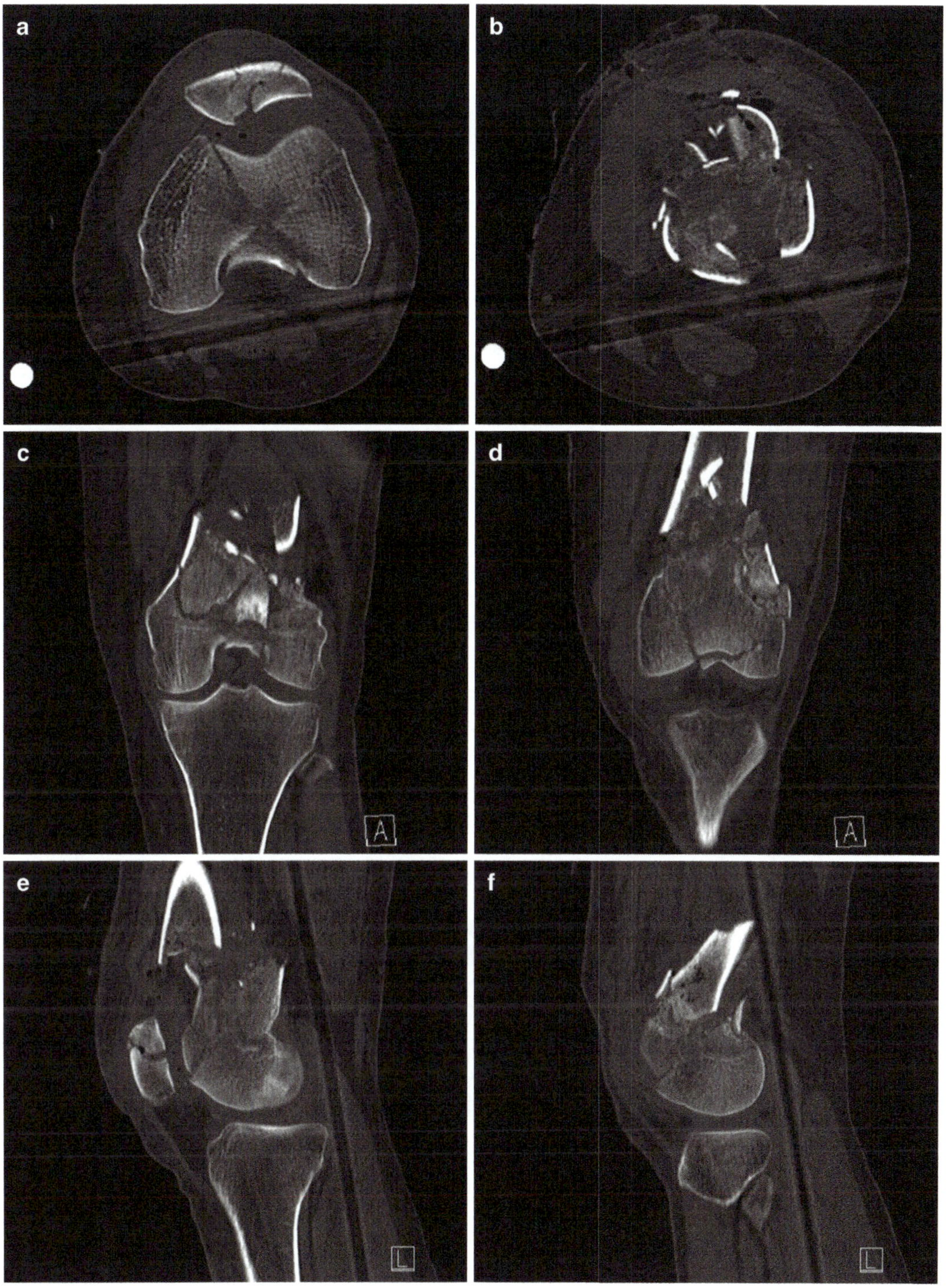

Fig. 19.7 Preoperative CT scan images of injured right knee. (**a** and **b**) Axial images showing intra-articular comminution with condylar widening/split with a large trochlear fragment. (**c** and **d**) Coronal images showing meta-diaphyseal dissociation and marked articular surface impaction, with meta-diaphyseal bone loss. (**e** and **f**) Sagittal images showing flexion deformity of the fracture pattern as well as a substantial lateral condyle Hoffa fragment

sides to achieve anatomic length and rotation. A long distal femoral locking plate was positioned appropriately on the distal segment. Care was taken not to position the plate too posterior, to ensure limited internal rotation of the articular block relative to the femoral shaft. As well, care was taken to limit medial displacement of the articular block, thus ensuring that no 'golf club deformity' was produced. After fixation of the distal femoral fracture the concomitant patellar fracture and extensor mechanism disruption was surgically addressed. As a final step, the bone void was filled with calcium sulphate resorbable beads (Osteoset®, Wright Medical) and vancomycin powder. The ligaments were deemed stable post fracture fixation, precluding the need for any further soft tissue procedures.

19.1.5.3 Outcome

Post-operatively, this patient continued on a 48-h course of IV antibiotics and the wound was monitored. She was initially made non-weight bearing with no range of motion for 2 weeks. Passive range of motion exercises in a hinged knee brace began thereafter. By the 6-week mark this patient began weight bearing with knee range of motion from 0° to 90°. Post-operative radiographs are shown in Fig. 19.8.

19.2 Acute Proximal Tibiofibular Injuries

19.2.1 Background and Mechanism of Injury

Acute proximal tibiofibular joint dislocations are rare injuries, accounting for less than 1% of all knee trauma [31]. Nonetheless, the majority of these injuries occur during sporting activities and may go unrecognized leading to prolonged pain and dysfunction [32].

The proximal tibiofibular joint is a synovial joint that has multiple normal anatomical variants in the population. In one of the earliest detailed descriptions of this joint, Ogden described two proximal tibiofibular joint anatomic variants: oblique and horizontal, with horizontal configuration being defined as <20° joint surface inclination relative to the horizontal plane [33]. In 10–12% of the population, the joint communicates directly with the knee joint [33–35]. The stabilizing structures around the joint include three broad ligamentous bands passing anteriorly, the posterior proximal tibiofibular ligament, and the structures of the posterolateral knee, including the popliteus and the lateral collateral ligament (LCL) [36].

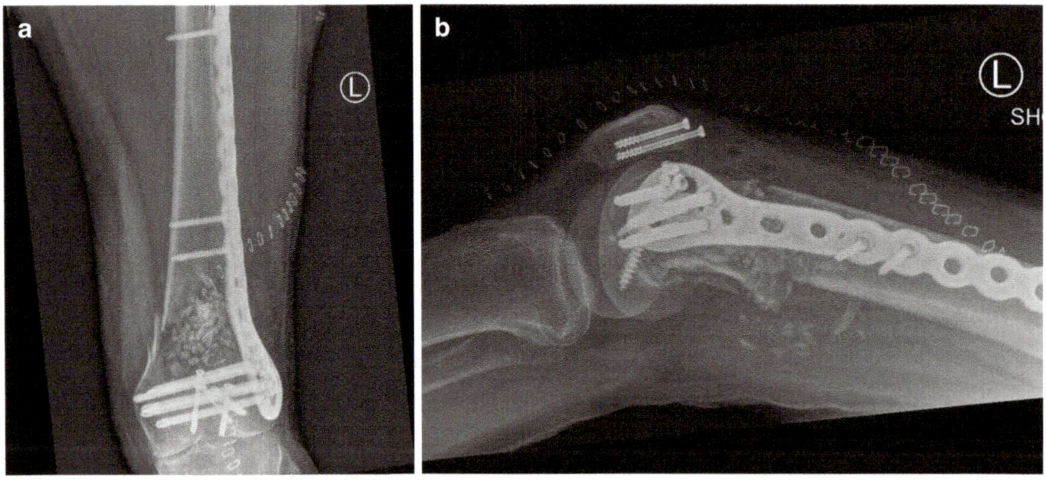

Fig. 19.8 Anteroposterior (**a**) and lateral (**b**) radiographs demonstrating immediate post-operative left knee fixation with a locked lateral plating along the diaphysis, augmented with calcium sulphate bone beads, with two cancellous screws providing fixation of the lateral Hoffa and intercondylar notch fragments

The most common mechanism of injury to the proximal tibiofibular joint is rotational, often through twisting of the knee in a flexed and externally rotated position, with concomitant inversion and plantar flexion of the foot [32, 37]. While sports injuries are the most common aetiology, high-energy polytraumatic injuries can also lead to proximal tibiofibular dislocations.

19.2.2 Clinical Presentation and Diagnostics

Diagnosis may be clinical or require imaging depending on the type and severity of the injury. In patients presenting with an isolated proximal tibiofibular joint disruption, localized pain and swelling over the fibular head is common. As well, prominence of the fibular head may be evident. Careful neurovascular examination should be undertaken as transient peroneal nerve palsy is common given its proximity to the proximal tibiofibular joint [36]. Plain radiographs may reveal the diagnosis, though not all cases are immediately evident. Contralateral radiographs can be helpful for direct comparison, with CT or MRI often being unnecessary for isolated injuries but may be indicated in polytrauma or cases with persistent posterolateral knee pain.

In his original case series, Ogden classified proximal tibiofibular joint dislocations into four types (Table 19.2). Type II injuries are the most common and usually sports-related, while Type III and IV injuries are more commonly related to high-energy mechanisms and direct trauma [38, 39].

Table 19.2 Ogden classification of proximal tibiofibular joint dislocations

Type I	Atraumatic subluxation
Type II	Anterolateral dislocation
Type III	Posteromedial dislocation
Type IV	Superior dislocation

19.2.3 Management Options and Evidence-Based Outcomes

19.2.3.1 Non-operative Management

Closed reduction should ideally be attempted under general anaesthesia with full muscle relaxation, which also allows conversion to open reduction if necessary. To facilitate reduction, the knee should be flexed between 80° and 110° to relax the biceps femoris and LCL [32, 36]. The foot can also be externally rotated, everted, and dorsiflexed to also relax the peroneals, extensor hallucis longus, and extensor digitorum longus (EDL) [36, 37, 39], though some authors argue that this is not necessary [40]. Direct pressure is then applied to the fibular head, with orientation of force depending on the direction and type of dislocation. Successful reduction is often accompanied by an audible and palpable "pop" [36].

19.2.3.2 Operative Management

In cases where closed reduction is unsuccessful, or surgery is required to address other injuries about the knee, open reduction internal fixation may be undertaken. To approach the proximal tibiofibular joint in isolation, a lateral curvilinear incision is made centred over the joint and the peroneal nerve is identified and protected just distally as is wraps around the fibular neck from posterolateral to anteromedial. Open reduction can then be attempted under full general anaesthesia with muscle relaxation. If still unsuccessful, muscular attachments of the proximal fibula including EDL, biceps femoris, and peroneus longus may need to be released to allow for complete reduction [36, 41, 42]. Successful reduction should be confirmed both by direct visualization and fluoroscopic confirmation following which fixation of the fibula to the tibia is needed to maintain alignment as the surrounding soft tissues heal. At least three different fixation techniques have been described: K-wire fixation, screw fixation, and dynamic suspensory suture button fixation. Tricortical fixation with a screw or K-wire have been demonstrated to be adequate, and are performed in a similar fashion, with either a screw or k-wire being placed perpendicular to the joint in a posterolateral to anteromedial direction, while taking care to protect the posterolateral structures of the knee [36, 42]. Alternatively, Warner et al. (2016) describe the treatment of chronic proximal tibiofibular joint instability using an anatomic reconstruction of the posterior ligamentous structures of the PTFJ with a semitendinosus autograft [43].

The dynamic yet powerful suture button may offer an option that more closely recreates the proximal tibiofibular anatomy, and has been described by Main et al. who used the Tightrope™ (Arthrex, Naples, Florida) device [41]. In their case report, the patient already had a history of mild degenerative joint disease in both knees, and was presenting with a chronic and recurrent case of proximal tibiofibular dislocation. It was felt that allowing micro-motion at the proximal tibiofibular joint would have a protective effect against accelerated OA for the patient. Two divergent sets of suture buttons were placed, one from anterolateral to posteromedial and the other from posterolateral to anteromedial. This was augmented by a bio-absorbable screw placed just below the level of the proximal syndesmosis. At 1 year post-operatively, the patient was asymptomatic from the perspective of her proximal tibiofibular joint [41].

Given the rarity of proximal tibiofibular dislocations, the body of literature on the topic is almost entirely composed of case reports. Ogden's original case series, circa 1974, may in fact be the largest case series on this condition, consisting of 43 patients. In that series, Ogden described a number of complications, specific to the dislocation type. Type I was associated with chronic subluxation and peroneal nerve injury leading to foot drop. Patients with Type II dislocations were all treated non-operatively, which was associated with chronic instability in some patients eventually leading to surgical fixation [39]. More recent reports generally attempt closed reduction, followed by immediate open reduction in cases of failed closed attempts. Unsurprisingly, these case reports generally demonstrate no complications, and full return to activity, including competitive sports, with both operative and non-operative management, but higher quality evidence is needed to confirm [36, 42, 44–46].

19.2.4 Case Presentation

The patient had been involved in a head-on motor vehicle collision and presented as a trauma team activation. The patient was diagnosed with a left olecranon fracture, bilateral femur fractures, a left proximal tibiofibular dislocation, and a left tibial shaft fracture. An anterolateral dislocation was noted on plain radiographs and confirmed on CT scan (Fig. 19.9) [32, 39].

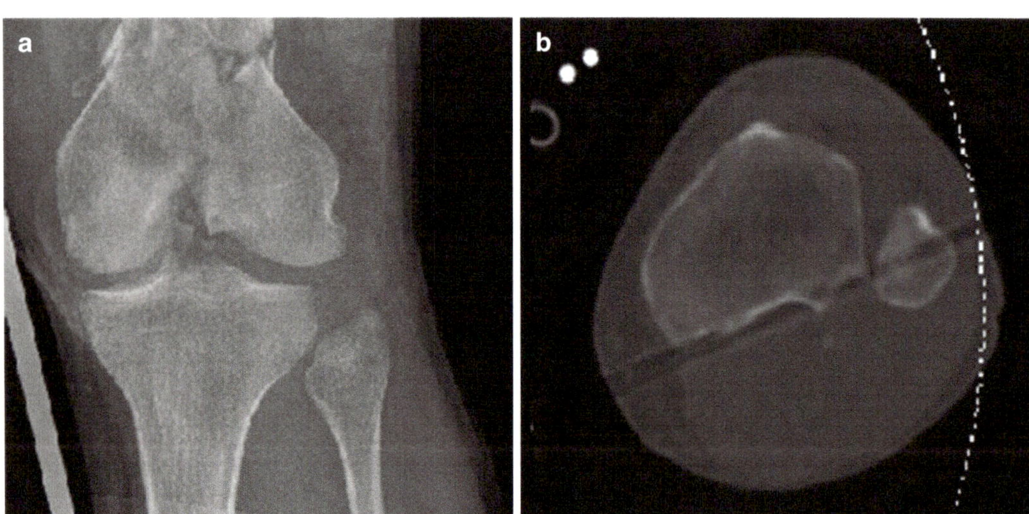

Fig. 19.9 (**a**) Anteroposterior radiograph demonstrating comminuted distal femur fracture and proximal tibiofibular dislocation; (**b**) Axial Computed Tomography scan confirming a Type II Anterolateral dislocation of the fibula

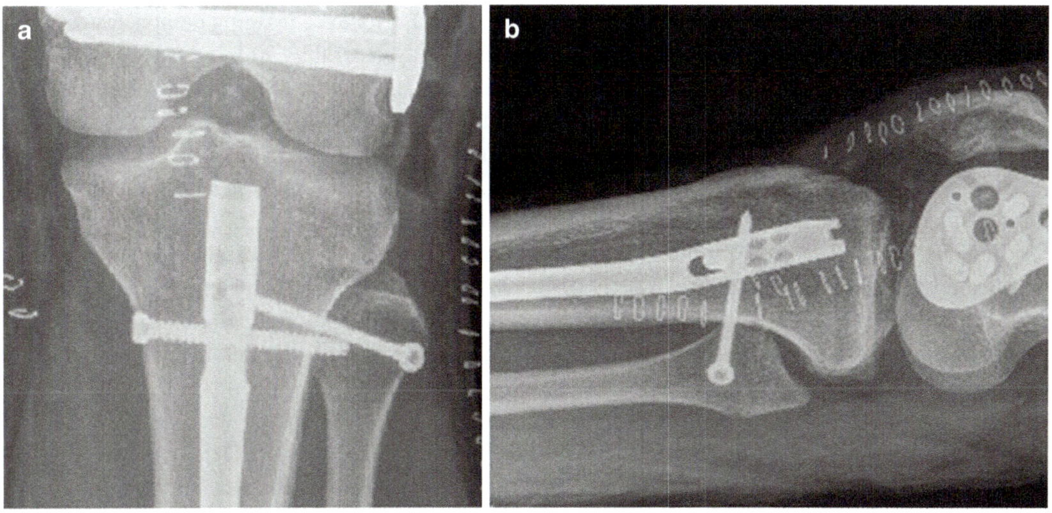

Fig. 19.10 Anteroposterior (**a**) and lateral (**b**) radiographs demonstrating fixation of the proximal tibiofibular joint with tricortical screw, intramedullary fixation of the tibia, and distal femoral locking plate

The patient was being taken to the operating room for their other injuries, and thus open reduction was performed. Minimally invasive-open reduction was achieved without any releases; however, a large enough incision was used to endure that the common peroneal nerve was intact and safe. A percutaneous tricortical screw was used to secure fixation of the proximal tibiofibular joint (Fig. 19.10). The patient was kept non-weight bearing in a long leg splint post-operatively. The patient will be monitored for symptomatology at the proximal tibiofibular joint at post-operative follow-ups to discern if hardware removal will be necessary.

19.3 Patellar Dislocation with Associated Osteochondral Fractures

19.3.1 Background and Mechanism of Injury

Lateral patellar dislocations are a common orthopaedic injury with a documented prevalence of 2.29–5.8 per 100,000 in the general population [47, 48]. The prevalence rises dramatically in adolescents to 11.9–29 per 100,000 and most commonly occurs during athletic endeavours [47, 48]. Several osseous and soft-tissue risk factors for dislocation have been identified including trochlear dysplasia, patella alta, tibial tubercle lateralization, generalized ligamentous laxity, and a history of previous dislocations [49–52].

Common associated injuries of patellar dislocations include chondral and osteochondral fractures. They are often found on the medial and central patellar facets and the lateral femoral condyle [53]. The prevalence of associated patellar chondral injuries is high, ranging from 38 to 95% [54–56]. Femoral-sided chondral injuries are less common and range from 5 to 32% [53, 54, 57, 58]. The high prevalence of osteochondral damage is thought to be due to the high prevalence in adolescents and differences in the properties of the chondral surface and subchondral bone [59]. Osteochondral lesions are more common in traumatic, high-energy dislocations when compared to low-energy recurrent dislocations in patients with underlying anatomic risk factors for dislocation [55]. Regardless of mechanism, however, the presence of osteochondral injuries in the setting of patellar dislocations significantly increases the rate of posttraumatic patellofemoral arthritis later in life [60, 61].

19.3.2 Clinical Presentation and Diagnostics

First-time patellar dislocations usually occur with a flexed knee and internal rotation of the tibia [62]. Acute patellar dislocations occur most commonly during athletics and tend to dislocate laterally [63]. Cartilage defects may present with ongoing pain and swelling, clicking and instability [64].

Plain radiographs of patellofemoral and tibiofemoral joints consisting of anteroposterior, lateral and skyline views should be obtained. Given that plain radiographs miss a large proportion of osteochondral injuries, they should primarily be utilized to assess for predisposing factors of patellar instability as well as concomitant injuries [49, 63]. Trochlear dysplasia can be assessed on plain radiographs utilizing the sulcus line, double contour sign and supratrochlear spurs [65, 66]. The Insall-Salvati, Caton-Deschamps and the Blackburn-Peel ratios are all measures of patellar height to assess for patella alta [67–69].

CT provides fine bony detail and three-dimensional reconstruction but comes with added radiation exposure. CT scans can be used to measure all of the same values as plain radiographs with the added benefit of accurately measuring the distance between the tibial tubercle and the trochlear groove (TT-TG) [70]. The TT-TG distance quantifies the lateralization of the tibial tubercle. Increased TT-TG distance increases the risk of recurrent patellar instability and is particularly important when tibial tubercle osteotomies are being considered in patients with predisposing malalignment [70].

MRI is considered the gold standard imaging modality for assessing both soft tissue, cartilaginous and bony injuries that occur with patellar dislocations [63, 71, 72]. Disruption of the medial ligamentous stabilizers, mainly the medial patellofemoral ligament (MPFL) and patellar retinaculum, are well visualized on MRI [57, 72–74]. Magnetic resonance imaging demonstrates a sensitivity of 81% when compared with arthroscopic evaluation of MPFL tears [72].

Bone oedema secondary to the contusion is seen on the medial patellar facet and the lateral femoral condyle after acute dislocations [71, 74]. Magnetic resonance imaging demonstrates a sensitivity of greater than 90% in assessing for chondral damage when compared to arthroscopy. Intra-articular loose bodies present as a separated fragment of chondral or osteochondral tissue and can be found in up to 33% of patients following patellar dislocations [57, 63, 75]. MRI should be obtained when there is clinical suspicion of an MPFL tear, osteochondral injury not elucidated on prior imaging, and recurrent patellar dislocations refractory to non-operative management.

19.3.3 Management Options and Evidence-Based Outcomes

The management of patellar dislocations with associated osteochondral lesions varies widely and is based largely on level IV evidence and expert opinion. Patients presenting with osteochondral lesions or loose intra-articular bodies are often excluded from clinical trials given the risk of further damage if left untreated [76–78]. Lesion size, location, chronicity, patient and surgeon preferences all play a role in the decision-making process. Given the lack of high-level evidence, there remains significant variation in the management of these injuries [79].

The presence of a loose intra-articular body following an episode of patellar instability is considered an indication for operative intervention in order to prevent symptoms and further chondral damage [76, 80–82]. Nikku et al. (2005) have performed the largest RCT to date examining the operative management of 127 primary patellar dislocations [76]. They did not find patellar realignment surgery to be beneficial, but they did find that the subset of patients presenting with loose bodies led to significantly poorer functional outcomes.

Surgical repair of unstable osteochondral fractures is the preferred method of management [83, 84]. Historically, these patients have had poor outcomes when treated non-operatively [85]. However, there is no consensus on the size, depth or location of a fragment that is considered amenable to fixation. Duthon et al. (2015) suggested that surgical fixation is favourable for fractures involving >10% of the articular surface [83]. Although limited to small case series and retrospective reviews, the outcomes of fixation after osteochondral fractures have been favourable for lesions of both the patella and femoral condyle [86–91]. Gesslin et al. (2019) retrospectively reviewed patients who underwent fixation compared to debridement for OCF lesions. Despite the fixation group presenting with larger fracture fragments, they had significantly better long-term clinical outcome scores and significantly fewer reoperations [86]. Kang et al. (2018) reviewed patients who underwent fixation compared to debridement for OCFs that did not involve the weight-bearing surface. They demonstrated that excision and debridement in this subgroup had improved clinical outcomes [92]. Should the fracture fragment be amenable to fixation, techniques for fixation vary widely and include bioabsorbable or nonabsorbable countersunk screw or pin fixation [86–91, 93]. The theoretical advantage of bioabsorbable implants is that they do not need to be removed if further revision surgery is required. Given the lack of comparative studies, method of fixation is left to the discretion of the treating surgeon.

Microfracture is a well-established technique aimed at marrow stimulation for chondral and osteochondral lesions [94]. Although short-term results have been favourable in younger patients, there is variable long-term efficacy particularly when examining older patients and microfractures of the patella and trochlea [95, 96]. Microfractures result in a fibrocartilaginous tissue that is biomechanically inferior when compared to the natural hyaline cartilage . Meta-analysis data has suggested that functional outcomes were improved if the lesions were <4 cm for all patients and <2 cm for the athletic subpopulation [97]. However, given the lack of literature that examines the efficacy of microfractures for the patellofemoral joint specifically, it is difficult to draw any conclusions about the size, depth and location that would benefit from microfracture [98]. It is the authors' experience that microfracture is rarely required or warranted by the time the patient seeks operative management following lateral patellar dislocations.

The MPFL is disrupted in the vast majority of acute patellar dislocations. However, the role of repair or reconstruction in the setting of an acute patellar dislocation remains controversial [79, 82]. Early randomized controlled trials focused on acute repair of the MPFL and demonstrated no differences in outcomes between surgical and conservative management [76, 99, 100]. The understanding of the anatomy and biomechanics of the MPFL has increased considerably in recent years, which has aided in the popularization of various reconstruction techniques [101, 102]. There is level I evidence that demonstrates lower dislocation rates and improved clinical outcomes in patients undergoing MPFL reconstruction compared to non-operative management in the setting of both acute and recurrent patellar dislocation [103–105]. However, these studies include both patients with normal anatomy and those who have anatomic risk factors for dislocation, making it challenging to apply these results to the individual patient. There remains a lack of data guiding the management of the MPFL in the setting of operative osteochondral lesions.

It is the senior author's recommendation that patients undergo a thorough preoperative assessment to assess for risk factors for patellar instability. In the absence of these risk factors, MPFL reconstruction is of questionable additional benefit in first time dislocators with osteochondral defects. However, if the patient presents with a history of recurrent instability and/or anatomic risk factors for instability, MPFL reconstruction is warranted. There are several proposed methods of MPFL reconstruction including single bundle vs. double bundle and various autografts or allografts [106].

19.3.4 Case Presentation

An otherwise healthy 13-year-old female presented to the orthopaedic outpatient clinic 4 days after a left knee injury. The patient reported that she was playing ball hockey in gym class and planted her foot when another player fell onto the outside of her knee. A "pop" was felt and the patient stated that she saw her knee cap dislocate laterally and reduce spontaneously. She had significant pain and swelling to the knee and was unable to ambulate. She presented to the Emergency Department where she was placed in a knee immobilizer. Initial plain radiographs demonstrated a fracture off the lateral femoral condyle with an intra-articular loose body (Fig. 19.11).

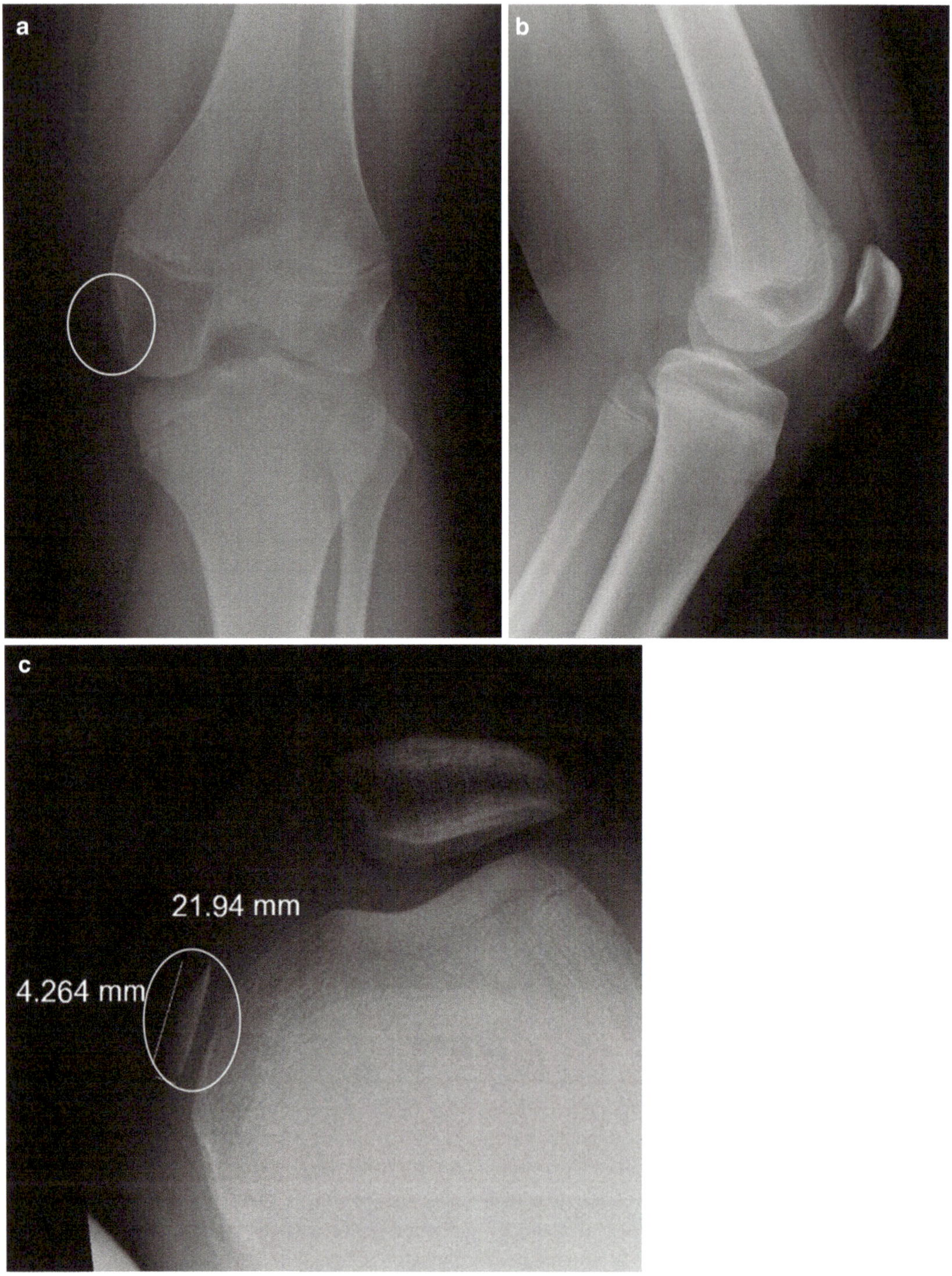

Fig. 19.11 Anteroposterior (**a**) lateral (**b**) and skyline (**c**) radiographs demonstrating acute fracture of the lateral femoral condyle with intra-articular loose body, circled in white

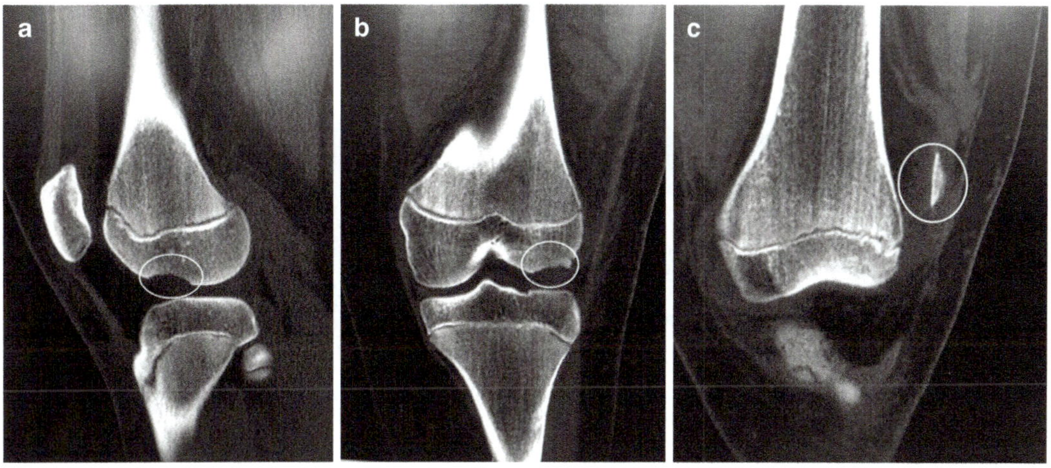

Fig. 19.12 Sagittal (**a**) and Coronal (**b**) cuts of the CT scan demonstrating osteochondral donor site on the lateral femoral condyle and associated cartilage fragment in the lateral joint recess on an anterior coronal slice (**c**)

The patient was seen in the orthopaedic clinic where a CT scan was triaged. The CT scan demonstrated an ossific fragment measuring 1.6 cm in its craniocaudal dimension × 0.5 cm in its transverse dimension × 1.5 cm in its AP dimension within the lateral aspect of the knee joint just superior to the lateral femoral condyle. The donor site involving the cortical and subcortical aspect of the inferior portion of the lateral femoral condyle measured 1.4 × 0.9 × 1.4 cm. The CT scan also demonstrated some lateral shift of the patella and subtle widening of the patellofemoral articulation in its medial aspect (Fig. 19.12).

Given the osteochondral fracture and associated loose bodies, we discussed the potential risks and benefits of undergoing operative intervention. The patient and his family consented to left knee arthroscopic loose body removal with possible open reduction and internal fixation of the osteochondral fracture. Given that the patient had no history of recurrent patellar instability and no risk factors on imaging, the decision was made to not perform an MPFL reconstruction at the index surgery.

19.3.4.1 Intraoperative Findings

Diagnostic arthroscopy identified a significant chondral defect at the lateral femoral condyle. This was subsequently debrided with the arthroscopic shaver. The loose osteochondral fragment was found in the lateral gutter and retrieved in one piece. It measured approximately 2.5 cm × 2 cm with a small piece of bone on the underside.

The operation was converted to an open procedure with the lateral vertical portal site extended proximally. A small lateral parapatellar approach was utilized to enter the knee joint. The defect was visualized and surrounding soft tissue and callous were removed. The osteochondral fragment was reduced and fixed with six 16 mm biodegradable SmartNail® implants (CONMED, Linvotec). Intraoperative images are shown in Fig. 19.13.

19.3.4.2 Outcome

The patient was placed in a hinged knee brace locked in full extension and instructed to be non-weight bearing for the first 6 weeks with progressive range of motion in the brace. At last follow-up this patient regained painless gait and range of motion of her knee with only 3° of terminal extension deficit and 90% quadriceps bulk compared to the contralateral knee. She continues with her athletic endeavours with a patellar brace. Post-operative radiographs have remained normal (Fig. 19.14).

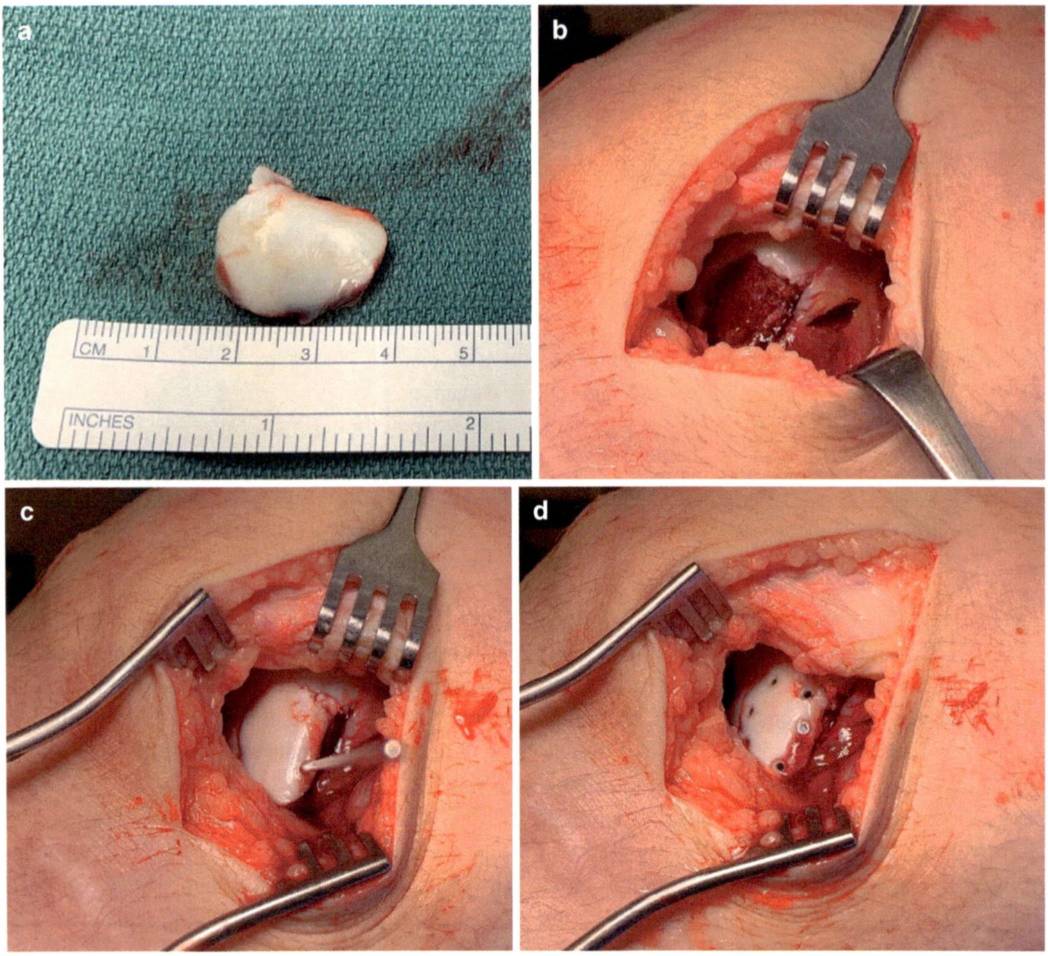

Fig. 19.13 Intraoperative images showing the 4 day-old lateral femoral condyle fracture fragment measuring approximately 2 cm in height (**a**) and the corresponding defect on the condyle (**b**). Provisional fixation (**c**) followed by definitive fixation with biodegradable SmartNail implants (**d**)

19.4 Tibia Physeal Fractures of the Knee in the Paediatric Population

19.4.1 Proximal Tibia Physeal Fractures

19.4.1.1 Background and Mechanism of Injury

Proximal tibia physeal fractures most commonly occur in adolescents 11–14 years of age. Given the stability of the proximal tibia via the medial collateral ligament (MCL), LCL, fibula, and tibial tubercle, displaced fractures of the proximal tibial physis requires a high-energy mechanism [107]. The mechanism of injury affects the degree and direction of the resulting displacement; hyperextension injuries result in anterior displacement of the epiphyseal fragment, and hyperflexion injuries result in anterior displacement of the metaphyseal fragment [108]. Given the location of the popliteal artery which runs along the posterior tibia and trifurcates just below the physis, these injuries present serious concern for laceration or thrombosis of the popliteal vessel in children [109].

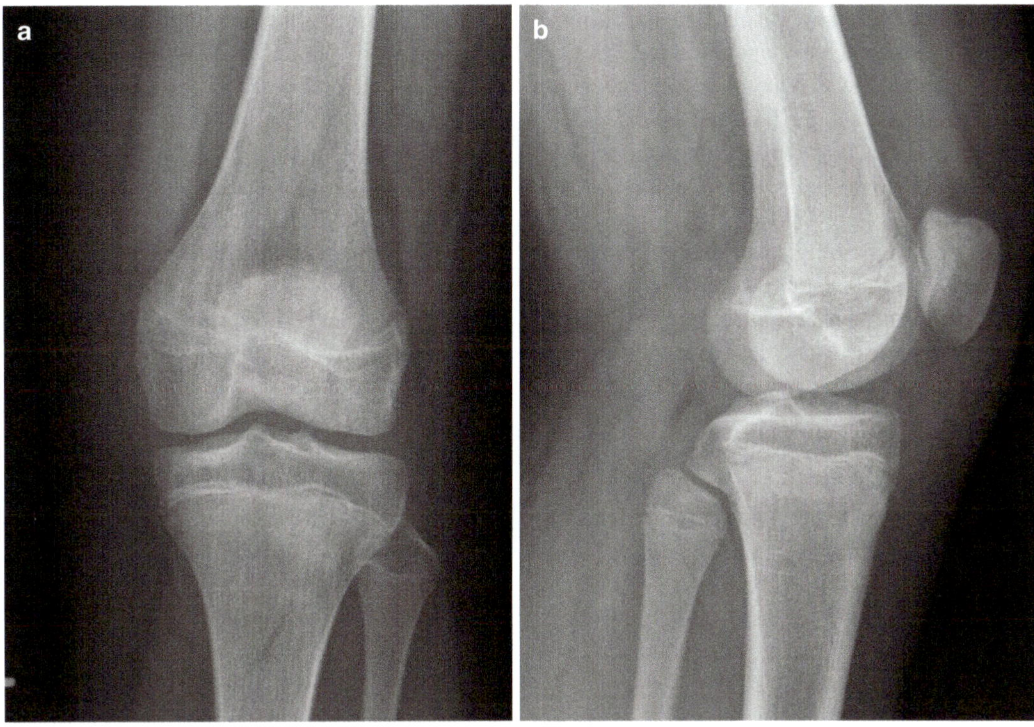

Fig. 19.14 Anteroposterior (**a**) and lateral (**b**) radiographs taken 6 weeks post-operatively demonstrating fracture healing and normal alignment

19.4.1.2 Clinical Presentation and Diagnostics

Patients with proximal tibial physeal fractures present with focal pain, soft tissue swelling and commonly a knee joint effusion. For all proximal tibial physeal fractures, a thorough neurological and particularly vascular examination of the leg is critical given that the incidence of vascular injuries are equivalent to that of multi-ligamentous knee dislocations [110]. Anteroposterior and lateral radiographs are required for initial diagnosis, with CT scans helpful as an adjunct to assess for the existence and degree of articular involvement. An MRI is another useful adjunct in displaced patterns to assess for ligamentous injuries that may be entrapped within the fracture gaps [110]. The widely used Salter-Harris classification for paediatric physeal fractures is the most commonly used system to classify proximal tibial physeal fractures [111].

19.4.1.3 Management Options and Evidence-Based Outcomes

For Salter-Harris types I and II fractures with displacement, an initial trial of closed reduction and long-leg casting may be acceptable if reduction achieves less than 2 mm of residual displacement [108, 110]. Residual displacement warrants open reduction to assess for soft tissue interposed between the fragments (MCL, LCL, pes anserinus, or periosteum), and pinning using transphyseal, smooth wires. Pins are typically placed in a crossed manner and can be inserted either anterograde or retrograde. Benefits of anterograde pinning include a less technically demanding procedure; however, the pins are often intra-articular leading to a higher risk of septic arthritis [110, 112].

Salter-Harris types III and IV fractures that are non- or minimally displaced can be managed with closed reduction and percutaneous screw fixation; however, any displacement warrants open reduction to achieve anatomic reduction of the articular surface under direct visualization [112]. The construct for fixation typically consists of screws or pins that are perpendicular to the physis within both the metaphysis and epiphysis.

The most commonly reported complications following physeal fractures of the proximal tibia are growth disturbances, vascular injury, neurological compromise, and less commonly nonunion. Growth disturbances have been reported to occur in up to 25% of proximal tibial physeal fractures, resulting in either unequal limb lengths or angular deformities [113]. Therefore, it is recommended that these patients be followed regularly until skeletal maturity with full leg-length films [113]. Vascular injuries of the limb occur in 10–15% of cases, and therefore, it is recommended that these patients be admitted for monitoring for at least 24 h post-operatively [109, 112].

19.4.2 Tibial Tubercle Fractures

19.4.2.1 Background and Mechanism of Injury

Fractures of the tibial tubercle most commonly occurs in adolescent males 12–17 years of age, comprising approximately 3% of all proximal tibia fractures [107, 114, 115]. The closure of the tubercle physis from proximal to distal during skeletal maturity leaves the distal aspect of the tubercle susceptible to injury [116]. The mechanism of injury is typically caused by jumping or forced flexion of the knee leading to a powerful contraction of the quadriceps muscle [107, 114, 117].

19.4.2.2 Clinical Presentation and Diagnostics

Patients with tibial tubercle fractures present with local soft tissue swelling and focal tenderness to palpation of the tubercle. When minimal swelling precludes an obvious diagnosis, pain with straight leg raise or resisted knee extension may provide a clue towards a possible diagnosis. Serial neurovascular examination is critical for any diagnosed or suspected tubercle fractures as damage to the anterior recurrent tibial artery may result in swelling and compression to the anterior compartment where the deep peroneal nerve and anterior tibial artery may be occluded [116].

Anteroposterior and lateral radiographs of the knee are required for initial diagnosis. In order to obtain a perfect lateral view of the tubercle, slight internal rotation of the leg provides a direct view of the apophysis, which is slightly lateral to the midline [116]. However, plain radiographs have been shown to underestimate the severity more than 50% of the time, and therefore, CT scans are useful to assess whether there is intra-articular or metaphyseal extension. The most commonly used classification is the Ogden modification of the Watson-Jones classification with grades I–III (relating to the location relative to the junction between the proximal tibia and the apophysis) each divided into subtypes A and B (for non-displaced or displaced/comminuted fractures, respectively) (Table 19.3) [118].

Table 19.3 Ogden modification of the Watson-Jones classification for tibial tubercle fractures

Type IA	Fracture line through ossification center of tibial tubercle with no displacement
Type IB	Anterior and proximal displacement of the fracture fragment
Type IIA	Fracture extends through the junction of proximal tibia and the tibial tubercle
Type IIB	Similar to IIA with comminuted tubercle fracture fragment and anterior displacement
Type IIIA	Fracture extends to the articular surface with associated discontinuity
Type IIIB	Intra-articular with comminution

19.4.2.3 Management Options and Evidence-Based Outcomes

Non-displaced fractures may be treated non-operatively with a long leg cast in extension [119]. Displaced fractures of the tubercle often require open reduction and internal fixation. A midline anterior approach is typically used, with intra-articular fractures commonly requiring arthroscopic assistance, or a parapatellar arthrotomy. The construct for fixation typically consists of two- to three cannulated, partially threaded screws perpendicular to the fracture, as screws have been shown to offer superior compression and fixation to percutaneous pins [120]. Washers may be used to prevent penetration into soft apophyseal bone [110, 117]. Given the anticipated significant anterior compartment swelling due to injury of the recurrent anterior tibial artery, intraoperative compartment pressure monitoring may be used if clinically indicated, necessitating possible decompression of the hematoma alone or in combination with a prophylactic anterior compartment fascia release distal to the surgical site [114]. Post-operative management includes admission to hospital for 24–48 h to monitor swelling of the anterior compartment, with the leg braced or splinted in extension for a minimum of 4 weeks [115].

The most common complications in order of acuity following tibial tubercle fractures are compartment syndrome, hardware prominence, bursitis, and growth disturbances. Compartment syndrome has been reported with incidence ranging from 2 to 20% [117, 118]. Hardware prominence resulting in bursitis is a problem primarily for thinner patients, and removal may be required in more than 50% of patients treated with open reduction and internal fixation [116, 120]. For patients younger than 13 years of age, long-term follow-up is suggested to monitor for growth arrest resulting in genu recurvatum [110, 114].

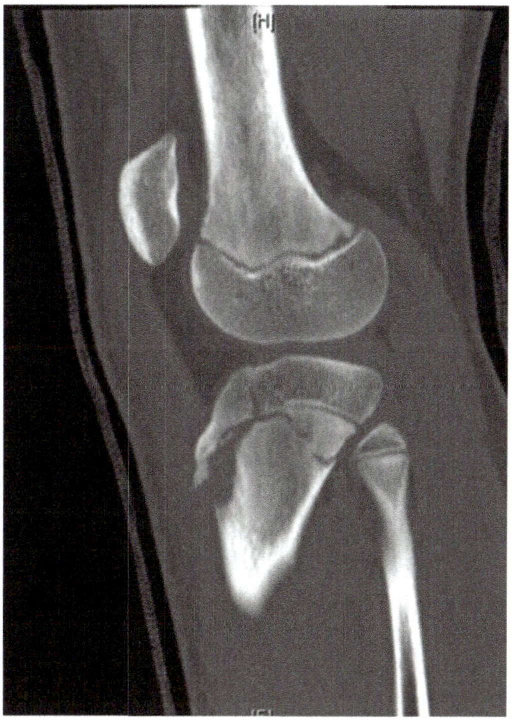

Fig. 19.15 Lateral sagittal CT image demonstrating a type IIA fracture of the tibial tubercle as well as a minimally displaced Salter-Harris type IV fracture of the proximal tibia physis

19.4.2.4 Case Presentation

A 12-year-old male presented to the Emergency Department with a left knee injury sustained during soccer, when early in the kick phase, the kicking leg was abruptly stopped and forced into eccentric contraction of the quadriceps, after hitting a section of raised playing surface. The patient was found to have no neurological compromise and compartments were monitored. Imaging, including plain radiographs and CT scan were performed, demonstrating a type IIA fracture of the tibial tubercle (Fig. 19.15) as well as a minimally displaced Salter-Harris type IV fracture of the proximal tibial physis.

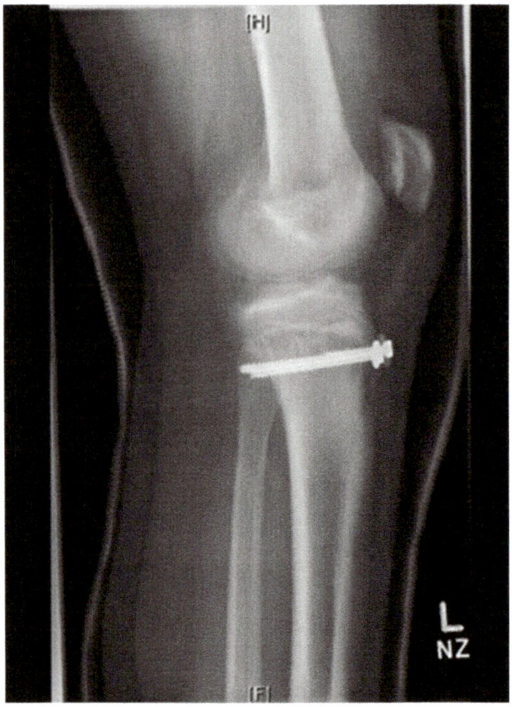

Fig. 19.16 Post-operative lateral radiographs. Two cannulated, partially threaded screws placed parallel to both the physis and perpendicular to the fracture. A washer was also used to prevent penetration into the bone

19.4.2.5 Intraoperative Findings

Operative management was undertaken using a midline incision. Given the swelling of the anterior compartment, the hematoma was evacuated, and a small fascial opening over the anterior compartment was made and left open. Two cannulated, partially threaded screws were placed parallel to both the physis and perpendicular to the fracture were placed using a washer to prevent penetration into the bone (Fig. 19.16). A long leg cast was applied, and the patient was admitted to hospital for monitoring of his compartments for 72-h post-operatively.

References

1. Schatzker J, McBroom R. The tibial plateau fracture. The Toronto experience 1968-1975. Clin Orthop. 1979;(138):94–104.
2. Moore TM. Fracture-dislocation of the knee. Clin Orthop. 1981;(156):128–40. https://doi.org/10.1097/00003086-198105000-00015.
3. Shao J, et al. Incidence and risk factors for surgical site infection after open reduction and internal fixation of tibial plateau fracture: a systematic review and meta-analysis. Int J Surg. 2017;41:176–82. https://doi.org/10.1016/j.ijsu.2017.03.085.
4. Porrino J, Richardson ML, Hovis K, Twaddle B, Gee A. Association of tibial plateau fracture morphology with ligament disruption in the context of multiligament knee injury. Curr Probl Diagn Radiol. 2018;47(6):410–6. https://doi.org/10.1067/j.cpradiol.2017.09.001.
5. Colletti P, Greenberg H, Terk MR. MR findings in patients with acute tibial plateau fractures. Comput Med Imaging Graph. 1996;20(5):389–94. https://doi.org/10.1016/S0895-6111(96)00054-7.
6. Bennett WF, Browner B. Tibial plateau fractures: a study of associated soft tissue injuries. J Orthop Trauma. 1994;8(3):183–8. https://doi.org/10.1097/00005131-199406000-00001.
7. Stannard JP, Lopez R, Volgas D. Soft tissue injury of the knee after tibial plateau fractures. J Knee Surg. 2010;23(4):187–92. https://doi.org/10.1055/s-0030-1268694.
8. Gardner MJ, et al. The incidence of soft tissue injury in operative tibial plateau fractures. A magnetic resonance imaging analysis of 103 patients. J Orthop Trauma. 2005;19(2):79–84. https://doi.org/10.1097/00005131-200502000-00002.
9. Warner SJ, et al. The effect of soft tissue injuries on clinical outcomes after tibial plateau fracture fixation. J Orthop Trauma. 2018;32(3):141–7. https://doi.org/10.1097/BOT.0000000000001042.
10. Caldas MTL, Malheiros DS, Lazzaroni AP, Avelino EA, Santos AJ. Injury of the knee ligaments associated with ipsilateral femoral shaft fractures. Rev Bras Ortop Engl Ed. 2013;48(5):438–40. https://doi.org/10.1016/j.rboe.2012.11.003.
11. Walker DM, Kennedy JC. Occult knee ligament injuries associated with femoral shaft fractures. Am J Sports Med. 1980;8(3):172–4. https://doi.org/10.1177/036354658000800305.
12. Dickson KF, et al. Magnetic resonance imaging of the knee after ipsilateral femur fracture. J Orthop Trauma. 2002;16(8):567–71. https://doi.org/10.1097/00005131-200209000-00005.
13. van Raay JJAM, Raaymakers ELFB, Dupree HW. Knee ligament injuries combined with ipsilateral tibial and femoral diaphyseal fractures: The 'floating knee. Arch Orthop Trauma Surg. 1991;110(2):75–7. https://doi.org/10.1007/BF00393877.
14. Dillin L, Slabaugh P. Delayed wound healing, infection, and nonunion following open reduction and internal fixation of tibial plafond fractures. J Trauma Inj Infect Crit Care. 1986;26(12):1116–9. https://doi.org/10.1097/00005373-198612000-00011.
15. Egol KA, Tejwani NC, Capla EL, Wolinsky PL, Koval KJ. Staged management of high-energy proximal tibia fractures (ota types 41): The results of a prospective, standardized protocol. J Orthop Trauma. 2005;19(7):448–55. https://doi.org/10.1097/01.bot.0000171881.11205.80.

16. Luo CF, Sun H, Zhang B, Zeng BF. Three-column fixation for complex tibial plateau fractures. J Orthop Trauma. 2010;24(11):683–92. https://doi.org/10.1097/BOT.0b013e3181d436f3.
17. Baker BJ, Escobedo EM, Nork SE, Henley MB. Hoffa fracture: a common association with high-energy supracondylar fractures of the distal femur. Am J Roentgenol. 2002;178(4):994. https://doi.org/10.2214/ajr.178.4.1780994.
18. Nork SE, et al. The association between supracondylar-intercondylar distal femoral fractures and coronal plane fractures. J Bone Jt Surg. 2005;87(3):564–9. https://doi.org/10.2106/JBJS.D.01751.
19. Jiang R, Luo CF, Wang MC, Yang TY, Zeng BF. A comparative study of Less Invasive Stabilization System (LISS) fixation and two-incision double plating for the treatment of bicondylar tibial plateau fractures. Knee. 2008;15(2):139–43. https://doi.org/10.1016/j.knee.2007.12.001.
20. McKee MD, et al. Open reduction and internal fixation compared with circular fixator application for bicondylar tibial plateau fractures: results of a multicenter, prospective, randomized clinical trial. J Bone Jt Surg Ser A. 2006;88(12):2613–23. https://doi.org/10.2106/JBJS.E.01416.
21. Young MJ, Barrack RL. Complications of internal fixation of tibial plateau fractures. Orthop Rev. 1994;23(2):149–54.
22. Griffin XL, Parsons N, Zbaeda MM, Mcarthur J. Interventions for treating fractures of the distal femur in adults. Cochrane Database Syst Rev. 2015;2015(8) https://doi.org/10.1002/14651858.CD010606.pub2.
23. Canadian Orthopaedic Trauma Society. Are locking constructs in distal femoral fractures always best? A prospective multicenter randomized controlled trial comparing the less invasive stabilization system with the minimally invasive dynamic condylar screw system. J Orthop Trauma. 2016;30(1):e1–6. https://doi.org/10.1097/BOT.0000000000000450.
24. Obremskey WT. LCP versus LISS in the treatment of open and closed distal femur fractures: does it make a difference? J Orthop Trauma. 2016;30(6):e212–6. https://doi.org/10.1097/BOT.0000000000000507.
25. Davis JT, Rudloff MI. Posttraumatic arthritis after intra-articular distal femur and proximal tibia fractures. Orthop Clin North Am. 2019;50(4):445–59. https://doi.org/10.1016/j.ocl.2019.06.002.
26. Moatshe G, Dornan GJ, Ludvigsen T, Løken S, LaPrade RF, Engebretsen L. High prevalence of knee osteoarthritis at a minimum 10-year follow-up after knee dislocation surgery. Knee Surg Sports Traumatol Arthrosc. 2017;25(12):3914–22. https://doi.org/10.1007/s00167-017-4443-8.
27. Wasserstein D, Henry P, Paterson JM, Kreder HJ, Jenkinson R. Risk of total knee arthroplasty after operatively treated tibial plateau fracture a matched-population-based cohort study. J Bone Jt Surg Ser A. 2014;96(2):144–50. https://doi.org/10.2106/JBJS.L.01691.
28. Abdel MP, von Roth P, Cross WW, Berry DJ, Trousdale RT, Lewallen DG. Total knee arthroplasty in patients with a prior tibial plateau fracture: a long-term report at 15 years. J Arthroplasty. 2015;30(12):2170–2. https://doi.org/10.1016/j.arth.2015.06.032.
29. Stevenson I, McMillan TE, Baliga S, Schemitsch EH. Primary and secondary total knee arthroplasty for tibial plateau fractures. J Am Acad Orthop Surg. 2018;26(11):386–95. https://doi.org/10.5435/JAAOS-D-16-00565.
30. Parratte S, Ollivier M, Argenson JN. Primary total knee arthroplasty for acute fracture around the knee. Orthop Traumatol Surg Res. 2018;104(1S):S71–80. https://doi.org/10.1016/j.otsr.2017.05.029.
31. Harvey GP, Woods GW. Anterolateral dislocation of the proximal tibiofibular joint: case report and literature review. Todays OR Nurse. 1992;14(3):23–7.
32. Sekiya JK, Kuhn JE. Instability of the proximal tibiofibular joint. J Am Acad Orthop Surg. 2003;11(2):120–8. https://doi.org/10.5435/00124635-200303000-00006.
33. Ogden JA. The anatomy and function of the proximal tibiofibular joint. Clin Orthop Relat Res. 1974;(101):186–91.
34. Eichenblat M, Nathan H. The proximal tibio fibular joint—an anatomical study with clinical and pathological considerations. Clin Orthop Relat Res. 1983;7(1):31–9. https://doi.org/10.1007/BF00267557.
35. Resnick D, Newell JD, Guerra J, Danzig LA, Niwayama G, Goergen TG. Proximal tibiofibular joint: anatomic-pathologic-radiographic correlation. Am J Roentgenol. 1978;131(1):133–8. https://doi.org/10.2214/ajr.131.1.133.
36. Nunes J, Direito-Santos B, Costa A, Tavares N, Varanda P, Duarte R. Acute proximal tibiofibular joint dislocation: a sports related injury?—two different cases. Ann Jt. 2019; https://doi.org/10.21037/aoj.2019.01.07.
37. Thomason PA, Linson MA. Isolated dislocation of the proximal tibiofibular joint. J Trauma Inj Infect Crit Care. 1986;26(2):192–5. https://doi.org/10.1097/00005373-198602000-00018.
38. Horan J, Quin G. Proximal tibiofibular dislocation. Emerg Med J. 2006;23(5):e33. https://doi.org/10.1136/emj.2005.032144.
39. Ogden JA. Subluxation and dislocation of the proximal tibiofibular joint. J Bone Jt Surg Ser A. 1974;56(1):145–54. https://doi.org/10.2106/00004623-197456010-00015.
40. Parkes JC, Zelko RR. Isolated acute dislocation of the proximal tibiofibular joint. Case report. J Bone Jt Surg Ser A. 1973;55(1):181–3. https://doi.org/10.2106/00004623-197355010-00020.
41. Main C, Norris B, Jensen R. Proximal tibiofibular fixation using a dynamic syndesmosis fixation device. Int J Orthop Surg. 2012;13(2) https://doi.org/10.5580/2222.
42. Robinson Y, Reinke M, Heyde CE, Ertel W, Oberholzer A. Traumatic proximal tibiofibular joint

dislocation treated by open reduction and temporary fixation: a case report. Knee Surg Sports Traumatol Arthrosc. 2007;15(2):199–201. https://doi.org/10.1007/s00167-006-0147-1.
43. Warner BT, Moulton SG, Cram TR, LaPrade RF. Anatomic reconstruction of the proximal tibiofibular joint. Arthrosc Tech. 2016;5(1):e207–10.
44. Hsieh C-H, Chen J-C. Acute dislocation of the proximal tibiofibular joint. J Orthop Sports Phys Ther. 2009;39(11):826. https://doi.org/10.2519/jospt.2009.0414.
45. Laing AJ, Lenehan B, Ali A, Prasad CVR. Isolated dislocation of the proximal tibiofibular joint in a long jumper. Br J Sports Med. 2003;37(4):366–7. https://doi.org/10.1136/bjsm.37.4.366.
46. Nieuwe Weme RA, Somford MP, Schepers T. Proximal tibiofibular dislocation: a case report and review of literature. Strateg Trauma Limb Reconstr. 2014;9(3):185–9. https://doi.org/10.1007/s11751-014-0209-8.
47. Fithian DC, et al. Epidemiology and natural history of acute patellar dislocation. Am J Sports Med. 2004;32(5):1114–21. https://doi.org/10.1177/0363546503260788.
48. Waterman BR, Belmont PJ, Owens BD. Patellar dislocation in the United States: role of sex, age, race, and athletic participation. J Knee Surg. 2011;25(1):51–7. https://doi.org/10.1055/s-0031-1286199.
49. Krause EA, Lin CW, Ortega HW, Reid SR. Pediatric lateral patellar dislocation: is there a role for plain radiography in the emergency department? J Emerg Med. 2013;44(6):1126–31. https://doi.org/10.1016/j.jemermed.2012.11.014.
50. Bollier M, Fulkerson JP. The role of trochlear dysplasia in patellofemoral instability. J Am Acad Orthop Surg. 2011;19(1):8–16. https://doi.org/10.5435/00124635-201101000-00002.
51. Koh JL, Stewart C. Patellar instability. Clin Sports Med. 2014;33(3):461–76. https://doi.org/10.1016/j.csm.2014.03.011.
52. Shirley ED, DeMaio M, Bodurtha J. Ehlers-Danlos syndrome in orthopaedics: etiology, diagnosis, and treatment implications. Sports Health. 2012;4(5):394–403. https://doi.org/10.1177/1941738112452385.
53. Vollnberg B, et al. Prevalence of cartilage lesions and early osteoarthritis in patients with patellar dislocation. Eur Radiol. 2012;22(11):2347–56. https://doi.org/10.1007/s00330-012-2493-3.
54. Nomura E, Inoue M, Kurimura M. Chondral and osteochondral injuries associated with acute patellar dislocation. Arthrosc J Arthrosc Relat Surg. 2003;19(7):717–21. https://doi.org/10.1016/S0749-8063(03)00401-8.
55. Farr J, Covell DJ, Lattermann C. Cartilage lesions in patellofemoral dislocations. Sports Med Arthrosc Rev. 2012;20(3):181–6. https://doi.org/10.1097/jsa.0b013e318259bc40.
56. Lee BJ, Christino MA, Daniels AH, Hulstyn MJ, Eberson CP. Adolescent patellar osteochondral fracture following patellar dislocation. Knee Surg Sports Traumatol Arthrosc. 2013;21(8):1856–61. https://doi.org/10.1007/s00167-012-2179-z.
57. Elias DA, White LM, Fithian DC. Acute lateral patellar dislocation at MR imaging: injury patterns of medial patellar soft-tissue restraints and osteochondral injuries of the inferomedial patella. Radiology. 2002;225(3):736–43. https://doi.org/10.1148/radiol.2253011578.
58. Luhmann SJ, Schoenecker PL, Dobbs MB, Gordon JE. Arthroscopic findings at the time of patellar realignment surgery in adolescents. J Pediatr Orthop. 2007;27(5):493–8. https://doi.org/10.1097/BPO.0b013e318093f4d8.
59. Buckwalter JA. Articular cartilage injuries. Clin Orthop Relat Res. 2002;(402):21–37. https://doi.org/10.1097/00003086-200209000-00004.
60. Sanders TL, Pareek A, Hewett TE, Stuart MJ, Dahm DL, Krych AJ. High rate of recurrent patellar dislocation in skeletally immature patients: a long-term population-based study. Knee Surg Sports Traumatol Arthrosc. 2018;26(4):1037–43. https://doi.org/10.1007/s00167-017-4505-y.
61. Salonen EE, Magga T, Sillanpää PJ, Kiekara T, Mäenpää H, Mattila VM. Traumatic patellar dislocation and cartilage injury: a follow-up study of long-term cartilage deterioration. Am J Sports Med. 2017;45(6):1376–82. https://doi.org/10.1177/0363546516687549.
62. Kramer J, White LM, Recht MP. MR imaging of the extensor mechanism. Semin Musculoskelet Radiol. 2009;13(4):384–401. https://doi.org/10.1055/s-0029-1242191.
63. Diederichs G, Issever AS, Scheffler S. MR imaging of patellar instability: injury patterns and assessment of risk factors. Radiographics. 2010;30(4):961–81. https://doi.org/10.1148/rg.304095755.
64. Mouzopoulos G, Borbon C, Siebold R. Patellar chondral defects: a review of a challenging entity. Knee Surg Sports Traumatol Arthrosc. 2011;19(12):1990–2001. https://doi.org/10.1007/s00167-011-1546-5.
65. Dejour H, Walch G, Nove-Josserand L, Guier C. Factors of patellar instability: an anatomic radiographic study. Knee Surg Sports Traumatol Arthrosc. 1994;2(1):19–26. https://doi.org/10.1007/BF01552649.
66. Amis AA, Firer P, Mountney J, Senavongse W, Thomas NP. Anatomy and biomechanics of the medial patellofemoral ligament. Knee. 2003;10(3):215–20. https://doi.org/10.1016/S0968-0160(03)00006-1.
67. Insall J, Salvati E. Patella position in the normal knee joint. Radiology. 1971;101(1):101–4. https://doi.org/10.1148/101.1.101.
68. Caton J. Method of measuring the height of the patella. Acta Orthop Belg. 1989;55(3):385–6.
69. Blackburn JS, Peel TE. A new method of measuring patellar height. J Bone Jt Surg Ser B. 1977;59(2):241–2. https://doi.org/10.1302/0301-620x.59b2.873986.

70. Tanaka MJ, Cosgarea AJ. Measuring malalignment on imaging in the treatment of patellofemoral instability. Am J Orthop (Belle Mead, N.J.). 2017;46(3):148–51.
71. Sanders TG, Paruchuri NB, Zlatkin MB. MRI of osteochondral defects of the lateral femoral condyle: incidence and pattern of injury after transient lateral dislocation of the patella. Am J Roentgenol. 2006;187(5):1332–7. https://doi.org/10.2214/AJR.05.1471.
72. Nomura E, Horiuchi Y, Inoue M. Correlation of MR imaging findings and open exploration of medial patellofemoral ligament injuries in acute patellar dislocations. Knee. 2002;9(2):139–43. https://doi.org/10.1016/S0968 0160(02)00002-9.
73. Kepler CK, Bogner EA, Hammoud S, Malcolmson G, Potter HG, Green DW. Zone of injury of the medial patellofemoral ligament after acute patellar dislocation in children and adolescents. Am J Sports Med. 2011;39(7):1444–9. https://doi.org/10.1177/0363546510397174.
74. Guerrero P, Li X, Patel K, Brown M, Busconi B. Medial patellofemoral ligament injury patterns and associated pathology in lateral patella dislocation: an MRI study. BMC Sports Sci Med Rehabil. 2009;1(1):17. https://doi.org/10.1186/1758-2555-1-17.
75. Kirsch MD, Fitzgerald SW, Friedman H, Rogers LF. Transient lateral patellar dislocation: diagnosis with MR imaging. Am J Roentgenol. 1993;161(1):109–13. https://doi.org/10.2214/ajr.161.1.8517287.
76. Nikku R, Nietosvaara Y, Aalto K, Kallio PE. Operative treatment of primary patellar dislocation does not improve medium-term outcome: a 7-year follow-up report and risk analysis of 127 randomized patients. Acta Orthop. 2005;76(5):699–704. https://doi.org/10.1080/17453670510041790.
77. Sillanpää PJ, M̈enp̈ HM, Mattila VM, Visuri T, Pihlajamäki H. Arthroscopic surgery for primary traumatic patellar dislocation: a prospective, nonrandomized study comparing patients treated with and without acute arthroscopic stabilization with a median 7-year follow-up. Am J Sports Med. 2008;36(12):2301–9. https://doi.org/10.1177/0363546508322894.
78. Camanho GL, Viegas AC, Bitar AC, Demange MK, Hernandez AJ. Conservative versus surgical treatment for repair of the medial patellofemoral ligament in acute dislocations of the patella. Arthrosc J Arthrosc Relat Surg. 2009;25(6):620–5. https://doi.org/10.1016/j.arthro.2008.12.005.
79. VandenBerg C, Sarkisova N, Pace JL, Rhodes J, Green DW. Current practice trends in the surgical management of patellofemoral instability; a survey of the Pediatric Research in Sports Medicine (PRiSM) Society. Orthop J Sports Med. 2019;7(3_suppl):2325967119S0009. https://doi.org/10.1177/2325967119s00090.
80. Gkiokas A, Morassi LG, Kohl S, Zampakides C, Megremis P, Evangelopoulos DS. Bioabsorbable pins for treatment of osteochondral fractures of the knee after acute patella dislocation in children and young adolescents. Adv Orthop. 2012;2012:249687. https://doi.org/10.1155/2012/249687.
81. Stefancin JJ, Parker RD. First-time traumatic patellar dislocation: a systematic review. Clin Orthop Relat Res. 2007;455:93–101. https://doi.org/10.1097/BLO.0b013e31802eb40a.
82. Liu JN, et al. Patellar instability management: a survey of the international patellofemoral study group. Am J Sports Med. 2018;46(13):3299–306. https://doi.org/10.1177/0363546517732045.
83. Duthon VB. Acute traumatic patellar dislocation. Orthop Traumatol Surg Res. 2015;101(1 Suppl):S59–67. https://doi.org/10.1016/j.otsr.2014.12.001.
84. Prince MR, King AH, Stuart MJ, Dahm DL, Krych AJ. Treatment of patellofemoral cartilage lesions in the young, active patient. J Knee Surg. 2015;28(4):285–95. https://doi.org/10.1055/s-0035-1549018.
85. Cash JD, Hughston JC. Treatment of acute patellar dislocation. Am J Sports Med. 1988;16(3):244–9. https://doi.org/10.1177/036354658801600308.
86. Gesslein M, Merkl C, Bail HJ, Krutsch V, Biber R, Schuster P. Refixation of large osteochondral fractures after patella dislocation shows better mid- to long-term outcome compared with debridement. Cartilage. 2019; https://doi.org/10.1177/1947603519886637.
87. Chotel F, Knorr G, Simian E, Dubrana F, Versier G. Knee osteochondral fractures in skeletally immature patients: French multicenter study. Orthop Traumatol Surg Res. 2011;97(8 Suppl):S154–9. https://doi.org/10.1016/j.otsr.2011.09.003.
88. Fuchs M, Vosshenrich R, Dumont C, Stürmer KM. Refixation of osteochondral fragments using absorbable implants. First results of a retrospective study. Chir Z Alle Geb Oper Medizen. 2003;74(6):554–61. https://doi.org/10.1007/s00104-003-0623-9.
89. Rüther H, Raschke D, Frosch S, Wachowski M, Seif A. Long-term clinical and MRI results after refixation of osteochondral fractures with resorbable implants. Orthop J Sports Med. 2017;5(4_suppl4):2325967117S0014. https://doi.org/10.1177/2325967117s00147.
90. Walsh SJ, Boyle MJ, Morganti V. Large osteochondral fractures of the lateral femoral condyle in the adolescent: outcome of bioabsorbable pin fixation. J Bone Jt Surg Ser A. 2008;90(7):1473–8. https://doi.org/10.2106/JBJS.G.00595.
91. Lidder S, Thomas M, Desai A, Skyrme A, Armitage A, Rajaratnam S. Osteochondral fractures of the knee in skeletally immature patients: short term results of operative fixation using Omnitech screws. Acta Orthop Belg. 2016;82(4):762–7.

92. Kang H, Li J, Chen XX, Wang T, Liu SC, Li HC. Fixation versus excision of osteochondral fractures after patellar dislocations in adolescent patients: a retrospective cohort study. Chin Med J (Engl). 2018;131(11):1296–301. https://doi.org/10.4103/0366-6999.232800.
93. Jehan S, Loeffler MD, Pervez H. Osteochondral fracture of the lateral femoral condyle involving the entire weight bearing articular surface fixed with biodegradable screws. J Pak Med Assoc. 2010;60(5):400–1.
94. Steadman JR, Rodkey WG, Rodrigo JJ. Microfracture: surgical technique and rehabilitation to treat chondral defects. Clin Orthop Relat Res. 2001;(391 Suppl):S362–9. https://doi.org/10.1097/00003086-200110001-00033.
95. Kreuz PC, et al. Is microfracture of chondral defects in the knee associated with different results in patients aged 40 years or younger? Arthrosc J Arthrosc Relat Surg. 2006;22(11):1180–6. https://doi.org/10.1016/j.arthro.2006.06.020.
96. Steadman JR, Briggs KK, Matheny LM, Guillet A, Hanson CM, Willimon SC. Outcomes following microfracture of full-thickness articular cartilage lesions of the knee in adolescent patients. J Knee Surg. 2015;28(2):145–50. https://doi.org/10.1055/s-0034-1373737.
97. Mithoefer K, Mcadams T, Williams RJ, Kreuz PC, Mandelbaum BR. Clinical efficacy of the microfracture technique for articular cartilage repair in the knee: an evidence-based systematic analysis. Am J Sports Med. 2009;37(10):2053–63. https://doi.org/10.1177/0363546508328414.
98. Smoak J, Kluczynski M, Marzo J. Systematic review of patient outcomes and associated predictors after microfracture in the patellofemoral joint. J Am Acad Orthop Surg Glob Res Rev. 2019;3(11):e10.5435.
99. Christiansen SE, Jakobsen BW, Lund B, Lind M. Isolated repair of the medial patellofemoral ligament in primary dislocation of the patella: a prospective randomized study. Arthrosc J Arthrosc Relat Surg. 2008;24(8):881–7. https://doi.org/10.1016/j.arthro.2008.03.012.
100. Sillanpää PJ, Mattila VM, Mäenpää H, Kiuru M, Visuri T, Pihlajamäki H. Treatment with and without initial stabilizing surgery for primary traumatic patellar dislocation: a prospective randomized study. J Bone Jt Surg Ser A. 2009;91(2):263–73. https://doi.org/10.2106/JBJS.G.01449.
101. Philippot R, Chouteau J, Wegrzyn J, Testa R, Fessy MH, Moyen B. Medial patellofemoral ligament anatomy: implications for its surgical reconstruction. Knee Surg Sports Traumatol Arthrosc. 2009;17(5):475–9. https://doi.org/10.1007/s00167-009-0722-3.
102. Stephen JM, Lumpaopong P, Deehan DJ, Kader D, Amis AA. The medial patellofemoral ligament: location of femoral attachment and length change patterns resulting from anatomic and nonanatomic attachments. Am J Sports Med. 2012;40(8):1871–9. https://doi.org/10.1177/0363546512449998.
103. Bitar AC, Demange MK, D'Elia CO, Camanho GL. Traumatic patellar dislocation: nonoperative treatment compared with MPFL reconstruction using patellar tendon. Am J Sports Med. 2012;40(1):114–22. https://doi.org/10.1177/0363546511423742.
104. Zheng X, et al. Surgical medial patellofemoral ligament reconstruction versus non-surgical treatment of acute primary patellar dislocation: a prospective controlled trial. Int Orthop. 2019;43(6):1495–501. https://doi.org/10.1007/s00264-018-4243-x.
105. Regalado G, Lintula H, Kokki H, Kröger H, Väätäinen U, Eskelinen M. Six-year outcome after non-surgical versus surgical treatment of acute primary patellar dislocation in adolescents: a prospective randomized trial. Knee Surg Sports Traumatol Arthrosc. 2016;24(1):6–11. https://doi.org/10.1007/s00167-014-3271-3.
106. Kyung H-S, Kim H-J. Medial patellofemoral ligament reconstruction: a comprehensive review. Knee Surg Relat Res. 2015;27(3):133–40. https://doi.org/10.5792/ksrr.2015.27.3.133.
107. Mubarak SJ, Kim JR, Edmonds EW, Pring ME, Bastrom TP. Classification of proximal tibial fractures in children. J Child Orthop. 2009;3(3):191–7. https://doi.org/10.1007/s11832-009-0167-8.
108. Vyas S, Ebramzadeh E, Behrend C, Silva M, Zionts LE. Flexion-type fractures of the proximal tibial physis: a report of five cases and review of the literature. J Pediatr Orthop B. 2010;19(6):492–6. https://doi.org/10.1097/BPB.0b013e32833cb764.
109. Burkhart SS, Peterson HA. Fractures of the proximal tibial epiphysis. J Bone Jt Surg Ser A. 1979;61(7):996–1002. https://doi.org/10.2106/00004623-197961070-00005.
110. Zionts LE. Fractures around the knee in children. J Am Acad Orthop Surg. 2002;10(5):345–55. https://doi.org/10.5435/00124635-200209000-00006.
111. Salter RB, Harris WR. Injuries involving the epiphyseal plate. J Bone Jt Surg. 1963;45A:587–622. https://doi.org/10.2106/00004623-196345030-00019.
112. Mayer S, Albright JC, Stoneback JW. Pediatric knee dislocations and physeal fractures about the knee. J Am Acad Orthop Surg. 2015;23(9):571–80. https://doi.org/10.5435/JAAOS-D-14-00242.
113. Gautier E, Ziran BH, Egger B, Slongo T, Jakob RP. Growth disturbances after injuries of the proximal tibial epiphysis. Arch Orthop Trauma Surg. 1998;118(1–2):37–41. https://doi.org/10.1007/s004020050307.
114. Frey S, Hosalkar H, Cameron DB, Heath A, David Horn B, Ganley TJ. Tibial tuberosity fractures in adolescents. J Child Orthop. 2008;2(6):469–74. https://doi.org/10.1007/s11832-008-0131-z.
115. Mosier SM, Stanitski CL. Acute tibial tubercle avulsion fractures. J Pediatr Orthop. 2004;24(2):181–4. https://doi.org/10.1097/01241398-200403000-00009.

116. Pandya NK, Edmonds EW, Roocroft JH, Mubarak SJ. Tibial tubercle fractures: complications, classification, and the need for intra-articular assessment. J Pediatr Orthop. 2012;32(8):749–59. https://doi.org/10.1097/BPO.0b013e318271bb05.
117. Zrig M, Annabi H, Ammari T, Trabelsi M, Mbarek M, Ben Hassine H. Acute tibial tubercle avulsion fractures in the sporting adolescent. Arch Orthop Trauma Surg. 2008;128(12):1437–42. https://doi.org/10.1007/s00402-008-0628-4.
118. Ogden JA, Tross RB, Murphy MJ. Fractures of the tibial tuberosity in adolescents. J Bone Jt Surg Ser A. 1980;62(2):205–15. https://doi.org/10.2106/00004623-198062020-00006.
119. McKoy BE, Stanitski CL. Acute tibial tubercle avulsion fractures. Orthop Clin North Am. 2003;34(3):397–403. https://doi.org/10.1016/S0030-5898(02)00061-5.
120. Pretell-Mazzini J, et al. Outcomes and complications of tibial tubercle fractures in pediatric patients: a systematic review of the literature. J Pediatr Orthop. 2016;36(5):440–6. https://doi.org/10.1097/BPO.0000000000000488.

Advances in Treating Arthrofibrosis

20

João V. Novaretti

20.1 Introduction

Arthrofibrosis typically involves a persistent and usually painful restriction of joint motion due to pathological proliferation of connective tissue. Knee joint arthrofibrosis is usually provoked by a complication of knee surgery, treatment of intra-articular fractures, ligamentous lesions, damage to the knee joint extensor muscle, or total arthroplasty of the knee [1]. The most commonly performed treatment options are limited to aggressive physical therapy, manipulation under anesthesia, and arthroscopic debridement. Yet, even after surgical treatment, the intraoperative range of motion achieved at the time of release is difficult to maintain. A significant number of patients develop residual or recurrent stiffness [2–4]. Therefore, several treatment options using biological interventions have been investigated to address such issues (Table 20.1). In this chapter, we discuss the advances in treating arthrofibrosis with some of the main emerging drug therapies and biological treatments.

J. V. Novaretti (✉)
Orthopaedics and Traumatology Sports Center (CETE), Department of Orthopaedics and Traumatology, Paulista School of Medicine (EPM), Federal University of São Paulo, São Paulo, Brazil

20.2 Sulfasalazine

Sulfasalazine is an anti-inflammatory drug that inhibits κB kinase, an activity that promotes myofibroblast apoptosis [5, 6]. Myofibroblasts, in fibrotic tissues, resist apoptosis and continue to synthesize and contract the extracellular matrix (ECM), leading to tissue stiffening [5, 7]. A recent study investigated the outcomes of intra-articular therapy with sulfasalazine in a rabbit model [8]. Drugs were encapsulated and implanted into the joints after fibrosis induction. Non-immobilized, untreated (normal) joint were compared to joints that were immobilized for 8 weeks with Kirschner wires. Joint stiffness was quantified by flexion-extension testing 8 weeks after removal of the Kirschner wires. Stiffness and intimal thickness were lower in joints treated with sulfasalazine compared with controls. Sulfasalazine also inhibited contractile activity in the cells. In conclusion, the results of the study demonstrate that sulfasalazine reduced stiffness by clearing myofibroblasts from fibrotic joints.

20.3 Rosiglitazone

Rosiglitazone is an antidiabetic drug in the thiazolidinedione class that has shown anti-fibrogenic effects [9]. The production of fibrotic connective tissue by activation of fibroblasts in response to Transforming Growth Factor beta

N. Nakamura et al. (eds.), *Advances in Knee Ligament and Knee Preservation Surgery*,
https://doi.org/10.1007/978-3-030-84748-7_20

Table 20.1 Drugs used for arthrofibrosis treatment

Drug	Type	Function on arthrofibrosis
Sulfasalazine	Anti-inflammatory—κB kinase inhibitor	Promotes myofibroblast apoptosis
Rosiglitazone	Antidiabetic—Thiazolidinedione class	Inhibits fibroblasts activation
Celecoxib	Nonsteroidal anti-inflammatory selective COX-2 inhibitor	Decreases myofibroblast activation
Recombinant antibodies	Non-modified anti-alpha2Ct antibody (ACA) and PEGylated ACA (P-ACA)	Blockage of collagen-collagen interaction mediated by the C-terminal telopeptide
Relaxin-2	Native antifibrotic hormone	Promotes MMP production and represses collagen production and expression of tissue inhibitors of metalloproteinases and TGF-β1
Interleukin-1 antagonist	Antagonism of interleukin-1, a mediator of the inflammatory response	Inhibits profibrotic mediators, fibroblast proliferation, and chemotaxis
Bevacizumab	Recombinant humanized monoclonal lg G1 antibody that targets the VEGF	Prevents deposition of fibrinous exudates, formation of the extracellular matrix with collagen deposition
Fosaprepitant	Neurokinin-1 receptor antagonist—Substance P inhibitor	Inhibits hypertrophic scarring, abdominal adhesions, and other types of fibrosis
Artesunate	Antimalarial drug	Inhibitory effect on cell proliferation and anti-fibrotic activity

COX-2 cyclooxygenase-2, *VEGF* vascular endothelial growth factor, *MMP* matrix metalloproteinase

proteins (TGF-beta) is related to development of arthrofibrosis. Previous study used hydrogels loaded with rosiglitazone intra-articular to evaluate the safety and efficacy of rosiglitazone in mitigating joint contracture in rabbits [10]. In vivo results after 8 weeks of immobilization showed a significantly improved contracture angle in animals that received rosiglitazone compared with controls. After 24 weeks from index surgery, benefits of the drug were still noticeable. Additionally, no adverse effects such as gross inflammation or arthritis were observed in the rabbits that received the intra-articular delivery of rosiglitazone. Another study investigated the use of rosiglitazone intramuscular and orally in rabbits that underwent contracture-forming surgery [11]. Eight weeks after, the animals underwent a surgical capsular release. One group received the drug and the other were controls. The animals were sacrificed after 16 weeks of free cage mobilization. No significant difference in post-traumatic contracture between groups was observed. Additionally, there was no difference in number or percentage of myofibroblasts between groups. However, there were 10 genes and 17 pathways that were significantly modulated by rosiglitazone in the posterior capsule. Thus, potentially antifibrotic genetic changes were observed after treatment with rosiglitazone in this animal model.

20.4 Celecoxib

Celecoxib is a nonsteroidal anti-inflammatory drug that selectively inhibits cyclooxygenase-2 (COX-2) thereby inhibiting prostaglandin E2 synthesis and downregulating a host of inflammatory cell types and growth factor activation [12, 13]. Decreased inflammatory cascade may decrease myofibroblast activation and, therefore, inhibit scar tissue formation.

A recent study investigated the efficacy of celecoxib in a rabbit model of arthrofibrosis by administering via intra-articular injection and a combination of intra-articular injection plus oral dosing [13]. Biomechanical and molecular analyses of contracted rabbit knee posterior capsule tissue after COX-2 inhibition revealed increased maximal passive extension and downregulation of collagen messenger RNA compared with controls. Histopathologic examination suggested a trend of decreased quantities of dense fibrous connective tissue with COX-2 inhibition. Thus, the study suggests that inhibiting the inflammatory cascade could potentially reduce pathologic myofibroblast activation, thereby reducing scar tissue formation and increasing the range of motion in arthrofibrotic joints. Another study investigated the properties of an collagen membrane as a potential celecoxib delivery scaffold intra-articular for the treatment of arthrofibrosis also in a rabbit model. Scaffolds exhibited celecoxib release through an initial burst release followed by sustained release of anti-fibrotic doses over 7 days. Therefore, the use of collagen scaffolds may be promising for treatment of arthrofibrosis [14].

20.5 Recombinant Antibodies

The arthrofibrosis formation in stiff joints consist mainly of fibril-forming collagen I and collagen III [15, 16]. A previous study used recombinant antibodies to interfere with the extracellular process of collagen fibril formation by blocking the critical collagen-collagen interaction mediated by the C-terminal telopeptide region of collagen I molecules in a rabbit model [17]. Two recombinant antibody variants of IgG type were used: non-modified anti-alpha2Ct antibody (ACA) and PEGylated ACA (P-ACA). The antibody was delivered directly to the cavities of injured knees in order to block the formation of collagen fibrils produced in response to injury. The authors observed that the antibody-treated knees demonstrated a significant reduction of flexion contracture in a mechanical test when compared with controls. Detailed microscopic and biochemical analyses verified that the reduction of flexion contracture in the treated group resulted from the antibody-mediated blocking of the assembly of collagen fibrils. Thus, the use of recombinant antibodies to target the formation of collagen fibrils may represent a valid treatment option for limiting joint stiffness.

20.6 Relaxin-2

Relaxin-2 is a native antifibrotic hormone upregulated during pregnancy to increase tissue laxity by promoting matrix metalloproteinase (MMP) production and by repressing collagen production and expression of tissue inhibitors of metalloproteinases and TGF-β1 [18, 19]. A recent study investigated the effects of relaxin-2 in a murine model of shoulder arthrofibrosis [19]. Multiple intra-articular injections of human relaxin-2, single intra-articular injections and multiple intravenous injections were tested. Multiple intra-articular injections of human relaxin-2 significantly improved range of motion, returning it to baseline measurements collected before limb immobilization. However, single intra-articular injection or multiple intravenous injections of relaxin-2 did not restore range of motion to baseline measurements. The histological hallmarks of contracture (e.g., fibrotic adhesions and reduced joint space) were absent in the animals treated with multiple intra-articular injections of relaxin-2 compared with the untreated controls and the single intra-articular injection and multiple intravenous injection-treated animals. Therefore, local delivery of relaxin-2 may be of value for treatment of arthrofibrosis.

20.7 Interleukin-1 Antagonist

Interleukin-1 is a crucial mediator of the inflammatory response and has an important role in promoting profibrotic mediators [20] and stimulating fibroblast proliferation and chemotaxis [21]. A previous study investigated the potential therapeutic effects of an intra-articular interleukin-1 receptor antagonist, anakinra, for treatment of arthrofibrosis in four patients with chronic refractory arthrofibrosis and four patients for limited arthrofibrosis of the knee [22]. All patients had improvements in range of motion and swelling while 75% of these patients were able to return to prior activity level. No adverse clinical reactions or infections were observed. These findings provide support for further study of interleukin-1 inhibition in the management of postoperative arthrofibrosis.

20.8 Bevacizumab

Bevacizumab is a recombinant humanized monoclonal lg G1 antibody that targets the vascular endothelial growth factor (VEGF). Apart from angiogenesis, VEGF facilitates deposition of fibrinous exudates, formation of the extracellular matrix with collagen deposition, and is released by fibroblasts and monocytes in the tumor stromal compartments [23–25]. Moreover, a previous study investigated the efficacy of intra-articular injection of bevacizumab in reducing arthrofibrosis and tested the effect of two injections in comparison to a single injection in a rabbit model [26]. The one-injection group had less microscopic aspects of arthrofibrosis (e.g., less inflammatory cells infiltration, less fibroblast number, less giant cell formation, less vascular density and less collagen deposition) than the control group. Yet, no statistically significant difference was found between the two groups in terms of range of joint motions and macroscopic adhesion score. The two-injection group had a better macroscopic adhesion score and a greater mean range of motion than the one-injection group. Additionally, all microscopic aspects except granulation tissue were significantly better in the two-injection group.

20.9 Fosaprepitant

Substance P is a neurotransmitter found predominantly in nerve tissues and is released after tissue injury. Increased substance P levels have been linked to hypertrophic scarring, abdominal adhesions, and other types of fibrosis [27–29]. Neurokinin-1 receptor is the cellular target of substance P. Fosaprepitant is a neurokinin-1 receptor antagonist and, therefore, a substance P inhibitor. A recent study investigated the effects of substance P inhibition using fosaprepitant to contracture formation after tissue injury in a rabbit model. Intra-articular injection of fosaprepitant was performed at four time points after the initial surgery (3, 6, 12, and 24 h) and compared with controls. Specimens were sacrificed at 72 h and at 24 weeks post-surgery. After sacrifice, all groups were analyzed by gene microarray expression profiling, bioinformatics, and biomechanical measurement. No changes in mean contracture angles between groups were observed. Yet, microarray gene expression analysis revealed that mRNA levels for proteins related to cell signaling, pro-angiogenic, pro-inflammatory and collagen matrix production were significantly different between control and fosaprepitant-treated rabbits. Thus, the study demonstrates that inhibition of substance P alters expression of pro-fibrotic gene in vivo and that targeting substance P may reduce the formation of post-traumatic joint contractures.

20.10 Artesunate

Artesunate (ART) is an antimalarial drug extracted from the Chinese medicinal herb *Artemisia annua* L that can exert inhibitory effect on cell proliferation and has shown anti-fibrotic activity [30–32]. A recent study investigated the effects of artesunate on post-traumatic knee arthrofibrosis in a rabbit model [33]. Sixty-four New Zealand white rabbits were randomly and equally divided into four groups: (1) low dose group administered with 15 mg/kg ART; (2) mediate dose group treated with 30 mg/kg ART; (3) high dose group fed with 60 mg/kg ART; (4) control group treated with the same volume of saline. After surgical removal of approximately 10×10 mm^2 area of cortical bone from both sides of the left femoral condyle, the surgical limb was then fixed in fully flexed position using Kirschner wires. All rabbits were administered intragastrically once a day for a total of 4 weeks. Artesunate induced cellular autophagy flux and inhibited cell proliferation in fibroblasts. Additionally, the severity of knee arthrofibrosis in the animals was alleviated with the use of intra-articular artesunate. Thus, artesunate may have a potential role for prevention of post-surgery knee arthrofibrosis.

20.11 Extracorporeal Shock Wave Therapy

Extracorporeal shock wave therapy (ESWT) has been used for treatment of musculoskeletal disorders. In particular, ESWT has been applied to treat fibrotic conditions such as adhesive capsulitis, plantar fibromatosis, and Dupuytren contracture [34, 35]. A recent study investigated the effects of ESWT on the arthrofibrosis formation immediately after the knee surgery in a rabbit model [36]. After surgery to induce knee arthrofibrosis with cortical bone resection of femoral condyle, knee joint of the animals was immobilized in full flexion with a fiberglass cast from the groin to the foot. On the cast, a circular opening in a diameter of 1.5 cm was made to facilitate administration of ESWT. In the ESWT group, 1000 shock waves at energy flux density of 0.2 mJ/mm^2 were delivered at the pulse repetition frequency of 3 Hz (about 5.5 min. For each session), 5 days per week for 4 weeks. After 4 weeks of surgery, contracture angle and macroscopic score of arthrofibrosis (i.e., the amount of adhesion) were significantly higher in the control group. Additionally, in the histological evaluation, the density of blood vessels was higher in the control group. No significant complication by the use of ESWT were observed. Therefore, ESWT was able to noninvasively and safely reduce formation of knee arthrofibrosis after surgery and may have a role in preventing arthrofibrosis in a clinical setting.

References

1. Czamara A, Kuzniecow M, Krolikowska A. Arthrofibrosis of the knee joint—the current state of knowledge. Literature review. Ortop Traumatol Rehabil. 2019;21:95–106.
2. Kornuijt A, Das D, Sijbesma T, de Vries L, van der Weegen W. Manipulation under anesthesia following total knee arthroplasty: a comprehensive review of literature. Musculoskelet Surg. 2018;102:223–30.
3. Stephenson JJ, Quimbo RA, Gu T. Knee-attributable medical costs and risk of re-surgery among patients utilizing non-surgical treatment options for knee arthrofibrosis in a managed care population. Curr Med Res Opin. 2010;26:1109–18.
4. Usher KM, Zhu S, Mavropalias G, Carrino JA, Zhao J, Xu J. Pathological mechanisms and therapeutic outlooks for arthrofibrosis. Bone Res. 2019;7:9.
5. Oakley F, Meso M, Iredale JP, et al. Inhibition of inhibitor of kappaB kinases stimulates hepatic stellate cell apoptosis and accelerated recovery from rat liver fibrosis. Gastroenterology. 2005;128:108–20.
6. Elsharkawy AM, Oakley F, Mann DA. The role and regulation of hepatic stellate cell apoptosis in reversal of liver fibrosis. Apoptosis. 2005;10:927–39.
7. Zhang J, Li Y, Bai X, Li Y, Shi J, Hu D. Recent advances in hypertrophic scar. Histol Histopathol. 2018;33:27–39.
8. Atluri K, Brouillette MJ, Seol D, et al. Sulfasalazine resolves joint stiffness in a rabbit model of arthrofi brosis. J Orthop Res. 2020;38(3):629–38.
9. Burgess HA, Daugherty LE, Thatcher TH, et al. PPARgamma agonists inhibit TGF-beta induced pulmonary myofibroblast differentiation and collagen production: implications for therapy of lung fibrosis. Am J Physiol Lung Cell Mol Physiol. 2005;288:L1146–53.
10. Arsoy D, Salib CG, Trousdale WH, et al. Joint contracture is reduced by intra-articular implantation of rosiglitazone-loaded hydrogels in a rabbit model of arthrofibrosis. J Orthop Res. 2018;36:2949–55.
11. Barlow JD, Morrey ME, Hartzler RU, et al. Effectiveness of rosiglitazone in reducing flexion contracture in a rabbit model of arthrofibrosis with surgical capsular release: a biomechanical, histological, and genetic analysis. Bone Jt Res. 2016; 5:11–7.
12. Davies NM, McLachlan AJ, Day RO, Williams KM. Clinical pharmacokinetics and pharmacodynamics of celecoxib: a selective cyclo-oxygenase-2 inhibitor. Clin Pharmacokinet. 2000;38:225–42.
13. Salib CG, Reina N, Trousdale WH, et al. Inhibition of COX-2 pathway as a potential prophylaxis against arthrofibrogenesis in a rabbit model of joint contracture. J Orthop Res. 2019;37:2609–20.
14. Trousdale WH, Salib CG, Reina N, et al. A drug eluting scaffold for the treatment of arthrofibrosis. Tissue Eng C Methods. 2018;24:514–23.
15. Hildebrand KA, Zhang M, Germscheid NM, Wang C, Hart DA. Cellular, matrix, and growth factor components of the joint capsule are modified early in the process of posttraumatic contracture formation in a rabbit model. Acta Orthop. 2008;79:116–25.
16. Steplewski A, Fertala J, Beredjiklian PK, et al. Auxiliary proteins that facilitate formation of collagen-rich deposits in the posterior knee capsule in a rabbit-based joint contracture model. J Orthop Res. 2016;34:489–501.
17. Steplewski A, Fertala J, Beredjiklian PK, et al. Blocking collagen fibril formation in injured knees reduces flexion contracture in a rabbit model. J Orthop Res. 2017;35:1038–46.
18. McDonald GA, Sarkar P, Rennke H, Unemori E, Kalluri R, Sukhatme VP. Relaxin increases ubiquitin-

dependent degradation of fibronectin in vitro and ameliorates renal fibrosis in vivo. Am J Physiol Renal Physiol. 2003;285:F59–67.
19. Blessing WA, Okajima SM, Cubria MB, et al. Intraarticular injection of relaxin-2 alleviates shoulder arthrofibrosis. Proc Natl Acad Sci U S A. 2019;116:12183–92.
20. Bonner JC. Regulation of PDGF and its receptors in fibrotic diseases. Cytokine Growth Factor Rev. 2004;15:255–73.
21. Gharaee-Kermani M, Phan SH. Role of cytokines and cytokine therapy in wound healing and fibrotic diseases. Curr Pharm Des. 2001;7:1083–103.
22. Brown CA, Toth AP, Magnussen B. Clinical benefits of intra-articular anakinra for arthrofibrosis. Orthopedics. 2010;33:877.
23. Bosch U, Zeichen J, Skutek M, Haeder L, van Griensven M. Arthrofibrosis is the result of a T cell mediated immune response. Knee Surg Sports Traumatol Arthrosc. 2001;9:282–9.
24. Ignjatovic D, Aasland K, Pettersen M, et al. Intra-abdominal administration of bevacizumab diminishes intra-peritoneal adhesions. Am J Surg. 2010;200:270–5.
25. Dong J, Grunstein J, Tejada M, et al. VEGF-null cells require PDGFR alpha signaling-mediated stromal fibroblast recruitment for tumorigenesis. EMBO J. 2004;23:2800–10.
26. Emami MJ, Jaberi FM, Azarpira N, Vosoughi AR, Tanideh N. Prevention of arthrofibrosis by monoclonal antibody against vascular endothelial growth factor: a novel use of bevacizumab in rabbits. Orthop Traumatol Surg Res. 2012;98:759–64.
27. Cohen PA, Gower AC, Stucchi AF, Leeman SE, Becker JM, Reed KL. A neurokinin-1 receptor antagonist that reduces intraabdominal adhesion formation increases peritoneal matrix metalloproteinase activity. Wound Repair Regen. 2007;15:800–8.
28. Lindstrom E, von Mentzer B, Pahlman I, et al. Neurokinin 1 receptor antagonists: correlation between in vitro receptor interaction and in vivo efficacy. J Pharmacol Exp Ther. 2007;322:1286–93.
29. Scott JR, Muangman P, Gibran NS. Making sense of hypertrophic scar: a role for nerves. Wound Repair Regen. 2007;15(Suppl 1):S27–31.
30. Xu Y, Liu W, Fang B, Gao S, Yan J. Artesunate ameliorates hepatic fibrosis induced by bovine serum albumin in rats through regulating matrix metalloproteinases. Eur J Pharmacol. 2014;744:1–9.
31. Wan Q, Chen H, Li X, Yan L, Sun Y, Wang J. Artesunate inhibits fibroblasts proliferation and reduces surgery-induced epidural fibrosis via the autophagy-mediated p53/p21(waf1/cip1) pathway. Eur J Pharmacol. 2019;842:197–207.
32. Liu Y, Huang G, Mo B, Wang C. Artesunate ameliorates lung fibrosis via inhibiting the notch signaling pathway. Exp Ther Med. 2017;14:561–6.
33. Wan Q, Chen H, Xiong G, et al. Artesunate protects against surgery-induced knee arthrofibrosis by activating Beclin-1-mediated autophagy via inhibition of mTOR signaling. Eur J Pharmacol. 2019;854:149–58.
34. Knobloch K, Vogt PM. High-energy focussed extracorporeal shockwave therapy reduces pain in plantar fibromatosis (Ledderhose's disease). BMC Res Notes. 2012;5:542.
35. Chen CY, Hu CC, Weng PW, et al. Extracorporeal shockwave therapy improves short-term functional outcomes of shoulder adhesive capsulitis. J Shoulder Elb Surg. 2014;23:1843–51.
36. Zhou Y, Yang K. Prevention of arthrofibrosis during knee repair by extracorporeal shock wave therapy: preliminary study in rabbits. Injury. 2019;50:633–8.

21 A View of Predisposing Factors by Novel 3D Imaging Techniques for the PF Joint

Yukiyoshi Toritsuka and Yuzo Yamada

Patellar dislocation or patellar instability (PI) usually occurs in patients with predisposing factors [1, 2]. Regarding treatments, medial patellofemoral ligament (MPFL) reconstruction has been widely performed with the intention of reconstructing the stabilizers against lateral dislocation of the patella [3–5]. In addition, trochleoplasties [6, 7] and tibial tubercle transfers [8] have been applied in order to reduce predisposing factors. Elimination of such factors by normalizing indicators is considered to be theoretically the best approach; however, various physical balances that had been maintained for years might be changed by the surgery and those changes might cause new problems. In addition, they are generally apt to be underestimated when compensating for prevention of dislocation [9].

Another question is whether current evaluation of predisposing factors is appropriate at all. The indicators for predisposing factors are defined on a plane obtained using diagnostic imaging modalities such as plain radiographs, computed tomography (CT), or magnetic resonance imaging (MRI). However, it is unclear whether they are sufficiently appropriate as indicators for evaluation of predisposing factors to decide indications for specific surgeries. Of course, there was no other option in times when evaluations were always carried out on films. But recently, progress in imaging technology has brought great benefits to the orthopedic field. Now, the technologies allow us to reconstruct images to create three-dimensional (3D) models from data obtained by those modalities leading to easier recognition of the real bones and joints [10–13]. If the conventional indicators are inadequate, can the new technologies bring new insights?

21.1 Limitation of Conventional Indicators for Predisposing Factors

Below are specific examples of limitations of conventional indicators on 2D images from the 3D point of view. The congruence angle is shown in Fig. 21.1 [14]. The reference points for measurement of the angle on the skyline view are shown in the figure. These points are actually not on one plane in the 3D space (Fig. 21.1). These show that the original sense of an indicator defined on an image projected onto a plane would become unclear in the 3D space.

Y. Toritsuka (✉)
Department of Health and Sports Sciences, Mukogawa Women's University, Nishinomiya, Hyogo, Japan
e-mail: torituka@mukogawa-u.ac.jp

Y. Yamada
Department of Orthopaedic Surgery, Yao Municipal Hospital, Yao-city, Osaka, Japan
e-mail: yuzo.yamada@hosp-yao.osaka.jp

N. Nakamura et al. (eds.), *Advances in Knee Ligament and Knee Preservation Surgery*,
https://doi.org/10.1007/978-3-030-84748-7_21

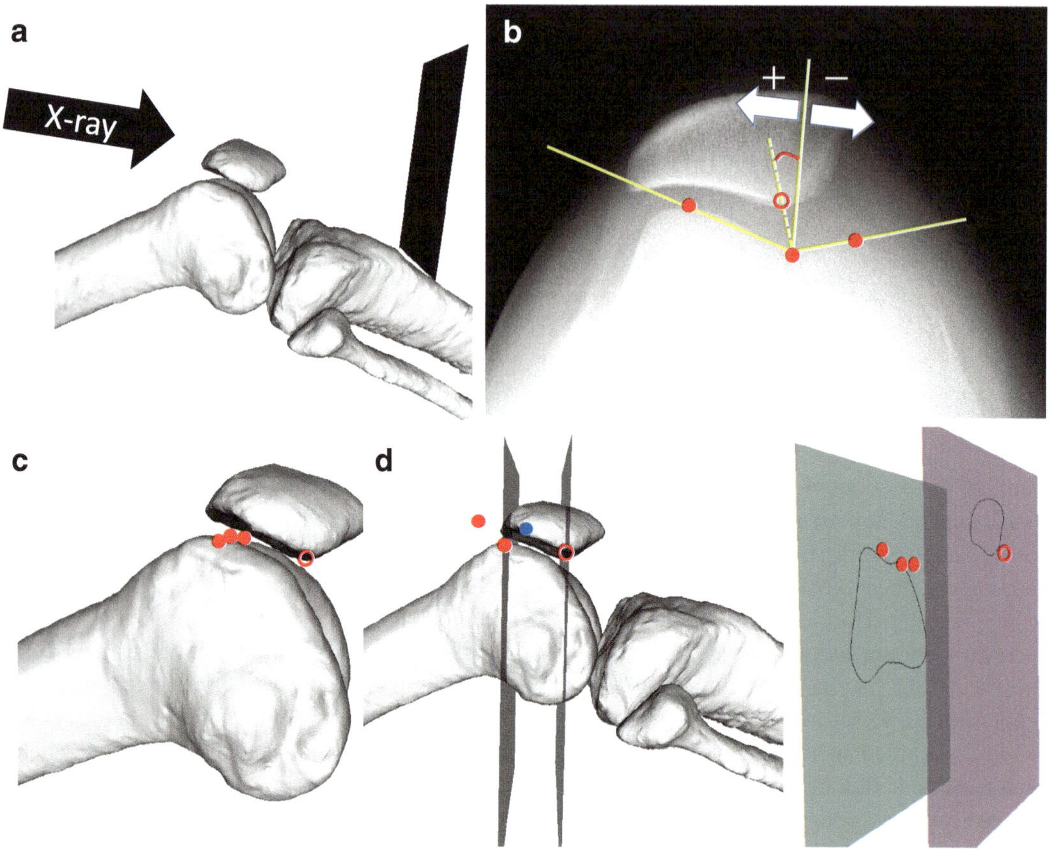

Fig. 21.1 Knee Merchant view. (**a**) A 3D image of X-ray projection. (**b**) The congruence angle. The open circle indicates the reference point on the patella and the three closed circles show the reference points on the trochlea used to determine the congruence angle. (**c**) The same reference points shown on the 3D image. (**d**) Image showing the planes including the reference point on the patella and those on the trochlea. There are at least two planes in the 3D space

The bisect offset index or the patellar tilt angle is typically evaluated by measuring the translation or rotation relative to the femoral reference lines on cross-sectional images (Fig. 21.2a) [15]. As the contour of the femur varies among images according to cross-sectional level or knee flexion angles, the reference lines on the femur become inconsistent. For example, when evaluated at the mid-patellar level, the cross section of the femur would be more proximally located in patients with patellar alta (Fig. 21.2b). When evaluated in the knee flexed position, the contours of the femur on the cross section might differ from those in knee extension (Fig. 21.2c). Hence, it is clear that the conventional indicators cannot help being influenced by the geometries of the femur on 2D images which differ depending on the location of cross sections and/or knee flexion angle, leading to unreliability when comparing the values among patients or among different knee positions even in the same patient.

However, 3D computer models could provide consistent references for direct evaluation, which are not affected by location of cross section or knee flexion angle. Thus, we believe that indicators should also be redefined for evalua-

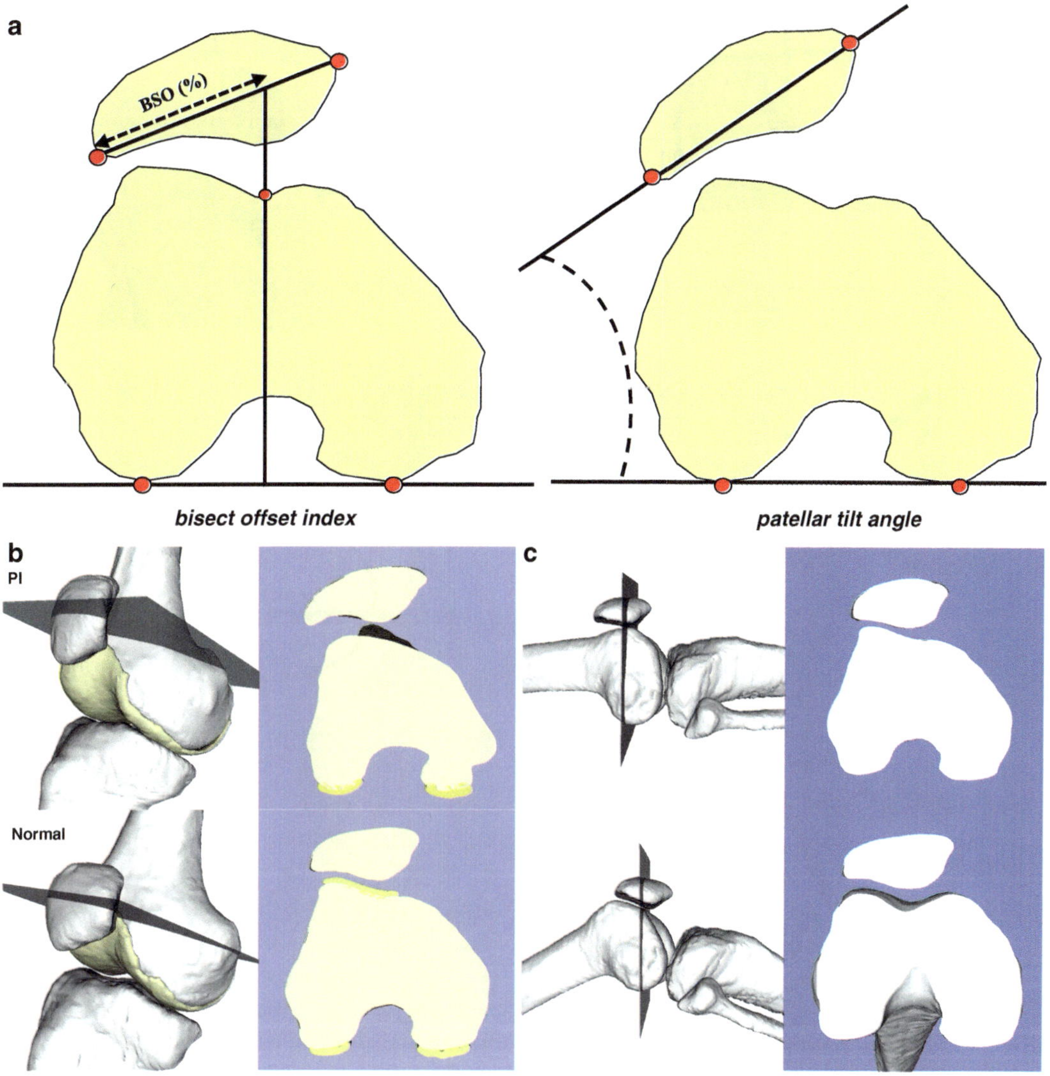

Fig. 21.2 (**a**) The bisect offset index (BSO) and the patellar tilt angle. These two images were created at the mid-patellar level. (**b**) Cross-sectional planes of a patient with PI and a normal control at the mid-patellar level in clinically used CT or MRI in relation to the location of the plane in each 3D computer model. These show that the locations of the two planes on which each indicator is calculated are different on each femur. *PI* patellar instability, *Normal* normal control. (**c**) Cross-sectional planes of a patient with PI at the mid-patellar level in clinically used CT or MRI in knee extension and in knee flexion in relation to the location of the plane in each 3D computer model. These show that the locations of the planes on which each indicator is calculated are different on each femur

tion of 3D images as a result of progress in imaging technology. We believe that new indicators for 3D images preserving the original senses of the conventional indicators would clearly describe anatomical characteristics in real bones and joints, or would accurately elucidate changes of the alignment with knee motion.

21.2 Method of Creating a 3D Knee Computer Model and the Coordinate System

In this chapter, we introduce our studies of the patellofemoral (PF) joint and provide a new view of predisposing factors recognized by a new imaging technology [12, 13, 16–18]. Our meth-

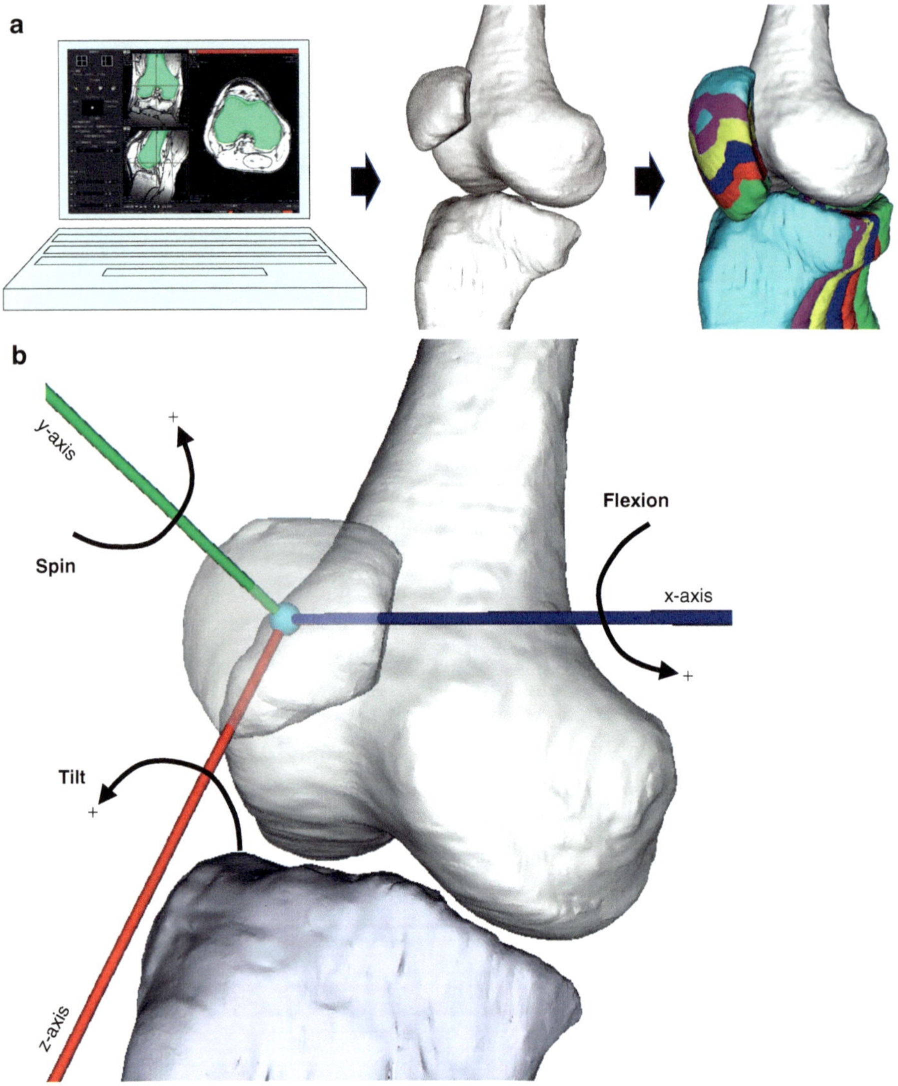

Fig. 21.3 (**a**) Creation of a 3D computer knee model. The positions of the patella and tibia are indicated at different degrees of knee flexion. The light blue indicates their position at 0°, the pink at 10°, the yellow at 20°, the blue at 30°, the red at 40°, and the green at 50°. (**b**) The global coordinate system. The line passing through the most proximal point and the most distal point was defined as the *z*-axis. The midpoint between these two points was defined as the patellar reference point (the light blue sphere). The line perpendicular to the *z*-axis and parallel to the line passing through the mediolateral osseous border of the patella was defined as the *x*-axis. The line perpendicular to the *z*-axis and *x*-axis was defined as the *y*-axis. Each motion around the *x*-axis, *y*-axis, and *z*-axis relative to the initial zero position was defined as *flexion*, *spin*, and *tilt* of the patella, respectively

ods are described briefly as follows: first, three-dimensional MR images were taken in the supine position at several knee flexion angles with the quadriceps muscle relaxed. Contours of the femur, patella, and tibia were semiautomatically extracted from 3D MRI data using the 3D motion analysis system. Three-dimensional computer models were then constructed and were automatically superimposed over images taken at each position by voxel-based registration (Fig. 21.3a) [13]. The coordination system for evaluation (the global coordination system) was established as follows: the patella at 0° of knee flexion was defined as the zero position, and the reference axes (x-, y-, and z-axis) were then set for each patella (Fig. 21.3b) [13]. Our results on normal patellar motion were closely comparable to previous cadaver studies despite the different methods employed [19–21]. Therefore, our method is reliable enough to assess patients' patellar motion.

21.3 New References for New Indicators in the 3D Space (the Anatomical Coordinate System)

Next, the anatomical coordinate system was established to reevaluate clinically relevant morphology or motions because the abovementioned global coordinate system, usually used in the field of basic science, is unsuitable to do so. The purpose of this coordinate system is to describe clinically relevant morphology or alignment instead of the conventional indicators for predisposing factors. In this coordinate system, predisposing factors were reevaluated in the 3D space using newly established indicators defined by newly established reference planes or lines in 3D space [12, 13]. Two new reference planes, the mid-sagittal plane and the femoral condylar planes (FCPs)/the femoral trochlear planes (FTPs) are explained here [12, 18].

21.3.1 The Mid-Sagittal Plane

The trans-epicondylar axis (TEA) was selected as the femoral reference axis because it forms not only an anatomical landmark but also the mechanical axis of the knee. The mid-sagittal plane was set on the femur defined as the plane perpendicular to the TEA and passing through the midpoint between the medial and lateral femoral epicondyles (Fig. 21.4) [13]. Thus, it formed a reference independent of the geometries of the femoral condyle and/or the femoral trochlea. In addition, since the mid-sagittal plane was set in 3D space, it forms a reproducible femoral reference that allows accurate description of patellar movement without the influence of knee flexion.

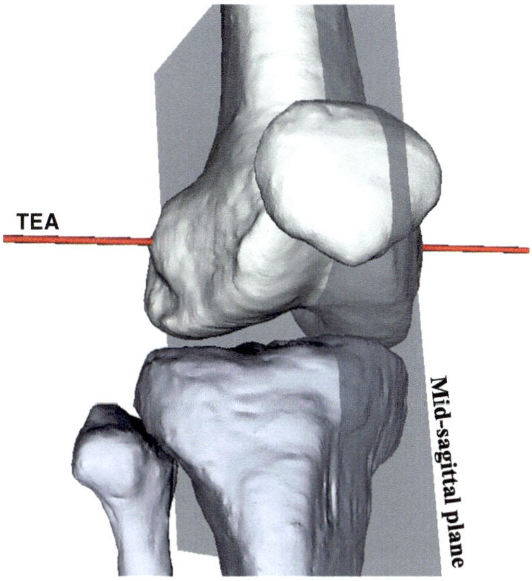

Fig. 21.4 The mid-sagittal plane was defined as the plane perpendicular to the TEA and passing through the midpoint between the medial and lateral femoral epicondyles. *TEA* trans-epicondylar axis

21.3.2 The Femoral Condylar Planes (FCPs) or the Femoral Trochlear Planes (FTPs)

Next, we considered whether it is appropriate to evaluate a 3D structure such as the femoral trochlea using planes perpendicular to the femoral axis, as in generally used CT or MRI evaluation. The femoral condylar planes (FCPs) or the femoral trochlear planes (FTPs) have been developed as another reproducible group of reference planes [12, 18]. The FCPs were established as virtual cross sections including the TEA [12]. FCP 0 was defined as the base plane including the superior bone–articular cartilage border of the intercondylar notch and FCP θ was also defined as the plane making optional angle θ to FCP 0 (Fig. 21.5a) [12]. These enable us to obtain reproducible cross sections which are as close as possible perpendicular to the articular surface of the trochlea and to evaluate the morphology of the femoral trochlea without the influence of the condylar/trochlear geometry. The femoral trochlear planes (FTPs) were similarly established [18]. Here, FTP 0 was defined as the base plane including the proximal edge of the femoral trochlea and FTP θ was also defined

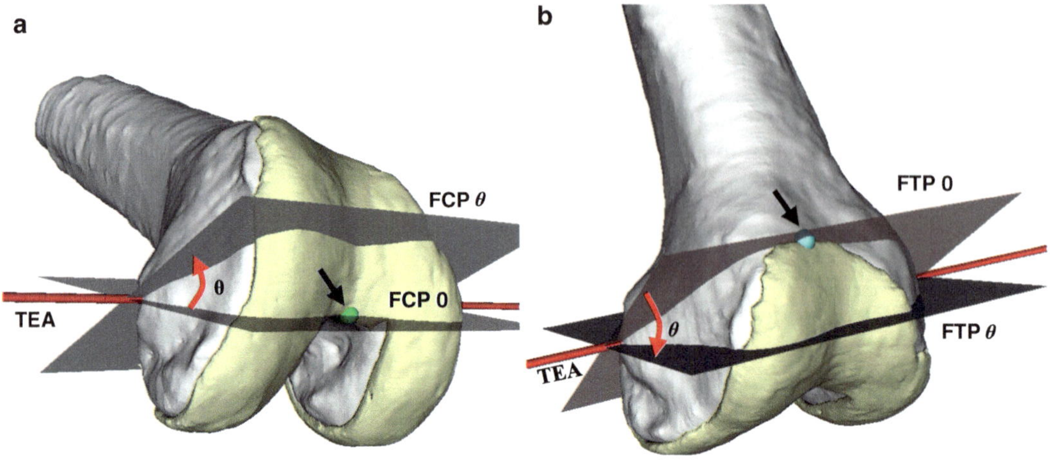

Fig. 21.5 (**a**) The femoral condylar planes (FCPs). The green sphere shows the superior bone–articular cartilage border of the intercondylar notch (black arrow). FCP θ was also defined as the plane making optional angle θ (red arrow) from FCP 0. *TEA* trans-epicondylar axis. (**b**) The femoral trochlear planes (FTPs). The light blue sphere shows the proximal edge of the femoral trochlea (black arrow). FTP θ was also defined as the plane making optional angle θ (red arrow) from FTP 0. *TEA* trans-epicondylar axis

as the plane making optional angle θ around the TEA to FTP 0 (Fig. 21.5b) [18]. These were essentially the same evaluation planes as the FCPs except that the base plane was set on the inlet of the trochlea, which is more appropriate to evaluate the morphological characteristics of the trochlea.

21.4 A New View of Predisposing Factors According to the New Indicators in the 3D Space

21.4.1 Patella Alta

Patella alta is one of the major predisposing factors for PI. As parameters for patella alta, measurements such as the ratio of patellar tendon length to the long axis of the patella (Insall-Salvati ratio), the ratio of the distance between the lower pole of the patella and the upper limit of the tibia, and the length of the patellar joint surface (Caton-Deschamps index) [22], and the patella:trochlea ratio of the cartilage baselines (the patellotrochlear index) [23] are commonly used; however, they do not directly indicate the patellar location relative to the femur. Thus, we have expressed how proximally the patella is located in comparison to the reference of the femur using the FCPs as the patellar center height [12]. The location of the patellar reference point was defined as the patellar center height and was expressed as the angle θ between FCP 0 and FCP θ including the patellar reference point (Fig. 21.6a) [12]. Our results indicated that the patellar center height in patients with PI was larger than that of normal controls, and showed a statistically significant correlation with articular cartilage height (Fig. 21.6b) [12]. The articular cartilage height was defined as the angle θ between FCP 0 and the FCP θ containing the proximal edge of the articular cartilage as an indicator expressing proximal distribution of articular cartilage (Fig. 21.6c) [12]. Thus, these measurements suggested that the patellae in patients with PI were located not simply proximally but maintained congruence with the more proximally distributed articular surface of the trochlea [12]. The distalization of the tibial tubercle is sometimes applied to patients with patella alta; therefore, distal transfer of the tibial tubercle might cause deterioration in the congruence between the patella and the trochlea leading to PF osteoarthrosis even if it would prevent dislocation of the patella.

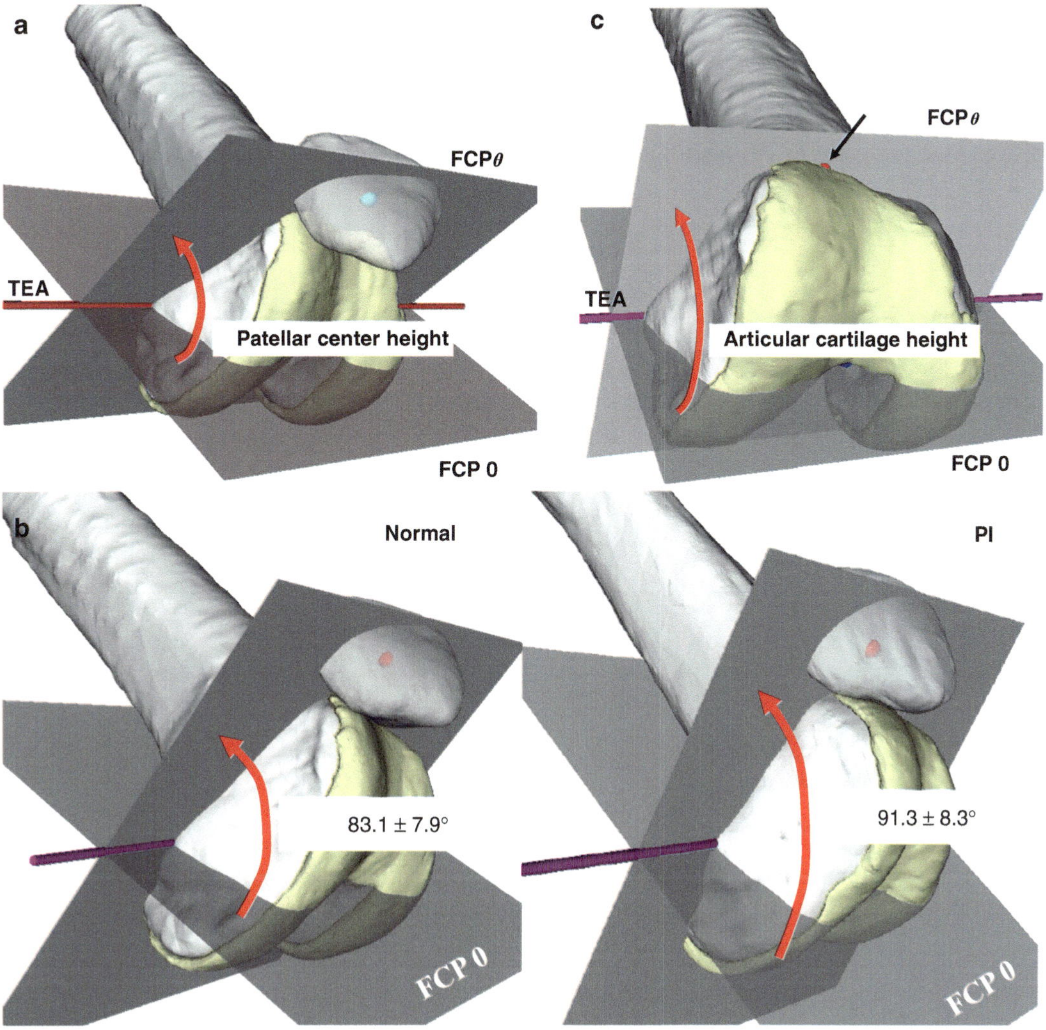

Fig. 21.6 (**a**) The patellar center height. The light blue sphere shows the patellar reference point. The patellar center height (red arrow) was expressed as the angle θ between FCP 0 and FCP θ including the patellar reference point. (**b**) Patellar center height in normal controls vs. patients with PI. The red spheres show the patellar reference points of a normal control and a patient with PI. The values are expressed as the mean ± standard deviation (SD) [12]. The mean values in patients with PI were significantly larger than those in normal controls ($P < 0.05$, Mann-Whitney U test) [12]. *PI* patellar instability. (**c**) The articular cartilage height. The red sphere shows the proximal edge of the articular cartilage of the femoral trochlea. The articular cartilage height (red arrow) was expressed as the angle θ between FCP 0 and FCP θ including the proximal edge of the articular cartilage

21.4.2 Lateral Shift of the Tibial Tubercle

Lateralization of the tibial tubercle is commonly expressed using tibial tubercle–trochlear groove (TT–TG) distance [22]. TT–TG distance is more advantageous than the Q angle in terms of independence from the location of the patella; however, the inter-observer and/or intra-observer reliability remains controversial due to difficulty in deciding on the reference point of the trochlear groove especially in cases with severe dysplasia. In brief, reliability of the TT–TG distance unfortunately depends on trochlear geometry. To compensate for this weak point, the tibial tubercle–posterior cruciate ligament (TT–PCL) distance would be another choice to express lateralization of the tibial tubercle [24]. However, using the reference plane in 3D space after constructing a 3D model of the knee joint is a more advantageous alternative. First, the location of the tibial tubercle was defined as the distance between the top point of the tibial tubercle and the mid-sagittal plane. Then, after dividing by each inter-epicondylar width, the individual values were expressed as *% tubercle shift* standardized by the knee size (Fig. 21.7a) [13]. This new indicator enables accurate presentation of the

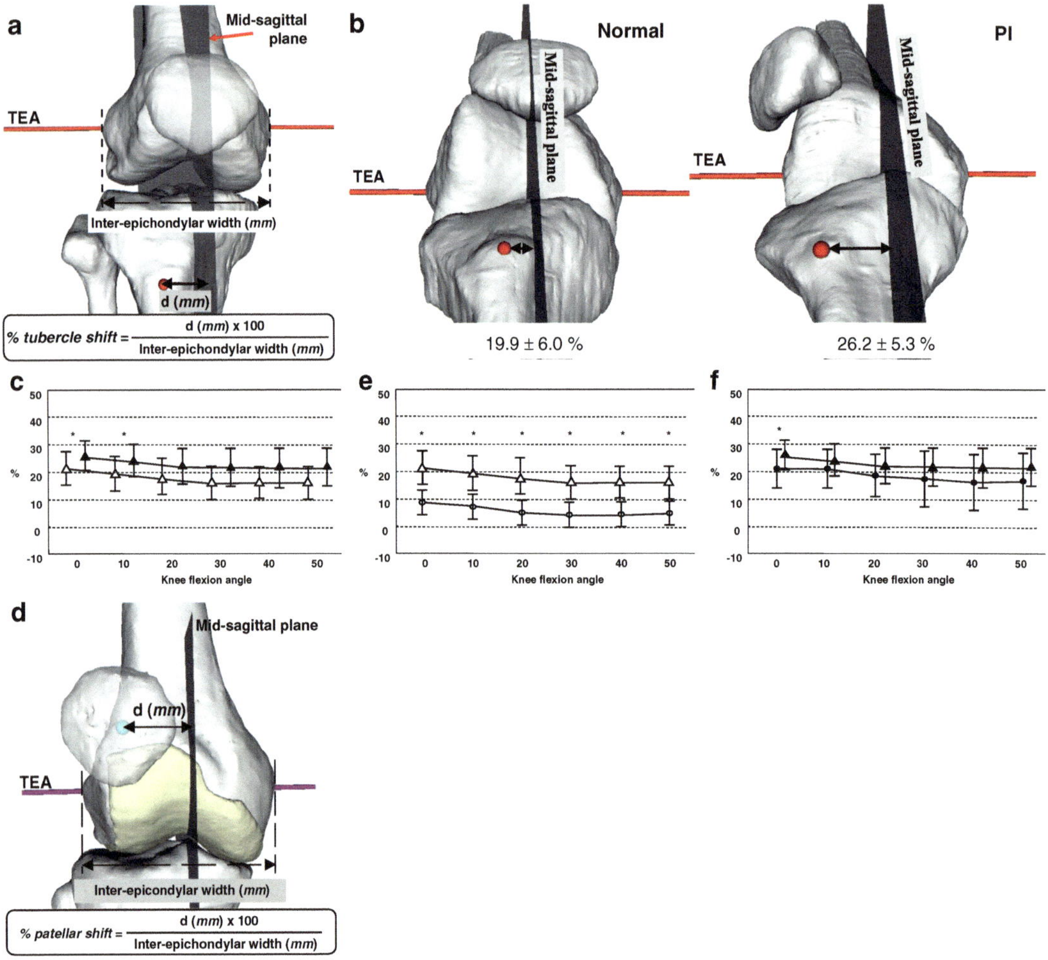

Fig. 21.7 (**a**) *% tubercle shift*. The red sphere shows the top point of the tibial tubercle. *TEA* trans-epicondylar axis. (**b**) *% tubercle shift* at 0° of knee flexion in normal controls vs. patients with PI. The red spheres show the top points of the tibial tubercle of a normal control and a patient with PI. The values are expressed as the mean ± standard deviation (SD) [13]. The mean values in patients with PI were significantly larger than those in normal controls ($P < 0.05$, Mann-Whitney U test) [13]. *PI* patellar instability, *TEA* trans-epicondylar axis. (**c**) Change of *% tubercle shift* of patients with PI and normal controls with knee flexion. The open triangles represent *% tubercle shift* of the normal controls and the closed triangles represent that of patients with PI. The values in the PI patients were significantly larger at 0° and 10° compared to those in the normal controls [13]. Values are mean ± standard deviation (SD). $^*P < 0.05$, Mann-Whitney U test. (**d**) *Percentage of patellar shift*. The light blue sphere shows the patellar reference point. *TEA* trans-epicondylar axis. (**e**) Change of *% tubercle shift* and *% patellar shift* of normal controls under knee flexion. The values are shown as the mean ± SD. The open triangles represent *% tubercle shift* and the open circles represent *% patellar shift* of the normal controls. Values of *% tubercle shift* were significantly larger at all angles than *% patellar shift* [13]. $^*P < 0.05$, Wilcoxon signed-rank test. (**f**) Change of *% tubercle shift* and *% patellar shift* in patients with PI under knee flexion. The values are shown as the mean ± SD. The closed triangles represent *% tubercle shift* and the closed circles represent *% patellar shift* of patients with PI. Values of *% tubercle shift* were significantly larger than *% patellar shift* only at 0° [13]. $^*P < 0.05$, Wilcoxon signed-rank test

location of the tibial tubercle relative to the femur even in cases with severe trochlear dysplasia. This indicator also identifies a more laterally located tibial tubercle in patients with PI (Fig. 21.7b) [13].

As TT–TG distance is originally an indicator in knee extension, changes cannot be seen with knee flexion. Expanded interpretation of TT–TG distance in the knee-flexed position might lead to inaccurate comparison among values in different positions due to changes of the contour of the trochlea on the evaluation planes caused by knee flexion. However, *% tubercle shift* enables accurate presentation of values even under changes of knee position. In addition, this enables us to describe the individual relationships between the lateralization of the tibial tubercle and the patella in the same flexion angle.

Our results showed that *% tubercle shift* showed a slight decrease with knee flexion both in the normal controls and the patients (Fig. 21.7c) [13]. The values in the patients were significantly greater at 0° and 10° than in the normal controls, indicating that the tibial tubercle was not more laterally located throughout the range of motion [13]. These findings suggested that tibial tubercle transfer might cause a potential risk of a too medially transferred tubercle in the knee flexed position if uniformly applied at 0° leading to increased pressure on the articular surface somewhere in the medial facet in the flexed position.

Next, the location of the patellar center was defined as *% patellar shift* expressed as the distance between the patellar reference point and the mid-sagittal plane divided by each inter-epicondylar width (Fig. 21.7d) [13]. When *% patellar shift* and *% tubercle shift* were compared in the same case, the latter showed significantly greater values at all angles in the normal controls but only at 0° in the patients (Fig. 21.7e, f) [13]. No case showed smaller *% tubercle shifts* in the normal control; however, smaller *% tubercle shifts* were found in the knees of half of the patients, indicating the patellae were located more laterally compared to the tibial tubercle in the patients (Table 21.1) [13]. These data suggest that the relationship between the patella and the tibial tubercle in patients with PI is different from that in normal controls because of the lateral location of the patella caused by disruption of the medial stabilizers, and that the relationship varied among cases and differed depending on knee flexion angles in patients with PI. Considering these observations, application for tibial tubercle transfer should be decided depending on the change of location of the tibial tubercle.

Table 21.1 Percentage of patellar shift vs. % tubercle shift in the patients

		Knee flexion angle (degree)					
Knee		0	10	20	30	40	50
P2 L	*% patellar shift*	***27.2***	***28.9***	***29.9***	***22.1***	15.4	12.3
	% tubercle shift	25.3	23.3	20.6	19.8	19.6	18.9
P3	*% patellar shift*	21.6	***20.5***	***14.4***	***14.2***	7.2	7.2
	% tubercle shift	23.8	20.0	14.2	13.2	11.4	11.9
P6	*% patellar shift*	***16.7***	***16.1***	***12.2***	***11.0***	8.4	***10.1***
	% tubercle shift	15.3	14.9	11.2	10.0	8.6	9.6
P7	*% patellar shift*	***29.2***	***30.5***	***29.2***	***41.9***	***42.5***	***43.9***
	% tubercle shift	25.9	25.2	23.9	27.4	26.7	26.5
P8L	*% patellar shift*	21.5	18.6	14.0	10.6	11.9	***13.2***
	% tubercle shift	23.7	21.2	17.1	15.1	15.1	12.0
P9	*% patellar shift*	26.0	26.3	***28.8***	***30.5***	***33.7***	***34.6***
	% tubercle shift	29.2	26.9	26.9	25.9	24.5	25.1
P11	*% patellar shift*	***25.7***	22.8	22.5	20.3	18.4	13.9
	% tubercle shift	24.3	23.6	22.9	22.4	22.9	20.3

Bold italic letters in *% patellar shift* indicate values larger than *% tubercle shift*

This table is cited from Yamada et al. JBJS Br 2007 [13]. Identification numbers of patients' knee were described in the first column. Fourteen knees from 12 patients with PI were analyzed in the study. A larger % patellar shift at more than one position was found in seven patients among them

Therefore, tibial tubercle transfer may be less applicable to patients than previously believed.

21.4.3 Patellar Shift and Tilt

Patellar shift and tilt have been widely used as indicators of malalignment of the patella. Thus, we tried to express these indicators in 3D space while retaining their original senses.

First, *3D shift* was defined as the same indicator as *% patellar shift*; location of the patellar reference point from 0° to 50° of knee flexion was defined as the distance from the mid-sagittal plane and was expressed as a percentage of the inter-epicondylar width standardizing the values according to individual knee size (Fig. 21.8a) [16, 17]. This can express lateralization of the patella throughout the range of knee flexion.

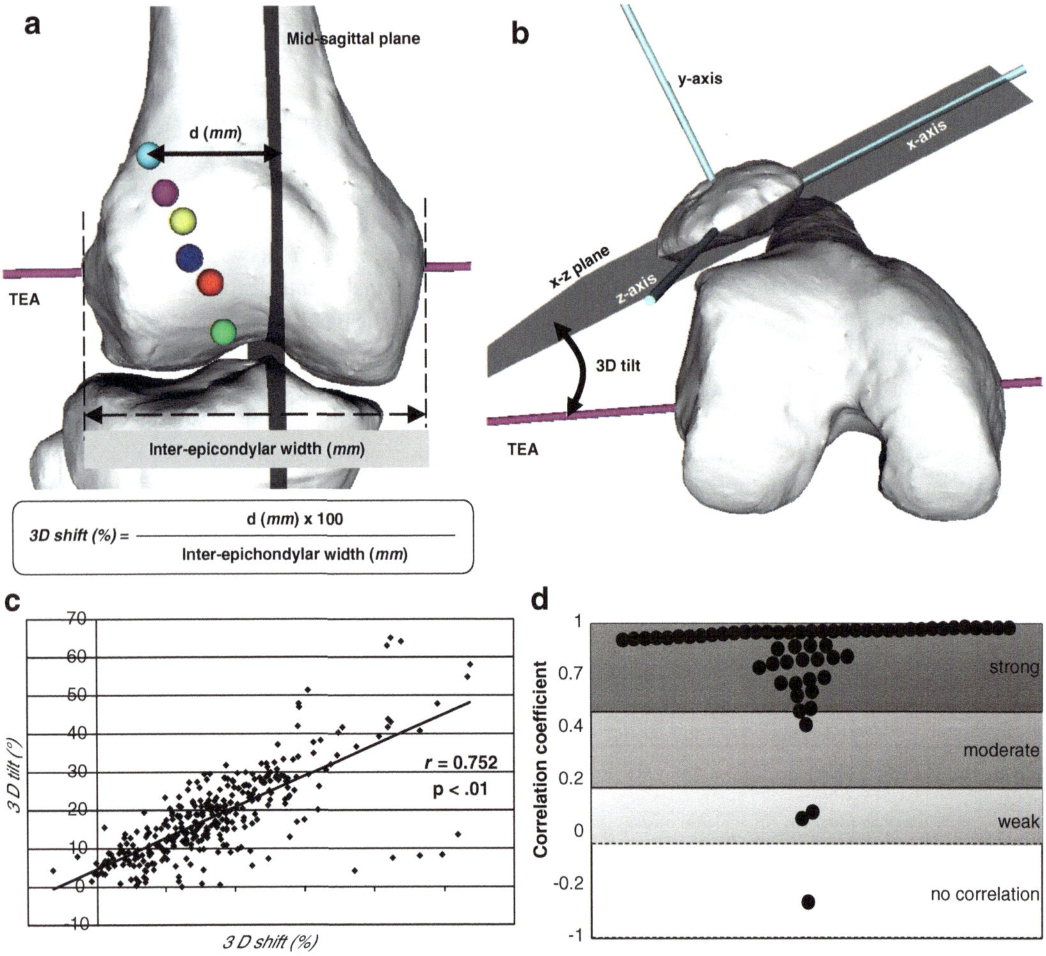

Fig. 21.8 (**a**) *3D shift. 3D shift* is the same indicator as *% patellar shift*. The light blue sphere represents the patellar reference point at 0°, the pink at 10°, the yellow at 20°, the blue at 30°, the red at 40°, and the green at 50°. *TEA* trans-epicondylar axis. (**b**) *3D tilt. 3D tilt* was defined as the angle between the *x-z* plane and the trans-epicondylar axis. *TEA* trans-epicondylar axis. (**c**) The relationship between *3D shift* and *3D tilt*. The data of 66 knees from 60 patients with PI at 0°, 10°, 20°, 30°, 40°, and 50° were used. Pearson's correlation coefficient between the *3D shift* and *3D tilt* was 0.752 ($P < 0.01$). (**d**) Distribution of individual correlation coefficients between *3D shift* and *3D tilt* in patients with PI. Pearson's correlation coefficient between the *3D shift* and *3D tilt* of the patella at 0°, 10°, 20°, 30°, 40°, and 50° was calculated for each knee. Correlation coefficient values >0.7 were defined as strong, 0.4–0.7 as moderate, 0.2–0.4 as weak, and 0–0.2 as no correlation

3D tilt was defined as the spatial angle between the TEA and the *x–z* plane of the patella in the abovementioned global coordinate system expressing inclination of the patella in 3D space (Fig. 21.8b) [13, 16]. This can also describe inclination of the patella through the range of knee flexion.

The analysis of *3D shift* and *3D tilt* at 0°, 10°, 20°, 30°, 40°, and 50° of knee flexion in patients with PI revealed a close relationship between the two indicators (Fig. 21.8c) (unpublished data). Then, Pearson's correlation coefficient between the *3D shift* and *3D tilt* of the patella of each knee was calculated. Correlation coefficient values >0.7 were defined as strong, 0.4–0.7 as moderate, 0.2–0.4 as weak, and 0–0.2 as no correlation. Overall, 95% of the knees showed a moderate/strong correlation (Fig. 21.8d) [16]. It may be natural from the 3D point of view because the patella is a sesamoid bone within the extensor mechanism of the knee. When the patella moves laterally, it inclines laterally under certain balancing conditions from the surrounding soft tissue before entering the femoral groove. It also tilts laterally on the convex trochlea after contacting the femoral groove and vice versa.

21.4.4 Classification of Patellar Tracking

These results suggested that it is not always necessary to use both indicators when evaluating or describing individual patellar alignment, at least for knees from PI patients in 0–50° of flexion [16]. Such a description may enable surgeons to describe patellar alignment more simply, which would lead to better and easier understanding of the individual characteristic pathology of each patient with PI. For better description of the patellar tracking, the maximum 3D shift *(Max-shift)* and the change of 3D shift from 0° to 50° *(*$Change_{0\text{–}50}$*)* were considered [17]. *Max-shift* was used as an indicator that represents the extent of lateral deviation of the patella and $Change_{0\text{–}50}$ for the movement direction of the patella with knee flexion [17]. First, the cutoff value (COV) of *Max-shift* was defined based on data from healthy volunteers. When a value was greater than the COV, it was defined as a *major subluxation* while it was defined as a *minor subluxation* when smaller than the COV (Fig. 21.9a) [17]. Next, the two COVs of $Change_{0\text{–}50}$ were similarly defined. When values were greater than the upper COV, they were defined as the *major-lateral type,* laterally moving patella with flexion, while they were defined as the *major-medial type,* medially moving patella with flexion when smaller than the lower COV [17]. When a value lay between the two COVs, it was defined as a *major-straight type* (Fig. 21.9b) [17].

This classification is very convenient for selecting surgical procedures. PI knees with a very similar tracking pattern to that of normal patellae (*minor-type* tracking) would be good candidates for conservative treatment, especially when they are first-time dislocators and would be safely able to undergo MPFL reconstruction. Patients with the *major-medial-type* of tracking would also be good candidates for MPFL reconstructions because a reconstructed MPFL tightens in knee extension, preventing lateral deviation of the patella. Patients with the *major-lateral-type* tracking might need lateral release in combination with MPFL reconstruction because the MPFL becomes lax with knee flexion. Tightness of the lateral retinaculum may cause lateral movement of the patella [17, 25, 26]. No further classification was performed on *the minor subluxations* because it is difficult to distinguish moving direction due to small mediolateral movement. Consequently, they were defined as *the minor type.*

Analysis using *3D shift* has enabled a new view of patellar tracking; however, we should be aware that there are not four separate types in PI because this classification is performed using cut-off values. Each parameter used in this classification showed continuous distribution. In other words, there are no clear borders between the types. However, these simultaneously imply a new concept in PI. As mentioned above, the results seem to present visually vague borders between habitual dislocators and the *major-lateral type*, and between the normal and the *minor type* in terms of patellar tracking. Thus, PI, in entirety, might be conceptually regarded as a continuous spectrum from normal to congenital patellar dislocation (Fig. 21.9c) [17, 25].

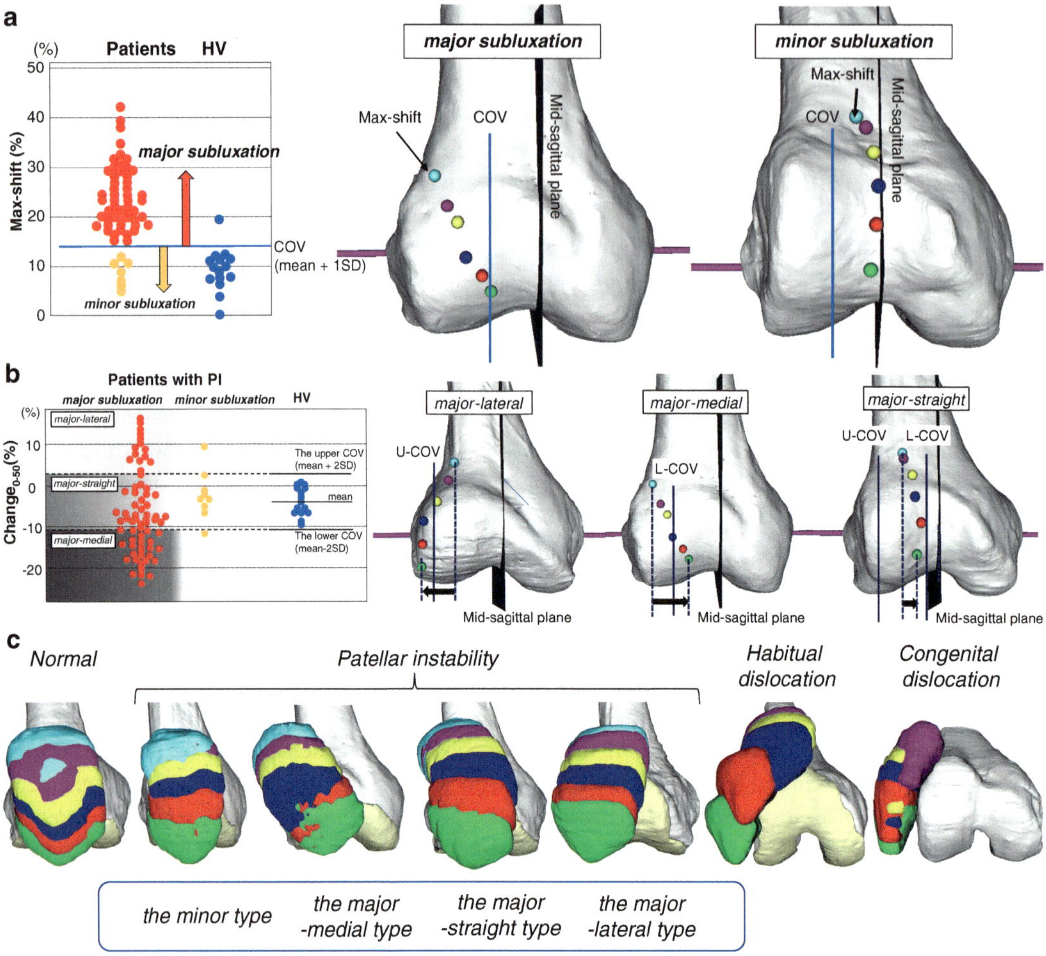

Fig. 21.9 (**a**) A *major subluxation* and a *minor subluxation* in patients with PI. Distribution of each value of *Max-shift* is shown in the chart on the left. The cutoff value (COV) of *Max-shift* was defined based on data from healthy volunteers (blue circles). The COV was the mean + 1SD of *Max-shift* of the healthy volunteers. When a value was greater than the COV, it was defined as a *major subluxation (red circles)* while it was defined as a *minor subluxation (yellow circles)* when smaller. The light blue sphere represents the patellar reference point at 0°, the pink at 10°, the yellow at 20°, the blue at 30°, the red at 40°, and the green at 50°. The blue lines on the 3D models represent the COV. *HV* healthy volunteers, *COV* cutoff value, *SD* standard deviation. (**b**) The *major-medial*, *major-lateral,* and *major-straight* types of *major subluxations.* The distribution of each value of $Change_{0-50}$ is shown in the chart on the left. The cutoff value (COVs) of $Change_{0-50}$ was defined based on data from healthy volunteers (blue circles). The COVs were the mean ± 2SD of $Change_{0-50}$ of the healthy volunteers. When greater than the upper COV, it was defined as a *major-lateral type* while it was defined as a *major-medial type* when smaller than the lower COV [17]. When a value fell between the two COVs, it was defined as a *major-straight type.* The red circles represent the major subluxations and the yellow circles, the minor subluxations in the chart on the left. The solid lines on the 3D models represent the COVs. The light blue sphere represents the patellar reference point at 0°, the pink at 10°, the yellow at 20°, the blue at 30°, the red at 40°, and the green at 50°. The dotted lines show the locations of the patellar reference points at 0° and at 50°. The arrows show the direction of movement of the patellae. HV: healthy volunteers. *COV* cutoff value, *U-COV* the upper COV, *L-COV* the lower COV, *SD* standard deviation. (**c**) A new concept of PI spectrum. PI, in entirety, might be conceptually regarded as a continuous spectrum from normal to congenital patellar dislocation

21.4.5 Morphology of the Trochlea

While trochlear dysplasia has also been evaluated by indicators on the 2D plane, it can be quite difficult to express the 3D structure of the femoral trochlea by simple cross-sectional CT or MR images, potentially resulting in misunderstandings. However, 3D models would visually present the real structural characteristics leading to much easier understanding of them. For example, we reported that articular cartilage on the trochlea is distributed more proximally and more laterally according to the patellar position in patients with PI compared to the normal position [12]. The use of 3D models could thus bring impressive images more directly to us.

From this point of view, there is still room for improvement in conventional evaluation of the trochlear morphology. The Dejour classification has been widely used but unfortunately the intra- and inter-rater agreements were still only fair [7, 27, 28]. This discrepancy might be caused by the 2D evaluation of the trochlea. While it originally consisted of a combination of a projected lateral view radiograph together with CT or axial MRI, the classification is sometimes performed based on an axial MR image alone. Thus, we have tried to three-dimensionally reevaluate the trochlear shape using reproducible cross sections as perpendicular to the articular surface of the trochlea as possible [21]. The trochlear types were evaluated on FTP 10, 20, 30, 40, 50, and 60 according to Dejour classification (Fig. 21.10a) [21]. This analysis revealed that the majority of patients with PI showed changes in their trochlear types on FTPs with various patterns (Fig. 21.10b) [21], and that low rates of agreement were found between the trochlear type on each FTP and the classification obtained in the clinical setting in patients with PI (Table 21.2) [21].

These findings suggested that types of trochlear dysplasia on one axial image provided by the Dejour classification could not represent the entire trochlear geometry. In other words, dysplastic trochleae would show morphological variation from a 3D point of view even when they were categorized within the same types according to Dejour classification. Thus, it should be noted that it is difficult to describe the entire trochlear shape using 2D images, especially for surgeons who intend to decide surgical indication for a trochleoplasty depending on the trochlear type.

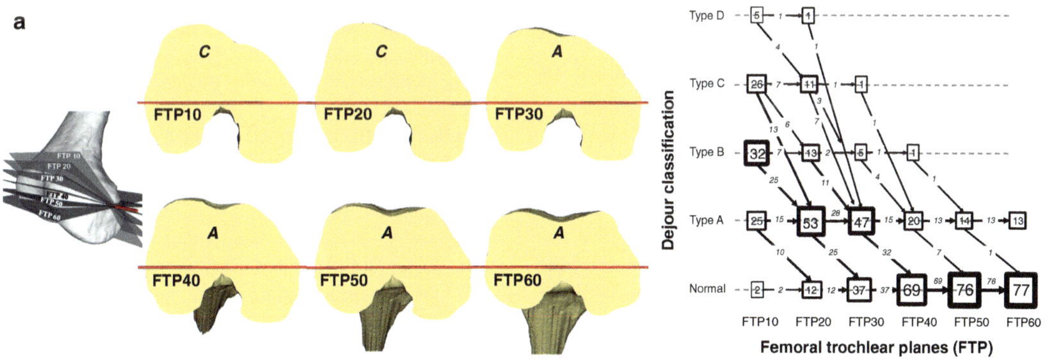

Fig. 21.10 (**a**) Change of trochlear type in patients with PI on FTPs according to Dejour classification. *FTP* femoral trochlear plane. (**b**) Change of trochlear type on FTPs according to Dejour classification in 90 knees of 81 patients with PI. The majority of patients with PI showed changes in their trochlear types on FTPs with various patterns [18]. The number of knees is shown in the squares or on the arrows

Table 21.2 Agreements of trochlear type on each FTP with that of clinically used Dejour classification according to the MRI axial plane

	Dejour classification	Total number	Number of knees showing agreement with the clinically used Dejour classification					
			FTP10	FTP20	FTP30	FTP40	FTP50	FTP60
PI	Normal	17	2 (12%)	10 (59%)	14 (82%)	16 (94%)	17 (100%)	17 (100%)
	A	25	9 (36%)	22 (88%)	13 (52%)	4 (16%)	3 (12%)	3 (12%)
	B	25	15 (60%)	4 (16%)	2 (8%)	0 (0%)	0 (0%)	0 (0%)
	C	20	11 (55%)	7 (35%)	1 (5%)	0 (0%)	0 (0%)	0 (0%)
	D	3	1 (33%)	1 (33%)	0 (0%)	0 (0%)	0 (0%)	0 (0%)
	Total	90	38 (42%)	44 (49%)	30 (33%)	20 (22%)	20 (22%)	20 (22%)
HV	Normal	15	3 (20%)	12 (80%)	15 (100%)	15 (100%)	15 (100%)	15 (100%)

PI patellar instability, *HV* healthy volunteers, *FTP* femoral trochlear plane
This table is cited from Yamada Y et al. JISAKOS 2019 [18]. The dysplastic trochlear types evaluated on the clinically used planes would not always match the ones on the virtual cross sections, closer to the perpendicular plane of the articular surface describing the contact surface to the patella

21.5 Summary

Anatomical characteristics in PI that could not be expressed by conventional indicators on 2D images have been revealed by analyses using 3D computer models as shown above. The new evaluation does not deny the conventional method but we believe it is time to introduce new indicators for 3D imaging to improve the understanding of PI. This approach is not applicable for daily practice now but we hope it can be a new standard for evaluation of PI as technology progresses.

References

1. Deutsch AL, Shellock FG, Mink JH. Imaging of the patellofemoral joint: emphasis on advanced techniques. In: Fox JM, Del Pizzo W, editors. The patellofemoral joint. New York: McGraw-Hill Inc; 1993. p. 75–103.
2. Halbrecht JL, Jackson DW. Acute dislocation of the patella. In: Fox JM, Del Pizzo W, editors. The patellofemoral joint. New York: McGraw-Hill Inc; 1993. p. 123–34.
3. Ellera-Gomes JL. Medial patellofemoral ligament reconstruction for recurrent dislocation of the patella: a preliminary report. Arthroscopy. 1992;8:335–40.
4. Nomura E, Inoue M, Kobayashi S. Long-term follow-up and knee osteoarthritis change after medial patellofemoral ligament reconstruction for recurrent patellar dislocation. Am J Sports Med. 2007 Nov;35(11):1851–8.
5. Toritsuka Y, Amano H, Mae T, Uchida R, Hamada M, Ohzono K, Shino K. Dual tunnel medial patellofemoral ligament reconstruction for patients with patellar dislocation using a semitendinosus tendon autograft. Knee. 2011;18(4):214–9.
6. Blønd L, Schöttle PB. The arthroscopic deepening trochleoplasty. Knee Surg Sports Traumatol Arthrosc. 2010;18(4):480–5.
7. Dejour D, Saggin P. The sulcus deepening trochleoplasty-the Lyon's procedure. Int Orthop. 2010;34:311–6.
8. Fulkerson JP, Becker GJ, Meaney JA, Miranda M, Folcik MA. Anteromedial tibial tubercle transfer without bone graft. Am J Sports Med. 1990;18(5):490–6; discussion 496–7.
9. Steiner T, Parker R. Subluxation and dislocation. In: Delee JC, Drez Jr D, Miller D, editors. DeLee & Drez's orthopaedic sports medicine. 3rd ed. Philadelphia: Saunders, an imprint of Elsevier Inc; 2010. p. 1548–72.
10. Ishii T, Mukai Y, Hosono N, Sakaura H, Fujii R, Nakajima Y, Tamura S, Iwasaki M, Yoshikawa H, Sugamoto K. Kinematics of the upper cervical spine in rotation: in vivo three-dimensional analysis. Spine. 2004;29(7):E139–44.
11. Moritomo H, Murase T, Goto A, Oka K, Sugamoto K, Yoshikawa H. Capitate-based kinematics of midcarpal joint during wrist radioulnar deviation: an in vivo

three-dimensional motion analysis. J Hand Surg Am. 2004;29(4):668–75.
12. Yamada Y, Toritsuka Y, Yoshikawa H, Sugamoto K, Horibe S, Shino K. Morphological analysis of the femoral trochlea in patients with recurrent dislocation of the patella using three-dimensional computer models. J Bone Joint Surg Br. 2007;89:746–51.
13. Yamada Y, Toritsuka Y, Horibe S, Sugamoto K, Yoshikawa H, Shino K. In vivo movement analysis of the patella using a three-dimensional computer model. J Bone Joint Surg Br. 2007;89:752–60.
14. Merchant AC, Mercer RL, Jacobsen RH, Cool CR. Roentgenographic analysis of patellofemoral congruence. J Bone Joint Surg Am. 1974;56-A:1391–6.
15. Brossmann J, Muhle C, Schroder C, et al. Patellar tracking patterns during active and passive knee extension: evaluation with motion triggered cine MR imaging. Radiology. 1993;187:205–12.
16. Yamada Y, Toritsuka Y, Nakamura N, Horibe S, Sugamoto K, Yoshikawa H, Shino K. Correlation of 3D shift and 3D tilt of the patella in patients with recurrent dislocation of the patella and healthy volunteers: an in vivo analysis based on 3-dimensional computer models. Am J Sports Med. 2017;45(13):3111–8.
17. Yamada Y, Toritsuka Y, Nakamura N, Horibe S, Sugamoto K, Yoshikawa H, Shino K. Patellar instability can be classified into four types based on patellar movement with knee flexion: a three-dimensional computer model analysis. JISAKOS. 2019;3(6):328–35.
18. Yamada Y, Toritsuka Y, Horibe S, Nakamura N, Sugamoto K, Yoshikawa H, Shino K. Classification of dysplasia of the femoral trochlea in patients with patellar instability depends on the evaluation plane. JISAKOS. 2019;4(6):290–5.
19. Heegaard J, Leyvraz PF, Van Kampen A, et al. Influence of soft structures on patellar three-dimensional tracking. Clin Orthop. 1994;299:235–43.
20. Hefzy MS, Jackson WT, Saddemi SR, Hsieh YF. Effects of tibial rotations on patellar tracking and patello-femoral contact areas. J Biomed Eng. 1992;14:329–43.
21. Patel VV, Hall K, Ries M, Lindsey C, Ozhinsky E, Lu Y, Majumdar S. Magnetic resonance imaging of patellofemoral kinematics with weight-bearing. J Bone Joint Surg Am. 2003;85-A:2419–24.
22. Dejour H, Walch G, Nove-Josserand L, Guier C. Factors of patellar instability: an anatomic radiographic study. Knee Surg Sports Traumatol Arthrosc. 1994;2(1):19–26.
23. Biedert RM, Albrecht S. The patellotrochlear index: a new index for assessing patellar height. Knee Surg Sports Traumatol Arthrosc. 2006;14(8):707–12.
24. Seitlinger G, Scheurecker G, Högler R, Labey L, Innocenti B, Hofmann S. Tibial tubercle-posterior cruciate ligament distance: a new measurement to define the position of the tibial tubercle in patients with patellar dislocation. Am J Sports Med. 2012;40(5):1119–25.
25. Toritsuka Y, Yamada Y, Nakamura N, Shino K. Lateral patellar dislocation: pathomechanism and treatment. In: Gobbi A, Espregueira-Mendes J, Nakamura N, editors. The patellofemoral joint-state of the art in evaluation and management. New York: Springer; 2014. p. 67–78.
26. Kita K, Tanaka Y, Toritsuka Y, Amano H, Uchida R, Takao R, Horibe S. Factors affecting the outcomes of double-bundle medial patellofemoral ligament reconstruction for recurrent patellar dislocations evaluated by multivariate analysis. Am J Sports Med. 2015;43(12):2988–96.
27. Lippacher S, Dejour D, Elsharkawi M, et al. Observer agreement on the Dejour trochlear dysplasia classification: a comparison of true lateral radiographs and axial magnetic resonance images. Am J Sports Med. 2012;40(4):837–43.
28. Tscholl PM, Wanivenhaus F, Fucentese SF. Conventional radiographs and magnetic resonance imaging for the analysis of trochlear dysplasia: the influence of selected levels on magnetic resonance imaging. Am J Sports Med. 2017;45(5):1059–65.

MPFL Reconstruction and Patellofemoral Chondral Status

22

Keisuke Kita, Shuji Horibe, Norimasa Nakamura, and Konsei Shino

22.1 Medial Patellofemoral Ligament Reconstruction

22.1.1 History

The MPFL was first described by Kaplan in 1957 as the transverse reinforcement extending from the base of the patella to the tendon of the medial head of the gastrocnemius [1]. In 1979, Warren and Marshall named it as the patellofemoral ligament [2]. In the 1990s, several authors revealed the importance of the MPFL as a primary restraint for the lateral deviation of the patella, which provides 50–60% of resistance to lateral displacement [3–5]. Nomura reported that MPFL injury was observed in 96% of the patients with acute patellar dislocation [6]. Thereafter, numerous surgical techniques to restore the function of the MPFL have been reported with excellent outcomes in terms of the prevention of re-dislocation.

K. Kita
Department of Sports Medicine, JCHO Osaka Hospital, Osaka, Japan
e-mail: keikita@hera.eonet.ne.jp

S. Horibe
Faculty of Comprehensive Rehabilitation, Osaka Prefecture University, Habikino, Osaka, Japan
e-mail: s-horibe@rehab.osakafu-u.ac.jp

N. Nakamura (✉)
Department of Orthopaedic Surgery, Osaka University Graduate School of Medicine, Suita, Osaka, Japan

Institute for Medical Science in Sports, Osaka Health Science University, Osaka, Japan

Global Center for Medical Engineering and Informatics, Osaka University, Suita, Osaka, Japan

Department of Rehabilitation Science, Osaka Health Science University, Osaka City, Osaka, Japan
e-mail: norimasa.nakamura@ohsu.ac.jp

K. Shino
Center for Sports Orthopaedics, Yukioka Hospital, Osaka, Japan

22.1.2 Anatomy

Warren and Marshall [2] described the MPFL as a collection of transverse fibers running across the plane of Layer II from the region of attachment of the medial collateral ligament to the patella. In 2002, Tuxøe [7] reported the MPFL as being 1.9 cm (1.0–3.0) in width and 5.3 cm (4.5–6.4) in length. Nomura [8] investigated the anatomy of the MPFL in detail and described that the total length of the MPFL was 58.8 ± 4.7 mm, the width and thickness being 12.0 ± 3.1 mm and 0.44 ± 0.19 mm in the middle. The center of the patellar attachment was located at 27 ± 10% from the upper end of the patella in the longitudinal patellar height, and the femoral attachment was superoposterior to the medial femoral epicondyle and just distal to the adductor tubercle. Thereafter, several researchers reported similar results [9–12]. Recently, Mochizuki [13] reported that the proximal fibers of the MPFL were mainly attached to the vastus

N. Nakamura et al. (eds.), *Advances in Knee Ligament and Knee Preservation Surgery*,
https://doi.org/10.1007/978-3-030-84748-7_22

intermedius tendon, without tight adhesion to the vastus medialis. Fulkerson [14] named it as the medial quadriceps tendon femoral ligament (MQTFL). Thereafter, the existence of the MQTFL has been supported by several authors [15, 16].

22.1.3 Repair or Reconstruction

Several attempts made to restore the function of impaired MPFL have been reported. Repair or reconstruction of the MPFL is a fundamental issue. Christiansen et al. [17] compared the effect of delayed surgical reinsertion of the MPFL with conservative treatment in patients with primary dislocation of the patella. They concluded that delayed primary repair did not reduce the risk of re-dislocation; nor did it produce any significantly better subjective functional outcome. Several authors reported similar results in chronic cases [18, 19]. Sillanpää et al. [20] reported that initial arthroscopic medial retinacular repair was neither followed by improved patellar stability nor reduced incidence of re-dislocations compared to nonoperative treatments. On the other hand, Schöttle et al. [21] reported that arthroscopic repair of the medial retinaculum was an effective technique in patients without trochlear dysplasia. Askenberger [22] described that operative repair of an MPFL injury in the acute phase in skeletally immature children with primary traumatic lateral patellar dislocation significantly reduced the rate of re-dislocation. A similar result was reported by Bryant [23] wherein MPFL repair in pediatric patients resulted in a low risk of recurrent instability, comparable to or better than that in allograft reconstruction. With regard to repair vs reconstruction, Puzzitiello [24] retrospectively compared the two and concluded that MPFL reconstruction may provide improved midterm clinical outcomes and a decreased recurrence rate compared with MPFL repair. More recently, a meta-analysis which compared repair with reconstruction was conducted by Previtali et al. [25]. This analysis concluded that MPFL reconstruction and medial patellofemoral soft-tissue surgical procedures were effective in restoring the medial restraining forces that prevent re-dislocation. But, MPFL reconstruction provided better functional outcomes, both at short-term and long-term follow-ups. Most patients with lateral patellar dislocation have underlying predisposing factors. To overcome these factors, the patella should be reinforced by a substance stronger than the native MPFL. Therefore, indications for MPFL repair should be limited to pediatric patients or patients without predisposing risk factors.

22.2 Chondral Damage After Patellar Dislocation

22.2.1 Osteochondral Fracture

Osteochondral fractures in association with acute patellar dislocation were first reported by Kroner in 1905 [27]. Since then, many authors have described similar cases [28, 29]. In most cases, the fracture is located in the inferior part of the medial facet of the patella; however, sometimes, the lateral edge of the femoral condyle is also involved. The presence of loose body formation lowers mid-term functional patient outcomes [30]. Several surgical techniques have been proposed for the reattachment of osteochondral fragments using metal screws, bioabsorbable pins, or sutures [31–33]. However, a controversy on the superiority of reattachment over only debridement still exists. Lee et al. [34] compared the surgical outcome of reattachment with that of loose body excision with microfracture. They concluded that while the excision group had smaller lesions, patients without fixation were less symptomatic. Gesslein et al. [35] retrospectively compared the results of refixation of large osteochondral fractures with those of debridement and concluded that refixation showed better clinical outcomes at mid- to long-term follow-up. While the medial reefing was performed for most of their patients and the rate of re-dislocation in the refixation group was significantly less than that of the debridement group, re-dislocation still occurred in 35.7% of the patients. There is another controversy regarding whether the function of the MPFL should be restored at the time of refixation of an osteochondral fragment. However, almost no data exists on this topic.

According to a survey by the International Patellofemoral Study Group [36], if an operative osteochondral fracture was identified, 89% of experts would choose to surgically address patellar instability at the same time as treating the osteochondral fracture with repair or excision. Sixty percent of experts would use MPFL or MQTFL reconstructions. However, 63% of the surgeons were more likely to perform MPFL repair if the ligament was avulsed off the patella.

22.2.2 Articular Cartilage Damage

A clear history of patellar dislocation or subluxation was noted in 28% of isolated PFOA diagnosed by X-ray that was reported by Iwano in 1990 [37]. With advancements in arthroscopy and MRI, intraarticular cartilage damage after patellar dislocation that could previously not be detected well by plain X-ray, has been mentioned in detail [38]. In 2003, Nomura [39] evaluated the frequency and precise pathology of articular cartilage injuries after acute patellar dislocation. He found 95% of the patients had articular cartilage injuries in the PF joint. All patients with cartilage injuries showed cartilage lesions on their patellar surfaces. Thirty-one percent of the patients had cartilage injury of the lateral femoral condyle. He also categorized the appearance of the cartilage lesions into three groups: cracks alone, cartilage defects caused by osteochondral or chondral fractures associated with cracks, and cartilage defects caused by osteochondral or chondral fractures. He also reported cartilage lesions in recurrent patellar dislocations as 96% of the patients had cartilage lesions of the patella. Fissuring was observed in 76% of the patients, commonly on the central dome, and erosive lesions were assumed as the extensions of the original acute injury [40]. In 2005, he suggested that the continuation of the patellar dislocation definitely made the patellar cartilage lesions worse [41]. Vollnberg [42] investigated MR images of 129 knees with patellar dislocation and concluded that cartilage defect size and prevalence of OA were correlated with the number of dislocations. Biomechanically, Stephen reported that peak lateral PF contact pressures significantly elevated the following MPFL transections [43]. Recently, Salonen [44] described that even single first-time or infrequently recurring traumatic lateral patellar dislocations seemed to be associated with gradual cartilage deterioration. Sanders [45] compared the risk of patellofemoral arthritis between patients who experienced a lateral patellar dislocation and matched individuals without a patellar dislocation and concluded that nearly half of the patients had symptoms and radiographic changes consistent with arthritis at 25 years after lateral patellar dislocation.

22.3 Conventional Surgical Procedure and Prevention of PFOA

Several surgical procedures for stabilizing the patella have been developed in the nineteenth century, including proximal and distal realignment. Transposition of the patellar ligament was first performed by Roux [46]. Since then, many authors have reported various techniques for distal realignment [47–50].

In 1976, Crosby reported that the incidence of late osteoarthritis was disturbingly high after tibial tubercle transfer (TTT) [51]. Juliusson [52] followed the patients who had undergone the modified Hauser procedure for 18 years, and revealed that more than two-thirds of the patients had some degree of OA. Similarly, PF cartilage deterioration after the Elmslie-Trillat procedure has been reported by several authors [53, 54]. On the other hand, more recently, Tscholl [55] reported that MPFL reconstruction with and without TTT is a reliable treatment option for recurrent patellar dislocation without deterioration of the PF joint at midterm follow-up when the indication for TTT was limited to patients with tibial tubercle-trochlear groove (TT-TG) distance >15 mm. Haj-Mirzaian et al. [56] revealed that TT-TG distance was associated with simultaneous lateral PF OA-related structural damage. Biomechanically, Kuroda [57] conducted a cadaveric study with regard to PF contact-pressure change after TTT and concluded that overmedialization of the tibial tuberosity tends to increase PF contact pressure, especially in patients with a normal Q angle. In addition, Stephen et al. [58] biomechanically investigated the ability of MPFL reconstruction to correct patellar kinematics and contact mechanics in the presence of a lateralized tibial tubercle and concluded that in patients with TT-TG greater than 15 mm, patellofemoral kinematics and contact mechanics could not be restored with isolated MPFL reconstruction. There is a possibility that past reports on PFOA after TTT could not accurately evaluate the lateral deviation of the tibial tubercle as they would have included patients with normal values due to the relatively ambiguous evaluation methods used (e.g., the Q angle). Recently, several authors reported good results in MPFL reconstruction when coupled with the TTT for patients with increased TT-TG distance. Neri [59] reported good clinical and radiological results in 133 MPFL reconstructions with or without TTT. Franciozi [60] reported that MPFL reconstruction with TTT yielded better clinical outcome than isolated MPFL reconstruction in patients with increased TT-TG distance. With regard to radiographic OA progression, Tscholl [55] reported no significant differences in postoperative clinical scores and osteoarthritis progression between patients with MPFL reconstruction with or without TTT at a mean follow-up time of 5.4 years. Therefore, when the indication for TTT was limited to increased TT-TG distance or patella alta, PFOA after TTT could be avoided. However, exact threshold values indicating TTT have never been addressed so far.

Advancement of the vastus medialis has been known as Insall's proximal realignment [61]. Zeichen [62] reported the midterm results after Insall's proximal realignment and PFOA was seen in 36.8% of the patients. Schüttler [63] described that plain radiographs showed a significant increase in PFOA in 43% of the patients at a mean follow-up period of 52 months. So far, there has been no promising surgical procedure preventing both further dislocation and future osteoarthritis.

22.4 MPFL Reconstruction and PF Chondral Status

In 1992, Ellera Gomes [64] first reported the results of MPFL reconstruction using artificial ligaments and 83.3% of the patients showed significant improvement in their initial knee-related complaints at an average follow-up of 39 months. Thereafter, numerous techniques to reconstruct the MPFL have been reported and they continue to gain popularity [65]. However, few reports exist on the long-term outcomes after MPFL reconstruction, especially with regard to the progression of PFOA. Long-term knee osteoarthritic changes after MPFL reconstruction were first reported by Nomura [66] in 2007. According to the paper, only 2 of the 22 knees showed definite progression from none to mild or moderate grade PFOA on plain X-ray at a mean follow-up of 11.9 years. Sillanpää [67] reported that no PFOA was seen on radiography after MPFL reconstruction at a median 10-year follow-up. Considering that previous published data revealed that PFOA after patellar stabilizing procedure was seen in 17–70% of the patients, MPFL reconstruction seems to reduce the risk of osteoarthrosis compared with the aforementioned procedures. However, factors that could overload the patellofemoral joint, including the usage of short grafts [68], over-tensioning the graft, and anteroproximal graft placement [69] must be taken care of.

Arthroscopic evaluation of the cartilaginous surface of the PF joint after MPFL reconstruction revealed that patellofemoral chondral status after isolated anatomical MPFL reconstruction was not altered at second-look arthroscopy in most parts of the patellofemoral joint [26]. Moreover, at the central ridge of the patella, significant improvement of the ICRS grading was observed and, in the lateral femoral condyle, recovery of chondral damage was seen in some cases (Fig. 22.1). This means that elevated lateral PF contact pressures after patellar dislocation [43] might decrease by correcting the alignment of the PF joint. These results suggest that MPFL reconstruction could change the nat-

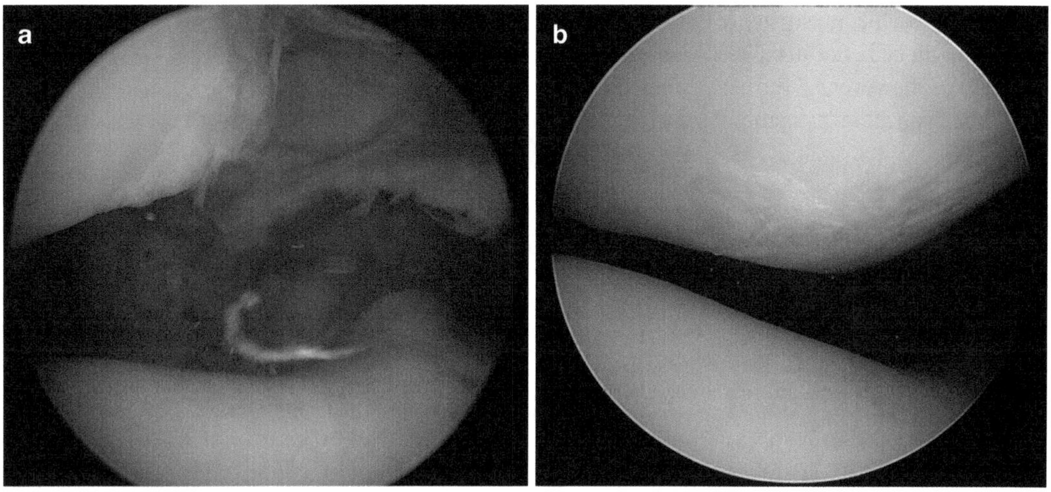

Fig. 22.1 Arthroscopic view of the chondral surface of the central ridge through the lateral suprapatellar portal at MPFL reconstruction (**a**) and at second-look arthroscopy (**b**). A gross articular cartilage fibrillation on the central ridge was seen, and the lesion was found to be deep to the subchondral bone on probing (**a**). A smooth fibrocartilaginous tissue covered the surface of the central ridge 6 months after MPFL reconstruction (**b**). (From [26], with permission)

ural course of PFOA for patients with recurrent patellar subluxation. On the other hand, half of the patients with improvement of chondral lesions in the central ridge had low-grade dysplasia, while half of the patients with deterioration of cartilage lesions in the femoral groove had high-grade trochlear dysplasia. It means that isolated MPFL reconstruction might increase PF contact pressure in high-grade trochlear dysplastic knees. However, there has been no biomechanical study addressing the effect of trochlear dysplasia on patellofemoral contact pressure after MPFL reconstruction.

Trochleoplasty is the only surgical procedure for correcting a dysplastic trochlea. Bereiter [70] first described the sulcus deepening trochleoplasty in 1994. He also reported good long-term clinical results of the procedure, but degenerative changes of the PF joint developed in 30% of the knees at a mean follow-up of 8.3 years [71]. Rouanet [72] reported similar results on sulcus deepening trochleoplasty, correcting PF stability even in patients with severe dysplasia, and the long-term functional outcome was better in this group; but it did not prevent PFOA. On the other hand, Ntagiopoulos [73] reported no radiological patellofemoral arthritis at a mean follow-up of 7 years after Dejour's sulcus deepening trochleoplasty. It is not clear whether trochleoplasty could prevent PFOA. Several short to midterm clinical results after MPFL reconstruction combined with trochleoplasty have been reported [74, 75]. However, there have been no data on whether trochleoplasty should be added to MPFL reconstruction.

22.5 Summary

- The prevalence of patellofemoral cartilage lesions is increased in conditions of patellar instability.
- Although there is little data on the long-term outcomes of MPFL reconstruction, it seems to be a more reliable surgery than other procedures, not only with regard to the prevention of further dislocation but also in the progression of PFOA, especially in patients with low-grade trochlear dysplasia.
- Coupling tibial tubercle osteotomy with MPFL reconstruction seems to be better for patients with increased TT-TG distance. However, its indications have not been described clearly.

- Adding trochleoplasty to MPFL reconstruction might be better in case of high-grade dysplasia; however, further clinical and biomechanical investigations are needed.

References

1. Kaplan EB. Factors responsible for the stability of the knee joint. Bull Hosp Joint Dis. 1957;18(1):51–9.
2. Warren LF, Marshall JL. The supporting structures and layers on the medial side of the knee: an anatomical analysis. J Bone Joint Surg Am. 1979;61:56–62.
3. Conlan T, Garth WP, Lemons JE. Evaluation of the medial soft-tissue restraints of the extensor mechanism of the knee. J Bone Joint Surg Am. 1993;75(5):682–93.
4. Desio SM, Burks RT, Buchus KN. Soft tissue restraints to lateral patellar translation in the human knee. Am J Sports Med. 1998;26(1):59–65.
5. Hautamaa PV, Fithian DC, Kaufman KR, Daniel DM, Pohlmeyer AM. Medial soft tissue restraints in lateral patellar instability and repair. Clin Orthop Relat Res. 1998;349:174–82.
6. Nomura E, Horiuchi Y, Inoue M. Correlation of MR imaging findings and open exploration of medial patellofemoral ligament injuries in acute patellar dislocations. Knee. 2002;2:139–43.
7. Tuxøe JI, Teir M, Winge S, Nielsen PL. The medial patellofemoral ligament: a dissection study. Knee Surg Sports Traumatol Arthrosc. 2002;10(3):138–40.
8. Nomura E, Inoue M, Osada N. Anatomical analysis of the medial patellofemoral ligament of the knee, especially the femoral attachment. Knee Surg Sports Traumatol Arthrosc. 2005;13(7):510–5.
9. Baldwin JL. The anatomy of the medial patellofemoral ligament. Am J Sports Med. 2009;37(12):2355–61.
10. Dirim B, Haghighi P, Trudell D, Portes G, Resnick D. Medial patellofemoral ligament: cadaveric investigation of anatomy with MRI, MR arthrography, and histologic correlation. AJR Am J Roentgenol. 2008;191(2):490–8.
11. LaPrade RF, Engebretsen AH, Ly VT, Johansen S, Wentorf FA, Engebretsen L. The anatomy of the medial part of the knee. J Bone Joint Surg Am. 2007;89(9):2000–10.
12. Philippot R, Chouteau J, Wegrzyn J, Testa R, Fessy MH, Moyen B. Medial patellofemoral ligament anatomy: implications for its surgical reconstruction. Knee Surg Sports Traumatol Arthrosc. 2009;17(5):475–9.
13. Mochizuki T, Nimura A, Tateishi T, Yamaguchi K, Muneta T, Akita K. Anatomic study of the attachment of the medial patellofemoral ligament and its characteristic relationships to the vastus intermedius. Knee Surg Sports Traumatol Arthrosc. 2013;21(2):305–10.
14. Fulkerson JP, Edgar C. Medial quadriceps tendon-femoral ligament: surgical anatomy and reconstruction technique to prevent patella instability. Arthrosc Tech. 2013;2(2):e125–8.
15. Placella G, Tei MM, Sebastiani E, Criscenti G, Speziali A, Mazzola C, Georgoulis A, Cerulli G. Shape and size of the medial patellofemoral ligament for the best surgical reconstruction: a human cadaveric study. Knee Surg Sports Traumatol Arthrosc. 2014;22(10):2327–33.
16. Tanaka MJ, Voss A, Fulkerson JP. The anatomic midpoint of the attachment of the medial patellofemoral complex. J Bone Joint Surg Am. 2016;98(14):1199–205.
17. Christiansen SE, Jakobsen BW, Lund B, Lind M. Isolated repair of the medial patellofemoral ligament in primary dislocation of the patella: a prospective randomized study. Arthroscopy. 2008;24(8):881–7.
18. Arendt EA, Moeller A, Agel J. Clinical outcomes of medial patellofemoral ligament repair in recurrent (chronic) lateral patella dislocations. Knee Surg Sports Traumatol Arthrosc. 2011;19(11):1909–14.
19. Camp CL, Krych AJ, Dahm DL, Levy BA, Stuart MJ. Medial patellofemoral ligament repair for recurrent patellar dislocation. Am J Sports Med. 2010;38(11):2248–54.
20. Sillanpää PJ, Mäenpää HM, Mattila VM, Visuri TT, Pihlajamäki H. Arthroscopic surgery for primary traumatic patellar dislocation: a prospective, nonrandomized study comparing patients treated with and without acute arthroscopic stabilization with a median 7-year follow-up. Am J Sports Med. 2008;36(12):2301–9.
21. Schöttle PB, Scheffler SU, Schwarck A, Weiler A. Arthroscopic medial retinacular repair after patellar dislocation with and without underlying trochlear dysplasia: a preliminary report. Arthroscopy. 2006;22(11):1192–8.
22. Askenberger M, Bengtsson Moström E, Ekström W, Arendt EA, Hellsten A, Mikkelsen C, Janarv PM. Operative repair of medial patellofemoral ligament injury versus knee brace in children with an acute first-time traumatic patellar dislocation: a randomized controlled trial. Am J Sports Med. 2018;46(10):2328–40.
23. Bryant J, Pandya N. Medial patellofemoral ligament repair restores stability in pediatric patients when compared to reconstruction. Knee. 2018;25(4):602–8.
24. Puzzitiello RN, Waterman B, Agarwalla A, Zuke W, Cole BJ, Verma NN, Yanke AB, Forsythe B. Primary medial patellofemoral ligament repair versus reconstruction: rates and risk factors for instability recurrence in a young, active patient population. Arthroscopy. 2019;35(10):2909–15.
25. Previtali D, Milev SR, Pagliazzi G, Filardo G, Zaffagnini S, Candrian C. Recurrent patellar dislocations without untreated predisposing factors: medial patellofemoral ligament reconstruction versus other medial soft-tissue surgical techniques-a meta-analysis. Arthroscopy. 2020;36(6):1725–34.
26. Kita K, Tanaka Y, Toritsuka Y, Yonetani Y, Kanamoto T, Amano H, Nakamura N, Horibe S. Patellofemoral

chondral status after medial patellofemoral ligament reconstruction using second-look arthroscopy in patients with recurrent patellar dislocation. J Orthop Sci. 2014;19(6):925–32.
27. Kroner M. Ein fall von flachenfraktur und luxation der patella. Deutache Med Wochenschr. 1905;31:996–7.
28. Krida A. Osteochondral fractures of the knee joint. Gynec Obstet. 1924;39:791–5.
29. Stewart SF. Frontal fractures of the patella. Ann Surg. 1925;81:536–9.
30. Nietosvaara Y, Aalto K, Kallio PE. Acute patellar dislocation in children: incidence and associated osteochondral fractures. J Pediatr Orthop. 1994;14(4):513–5.
31. Dhawan A, Hospodar PP. Suture fixation as a treatment for acute traumatic osteochondral lesions. Arthroscopy. 1999;15(3):307–11.
32. Gkiokas A, Morassi LG, Kohl S, Zampakides C, Megremis P, Evangelopoulos DS. Bioabsorbable pins for treatment of osteochondral fractures of the knee after acute patella dislocation in children and young adolescents. Adv Orthop. 2012; https://doi.org/10.1155/2012/249687.
33. Hammerle CP, Jacob RP. Chondral and osteochondral fractures after luxation of the patella and their treatment. Arch Orthop Trauma Surg. 1980;97(3):207–11.
34. Lee BJ, Christino MA, Daniels AH, Hulstyn MJ, Eberson CP. Adolescent patellar osteochondral fracture following patellar dislocation. Knee Surg Sports Traumatol Arthrosc. 2012;21(8):1856–61.
35. Gesslein M, Merkl C, Bail HJ, Krutsch V, Biber R, Schuster P. Refixation of large osteochondral fractures after patella dislocation shows better mid- to long-term outcome compared with debridement. Cartilage. 2019; https://doi.org/10.1177/1947603519886637.
36. Liu JN, Steinhaus ME, Kalbian IL, Post WR, Greeen DW, Strickland SM, Shubin Stein BE. Patellar instability management: a survey of the International Patellofemoral Study Group. Am J Sports Med. 2018;46(13):3299–306.
37. Iwano Y, Kurosawa H, Tokuyama H, Hoshikawa Y. Roentgenographic and clinical findings of patellofemoral osteoarthrosis. With special reference to its relationship to femorotibial osteoarthrosis and etiologic factors. Clin Orthop Relat Res. 1990;252:190–7.
38. Stanitski CL, Paletta GA Jr. Articular cartilage injury with acute patellar dislocation in adolescents. Arthroscopic and radiographic correlation. Am J Sports Med. 1998;26(1):52–5.
39. Nomura E, Inoue M, Kurimura M. Chondral and osteochondral injuries associated with acute patellar dislocation. Arthroscopy. 2003;19(7):717–21.
40. Nomura E, Inoue M. Cartilage lesions of the patella in recurrent patellar dislocation. Am J Sports Med. 2004;32(2):498–502.
41. Nomura E, Inoue M. Second-look arthroscopy of cartilage changes of the patellofemoral joint, especially the patella, following acute and recurrent patellar dislocation. Osteoarthr Cartil. 2005;13(11):1029–36.
42. Vollnberg B, Koehlitz T, Jung T, Scheffler S, Hoburg A, Khandker D, Hamm B, Wiener E, Diederichs G. Prevalence of cartilage lesions and early osteoarthritis in patients with patellar dislocation. Eur Radiol. 2012;22(11):2347–56.
43. Stephen JM, Kader D, Lumpaopong P, Deehan DJ, Amis AA. Sectioning the medial patellofemoral ligament alters patellofemoral joint kinematics and contact mechanics. J Orthop Res. 2013;31(9):1423–9.
44. Salonen EE, Magga T, Sillanpää PJ, Kiekara T, Mäenpää H, Mattila VM. Traumatic patellar dislocation and cartilage injury: a follow-up study of long-term cartilage deterioration. Am J Sports Med. 2017;45(6):1376–82.
45. Sanders TL, Pareek A, Johnson NR, Stuart MJ, Dahm DL, Krych AJ. Patellofemoral arthritis after lateral patellar dislocation: a matched population-based analysis. Am J Sports Med. 2017;45(5):1012–7.
46. Roux D. Luxation Habituelle de la Rotule. Rev Chir Paris. 1888;8:682–9.
47. Goldthwait JE. Slipping or recurrent dislocation of the patella: with the report of eleven cases. Boston Med Surg J. 1904;150:169–74.
48. Goutallier D, Debeyre J. Le recentrage rotulien dans les arthroses femoro patellaires lateralisees. Rev Chir Orthop. 1974;60:377–86.
49. Hauser EW. Total tendon transplant for slipping patella. Surg Gynecol Obstet. 1938;66:199.
50. Trillat A, Dejour H, Coutette A. Diagnostic et traitement des subluxations recidivantes de la rotule. Rev Chir Orthop. 1964;50:813–24.
51. Crosby EB, Insall J. Recurrent dislocation of the patella. Relation of treatment to osteoarthritis. J Bone Joint Surg Am. 1976;58(1):9–13.
52. Juliusson R, Markhede G. A modified Hauser procedure for recurrent dislocation of the patella. A long-term follow-up study with special reference to osteoarthritis. Arch Orthop Trauma Surg. 1984;103(1):42–6.
53. Farr S, Huyer D, Sadoghi P, Kaipel M, Grill F, Ganger R. Prevalence of osteoarthritis and clinical results after the Elmslie-Trillat procedure: a retrospective long-term follow-up. Int Orthop. 2014;38(1):61–6.
54. Nakagawa K, Wada Y, Minamide M, Tsuchiya A, Moriya H. Deterioration of long-term clinical results after the Elmslie-Trillat procedure for dislocation of the patella. J Bone Joint Surg Br. 2002;84(6):861–4.
55. Tscholl PM, Wanivenhaus F, Centmaier-Molnar V, Camenzind RS, Fucentese SF. Clinical and radiological results after one hundred fifteen MPFL reconstructions with or without tibial tubercle transfer in patients with recurrent patellar dislocation-a mean follow-up of 5.4 years. Int Orthop. 2020;44(2):301–8.
56. Haj-Mirzaian A, Guermazi A, Pishgar F, Roemer FW, Sereni C, Hakky M, Zikria B, Demehri S. Patellofemoral morphology measurements and their associations with tibiofemoral osteoarthritis-related structural damage: exploratory analysis on the osteoarthritis initiative. Eur Radiol. 2020;30(1):128–40.

57. Kuroda R, Kambic H, Valdevit A, Andrish J. Distribution of patellofemoral joint pressures after femoral trochlear osteotomy. Knee Surg Sports Traumatol Arthrosc. 2002;10(1):33–7.
58. Stephen JM, Dodds AL, Lumpaopong P, Kader D, Williams A, Amis AA. The ability of medial patellofemoral ligament reconstruction to correct patellar kinematics and contact mechanics in the presence of a lateralized tibial tubercle. Am J Sports Med. 2015;43(9):2198–207.
59. Neri T, Parker DA, Beach A, Gensac C, Boyer B, Farizon F, Philippot R. Medial patellofemoral ligament reconstruction with or without tibial tubercle transfer is an effective treatment for patellofemoral instability. Knee Surg Sports Traumatol Arthrosc. 2019;27(3):805–13.
60. Franciozi CE, Ambra LF, Albertoni LJB, Debieux P, Granata GSM Jr, Kubota MS, Carneiro M, Abdalla RJ, Luzo MVM, Cohen M. Anteromedial tibial tubercle osteotomy improves results of medial patellofemoral ligament reconstruction for recurrent patellar instability in patients with tibial tuberosity-trochlear groove distance of 17 to 20 mm. Arthroscopy. 2019;35(2):566–74.
61. Insall J, Falvo KA, Wise DW. Chondromalacia patellae. A prospective study. J Bone Joint Surg Am. 1976;58:1–8.
62. Zeichen J, Lebenhoffer P, Gerich T, Tsheme H, Bosch U. Medium-term results of the operative treatment of recurrent patellar dislocation by Insall proximal realignment. Knee Surg Sports Traumatol Arthrosc. 1999;7(3):173–6.
63. Schüttler KF, Struewer J, Roessler PP, Gesslein M, Rominger MB, Ziring E, Efe T. Patellofemoral osteoarthritis after Insall's proximal realignment for recurrent patellar dislocation. Knee Surg Sports Traumatol Arthrosc. 2014;22(11):2623–8.
64. Ellera Gomes JL. Medial patellofemoral ligament reconstruction for recurrent dislocation of the patella: a preliminary report. Arthroscopy. 1992;8(3):335–40.
65. Stupay KL, Swart E, Shubin Stein BE. Widespread implementation of medial patellofemoral ligament reconstruction for recurrent patellar instability maintains functional outcomes at midterm to long-term follow-up while decreasing complication rates: a systematic review. Arthroscopy. 2015;31(7):1372–80.
66. Nomura E, Inoue M, Kobayashi S. Long-term followup and knee osteoarthritis change after medial patellofemoral ligament reconstruction for recurrent patellar dislocation. Am J Sports Med. 2007;35(11):1851–8.
67. Sillanpää P, Mattila VM, Visuri T, Mäenpää HH, Pihlajamäki H. Ligament reconstruction versus distal realignment for patellar dislocation. Clin Orthop Relat Res. 2008;466(6):1475–84.
68. Elias JJ, Cosgarea AJ. Technical errors during medial patellofemoral ligament reconstruction could overload medial patellofemoral cartilage: a computational analysis. Am J Sports Med. 2006;34(9):1478–85.
69. Stephen JM, Kaider D, Lumpaopong P, Deehan DJ, Amis AA. The effect of femoral tunnel position and graft tension on patellar contact mechanics and kinematics after medial patellofemoral ligament reconstruction. Am J Sports Med. 2014;42(2):364–72.
70. Bereiter H, Gautier E. Die trochleaplastik als chirurgische Therapie der rezidivierenden Patellaluxation bei Trochleadysplasie des Femurs. Arthroskopie. 1994;7:281–6.
71. von Knoch F, Böhm T, Bürgi ML, von Knoch M, Bereiter H. Trochleaplasty for recurrent patellar dislocation in association with trochlear dysplasia. A 4- to 14-year follow-up study. J Bone Joint Surg Br. 2006;88(10):1331–5.
72. Rouanet T, Gougeon F, Fayard JM, Rémy F, Migaud H, Pasquier G. Sulcus deepening trochleoplasty for patellofemoral instability: a series of 34 cases after 15 years postoperative follow-up. Orthop Traumatol Surg Res. 2015;101(4):443–7.
73. Ntagiopoulos PG, Byn P, Dejour D. Midterm results of comprehensive surgical reconstruction including sulcus-deepening trochleoplasty in recurrent patellar dislocations with high-grade trochlear dysplasia. Am J Sports Med. 2013;41(5):998–1004.
74. Banke IJ, Kohn LM, Meidinger G, Otto A, Hensler D, Beitzel K, Imhoff AB, Schöttle PB. Combined trochleoplasty and MPFL reconstruction for treatment of chronic patellofemoral instability: a prospective minimum 2-year follow-up study. Knee Surg Sports Traumatol Arthrosc. 2014;22(11):2591–8.
75. Nelitz M, Dreyhaupt J, Lippacher S. Combined trochleoplasty and medial patellofemoral ligament reconstruction for recurrent patellar dislocations in severe trochlear dysplasia: a minimum 2-year follow-up study. Am J Sports Med. 2013;41(5):1005–12.

Osteotomy: Coronal and Axial Plane Deformity

23

Humza Shaikh, Rajiv Reddy, Christopher M. Gibbs, Ryan Murray, and Volker Musahl

23.1 Introduction

The knee is a complex joint that is subject to a high degree of mechanical stress throughout weight bearing and range of motion. The alignment of the lower extremity has a substantial impact on the biomechanical function of the knee joint. Deviations from normal alignment of the lower extremity subject intra- and extra-articular structures to abnormal forces, predisposing the knee joint to a variety of pathologies [1–4]. Furthermore, intra-articular pathologies such as cartilage injuries or meniscal deficiency can be further affected by limb alignment deviations. Fortunately, malalignment of the lower extremity can be addressed through surgical procedures aimed at correcting alignment – either in isolation or in the setting of concomitant intra-articular procedures. In general, varus deformity of the lower extremity results in overload of the medial compartment and can be corrected with osteotomies about the proximal tibia such as medial opening wedge or lateral closing wedge procedures. Tibial osteotomies are also useful to correct sagittal plane malalignment, such as excess posterior tibial slope. In contrast, valgus limb alignment results in lateral compartment overload and can be corrected with procedures about the distal femur, including medial closing wedge and lateral opening wedge osteotomies. These procedures are commonly performed in conjunction with articular cartilage procedures or meniscal transplantation, to treat symptomatic lateral compartment osteoarthritis in the setting of lower extremity deformity. Furthermore, rotational malalignment of the lower extremity can have a significant effect on not only the knee, but also the hip and ankle as well. Rotational abnormalities most commonly occur with derangement of the patellofemoral joint, particularly patellofemoral instability. Fortunately, derotational osteotomies to correct femoral anteversion and external tibial torsion are available to treat these abnormalities. Correcting the underlying deformity can treat symptomatic pathology about the knee, such as patellofemoral instability. This chapter will provide an overview of the assessment of limb alignment as well as the indications and techniques to correct malalignment about the knee in the treatment of lower extremity deformity.

H. Shaikh · R. Reddy · C. M. Gibbs · R. Murray
Department of Orthopaedic Surgery, University of Pittsburgh Medical Center, Pittsburgh, PA, USA
e-mail: shaikhh@upmc.edu; rpr17@pitt.edu

V. Musahl (✉)
Department of Orthopaedic Surgery, University of Pittsburgh Medical Center, Pittsburgh, PA, USA

UPMC Center for Sports Medicine, Pittsburgh, PA, USA
e-mail: musahlv@upmc.edu

N. Nakamura et al. (eds.), *Advances in Knee Ligament and Knee Preservation Surgery*,
https://doi.org/10.1007/978-3-030-84748-7_23

23.2 Limb Alignment and Preoperative Planning

The knee joint is the largest and among the most complex joints in the human body. Due to its location and weight-bearing function, the knee joint encounters substantial axial and mechanical load. Thus, malalignment of the knee joint can cause significant harm over time. Proper understanding of lower limb anatomic and mechanical axes, and knee joint angles, is crucial for surgical decision-making [5, 6].

The anatomic axes of the femur and tibia refer to a line drawn along the diaphysis of these long bones, while the mechanical axis of the lower extremity is defined as a line connecting the center of the femoral head to the center of the tibial plafond [7], Fig. 23.1.

Due to the greater distance between the femoral heads compared with the centers of the knee and ankle joints, the mechanical axis of the lower extremity runs craniolateral to caudomedial. In the coronal plane, the distance between the mechanical axis and center of the knee joint is termed the mechanical axis deviation (MAD), and is commonly used to define coronal alignment [8], Fig. 23.2. More commonly used is the mechanical femorotibial angle (mFTA), the angle between the mechanical axes of the femur and tibia [9], Fig. 23.1.

Lines tangent to the distal femoral condyles and tibial plateau are used to determine joint orientation angles and the joint line convergence angle (JLCA). The medial proximal tibial angle (MPTA) and lateral distal femoral angle (LDFA) are most commonly used, and are defined as the angle between the joint line and corresponding long bone anatomic axis [9], Fig. 23.3.

Additionally, sagittal and axial plane deformities effect kinematics and can mimic coronal malalignment, so both clinical and radiographic evaluation is imperative. Posterior tibial slope (PTS) and femoral and tibial torsion must also be evaluated.

Physiologic ranges for all parameters, for purposes of deformity analysis, are listed in Table 23.1.

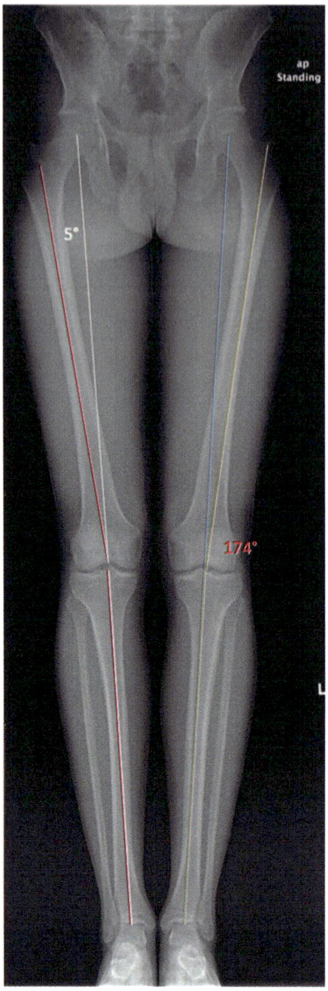

Fig. 23.1 Full-length standing lower extremity x-ray demonstrating the mechanical (white) and anatomic (red) axes of the femur and tibia of the left lower extremity. The anatomical mechanical femoral angle is 5°. The mechanical axis of the left lower extremity is marked with a blue line. The anatomical axis of the femur and tibia of the left leg are marked with a yellow line and forms a 174° anatomical femorotibial angle

23.3 Varus Malalignment

Although physiologic varus can exist, in varus malalignment, the lateral angle between the anatomic axes of the femur and tibia is greater than 173–175°. The mechanical axis is more than 4 ± 2 mm medial to the center of the knee joint, resulting in increased medial MAD and decreased mFTA [8].

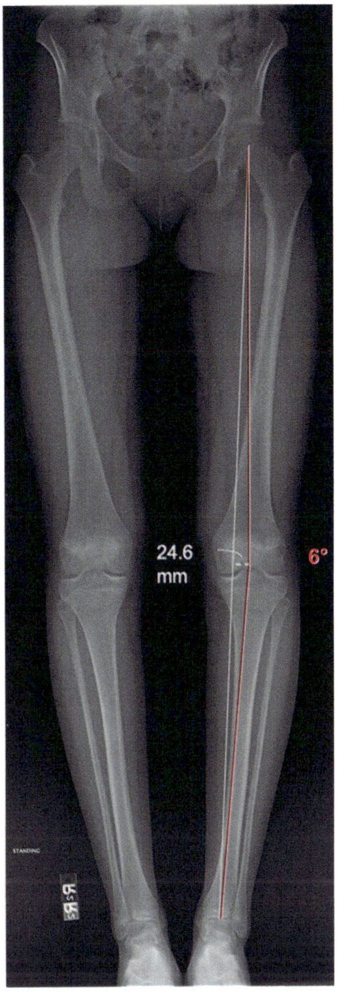

Fig. 23.2 Full-length standing lower extremity x-ray demonstrating varus alignment of the left lower extremity. The femorotibial angle is 6° varus (red line). The mechanical axis of the left lower (white line) extremity is 24.6 mm medial to the center of the knee joint

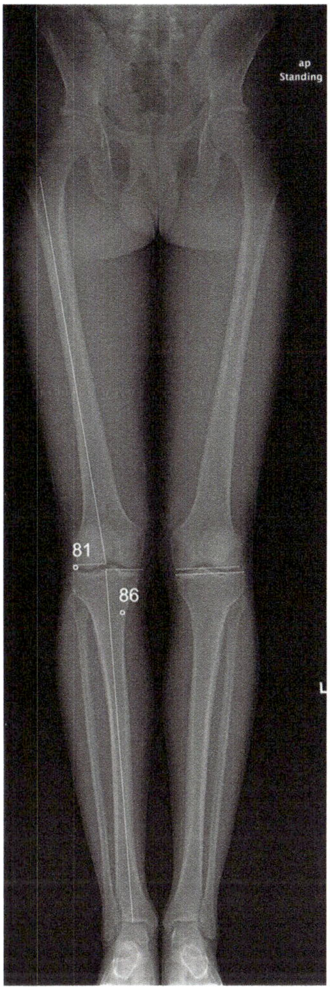

Fig. 23.3 Full-length standing lower extremity x-ray demonstrating the anatomic medial proximal tibial angle of 86° and anatomic lateral distal femoral angle of 81° of the right lower extremity. Femoral and tibial joint lines are demonstrated on the left lower extremity

In the presence of varus malalignment, more load bearing occurs in the medial compartment, which correlates with increased medial tibiofemoral contact pressures and increased medial meniscal extrusion [10]. This may result in accelerated degeneration of the medial compartment and subsequent symptoms. Symptomatic degeneration of the medial compartment can be addressed in the setting of malalignment with osteotomies which redirect the mechanical, weight-bearing axis toward the midline, offloading the symptomatic compartment. Often, valgus-producing HTOs can be done in conjunction with medial compartment cartilage restoration or meniscal transplant procedures.

Table 23.1 Coronal plane parameters and physiologic ranges

Parameter	Physiologic range
mFTA [°]	177–181
mMPTA [°]	85–90
mLDFA [°]	85–90
JLCA [°]	0–3
MAD [mm]	3–17 (medial)

For coronal plane deformities, the aim is typically to restore native valgus, with a correction to

3–5° valgus mechanical axis, largely following methodology described by Dugdale and Noyes [11]. In short, a femoral weight-bearing line is defined as the line from the center of the femoral head to the point on the tibial plateau that is 62.5% of the coronal width, measured from medial to lateral. The tibial weight-bearing line is measured similarly, from this point on the tibial plateau to the center of the tibial plafond. The angle formed by these two lines defines the angle of correction, and an equivalent wedge of bone is either resected (LCW) or gap is created (MOW).

23.3.1 Medial Opening High Tibial Osteotomy (HTO)

The purpose of the medial open-wedge high tibial osteotomy (MOW-HTO) is to treat symptomatic medial compartment pathology in the setting of varus deformity [12, 13], Fig. 23.4a. The procedure shifts the mechanical weight-bearing axis laterally to decompress the medial compartment, thereby reducing medial compartment load, which is particularly useful in young, active patients. It can also be protective for concurrent ligament reconstructions. With PCL reconstructions, increased PTS has been shown to limit posterior tibial translation, off-loading the translational forces on the reconstructed ligament [14]. Similarly, decreased PTS limits anterior tibial translation, off-loading ACL reconstructions and decreasing failure rates [3, 15]. In contrast to the lateral closing-wedge technique, the MOW-HTO requires only one osteotomy, can be fine-tuned intraoperatively, does not shorten the lower extremity, spares the proximal tibia-fibular joint, and obviates the need for dissection of the peroneal nerve.

Indications for MOW-HTO include symptomatic medial compartment cartilage defects, meniscal deficiency, medial compartment osteoarthritis with varus deformity of the lower extremity in young, active patients, proximal tibial vara, and ligamentous deficiency that may benefit from change in tibial slope. The preoperative range of motion of the knee should be at least 0–10° of extension and 120° of flexion. Patients should be younger than 50–60 years old.

Contraindications include severe obesity, inflammatory arthropathy, femoral-sided deformity, poor bone quality, lateral compartment degeneration or meniscal deficiency, and limited range of motion, especially flexion contractures.

23.3.1.1 Surgical Technique

The patient is positioned in the supine position. A longitudinal 6–8 cm anteromedial incision is made between the tibial tubercle and posteromedial border of the tibia. Dissection is carried down to the sartorial fascia. The medial collateral ligament, sartorial fascia, and patellar tendon are identified and protected. The sartorial fascia is incised in line with the gracilis tendon, and the tibial attachments of the sMCL and the pes anserine are mobilized. Retractors are placed deep to

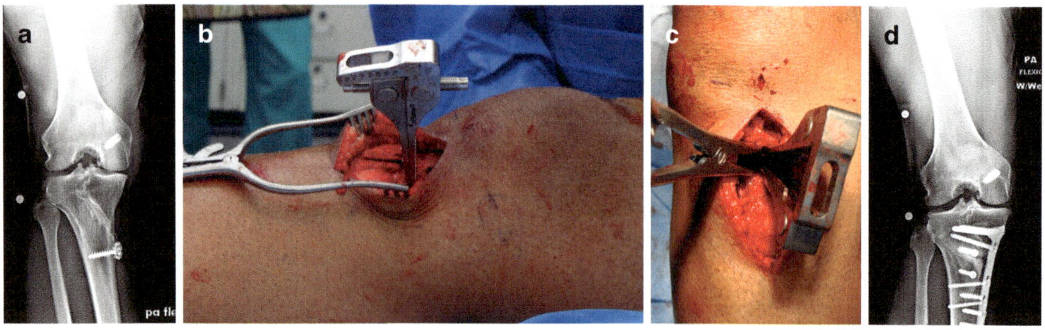

Fig. 23.4 Medial opening high tibial osteotomy (HTO). (**a**) Preoperative radiographs of a 40 yo with symptomatic medial compartment OA. (**b**, **c**) Once the osteotomy is complete, a lamina spreader is used to gently distract the osteotomy. (**d**) 2y follow-up XRs showing preserved medial compartment joint space with mild OA progression

the patellar tendon and the medial collateral ligament and popliteus muscle following subperiosteal dissection. The posteromedial retractor is passed safely behind the tibia to protect the neurovascular structures.

Several techniques are described for the planning and execution of the osteotomy. In general, a pair of reference pins may be placed parallel, and approximately 5 cm distal, to the medial joint line, from distal-medial to proximal-lateral as a guide for saw trajectory during axial osteotomy. Two additional parallel pins are inserted anterior to posterior, parallel to the posterior tibial slope, to monitor sagittal plane alignment. Fluoroscopy is used to confirm the level and trajectory of the saw cut. An oscillating saw is then used to perform the osteotomy to achieve the desired correction. Beginning at the medial cortex, the osteotomy should aim for the level of the proximal tibiofibular joint. Depending patellofemoral joint condition, the initial frontal osteotomy, is either made distal or proximal to the tibial tubercle, as supratubercle osteotomies can induce patella baja and increase patellofemoral contact pressures [16]. The osteotomy should stop at least 1 cm medial to the lateral cortex to avoid violating the lateral hinge and it should not terminate less than 1.5 cm inferior to the tibial plateau to avoid iatrogenic intra-articular extension. The osteotomy is often completed with osteotomes and with attention paid to the posterior cortex as this is a common site for incomplete osteotomy. Once the osteotomy is complete, stacked osteotomes or a lamina spread may be used to begin to gently distract the osteotomy to a predetermined level based on the desired correction, Fig. 23.4b, c. Commercially available wedges and sizing devices may also be used for this purpose. Once the desired amount of distraction has been achieved, a medial plate is selected and fixed distally with a non-locking screw, then proximally with locking screws, followed by at least two to three locking screws distally. Fluoroscopy is used to verify screw positioning and to ensure that the lateral cortex and tibial plateau remain intact. Allograft or autologous bone graft is used to pack into the osteotomy site. The wound is irrigated and closed in layers. The patient is placed in a hinged knee brace [17].

23.3.1.2 Postoperative Management

Cryotherapy accompanied by intermittent pneumatic compression for venous thromboembolism (VTE) prophylaxis should begin immediately after the operation. Pharmacologic VTE chemoprophylaxis should be employed postoperatively based on patient risk factors. Patients are kept non-weight bearing for 6 weeks and are advanced as tolerated thereafter.

Skin sutures are removed on postoperative day 10–12. Radiological follow-up is done on postoperative day 3 and postoperative week 6.

23.3.1.3 Results

MOW-HTO performed for medial compartment varus osteoarthritis of the knee has good clinical outcomes with high patient satisfaction. Cumulative survival rate is 97% at 5 years, 87% at 10 years, and 85% at 13 years [18]. In a study by Schuster, subjective IKDC scores significantly improved from 44 ± 11 preoperatively to 70 ± 13, 66 ± 15, 66 ± 15, and 65 ± 17 at 1, 3, 5, and 10 years, respectively. Poor cartilage regeneration and low preoperative IKDC score (<40) were associated with decreased survival [19]. Another study found that at follow-up 95% of patients reported an excellent or good score according to the IKDC and HSS scoring systems [20].

23.3.2 Lateral Closing HTO

The closing-wedge procedure was once the gold standard osteotomy in the treatment of arthritis [21]. Direct cortical contact is believed to improve osteotomy union and decrease risk of implant failure [22]. However, with the advent of implant designs such as plate fixators with angular stable locking head screws, the medial open-wedge technique has now gained popularity over the closed-wedge technique [23–27]. Although the medial open-wedge technique offers increased precision, decreased surgical time, and reduced risk of peroneal nerve injury, there are still certain indications for lateral closing-wedge osteotomy. In the setting of revision ACL reconstruction, with associated proximal tibia vara and increased posterior tibial slope, LCW-HTO could help protect the reconstruction. Additionally, LCW may be preferable in obese patients who would benefit from direct cortical contact, in patients with existing patella baja, and those with leg length discrepancies in whom the operative leg is longer. If a lateral approach to the knee is planned or has previously been done, LCW also obviates the need for a second medial incision.

23.3.2.1 Surgical Technique

Patient is positioned in a supine position. A straight longitudinal incision is made on the anterolateral aspect of the proximal tibia. The extensor muscles are then detached from the tibia and fibula. The peroneal nerve is exposed, and the neck of the fibula is prepared. A proximal fibular osteotomy or partial fibular head resection is then performed to allow for closure for osteotomy gap. Fluoroscopy is now used to determine the level of the oblique tibial osteotomy. Reference pins are then placed to guide the coronal correction, as determined preoperatively, Fig. 23.5. A partial osteotomy of the tuberosity is performed in the coronal plane with an osteotome in place, in order to protect the tuberosity during the horizontal osteotomy. The horizontal osteotomy is performed between the K-wires using a saw blade. After wedge resection, the osteotomy gap is closed, and the osteotomy is fixed with a locking compression plate. The extensor muscles are reattached, and the wound is closed [28].

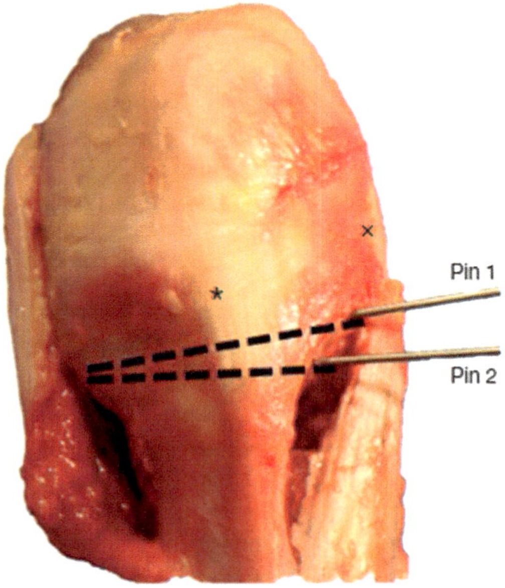

Fig. 23.5 Cadaver model. *patellar ligament. x joint line. The position of the two pins shows where the osteotomy will be performed

23.3.2.2 Postoperative Management

The patient is immediately allowed to be partially weight bearing with the knee extended in a removable splint. Passive physical therapy should also begin on the first postoperative day. Follow-up radiography is performed at 6 and 12 weeks postoperatively. If bone consolidation is present radiographically at 6 weeks postoperatively, the patient is allowed full weight bearing [7, 29].

23.3.2.3 Results

Closing-wedge osteotomy performed for medial varus osteoarthritis of the knee has good clinically reported outcomes with high patient satisfaction. According to the Crosby–Insall grading system, 97% of patients reported excellent and good results. Mean survivorship is reported at 12.6 ± 7.1 years, with rates of 92% at 10 years, 82% at 15 years, and 80% at 20 years [30]. Adverse events are reported at 5% and revision rate is 10% at a mean period of 12.8 years [31]. Unsatisfactory results are typically due to unstable implants, poor patient selection, or inadequate preoperative planning.

23.3.2.4 Complications

Rare complications unrelated to limb alignment, such as neurovascular injuries, are possible and require an intraoperative multidisciplinary team to acutely manage injuries to the blood vessels or nerves with appropriate procedures.

Common complications include failure to achieve proper correction as well as overcorrection. Preoperative planning is essential in order to avoid errors in correction. Intraoperative verification of limb alignment prior to fixation can help avoid potential errors. With the medial opening wedge technique, lateral cortical hinge fractures and intra-articular fractures are also possible if the saw is taken too deep laterally or proximally. Additionally, nonunion is a risk. Delayed consolidation of the osteotomy gap presents as persistent pain with ambulation 6–12 weeks postoperatively without evidence of radiographic healing. Oftentimes, watchful waiting will result in union on a delayed basis; however, secondary bone grafting may be required.

Complications with lateral closing wedge osteotomy are similar, with the exception of increased risk of damage to the peroneal nerve, given the surgical approach and need for high fibular osteotomy.

Additional complications, common to many orthopedic procedures, including infection, VTE, postoperative hematoma, or compartment syndrome require vigilant monitoring and proper treatment. Compartment syndrome, in particular, requires diagnostic suspicion and vigilance to ensure early diagnosis and treatment with fasciotomies when indicated.

23.4 Valgus Malalignment

In valgus malalignment, the lateral angle between the anatomic axes of the femur and tibia is less than 173–175°, the deviation of the mechanical axis from the center of the knee joint is greater than 10 mm laterally, and the distance between the medial malleoli is increased [8], Fig. 23.6. In the presence of such a deviation, the mechanical transfer of stress in the joint is no longer uniform because it is distributed more to the lateral compartment. Thus, valgus malalignment is also regarded as a pre-arthritic deformity.

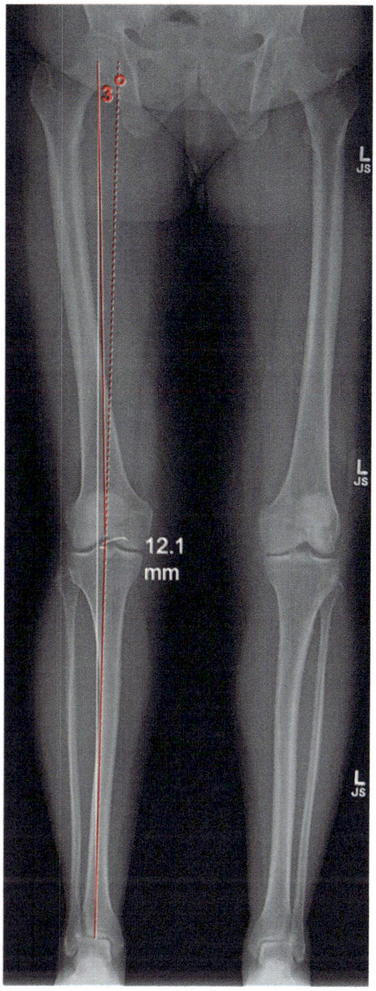

Fig. 23.6 Full-length standing lower extremity x-ray demonstrating valgus alignment of the right lower extremity. The femorotibial angle is 3° valgus (red lines). The mechanical axis of the right lower (white line) extremity is 12.1 mm lateral to the center of the knee joint

23.4.1 Medial Closing Distal Femoral Osteotomy (DFO)

The indications for varus distal femoral osteotomy are lateral compartment degeneration including cartilage lesions, or meniscal deficiency with valgus alignment. The ideal patient desires an active lifestyle and is under 60 years of age, although no strict age cutoffs are followed. The preoperative range of motion in the knee should be at least 90°of flexion-extension without flexion contracture. This osteotomy is particularly suited for larger angular corrections, >17.5°, patients with limb length discrepancies (longer operative limb), those at higher risk for nonunion, and in patients requiring earlier postoperative weight bearing [32].

Contraindications for this procedure include restricted movement at the knee (especially an extension deficit greater than 10°) and medial compartment cartilage lesions or meniscal deficiency.

23.4.1.1 Surgical Technique

Patient is positioned supine. A 10 cm longitudinal skin incision extends proximally in line with the medial epicondyle of the femur. Dissection is performed to the level of the vastus medialis, which is then elevated anteriorly from the intermuscular septum, taking care to identify and protect the femoral artery proximally in the adductor canal.

A pre-contoured plate is placed on the medial femur. The site of the osteotomy between the screw holes at the metadiaphyseal region is marked. Using cautery, a transverse medial incision is made through the periosteum and extended along the femur. A blunt retractor is placed posteriorly to protect the neurovascular structures and anteriorly to protect the quadriceps.

The osteotomy is guided with two bicortical K-wires from medial to lateral according to the planned osteotomy angle and cortical length of the triangle base. The medial cortex of the femur is cauterized longitudinally to serve as a perpendicular referencing guide to avoid malrotation. An oscillating saw is used between the two K-wires to create the wedge osteotomy, while preserving 5 mm of lateral femoral cortex. Irrigation is essential in order to avoid thermal injury from the saw. The wedge is resected, K-wires are removed, and the osteotomy is closed with a varus force on the lower leg.

The pre-contoured medial distal femoral locking plate is placed and pinned distally and proximally. A locking screw is inserted distally into the plate. Fluoroscopy is used to confirm alignment, and a long rod is used to check that the weight-bearing line from the femoral head to the talus passes through the center of the knee. The remaining distal locking holes are now filled with screws. The osteotomy site is compressed using an eccentrically drilled bicortical non-locking screwing just proximal to the osteotomy. Unicortical locking screws are used on the remaining three holes. Fluoroscopy is used to verify plate positioning. The wound is irrigated and the fascia overlying the vastus medialis and skin are closed [33].

23.4.1.2 Postoperative Management

Gentle range of motion can be performed on the day of the surgery as long as the compression bandage is in place. Swelling will persist but can be reduced with cryotherapy and an intermittent pneumatic pump. The bandage must be changed, and the soft tissue must be evaluated on postoperative day 1. Mobilization with partial weight bearing starts on postoperative day 1. Partial loading should continue for 6 weeks. If the patient has unrestricted range of motion and is radiographically healing the osteotomy, loading should be increased up to the pain threshold in postoperative week 7. If the osteotomy is not fully closed or the fixation stability is suboptimal, loading should be increased slowly over 2–4 weeks. Sutures are removed on postoperative day 10–12. Radiological assessment of the osteotomy should take place immediately after the procedure, after mobilization of the patient, 6 weeks postoperatively, and 3 months postoperatively. Full weight bearing is permitted if radiological assessment shows a fully consolidated osteotomy.

During the rehabilitation period, the patient should not apply torsional loading to the leg, as the implant is particularly sensitive to torsional weight bearing. The physical therapy regimen should include active and passive exercises. Prophylactic anticoagulation should be continued until full weight bearing is allowed. Electrotherapy with an EMS device is recommended for muscle stimulation, especially for the vastus medialis.

23.4.1.3 Results

There is a high reoperation rate ranging from 25–40% in patients who undergo medial closing-wedge osteotomy, often due to the need for hardware removal or conversion to total knee arthroplasty. Survival rate is 98% at 5 years and 92% at 10 years, and patients requiring conversion to arthroplasty tend to be older [34].

23.4.2 Lateral Opening DFO

Though the medial closing wedge has historically been more popular, lateral opening wedge osteotomies are becoming increasingly common due to ease of surgical exposure, need for single bone cut, and improved accuracy of correction [32]. The ideal candidate is an active patient under 65 years of age with a valgus malalignment of less than 20° with lateral compartment osteoarthritis, Fig. 23.7a. The procedure is indicated in valgus knees with lateral meniscal deficiency and can be an isolated procedure or combined with a lateral meniscal transplant. It is also indicated in osteochondral lesions of the lateral compartment with valgus malalignment, Fig. 23.7b. Lastly, it can also be performed in case of chronic MCL deficiency and can be done in conjunction with MCL reconstruction [35].

Contraindications include obesity, inflammatory arthropathy, impaired range of motion, large corrections requiring greater than 15 mm of lateral opening, and advanced bi/tricompartmental arthrosis [35].

23.4.2.1 Surgical Technique

The patient is positioned supine. A 12 cm incision is made over the lateral portion of the femur starting from the lateral epicondyle and extending proximally. The iliotibial band is split and the vastus lateralis is elevated anteriorly from the intermuscular septum. The lateral femoral cortex is now exposed. The starting point of the guide wire is the mid antero-posterior position on the lateral femur, three fingers proximal to the lateral epicondyle. The trajectory is 20 degrees caudal, aiming towards the medial epicondyle while making sure the osteotomy plane is proximal to the medial femoral condyle. A mark perpendicular to the osteotomy is made on the lateral femoral cortex to reduce the risk of malrotation. The knee must be flexed to minimize injuries and provide slack for the neurovascular structures. The osteotomy is started with an oscillating saw proximal to the guide wire to prevent distal migration. The blade should be on a plane perpendicular to the coronal plane. The cut is made with thin osteotomes within 1 cm from the medial cortex. Gentle varus forces should be applied to check mobility of the osteotomy. The osteotomy is gently and gradually opened using a lamina spreader until the proper correction is attained. A long alignment rod is now placed from the center of the femoral head to the center of the ankle to assess the alignment correction with fluoroscopy. The rod should pass through the center of the knee for neutral alignment. The osteotomy is fixed with a femoral plate, Fig. 23.7c. After plating, the osteotomy gap is bone grafted. Although different options are available, the preferred option is the femoral head allograft in which two wedges are obtained [35].

23.4.2.2 Postoperative Management

In the postoperative period, the patient is kept in a sterile cotton dressing and ace wrap bandage. Mechanical and chemical DVT prophylaxis are provided based on preoperative risk factors. The leg is placed in a knee immobilizer and locked in full extension for ambulation for 6 weeks postoperatively. Knee range-of-motion exercises may be performed. The patient is partial weight bear-

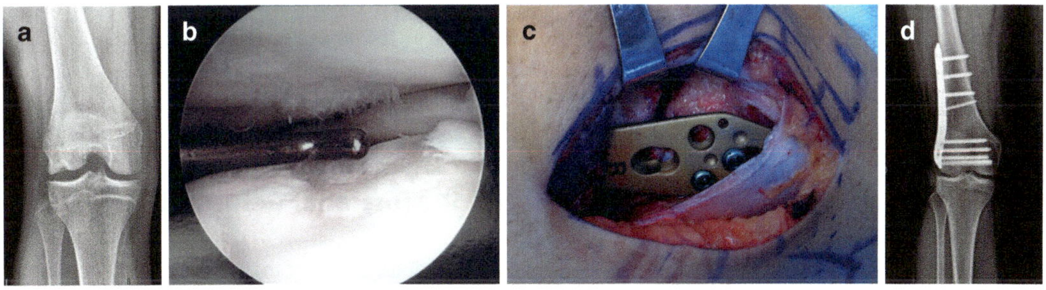

Fig. 23.7 Lateral opening distal femoral osteotomy (DFO). (**a**) Preoperative radiographs of a 19 yo runner with symptomatic lateral compartment OA. (**b**) Osteochondral lesion of lateral compartment. (**c**) A precontoured medial distal femoral locking plate. (**d**) 3y follow-up XRs showing preserved lateral compartment joint space with mild OA progression

ing for 6 weeks postoperatively but may subsequently progress to full weight bearing thereafter. Low-impact strength and aerobic exercises are permitted. Radiographic follow-up occurs at postoperative month 3 [36].

23.4.2.3 Results

Opening-wedge distal femoral osteotomy has shown good results with a low rate of conversion to arthroplasty. Reported survival rates vary from study to study and range from 79% to 91% at 5 years. Conversion to total knee arthroplasty is 12% at 45–78 months [35]. Biomechanical studies have similarly shown that opening-wedge DFOs decrease lateral compartment pressures throughout knee range of motion [37].

23.4.2.4 Complications

Preoperative planning is essential in order to avoid errors in the degree of correction. Given that the osteotomy is more proximal, angular corrections are affective over a longer length, accentuating errors. Intraoperative verification of the corrected mechanical axis prior to fixation can help avoid these errors. With MCW-DFO, repairs of the incised medial patellofemoral ligament and the distal insertion of the vastus medialis prevents lateral patellar instability. Many potential complications specific to medial closing DFO arise from the application of the plate. The fixed-angle locking must be performed precisely with the orientation of the locking head screws dictated by the plate-hole design. Proper positioning of the implant is essential, with the plate shaft aligned parallel to the longitudinal axis of the femur and the distal plate head anteromedial to the medial femoral condyle. In order to prevent the locking head screws from protruding posteriorly from the condylar block, they should be angled in the frontal plane.

The procedure places the femoral artery, femoral vein, sciatic nerve, and vascular bundle at risk of injury at the posterior aspect of the femur. Therefore, the posterior femoral cortex should only be cut if the soft tissues beyond it are protected by a retractor. Furthermore, the genicular arteries and veins are susceptible to bleeding if the intermuscular septum is inadvertently divided. In the case of injury to large blood vessels, repair by vascular surgery is required.

Delayed bone healing presents as persistent pain upon loading and an absence of callus on radiographs. However, given the biomechanical strength of this closing wedge construct fixed in compression, an absence of callus may result from primary bone healing. Therefore, careful attention should be paid to the resolution of the osteotomy line on radiographs, particularly in cases without a gap in the osteotomy. If no signs of bone healing are present after more than 3 months, secondary bone grafting procedures should be considered after a thorough evaluation with advanced imaging.

With LOW-DFO, intra-articular fractures and breaches of the medial hinge are the most common intraoperative complications. This is often due to incomplete osteotomy or guide pin placement too near the joint line. Hardware irritation is also a particular issue with LOW-DFO. Neurovascular injuries can also occur and may be prevented by using the lateral transverse artery as a landmark for finding proper osteotomy height. Nonunion occurs at a rate of 2–5%, with significantly increased risk associated with smoking and obesity [38]. Implant failure or osteotomy collapse may occur due to inadequate fixation or aggressive rehabilitation [35].

Complications common to many orthopedic procedures include postoperative tissue swelling, lymphedema, VTE, compartment syndrome. Postoperative infection remains a possible complication, but if early, may be treated by surgical revision with debridement and systemic antibiotics. The plate fixator does not need to be removed if the soft tissue cover is intact and if the osteosynthesis is stable. If this is not the case, an external fixator should be applied.

23.5 Torsional Deformities

Torsion deformities of the leg often present as anterior knee pain due to patellar maltracking or patellofemoral instability, gait disturbance, and hip impingement. In the absence of clinical symptoms, torsional deformities do not require correction. If

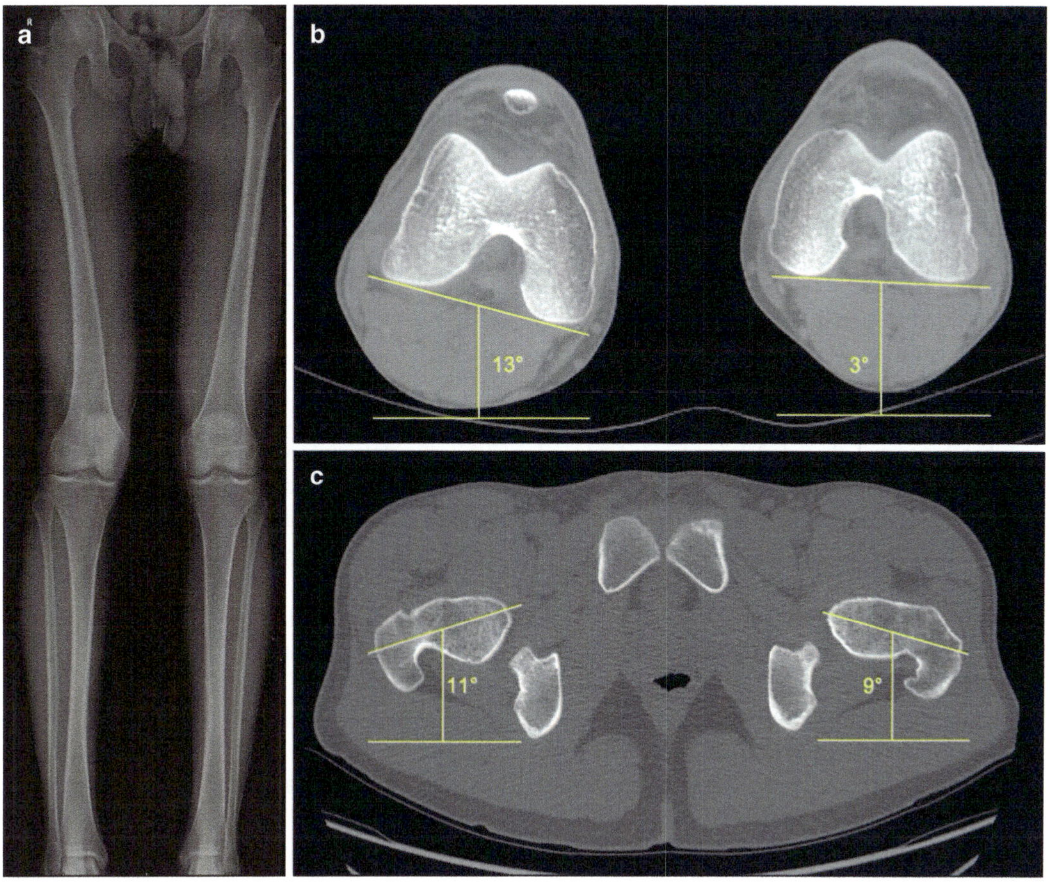

Fig. 23.8 Preoperative planning for femoral derotational osteotomy. (**a**) Full-length standing lower extremity x-ray of a 24-year-old male, demonstrating right lower extremity valgus with increased mFTA and mLDFA after ACL reconstruction. (**b**) On axial CT cuts through the distal femur, angle of posterior femoral condyles to horizontal is measured and similarly (**c**) femoral neck anteversion is calculated. The difference between these two measurements demonstrates femoral torsion. This patient femoral internal rotation of 24° (11° + 13°) on the right, as compared to 6° (9° + (−3°)) of internal rotation on the left

clinical symptoms are present but pathological torsion is only slight, conservative treatment with physical therapy and corrective insoles is usually successful. For more severe symptoms refractory to conservative management, derotational osteotomies may be indicated. In the setting of concomitant axial and coronal plane deformity, multiplanar corrections are possible [39].

23.5.1 Femoral Rotational Osteotomy

Acetabular anteversion angle typically ranges from 15–20°, while femoral torsion angle is roughly 15 ± 5°. Femoral torsion (or version) is measured as the angle between a line drawn tangential to the posterior femoral condyles and one in line with the femoral neck, based on axial CT images, Fig. 23.8. Though there is no clear threshold, for excess, symptomatic femoral torsion, femoral rotational osteotomy may be considered.

23.5.1.1 Surgical Technique

The patient is positioned supine. A 10–15 cm skin incision is made beginning at the greater trochanter and running distally. The fascia lata is incised longitudinally and the vastus lateralis is detached in an L-shape from the innominate tubercle. Two Hohmann elevators are inserted to expose the femur and intertrochanteric region. Using an osteotome or cautery, the location of the transverse osteotomy is marked cranial to the lesser trochanter. Image intensification is used to plan the position of the angled-blade plate on the femur (Synthes). The lower bend in the plate should be at the level of the planned osteotomy so that the plate blade is in the femoral neck. The insertion site of the blade is also marked using a chisel or cautery at the greater trochanter. K-wire is now inserted bicortically cranial to the osteotomy mark. Another bicortical threaded wire is also placed caudal to the mark and is positioned to form the calculated correction angle with the first wire. The wires must lie perpendicularly to the mechanical femoral axis. Using image intensification, another wire is advanced into the femoral neck in contact with the bone in order to mark the position and anteversion of the femoral neck. The insertion site and blade bed are now prepared by hammering a seating chisel with U-profile parallel to this wire. Advancing the chisel into the bone using a slotted hammer will allow controlled rotation. For proper orientation of the chisel blade in the frontal and sagittal planes, the tongue of the guiding angle must be aligned parallel to the femoral shaft in both planes. The osteotome is then left in position to act as a guide. The leg is positioned in full extension. The blade should now be hammered back out 1–2 cm to help extract the seating chisel after the osteotomy. The oscillating saw is used under cooling to perform the osteotomy under the protection of two Hohmann retractors. The osteotomy is performed at 90° to the mechanical axis. The saw blade must be in the same plane as the first and second wires. The chisel is now withdrawn, and the blade plate is hammered into the implant bed. The planned correction angle is achieved by rotating the distal segment until the second K-wire lies parallel to the first K-wire. The angled blade plate is stabilized temporarily with Verbrugge clamps. Compression is applied to the osteotomy gap by a plate compression device. After inserting the bicortical screws, radiographic documentation is recorded, and drains are inserted followed by wound closure [40].

23.5.1.2 Postoperative Management

Dressings are changed and the drains are removed on the first postoperative day. Mobilization should begin with partial weight bearing. Active hip motion against resistance should be avoided although passive hip motion is permitted. Radiological follow-up should be performed on postoperative day 3 and after postoperative week 6. Weight bearing can be increased after postoperative week 7 based on clinical and radiological findings [7].

23.5.1.3 Results

Femoral rotational osteotomy has shown good results clinically. Good to excellent outcomes are reported at a range of 93–98%. In one study, the modified Harris Hip Score improved by 29 points at an average follow-up of 6.5 years. Seventy-eight percent of patients required a subsequent surgery, of which 91% were implant removals [41].

23.5.2 Tibial Rotational Osteotomy

The standard tibial torsion angle is 23 ± 5.1°, measured on axial CT as the angle between the knee joint axis and transmalleolar axis. Again, though no clear threshold exists, if the patient's torsion angle exceeds two standard deviations from the norm (>35° or < 1°), and the patient has symptomatic gait disturbance or pain, tibial rotational osteotomy may be considered.

23.5.2.1 Surgical Technique

Patient is placed in a supine position. The knee is in near full extension with the knee slightly flexed on a towel. A 5–7 cm longitudinal skin incision is made anterior to the tuberosity. The tibialis anterior fascia is dissected longitudinally from 1 cm lateral to the tibial attachment. A 5 cm portion of the tibialis anterior muscle is detached in order to expose the lateral tibia. A longitudinal incision is now made on the periosteum medial to the tuberosity in order to pass a bone rasp beneath it. The medial collateral ligament and tendons of the pes anserinus must now be protected. Image intensification is used with a measuring rod to identify the tibial axis and to mark the direction of the saw cut with a K-wire. Another mark is made with cautery perpendicular to the tibial mechanical axis 1 cm distal to the cranial margin of the tuberosity. This marks the level of osteotomy. Image intensification is used to position the internal plate fixator on the lateral tibia. The two proximal screw holes are pre-drilled. The plate is removed, and a mark is made for the osteotomy. If the tuberosity needs to be rotated with the distal segment, the anterior osteotomy should be cranial. If the tuberosity need not be rotated, the saw cuts may run distally. For valgus opening-wedge tibial correction osteotomy, the angle between the oblique osteotomy and the transverse osteotomy should be 100° and the tuberosity segment should be 10–15 mm wide.

A thin saw blade is used to make the anterior oblique osteotomy. The saw cut should be strictly in the frontal plane and the patellar tendon should be protected. K-wires are inserted bicortically parallel to each other with one proximal to the osteotomy site and one distal to it. Both wires should be perpendicular to the tibial axis. Transverse tibial osteotomy is performed using image intensification from lateral to medial, as well as perpendicular to the tibial mechanical axis. An angled protection plate should be left in the osteotomy cut while the transverse tibial saw cut is made in order to protect the tibial tuberosity. After removing the protective saw blade from the tuberosity cut, a small portion of bone should be removed from behind the tibial tuberosity to allow rotation. The segment should be verified to see if rotation without resistance is possible, and if not, any residual bone bridges are chiseled. The distal segment is rotated relative to the proximal segment until the proper correction is achieved. Proper alignment of the foot relative to the patella and tibial tuberosity is essential. Before applying the fixed-angle plate fixator, the correction should be temporarily stabilized with two K-wires. The plate should be positioned so that the two proximal plate holes are aligned with the two holes in the proximal tibial segment. Two locking head screws are inserted bicortically. Three monocortical locking screws should be inserted for stable fixation distal to the osteotomy [42].

If the derotation is less than 20°, decompression of the peroneal nerve or osteotomy of the fibula are not necessary. It is recommended, however, to split the fascia of the anterior compartment to prevent pressure increase. The muscles should now be reattached, and drains inserted. Wound closure should happen in layers.

23.5.2.2 Postoperative Management

Dressings are changed and the drains are removed on the first postoperative day. Mobilization starts with partial weight bearing. Active knee motion against resistance should be avoided although unrestricted passive and active motion is permitted. Radiological follow-up should be performed on postoperative day 3 and after postoperative week 6. Full weight bearing is permitted postoperative week 7 based on clinical and radiological findings.

23.5.2.3 Results

A study done by Fouilleron shows good clinical results. Ninety-four percent of patients in the study who underwent proximal tibial derotational osteotomy were satisfied or very satisfied with the outcome. Seventy-five percent of the patients in the study had good to excellent results based on the Lille test. The functional results were significantly improved on the Lille score, increasing from 54.8 ± 16.9 to 85.2 ± 14. Patellofemoral pain improved significantly in all patients. Mean postoperative tibial torsion measured at clinical follow-up was 8.6°, with a mean decrease of 25.2° [43].

23.5.2.4 Complications

Complications associated with femoral rotation surgery include overcorrection or under correction due to poor preoperative planning, postoperative infections, malalignment of the frontal or sagittal planes, fractures, secondary loss of correction, implant fatigue failures, and alteration to the foot progression angle. In addition, femoral head necrosis has been reported after intertrochanteric femoral osteotomy. Lastly, if extensive torsional correction is required, internal rotational osteotomy of the femur may damage the sciatic nerve [7].

With tibial osteotomies, distal external rotation of the tibia may also damage the peroneal and tibial nerve. Thus, intracompartmental decompression by fasciotomy is recommended.

23.6 Take Home Message

Osteotomies about the knee are useful, often forgotten procedures, that have proven success with correction of malalignment in the setting of early degenerative changes, ligament reconstruction procedures, and augmentation of joint preserving procedures such as osteochondral and meniscus reconstruction. Precise preoperative planning is crucial to determining appropriate correction.

Computed tomography can render models that allow surgeons to improve their multiplanar preoperative planning [44]. With careful planning and execution, osteotomies and concomitant procedures allow for return to work/play for patients [45–47].

The reported physiological ranges are based on [10]. JLCA, joint line convergence angle (positive values indicate medial convergence); MAD, mechanical axis deviation (positive values indicate medial MAD); mFTA, mechanical femorotibial angle (values >180° indicate valgus alignment, values <180° indicate varus alignment); mLDFA, mechanical lateral distal femoral angle; mMPTA, mechanical medial proximal tibial angle.

References

1. Samuelsen BT, et al. Posterior medial meniscus root tears potentiate the effect of increased Tibial slope on anterior cruciate ligament graft forces. Am J Sports Med. 2020;48(2):334–40.
2. Mehl J, et al. Osseous valgus alignment and posteromedial ligament complex deficiency lead to increased ACL graft forces. Knee Surg Sports Traumatol Arthrosc. 2019;28:1119–29.
3. Bernhardson AS, et al. Tibial slope and its effect on force in anterior cruciate ligament grafts: anterior cruciate ligament force increases linearly as posterior Tibial slope increases. Am J Sports Med. 2019;47(2):296–302.
4. Yamaguchi KT, et al. Effects of anterior closing wedge Tibial osteotomy on anterior cruciate ligament force and knee kinematics. Am J Sports Med. 2018;46(2):370–7.
5. Morrison J. Bioengineering analysis of force actions transmitted by the knee joint. Biomed Eng. 1968;3:164–70.
6. Morrison J. Function of the knee in various activities. Bio-Med Eng. 1969;4:573–580s.
7. Lobenhoffer P, et al. Osteotomies around the knee: indications, planning, surgical techniques using plate fixators. Stuttgart: AO Publishing; 2008.
8. Paley D, Pfeil J. Principles of deformity corrections around the knee. Orthopade. 2000;29(1):18–38.
9. Paley D, et al. Deformity planning for frontal and sagittal plane corrective osteotomies. Orthop Clin North Am. 1994;25(3):425–65.
10. Willinger L, et al. Varus alignment increases medial meniscus extrusion and peak contact pressure: a biomechanical study. Knee Surg Sports Traumatol Arthrosc. 2020;28(4):1092–8.
11. Dugdale TW, Noyes FR, Styer D. Preoperative planning for high tibial osteotomy. The effect of lateral tibiofemoral separation and tibiofemoral length. Clin Orthop Relat Res. 1992;274:248–64.
12. Coventry MB. Upper tibial osteotomy for gonarthrosis. The evolution of the operation in the last 18 years and long term results. Orthop Clin North Am. 1979;10(1):191–210.
13. Lobenhoffer P, De Simoni C, Staubli AE. Opening wedge high-tibial osteotomy with rigid plate fixation. Tech Knee Surg. 2002;1:93–105.
14. Novaretti JV, et al. The role of osteotomy for the treatment of PCL injuries. Curr Rev Musculoskelet Med. 2018;11(2):298–306.
15. Yoon KH, et al. Influence of posterior Tibial slope on clinical outcomes and survivorship after anterior cruciate ligament reconstruction using hamstring autografts: a minimum of 10-year follow-up. Arthroscopy. 2020;36(10):2718–27.
16. LaPrade RF, et al. Patellar height and tibial slope after opening-wedge proximal tibial osteotomy: a prospective study. Am J Sports Med. 2010;38(1):160–70.
17. Chahla J, et al. Medial opening wedge proximal Tibial osteotomy. Arthrosc Tech. 2016;5(4):e919–28.
18. Jin C, et al. Survival and risk factor analysis of medial open wedge high Tibial osteotomy for unicompartment knee osteoarthritis. Arthroscopy. 2020;36(2):535–43.
19. Schuster P, et al. Ten-year results of medial openwedge high Tibial osteotomy and chondral resurfacing in severe medial osteoarthritis and Varus malalignment. Am J Sports Med. 2018;46(6):1362–70.
20. Puddu G. Outcomes of opening wedge high tibial osteotomy. Orthopaedic Proc. 2018;91.
21. Coventry MB. Osteotomy of the upper potion of the tibia for degenerative arthritis of the knee. A preliminary report. J Bone Joint Surg Am. 1965;47:894–990.
22. Luites JW, et al. Fixation stability of opening- versus closing-wedge high tibial osteotomy: a randomised clinical trial using radiostereometry. J Bone Joint Surg Br. 2009;91(11):1459–65.

23. Franco V, et al. Open wedge osteotomy of the distal femur in the valgus knee. Orthopade. 2004;33(2):185–92.
24. Jacobi M, Jakob RP. Open wedge osteotomy in the treatmen of medial osteoarthritis of the knee. Tech Knee Surg. 2005;4(2):70–8.
25. Lobenhoffer P, Agneskirchner JD. Improvements in surgical technique of valgus high tibial osteotomy. Knee Surg Sports Traumatol Arthrosc. 2003;11(3):132–8.
26. Staubli AE, et al. TomoFix: a new LCP-concept for open wedge osteotomy of the medial proximal tibia--early results in 92 cases. Injury. 2003;34(Suppl 2):B55–62.
27. Stoffel K, Stachowiak G, Kuster M. Open wedge high tibial osteotomy: biomechanical investigation of the modified Arthrex Osteotomy Plate (Puddu Plate) and the TomoFix Plate. Clin Biomech (Bristol, Avon). 2004;19(9):944–50.
28. Mattei L, et al. Closing wedge tibial osteotomy: is it an actual procedure nowadays? Annals of Joint. 2017;2(6):6.
29. Lee DC, Byun SJ. High tibial osteotomy. Knee Surg Relat Res. 2012;24(2):61–9.
30. Berruto M, et al. Closing-wedge high tibial osteotomy, a reliable procedure for osteoarthritic varus knee. In: Knee Surgery, Sports Traumatology, Arthroscopy; 2020. p. 1–7.
31. Berruto M, et al. Closing-wedge high tibial osteotomy, a reliable procedure for osteoarthritic varus knee. Knee Surg Sports Traumatol Arthrosc. 2020;28:1–7.
32. Sherman SL, Thompson SF, Clohisy JCF. Distal femoral Varus osteotomy for the Management of Valgus Deformity of the knee. JAAOS - J Am Acad Orthopaedic Surg. 2018;26(9):313–24.
33. Duethman NC, et al. Medial closing wedge distal femoral osteotomy. Clin Sports Med. 2019;38(3):361–73.
34. Wylie JD, Maak TG. Medial closing-wedge distal femoral osteotomy for genu Valgum with lateral compartment disease. Arthrosc Tech. 2016;5(6):e1357–66.
35. Pilone C, et al. Lateral opening wedge distal femoral osteotomy for lateral compartment arthrosis/overload. Clin Sports Med. 2019;38(3):351–9.
36. O'Malley MP, et al. Distal femoral osteotomy: lateral opening wedge technique. Arthrosc Tech. 2016;5(4):e725–30.
37. Wylie JD, et al. The effect of lateral opening wedge distal femoral Varus osteotomy on tibiofemoral contact mechanics through knee flexion. Am J Sports Med. 2018;46(13):3237–44.
38. Liska F, et al. Smoking and obesity influence the risk of nonunion in lateral opening wedge, closing wedge and torsional distal femoral osteotomies. Knee Surg Sports Traumatol Arthrosc. 2018;26(9):2551–7.
39. Imhoff FB, et al. Derotational osteotomy of the distal femur for the treatment of patellofemoral instability simultaneously leads to the correction of frontal alignment: a laboratory cadaveric study. Orthop J Sports Med. 2018;6(6):2325967118775664.
40. Nelitz M. Femoral Derotational osteotomies. Curr Rev Musculoskelet Med. 2018;11(2):272–9.
41. Buly RL, et al. Femoral derotation osteotomy in adults for version abnormalities. J Am Acad Orthop Surg. 2018;26(19):e416–25.
42. Walton DM, et al. Proximal tibial derotation osteotomy for torsion of the tibia: a review of 43 cases. J Child Orthop. 2012;6(1):81–5.
43. Fouilleron N, et al. Proximal tibial derotation osteotomy for torsional tibial deformities generating patello-femoral disorders. Orthop Traumatol Surg Res. 2010;96(7):785–92.
44. Yan J, et al. Outcome reporting following navigated high tibial osteotomy of the knee: a systematic review. Knee Surg Sports Traumatol Arthrosc. 2016;24(11):3529–55.
45. Kunze KN, et al. Return to work and sport after proximal Tibial osteotomy and the effects of opening versus closing wedge techniques on adverse outcomes: a systematic review and meta-analysis. Am J Sports Med. 2019:363546519881638.
46. Hoorntje A, et al. High rates of return to sports activities and work after osteotomies around the knee: a systematic review and meta-analysis. Sports Med. 2017;47(11):2219–44.
47. Ekhtiari S, et al. Return to work and sport following high Tibial osteotomy: a systematic review. J Bone Joint Surg Am. 2016;98(18):1568–77.

Patient-Specific Instrumentation and 3-D Osteotomy

24

Wouter Van Genechten, Annemieke van Haver, and Peter Verdonk

24.1 Introduction

Osteotomies around the knee are well-established, joint-preserving surgical interventions which primarily aim to correct the mal-aligned lower limb in the coronal plane, hereby inducing mechanical unloading of either the medial or lateral arthritic knee compartment [1]. In neutral alignment, the medial compartment bears up to 55–70% of a person's weight during the stance phase of gait, which increases with 5% for every 1° of additional varus deformity [2]. The fact that a constitutional varus alignment of 3° or more is found in a significant number of adults, contributes to the overall high prevalence of medial relative to lateral knee osteoarthritis (OA) [3–5]. Consequently, osteotomies towards valgus are most commonly performed, and since varus deformities are frequently found in the proximal tibia (mechanical medial proximal tibial angle (mMPTA) < 85°), surgical corrections are preferred at this level. Both the medial opening-wedge and the lateral closing-wedge high tibial osteotomy (HTO) have shown to be effective for unloading the diseased medial compartment [6]. When performed in a timely fashion, it can delay or even prevent the development to end-stage knee OA [7]. For several reasons such as the need for a fibular osteotomy, risk of peroneal nerve damage, and extended soft tissue dissections, the lateral closing-wedge approach has fallen into disuse [2, 6, 8]. Therefore, modern opening-wedge HTO forms currently the standard with reported survival rates of >90% at 10 years in young (<65 years) and physically active patients [8–10]. Nevertheless, conventional opening-wedge HTO remains a technically demanding procedure with a considerable risk of complications, including (unstable) lateral hinge fractures, delayed or non-union of the gap, over/under-correction, and unintended increase of the tibial slope [11–14].

Considering the accuracy of conventional HTO procedures in the coronal plane, Van den Bempt et al. uncovered a surprisingly low achievement of the planned correction [15]. Eight out of 11 conventional HTO cohorts were unable to reach a threshold of 75% accurate corrections

W. Van Genechten (✉)
More Institute, Antwerp, Belgium

University of Antwerp, Antwerp, Belgium

A. van Haver
More Institute, Antwerp, Belgium

P. Verdonk
More Institute, Antwerp, Belgium

University of Antwerp, Antwerp, Belgium

ORTHOCA, Antwerp, Belgium

N. Nakamura et al. (eds.), *Advances in Knee Ligament and Knee Preservation Surgery*,
https://doi.org/10.1007/978-3-030-84748-7_24

within a self-defined accuracy interval. Since realignment surgery is a highly individualized intervention associated with a small tolerance for error, these results are posing a major concern regarding intervention durability [16]. Both unprecise preoperative osteotomy planning and subsequent challenging translation into surgery are considered to form the basis of inaccurate osteotomy corrections [17]. The introduction of computer navigation in the field of knee osteotomies has certainly been a step towards more accurate surgical outcomes, mainly due to the real-time visualization of the corrected limb [18]. However, expensive equipment, a long learning-curve with prolonged surgical duration and unpredicted technology failure have constrained this approach from becoming widespread among orthopaedic knee surgeons [15, 19].

Since modern volumetric imaging modalities such as very low-dose computer tomography (CT) scans and magnetic resonance imaging (MRI) became available on large-scale, several attempts have been made to virtually simulate surgeries in suitable medical software and to print 3-D anatomical models [20]. Shortly afterwards, the intra-operative use of 3-D-printed patient-specific instrumentation (PSI) was introduced, first in maxillofacial surgery which was later successfully translated to surgical corrections of the spine and mal-union fractures of the forearm [21–23]. The implementation of PSI in realignment surgery of the lower limb, however, is relatively new [24]. The thought of having customized surgical tools available during surgery, which instantly determine the osteotomy plane together with the intended correction in both the coronal and sagittal plane, sounded very appealing and led to the development of a handful innovative PSI approaches for knee osteotomies [16, 25–28]. Therefore, this chapter provides an overview about the clinical use of 3-D osteotomy planning, customized guide printing and PSI in the operating room (OR) with accuracy outcomes of several techniques developed for knee osteotomy surgery. Further, the author's onsite preferred PSI approach is discussed, together with general considerations and concerns about the topic.

24.2 Osteotomy Planning

A proper full-leg bipodal standing radiograph has always been the benchmark both for determining mal-alignment of the lower limb and for osteotomy planning [29]. However, questions have been raised about the reliability and effect of slight knee flexion and limb rotation on 2-D image measurements [30–32]. Moreover, the factor weight-bearing might cause an overestimation of the preoperative varus alignment, which should theoretically result in high numbers of overcorrected osteotomies [33–35]. Finally, full-leg radiographs only allow osteotomy planning in a single plane (coronal), while most HTO surgeries consist of a biplanar bone cut.

Despite the imperfections, a full-leg standing radiograph still forms a cornerstone in the planning phase, even in the majority clinical PSI studies (Table 24.1) [14, 25–27, 36, 37]. Now, considering 3-D bone modelling for osteotomy planning, a baseline CT-scan appears to be the better option over MRI because it is less expensive, the imaging waiting times are shorter, and it provides clearer spatial resolution to segment the bones [38]. A scan of the knee joint, or at minimum of the proximal tibia is obligatory to perform a multiplanar osteotomy simulation and to design PSI. The obtained imaging DICOM files from the scan are easily loaded into the dedicated segmentation software after which the anatomical bone models are exported as STL-files to maintain scale and composition. Finally, the bone models are transferred to 3-D medical planning software to virtually pre-plan the correction size and define the bone cut (plane, depth and starting point) which is ultimately followed by PSI design and printing. [25, 26, 36, 37].

Some authors have recently implemented the mechanical medial proximal tibial angle (mMPTA) as primary planning angle. [25, 37] The mMPTA strictly limits the correction change to the tibial bone in contrast to the mechanical femorotibial angle (mFTA) or weight-bearing line (WBL%) which might be prone to variation by a patient's position during preoperative imaging. Moreover, this angle has proven to be the only predictor for alignment errors after opening-

Table 24.1 Overview of laboratory and clinical studies using patient-specific instrumentation (PSI) for osteotomy surgery around the knee joint

Author (Year)	# PSI cases (HTO/DFO)	Planning	Target (planning)	PSI Technique	Accuracy with PSI		Conventional controls	Postop measurements (2D/3D)
					Coronal plane	Sagittal plane		
Laboratory studies								
Kwun et al. (2017)	10 Porcine HTO	2-D and 3-D simulation	62.5%	Printing of gap volume (wedge)	Postop: 61.8% ± 1.5	Pre: 11.2° ± 2.2 Post: 11.4° ± 2.5	No	2-D
Donnez et al. (2018)	10 Human HTO	3-D simulation	Random	Cutting guide with pre-drilled matching holes for final plate	ΔmMPTA: 0.2° ± 0.3 (−0.3° to 0.5°)	ΔTS: −0.1° ± 0.5 (−0.7° to 0.8°)	No	3-D
Clinical studies								
Victor et al. (2013)	4 HTO/10 DFO	3-D simulation	Variable	Cutting guide with pre-drilled matching holes for final plate	ΔmFTA: 0° ± 0.72 (−1° to 1°)	ΔTS: 0.3° ± 1.14 (−0.9° to 3°)	No	2-D
Perez-Mananez et al. (2016)	8 HTO	2-D planning and 3-D simulation knee	62%	Cutting guide with 3 spacer wedges	ΔmFTA: 0.5° (0° to 1.2°)	Not mentioned	Yes (*n* = 20)	2-D
Arnal-burro et al. (2017)	12 DFO	2-D planning and 3-D simulation knee	62%	Cutting guide with 3 spacer wedges	ΔmFTA: 0.28° (0° to 1°)	Not mentioned	Yes (*n* = 20)	2-D
Munier et al. (2017)	10 HTO	Full leg 2-D and full-leg 3-D simulation	HKA: 2-4° valgus	Cutting guide with pre-drilled matching holes for final plate	100% within [−2°;+2°] mFTA	90% within [−2°;+2°]	No	2-D and 3-D
Yang et al. (2018)	10 HTO	2-D planning and 3-D simulation knee	62.5%	Biplanar cutting guide with holes for rod matching	Postop: 60.2% ± 2.8%	Preop: 9.9° ± 0.47 Postop: 10.1° ± 0.36	No	2-D
Kim et al. (2018)	20 HTO	2-D planning and 3-D simulation knee	62.5%	Printing of gap volume (wedge),	ΔWBL: 3.9% ± 4.5	Preop: 9.6° ± 3.3 Postop: 9.8° ± 3.2	No	2-D

(continued)

Table 24.1 (continued)

Author (Year)	# PSI cases (HTO/DFO)	Planning	Target (planning)	PSI Technique	Accuracy with PSI		Conventional controls	Postop measurements (2D/3D)
					Coronal plane	Sagittal plane		
Kim et al. (2018)	20 HTO	2-D planning and 3-D simulation knee	62.5%	Printing of gap volume (wedge), no slot for bone cut	ΔWBL: 2.3% ± 2.5	Preop: 8.6° ± 3.3 Postop: 8.9° ± 3.1	Yes (*n* = 20)	2-D
Jones et al. (2018)	18 HTO	3-D simulation	Not mentioned	Cutting guide position based on distant landmarks and 'correction block'	100% within [-3°;+3°] mFTA	100% within [-3°;+3°] TS	No	3-D
Chaouche et al. (2019)	100 HTO	Full leg 2-D and full-leg 3-D simulation	Not mentioned	Cutting guide with pre-drilled matching holes for final plate	ΔmFTA: 1.0° ± 0.9 ΔmMPTA: 0.5° ± 0.6	ΔTS: 0.4° ± 0.8	No	3-D
Fucentese et al. (2020)	23 HTO	Full leg 2-D and full-leg 3-D simulation	Majority on 62.5%	Cutting guide with pre-drilled matching holes for final plate	ΔmFTA: 0.8° ± 1.5 74% within [−2°;+2°] mFTA	ΔTS: 1.7° ± 2.2 61% within [−2°;+2°] TS	No	2-D and 3-D
Authors (2020)	10 HTO	3-D simulation (full-leg)	Lateral spine	Customized wedge and cast for structural bone graft	ΔmFTA: -0.4° ± 1.0 100% within [−2°;+2°] mFTA	ΔTS: 2.1° ± 2.6	No	2–D and 3-D

HTO high tibial osteotomy, *DFO* distal femur osteotomy, *Δ* difference, *mMPTA* mechanical medial proximal tibial angle, *mFTA* mechanical femorotibial angle, *WBL* weight-bearing line, *TS* tibial slope wedge HTO and makes its inclusion in modern HTO planning recommendable to improve correction accuracy [39]. In addition, the authors support the conduction of mMPTA measurements in order to control joint line orientation ((JLO) <5°) after HTO and to maintain the conversion option to arthroplasty in a later stadium. The planned mMPTA should not exceed 95° as this might induce excessive joint line obliquity with increased shear stress on the articular cartilage [40]. A double-level osteotomy might be indicated in large varus corrections which can, on their turn, be planned more precisely in 3-D imaging software. In one PSI strategy, final plate type and positioning are already included in the 3-D planning by determining the predrilled screw holes in the PSI guide (Fig. 24.1). [14, 24, 25, 37] This eventually facilitates immediate and correct implant positioning intraoperatively but leaves a small margin for unexpected alternations during surgery.

Chernchujit et al. recently reported on a planning technique to correct the non-weight bearing component of a full-leg supine CT-scan by using a 2-D full-leg standing radiograph [41]. Accordingly, a full-leg 3-D model under 'weight-bearing circumstances' was created to simulate the intended osteotomy; however, no PSI was printed or used intraoperatively. Despite precise 3-D planning, only 79% of cases ($n = 19$) fell into a wide ±3° range around target, which emphasizes the actual need for customized surgical tools during surgery on top of preoperative 3-D simulation [41].

Overall, the main advantages of executing a preoperative 3-D osteotomy planning are (1) the reliable angle measurements based on exact identification of unique bony landmarks, (2) the multiplanar and multilevel simulation of the surgery and finally (3) the ideal tool for designing PSI and tailor-made anatomical models [42]. With the availability of 3-D bone models, the intended correction size can be planned very precisely in a way that even the thickness of the sawblade can be taken into account [16].

24.3 3-D Printing of PSI: Materials and Equipment

The availability of 3-D planning software, medical grade resin, a 3-D printer and most importantly, trained personnel are mandatory for streamlining an in-hospital preoperative planning and printing process of PSI. If one of these requirements onsite is missing, external companies can be involved; however, this may result in an increased cost per case, a longer manufacturing process and more complex logistics. Therefore, it can be recommended for certain hospitals/orthopaedic departments to invest in a 3-D core facility, especially in case of high surgical turnovers and short waiting lists. Moreover, 3-D planning and PSI is far from only reserved

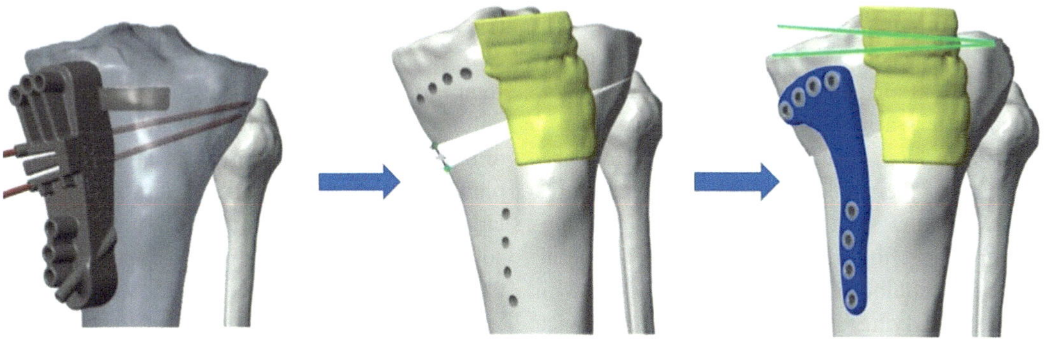

Fig. 24.1 Design of a customized osteotomy guide equipped with drill holes which eventually match with the screw holes of the plate during optimal (planned) gap distraction in opening-wedge HTO. Final plate type and positioning are included in the 3-D osteotomy planning. (Donnez et al. [45], Munier et al. [25], Chaouche et al. [37]; permission from the authors was obtained to publish illustrations)

for knee osteotomies. PSI has proven its value in multiple disciplines and operations such as maxillofacial/craniofacial surgery, bone tumour resections, osteotomies for mal- or non-union fractures and corrections of forearm deformities [43]. So theoretically, a 3-D core facility can supply several departments of interest, hereby sharing the costs of its own establishment and maintenance.

When used in the OR, anatomical patient models and PSI are printed in medical grade resin. Polyamide (or nylon) is the most commonly used material for guide manufacturing because of its biocompatibility and good mechanical properties [14, 16]. When devices are printed with selected laser sintering (SLS), the polyamide powder is fused into a solid model, which does not need structural support. Further, acrylonitrile butadiene styrene (ABS), a thermoplastic polymer, forms another choice and has been frequently used to print PSI guides for knee osteotomies [26, 27, 44]. Using this material, Perez-mananez was able to print PSI for less than €5 euro per patient, based on an ABS purchase price of €0.04/gram [26]. Arnal-burro et al. used polylactic acid (PLA), a thermoplastic polyester, and was able to print the required PSI per patient for even half of this price [36]. His group proposed a reasonable price range of €500–2000 for purchasing a suitable 3-D printer compatible with this material. The drawback of these inexpensive fused deposition modelling (FDM) 3-D printers, however, is the lower printing accuracy and the obvious layer lines which are inherent to filament printing. Since 2016, it is also possible to 3-D print medical grade photopolymer resins with a desktop stereolithography (SLA) printer, which offers a high resolution, accuracy and a smooth surface finish. A drawback of SLA printing is that the models need support structures which require manual removal after printing. Nevertheless, the authors have been using this printing technique for several years onsite with overall satisfying outcomes. Finally, the device should be safely sterilized in a standardized steam pressure autoclave, gamma ray sterilization or low temperature hydrogen peroxide sterilization (STERRAD sterilization) according to the instructions on the technical data sheet of the used material [14, 16, 28].

24.4 PSI Techniques and Accuracy

In a recent controlled laboratory study, the importance of PSI cutting guides was highlighted for improving osteotomy accuracy [17]. Customized slot guides (closed) were compared to open guides and free-hand sawing on a mid-shaft femur model. The closed guides had favorable outcomes in both precision of the osteotomy cut and translation of the preoperative 3-D planning. The authors concluded that the use of PSI guides (open and closed) leads to more predictable outcomes in osteotomy surgery and bony resections and can be recommended especially in multiplanar and rotational corrections [17].

In the context of osteotomies around the knee joint, PSI guides can be beneficial in two ways: first by defining the starting point, inclination angle and plane for the actual bone cut(s) and secondly by determining the planned gap opening at the medial cortex. Victor et al. designed and clinically tested the first PSI prototype for knee osteotomies (HTO and distal femur osteotomies (DFO)) which included a robust frame for fitting patient's bony landmarks to assure proper positioning (Table 24.1) [24]. This guide was equipped with a cutting slot and drill holes which would later match with the screw holes of the fixation plate as under optimal gap distraction (Fig. 24.2). After 14 cases, an accuracy outcome of $0° \pm 0.72$ ΔmFTA relative to the planning was found in the coronal plane with all cases falling within [−1°; +1°] around the target. Overall, minor changes were observed in the sagittal plane. Despite these highly accurate results, a large incision (13 cm femur and 12 cm tibia) and soft tissue dissection was required to properly fit the guide, inducing higher risk for wound infections and delayed or non-union of the gap [16]. Nevertheless, this pioneer technique was later adopted by several research groups developing their own PSI technique for opening-wedge

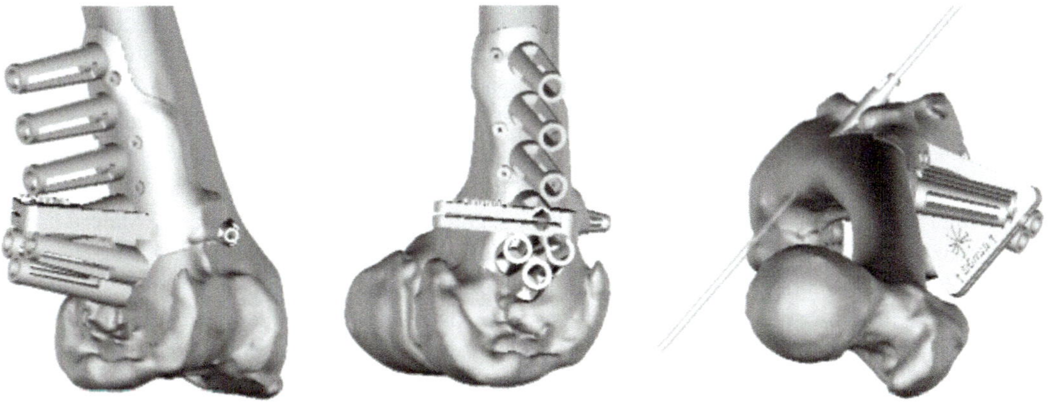

Fig. 24.2 Design of the first PSI guide for osteotomies around the knee (distal femur osteotomy (DFO)). (From Victor et al. [24]; permission from the corresponding author was obtained for illustration reprinting)

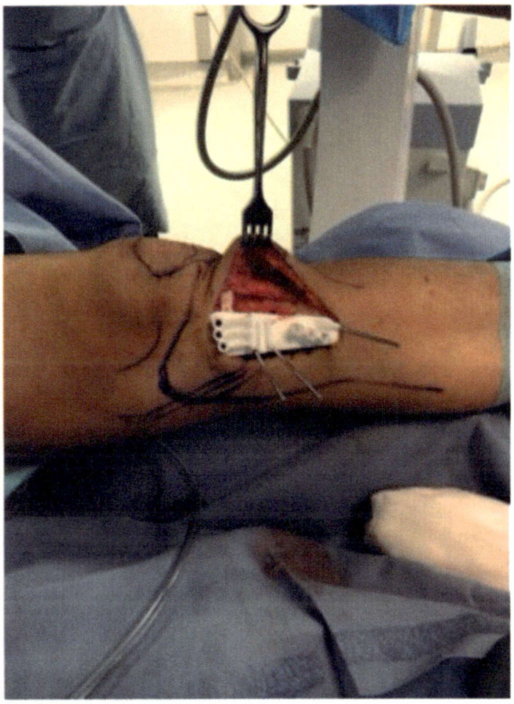

Fig. 24.3 Intraoperative positioning and fixation of a PSI cutting guide in opening-wedge HTO. (Donnez et al. [45], Munier et al. [25], Chaouche et al. [37]; permission from the authors was obtained to publish illustrations)

HTO. [14, 25, 37, 45] The largest case series with PSI was recently published by Chaouche et al., who included 100 opening-wedge HTO cases (Figs. 24.1 and 24.3) [37]. In the coronal plane, an accuracy of 1.0° ± 0.9 ΔmFTA and 0.5° ± 0.6 ΔmMPTA was established, while the planned and postoperative tibial slope differed with 0.4° ± 0.8. The authors concluded that by applying this PSI technique, predictable correction outcomes can be delivered, without increasing (non-)specific HTO complications [37].

To avoid large skin incisions for robust PSI guides, Jones et al. developed an external device to align the osteotomy cutting guide based on distant superficial bony landmarks including the fibular head and maleolli [16]. His group suggested to use a customized 'correction block' fixed with 3 k-wires to determine and maintain the intended gap opening during surgery. Preliminary results with this technique ensure an accuracy within 3° around the target after 18 HTO cases [16]. In this way, an HTO can be performed minimally invasive while maintaining freedom for the surgeon to choose the fixation device and plate positioning. However, the authors admit to a longer multi-step procedure which is in conflict with a principal advantage of PSI, namely, reducing the time and complexity of the operation [17, 26, 36].

Another way to obtain the planned limb realignment is simply to print the complementary wedge spacers needed to fill the osteotomy gap [26, 27, 44]. Perez-Mananez et al. described this approach by exchanging the spacers for structural bone autograft derived from the iliac crest in 8 HTO cases [26]. In combination with a customized positioning guide, an average accuracy of 0.5° ΔmFTA (ranging 0°–1.2°) was demonstrated. Twenty conventional control HTOs were performed, and although showing lower accuracy (average 1.1° ΔmFTA (ranging 0°–2.8°)), both groups were not significantly different. Interestingly, an additional 3-D anatomical model of the proximal tibia was always available intra-operatively to confirm fitting of the cutting guide. Shortly thereafter, the exact same PSI approach was evaluated for 12 DFOs and compared to the conventional technique [36]. Mechanical axis deviation in the coronal plane was on average 0.28° ΔmFTA (ranging 0°–1°) for PSI and 1.8° ΔmFTA (ranging 0°–4°) in controls, which was significantly different.

Similarly, but without the inclusion of an osteotomy cutting guide and the implementation of bone autograft, Kim et al. demonstrated a lower absolute difference from the correction target of 62.5% in 20 PSI HTO cases (2.3% ± 2.5 ΔWBL) compared to 20 conventional controls (6.2% ± 5.1 ΔWBL) [27]. The tibial slope remained almost unchanged in the PSI cases, while for the conventional approach, a statistically significant increase was observed. Finally, Yang et al. found an alternative way to obtain the desired wedge opening by designing a biplanar cutting guide consisting of a proximal and distal part, each equipped with an aligning hole [28]. While distracting the osteotomy, a metal rod was placed in the proximal hole and only fitted in the second distal hole of the guide when the planned osteotomy gap was obtained. A pilot study of 10 HTOs yielded a postoperative alignment of 60.2% ± 2.8 while aiming for 62.5% and a tibial slope that barely increased relative to the preoperative status.

24.5 PSI Technique of the Authors

24.5.1 3-D Planning

Preoperatively, the patient receives a full-leg bipodal standing radiograph and a supine CT-scan of the affected limb according to the Trumatch knee scanning protocol [38]. This low-dose protocol is specially designed for creating 3-D models by scanning the anatomical reference points, including hip and ankle joint at 5 mm slice thickness and spacing and the knee joint at 0.5 mm slice thickness and spacing in a 150 mm range. The resultant Digital Imaging and Communications in Medicine (DICOM) files are loaded into the segmentation software Mimics® (Materialise®, Heverlee, Belgium) to separate bony structures from surrounding soft tissue. The 3-D model of the lower limb is then transferred to the planning software 3-matic® (Materialise®, Heverlee, Belgium), in which the desired osteotomy cut and wedge opening are simulated, aiming for the postoperative mechanical axis to pass through the lateral spine (Fig. 24.4). All osteotomies are simulated using the mMPTA as main planning angle. At the end of the planning process, a personalized fitting wedge and cast are designed and 3-D printed in certified biocompatible photopolymers and sterilized by hydrogen peroxide gas plasma (Fig. 24.5). For safety reasons, the printed cast is labeled with the surgery side, the amount of correction (°) and the patient's initials. To ensure proper positioning of the printed wedge in the osteotomy gap, two grooves are created which should match with the medial cortex of the proximal and distal tibial fragment. Although this planning method looks seemingly time-consum-

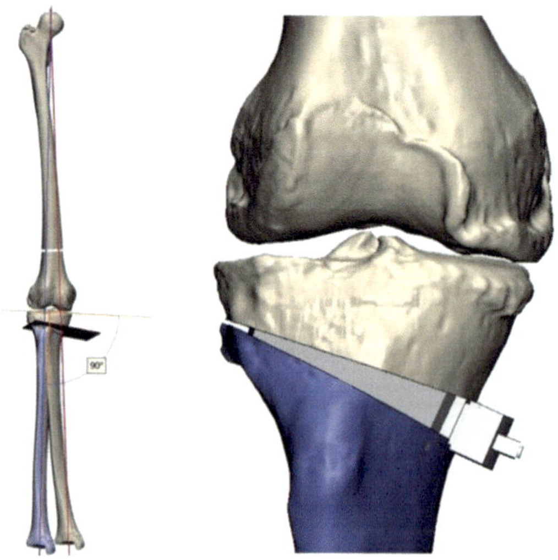

Fig. 24.4 Alignment determination on a 3-D bone model of the lower limb with virtual 3-D HTO planning and required gap opening/spacer size

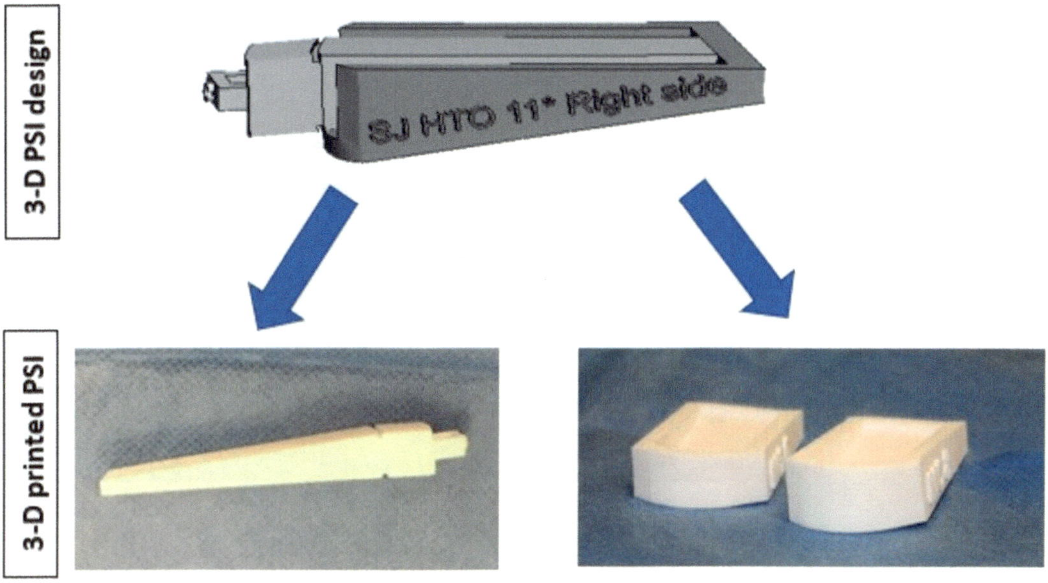

Fig. 24.5 Final design and 3-D printed models of the wedge spacer with complementary cast to trim customized bone allograft

ing, the time from scanning the lower extremity up to the availability of sterilized PSI in the OR can be fit in a 48 h streamlined flow due to the onsite availability of the required software, resin and 3-D printing equipment.

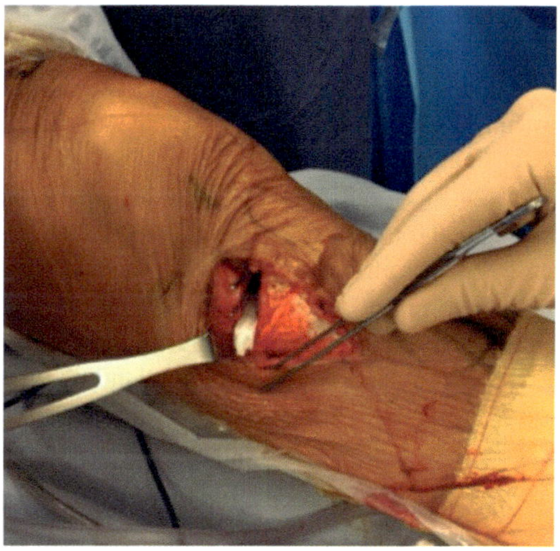

Fig. 24.6 Intraoperative introduction of the personalized wedge spacer which instantly provides the intended correction, while the identical structural bone graft is prepared

24.5.2 Surgical Technique for MOW-HTO

A vertical medial skin incision is made on the tibia. Under fluoroscopic control, two parallel K-wires are introduced horizontally, starting 3–4 cm below the medial tibial joint line on the medial cortex and aimed laterally, proximally of the tibiofibular joint and 1 cm below to the lateral joint line. The horizontal osteotomy is performed distal in contact with the 2 K-wires on the medial side using an oscillating saw, followed by an oblique step osteotomy at the level of the tibial tubercle, as planned in 3-D. The horizontal osteotomy is gently opened by inserting five chisels in a progressive manner posteriorly, without full engagement. The personalized wedge spacer is now introduced in the gap while giving mild valgus stress (Fig. 24.6). The two grooves on the printed wedge are checked for matching the medial cortices. The bone graft preparations are started, while the customized spacer remains in the osteotomy gap, keeping the tibia in the intended corrected position. The printed negative cast is used as a box in which the bone allograft (half femoral head) can be

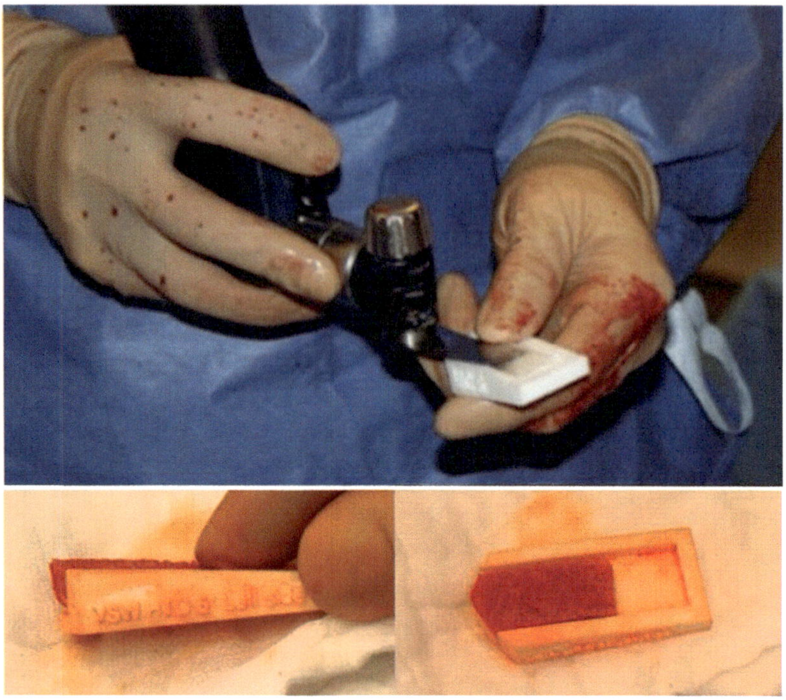

Fig. 24.7 Precise trimming of the structural bone allograft derived from half of a femur head in the dedicated cast of the patient

precisely customized. The bone allograft is trimmed triangularly by a sawblade until the size matches the original printed wedge (Fig. 24.7). When ready, the printed spacer is exchanged for the wedge-shaped structural bone graft which ultimately provides an identical alignment correction. The osteotomy is finally fixed with a TomoFix® locking plate (Depuy-Synthes GmbH, Solothurn, Switzerland).

24.5.3 Accuracy Outcome

For study purposes, ten patients that were operated according to this novel PSI technique received a full leg CT-scan and radiograph at 3 months postoperatively to assess accuracy outcomes in the coronal and sagittal plane (Table 24.2). Accuracy results showed that 90% (9/10) were within an accuracy range of [−1.5°;+1.5°] mFTA around the target, while all cases were within [−2°;+2°]. In the sagittal plane, an absolute ΔTS of 2.7° ± 1.8 was observed with an effective average slope increase of 2.1°. In comparison to previous PSI osteotomy studies (Table 24.1), our pilot study showed highly accurate and therefore similar results in the coronal plane, while assessment was performed on more reliable 3-D imaging postoperatively. However, in the sagittal plane, an unintended slight increase of the posterior slope was observed. The authors hypothesize that this might have been due to the

Table 24.2 Accuracy outcomes of the authors personal PSI technique for opening-wedge HTO

Angle	Outcome	3-D imaging (mean ± SD)	2-D imaging (mean ± SD)
mFTA (°)	Relative Δ	−0.4 ± 1.0	-0.5 ± 1.3
	Absolute Δ	0.9 ± 0.6	1.2 ± 0.7
mMPTA (°)	Relative Δ	−1.0 ± 1.4	0.3 ± 2.2
	Absolute Δ	1.3 ± 1.1	1.7 ± 1.3
TS (°)	Relative Δ	2.1 ± 2.6	0.0 ± 3.2
	Absolute Δ	2.7 ± 1.8	2.2 ± 2.2

Δ difference, *mMPTA* mechanical medial proximal tibial angle, *mFTA* mechanical femorotibial angle, *WBL* weight-bearing line, *TS* tibial slope, *SD* standard deviation limited width of the printed wedge and structural graft (1 cm) which allowed for tibia plateau tilting in the sagittal plane. Therefore, a larger case series is currently ongoing to investigate a resized model of this PSI technique.

24.6 General Factors to Consider in 3-D Planning and PSI Osteotomy

Besides accurately obtaining the planned osteotomy correction, some practical and logistical factors need to be considered when applying 3-D planning and printing of PSI in clinical practice. Firstly, 3-D imaging in any form (CT or MRI) of the proximal tibia is minimally required to simulate the bone cut and plan the osteotomy opening in a multiplanar fashion. This might be associated with an additional cost and in case of CT-scan, with increased radiation exposure on top of a standard preoperative full-leg radiograph. The effective radiation dose of a CT-scan is largely dependent on the applied slice thickness, spacing and scanned area. Therefore, very low-dose protocols for scanning the lower limb have been established, only targeting a centred range of the hip, knee and ankle joint resulting in reliable 3-D anatomic models for planning realignment and arthroplasty surgery [20]. In this way, the effective radiation dose can be reduced to the equivalent of one full-leg standing radiograph. Altogether, the slight increase in radiation dose for 3-D planning purposes should be put in perspective to the reduced need for fluoroscopy intraoperatively when applying PSI [16, 26, 28, 36].

Primary goals of PSI are to facilitate technically demanding osteotomy surgeries, leading to reduced operating times while minimizing human correction errors [17, 36]. Perez-Mananes recorded the tourniquet time in HTO cases with and without PSI which was on average 61 and 92 minutes, respectively [26]. Similar for DFO operations, significantly reduced surgery times were observed in favour of the PSI technique [36]. In addition, the saved OR time was financially translated and yielded €522/procedure, which ultimately appeared to cover the cost of a new 3-D printer. Nevertheless, preoperative 3-D planning and printing is obviously more time-consuming relative to conventional methods and often requires the collaboration with a biomedical engineer. So, in short, the time and associated cost saved during PSI surgery can be directly reinvested in the preoperative planning and production phase of the next osteotomy patient, resulting in a sustainable and economically healthy feedback system. This is in contrast to the use of computer navigation, which is, despite delivering highly accurate corrections in lower limb realignment, prolonging the operation time, technically more demanding and very expensive on top [18, 19].

A legitimate concern, however, is the effect of PSI mal-positioning as this might potentially increase the risk of tibia plateau fractures, intra-articular screw positioning, inaccurate translation of the planning and poor clinical outcomes [46]. To assess the potential consequences, Jud et al. simulated guide mal-positioning (cutting slot with predrilled screw holes for matching plate fixation) by stepwise translation (5 mm) and rotation (2.5°) on the proximal tibia in 3-D medical software [46]. Although a proximal 5 mm translation of the guide resulted in surgical failure, the authors concluded that PSI mal-positioning was safe within the possible 'degrees of freedom' and had low impact on coronal accuracy. Tibial slope changes, however, were not assessed.

Finally, the experience of the surgeon should be taken into account when the accuracy and potential advantages of PSI and conventional HTO studies are investigated. The authors hypothesize that the implementation of PSI might be most beneficial in case of young or unexperienced orthopaedic surgeons performing standard knee osteotomies, since a short learning curve can be expected with most PSI guides. However, for the experienced senior surgeon, satisfying accuracy levels can potentially be obtained with conventional HTO approaches, but PSI might still be valuable in more complex surgeries such as large or rotational corrections, multiplanar deformities and double-level osteotomies.

In future perspectives, technological development might further reduce the radiation exposure and advance required imaging such as EOS weight-bearing full-leg CT-scan and cone-beam. Further, the automation of the segmentation/ planning process should be stimulated and the cost of 3-D software and printers decreased to enhance the onsite accessibility of medical 3-D technology. Additionally, advanced technology with biomechanical finite element analysis will evolve, attempting to customize the fixation hardware and improve implant size and fit to the 'post-distraction' medial cortex [47, 48]. This approach might potentially result in less postoperative skin irritation and subsequently lowering the reoperation rate for hardware removal after knee osteotomies.

24.7 Conclusion

Three-dimensional osteotomy planning and PSI printing have successfully found their way into the field of knee osteotomy surgery. A handful of PSI techniques have been developed and clinically tested over the past decade, showing overall highly accurate outcomes in the coronal plane, while the tibial slope can be well-controlled. Despite these promising preliminary results, the biplanar accuracy and long-term clinical advantage over conventional HTO surgery remains to be determined in large comparative, and preferably randomized, trials. In the meantime, technological development might further (1) reduce the radiation exposure and advance required imaging, (2) stimulate the automation of the segmentation processes and (3) decrease the cost of 3-D software and printers to make medical 3-D technology accessible for the majority of hospitals. In addition, radiation exposure, costs for equipment, time-intensive preoperative planning and experience of the surgeon are factors that need to be outbalanced with the relative benefits associated with surgical accuracy. Nevertheless, in complex osteotomy cases, the authors advocate the use 3-D planning and PSI. It can guide the surgeon through the operation, leading to satisfying accuracy outcomes, as this remains one of the most important factors in the durability of joint-preserving osteotomies around the knee.

References

1. Brinkman JM, Lobenhoffer P, Agneskirchner JD, Staubli AE, Wymenga AB, Van Heerwaarden RJ. Osteotomies around the knee: patient selection, stability of fixation and bone healing in high tibial osteotomies. J Bone Jt Surg - Ser B. 2008;90(12):1548–57.
2. Liu X, Chen Z, Gao Y, Ztthang J, Jin Z. High Tibial Osteotomy: Review of Techniques and Biomechanics. J Healthc Eng. 2019;2019:8363128.
3. Bellemans J, Colyn W, Vandenneucker H, Victor J. The chitranjan ranawat award: is neutral mechanical alignment Normal for all patients? The concept of constitutional Varus. Clin Orthop Relat Res. 2012;470(1):45–53.
4. Sharma L, Song J, Felson DT, Cahue S, Shamiyeh E, Dunlop DD. The role of knee alignment in disease progression and functional decline in knee osteoarthritis. J Am Med Assoc. 2001;286(2):188–95.
5. Brouwer GM, Van Tol AW, Bergink AP, Belo JN, Bernsen RMD, Reijman M, et al. Association between valgus and varus alignment and the development and progression of radiographic osteoarthritis of the knee. Arthritis Rheum. 2007;56(4):1204–11.
6. Duivenvoorden T, Brouwer RW, Baan A, Bos PK, Reijman M, Bierma-Zeinstra SMA, et al. Comparison of closing-wedge and opening-wedge high tibial osteotomy for medial compartment osteoarthritis of the knee: a randomized controlled trial with a six-year follow-up. J Bone Joint Surg Am. 2014;96:1425–32.
7. Brouwer RW, Bierma-Zeinstra SM. A, van Raaij TM, Verhaar J a N. osteotomy for medial compartment arthritis of the knee using a closing wedge or an opening wedge controlled by a Puddu plate. A one-year randomised, controlled study. J Bone Joint Surg Br. 2006;88(11):1454–9.
8. Sabzevari S, Ebrahimpour A, Khalilipour Roudi M, Kachooei AR. High Tibial osteotomy: a systematic review and current concept. Arch Bone Jt Surg [Internet]. 2016;204(43):204–12. Available from: http://abjs.mums.ac.ir
9. Hantes ME, Natsaridis P, Koutalos AA, Ono Y, Doxariotis N, Malizos KN. Satisfactory functional and radiological outcomes can be expected in young patients under 45 years old after open wedge high tibial osteotomy in a long-term follow-up. Knee Surg Sport Traumatol Arthrosc [Internet]. 2018;26(11):3199–205. https://doi.org/10.1007/s00167-017-4816-z.
10. Schallberger A, Jacobi M, Wahl P, Maestretti G, Jakob RP. High tibial valgus osteotomy in unicompartmental medial osteoarthritis of the knee: a retrospective follow-up study over 13-21 years. Knee Surg Sports Traumatol Arthrosc. 2011 Jan;19(1):122–7.

11. Takeuchi R, Ishikawa H, Kumagai K, Yamaguchi Y, Chiba N, Akamatsu Y, et al. Fractures around the lateral cortical hinge after a medial opening-wedge high tibial osteotomy: a new classification of lateral hinge fracture. Arthrosc - J Arthrosc Relat Surg. 2012;28(1):85–94.
12. Woodacre T, Ricketts M, Evans JT, Pavlou G, Schranz P, Hockings M, et al. Complications associated with opening wedge high tibial osteotomy - a review of the literature and of 15 years of experience. Knee [Internet]. 2016;23(2):276–82. https://doi.org/10.1016/j.knee.2015.09.018.
13. Nha KW, Kim HJ, Ahn HS, Lee DH. Change in posterior Tibial slope after open-wedge and closed-wedge high Tibial osteotomy. Am J Sports Med. 2016;44(11):3006–13.
14. Fucentese SF, Meier P, Jud L, Köchli GL, Aichmair A, Vlachopoulos L, et al. Accuracy of 3D-planned patient specific instrumentation in high tibial open wedge valgisation osteotomy. J Exp Orthop. 2020;7(1):1–7.
15. Van den Bempt M, Van Genechten W, Claes T, Claes S. How accurately does high tibial osteotomy correct the mechanical axis of an arthritic varus knee? A systematic review. Knee [Internet]. 2016;23(6):925–35. https://doi.org/10.1016/j.knee.2016.10.001.
16. Jones GG, Jaere M, Clarke S, Cobb J. 3D printing and high tibial osteotomy. EFORT Open Rev [Internet]. 2018;3(5):254–9. Available from: http://online.boneandjoint.org.uk/doi/10.1302/2058-5241.3.170075
17. Sys G, Eykens H, Lenaerts G, Shumelinsky F, Robbrecht C, Poffyn B. Accuracy assessment of surgical planning and three-dimensional-printed patient-specific guides for orthopaedic osteotomies. Proc Inst Mech Eng Part H J Eng Med. 2017;231(6):499–508.
18. Saragaglia D, Roberts J. Navigated osteotomies around the knee in 170 patients with osteoarthritis secondary to genu varum. Orthopedics. 2005;28(10 Suppl):s1269–74.
19. Iorio R, Pagnottelli M, Vadalà A, Giannetti S, Di Sette P, Papandrea P, et al. Open-wedge high tibial osteotomy: comparison between manual and computer-assisted techniques. Knee Surg Sport Traumatol Arthrosc. 2013;21(1):113–9.
20. Henckel J, Richards R, Lozhkin K, Harris S, Baena FMR, Barrett ARW, et al. Very low-dose computed tomography for planning and outcome measurement in knee replacement: the imperial knee protocol. J Bone Jt Surg - Br Vol [Internet]. 2006;88-B(11):1513–8. Available from: http://www.bjj.boneandjoint.org.uk/cgi/doi/10.1302/0301-620X.88B11.17986
21. Sarment DP, Al-Shammari K, Kazor CE. Stereolithographic surgical templates for placement of dental implants in complex cases. Int J Periodontics Restorative Dent [Internet]. 2003;23(3):287–95. Available from: http://www.ncbi.nlm.nih.gov/pubmed/12854779
22. Lu S, Zhang YZ, Wang Z, Shi JH, Chen YB, Xu XM, et al. Accuracy and efficacy of thoracic pedicle screws in scoliosis with patient-specific drill template. Med Biol Eng Comput [Internet]. 2012;50(7):751–8. Available from: http://www.ncbi.nlm.nih.gov/pubmed/22467276
23. Byrne A-M, Impelmans B, Bertrand V, Van Haver A, Verstreken F. Corrective osteotomy for Malunited Diaphyseal forearm fractures using preoperative 3-dimensional planning and patient-specific surgical guides and implants. J Hand Surg Am [Internet]. 2017;42(10):836.e1–836.e12. Available from: http://www.ncbi.nlm.nih.gov/pubmed/28709790
24. Victor J, Premanathan A. Virtual 3D planning and patient specific surgical guides for osteotomies around the knee: a feasibility and proof-of-concept study. Bone Joint J. 2013;95 B(11 Suppl A):153–8.
25. Munier M, Donnez M, Ollivier M, Flecher X, Chabrand P, Argenson JN, et al. Can three-dimensional patient-specific cutting guides be used to achieve optimal correction for high tibial osteotomy? Pilot study. Orthop Traumatol Surg Res [Internet]. 2017;103(2):245–50. https://doi.org/10.1016/j.otsr.2016.11.020.
26. Pérez-Mañanes R, Burró JA, Manaute JR, Rodriguez FC, Martín JV. 3D surgical printing cutting guides for open-wedge high Tibial osteotomy: do it yourself. J Knee Surg. 2016;29(8):690–5.
27. Kim H-J, Park J, Shin J-Y, Park I-H, Park K-H, Kyung H-S. More accurate correction can be obtained using a three-dimensional printed model in open-wedge high tibial osteotomy. Knee Surg Sport Traumatol Arthrosc [Internet]. 2018;26:3452–8. Available from: http://link.springer.com/10.1007/s00167-018-4927-1.
28. Yang JCS, Chen CF, Luo CA, Chang MC, Lee OK, Huang Y, et al. Clinical experience using a 3D-printed patient-specific instrument for medial opening wedge high Tibial osteotomy. Biomed Res Int. 2018;2018:1–9.
29. Schröter S, Ihle C, Mueller J, Lobenhoffer P, Stöckle U, van Heerwaarden R. Digital planning of high tibial osteotomy. Interrater reliability by using two different software. Knee Surgery, Sport Traumatol Arthrosc. 2013;21(1):189–96.
30. Swanson KE, Stocks GW, Warren PD, Hazel MR, Janssen HF. Does axial limb rotation affect the alignment measurements in deformed limbs? Clin Orthop Relat Res. 2000;371:246–52.
31. Koshino T, Takeyama M, Jiang LS, Yoshida T, Saito T. Underestimation of varus angulation in knees with flexion deformity. Knee. 2002;9(4):275–9.
32. Kawakami H, Sugano N, Yonenobu K, Yoshikawa H, Ochi T, Hattori A, et al. Effects of rotation on measurement of lower limb alignment for knee osteotomy. J Orthop Res. 2004;22(6):1248–53.
33. Specogna AV, Birmingham TB, Hunt MA, Jones IC, Jenkyn TR, Fowler PJ, et al. Radiographic measures of knee alignment in patients with varus gonarthrosis: effect of weightbearing status and associations with dynamic joint load. Am J Sports Med. 2007;35(1):65–70.
34. Ogawa H, Matsumoto K, Ogawa T, Takeuchi K, Akiyama H. Preoperative varus laxity correlates with overcorrection in medial opening wedge

high tibial osteotomy. Arch Orthop Trauma Surg. 2016;136(10):1337–42.
35. Lee YS. Eu., Kim MG y., Byun HW o., Kim SB u., Kim JG o. reliability of the imaging software in the preoperative planning of the open-wedge high tibial osteotomy. Knee Surg Sports Traumatol Arthrosc. 2015;23(3):846–51.
36. Arnal-Burró J, Pérez-Mañanes R, Gallo-del-Valle E, Igualada-Blazquez C, Cuervas-Mons M, Vaquero-Martín J. Three dimensional-printed patient-specific cutting guides for femoral varization osteotomy: do it yourself. Knee. 2017;24(6):1359–68.
37. Chaouche S, Jacquet C, Fabre-Aubrespy M, Sharma A, Argenson JN, Parratte S, et al. Patient-specific cutting guides for open-wedge high tibial osteotomy: safety and accuracy analysis of a hundred patients continuous cohort. Int Orthop. 2019;43(12):2757–65.
38. DePuy Orthopaedics. Radiation exposure considerations with the use of TruMatch personalized solutions. 2010.
39. Kubota M, Ohno R, Sato T, Yamaguchi J, Kaneko H, Kaneko K. The medial proximal tibial angle accurately corrects the limb alignment in open-wedge high tibial osteotomy. Knee Surg Sport Traumatol Arthrosc [Internet]. 2018;27:1–7. https://doi.org/10.1007/s00167-018-5216-8.
40. Nakayama H, Schröter S, Yamamoto C, Iseki T, Kanto R, Kurosaka K, et al. Large correction in opening wedge high tibial osteotomy with resultant joint-line obliquity induces excessive shear stress on the articular cartilage. Knee Surg Sport Traumatol Arthrosc. 2018;26(6):1873–8.
41. Chernchujit B, Tharakulphan S, Prasetia R, Chantarapanich N, Jirawison C, Sitthiseripratip K. Preoperative planning of medial opening wedge high tibial osteotomy using 3D computer-aided design weight-bearing simulated guidance: technique and preliminary result. J Orthop Surg. 2019;27(1):1–8.
42. Victor J, Van Doninck D, Labey L, Innocenti B, Parizel PM, Bellemans J. How precise can bony landmarks be determined on a CT scan of the knee? Knee [Internet]. 2009;16(5):358–65. https://doi.org/10.1016/j.knee.2009.01.001.
43. Auricchio F. Instructional lecture : general Orthopaedics 3D printing : clinical applications in orthopaedics and traumatology. EFORT Open Rev. 2016;1(May):121–7.
44. Kwun JD, Kim HJ, Park J, Park IH, Kyung HS. Open wedge high tibial osteotomy using three-dimensional printed models: experimental analysis using porcine bone. Knee [Internet]. 2017;24(1):16–22. https://doi.org/10.1016/j.knee.2016.09.026.
45. Donnez M, Ollivier M, Munier M, Berton P, Podgorski JP, Chabrand P, et al. Are three-dimensional patient-specific cutting guides for open wedge high tibial osteotomy accurate? An in vitro study. J Orthop Surg Res. 2018;13(1):1–8.
46. Jud L, Fürnstahl P, Vlachopoulos L, Götschi T, Leoty LC, Fucentese SF. Malpositioning of patient-specific instruments within the possible degrees of freedom in high-tibial osteotomy has no considerable influence on mechanical leg axis correction. Knee Surgery, Sport Traumatol Arthrosc [Internet]. 2019;28(5):1356–64. https://doi.org/10.1007/s00167-019-05432-3.
47. Koh YG, Lee JA, Lee HY, Chun HJ, Kim HJ, Kang KT. Design optimization of high tibial osteotomy plates using finite element analysis for improved biomechanical effect. J Orthop Surg Res. 2019;14(1):1–10.
48. Yoo OS, Lee YS, Lee MC, Park JH, Kim JW, Sun DH. Morphologic analysis of the proximal tibia after open wedge high tibial osteotomy for proper plate fitting. BMC Musculoskelet Disord [Internet]. 2016;17(1):1–9. https://doi.org/10.1186/s12891-016-1277-3.

Save the Meniscus: Advances in Meniscal Repair Techniques

25

Johannes Zellner and Peter Angele

25.1 Introduction

Meniscal lesions represent one of the most common intra-articular knee injury and are therefore one of the most frequent cause of surgical procedures in orthopedic surgery. The mean annual incidence of meniscal lesions has been reported to be 66 per 100,000 inhabitants, 61 of which result in partial or subtotal meniscectomy [1]. The changes in "pivoting" sports activities in the past few decades have resulted in increased injury rates of the meniscus [2]. Especially in combination with anterior cruciate ligament injuries, a high incidence of acute meniscal lesions (40–80%) can be detected.

Additionally, an increasing number of degenerative meniscus lesions have been detected over the last few decades. Although it is still under debate whether these meniscus lesions are better treated conservatively or operatively, there is no discussion about the fact that such degenerative meniscal changes are associated with the development of osteoarthritis of the knee.

Meniscus integrity is the key for joint health of the knee. Untreated meniscus tears cause intermittent pain, joint swelling, recurrent mechanical symptoms (clicking, catching, giving way) and, therefore, significant reduction in quality of life in predominately young and active patients [3].

In the long-term, meniscus tears can result in the onset of joint degeneration and, finally, knee osteoarthritis with all its consequences including pain, immobility, and the need of knee arthroplasty [2, 4–7]. In a recent published case-control study, specific meniscus tear morphologies (meniscus extrusion, complex tears, tears with large radial involvement) have shown to be significantly more common in patients with progressive development of osteoarthritic changes in a 2-year follow-up indicating that these meniscus tears represent a negative prognostic risk factor for later development of osteoarthritis [7].

Removal of meniscus tears leads to short-term relief of clinical symptoms but also to knee osteoarthritis in long term [6, 8–10]. Especially, the amount meniscus removed, lateral meniscectomy, concomitant injuries such as ACL ruptures, malalignment, high BMI, and longer duration of clinical symptoms preoperatively have been identified as negative prognostic risk factors for

J. Zellner · P. Angele (✉)
Sporthopaedicum Regensburg, Hildegard von Bingen Strasse 1, Regensburg, Germany

University Medical Centre Regensburg, Department of Trauma Surgery, Franz Josef Strauss Allee 11, Regensburg, Germany
e-mail: zellner@sporthopaedicum.de; angele@sporthopaedicum.de

N. Nakamura et al. (eds.), *Advances in Knee Ligament and Knee Preservation Surgery*, https://doi.org/10.1007/978-3-030-84748-7_25

the onset of osteoarthritis in systematic reviews [6, 10]. Elevated expression levels of arthritis-related markers in meniscus tears in patients under 40 years old, compared to patients over 40 years, and in patients with meniscus and anterior cruciate ligament tears, compared to patients with isolated meniscus tears, indicate an increased catabolic response suggesting a higher risk for progression of osteoarthritis following partial meniscectomy [11].

Knowing the risk for the onset of osteoarthritis after meniscectomy, the majority of meniscus tears are still treated with partial meniscectomy as shown in a cohort of more than 1000 young patients undergoing anterior cruciate ligament reconstruction [12].

Therefore, the main goal in the care of meniscal tears should be the maintenance of as much meniscus tissue as possible [2, 12–14]. This includes repair of meniscus tears and regeneration of meniscus defects with regenerative treatment approaches like biological augmentation.

25.2 Anatomical, Physiological and Biomechanical Considerations on Meniscus Regeneration Capacity

The meniscus plays a decisive role for the integrity of the knee joint. This includes shock absorption and transmission but also joint stabilization, proprioception, lubrication, and nutrition of the articular cartilage [1]. Biomechanical studies have shown that a loss of meniscus integrity leads to remarkable changes in kinematics and load distribution in the knee joint. The pressure on the surrounding native articular cartilage subsequently increases. Even a resection of only 15–34% of meniscus tissue enhances the load on the surrounding hyaline cartilage up to 350% [15].

In accordance to that, osteoarthritis of the knee, as a resulting effect of meniscectomy, has already been described a long time ago [16]. According to the current literature, partial meniscectomy is also well known to predispose the knee for the development and an early onset of osteoarthritis [17, 18]. Especially, the following criteria are defined as risk factors for the development of degenerative changes in context to meniscus injuries (according to Mordecai) [19]:

- Partial meniscectomy of the lateral meniscus.
- Resection of larger portions of meniscus tissue.
- Radial tears reducing or cancelling the meniscus ring tension (functional meniscectomy).
- Preexisting cartilage lesions.
- Persisting ligamentous joint instability.
- Axis deviation (varus-medial, valgus-lateral).
- Obesity.
- Age > 40 years.
- Low activity level.

According to the increasing knowledge concerning the biology and function of the meniscus, there is a consensus to preserve as much meniscus tissue as possible in the treatment of meniscus injuries. Thus, different techniques for the therapy of meniscus tears have been developed over time. Today, meniscus repair is the gold standard for the treatment of meniscus lesions particularly in vascularized portion. Whereas initially this procedure was performed as an open procedure, up to now it is almost exclusively performed arthroscopically. Different techniques for meniscus suturing have to be distinguished: all-inside, outside-in, and inside-out.

The vascularization and nutritional situation of the injured meniscus area as well as the type of meniscus tear are important for the success of a meniscus reconstruction.

While the inner 2/3 of the meniscus ("white-white") is nourished by diffusion from the synovial fluid, the periphery in the so-called red-red zone has a vascular supply. Between the white-white zone and the vascularized portion, a red-white transition zone is located. Especially, the outer third and, to a lesser extent, the red-white transition zone show a regenerative potential with good conditions for a successful meniscus suturing [20] (Fig. 25.1).

However, the meniscus still remains a challenging structure for repair and restoration. The question arises whether the limited healing capacity mainly in the inner thirds of the menis-

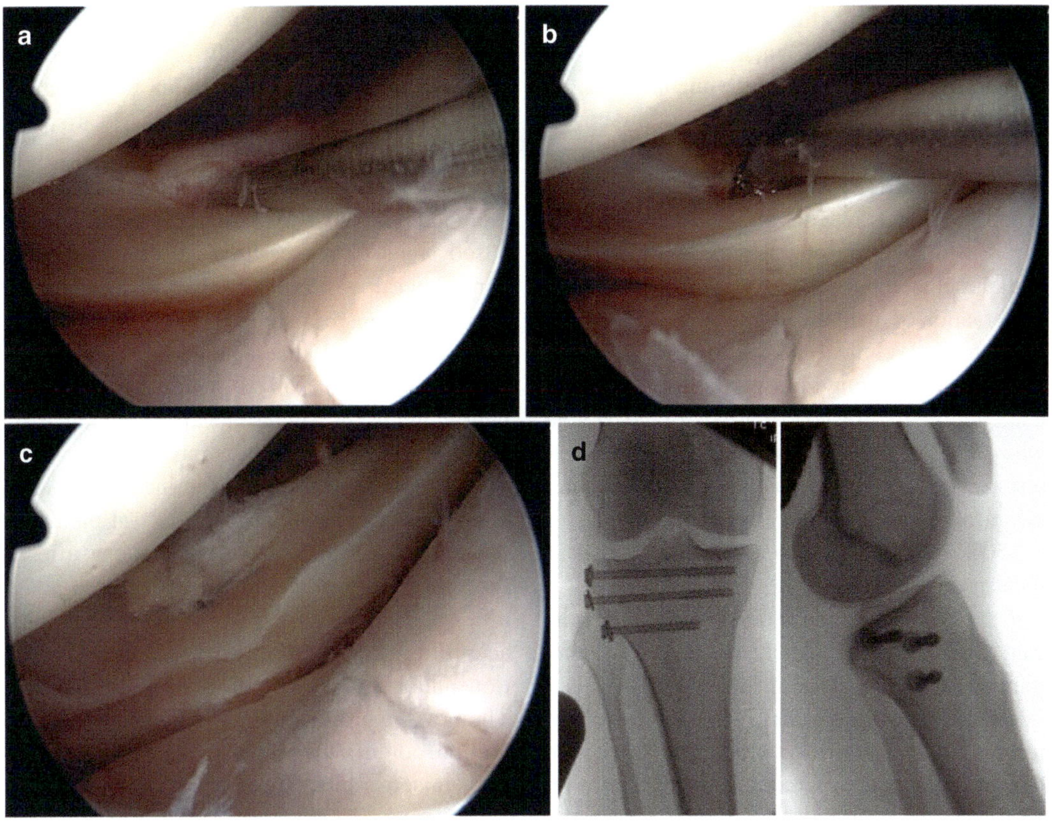

Fig. 25.1 Meniscus suturing of a meniscal tear in vascularized portion of the lateral meniscus in case of lateral tibial plateau fracture. (**a**) Arthroscopically assisted anatomical fracture reduction and evaluation of the meniscus tear. (**b**) Rasping of the meniscus tear. (**c**) All-inside suture at the posterior horn of the lateral meniscus D intraoperative X-ray

cus can be overcome by innovative treatment strategies like, for example, biological treatment augmentation. Additionally, over the last decades, different tissue engineering approaches came in the focus of research to enhance the healing potential in order to save or to rebuild as much meniscus tissue as possible to improve long-term outcome after meniscus treatment and to prevent the onset of osteoarthritis.

25.3 Meniscus Reconstruction Improves the Knee Function in Long-Term

The first description of a meniscus suture technique was published by Annandale in 1885 [21]. Since then, the treatment options for the reconstructive therapy of meniscus lesions have been significantly advanced, especially by the development of arthroscopic techniques. Regarding studies and meta-analysis describing the long-term outcome after meniscal reconstructive therapy, the technical development of the treatment options (open versus arthroscopic procedures) have to be considered.

Tengrootenhuysen et al. retrospectively compared the clinical outcome after successful and failed meniscus suture in 119 patients after a mean follow-up of more than 5 years [22]. The successful reconstruction of the meniscus was associated with a significant improvement of the knee function according to the IKDC and Lysholm score.

Xu et al. evaluated the long-term outcome of meniscus reconstruction in comparison to the long-term outcome after partial meniscectomy [23]. According to the inclusion criteria, 367 patients of seven studies were included in this meta-analysis. After a mean follow-up of 84 months, they detected a significant improvement of the IKDC and Lysholm score in the group of patients receiving a meniscus suture in contrast to the group of patients who have had partial meniscectomy. They summarized that the preservation of meniscus tissue is associated with an improved clinical and functional outcome over a mid- and long-term period. In addition to that, Stein et al. showed that 96.2% of the patients who had a meniscus reconstruction were able to restore their pre-injury activity level within a mean follow-up of almost 9 years in comparison to 50% of the patients who had a partial meniscectomy.

Overall, the current literature shows a significant positive effect of meniscus suture on knee joint function in long-term. However, the question remains to what extent meniscus preserving techniques are able to positively influence the development of degenerative changes within the knee joint.

25.4 Prevention of Osteoarthritis by Meniscus Suturing in Long Term

The integrity of the meniscus is of impact for the prevention of osteoarthritis, such as shown by (partial) meniscectomy. It usually goes along with a loss of symptoms and functional improvement in short-term [24]. However, the long-term outcome after (partial) meniscectomy shows a trend toward degenerative effects. Englund et al. described an association between the degenerative effect and the amount of lost meniscus tissue [25]. Even if the partial meniscectomy does not show that extended destructive effect, osteoarthritic changes are also documented after a follow up of 16 years after partial meniscectomy [26]. So, Papalia et al. defined the amount of resected meniscus tissue as a predictive factor for the development of osteoarthritis [10].

In a systematic review concerning the outcome after arthroscopically performed partial meniscectomy with a minimum follow-up of 8 years and a mean age of 36 years, satisfying results concerning the functional outcome were found by Petty et al. [18]. Nevertheless, all included studies evaluating radiologically based signs of osteoarthritis in the index and contralateral site detected significantly enhanced signs of osteoarthritis in the partially meniscectomized knee. Comparing medial and lateral, especially partial meniscectomy of the lateral meniscus shows negative influence on the development of degenerative changes [8]. In this context Lee et al. examined 49 patients after subtotal resection of the lateral meniscus having lateral meniscus replacement after a mean of 4.5 years. The authors observed a significant development of signs of osteoarthritis according to the Kellgren-Lawrence classification and a progressive loss of the joint line. Though, the process of progressive joint degeneration could have positively been influenced by meniscus replacement [27].

In contrast to (partial) meniscectomy meniscus preserving techniques such as meniscus suturing show a cartilage protective effect in long-term. Noyes et al. evaluated the meniscal status of 33 patients having meniscus suture after a mean follow-up of 16.8 years by MRI scan. No degenerative changes in the operated compartment or differences concerning the status of degeneration in comparison to the healthy, contralateral site were found in patients after having successful meniscus reconstruction [28]. Johnson et al. compared the injured and contralateral knee joint 10 years after meniscus suture on X-rays [29]. Only 8% of these patients developed osteoarthritic signs on the operated site, while degenerative changes were also found in even 3% of the contralateral, intact knee joints. Furthermore, Tengrootenhuysen et al. analyzed differences between patients after a successful meniscus suture and patients in whom the meniscus suture failed [22]. In 14% of the patients having a successful reconstruction of the meniscus, signs of osteoarthritis were documented in X-ray. In contrast to that, in more than 80% of the patients with a failed meniscus preserving therapy, signs of osteoarthritis were seen.

Regarding the development of osteoarthritis of the knee, techniques preserving a functional intact meniscus tissue show advantages in comparison to partial meniscectomy. Stein et al. showed no progress of radiological signs of osteoarthritis in 81% of the evaluated patients after almost 9 years after meniscus suturing, whereas a reduction of a degenerative progress was seen in 40% of the patients after partial meniscectomy [2]. Similar results were found by Paxton et al. [9]. While 78% of the patients had no progress of the osteoarthritic status according to the X-ray after having reconstruction of the meniscus, just in 64% of the patients, who had partial meniscectomy, no further development of osteoarthritis was detected. Especially in younger patients, further studies showed also clear advantages of the meniscus preserving techniques in contrast to the partial meniscectomy regarding osteoarthrosis preventing qualities [30].

25.5 Higher Revision Rate after Meniscus Suture Compared to Partial Meniscectomy

The current literature contains different studies about the failure rate after meniscus suture. Johnson et al. report a secondary meniscectomy rate of 24% within 10 years after meniscus repair [29]. Also Nepple et al. documented a meniscus suture failure rate of 23% regarding an observation period of at least 5 years [31]. However, most of these long-term outcome studies refer to older meniscus suture techniques that were predominantly performed in an open procedure.

Regarding recent arthroscopic meniscus suture techniques, a further improvement of long-term outcome and a reduced failure rate are described. Lozano et al. reviewed the outcome after all-inside meniscus suture and found a mean failure rate of 15% [32].

When looking at the literature, the question which arises is what should be considered as a treatment failure, as MRI follow up is not always meaningful. Pujol et al. analyzed MRI examinations of the knee joint 10 years after meniscus suturing and found a hyperintense signal in the treated meniscus in 87% of the cases. The authors concluded that MRI is not suitable for the analysis of the healing status of the meniscus after meniscus suture [33].

If a treatment failure is defined as a necessary re-operation, there is a clear advantage for the partial meniscectomy compared to meniscus reconstruction. In a review, Paxton et al. analyzed 95 studies regarding the outcome and re-operation rate after meniscus treatment [9]. For the period of 0 to 4 years after the first meniscus surgery, they found a re-operation rate of 1.4% in the meniscectomy group compared to 16.5% in the meniscus suture group. In the observation period longer than 10 years, a ratio of 3.9% for meniscectomy to 20.7% for meniscus suture was detected. However, the re-operation was defined as a further meniscus treatment. Whether and how many patients in which group had to be converted to, for example, an arthroplasty remains unclear. Nevertheless, meniscus repair has a higher revision rate over time. This is a fact that has to be explicitly discussed with the patient before meniscus treatment.

On the other hand, revision surgery has not necessarily to be classified as a complete failure of the meniscus suture. Pujol et al. showed that a partial restoration of meniscus is also possible [34]. In 37 patients, the amount of meniscal substance resected during the revision was compared with the initial rupture. They found that in 52% approximately the same amount, but in 35% of the cases even less meniscus tissue had to be removed during the revision surgery. Regarding the fact that more meniscus tissue also means enhanced protection for the surrounding cartilage, this could also have a positive effect on the long-term outcome.

Despite promising results for successful meniscus repair regarding functional outcome and prevention of osteoarthritis, there is still the need to improve the healing rate after meniscus suture.

25.6 Stimulation of the Regenerative Potential of the Meniscus Tissue

In daily clinical practice, the regenerative potential of the meniscus can be further supported by various measures.

Refreshing of the margins of the meniscus tears is an obligate procedure before each meniscus suture. Different further techniques, such as the trephination of the meniscus margins by awls or K-wires as well as roughening of the defect sites by special meniscus tissue rasps, are available. In a comparative study, Zhang et al. analyzed the effect of such a refreshment of the meniscus defect site by trephination before meniscus suturing [35]. They found a significantly lower failure rate of the meniscus suture when, in addition to the suture, a trephination was performed before.

In addition to that, a beneficial joint milieu can positively influence the meniscus regeneration. Cannon et al. detected an increased healing rate of 93% in patients after meniscus suture and simultaneous anterior cruciate ligament (ACL) reconstruction in comparison to a healing rate of 50% in patients, who had an isolated meniscus suture without simultaneous ACL replacement [36]. This fact has led to a marked increase of the number of meniscus sutures in combination with an ACL replacement in recent years. The positive effect presumably can be ascribed to the opening of the bone marrow space by drilling the femoral and tibial tunnels for the ACL reconstruction. Via these medullary tunnels, mesenchymal stem cells as well as bioactive substances, which support the meniscus regeneration, may arrive to the meniscus defect site and influence the joint milieu. To imitate this effect, some authors also recommend a trephination of the notch before meniscus suture to support meniscus healing [19].

Although improvements regarding successful regeneration after meniscus reconstruction could be seen in recent years, there is still a lack of treatment options for meniscal injuries particularly in the avascular zone and for critical size defects. Biological augmentation approaches might be possible future perspective for these kinds of meniscal pathologies.

A successful meniscus repair requires a whole joint approach of the knee. The key factor for a positive outcome after meniscus repair is to address all comorbidities. Therefore, a detailed analysis of the affected knee regarding leg alignment, stability, and status of degeneration is essential to detect all pathologies. All comorbidities should be considered for the decision of a partial meniscectomy or for meniscal repair potentially in combination with correction of the alignment or ligamentous stabilization which can be planned in a one or two step procedure (Fig. 25.2).

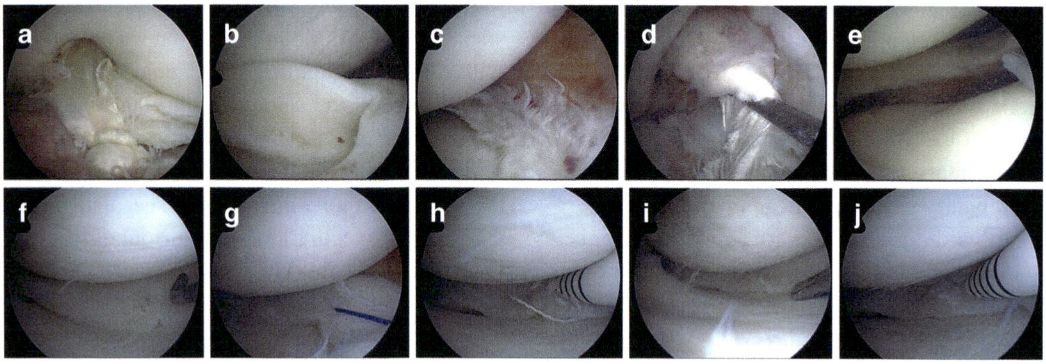

Fig. 25.2 23-year-old man with a bucket handle tear of his right medial meniscus (**a**–**c**) in combination with a rerupture of his ACL reconstruction (**d**) and a primarily non-addressed posterolateral instability (**e**); reposition of the bucket handle tear with a probe (**f**) and repair of the medial meniscus with an outside-in suture of the pars intermedia (**g**) and two all-inside sutures for the posterior horn (**h**–**j**). After bone augmentation of the tibial tunnel together with the meniscus reconstruction, the ACL and posterolateral stabilization is planned in a second step

25.7 Potential Ways for Meniscus Healing Enhancement by Suture Augmentation

25.7.1 Augmentation of Meniscus Suture with Bioactive Substances/Growth Factors

In preclinical trials and in vitro studies, various factors have been identified to have therapeutic positive effect and the potential to enhance meniscal repair. PDGF, FGF-2, IGF-I, and TGF-ß have shown positive effect on activation of cell proliferation and survival. TGF-ß and SDF-I also revealed influence and cell migration. In different studies, growth factors like PDGF, TGF-ß, BMP7 HGF, FGF-2, and IGF-I stimulated anabolic pathways, while IL-I receptor antagonist, TNF antibody, inhibitors of MMPs, and TGF-ß inhibit inflammatory and catabolic pathways. The activation of biomechanical signaling pathways are also pro-anabolic or anti-catabolic [37].

In daily clinical practice, a single-stage regenerative treatment would be preferable for meniscus injuries. Especially, clinically applicable bioactive substances or isolated growth factors like platelet-rich plasma (PRP) as a cocktail of bioactive substances or bone morphogenetic protein 7 (BMP7) are in the focus of interest. However, in literature, results for the use of, e.g., PRP or isolated growth factors are ambiguous in preclinical and first clinical studies.

Another study evaluated the effects of PRP and BMP7 on the regeneration of avascular meniscal defects. In vitro analysis showed that PRP secretes multiple growth factors over a period of 8 days. BMP7 enhances the collagen II deposition in an aggregate culture model of MSCs. However, applied to different shaped meniscal defects in vivo, PRP or BMP7 in combination with a composite matrix failed to improve meniscus healing in the avascular zone in a rabbit model [38–40]. In a similar model, Koch et al. saw no effect by additional application of PRP to the suture for repair of a vascular meniscal tear. However, the augmentation of a meniscal suture with autologous bone marrow concentrate showed improved healing of tears in the avascular zone of the meniscus in a rabbit model [41].

Theoretically, a highly angiogenic growth factor like VEGF might have a positive effect on the regeneration of an avascular tissue like the inner zone of meniscus. However, there are reports that VEGF-coated PDLLA sutures failed and showed even worse results than uncoated sutures when meniscal tears in the avascular zone of meniscus were reconstructed in a rabbit model [42].

Further information of the repair mechanism at the defect site is needed to develop special release systems or carriers for the appropriate application of growth factors to support biological augmentation of meniscus regeneration.

25.7.2 Augmentation of Meniscus Suture with Mesenchymal Stem Cells

Repair cells of meniscus injury can either be located in the meniscus tissue itself or entering the meniscus predominately via circulation. Endogenous repair of meniscus injury seems to be dependent of the different vascularization of the outer and the inner zone of the meniscus [43]. Repair in the vascularized outer zone can be achieved but fail to encourage healing in the avascular inner zone of the meniscus. However, in several studies also regeneration could be seen in the inner zone of the meniscus indicating regenerative potential independently from the vascularization [44]. Hennerbichler et al. have shown in an experimental setup that punch defects, which were directly filled with the removed punches, showed no significant difference in healing potential between the vascularized and avascularized meniscus zone [45]. Croutze and coworkers could demonstrate equivalent differentiation potential toward chondrogenic phenotype and extracellular matrix production of isolated human meniscus cells from the inner and the outer zone [46].

From a clinical standpoint, meniscus cells from meniscectomized tissue would be the ideal source for cell-based repair. Meniscal cells with regional multilineage differentiation described by Mauck et al. [47] might promote this meniscal regeneration capacity. In a rabbit model, Zellner et al. could achieve successful meniscal repair in the avascular zone [48]. As the resection of the complete medial meniscus was necessary to gain a sufficient number of cells and due to a reduced potential of these meniscal debris cells for differentiation, an alternative cell source has to be found.

Matsukura et al. found elevated levels of mesenchymal stem cells in the synovium fluid after meniscus injury compared to normal knee joints suggesting that mesenchymal stem cells in the fluid may play a role in regeneration of meniscus [49].

Preclinical trials have shown enhanced healing of meniscal lesions with the application of mesenchymal based cells [1, 39, 40, 50–52]. In an animal model, locally applied expanded mesenchymal stem cells from the bone marrow have achieved regeneration of longitudinal meniscus tears in the avascular zone in the lateral meniscus with differentiated meniscus-like tissue detected by histology, immunohistochemistry, and biomechanical analysis. In contrast control groups with untreated tears, treatment with meniscus suture alone, or meniscus suture in combination with implanted cell-free biomaterials revealed no recognizable healing. Ischimura et al. showed a faster and improved healing of avascular meniscal defects in a rabbit model by using bone marrow fibrin clot constructs compared to fibrin clot alone [53].

Treatment of meniscal full size defects with a hyluronan-collagen scaffold seeded with autologous mesenchymal stem cells after resection of the pars intermedia of the medial meniscus in a rabbit model resulted in a complete defect filling after 3 months in vivo. Only treatment with mesenchymal stem cells was able repair this critical size meniscal defects with stable differentiated meniscus-like tissue compared to untreated defects or the treatment with a cell-free hyaluronan collagen scaffold [52]. Similar results were detected for treatment of isolated avascular meniscal punch defects in the pars intermedia of the lateral meniscus in a rabbit model [40].

It is not clear whether this is a direct action of the mesenchymal-based cells or is rather mediated by secretion of certain stimulating factors [54]. Despite the fact that meniscus regeneration seems to be feasible by growth factors and mononucleated cells, not many of the cell-based strategies has entered clinical practice to date [43]. The implementation of cell-based strategies is mainly limited by the necessity to expand cells prior to transplantation resulting in high treatment costs.

Whitehouse et al. conducted a first in human safety study of five patients with a critical avascular meniscal tear. Autologous MSCs were taken from the iliac crest, expanded, cultured, and seeded on a collagen scaffold. These MSC-scaffold constructs were implanted in the meniscal tears and secured in the defect with sutures. At 2 years post-op, three patients were asymptomatic with functional improvement, and no signs of a re-tear in the MRI. Two patients required subsequent meniscectomy due to non-healing after approximately 15 months [55].

In summary, local or systemic stem cells seem to play a fundamental and essential role in the regeneration of meniscus injury, either as direct repair cells or as a source for secretion of bioactive modulators or immunomodulation. However, their role and potential to improve meniscus regeneration in daily clinical care requires further study.

25.8 Conclusion

Meniscus integrity is the key for joint health. Therefore, the main goal of every meniscus treatment should be the maintenance of as much meniscus tissue as possible. Meniscus preserving techniques to obtain a functional intact meniscus after meniscus injury in long-term are of great importance for the prevention of the development of osteoarthritis in the knee joint. Healing rates for meniscus repair have been improved. However, strategies for a successful meniscus repair should be developed for every meniscal zone and all meniscal defect situations. Due to growing knowledge in recent years and improved techniques for application, the authors have the chance to implement biological meniscus augmentation in daily clinical practice. Efforts in all research fields should be taken to translate these approaches in clinical practice as the standard of care for meniscus regeneration where needed.

References

1. Makris EA, Hadidi P, Athanasiou KA. The knee meniscus: structure-function, pathophysiology, current repair techniques, and prospects for regeneration. Biomaterials. 2011;32(30):7411–31.
2. Stein T, Mehling AP, Welsch F, von Eisenhart-Rothe R, Jager A. Long-term outcome after arthroscopic meniscal repair versus arthroscopic partial meniscectomy for traumatic meniscal tears. Am J Sports Med. 2010;38(8):1542–8.
3. McDermott I. Meniscal tears, repairs and replacement: their relevance to osteoarthritis of the knee. Br J Sports Med. 2011;45(4):292–7.
4. Lohmander LS, Englund PM, Dahl LL, Roos EM. The long-term consequence of anterior cruciate ligament and meniscus injuries: osteoarthritis. Am J Sports Med. 2007;35(10):1756–69.
5. Borchers JR, Kaeding CC, Pedroza AD, Huston LJ, Spindler KP, Wright RW. Intra-articular findings in primary and revision anterior cruciate ligament reconstruction surgery: a comparison of the MOON and MARS study groups. Am J Sports Med. 2011;39(9):1889–93.
6. Jeong HJ, Lee SH, Ko CS. Meniscectomy. Knee Surg Relat Res. 2012;24(3):129–36.
7. Badlani JT, Borrero C, Golla S, Harner CD, Irrgang JJ. The effects of meniscus injury on the development of knee osteoarthritis: data from the osteoarthritis initiative. Am J Sports Med. 2013;41(6):1238–44.
8. Salata MJ, Gibbs AE, Sekiya JK. A systematic review of clinical outcomes in patients undergoing meniscectomy. Am J Sports Med. 2010;38(9):1907–16.
9. Paxton ES, Stock MV, Brophy RH. Meniscal repair versus partial meniscectomy: a systematic review comparing reoperation rates and clinical outcomes. Arthroscopy. 2011;27(9):1275–88.
10. Papalia R, Del Buono A, Osti L, Denaro V, Maffulli N. Meniscectomy as a risk factor for knee osteoarthritis: a systematic review. Br Med Bull. 2011;99:89–106.
11. Brophy RH, Matava MJ. Surgical options for meniscal replacement. J Am Acad Orthop Surg. 2012;20(5):265–72.
12. Fetzer GB, Spindler KP, Amendola A, Andrish JT, Bergfeld JA, Dunn WR, et al. Potential market for new meniscus repair strategies: evaluation of the MOON cohort. J Knee Surg. 2009;22(3):180–6.
13. Starke C, Kopf S, Petersen W, Becker R. Meniscal repair. Arthroscopy. 2009;25(9):1033–44.
14. Abrams GD, Frank RM, Gupta AK, Harris JD, McCormick FM, Cole BJ. Trends in meniscus repair and meniscectomy in the United States, 2005-2011. Am J Sports Med. 2013;41(10):2333–9.
15. Radin EL, de Lamotte F, Maquet P. Role of the menisci in the distribution of stress in the knee. Clin Orthop Relat Res. 1984;185:290–4.
16. Fairbank TJ. Knee joint changes after meniscectomy. J Bone Joint Surg. 1948;30B(4):664–70.
17. McDermott ID, Amis AA. The consequences of meniscectomy. J Bone Joint Surg. 2006;88(12):1549–56.
18. Petty CA, Lubowitz JH. Does arthroscopic partial meniscectomy result in knee osteoarthritis? A systematic review with a minimum of 8 years' follow-up. Arthroscopy. 2011;27(3):419–24.
19. Mordecai SC, Al-Hadithy N, Ware HE, Gupte CM. Treatment of meniscal tears: an evidence based approach. World J Orthop. 2014;5(3):233–41.
20. Arnoczky SP. Building a meniscus. Biologic considerations. Clinical Orthopaed Relat Res. 1999;367(Suppl):S244–53.
21. Di Matteo B, Tarabella V, Filardo G, Vigano A, Tomba P, Marcacci M. Thomas Annandale: the first meniscus repair. Knee Surg Sports Traumatol Arthrosc. 2013;21(9):1963–6.
22. Tengrootenhuysen M, Meermans G, Pittoors K, van Riet R, Victor J. Long-term outcome after meniscal repair. Knee Surg Sports Traumatol Arthrosc. 2011;19(2):236–41.
23. Xu C, Zhao J. A meta-analysis comparing meniscal repair with meniscectomy in the treatment of meniscal tears: the more meniscus, the better outcome? Knee surg Sports Traumatol Arthrosc. 2015;23(1):164–70.
24. Mezhov V, Teichtahl AJ, Strasser R, Wluka AE, Cicuttini FM. Meniscal pathology - the evidence for treatment. Arthritis Res Ther. 2014;16(2):206.
25. Englund M, Lohmander LS. Risk factors for symptomatic knee osteoarthritis fifteen to twenty-two years after meniscectomy. Arthritis Rheum. 2004;50(9):2811–9.
26. Englund M, Roos EM, Lohmander LS. Impact of type of meniscal tear on radiographic and symptomatic knee osteoarthritis: a sixteen-year followup of meniscectomy with matched controls. Arthritis Rheum. 2003;48(8):2178–87.
27. Lee BS, Bin SI, Kim JM. Articular cartilage degenerates after subtotal/Total lateral meniscectomy but radiographic arthrosis progression is reduced after meniscal transplantation. Am J Sports Med. 2016;44(1):159–65.
28. Noyes FR, Chen RC, Barber-Westin SD, Potter HG. Greater than 10-year results of red-white longitudinal meniscal repairs in patients 20 years of age or younger. Am J Sports Med. 2011;39(5):1008–17.
29. Johnson MJ, Lucas GL, Dusek JK, Henning CE. Isolated arthroscopic meniscal repair: a long-term outcome study (more than 10 years). Am J Sports Med. 1999;27(1):44–9.

30. Sommerlath KG. Results of meniscal repair and partial meniscectomy in stable knees. Int Orthop. 1991;15(4):347–50.
31. Nepple JJ, Dunn WR, Wright RW. Meniscal repair outcomes at greater than five years: a systematic literature review and meta-analysis. J Bone Joint Surg Am. 2012;94(24):2222–7.
32. Lozano J, Ma CB, Cannon WD. All-inside meniscus repair: a systematic review. Clin Orthop Relat Res. 2007;455:134–41.
33. Pujol N, Tardy N, Boisrenoult P, Beaufils P. Magnetic resonance imaging is not suitable for interpretation of meniscal status ten years after arthroscopic repair. Int Orthop. 2013;37(12):2371–6.
34. Pujol N, Barbier O, Boisrenoult P, Beaufils P. Amount of meniscal resection after failed meniscal repair. Am J Sports Med. 2011;39(8):1648–52.
35. Zhang Z, Arnold JA. Trephination and suturing of avascular meniscal tears: a clinical study of the trephination procedure. Arthroscopy. 1996;12(6):726–31.
36. Cannon WD Jr, Vittori JM. The incidence of healing in arthroscopic meniscal repairs in anterior cruciate ligament-reconstructed knees versus stable knees. Am J Sports Med. 1992;20(2):176–81.
37. Cucchiarini M, McNulty AL, Mauck RL, Setton LA, Guilak F, Madry H. Advances in combining gene therapy with cell and tissue engineering-based approaches to enhance healing of the meniscus. Osteoarthritis Cartilage. 2016;24(8):1330–9.
38. Zellner J, Taeger CD, Schaffer M, Roldan JC, Loibl M, Mueller MB, et al. Are applied growth factors able to mimic the positive effects of mesenchymal stem cells on the regeneration of meniscus in the avascular zone? Biomed Res Int. 2014;2014:537686.
39. Zellner J, Hierl K, Mueller M, Pfeifer C, Berner A, Dienstknecht T, et al. Stem cell-based tissue-engineering for treatment of meniscal tears in the avascular zone. J Biomed Mater Res B Appl Biomater. 2013;101(7):1133–42.
40. Zellner J, Mueller M, Berner A, Dienstknecht T, Kujat R, Nerlich M, et al. Role of mesenchymal stem cells in tissue engineering of meniscus. J Biomed Mater Res A. 2010;94(4):1150–61.
41. Koch M, Hammer S, Fuellerer J, Lang S, Pfeifer CG, Pattappa G, et al. Bone marrow aspirate concentrate for the treatment of avascular meniscus tears in a one-step procedure-evaluation of an in vivo model. Int J Mol Sci. 2019;20(5):1120.
42. Petersen W, Pufe T, Starke C, Fuchs T, Kopf S, Neumann W, et al. The effect of locally applied vascular endothelial growth factor on meniscus healing: gross and histological findings. Arch Orthop Trauma Surg. 2007;127(4):235–40.
43. Scotti C, Hirschmann MT, Antinolfi P, Martin I, Peretti GM. Meniscus repair and regeneration: review on current methods and research potential. Eur Cell Mater. 2013;26:150–70.
44. Pabbruwe MB, Kafienah W, Tarlton JF, Mistry S, Fox DJ, Hollander AP. Repair of meniscal cartilage white zone tears using a stem cell/collagen-scaffold implant. Biomaterials. 2010;31(9):2583–91.
45. Hennerbichler A, Moutos FT, Hennerbichler D, Weinberg JB, Guilak F. Repair response of the inner and outer regions of the porcine meniscus in vitro. Am J Sports Med. 2007;35(5):754–62.
46. Croutze R, Jomha N, Uludag H, Adesida A. Matrix forming characteristics of inner and outer human meniscus cells on 3D collagen scaffolds under normal and low oxygen tensions. BMC Musculoskelet Disord. 2013;14:353.
47. Mauck RL, Martinez-Diaz GJ, Yuan X, Tuan RS. Regional multilineage differentiation potential of meniscal fibrochondrocytes: implications for meniscus repair. Anat Rec. 2007;290(1):48–58.
48. Zellner J, Pattappa G, Koch M, Lang S, Weber J, Pfeifer CG, et al. Autologous mesenchymal stem cells or meniscal cells: what is the best cell source for regenerative meniscus treatment in an early osteoarthritis situation? Stem Cell Res Ther. 2017;8(1):225.
49. Matsukura Y, Muneta T, Tsuji K, Koga H, Sekiya I. Mesenchymal stem cells in synovial fluid increase after meniscus injury. Clin Orthop Relat Res. 2014;472(5):1357–64.
50. Hasan J, Fisher J, Ingham E. Current strategies in meniscal regeneration. J Biomed Mater Res B Appl Biomater. 2013;102(3):619–34.
51. Izuta Y, Ochi M, Adachi N, Deie M, Yamasaki T, Shinomiya R. Meniscal repair using bone marrow-derived mesenchymal stem cells: experimental study using green fluorescent protein transgenic rats. Knee. 2005;12(3):217–23.
52. Angele P, Johnstone B, Kujat R, Zellner J, Nerlich M, Goldberg V, et al. Stem cell based tissue engineering for meniscus repair. J Biomed Mater Res A. 2008;85(2):445–55.
53. Ishimura M, Tamai S, Fujisawa Y. Arthroscopic meniscal repair with fibrin glue. Arthroscopy. 1991;7(2):177–81.
54. Caplan AI, Dennis JE. Mesenchymal stem cells as trophic mediators. J Cell Biochem. 2006;98(5):1076–84.
55. Whitehouse MR, Howells NR, Parry MC, Austin E, Kafienah W, Brady K, et al. Repair of torn avascular meniscal cartilage using undifferentiated autologous mesenchymal stem cells: from in vitro optimization to a first-in-human study. Stem Cells Transl Med. 2017;6(4):1237–48.

26 Meniscus Root Tears

Jin Goo Kim, Dhong Won Lee,
and Kyu Sung Chung

26.1 Introduction

The roles of menisci are shock absorption, load transmission, joint stabilization, contribution to proprioception, and lubrication [1]. Meniscal functions are enabled by robust anterior and posterior root attachments of the menisci to the tibial plateau. The most important function for the protection of articular cartilage is the maintenance of hoop tension which allows appropriate load transmission. Meniscal root tear has been reported to result in loss of hoop tension and biomechanical conditions similar to total meniscectomy [2]. Ultimately, these altered biomechanics of the knee leads to accelerated osteoarthritis (OA) progression and malalignment [2, 3]. Medial meniscus posterior root tears (MMPRTs) are commonly degenerative and seen in middle-aged females, while lateral meniscus posterior root tears (LMPRTs) are more likely to occur concurrently with an anterior cruciate ligament (ACL) tear [4, 5]. The diagnosis of root tears is sometimes challenging because no clinical test or definitive symptoms exist. Magnetic resonance imaging (MRI) is the best modality to diagnose meniscal root tears and concomitant pathologies such as medial meniscal extrusion (MME), status of articular cartilage, and subchondral lesions [6]. It is important to identify root tears early and not to delay appropriate treatment to restore tibiofemoral mechanics based on biomechanical studies [2, 7]. In this chapter, we present anatomy and structural properties of the root attachments and diagnosis and treatment based on recent literature.

J. G. Kim (✉)
Hanyang University Myong Ji Hospital,
Goyang-si, Republic of Korea
e-mail: jgkim@mjh.or.kr

D. W. Lee
Konkuk University Medical Center,
Seoul, Republic of Korea
e-mail: osdoctorknee@kuh.ac.kr

K. S. Chung
Inje University Seoul Paik Hospital,
Seoul, Republic of Korea

26.2 Anatomy and Function

The roles of the menisci are dependent upon the anterior and posterior roots of the menisci being firmly attached to the bone, with secure meniscal attachments preventing meniscal extrusion from the joint surface when an axial load is applied [1, 8]. The insertion of posterior horn of the menisci has a significantly greater thickness of interdigitations and has three zones including subchondral bone, calcified fibrocartilage, and uncalcified fibrocartilage [9, 10]. This enables the posterior horn to play a principal role in the stability of the menisci [1, 11].

Andrews et al. [12] reported that near the tibial insertion, the root contains large ligament like collagen fascicles, and the root may continue into

N. Nakamura et al. (eds.), *Advances in Knee Ligament and Knee Preservation Surgery*,
https://doi.org/10.1007/978-3-030-84748-7_26

the outer portion of the meniscus where it then blend with the more fibrocartilage-like inner portions of the tissue. The histological study reported by Hino et al. [13] showed that medial meniscal posterior root attachment is mainly localized in the anterior one-third.

Johannsen et al. [14] reported that the posterior root of the medial meniscus inserts 9.6 (±0.8) mm posterior and 0.7 (±0.4) mm lateral to the medial tibial eminence apex, and average distance from posterior cruciate ligament (PCL) is 8.2 (±0.7) mm. The average area of the medial meniscal posterior root attachment is 30.4 (±2.9) mm^2, whereas the average area of the transverse shiny white fibers is 47.3 (±4.4) mm^2. The posterior root of the lateral meniscus inserts 4.2 (±0.4) mm medial and 1.5 (±0.7) mm posterior to the apex of the lateral tibial eminence, and distance from PCL is 12.7(±1.1) mm. The average area of the lateral meniscal posterior root attachment is 39.2 (±2.4) mm^2 (Fig. 26.1). Ellman [15] revealed that three (posterior medial, posterior lateral, and anterior lateral) of four meniscal attachments contain supplemental fibers (shiny white fibers) that make up a significant percentage of the native root attachments, and the supplemental fibers significantly contribute to the failure strengths of the meniscal roots. Ultimate failure strength and stiffness of the posterior medial attachment are 513.8 (388.4–639.1) N

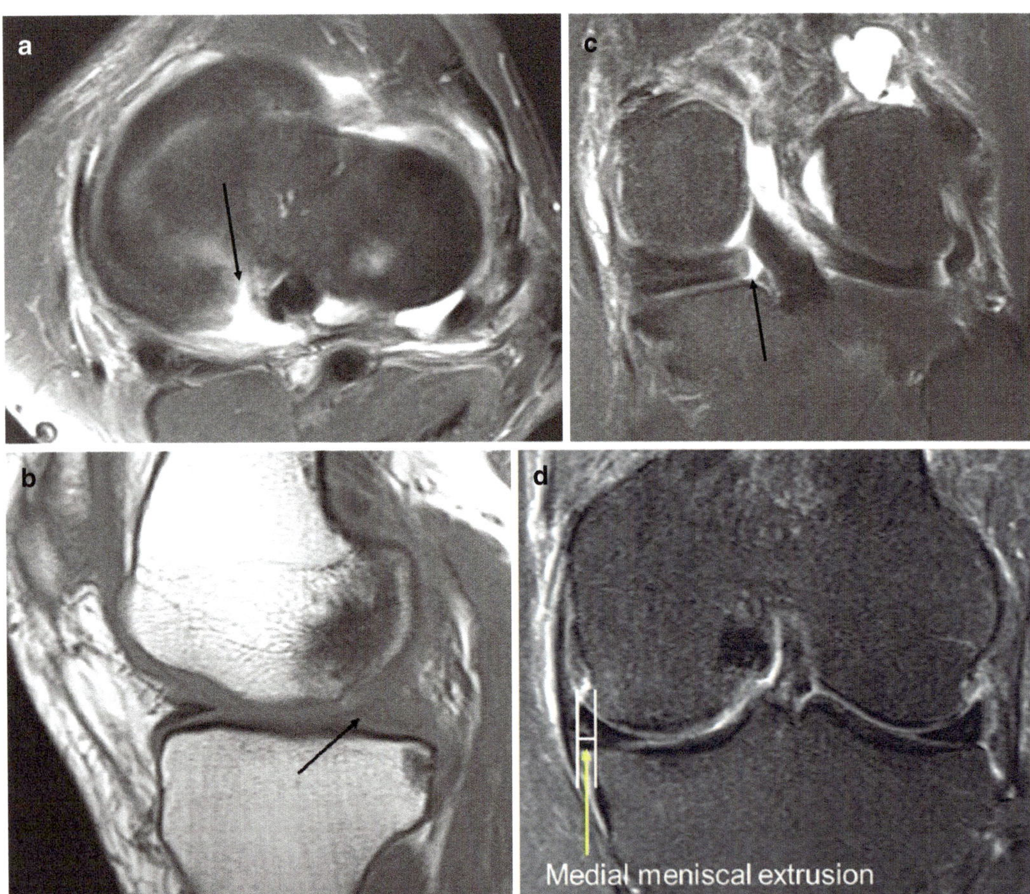

Fig. 26.1 Illustration presenting the medial and lateral meniscal posterior roots and relevant anatomy. (A) Superior view and (B) posterior view. (From Adam M Johannsen, David M Civitarese, Jeffrey R Padalecki, Mary T Goldsmith, Coen A Wijdicks, Robert F LaPrade. Qualitative and quantitative anatomic analysis of the posterior root attachments of the medial and lateral menisci. Am J Sports Med. 2012 Oct;40(10):2342–2347. Reprinted with permission)

and 122.7 (95.1–150.3) N/mm, respectively, and the shiny white fibers of posterior medial root account 47.8% and 34.2% of native root strength and stiffness.

Arnoczky and Warren [16] revealed that the meniscal root attachment has better vesicular supply via the perimeniscal capillary plexus through the bony attachment compared to the middle horn of meniscus.

26.3 Biomechanical Effects of Meniscal Root Tears

The meniscal roots act as anchors for the anterior and posterior horns, and the posterior horn of the medial meniscus moves least [17]. The relative immobility of the posterior part of the medial meniscus contribute to joint stability and shock absorption; however, the posterior root of medial meniscus is more susceptible to be damaged by compressive and shearing forces than other parts [18]. The loss of a "hoop stress" can be caused by disruption of the root attachments, similar to a radial tear in the body of the meniscus.

Previous studies have concluded that medial meniscal posterior root tears (MMPRTs) lead to deleterious alterations in tibiofemoral contact mechanics [2]. This includes failure of load distribution accompanied by decreased tibiofemoral contact area and increased tibiofemoral contact pressure because of meniscal extrusion. As a result, loss of articular cartilage, joint space narrowing, and progressive osteoarthritis occur similar to that observed following total meniscectomy. A cadaveric study by Allaire et al. [19] revealed that MMPRTs caused a 25% increase in peak contact pressure compared to that shown in the intact condition ($p < 0.001$), and no difference was found in peak contact pressure between the total meniscectomy and root tear. However, when the MMPRTs were repaired, the loading profiles returned to normal. In a similar study, Marzo and Gurske-DePeri [20] showed that avulsion of the posterior root of medial meniscus resulted in a significant increase in medial peak contact pressure and a significant decrease in contact area compared with controls (5084 ± 1087 kPa vs. 3841 ± 1240 kPA and 474 ± 79 mm^2 vs. 594 ± 59 mm^2). When the avulsion of the posterior root was repaired, the loading profiles were restored to values equal to the control [20]. Kim and colleagues [21] reported that the contact pressure differed significantly between root tear and root repair specimens at 30° and 60° of flexion ($p = 0.04$ and 0.03, respectively), and they concluded that meniscal root repair improved tibiofemoral contact mechanics, although full restoration to the level of normal meniscal function was not possible. Chung et al. [22] showed that the peak contact pressure was significantly higher and contact surface area significantly lower in MMPRTs than in normal porcine knees, respectively. They also showed that the peak contact pressure and contact surface area improved significantly after meniscal root repair, especially if the repair was performed using a locking mechanism.

Hein et al. [3] reported that medial displacement following the avulsion of the medial meniscal posterior root (3.28 mm) was significantly greater than the native knee (1.60 mm) and a repaired posterior root (1.46 mm). Gap formation was also significantly larger in the avulsed posterior root compared to the repaired state at 0 ($p < 0.02$) and 1800 N ($p < 0.02$). Marsh and colleagues [23] revealed that MMPRTs increased lateral tibial translation and medial compartment anteroposterior excursion in dynamic activities. They suggest that MMPRTs lead to significant changes to in vivo kinematics, and the magnitude of these changes are influenced by dynamic task difficulty [23]. Due to these biomechanical and kinematic changes observed following MMPRTs, an increase in varus limb alignment is ultimately seen [3]. To counteract these changes, medial meniscal root repair theoretically reduces negative effects on knee biomechanics and kinematics, and it can allow the profiles to restore to the native state as close as normal [2, 6, 19, 21, 24–27].

Contrary to MMPRTs, the biomechanical effects of lateral meniscal posterior root tears depend on the state of the meniscofemoral ligaments. Forkel et al. [28] demonstrated that posterior lateral attachment has a better prognosis in terms of OA progression when the meniscofemoral ligament is intact. However, long-term outcomes of lateral meniscal root tears have been shown to significantly affect the lateral compartment, because the lateral meniscus is responsible for the distribution of 70% of load within the lateral compartment [7, 29]. The posterior root of the lateral meniscus also importantly acts as a secondary stabilizer of tibial anterior translation at lower flexion angles and the secondary stabilizer of internal rotation at higher flexion angles [30]. Based on these biomechanical studies, lateral meniscal posterior root repair should be considered during anterior cruciate ligament (ACL) reconstruction to avoid residual instability and increased stress on the ACL graft [30].

26.4 Diagnosis of Meniscal Root Tears

26.4.1 Clinical Presentation

The clinical feature of medial and lateral posterior root tears is different. The lateral meniscus posterior root tears (LMPRTs) are more likely to have a ligament injury (mainly, ACL tear). However, most of MMPRTs occur in degenerative conditions following minor traumatic event such as squatting. This is due to an age-related degeneration of the posterior root, which shows increased fibrocartilage formation and leads to decreased ability to withstand tensile stress [31].

Matheny et al. [32] showed that patients with LMPRTs were 10.3 times (95% CI 2.6–42.5) more likely to have ACL injuries than patients with MMPRTs ($p = 0.012$), while patients who had MMPRTs were 5.8 times (95% CI 1.6–20.5) more likely to have cartilage lesion with an Outerbridge grade 2 or higher defect than patients who had LMPRTs ($p = 0.044$). Krych et al. [5] demonstrated that patients with MMPRTs had a significantly higher age (MMPRTs = 51.4 years vs. LMPRTs = 24.6 years, $p < 0.0001$), higher Kellgren-Lawrence (K-L) score (MMPRTs = 1.3 vs. LMPRT = 0.6, $p < 0.0001$), and higher rate of meniscal extrusion (MMPRTs = 72% vs. LMPRTs = 20%, $p < 0.0001$).

Since most MMPRTs have no inciting trauma, the clinical diagnosis of it is generally difficult. The most common symptom and sign of MMPRTs are posterior knee pain on deep knee flexion and joint line tenderness. A high degree of suspicion is required in the presence of other risk factors including increased age, female sex, high body mass index, and varus alignment [2, 7].

Habata et al. [33] described the main symptom of a MMPRT as a "click" or "feeling of shock" that is often associated with a minor traumatic event in older patients. Bae et al. [34] showed that a single event of painful popping is a highly predictive clinical sign (a positive predictive value of 96.5%, a negative predictive value of 81.8%, a sensitivity of 35.0%, a specificity of 99.9%, and a diagnostic accuracy of 77.9%) of MMPRTs in middle-aged to older Asian people. Traditionally, Asian lifestyles which have more lotus or squatting positions than Western lifestyles can result in greater impingement of the less mobile medial meniscus posterior root, as the meniscus moves posteriorly with deep flexion with this repetitive impingement eventually leading to subsequent degeneration of medial compartment and MMPRT [2, 17, 35]. Lee et al. [36] revealed that more than 80% of patients who had a MMPRT ($n = 38$) with an event of painful popping within 3 weeks showed that medial compartment degeneration (K-L ≥ grade 2) preceded the event of painful popping. Therefore, a thorough questioning of the patient is advised at initial examination, especially for the presence of painful popping which has diagnostic importance. Chung and colleagues [37] reported that regional geometry such as larger dimension of medial femoral condyle (MFC) to medial tibial condyle (MTC) can be one of the direct contributors to the genesis of MMPRTs which sometimes accompanied painful popping event.

26.4.2 Imaging

Magnetic resonance imaging (MRI) is the best non-invasive modality to diagnose meniscal root tears and concomitant pathologies [6]. T2-weighted sequences (especially, coronal planes) are generally considered to be the most useful images for detection of root tears given their optimum sensitivity and specificity [38]. Root tears are defined as complete tears within 9–10 mm from the root attachment, which significantly alter the biomechanics of the posterior roots [4, 24, 39].

Lee et al. [40] reported that all 36 MMPRTs were correctly diagnosed by MRI, with findings showing a ghost sign on sagittal planes in 100% (36/36), radial linear defect on axial planes in 94% (34/36), and vertical linear defect on coronal planes in 100% (36/36). They found that MMPRTs were associated with cartilage defects of the medial femoral condyle (89%) and medial meniscal extrusion ≥3 mm (67%) (Fig. 26.2). Lee et al. [36] found that the mean medial meniscal extrusion (MME), relative percentage of extrusion (RPE), and the ratio between MME and MFC/MTC at 0° were 2.9 ± 1.2 mm, 22.0 ± 10.3%, and 3.2 ± 1.3, in 38 MMPRTs. Of 38 patients with MMRPTs, 20 (52.6%) presented with a MME ≥ 3 mm, 33 (86.8%) had modified Outerbridge scale ≥2, and 32 (84.2%) showed K-L grade ≥ 2 [36]. These results agree with the study by Magee et al. [41] who observed that MME in patients >50 years may be associated with a meniscal root "stretch injury" due to degeneration of the meniscus without a tear detectable on arthroscopy. These menisci may have increased laxity due to loss of function of meniscal collagen fibers. This predisposes the patients to premature OA and vulnerable to MMPRTs [31, 41]. (Fig. 26.3).

Choi et al. [42] demonstrated that MMPRTs were found in 28.6% (120/419), and they are significantly associated with age, medial meniscal tears, MME, osseous change at the root attachment, cartilage lesion, and insufficiency fracture. Pathological MME is defined as ≥3 mm extrusion on midcoronal imaging, and it has been reported that pathological MME are strongly correlated with MMPRTs [41, 43]. Choi et al. [44] showed that MMPRTs were found in 26.6% (66/248) of medial meniscal tears, and mean MME in MMPRTs was 3.8 ± 1.4 mm (vs. 2.7 ± 1.3 mm in non MMPRTs, $p < 0.001$). In addition, MME was significantly correlated with the severity of chondral lesions of the MFC ($p < 0.001$). Furumatsu and colleagues [45] found that MME increased progressively within the short duration after the onset of painful popping in MMPRTs and recommended that preoperative MME assessment is important in determining

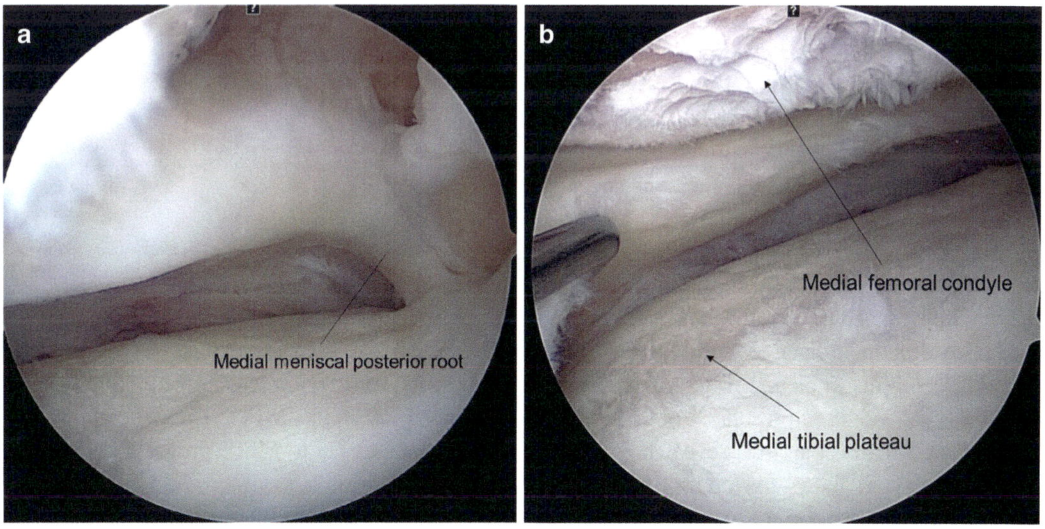

Fig. 26.2. (**a**) The medial meniscal posterior root of left knee shows increased laxity due to loss of function of meniscal collagen fibers. (**b**) In this situation, osteoarthritic change of the medial compartment precedes the medial meniscal posterior root tear

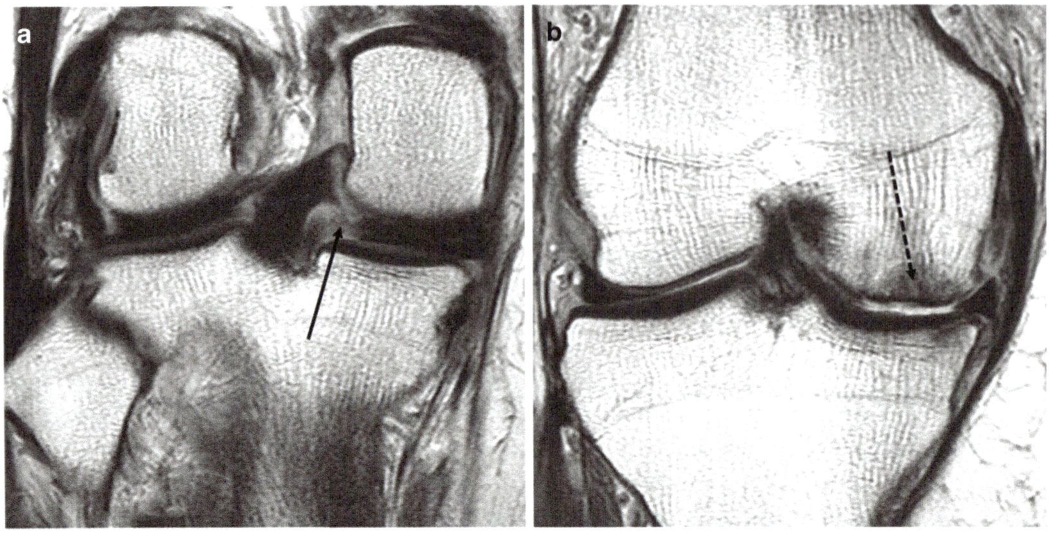

Fig. 26.3 (**a**) Black arrow indicates the medial meniscal posterior root tear of the right knee. (**b**) Dotted black arrow shows subchondral lesion of medial compartment, and it can be an indirect expression for a meniscal dysfunction after the posterior root tear

disease duration and treatment strategy. Krych et al. [46] showed that MME and modified Outerbridge scale worsened at a rapid rate after the diagnosis of MMPRTs on MRI. In less than 5 months in the subacute group (follow-up MRI within 12 months of initial imaging), MME progression without treatment was positively correlated with MFC cartilage degeneration. Guermazi and colleagues [47] reported that adjusted relative risk of cartilage degeneration was 2.03 (95% confidence interval [CI]: 1.18, 3.48) for the MMPRTs and 1.84 (95% CI: 1.32, 2.58) for the other meniscal tears and concluded that isolated MMRPTs are associated with incident and progressive medial compartment OA.

Subchondral lesions of the medial compartment can be an indirect expression of meniscal dysfunction related to posterior root tears, and medial compartment bone marrow edema, osteonecrosis, and insufficiency fracture are commonly seen in the presence of MMPRTs (Fig. 26.4). Umans et al. [48] revealed that bone marrow edema at the posterior root attachment may result from abnormal stress related to root degeneration and be the prequel of MMPRTs. They found that bone marrow edema at the posterior root attachment resolved on post-tear MRI and the osseous lesion was moved to the weight-bearing area of the MFC. These bone marrow signal changes led to MME progression and cartilage degeneration in 40% of 15 patients with MRI diagnosed MMPRTs and a MRI antecedent to MMTPTs [48]. Sung et al. [49] showed that osteonecrosis was observed in 12 of 36 knees (33.3%) with MMPRTs and 4 of 27 (14.8%) with

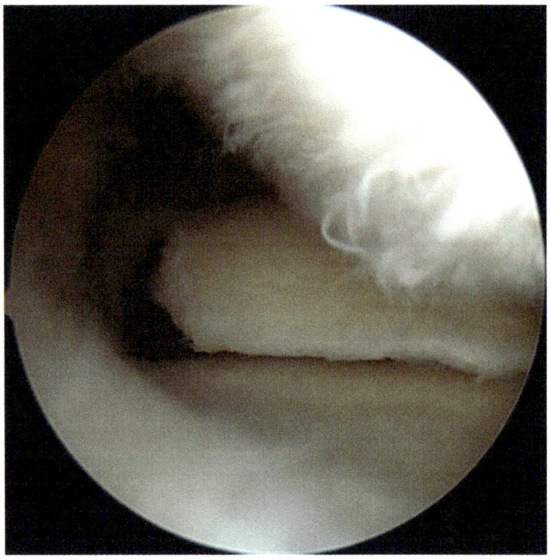

Fig. 26.4 Confirming of root tear

medial meniscal horizontal tears. Furthermore, the mean RPE of MMPRTs was greater than that of medial meniscal horizontal tears (46.1 ± 9.0% vs. 35.3 ± 13.2, p = 0.01). A recent systematic review concluded that MMPRTs increase contact pressure and create an environment within the knee from which insufficiency fractures can occur, and clinicians should be cognizant of the high prevalence of MMPRTs in patients with medial compartment osteonecrosis [50]. The authors' systematic review recommended the term "subchondral insufficiency fractures of the knee (SIFK)" instead of "spontaneous osteonecrosis of the knee (SONK)", because the pathology of SONK seems to be more like a traditional ischemia-induced osteonecrosis and therefore is likely to be multifactorial. In this context, Yao and colleagues [51] have used the term "presumptive subarticular stress reactions (PSSR)" and showed that PSRR are significantly associated with older age and MMPRTs. They postulated that abnormal mechanical stress such as loss of hoop tension is important in the pathogenesis of PSRR detected by MRI [51].

26.5 Clinical Results of Root Repair

The following demonstrates specific clinical and radiological results of root repair focusing on several factors:

1. *Clinical scores*: Table 26.1 summarizes the clinical outcomes after root repair. Based on these previous studies, root repair results in significant improvements in the postoperative clinical subjective scores compared with the preoperative status. However, most of them are based on short-term follow-up periods. Although follow-up period was extended (minimum 5-year follow-up), the postoperative clinical scores (Lysholm and IKDC subjective score) significantly improved compared to preoperative scores [59]. In a 5–10-year follow-up study, the mean Lysholm score improved significantly from 51.8 preoperatively to 83.0 at the final follow-up [60]. Thus, root repair can result in significantly improved clinical scores in both short-term and mid- to long-term follow-up periods.
2. *Progression of arthritis*: As demonstrated in Table 26.1, root repair does not completely prevent the progression of osteoarthritis. Kellgren-Lawrence (KL) grade progression is observed in 5–30% of patients postoperatively in short-term follow-up examinations. In mid-term follow-up, 68% of patients had K-L grade progression postoperatively; thus, the risk of progression of arthritis seems to increase as time goes on [59]. In terms of progression of cartilage grade, 7–24% of patients had worsened cartilage grade postoperatively. It is assumed that this problem would be associated with persistent meniscus extrusion and incomplete healing after root repair.
3. *Meniscus extrusion*: According to a meta-analysis, [65] meniscus extrusion was not reduced completely, although extrusion was likely to decrease postoperatively. In Table 26.1, studies of Kim JH et al. [54], Kim et al. [56], and Lee et al. [58] showed decreased meniscus extrusion, postoperatively, whereas Moon et al. [53] showed increased postoperative extrusion.

Persistent meniscus extrusion following root repair is an ongoing issue. Greater meniscus extrusion is a significant predictor of the progression of arthritic changes in osteoarthritic knees [66]. Therefore, it seems logical that if meniscus extrusion can be eliminated or reduced after root repair, the chance of subsequent degenerative arthritis will be reduced. Chung et al. investigated the correlation between meniscus extrusion and the quality of the result after root repair (increased extrusion group versus decreased extrusion group) [67]. Transtibial pullout repair using simple stitches led to favorable midterm clinical scores, regardless of meniscus extrusion status confirmed by 1-year follow-up MRI. However, patients with decreased meniscus extrusion have more favorable clinical scores and radiographic findings at midterm follow-up than those with increased extrusion (Lysholm score, 81 vs 88; IKDC score, 71 vs 79; per-

Table 26.1 The clinical characteristics and outcomes of the eligible studies

Study	Study design (level of evidences)	Repair technique	Tear location	No. of patients	Mean age(years)	Mean follow-up (months)	Mean Lysholm score		Progression of KL grade	Progression of cartilage grade method, %	Mean meniscus extrusion (mm)		Healing method, %
							Before	After			Before	After	
Jung et al. 2012 [52]	Case series (IV)	Suture anchor	Medial root	13	53.2	30.8	69.1	90.3	No data	No data No data	3.9	3.5*	MRI complete: 50% (5/10) partial: 40% (4/10) no: 10% (1/10)
Moon HK et al. 2012 [53]	Case series (IV)	Pull-out (simple stitch) PDS	Medial root	51	59	33	48.3	83.2	No data	MRI 7% (2/31)	3.6	5.9	MRI complete: 90.3% (28/31) partial: 9.7% (3/31) no: 0%
Kim JH et al. 2011 [54]	Comparative study (III) (pull-out vs suture anchor)	Pull-out (simple stitch) Ethibond	Medial root	22	53.2	25.9	54.3	92.5	14% (3/22)	MRI 18% (3/17)	4.3	2.1	MRI complete: 65% (11/17) partial: 35% (6/17) no: 0%
		Suture anchor	Medial root	23	52.8	26.8	55.4	93.2	9% (2/23)	MRI 21% (3/14)	4.1	2.2	MRI complete: 85% (12/14) partial: 15% (2/14) no: 0%
Seo HS et al. 2010 [55]	Case series (IV)	Pull-out (simple stitch) PDS	Medial root	21	55.4	13.4	56.1	83	5% (1/21)	Second-look arthroscopy 9% (1/11)	No data		Second-look arthroscopy complete: 0% lax: 45% (5/11) scar: 36% (4/11) no: 19% (2/11)
Kim SB et al. 2011 [56]	Comparative study (III) (vs meniscectomy)	Pull-out (simple stitch) PDS	Medial root	30	55.2	48.5	56.8	85.1	30% (9/30)	MRI 20% (6/30)	3.13	2.94	MRI complete: 56.7% (17/30) partial: 36.7% (11/30) no: 6.7% (2/30)

Lee JH et al. 2009 [57]	Case series (IV)	Pull-out (simple stitch) Ethibond	Medial root	20	51.2	31.8	57	93.1	5% (1/21)	Second-look arthroscopy 0% (0/10)	No data		Second-look arthroscopy complete: 100% (10/10) partial: 0% no: 0%
Lee DW et al. 2014 [58]	Comparative study (III) (vs simple stitch)	Pull-out (Mason-Allen stitch) PDS	Medial root	25	55.7	24.1	57.4	87.6	8% (2/25)	MRI 24% (6/25)	4.7	4.1	MRI complete: 60% (15/25) partial: 36% (9/25) no: 4% (1/25)
Chung KS et al. 2015 [59]	Comparative study (III) (vs meniscectomy)	Pull-out (simple stitch) PDS	Medial root	37	55.5	67.5	52.3	84.3	68% (25/37)	No data	No data		No data
Chung KS et al. 2018 [60]	Case series (IV)	Pull-out (simple stitch) PDS	Medial root	91	58.7	84.8	51.8	83	No data	No data	No data		No data
Lee SS et al. 2018 [61]	Case series (IV)	Pull-out (simple stitch) PDS	Medial root	56	53.3	40.6	48.7	81.5	41% (23/56)	No data	No data		Second-look arthroscopy stable: 69.7% (23/33) no: 30.3% (10/33)
Kim CW et al. 2019 [62]	Comparative study (III) (vs meniscectomy)	Pull-out (simple or modified Mason-Allen stitch) PDS or Tigerwire	Medial root	21	55.9	39.2	51.7	80.9	52% (11/21)	No data	No data	2.2	Second-look arthroscopy lax healing: 100% (21/21)
LaPrade et al. 2016 [63]	Case series (IV)	Pull-out (simple stitch) FiberWire	Medial root	35	41.0	30	54	84	No data	No data	No data		No data
			Lateral root	15	32.2		35	75					
Ahn JH et al. 2010 [64]	Case series (IV)	All-inside (simple stitch) PDS	Lateral root	25	28.8	18	62.3	92.9	No data	No data	No data		Second-look arthroscopy stable: 88% (8/9) no: 12% (1/9)

centage of K-L grade progression, 87% vs 50%; progression of joint space narrowing, 1.1 mm vs 0.6 mm). There are several factors to consider to reduce meniscus extrusion. First, locking mechanism sutures such as the modified Mason-Allen sutures is recommended because it has superior holding power and large meniscus-bone contact area that improves healing potential [24, 50]. The modified Mason-Allen sutures repair provides a superior contact surface area compared with that noted after fixation using a simple suture repair [19]. In a clinical study, the modified Mason-Allen suture repair showed reduced meniscal extrusion and more favorable radiological outcomes than simple suture repair [58]. Secondly, anatomic root repair which restores the bone bed in the native root attachment area is very important. In patients with a narrow compartment with a tight knee, it is difficult to access the native root attachment area. Non-anatomic repair can increase meniscus extrusion as shown in a biomechanical study, where non-anatomic repair did not restore the contact area or mean contact pressures to that of anatomic repair, whereas the anatomic repair produced near-intact contact area and peak contact pressures compared with the intact knee [68]. Thus, anatomic root repair is critical factor to reduce meniscus extrusion.

One of the additional procedures to help in reducing extrusion is centralization technique [69–71]. The centralization technique is where the midbody of the meniscus is centralized and stabilized onto the rim of the tibial plateau to reduce extrusion. Sutures for the centralization can share the load with those for the pullout repair, so the failure risk of the pullout sutures at the torn edge can be reduced. However, centralization can present a risk to limit the normal motion of the meniscus during knee extension-flexion, and there is no specific report of clinical results after centralization in root repair. Consequently, efforts to reduce meniscus extrusion during root repair can be rewarded with improved results, thus, one of the main goals of the root repair is to reduce meniscus extrusion as much as possible.

4. *Meniscus healing:* The healing condition of the repaired root is a critical factor because it is associated with postoperative meniscal extrusion status and progression of arthritis; however, healing status after root repair is still debatable. The results of the healing status after root repair are shown in Table 26.1. MRI and 2nd look arthroscopy were used to confirm healing status, with 2nd look arthroscopy being a more reliable method to accept healing status because it can evaluate actual restoration of meniscus attachment.

   MRI has demonstrated 56.7–90.3% of complete healing, 9.7–36.7% of partial healing, and 0–6.7% of non-healing postoperatively [52–54, 56, 58]. Interestingly, second-look arthroscopy results are debatable. Seo et al. [55] reported that in their series that there was no case with complete healing. Only five cases of lax healing (45%), four cases of scar tissue healing (36%), and two cases of non-healing (19%) were observed. However, they did not make a bleeding bone bed which is essential to get bone-to-meniscus healing [55]. Kim et al. [62] reported they found lax healing in all cases from second-look arthroscopy. In contrast, Lee SS et al. [61] reported that 69.7% of patients were classified into a stable healed group as judged by second-look arthroscopy. Importantly, they made a bone bed to promote healing. Lee et al. [57] reported complete healing in all cases. Consequently, surgeons can get favorable healing results after root repair by appropriate surgical technique.

In lateral root tears, Ahn et al. [64] reported complete healing was shown in 88% of patients (8/9) by second-look arthroscopy, although they performed concomitant ACL reconstruction and all-inside root repair.

5. *Mid- and long-term survivorship:* The mid- and long-term results are valuable because the primary aim of root repair is the prevention or delay of arthritis progression following meniscus root repair. Unfortunately, little evi-

dence is available for assessing mid- and long-term survivorship in patients undergoing pullout repair in MMPRTs. In a comparative study between partial meniscectomy and pullout repair in patients with MMPRTs at a minimum 5-year follow-up,[59] the repair group had significantly better Lysholm (84.3 vs 62.8) and IKDC (73.7 vs 49.3) scores than the meniscectomy group. In terms of radiological results, the repair group showed less K-L grade progression (percentage of patients with K-L grade progression; 68% vs 100%) and less medial joint space narrowing (0.8 mm vs 2.3 mm) than the meniscectomy group. The rate of conversion to total knee arthroplasty was 35% in meniscectomy group, whereas there was no conversion to total knee arthroplasty in repair group. The 5-year survival rates in repair and meniscectomy group were 100% and 75%, respectively ($p < 0.001$). Chung et al[60] reported mid- to long-term survival rates in patients with pullout repair of MMPRTs. Clinical failures were defined as cases requiring conversion to total knee arthroplasty (TKA) or having **a** final Lysholm score <65 or less than their preoperative scores. Among 91 patients, four patients failed due to conversion to TKA ($n = 1$) or having final Lysholm scores <65 or less than the preoperative scores (n=3) during mean follow-up duration of 84.8 months. Thus, the overall Kaplan-Meier probabilities of survival after root repair were 99% at 5 years, 98% at 6 years, 95% at 7 years, and 92% at 8 years.

In comparative study between meniscectomy and pullout repair in patients with MMPRTs at a minimum 10-year follow-up, [72] the repair group had significantly better Lysholm (77.1 vs 58.2) and IKDC (63.7 vs 44.4) scores than the meniscectomy group. Fifty-six percent of patients in the meniscectomy group and 22% of patients in the repair group were converted to TKA in the follow-up period ($p = 0.016$). According to Kaplan-Meier analysis, the 10-year survival rate between meniscectomy and repair groups was 44.4% versus 79.6% ($p = 0.004$). Consequently, pullout repair demonstrated a high clinical survival rate in mid- and long-term follow-up examinations, and it is an effective treatment to prevent or delay progression of arthritis in patients with a medial meniscus posterior root tear.

6. *Prognostic factors:* Before performing a root repair, identifying preoperative prognostic factors is critical to selecting the most appropriate treatment and predicting postoperative results. As we mentioned above, Chung KS et al. reported that patients with decreased meniscus extrusion have more favorable clinical scores and radiographic findings at mid-term follow-up than those with increased extrusion [67]. This study indicates that one of the main goals of the repair of MMPRTs is to reduce meniscus extrusion as much as possible. In short-term follow-up, patients with Outerbridge grade 3 or 4 chondral lesions had poorer results than those with grade 1 or 2 lesions in terms of clinical scores (American knee society score and Lysholm score) and patients with varus alignment greater than 5° had poorer results than those with varus alignment less than 5° [53].

   The predictors of unfavorable clinical and radiologic outcomes at a minimum of 5 years after root repair were investigated by Chung KS et al. [73] Unfavorable prognostic factors of the Lysholm score were grade 3 or 4 chondral lesions (odds ratio OR = 5.993; $p = 0.028$) and varus mechanical alignment (odds ratio = 1.644; $p = 0.017$) and for IKDC score were $\geq$ grade 3 chondral lesions (odds ratio = 11.146; $p = 0.038$) and older age (odds ratio = 1.200; $p = 0.017$). Preoperative higher chondral lesion (grade 3 or 4) significantly increased the risk of K-L grade progression (odds ratio = 11.000; $p = 0.031$).

Clinically, Outerbridge grade 3 or 4 chondral lesions, more varus alignment, and older age were found to predict a poor prognosis after root repair. These poor prognostic factors should be taken into consideration during surgical decision making. If the patients with those factors need root repair, the possibility of poor outcomes should be discussed when obtaining informed consent.

Malalignment can cause abnormal pressure and have a negative effect on fixed roots. Thus, concomitant surgical procedure such as corrective osteotomy may be considered in MMPRT patient with significant varus alignment in conjunction with root repair.

26.6 Options of Surgical Procedures of Root Repair

There are several options of how to perform a posterior root repair. Based on a meta-analysis in MMPRTs, the most common technique is arthroscopic transtibial pullout fixation with non-locking mechanism through an anterior portal [65].

26.6.1 Transtibial Pullout Repair Versus suture Anchor Repair

Most root repair studies are based on transtibial pullout fixation. There has not been a prospective randomized comparative study between the two techniques. In a comparative study between suture anchor repair and transtibial pullout repair in MMPRTs, [54] both techniques showed symptomatic improvement and no significant differences in Kellgren-Lawrence (K-L) grade in mean follow-up duration of 25.9 months. In follow-up MRI, complete structural healing was observed in 50% of the pullout fixation group and 52% of the suture anchor fixation group. Mean meniscal extrusion of 4.3 mm in the pullout fixation group and 4.1 mm in the suture anchor fixation group preoperatively was significantly decreased to 2.1 mm and 2.2 mm postoperatively, respectively. Consequently, there were no significant differences between transtibial pullout repair and suture anchor repair clinically and radiologically.

Jung YH et al. demonstrated root repair with suture anchor fixation through posterior portal and showed symptomatic improvement (Lysholm score: 69.1 preoperatively, 90.3 postoperatively) at a mean follow-up duration of 30.8 months [52]. On MRI, 50% of patients showed complete healing, 40% of patients showed partial healing, and 10% of patients showed no healing. However, mean extrusion of the midbody of the medial meniscus was 3.9 mm preoperatively and 3.5 mm postoperatively; thus, extrusion was not significantly decreased.

26.6.2 Anterior Portal Versus Posterior Portal

Most studies have also used an anterior portal. The anterior portal approach is more commonly used because it is easier to approach in the posterior root attachment area than the posterior portal approach. However, especially in patients with a tight medial compartment, it is difficult to visualize and use instruments to address meniscal pathologies. The aggressive force needed to open the medial compartment in tight knee may result in unwanted complications such as rupture of the MCL or fracture. Thus, periosteal detachment of the distal sMCL or pie-crust release of the sMCL is needed to overcome the tight medial compartment. Chung KS et al. reported that the release of the distal attachment of the sMCL during root repair did not result in residual instability and complications [51]. However, some surgeons may have concerns of sMCL injury and hesitate to perform sMCL release. In this situation, posterior transseptal portal approach is an alternative method to approach and visualize the posterior root area without sMCL release procedures [61].

26.6.3 Non-locking Versus Locking Mechanism Sutures

Complex suture patterns with locking mechanisms illustrate higher maximum failure loads and lower displacement during cyclic loading [74]. Among the four different suture techniques (simple sutures, horizontal mattress sutures, modified Mason-Allen sutures, and modified loop stitches), the MMA technique provided the best biomechanical properties with regard to cyclic loading and load-to-failure testing [74].

In a biomechanical study comparing tibiofemoral contact mechanics between simple sutures and modified Mason-Allen sutures in transtibial pullout repair, the peak contact pressure and contact surface area improved significantly after fixation, regardless of the fixation method. However, modified Mason-Allen sutures provided a superior contact surface area compared with simple sutures [19].

In a clinically comparative study between simple sutures and modified Mason-Allen sutures in root repair, [58] Lysholm score and IKDC score improved significantly in both groups in mean follow-up duration of 24 months. Although the clinical outcomes did not differ between the groups at final follow-up, postoperative meniscus extrusion decreased 0.6 mm in the modified Mason-Allen suture group, whereas extrusion increased 1 mm in the simple repair group on follow-up MRI. Regarding radiological outcomes, the modified Mason-Allen suture group did not show significant progression in the K-L grade and cartilage degeneration, whereas both measures increased significantly in the simple sutures group. Thus, the modified Mason-Allen repair showed reduced meniscal extrusion and more favorable radiological outcomes [58].

26.6.4 Non-absorbable Versus Absorbable Sutures

In a biomechanical study comparing biomechanical properties (cyclic loading and load-to-failure testing) of four different suture materials [No. 2 PDS™ (Ethicon, Somerville, NJ, USA), No. 2 Ethibond™ (Ethicon, Somerville, NJ, USA), No. 2 FiberWire™ (Arthrex, Naples, FL, USA), 2-mm Fiber-Tape™ (Arthrex, Naples, FL, USA)] for transtibial pullout repair of MMPRTs, [75] PDS™ showed the lowest values for maximum load and stiffness, whereas FiberWire™ showed the highest values for maximum load and stiffness. Thus, FiberWire™ may improve healing rates and avoid progressive extrusion of the meniscus after transtibial pullout repair. However, non-absorbable suture materials can damage meniscus tissue when pulling out the sutures under maximum force; thus, surgeons need to be careful during fixation procedures.

26.7 Senior Authors' Preferred Approach for Medial Meniscus Posterior Root Tears Repair

Recently, my preferred technique is arthroscopic transtibial root repair using a modified Mason-Allen stitch with locking mechanism [24, 50]. The arthroscope is introduced through the anterolateral (AL) portal, and the working instruments are introduced through the anteromedial (AM) portal. Arthroscopic examination is routinely performed to confirm the presence of a root tear or abnormality of other intra-articular structures.

If MMPRT is confirmed on arthroscopic examination, the superficial medial collateral ligament (sMCL) is released to get a sufficient working space. The release of the sMCL is achieved using a periosteal elevator directed toward the distal attachment area of the sMCL via a 3 cm longitudinal skin incision made at the anteromedial aspect of the proximal tibia [51]. The distal attachment of the sMCL is released completely via subperiosteal stripping in two directions, distally (from just inferior of the pes anserinus attachment to the distal tibial attachment of the sMCL) and posteromedially (the posteromedial crest of the proximal tibia beneath the tibial attachment of the posterior oblique ligament and the proximal attachment of the sMCL), while preserving the deep medial collateral ligament, proximal sMCL, and posterior oblique ligament.

After getting a larger working space and more clear visualization by the sMCL release, the root tear (Fig. 26.4) and landmarks relevant to the insertion of the medial meniscus, including the PCL insertion point, medial tibial spine, and articular margin of the tibial plateau, should be identified by arthroscopy. A meniscus resector and shaver are used to remove fibrous tissue and get fresh meniscal tissue.

Next, an 8 mm large curette is inserted through the AM portal to make a bone bed at the native root insertion site (Fig. 26.5). The bone bed is positioned just on the medial side of posterior cruciate ligament and just to the posterior side of the medial eminence of the tibia [76]. This is an important procedure to get meniscus to bone healing – a larger bony bed is therefore recommended to improve healing potential.

Next, a crescent-shaped suture hook (Linvatec; Largo, FL, USA) loaded with No. 1 polydioxanone (PDS; Ethicon; Somerville, NJ, USA) is then passed through the AM portal. The detached portion of the medial meniscus posterior horn is penetrated by the sharp tip of the suture hook at a point 5 mm medial to the torn edge in a vertical direction from the femoral side to the tibial side (Fig. 26.6). Then, the No. 1 PDS suture is advanced through the suture hook to the tibial side and taken out through the AM portal using a suture retriever. Another strand of suture, this time with Maxon ™ (Covidien; Minneapolis, MN, USA) to differentiate from the previous PDS, is placed in a position inside that of the first suture, in an identical manner via the same portal (Fig. 26.7). The superior ends of the two sutures are then tied outside the portal, and the inferior end of the Maxon™ (Covidien; Minneapolis, MN, USA) suture is pulled out. Using the shuttle relay method, the Maxon™ (Covidien; Minneapolis, MN, USA) suture is exchanged

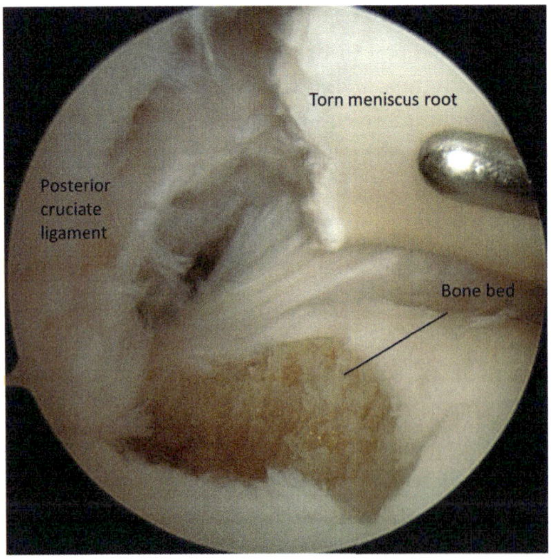

Fig. 26.5 Making a bone bed at the native root insertion site

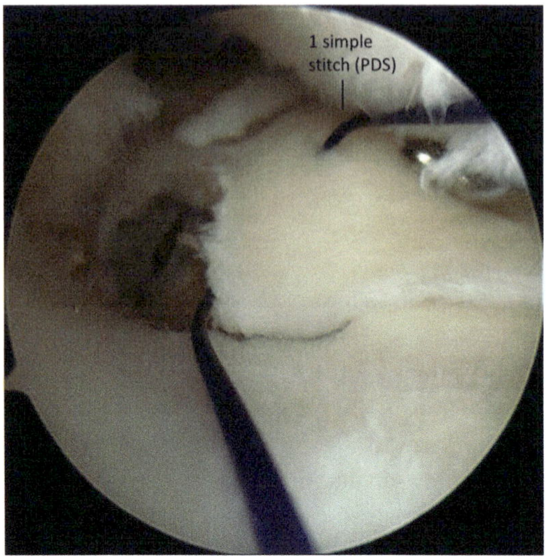

Fig. 26.6 Inserting PDS suture by crescent-shaped suture hook

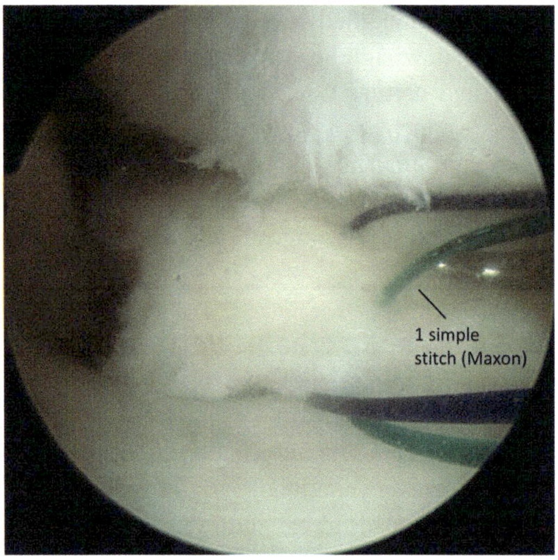

Fig. 26.7 Inserting Maxon™ suture by crescent-shaped suture hook

with the PDS suture so that the horizontal loop is completed (Fig. 26.8). A crescent-shaped suture hook loaded with No.1 PDS is again passed through the AM portal, and a simple vertical stitch is made that overlays and crosses the horizontal suture (Fig. 26.9). Both ends of the suture are then taken out through the AM portal; the resulting cruciate-shaped stitch constitutes a modified Mason-Allen stitch. If the quality of the

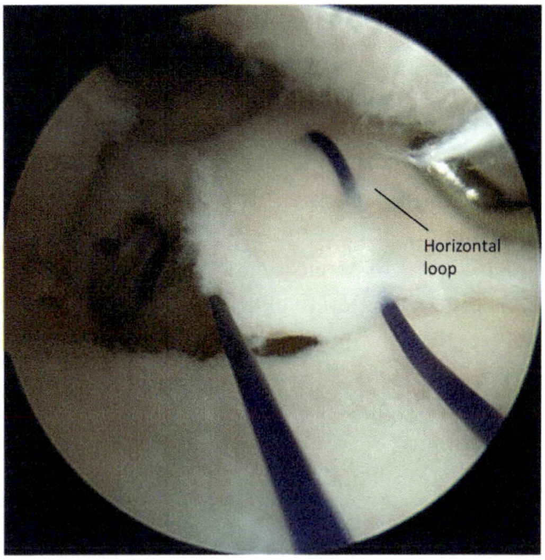

Fig. 26.8 Using the shuttle relay method, the Maxon™ suture is exchanged with the PDS suture so that the horizontal loop is completed

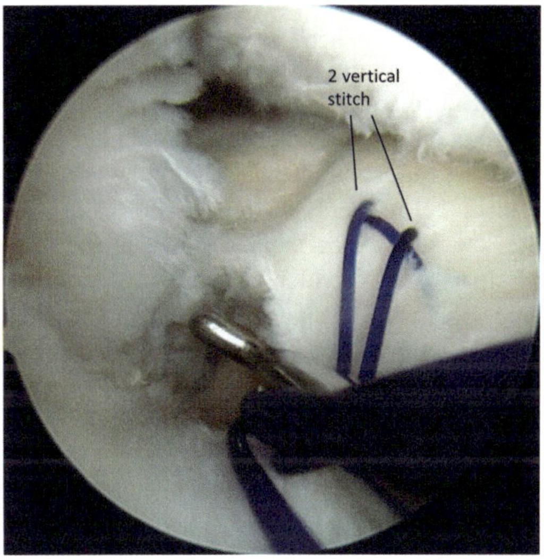

Fig. 26.9 Inserting simple vertical stitch (overlaying and crossing the horizontal suture)

root tissue is not good or the horizontal loop is too close to the torn edge of the root, additional vertical suturing may be performed.

Next, the soft tissue is detached from the previously incised area to allow for sMCL release to make a tibial tunnel. An anterior cruciate ligament reconstruction tibial tunnel guide (Linvatec; Largo, FL, USA) is inserted through

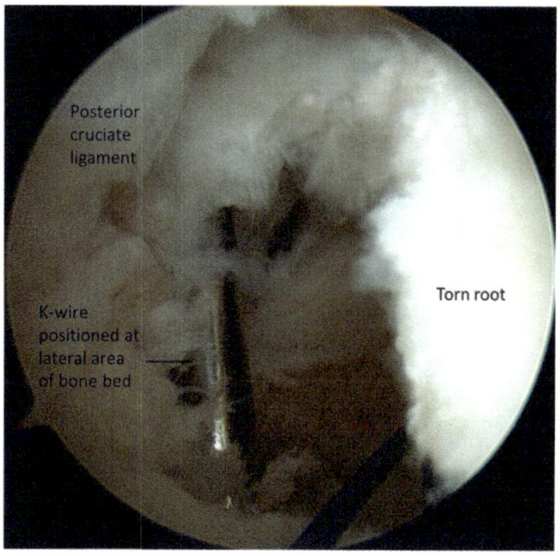

Fig. 26.10 K-wire is visualized directly using an arthroscope. The K-wire tip should be positioned at far lateral area of the bone bed and just medial area of the posterior cruciate ligament. After confirming suitable tunnel position, the K-wire is pulled back through

the AM portal, with its tip placed in contact with normal attachment site of the meniscal root. A Kirschner wire (K-wire) is then passed through the guide, with the K-wire visualized directly using the arthroscope (Fig. 26.10). The K-wire tip should be positioned at a far lateral area of the bone bed and just medial to the area of the posterior cruciate ligament. After confirming suitable tunnel position, the K-wire is pulled back through.

Next, a metal wire with a loop is then inserted into the tibial tunnel (from the anterior opening of the tibial tunnel) until its tip can be seen; it is then taken out through the AM portal using a suture grasper.

In the next step, the metal wire is taken out from the tibial tunnel together with the ends of the PDS strands after properly engaging PDS strands within the metal wire loop. The meniscus is reduced and stabilized when the ends of the sutures are pulled through the tibial tunnel under adequate tension (Fig. 26.11).

The suture ends are then tied over an Endobutton (Smith & Nephew; Andover, MA, USA), which is placed under the periosteum overlying the anteromedial tibial cortex with the knee at 0° flexion.

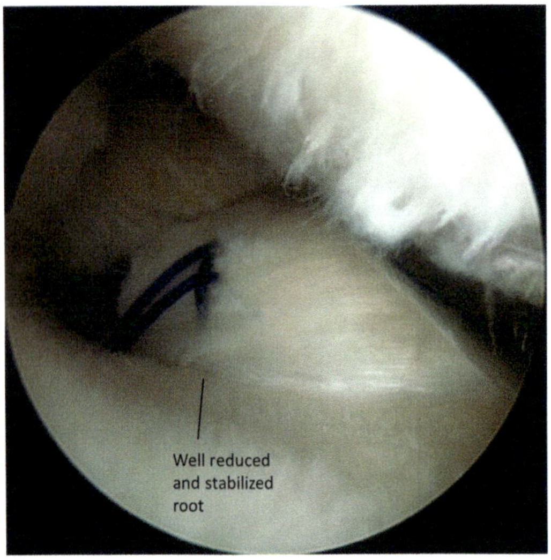

Fig. 26.11 The meniscus is reduced and stabilized when the ends of the sutures are pulled through the tibial tunnel under adequate tension

Finally, an arthroscopic evaluation is performed to confirm condition of the torn meniscal root and the entire medial meniscus.

Regarding to concomitant corrective osteotomy, it is considered in patient with mechanical varus alignment greater than 5 degree. In a previously reported study by Moon and colleagues [53], varus alignment greater than 5 degree was found to be independent negative prognostic factors after root repair. When performing concomitant osteotomy, PDS sutures are inserted via same manners before osteotomy procedures. Then, corrective osteotomy and internal fixation procedures are performed completely. The most critical problem of concomitant osteotomy procedures is that metal screws can interfere with making a bone tunnel for transtibial pullout repair. In order to solve this problem, when making a bone tunnel, if the starting point of bone tunnel is just behind the locking plate (not in front of the plate), there is little chance that tunnel and screw overlap. Another way is to use a smaller plate than usual. After that, the rest can be done in the same way as described above.

26.8 Postoperative Rehabilitation

After 3 weeks of immobilization, range of motion (ROM) exercises are started and progressed up to 90° flexion until 6 weeks postoperatively. Toe touch weight-bearing using crutches commences immediately after surgery, with the brace locked to allow for full extension of the knee joint in the first three postoperative weeks. Progressive partial weight-bearing exercises commence 3 weeks postoperatively. Full weight-bearing and progressive closed kinetic chain strengthening exercises are permitted 6 weeks after surgery. Light running is permitted after 3 months, and sports participation is allowed at 6 months. Lifestyle modifications aimed at avoiding deep knee flexion should be recommended for all patients.

26.9 Conclusions

Posterior meniscus root tears completely disrupt the continuity of the circumferential fibers and lead to loss of hoop tension, loss of load sharing ability, and an unacceptable increase in peak pressures, which can lead to degenerative arthritis. However, root repair can restore the hoop tension of the meniscus and its ability to dissipate forces, which can delay the progression of arthritis significantly compared with meniscectomy.

Encouraging results from root repair over the last decade have led to increased interest in this procedure.

However, there are several challenges that need to be addressed in the future including how to get complete healing, how to reduce meniscal extrusion, how to manage concomitant cartilage problems, and which degree of mechanical alignment is acceptable to achieve favorable outcomes. In facing these challenges, we should continue to improve surgical and perioperative management of patients undergoing root repair, so as to save the meniscus and ultimately restore normal knee function.

References

1. Villegas DF, Hansen TA, Liu DF, Donahue TL. A quantitative study of the microstructure and biochemistry of the medial meniscal horn attachments. Ann Biomed Eng. 2008;36(1):123–31.
2. Lee DW, Ha JK, Kim JG. Medial meniscus posterior root tear: a comprehensive review. Knee Surg Relat Res. 2014;26(3):125–34.
3. Hein CN, Deperio JG, Ehrensberger MT, Marzo JM. Effects of medial meniscal posterior horn avulsion and repair on meniscal displacement. Knee. 2011;18(3):189–92.
4. LaPrade CM, James EW, Cram TR, Feagin JA, Engebretsen L, LaPrade RF. Meniscal root tears: a classification system based on tear morphology. Am J Sports Med. 2015;43(2):363–9.
5. Krych AJ, Bernard CD, Kennedy NI, Tagliero AJ, Camp CL, Levy BA, et al. Medial vs. Lateral Meniscus Root Tears: Is There a Difference in Injury Presentation, Treatment Decisions, and Surgical Repair Outcomes? Arthroscopy. 2020;36(4):1135–41.
6. Pache S, Aman ZS, Kennedy M, Nakama GY, Moatshe G, Ziegler C, et al. Meniscal root tears: current concepts review. Arch Bone Jt Surg. 2018;6(4):250–9.
7. Kennedy MI, Strauss M, LaPrade RF. Injury of the meniscus root. Clin Sports Med. 2020;39(1):57–68.
8. Johnson DL, Swenson TM, Livesay GA, Aizawa H, Fu FH, Harner CD. Insertion-site anatomy of the human menisci: gross, arthroscopic, and topographical anatomy as a basis for meniscal transplantation. Arthroscopy. 1995;11(4):386–94.
9. Holmes SW Jr, Huff LW, Barnes AJ, Baier AJ. Anatomic reinforced medial meniscal root reconstruction with Gracilis autograft. Arthrosc Tech. 2019;8(3):e209–e13.
10. Lee DW, Haque R, Chung KS, Kim JG. Arthroscopic medial meniscus posterior root reconstruction using auto-Gracilis tendon. Arthrosc Tech. 2017;6(4):e1431–e5.
11. Kan A, Oshida M, Oshida S, Imada M, Nakagawa T, Okinaga S. Anatomical significance of a posterior horn of medial meniscus: the relationship between its radial tear and cartilage degradation of joint surface. Sports Med Arthrosc Rehabil Ther Technol. 2010;2:1.
12. Andrews SH, Rattner JB, Jamniczky HA, Shrive NG, Adesida AB. The structural and compositional transition of the meniscal roots into the fibrocartilage of the menisci. J Anat. 2015;226(2):169–74.
13. Hino T, Furumatsu T, Miyazawa S, Fujii M, Kodama Y, Kamatsuki Y, et al. A histological study of the medial meniscus posterior root Tibial insertion. Connect Tissue Res. 2019;61:1–8.
14. Johannsen AM, Civitarese DM, Padalecki JR, Goldsmith MT, Wijdicks CA, LaPrade RF. Qualitative and quantitative anatomic analysis of the posterior root attachments of the medial and lateral menisci. Am J Sports Med. 2012;40(10):2342–7.
15. Ellman MB, LaPrade CM, Smith SD, Rasmussen MT, Engebretsen L, Wijdicks CA, et al. Structural properties of the meniscal roots. Am J Sports Med. 2014;42(8):1881–7.
16. Arnoczky SP, Warren RF. Microvasculature of the human meniscus. Am J Sports Med. 1982;10(2):90–5.
17. Kim JG, Lee YS, Bae TS, Ha JK, Lee DH, Kim YJ, et al. Tibiofemoral contact mechanics following posterior root of medial meniscus tear, repair, meniscectomy, and allograft transplantation. Knee Surg Sports TraumatolArthrosc. 2013;21(9):2121–5.
18. Allaire R, Muriuki M, Gilbertson L, Harner CD. Biomechanical consequences of a tear of the posterior root of the medial meniscus. Similar to total meniscectomy. J Bone Joint Surg Am. 2008;90(9):1922–31.
19. Chung KS, Choi CH, Bae TS, Ha JK, Jun DJ, Wang JH, et al. Comparison of tibiofemoral contact mechanics after various Transtibial and all-inside fixation techniques for medial meniscus posterior root radial tears in a porcine model. Arthroscopy. 2018;34(4):1060–8.
20. Marsh CA, Martin DE, Harner CD, Tashman S. Effect of posterior horn medial meniscus root tear on in vivo knee kinematics. Orthop J Sports Med. 2014;2(7):2325967114541220.
21. Padalecki JR, Jansson KS, Smith SD, Dornan GJ, Pierce CM, Wijdicks CA, et al. Biomechanical consequences of a complete radial tear adjacent to the medial meniscus posterior root attachment site: in situ pull-out repair restores derangement of joint mechanics. Am J Sports Med. 2014;42(3):699–707.
22. Okazaki Y, Furumatsu T, Masuda S, Miyazawa S, Kodama Y, Kamatsuki Y, et al. Pullout repair of the medial meniscus posterior root tear reduces proton density-weighted imaging signal intensity of the medial meniscus. Acta Med Okayama. 2018;72(5):493–8.
23. Faucett SC, Geisler BP, Chahla J, Krych AJ, Kurzweil PR, Garner AM, et al. Meniscus root repair vs meniscectomy or nonoperative management to prevent knee osteoarthritis after medial meniscus root tears: clinical and economic effectiveness. Am J Sports Med. 2019;47(3):762–9.
24. Lee DW, Jang SH, Ha JK, Kim JG, Ahn JH. Meniscus root refixation technique using a modified Mason-Allen stitch. Knee Surg Sports Traumatol Arthrosc. 2013;21(3):654–7.
25. Forkel P, Herbort M, Schulze M, Rosenbaum D, Kirstein L, Raschke M, et al. Biomechanical consequences of a posterior root tear of the lateral meniscus: stabilizing effect of the meniscofemoral ligament. Arch Orthop Trauma Surg. 2013;133(5):621–6.
26. Frank JM, Moatshe G, Brady AW, Dornan GJ, Coggins A, Muckenhirn KJ, et al. Lateral meniscus posterior root and Meniscofemoral ligaments as stabilizing structures in the ACL-deficient knee: a biomechanical study. Orthop J Sports Med. 2017;5(6):2325967117695756.
27. Park DY, Min BH, Choi BH, Kim YJ, Kim M, Suh-Kim H, et al. The degeneration of meniscus roots is accompanied by fibrocartilage formation, which may precede meniscus root tears in osteoarthritic knees. Am J Sports Med. 2015;43(12):3034–44.

28. Matheny LM, Ockuly AC, Steadman JR, LaPrade RF. Posterior meniscus root tears: associated pathologies to assist as diagnostic tools. Knee Surg Sports Traumatol Arthrosc. 2015;23(10):3127–31.
29. Messner K, Gao J. The menisci of the knee joint. Anatomical and functional characteristics, and a rationale for clinical treatment. J Anat. 1998;193(Pt 2):161–78.
30. Habata T, Uematsu K, Hattori K, Takakura Y, Fujisawa Y. Clinical features of the posterior horn tear in the medial meniscus. Arch Orthop Trauma Surg. 2004;124(9):642–5.
31. Bae JH, Paik NH, Park GW, Yoon JR, Chae DJ, Kwon JH, et al. Predictive value of painful popping for a posterior root tear of the medial meniscus in middle-aged to older Asian patients. Arthroscopy. 2013;29(3):545–9.
32. Bin SI, Kim JM, Shin SJ. Radial tears of the posterior horn of the medial meniscus. Arthroscopy. 2004;20(4):373–8.
33. Vedi V, Williams A, Tennant SJ, Spouse E, Hunt DM, Gedroyc WM. Meniscal movement. An in-vivo study using dynamic MRI. J Bone Joint Surg. 1999;81(1):37–41.
34. Lee DW, Moon SG, Kim NR, Chang MS, Kim JG. Medial knee osteoarthritis precedes medial meniscal posterior root tear with an event of painful popping. Orthop Traumatol Surg Res. 2018;104(7):1009–15.
35. Chung JY, Song HK, Jung MK, Oh HT, Kim JH, Yoon JS, et al. Larger medial femoral to tibial condylar dimension may trigger posterior root tear of medial meniscus. Knee Surg Sports Traumatol Arthros. 2016;24(5):1448–54.
36. Lee SY, Jee WH, Kim JM. Radial tear of the medial meniscal root: reliability and accuracy of MRI for diagnosis. AJR Am J Roentgenol. 2008;191(1):81–5.
37. Sharif B, Ashraf T, Saifuddin A. Magnetic resonance imaging of the meniscal roots. Skeletal Radiol. 2020;49:661–76.
38. Lee YG, Shim JC, Choi YS, Kim JG, Lee GJ, Kim HK. Magnetic resonance imaging findings of surgically proven medial meniscus root tear: tear configuration and associated knee abnormalities. J Comput Assist Tomogr. 2008;32(3):452–7.
39. Magee T. MR findings of meniscal extrusion correlated with arthroscopy. J Magn Reson Imaging. 2008;28(2):466–70.
40. Lerer DB, Umans HR, Hu MX, Jones MH. The role of meniscal root pathology and radial meniscal tear in medial meniscal extrusion. Skelet Radiol. 2004;33(10):569–74.
41. Choi CJ, Choi YJ, Lee JJ, Choi CH. Magnetic resonance imaging evidence of meniscal extrusion in medial meniscus posterior root tear. Arthroscopy. 2010;26(12):1602–6.
42. Choi JY, Chang EY, Cunha GM, Tafur M, Statum S, Chung CB. Posterior medial meniscus root ligament lesions: MRI classification and associated findings. AJR Am J Roentgenol. 2014;203(6):1286–92.
43. Furumatsu T, Kamatsuki Y, Fujii M, Kodama Y, Okazaki Y, Masuda S, et al. Medial meniscus extrusion correlates with disease duration of the sudden symptomatic medial meniscus posterior root tear. Orthop Traumatol Surg Res. 2017;103(8):1179–82.
44. Krych AJ, Johnson NR, Mohan R, Hevesi M, Stuart MJ, Littrell LA, et al. Arthritis progression on serial MRIs following diagnosis of medial meniscal posterior horn root tear. J Knee Surg. 2018;31(7):698–704.
45. Guermazi A, Hayashi D, Jarraya M, Roemer FW, Zhang Y, Niu J, et al. Medial posterior meniscal root tears are associated with development or worsening of medial tibiofemoral cartilage damage: the multicenter osteoarthritis study. Radiology. 2013;268(3):814–21.
46. Umans H, Morrison W, DiFelice GS, Vaidya N, Winalski CS. Posterior horn medial meniscal root tear: the prequel. Skelet Radiol. 2014;43(6):775–80.
47. Sung JH, Ha JK, Lee DW, Seo WY, Kim JG. Meniscal extrusion and spontaneous osteonecrosis with root tear of medial meniscus: comparison with horizontal tear. Arthroscopy. 2013;29(4):726–32.
48. Hussain ZB, Chahla J, Mandelbaum BR, Gomoll AH, LaPrade RF. The role of meniscal tears in spontaneous osteonecrosis of the knee: a systematic review of suspected etiology and a call to revisit nomenclature. Am J Sports Med. 2019;47(2):501–7.
49. Yao L, Stanczak J, Boutin RD. Presumptive subarticular stress reactions of the knee: MRI detection and association with meniscal tear patterns. Skelet Radiol. 2004;33(5):260–4.
50. Chung KS, Ha JK, Ra HJ, Kim JG. Arthroscopic medial meniscus posterior root fixation using a modified Mason-Allen stitch. Arthrosc Tech. 2016;5(1):e63–6.
51. Chung KS, Ha JK, Ra HJ, Kim JG. Does release of the superficial medial collateral ligament result in clinically harmful effects after the fixation of medial meniscus posterior root tears? Arthroscopy. 2017;33(1):199–208.
52. Jung YH, Choi NH, Oh JS, Victoroff BN. All-inside repair for a root tear of the medial meniscus using a suture anchor. Am J Sports Med. 2012;40(6):1406–11.
53. Moon HK, Koh YG, Kim YC, Park YS, Jo SB, Kwon SK. Prognostic factors of arthroscopic pull-out repair for a posterior root tear of the medial meniscus. Am J Sports Med. 2012;40(5):1138–43.
54. Kim JH, Chung JH, Lee DH, Lee YS, Kim JR, Ryu KJ. Arthroscopic suture anchor repair versus pullout suture repair in posterior root tear of the medial meniscus: a prospective comparison study. Arthroscopy. 2011;27(12):1644–53.
55. Seo HS, Lee SC, Jung KA. Second-look arthroscopic findings after repairs of posterior root tears of the medial meniscus. Am J Sports Med. 2011;39(1):99–107.
56. Kim SB, Ha JK, Lee SW, Kim DW, Shim JC, Kim JG, et al. Medial meniscus root tear refixation: comparison of clinical, radiologic, and arthroscopic findings with medial meniscectomy. Arthroscopy. 2011;27(3):346–54.

57. Lee JH, Lim YJ, Kim KB, Kim KH, Song JH. Arthroscopic pullout suture repair of posterior root tear of the medial meniscus: radiographic and clinical results with a 2-year follow-up. Arthroscopy. 2009;25(9):951–8.
58. Lee DW, Kim MK, Jang HS, Ha JK, Kim JG. Clinical and radiologic evaluation of arthroscopic medial meniscus root tear refixation: comparison of the modified Mason-Allen stitch and simple stitches. Arthroscopy. 2014;30(11):1439–46.
59. Chung KS, Ha JK, Yeom CH, Ra HJ, Jang HS, Choi SH, et al. Comparison of clinical and radiologic results between partial meniscectomy and Refixation of medial meniscus posterior root tears: a minimum 5-year follow-up. Arthroscopy. 2015;31(10):1941–50.
60. Chung KS, Noh JM, Ha JK, Ra HJ, Park SB, Kim HK, et al. Survivorship analysis and clinical outcomes of Transtibial pullout repair for medial meniscus posterior root tears: a 5- to 10-year follow-up study. Arthroscopy. 2018;34(2):530–5.
61. Lee SS, Ahn JH, Kim JH, Kyung BS, Wang JH. Evaluation of healing after medial meniscal root repair using second-look arthroscopy, clinical, and radiological criteria. Am J Sports Med. 2018;46(11):2661–8.
62. Kim CW, Lee CR, Gwak HC, Kim JH, Park DH, Kwon YU, et al. Clinical and radiologic outcomes of patients with lax healing after medial meniscal root repair: comparison with subtotal meniscectomy. Arthroscopy. 2019;35(11):3079–86.
63. LaPrade RF, Matheny LM, Moulton SG, James EW, Dean CS. Posterior meniscal root repairs: outcomes of an anatomic Transtibial pull-out technique. Am J Sports Med. 2017;45(4):884–91.
64. Ahn JH, Lee YS, Yoo JC, Chang MJ, Park SJ, Pae YR. Results of arthroscopic all-inside repair for lateral meniscus root tear in patients undergoing concomitant anterior cruciate ligament reconstruction. Arthroscopy. 2010;26(1):67–75.
65. Chung KS, Ha JK, Ra HJ, Kim JG. A meta-analysis of clinical and radiographic outcomes of posterior horn medial meniscus root repairs. Knee Surg Sports Traumatol Arthrosc. 2016;24(5):1455–68.
66. Emmanuel K, Quinn E, Niu J, Guermazi A, Roemer F, Wirth W, et al. Quantitative measures of meniscus extrusion predict incident radiographic knee osteoarthritis--data from the osteoarthritis initiative. Osteoarthr Cartil. 2016;24(2):262–9.
67. Chung KS, Ha JK, Ra HJ, Nam GW, Kim JG. Pullout fixation of posterior medial meniscus root tears: correlation between meniscus extrusion and midterm clinical results. Am J Sports Med. 2017;45(1):42–9.
68. LaPrade CM, Foad A, Smith SD, Turnbull TL, Dornan GJ, Engebretsen L, et al. Biomechanical consequences of a nonanatomic posterior medial meniscal root repair. Am J Sports Med. 2015;43(4):912–20.
69. Koga H, Watanabe T, Horie M, Katagiri H, Otabe K, Ohara T, et al. Augmentation of the pullout repair of a medial meniscus posterior root tear by arthroscopic centralization. Arthrosc Tech. 2017;6(4):e1335–e9.
70. Koga H, Muneta T, Watanabe T, Mochizuki T, Horie M, Nakamura T, et al. Two-year outcomes after arthroscopic lateral meniscus centralization. Arthroscopy. 2016;32(10):2000–8.
71. Chernchujit B, Prasetia R. Arthroscopic direct meniscal extrusion reduction: surgical tips to reduce persistent meniscal extrusion in meniscal root repair. Eur J Orthopaed Surg Traumatol. 2018;28(4):727–34.
72. Chung KS, Ha JK, Ra HJ, Yu WJ, Kim JG. Root repair versus partial meniscectomy for medial meniscus posterior root tears: comparison of long-term survivorship and clinical outcomes at minimum 10-year follow-up. Prognostic factors in the midterm results of pullout fixation for posterior root tears of the medial meniscus. Am J Sports Med. 2020;48(8):1937–44.
73. Chung KS, Ha JK, Ra HJ, Kim JG. Prognostic factors in the midterm results of pullout fixation for posterior root tears of the medial meniscus. Arthroscopy. 2016;32(7):1319–27.
74. Feucht MJ, Grande E, Brunhuber J, Burgkart R, Imhoff AB, Braun S. Biomechanical evaluation of different suture techniques for arthroscopic transtibial pull-out repair of posterior medial meniscus root tears. Am J Sports Med. 2013;41(12):2784–90.
75. Feucht MJ, Grande E, Brunhuber J, Rosenstiel N, Burgkart R, Imhoff AB, et al. Biomechanical evaluation of different suture materials for arthroscopic transtibial pull-out repair of posterior meniscus root tears. Knee Surg Sports Traumatol Arthrosc. 2015;23(1):132–9.
76. Smigielski R, Becker R, Zdanowicz U, Ciszek B. Medial meniscus anatomy-from basic science to treatment. Knee Surg Sports Traumatol Arthrosc. 2015;23(1):8–14.

Meniscus Substitution

27

Francesca de Caro, Jonas Grammens, Wouter Van Genechten, Rene Verdonk, and Peter Verdonk

27.1 Meniscal Allograft

Meniscal allograft transplantation (MAT) was first performed in the 1984 [1].

Nowadays, even if this procedure is estimated to be performed only on 1/1000000 patients, it cannot be considered an experimental surgery anymore and recent guidelines on the use of MAT have been compiled [2, 3].

The aim of meniscal allograft transplantation is to limit the negative effects of a meniscal loss [4–6].

Main indication for this kind of treatment is a painful meniscus deficient patient who has not yet developed advanced osteoarthritis (OA). Recent studies by Stone et al. [7] and Lee et al. re[8] have shown that good mid-term results can be achieved also in patients with and advanced grade of OA treated by meniscal allograft transplantation as a salvage procedure. Unfortunately, while post-operative clinical scores are not different to those of patients treated with standard indications, in patients with advance OA, graft survival rate is significantly lower in the salvage procedure group. Of importance is the absence of squaring to the femoral condyle. These anatomical changes obliterate the triangular space in between tibial plateau and femoral condyle and cause extrusion of the meniscus graft. Meniscal allograft transplantation is moreover not recommended in asymptomatic patients since the chondroprotective effect of this procedure is still debated and the reoperation rate can be as high as 32% [9]. However, it is advised to provide strict clinical and/or imaging follow-up in young asymptomatic patients with critical meniscus loss, to identify early onset of osteoarthrosis. Other contraindications to meniscal transplantation include: obesity, skeletal immaturity, synovial disease, inflammatory arthritis, and previous joint infection.

Another main indication to a meniscal allograft in conjunction with an ACL reconstruction is a deficient medial meniscal instable knee. Menisci and anterior cruciate ligament (ACL) play a synergic role in knee biomechanics, with medial and lateral menisci acting as secondary restrains for antero-posterior and rotatory laxity [10, 11]. Many clinical studies have demon-

F. de Caro
Department Of Orthopaedics, Division of Minimally Invasive Surgery, Istituto di Cura Città di Pavia, Pavia, Italy

J. Grammens · W. Van Genechten
Antwerp Surgical Training, Anatomy and Research Centre, Antwerp University, Antwerp, Belgium
e-mail: jonas.grammens@uantwerpen.be

R. Verdonk
Department of Orthopaedic Surgery, University Libre de Bruxelles, Brussels, Belgium

P. Verdonk (✉)
ORTHOCA, Antwerp, Belgium

Department of Orthopaedic Surgery, Antwerp University, Antwerp, Belgium

N. Nakamura et al. (eds.), *Advances in Knee Ligament and Knee Preservation Surgery*,
https://doi.org/10.1007/978-3-030-84748-7_27

strated the importance of the menisci in achieving good outcomes following ACL reconstruction [12]. The addition of a MAT, particularly of a medial graft, to an ACL or revision ACL reconstruction in a patient who lacks the meniscus, may aid in improving outcome [13].

Besides stability, it is very important to address any malalignment, if present [14] and isolated cartilage lesions which could benefit each other in terms of healing and outcome. Good results have been obtained also by treating young patients with a distal femoral osteotomy and lateral meniscal transplantation, with the aim to procrastinate as much as possible a more invasive surgery like a knee replacement. Thus, to ensure an optimal outcome, patient selection and preoperative counseling and expectation setting are of the utmost importance [15].

A recent meta-analysis from Amendola et al. [16] suggest that even active patients can undergo this kind of surgery, with 77% of athletes and physically active patients able to return to some level of sports after MAT especially those involved in low-impact sports. For what concerns high-impact and strenuous activities, these should be considered and discussed with caution only in selected patients.

Concerning graft procurement and preparation, the International Meniscus Reconstruction Experts Forum recommends nonirradiated frozen or viable meniscal allografts to be provided with the peripheral meniscus-tibial ligaments remaining intact [2]. Fresh viable meniscal grafts may be the most valid option maintaining the cells viability and extracellular matrix integrity. Verdonk et al. reported good clinical results at long-term follow-up with viable meniscal grafts with a mean survivor rate of 72% for both the medial and lateral grafts at 10 years follow-up [17] (Fig. 27.1). Unfortunately, due to the high costs and difficulty to obtain the viable grafts, some authors use cryopreservation, even knowing that of all the storage procedures, this is the one that mostly alters the properties of the tissue [3]. Lyophilization is abandoned due to a higher incidence of graft shrinking.

One of the most important preoperative evaluations is the correct sizing of the graft. A small graft will probably result in an early failure due to the increased biomechanical load [18]. On the other hand, a bigger graft with its extruded position will result in a continued overload of the articular cartilage. Nevertheless, it is advised to oversize rather than to undersize, since oversizing can be partially corrected surgically.

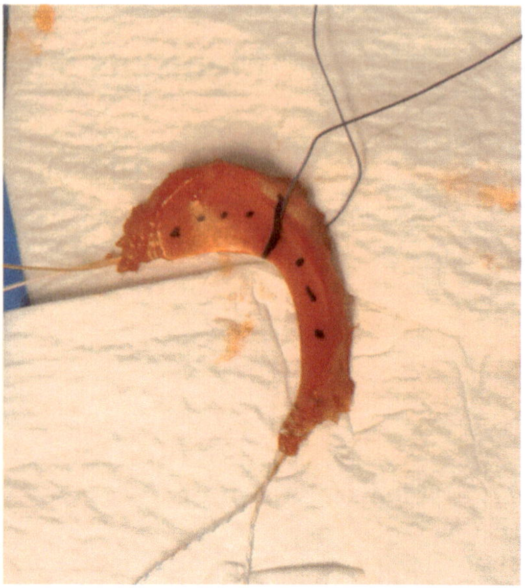

Fig. 27.1 NUsurface® non-anchored meniscal spacer with a discoid deformable structure

No sizing method has been identified as being most reliable. The most used one is the Pollard method, which relies on calibrated AP and lateral radiographs [19]. Although this method is widely used, it incorporates a number of flaws mainly affecting lateral allograft sizing and new 3D MRI based measures on contralateral knee are being proposed [20]. The authors recommend MRI measurements in the antero-posterior and medio-lateral direction from rim to rim and from rim to mid spine, respectively, of the affected compartment.

Besides sizing, also the anatomy of the medial anterior horn, both the recipient and the graft must be taken into account. Implanting a meniscus allograft posterior to the tibial edge (type 1 or 2) in a knee with type 3 meniscus might result in overstuffing of the anterior compartment, extension loss and increased stress in the graft [21]. Finally, the use of a bone block during the surgical technique will require a higher sizing accuracy than that needed for the soft tissue fixation.

No superiority of one surgical technique over the other (bone block vs soft tissue) is at the moment accepted [2], with 74% of the surgeons preferring to use bone fixation compared with 26% preferring soft tissue. More specifically, the preference is for a slot/bone bridge technique on the lateral side, and bone plugs for the medial side.

Since the demand for meniscus allografts increase, limited availability of grafts, more specifically larger lateral grafts and smaller medial grafts, might become a real problem.

27.2 Meniscal Scaffolds

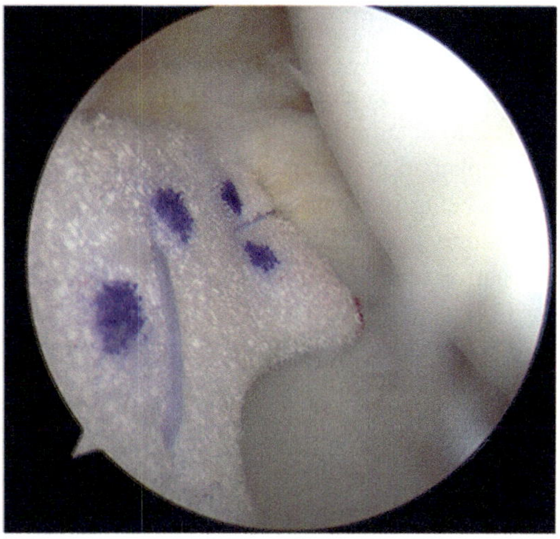

Fig. 27.2 Actifit® porous polycaprolactone and urethane meniscal scaffold

As MAT is now an established and effective procedure, some considerations still need to be made about meniscal scaffolds [22–24].

A meniscus scaffold is a biocompatible structure that provides a 3-dimensional support to fibrocartilaginous tissue regeneration when a segmental meniscal defect is present. Up to day, two acellular meniscal scaffolds have been available in Europe. The first, the collagen meniscus implant (CMI; Stryker, Kalamazoo, MI) is a bovine collagen derived construct that consists entirely of type 1 collagen from bovine purified Achilles tendon. This is a highly bioresorbable (12–18 months) and highly porous scaffold [25].

The second, more recent scaffold (Actifit®, Orteq Ltd., London, UK) consists of porous polycaprolactone and urethane segments (Fig. 27.2), and its purpose is to restore the lost meniscal tissue and function by providing a structure for new tissue ingrowth, being slowly biodegradable with an estimated degradation time of 4–6 years [26].

In contrast to MAT, the indication for a meniscus scaffold is limited to segmental meniscus loss with an intact meniscal rim and intact horn, since the scaffold is not intended to treat total or subtotal meniscal defects.

The ideal indication for the implantation of a meniscal scaffold is a segmental meniscal loss with intact stable rim, in young patients with international cartilage repair society (ICRS) classification <3, no malalignment or instability of the knee. A body mass index of 35 is a clear contraindication. The typical squaring of the femoral condyle that can appear after a meniscectomy is another strict contraindication [27].

The surgical technique, for both devices, consists of an arthroscopic debridement of the meniscal lesion, up to the vascularized zone, being aware of not damaging the meniscal rim. This can be punctured with a microfracture pick to open up vascular channels or rasped to try to promote healing. When the debridement is complete, the lesion is measured using a meniscus ruler provided with the scaffold. The scaffold is then cut to fit, being aware of cutting 3–5 mm more than measured because a natural shrinkage due to suturing can be expected. Cranial and caudal surfaces can be marked and the device is then introduced through the antero-medial or antero-lateral portal using a blunt-nosed grasper. Fixation is started with a horizontal all-inside suture from the posterior edge of the implant to the native meniscus [28].

Short- and mid-term clinical results of polyurethane and collagen meniscal scaffolds are now established, even if the number of scientific documentations is limited [27–29]. Significant improvement of the clinical scores, both of pain and knee function, are reported by most authors, demonstrating the effectiveness of the scaffold and its durability. These clinical results remain stable over time, although few studies report long-term clinical outcomes [30–32].

Preoperative cartilage status is found to be a predictor of clinical outcomes. It has been proven that cartilage damage should not exceed ICRS grade 2 to obtain predictable results after meniscal implantation, independently of which scaffold is being implanted. Moreover, concomitant procedures seem to influence clinical outcome in a negative sense, compromising a slower recovery [33].

It is still debated whether there is an indication to implant these scaffolds in an acute setting, during the primary meniscectomy. These controversial findings can be explained by the rapid clinical improvement offered by a simple meniscectomy, but current evidence tends to support the use of meniscal scaffolds for chronic lesions, with better clinical outcomes at longer follow-up [34].

In contrast to the overall good clinical results, MRI findings are disappointing, showing a reduction in the size of the scaffold, with hyperintense signal that tends to diminish over time, but never reaches the level of normal meniscus tissue. These mediocre results, however, are not correlated to the good clinical results [35]. For what concerns the radiographic evaluation, only Zaffagnini et al. [32] report a significantly less medial joint space narrowing in patients treated by CMI implantation. Based on the available imaging data, the chondroprotective effect of meniscus scaffolds cannot be established [36].

Only few studies report histologic data. Essential structural elements of the human meniscus, evidence of tissue ingrowth, and regrowth have been documented [26, 31].

Although meniscus scaffolds have been proved to be a safe, easy technique for meniscal tissue regeneration, further improvement on scaffold technology, surgical technique, and enhancing biology could be achieved [37].

27.3 Meniscal Implants

The latest advanced treatment for patients suffering painful medial meniscus insufficiency is the NUsurface® (Active Implants Ltd., Netanya) meniscal spacer (Fig. 27.3).

The NUsurface® implant is a non-anchored meniscal spacer with a discoid deformable structure. It is designed for the medial compartment only and comes in 7 discrete sizes. The implant has some creep characteristics and can therefore adapts to the space in the knee of each individual patient. This implant is conceived as the missing dowel between minimally invasive meniscal repair and knee replacement.

Fig. 27.3 Preparation of a fresh viable meniscal graft

The implant is made of urethane polycarbonate (PCU) which offers exceptional abrasion resistance, flexibility, and high tensile strength. Moreover, the implant is reinforced at the periphery by ultra-high molecular weight polyethylene fibers. This complex structure permits a better distribution of the contact forces in the knee in a similar pattern to that of the natural meniscus. The implant is biologically inert and is neither absorbed nor degraded in the body [38].

The main indication for the use of this spacer is medial knee pain caused by a degenerative, irreparable lesion of the medial meniscus. Moreover, it is indicated in those patients with chronic knee pain due to a previous partial meniscectomy, aged from 50 to 70 years old, with minimal cartilage wear, where there are limited treatment options. Even in these cases, the purpose of its use, demonstrated by clinical studies, is to delay the progression of osteoarthritis and improve clinical function.

It is important to verify pre-operatively that a minimum of 2 mm intact medial meniscal rim is present. Active infections, obesity, malalignment, unstable knees, and radiographic evidence of rapid joint degradation or bone absorption are clear contraindications. The device is also contraindicated in patients with insufficient synovial fluid in the knee, for example, patients with Sjögren syndrome. The implant should not be used if femoral squaring, larger ICRS grade 4 focal cartilage lesions or bone is exposed at the tibial spine level as these can induce abrasion.

The surgical technique provides a first arthroscopic moment where it is of fundamental importance to resect the lesioned and unstable meniscal tissue until a stable margin is reached, without damaging the healthy articular cartilage. The aim is to create a nice vertical wall of the meniscal edge about 2–4 mm thick, which allows to sustain the implant in the absence of any anchoring. A notchplasty can be performed to make sure that there is room for the lateral wall of the implant, so that this can slide freely during knee range of movement.

When the arthroscopy is over, with the knee flexed, a longitudinal incision of 4–7 cm is made along the medial edge of the patellar tendon to open the capsule of the knee and allow access to the medial compartment. At this point the meniscectomy of the anterior portion of the medial meniscus can be completed, leaving the intermeniscal connection intact. A trial implant is inserted to check sizing and potential bony conflicts. A plasty of the lateral wall of the medial condyle and the roof of the notch are indicated to shape the medial compartment to the implant. The knee is mobilized through the entire range of flexion-extension movements, observing through arthrotomy or with arthroscopy from the external access. Moreover, the trial implant is radio-opaque and can be observed dynamically with an image intensifier if necessary. This allows to evaluate the dimensions of the trial implant with respect to the tibial edge and the notch to exclude any "impingement" with the joint structures. When all the measures are done and the size is confirmed, the trial implant can be taken out and the final device implanted.

The main advantages of the NUsurface® meniscus prosthesis is the absence of any fixation to bone or meniscal tissue. This implies that no need of vital meniscal tissue necessary, and no laceration of endogenous tissue can occur. Moreover, no partial resection to the femur or tibia is made, thus not limiting future knee surgical procedures. In case of the failure of the implant, for example, for its breakage, the implant can easily be replaced and it is not a big issue to perform alternative surgical treatments.

Different animal studies have demonstrated how this device protects further cartilage wear in case of medial meniscal insufficiency [39]. A pilot study from Verdonk et al. has assessed the static kinematics and the motion pattern of three knees implanted with an artificial polycarbonate-urethane prosthesis by an open-MRI. The device did not affect femoral roll-back and tibio-femoral contact points, though antero-posterior movement was demonstrated to be slightly different [40].

Finally, preliminary results of the first clinical evaluation of a PCU meniscal implant have shown a considerable reduction of knee pain and increased activity levels after 1 year of follow-up in 61 patients [40]. The device is marked CE and it is nowadays currently used in clinical practice in the European space. A large FDA IDE trial is currently in its final stage.

References

1. Wirth CJ, Milachowski KA, Weismeier K. Meniscus transplantation in animal experiments and initial clinical results [in German]. Z Orthop Ihre Grenzgeb. 1986;124(4):508–12.
2. Getgood A, LaPrade RF, Verdonk P, Gersoff W, Cole B, Spalding T. IMREF group international meniscus reconstruction experts forum (IMREF) 2015 consensus statement on the practice of meniscal allograft transplantation. Am J Sports Med. 2017;45(5):1195–205. Epub 2016 Aug 25
3. Figueroa F, Figueroa D, Calvo R, Vaisman A, Espregueira-Mendes J. Meniscus allograft transplantation: indications, techniques and outcomes. EFORT Open Rev. 2019;4(4):115–20. https://doi.org/10.1302/2058-5241.4.180052.
4. Masouros SD, McDermott ID, Amis AA, Bull AMJ. Biomechanics of the meniscus-meniscal ligament construct of the knee. Knee Surg Sports Traumatol Arthrosc. 2008;16:1121–32.
5. King D. The function of semilunar cartilages. J Bone Joint Surg. 2020;18:1069–76.
6. Heijink A, Gomoll AH, Madry H, Drobnic M, Filardo G, Espregueira-Mendes J, Van Dijk CN. Biomechanical considerations in the pathogenesis of osteoarthritis of the knee. Knee Surg Sports Traumatol Arthrosc. 2020;3:423–35.

7. Stone KR, Pelsis JR, Surrette ST, Walgenbach AW, Turek TJ. Meniscus transplantation in an active population with moderate to severe cartilage damage. Knee Surg Sports Traumatol Arthrosc. 2015;23:251–7.
8. Lee BA, Bin SI, Kim JM, Kim WK, Choi JW. Survivorship after meniscal allograft transplantation according to articular cartilage status. Am J Sports Med. 2017;45:1095–101.
9. Harris JD, Abrams GD, Hussey KE, Wilson H, Frank R, Gupta AK, Bach BR Jr, Cole BJ. Survival and reoperation rates after meniscal allograft transplantation: analysis of failures for 172 consecutive transplants at a minimum 2-year follow-up. Am J Sports Med. 2014;42(4):892–7. https://doi.org/10.1177/0363546513520115. Epub 2014 Feb 14
10. Novaretti JV, Lian J, Patel NK, Chan CK, Cohen M, Musahl V, Debski REJ. Partial lateral meniscectomy affects knee stability even in anterior cruciate ligament-intact knees. Bone Joint Surg Am. 2020; https://doi.org/10.2106/JBJS.19.00712. [Epub ahead of print].
11. Elmansori A, Lording T, Dumas R, Elmajri K, Neyret P, Lustig S. Proximal tibial bony and meniscal slopes are higher in ACL injured subjects than controls: a comparative MRI study. Knee Surg Sports Traumatol Arthrosc. 2017;25(5):1598–605. https://doi.org/10.1007/s00167-017-4447-4.
12. Robb C, Kempshall P, Getgood A, Standell H, Sprowson A, Thompson P, Spalding T. Meniscal integrity predicts laxity of anterior cruciate ligament reconstruction.Knee. Surg Sports Traumatol Arthrosc. 2015;23(12):3683–90. https://doi.org/10.1007/s00167-014-3277-x. Epub 2014 Sep 13
13. Zaffagnini S, Grassi A, Romandini I, Marcacci M, Filardo G. Meniscal allograft transplantation combined with anterior cruciate ligament reconstruction provides good mid-term clinical outcome. Knee Surg Sports Traumatol Arthrosc. 2019;27(6):1914–23. https://doi.org/10.1007/s00167-018-5078-0. Epub 2018 Aug 6
14. De Bruycker M, Verdonk P, Verdonk R. Meniscal allograft transplantation: a meta-analysis. SICOT J. 2017;3:33.
15. Leong NL, Southworth TM, Cole BJ. Distal femoral osteotomy and lateral meniscus allograft transplant. Clin Sports Med. 2019;38(3):387. https://doi.org/10.1016/j.csm.2019.02.007.
16. Grassi A, Bailey JR, Filardo G, Samuelsson K, Zaffagnini S, Amendola A. Sports health return to sport activity after meniscal allograft transplantation: at what level and at what cost? A Syst Rev Meta-Anal. 2019;11(2):123–33. https://doi.org/10.1177/1941738118819723. Epub 2019 Jan 14.
17. Verdonk PC, Demurie A, Almqvist KF, Veys EM, Verbruggen G, Verdonk R. Transplantation of viable meniscal allograft: surgical technique. J Bone Joint Surg Am. 2006;88(suppl 1, pt 1):109–18.
18. Stevenson C, Mahmoud A, Tudor F, Myers P. Meniscal allograft transplantation: undersizing grafts can lead to increased rates of clinical and mechanical failure. Knee Surg Sports Traumatol Arthrosc. 2019;27(6):1900–7. https://doi.org/10.1007/s00167-019-05398-2. Epub 2019 Feb 14
19. Pollard ME, Kang Q, Berg EE. Radiographic sizing for meniscal transplantation. Arthroscopy. 1995;11:684–7.
20. Beeler S, Jud L, Von Atzigen M, Sutter R, Fürnstahl P, Fucentese SF, Vlachopoulos L. Three-dimensional meniscus allograft sizing-a study of 280 healthy menisci. J Orthop Surg Res. 2020;15(1):74. https://doi.org/10.1186/s13018-020-01591-z.
21. De Coninck T, Vanrietvelde F, Seynaeve P, Verdonk P, Verstraete K. MRI imaging of the anatomy of the anterior horn of the medial meniscus. Acta Radiol. 2017;58(4):464–71.
22. Espejo-Reina A, Aguilera J, Espejo-Reina MJ, Espejo-Reina MP, Espejo-Baena A. One-third of meniscal tears are repairable: an epidemiological study evaluating meniscal tear patterns in stable and unstable knees. Arthroscopy. 2019;35:857–63.
23. Abrams GD, Frank RM, Gupta AK, Harris JD, McCormick FM, Cole BJ. Trends in meniscus repair and meniscectomy in the United States, 2005-2011. Am J Sports Med. 2013;41:2333–9.
24. de Caro F, Perdisa F, Dhollander A, Verdonk R, Verdonk P. Meniscus scaffolds for partial meniscus defects. Clin Sports Med. 2020 Jan;39(1):83–92. https://doi.org/10.1016/j.csm.2019.08.011.
25. Stone KR, Steadman JR, Rodkey WG, Li ST. Regeneration of meniscal cartilage with use of a collagen scaffold. Analysis of pre- liminary data. J Bone Joint Surg Am. 1997;79(12):1770–7.
26. Verdonk R, Verdonk P, Huysse W, Forsyth R, Heinrichs EL. Tissue ingrowth after implantation of a novel, biodegradable poly- urethane scaffold for treatment of partial meniscal lesions. Am J Sports Med. 2011;39(4):774–82. https://doi.org/10.1177/0363546511398040.
27. de Caro F, Perdisa F, Dhollander A, Verdonk R, Verdonk P. Meniscus scaffolds for partial meniscus defects. Clin Sports Med. 2020;39(1):83–92. https://doi.org/10.1016/j.csm.2019.08.011.
28. Moran CJ, Withers DP, Kurzweil PR, Verdonk PC. Clinical application of scaffolds for partial meniscus replacement. Sports Med Arthrosc Rev. 2015;23(3):156–61. https://doi.org/10.1097/JSA.0000000000000072.
29. Pereira H, Cengiz IF, Gomes S, Espregueira-Mendes J, Ripoll PL, Monllau JC, Ris RL, Oliveira JM. Meniscal allograft transplants and new scaffolding techniques. Effort open Rev EOR. 2019;4 https://doi.org/10.1302/2058-5241.4.180103.
30. Monllau JC, Gelber PE, Abat F, et al. Outcome after partial medial meniscus substitution with the collagen meniscal implant at a minimum of 10 years' follow-up. Arthroscopy. 2011;27(7):933–43.
31. Bulgheroni E, Grassi A, Bulgheroni P, et al. Long-term outcomes of medial CMI implant versus partial medial meniscectomy in patients with concomitant ACL reconstruction. Knee Surg Sports Traumatol Arthrosc. 2015;23:3221–7.

32. Zaffagnini S, Marcheggiani Muccioli GM, Lopomo N, et al. Prospective long-term outcomes of the medial collagen meniscus implant versus partial medial meniscectomy. A minimum 10-year follow-up study. Am J Sports Med. 2011;39(5) https://doi.org/10.1177/0363546510391179.
33. Gelber PE, Isart A, Erquicia JI, et al. Partial meniscus substitution with a polyurethane scaffold does not improve outcome after an open-wedge high tibial osteotomy. Knee Surg Sports Traumatol Arthrosc. 2015;23:334–9.
34. Rodkey WG, DeHaven KE, Montgomery WH 3rd, Baker CL Jr, Beck CL Jr, Hormel SE, Steadman JR, Cole BJ, Briggs KK. Comparison of the collagen meniscus implant with partial meniscectomy. A prospectiverandomized trial. J Bone Joint Surg Am. 2008;90(7):1413–26. https://doi.org/10.2106/JBJS.G.
35. Kovacs BK, Huegli R, Harder D, Cedro L, Berbig R, Amsler F, Bensler S, Hirschmann MT, Hirschmann A. MR variability of collagen meniscal implant remodelling in patients with good clinical outcome. Knee Surg Sports Traumatol Arthrosc. 2019 Sep 28; https://doi.org/10.1007/s00167-019-05715-9.
36. Shin YS, Lee HN, Sim HB, Kim HJ, Lee DH. Polyurethane meniscal scaffolds lead to better clinical outcomes but worse articular cartilage status and greater absolute meniscal extrusion. Knee Surg Sports Traumatol Arthrosc. 2018;26(8):2227–38. https://doi.org/10.1007/s00167-017-4650-3. Epub 2017 Jul 26
37. Koch M, Achatz FP, Lang S, et al. Tissue engineering of large full-size meniscus defects by apolyurethane scaffold: accelerated regeneration by mesenchymal stromal cells. Stem Cells Int. 2018;2018:8207071.
38. Elsner JJ, Portnoy S, Zur G, Guilak F, Shterling A, Linder-Ganz E. Design of a free-floating polycarbonate-urethane meniscal implant using finite element modeling and experimental validation. J Biomech Eng. 2010;132(9):095001.
39. Zur G, Linder-Ganz E, Elsner JJ, Shani J, Brenner O, Agar G, Hershman EB, Arnoczky SP, Guilak F, Shterling A. Chondroprotective effects of a polycarbonate-urethane meniscal implant: histopathological results in a sheep model. Knee Surg Sports Traumatol Arthrosc. 2011;19(2):255–63. https://doi.org/10.1007/s00167-010-1210-5. Epub 2010 Jul 16
40. De Coninck T, Elsner JJ, Linder-Ganz E, Cromheecke M, Shemesh M, Huysse W, Verdonk R, Verstraete K, Verdonk P. In-vivo evaluation of the kinematic behavior of an artificial medial meniscus implant: a pilot study using open-MRI. Clin Biomech. 2014;29(8):898–905. https://doi.org/10.1016/j.clinbiomech.2014.07.001. Epub 2014 Jul 17

28 Update on Indications, Techniques, and Outcomes of Meniscal Allograft Transplantation (MAT)

Trevor R. Gulbrandsen, Alan G. Shamrock, and Seth L. Sherman

28.1 Introduction

The meniscus is a critical intra-articular structure that plays a significant role in knee performance including lower limb kinetics and overall joint health. The knee menisci are dynamically involved in multiple joint functions including load distribution, joint lubrication, chondroprotection, articular nutrition, secondary stabilization, and proprioception. Meniscal injury results in altered mechanics and progressive joint dysfunction that may ultimately result in premature osteoarthritis of the knee [1–5].

There has been a significant paradigm shift in the management of meniscal pathology since the misconception of the 1970s that the meniscus is a useless structure [6]. Our current understanding has led us toward a strategy of meniscus preservation if at all possible [7, 8]. For meniscal deficient patients who continue to be symptomatic, replacement of the entire meniscus through meniscal allograft transplantation (MAT) has evolved as a reliable surgical solution for this challenging clinical problem.

T. R. Gulbrandsen · A. G. Shamrock
Department of Orthopaedic Surgery, University of Iowa Hospitals and Clinics, Iowa City, IA, USA

S. L. Sherman (✉)
Orthopaedic Surgery, Stanford University, Pao Alto, CA, USA
e-mail: shermans@stanford.edu

The ultimate goal of the MAT procedure is to restore native biomechanical properties of the native meniscus, thereby reestablishing knee function, reducing pain, improving the patient's quality of life, and possibly delaying osteoarthritis [9–11]. However, the surgeon and patient need to be prepared as a successful MAT requires careful preoperative planning and postoperative management. It is necessary that the surgeon understands the proper surgical technique including how and why to perform each surgical step. Additional key aspects of management include appropriate patient selection, preoperative discussion of patient expectations (especially long term), thorough preoperative planning, and firm adherence to postoperative rehabilitation protocols. Through strict preparation and thorough planning, the surgeon can provide the patient a reliable operation with a minimized risk of complication. Although controversy still exists, there have been substantial advancements since the first meniscal transplantation in 1984, [12] with more evidence-based indications and techniques contributing to improved long-term outcomes of MAT.

28.2 Indications

Despite the recent trend and advanced understanding of the importance in meniscus preservation, meniscus deficiency is still common [13]. It

N. Nakamura et al. (eds.), *Advances in Knee Ligament and Knee Preservation Surgery*,
https://doi.org/10.1007/978-3-030-84748-7_28

occurs in the setting of failed repairs, irreparable tear patterns, or aggressive subtotal meniscectomy. The clinical presentation of post meniscectomy syndrome includes unicompartmental pain and recurrent painful effusions following a previously sustained injury or meniscectomy. Initial treatment includes conservative management (i.e., NSAIDs, biologic injections, unloader brace, activity modification, rehabilitation programs). However, MAT should be considered when symptoms persist [9, 14–16].

Meniscal allograft transplant has historically been indicated for younger (ages 15–50) patients with symptomatic meniscal deficiency following previous subtotal or functional meniscectomy. Contraindications have included obesity, current tobacco use, skeletal immaturity, advanced osteoarthritis, inflammatory arthropathy, and history of recent or remote septic arthritis [17–19]. Any concurrent pathology will need to be addressed when considering MAT.

However, recent literature has continued the debate on specific indications and contraindications for MAT. Currently, symptomatic cartilage lesions in the setting of meniscus deficiency is an important indication for MAT. The benefits of the offloading aspects of MAT may be beneficial to optimize the outcome of the symptomatic cartilage lesion [20, 21]. Lee and colleagues investigated survivorship of MAT according to the cartilage defect grade present. Patients were grouped into low-grade lesions (ICRS [International Cartilage Repair Society] Grade II) on both the femoral and tibial articular surface, high-grade lesions (ICRS Grade II or IV) on one articular surface (femur or tibia), and high-grade lesions on both the femoral and tibial articular surface. Diffuse exposed subchondral bone was not included. Malalignment was corrected prior to the MAT procedure. The study reported that the postoperative PRO scores were not significantly different between the three groups. However, the overall graft survival at 5 years was significantly lower in the group with high-grade chondral lesions on both articular surfaces (62.2%) when compared to the other two groups (low-grade, 93.8%; high-grade lesions (tibia or femur), 90.9%). There was no association with sex, affected compartment, or malalignment and concomitant procedures [22]. Stone et al. investigated the impact of chondral defects and return to sport. Out of 49 MATs in patients with Grade III or IV Outerbridge defects, who were followed for a mean of 8.6 ± 4.2 years, 73.5% were able to resume sporting activities [23].

The role of prophylactic MAT continues to remain a controversial subject. It has been demonstrated that meniscal deficiency has been associated with progressive radiographic joint space narrowing [24]. However, clinical symptoms infrequently correlate with radiographic findings [5, 25]. Due to the unpredictable aspect of these findings, MAT is not routinely indicated for asymptomatic meniscal deficiency. The International Meniscus Reconstruction Experts Forum (IMREF) published a report after surveying orthopedic surgeons. In this report, 42% stated that they would not perform MAT for asymptomatic reasons. However, 18% reported that they would for the lateral meniscus [26]. One specific case in which prophylactic MAT may be considered is an active female with valgus deformity in the setting of total or subtotal lateral meniscectomy. Due to the anatomic and biomechanical characteristics of the lateral compartment, valgus deformity in the setting of lateral meniscal deficiency can lead to the rapid progression of lateral compartment chondral disease and worsening valgus malalignment [15, 27]. Therefore, early surgical intervention including alignment correction (varus producing osteotomy) combined with MAT is strongly considered to help mitigate the devastating effects of disease progression.

Chronic anterior cruciate ligament (ACL) instability or failed ACL reconstruction in the setting of meniscus deficiency is another indication for MAT [28]. The medial meniscus plays a role as a secondary stabilizer to anterior tibial translation, while the lateral meniscus provides rotational stability during pivot shift. As such, MAT can be implemented as an adjunct during ACL reconstruction for patients with significant meniscal deficiency to improve stability and reduce risk of ACL graft failure [2, 28]. This is particularly true for those with high-grade sagittal instability (Lachman Grade IIIB) and/or explosive Grade III pivot shift. It is important to note that these patients may present with chronic functional instability and not necessarily painful effusions localized to the compartment where the meniscus is deficient.

Overall, the decision to perform MAT should be made on a case by case basis, with a lower threshold to perform the procedure in young, symptomatic patients. However, it is critical that the patient has a thorough understanding of the risks and benefits of the procedure, including the risk of failure and need for additional or revision procedures in the future.

28.3 Applied Surgical Anatomy

The medial and lateral meniscus have distinct anatomic and biomechanical characteristics that influence decision making including timing of procedure, need for concomitant procedures, and the specific surgical technique utilized (Fig. 28.1).

The medial compartment consists of a concave tibial platform with the meniscus covering approximately 60–65% of the articular surface. The medial meniscus bears on average 50% of the load, and therefore meniscectomy is often tolerated in this compartment as it delays symptoms and decreases the need for concomitant cartilage restoration [29, 30]. Additionally, the medial meniscus is inherently more stable, when compared to the lateral meniscus, as it has the associated attachments of the meniscotibial (coronary) ligaments to the deep medial collateral ligament. Therefore, consideration to reproduce these important attachments during medial MAT is strongly encouraged. The medial meniscus is a secondary stabilizer to anterior tibial translation; therefore, medial MAT is often indicated with revision or chronic ACL reconstruction with concomitant meniscal deficiency.

The medial meniscus root insertions are widespread and oriented obliquely in the axial plane. The anatomy of the anterior horn of the medial meniscus is variable, with three types described (Fig. 28.2). Type 1 involves insertion lateral to the tibial spine and posterior to the anterior tibial edge, Type 2 with insertion medial to the tibial spine, and Type 3 with the insertion anterior to the anterior tibial edge [31]. It has been theorized that securing the anterior horn medial meniscus allograft posterior to the tibial edge in Type 3 medial menisci results in overcrowding of the compartment and increased risk of failure [19]. Furthermore, coronal malposition of root insertions has been shown to have greater influence on meniscal extrusion following MAT compared to sagittal malposition [32]. Due to these unique anatomic characteristics of the medial meniscus, bone plug or soft tissue only fixation is commonly utilized in the medial compartment [27,

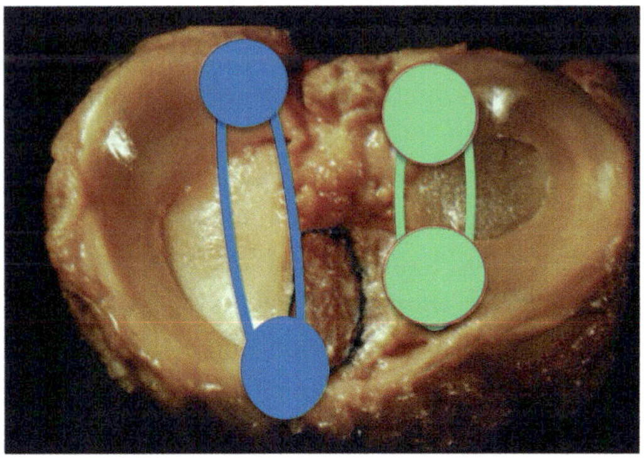

Fig. 28.1 Comparison of the medial and lateral menisci. The medial meniscus root insertions are widespread and oriented obliquely in the axial plane (blue), while the lateral meniscus root insertions are narrow and in a vertical orientation in the axial plane (green)

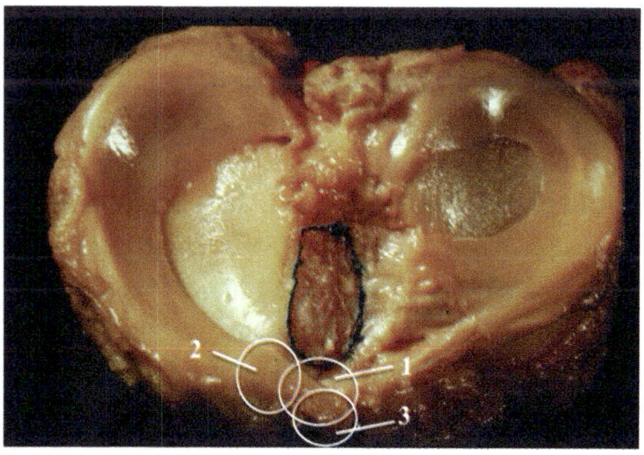

Fig. 28.2 The variation in attachments of the anterior horn of the medial meniscal. Type 1: flat anterior intercondylar region. Type 2: downward slope of the anteromedial articular plateau. This is medial to the intercondylar region (Type 1). Type 3: downward slope of the medial tibial plateau anterior to the intercondylar region (Type 1)

29, 30, 33]. This is due to the advantage of maintained bone stock, ease of concomitant ACL reconstruction, and the avoidance of disrupting the native ACL footprint [16, 27, 34].

The lateral tibial articular surface is convex with the lateral meniscus covering approximately 80–85% of the articular surface and bearing 70% of the load [33]. These distinct anatomic characteristics result in a higher incidence of symptoms following lateral meniscectomy. Additionally, there is a higher risk of chondral degeneration with associated clinical symptoms when comparted to medial meniscectomy [5]. Therefore, concomitant cartilage restoration is often required in the setting of symptomatic lateral meniscus deficiency. The lateral meniscus is O-shaped with narrow, vertically (axial plane) oriented roots. Many surgeons prefer to utilize the bone-in-slot technique for the lateral MAT due to the fixed distance between the roots and increased fixation strength at time zero [16, 27, 34]. However, soft tissue only or bone plug strategies can be utilized on the lateral side as well.

28.4 Treatment of Concomitant Pathology

The knee joint is a highly integrated organ that functions as one mechanical unit. Therefore, optimization of the knee joint environment is critical to achieve favorable outcomes following MAT. Conditions including high-grade cartilage defects, ligament insufficiency, and limb malalignment must be recognized and addressed simultaneously or in a staged fashion. Staged or simultaneous treatment of these concomitant issues has resulted in superior or equivocal outcomes when compared to isolated MAT [1, 21, 35, 36].

Staged arthroscopy is a beneficial procedure for accurate identification of meniscal and articular pathology, lysis of adhesions, and can be helpful when planning for a complex salvage procedure. Additionally, corrective procedures for limb malalignment including extra-articular osteotomy can be performed during staging arthroscopy. ACL tunnel bone grafting for tunnel widening is also performed in stage one. Intra-articular procedures including ligament reconstruction and cartilage restoration should be performed during the time of MAT.

Ligament insufficiency including ACL, posterior cruciate ligament (PCL), medial collateral ligament (MCL), and posterolateral corner (PLC) are risk factors for meniscal repair and MAT failure. Ligament procedures are traditionally performed concomitantly with MAT with concomitant ACL reconstruction having no negative effect on the postoperative outcome following MAT [16, 37, 38].

Similarly, cartilage restoration is symbiotic with MAT and should be performed in the same setting (Fig. 28.3) [39]. Specific technique and graft choice for cartilage restoration is lesion specific but can be performed during an open approach after arthroscopic or arthroscopic-assisted MAT. Untreated meniscal deficiency is a contraindication to isolated cartilage restoration, and the presence of untreated Grade III–IV chondral lesions is a contraindication to isolated MAT. However, concomitant procedures have been shown to produce favorable outcomes [21, 39]. Getgood et al. published a case series of patients who underwent combined MAT and tibial osteochondral allografting. They reported favorable outcomes with 5- and 10-year survivorship at 78% and 69%, respectively, for MAT component and 73% and 68%, respectively, for the osteochondral allograft [40].

It is beneficial to correct the mechanical axis to neutral or slightly varus in the presence of an axis deviation of more than 3 degrees [41]. Special care should be taken to avoid overcorrection. Lateral opening or medial closing wedge distal femoral osteotomy (DFO) is often the treatment of choice for valgus malalignment, while medial opening wedge high tibial osteotomy (HTO) is utilized to correct varus malalignment (Fig. 28.4). Additionally, mechanical axis correction can provide significant pain relief and MAT, and other intra-articular procedures may be avoided in patients who become asymptomatic following realignment.

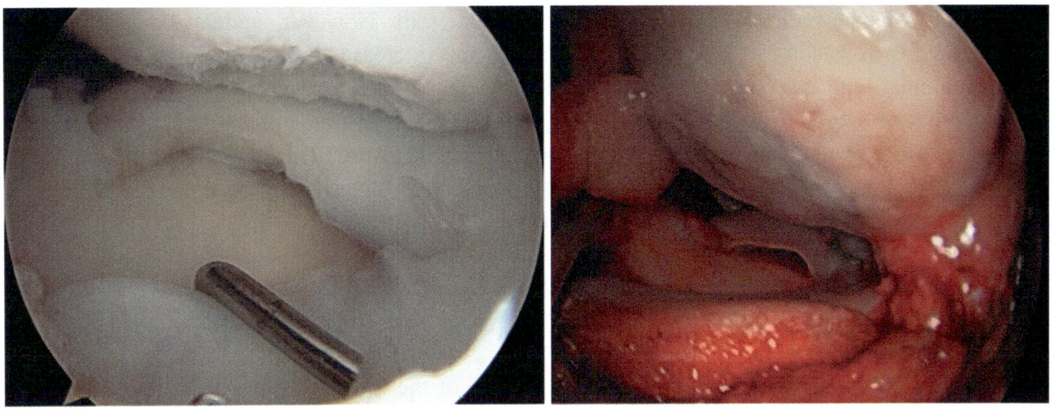

Fig. 28.3 Concomitant chondral restoration with arthroscopic MAT

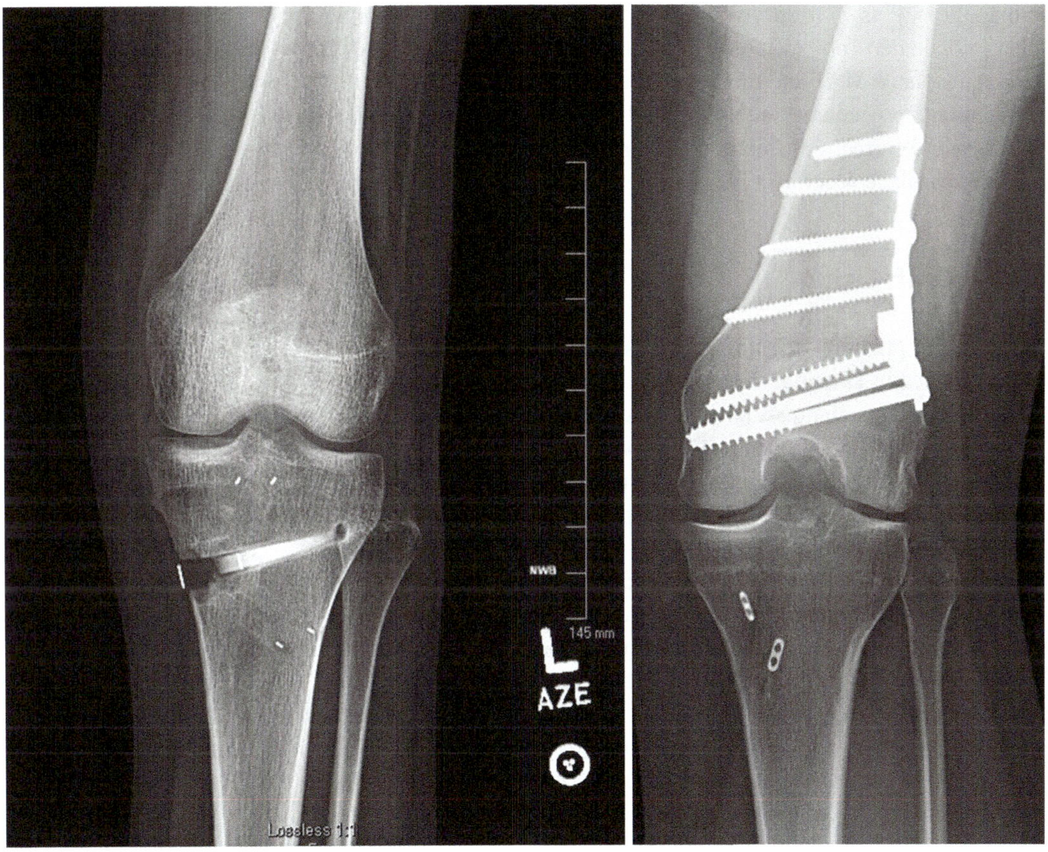

Fig. 28.4 Examples of malalignment correction prior to MAT. The left radiograph demonstrates correction of varus deformity utilizing the high tibial osteotomy (HTO). The right radiograph demonstrates lateral opening wedge distal femoral osteotomy (DFO) to correct valgus malalignment

However, debate continues regarding the overall benefits of osteotomy, including whether MAT combined with osteotomy provides superior outcomes compared to an isolated osteotomy procedure. Bloch et al. recently performed a prospective case series investigating 240 patients who underwent MAT. This cohort consisted of five groups: (A) patient with good chondral surfaces, (B) good chondral surfaces who underwent concomitant osteotomy, (C) good chondral surfaces with concomitant ACL reconstruction, (D) unipolar full thickness chondral wear, and (E) bipolar full-thickness chondral wear. They reported an overall better survivorship with groups A–C (no significant chondral damage) compared to groups D–E (significant chondral damage) (95% vs 77% survivorship at 5 years). They concluded no difference in outcomes between isolated MAT and those who underwent concomitant procedures including ACL reconstruction and corrective osteotomy, as long as the articular cartilage was intact [42].

28.5 Graft Sizing

Preoperative planning, including graft sizing, is biomechanically essential for proper outcomes. Several different techniques have been suggested for meniscus sizing with the utilization radiographs, computed tomography (CT), magnetic resonance imaging (MRI), and anthropometric data [19]. For medial MAT, the Pollard radiographic method is utilized to obtain length and width of the meniscus [43]. The Yoon equation for length and the anthropometric method for width is preferred for lateral MAT. The mediolateral sizing is more important when compared to anteroposterior sizing [44]. Utilizing MRI obtained from the contralateral knee may be beneficial in select cases. Inaccurate sizing can result in continued symptoms, meniscus extrusion, and ultimate failure [17]. Graft size should be within 10% of native meniscus. The risk for graft extrusion or early failure is increased with oversized grafts; therefore, if given the choice, an undersized graft is preferred over an oversized graft. However, undersized graft experiences increased biomechanical load with increased shear forces resulting in risk of allograft disruption and tear [17, 19]. Therefore, correct measurements and the utilization of a reliable tissue bank is critical.

28.6 Meniscus Transplant: Surgical Options

One of the most relevant topics for MAT is the fixation type utilized. Surgical techniques have dramatically improved over the last two decades. The ideal MAT technique theoretically would consist of being minimally invasive, bone stock preserving, providing options to handle graft mismatch, providing anatomic meniscus footprint restoration, securing adequate time zero fixation, and ultimately be reproducible. Currently, there are three standard techniques implemented in MAT fixation: suture-only, double bone plug, and bone-in-slot (bone bridge/trough). Novel hybrid techniques have increasing evidence for use. Controversy surrounds that MAT fixation method provides the best long-term results.

28.7 Bone Plug Technique

The traditional MAT bone plug technique involves preparing a graft with two 7–9 mm bone plugs attached to each root. After thorough debridement of any native meniscus remnants, two 8–10 mm tibial sockets are made. The plugs are secured with a bone-to-bone fixation at the roots, while the meniscal horns and body are fixed with sutures [45]. This technique is usually preferred in the medial compartment. (Fig. 28.5).

This technique significantly preserves bone stock. Additional advantages include decreased surgical time, the ability to accommodate graft size mismatch, no required "flipping" of the meniscal transplant (as there is in bone-in-slot technique), and avoidance from the risk of dis-

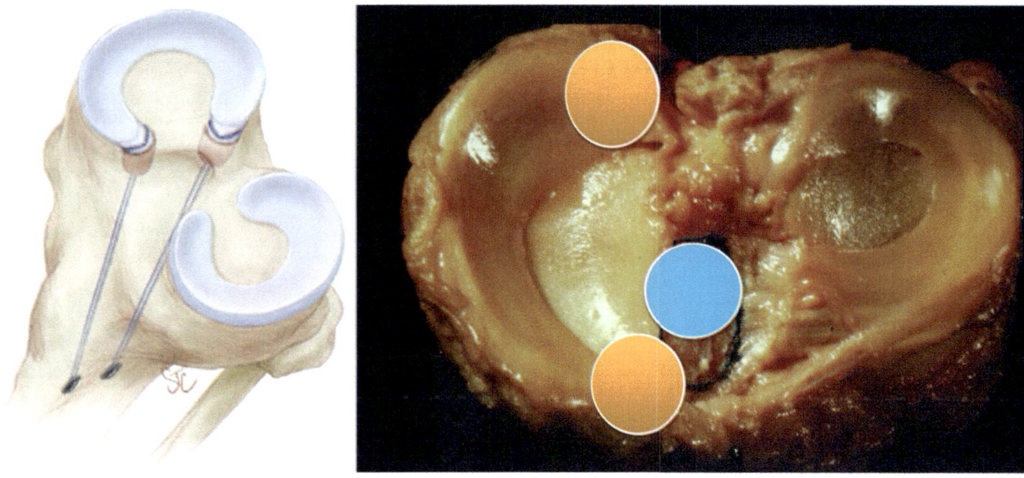

Fig. 28.5 (**a**) The bone plug technique is ideal for medial MAT procedures due to the oblique orientation (axial plane) of the medial meniscus roots (orange). Additionally, it does not disrupt the ACL (blue)

rupting the ACL footprint. The technique also uses familiar arthroscopic methods of socket preparation and peripheral meniscal repair. The disadvantages include the challenge in fully seating the bone plugs with an 8–10 mm depth, lack of fixation strength at time zero that may complicate early postoperative rehabilitation, and risk of tunnel coalescence in the lateral compartment due to the close proximity of the meniscus root insertions.

The importance of the fixation distance as well as the proper placement of the anterior and posterior horn attachments has been reported [28, 46–48]. The bone plug technique allows for proper anatomic placement of the meniscal horn attachments and allows the ability to handle graft mismatch if needed while also possibly making revision scenarios less challenging.

Studies continue to demonstrate good outcomes with MAT while using the bone plug technique, and many patients are able to return to pre-injury activity level [26, 49, 50].

28.8 Bone-in-Slot Technique

The bone-in-slot (trough, bone bridge) technique is preferred for the lateral compartment. In this procedure, a 10 mm high and 10 mm wide bone bridge is created on the meniscus allograft. A trough is prepared on the tibia for the allograft to "key" in place (Fig. 28.6) [51]. This technique has advantages, including a strong time zero fixation as well as the ability to maintain the anatomical proximity of the anterior and posterior meniscal root. However, the bone-in-slot technique has several disadvantages including it being a technically challenging technique ("flipping" the meniscus into place), decreased ability in dealing with graft mismatch/sizing issues, loss of substantial tibial bone stock, and decreased ease of revision MAT procedures. Avoidance of excessive constraint on the lateral meniscus graft during transplantation is essential as the native lateral meniscus is more mobile with no additional attachments to the lateral collateral ligament (LCL) or popliteal hiatus. Non-physiological tension through unnecessary suture fixation on the rim can lead to graft extrusion and failure. Due to the need for an accurate matched graft, this technique involves strict preoperative sizing as well as a reliable allograft bank. Recently, Kim et al. investigated the risk associated with non-anatomic horn position in the bone-in-slot technique for lateral MAT [52]. This study demonstrated that in the 214 isolated lateral bone-in-slot MAT

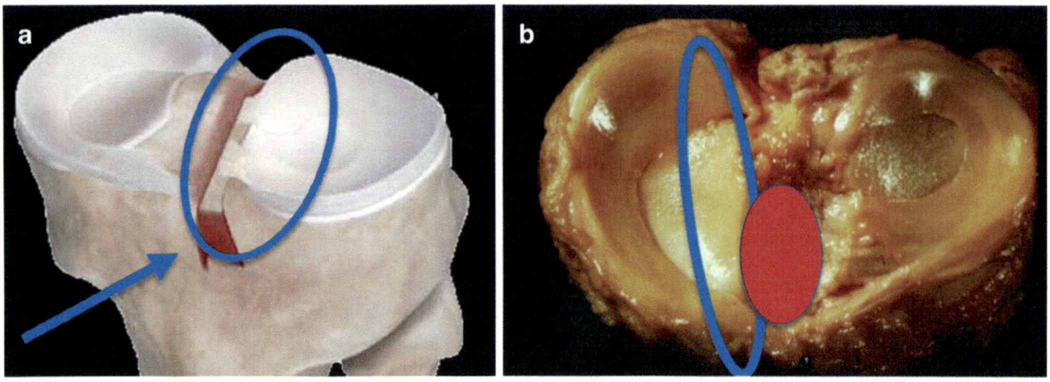

Fig. 28.6 (**a**) The bone-in-slot technique is preferred on the lateral compartment due to the vertical orientation (axial plane) of the lateral meniscus roots and the limited disruption of adjacent structures. (**b**) If performed on the medial compartment, it would interfere with the anatomic orientation of the medial meniscus root insertions and disrupt the medial portion of the ACL footprint (red)

cohort included, 11.1% of the non-anatomic group experienced early failure (within 1 year of MAT), while only 3.1% in anatomic group experienced early failure (odds ratio = 3.88; 95% CI, 1.13–13.26).

28.9 Suture-Only Technique

The suture-only technique involves only a soft tissue graft. Stay sutures are utilized to secure the body and meniscal horns, while the roots are fixed via a transtibial tunnel in a meniscal root repair fashion. It is technically the simplest option, and preoperative graft sizing is less demanding than techniques involving bony fixation. Nevertheless, graft extrusion has been a concern with suture-only fixation. Recent studies have shown that MATs fixed with bone plugs had a reduced incidence of extrusion compared to those with suture fixation [53–55]. Abat et al. found that meniscal allografts fixed with suture-only technique not only showed a significantly higher degree of extruded meniscal body than that fixed with bony fixation but also had a higher graft tear rate [56]. However, studies have suggested that although graft extrusion is common, it has not proven to have a relevant effect on the clinical and functional results [57]. Even so, suture-only fixation is currently performed with more frequency outside of the United States.

28.10 Hybrid Technique

An emerging technique utilizing an adjustable loop cortical fixation construct has recently been performed with positive results. The adjustable loop suspensory cortical suture is a #5 ultra-high molecular weight polyethylene (UHMWPE) suture loop that provides a four-point knotless locking system that does not rely on fixation distance. The graft has smaller bone plugs (9 mm in diameter and 3 mm in depth) located in anatomic sockets, therefore preserving bone stock whether used in the medial or lateral compartments (Fig. 28.7). Only one adjustable loop cortical fixation device per root is needed.

This technique also uses familiar arthroscopic methods for socket preparation. A well-known drilling technique is performed utilizing RetroCutter posteriorly or low-profile reamer anteriorly to create the tibia bone sockets (9.5 mm diameter by 5–10 mm deep) at the anterior and posterior root insertions. The adjustable loop is easily passed through the sockets and is secured over tibial cortical bone with an 8 × 12 mm slotted metallic attachable button system (Arthrex). After the MAT is shuttled into the joint arthroscopically and the roots are provisionally fixed, peripheral meniscal repair (6–8 total points of fixation) is performed with an all-inside technique on the posterior horn (1–2 fixation points), inside-out technique for the middle meniscus (3–4 fixation

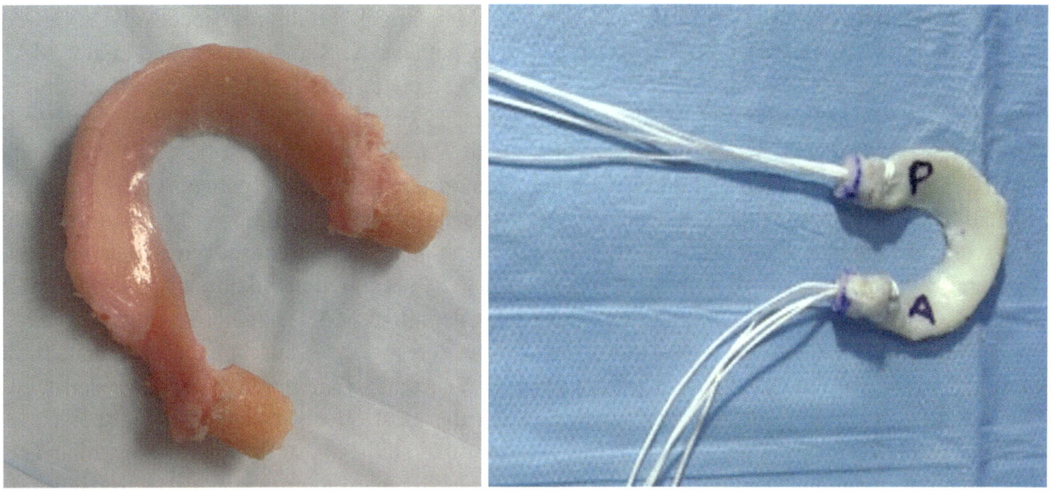

Fig. 28.7 MAT allograft that is prepared (9 mm in diameter and 5 mm in depth) for the arthroscopic MAT using anatomic bone sockets and suspensory cortical fixation technique

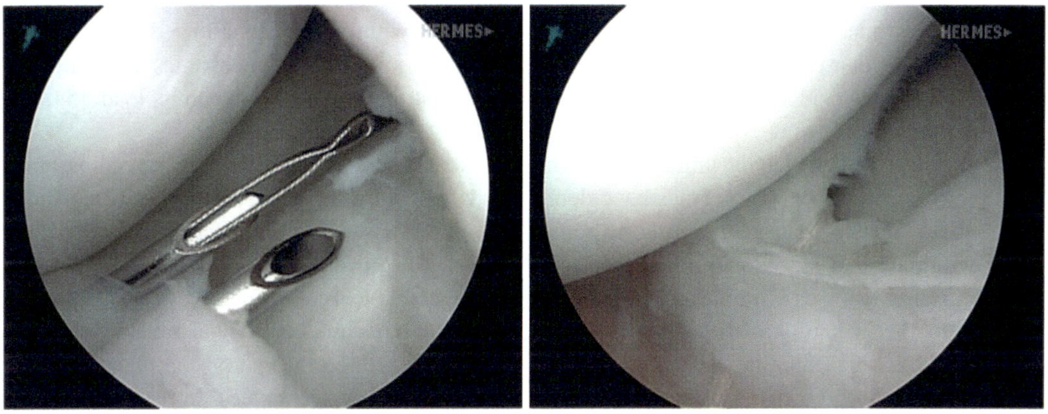

Fig. 28.8 A hybrid meniscal fixation is performed. There are 6–8 total points of fixation with an all-inside technique for the posterior horn (1–2 fixation points), inside-out technique for the middle meniscus (3–4 fixation points), and the outside-in technique for the anterior horn (1–2 fixation points)

points), and the outside-in technique for the anterior horn (1–2 fixation points) (Fig. 28.8).

This hybrid technique incorporates advantages of the other three techniques. It is all arthroscopic and easier to perform (similar to soft tissue only) but has strong fixation (similar to trough) without losing bone stock and can accommodate for graft mismatch (similar but easier than traditional bone plug technique). It provides other benefits including decreased surgical time and ability to manage graft extrusion following provisional peripheral fixation (adjustable cortical fixation) (Fig. 28.9).

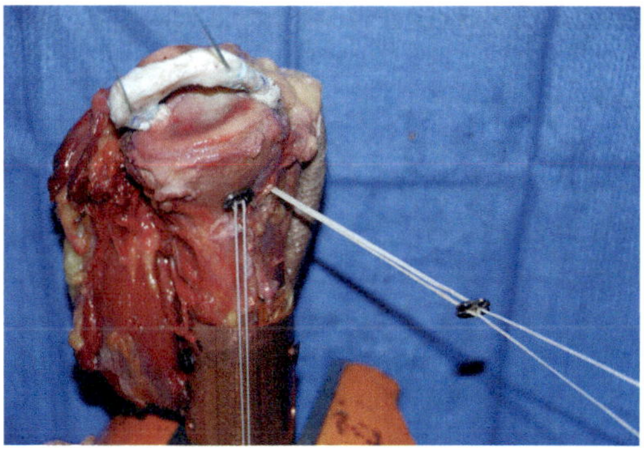

Fig. 28.9 Cadaveric representation of the adjustable loop cortical fixation construct in a medial MAT

28.11 Considerations in the Adolescent Population

There has been an increased rate of adolescents participating in competitive sports, which has resulted in an increase of meniscal injuries in this population [58]. Premature meniscal deficiency in adolescents results in a significant risk for early degenerative joint disease and future complications. Discoid meniscus tears provide an additional problem due to the possibility of unavoidable lateral meniscectomy [59, 60]. Determining appropriate indications and management can be challenging in this age group [61].

If treating meniscal deficiency with close observation, adolescent patients with open physes and meniscal deficiency will require at minimum annual clinical evaluation looking for painful effusion, x-rays including alignment views, and consideration for MRI assessment to monitor the articular cartilage [19]. If symptoms develop or there is evidence of progressive joint deterioration, MAT may be indicated [61]. Although MAT in a skeletally immature patient continues to be controversial, studies have demonstrated low revision rates in the pediatric population with safe and reproducible outcome [62, 63]. A current MRI is essential, as MAT sizing is key due to the lack of growth of the implant. However, MAT should not be delayed due to the potential chondroprotective affects. Practice guidelines and recommendations will continue to change as more skeletally immature patients undergo MAT [19, 64].

28.12 Updated Outcomes

MAT is an effective procedure for symptomatic post-meniscectomy syndrome and provides improved patient satisfaction with good survivorship and outcomes [63, 65–67]. In a prospective study, LaPrade et al. found that MAT significantly reduced pain, decreased activity-related effusion, and improved function in patients with previous meniscectomy. At 2-year follow-up, 91% of patients showed significant improvements in both pain and function [10].

McCormick et al. retrospectively reviewed 178 patients who underwent bone-in-slot MAT (41% isolated MAT, 60% concomitant procedure; 127 medial, 71 lateral, 2 bicompartmental) with a minimum 2-year follow-up. This study demonstrated that only 8 patients (4.7%) required subsequent MAT revision or knee arthroplasty indicating a 95% allograft survival rate. Although 64 (32%) required an additional procedure after the definitive MAT, a majority (59%) of these procedures were arthroscopic debridements [68].

A current area of research interest is determining the most effective method of graft fixation. Multiple studies have demonstrated similar outcomes in regard to graft survival with the different fixation options [9, 48]. Several studies have shown the superiority of bone plug fixation compared to suture-only fixation for medial meniscus transplantations and that the secure fixation achieved with bone plugs allowed restoration of optimum joint contact mechanics with superior load distribution under dynamic load [54, 69, 70].

Abat et al. compared clinical and radiographic outcomes of MAT with suture-only versus bony fixation. They found no difference in clinic outcomes but did report a significantly higher percentage of extruded meniscal tissue in the suture-only technique. They also demonstrated a significant difference in complication rate between the two techniques. In patients with the suture-only technique, there was a 33.3% overall complication rate including arthrofibrosis, infections, and graft tears compared to only 16.4% with bony fixation [71].

Outcomes depend on proper patient selection and thorough preoperative planning [9]. One important aspect of patient selection and expected outcomes includes the evaluation of the amount of chondral damage. As previously mentioned, Bloch et al. performed a prospective study that included 240 knees [42]. While MAT without chondral wear demonstrated 95% survivorship at 5 years, knees with full thickness chondral wear demonstrated a survivorship rate of only 77%. However, Frank and colleagues evaluated a cohort of 100 patients indicated for isolated osteochondral allograft transplantation (OCA).

Of these patients, 50 underwent concomitant MAT. They reported no difference in reported clinical outcomes, complications, or failure rate, concluding favorable outcomes of OCA with concomitant MAT [72].

In the current climate of cost-effective, value-based outcomes, financial factors should be considered. Bendich and colleagues evaluated factors that make MAT cost-effective in delaying the progression of osteoarthritis [73]. Using a Markov model, cost-effectiveness was assessed by comparing MAT to non-operative management in patients with previous meniscectomy. This study concluded that compared to non-operative management, MATT needs to be 31% more effective in delaying osteoarthritis in order to be cost-effective. This study also demonstrated that MAT is most cost-effective in patients who are 20–29 year-olds but is less cost-effective in obese patients (BMI 30–35).

There is a paucity of literature regarding return to sport after MAT. Several low-quality studies suggest that return to pre-injury activity is possible. Zaffagnini et al. recently reported that 66 out of 89 (74%) of patients who underwent MAT returned to sports after 8 months of strict postoperative rehabilitation. This particular cohort was able to return to rather high-demand sports including basketball, soccer, rugby, and volleyball. However, only 44 (49%) returned to the same level of play as pre-injury [74]. Similar results have been reported, with case series demonstrating 75–85% return to play after MAT [75, 76]. However, long-term durability of the allograft with repetitive loads is unknown. At present, MAT is not recommended by IMREF for athletes participating in contact/collision sport [26]. Therefore, sufficient patient education and discussion on postoperative expectation is necessary.

There are no standard evidence-based rehabilitation protocol following MAT to guide use of a brace, weightbearing, and ROM. Some surgeons initiate full weightbearing and range of motion much early on in the postoperative period. However, other surgeons recommend early weightbearing restrictions with a brace and graduated or delayed range of motion. Lee et al. investigated the impact of delayed rehabilitation on graft extrusion in a cohort of 53 patients who underwent lateral MAT, 25 with a standard rehabilitation, and 28 who underwent delayed rehabilitation, which included 3 weeks of immobilization followed by the utilization of unloading braces for 9 weeks [77]. At a mean follow-up of 2 years, the delayed rehabilitation cohort demonstrated decreased graft extrusion and relative percent of extrusion. Additionally, the delayed group had less joint space narrowing and progression of arthritis, providing additional evidence for the possible benefit of delayed rehabilitation. In general, rehabilitation will also depend on the need for concomitant procedures, time zero fixation type, and can be influenced by other patient factors.

The chondroprotective impact of MAT is undetermined. Historical reports demonstrate a lack of evidence that MAT decreases the advancement of osteoarthritis and joint space narrowing [9]. However, Jiang et al. published findings on a four- to six-year follow-up study comparing immediate and delayed MAT. Eight patients underwent MAT immediately after meniscectomy, while ten patients underwent a delayed MAT procedure (mean delay of 35 months). Although there was no different in PRO, patients who underwent MAT immediately after meniscectomy demonstrated less cartilage degeneration on radiographs and MRI [78]. Nevertheless, due to the small sample sizes and limited long-term outcome, conclusions cannot be determined, and the orthopedic surgeon must determine operative management on a case by case basis [79].

28.13 Conclusion

MAT has proven to be a safe and effective procedure for symptomatic meniscal deficiency or following previous failed procedures. There continues to be advancement in this field as we continue to investigate proper indications and techniques. However, there is still much debate, and further research is needed to clarify indications, proper surgical techniques, and overall long-term outcomes of MAT. Proper patient selection and preoperative planning are key to a successful MAT. As a salvage intervention, the main goal is improvement of quality of life. Return to preinjury sporting activity level is a secondary goal that may not be achievable. Therefore, adequate patient education and managing patient expectations are crucial. The patient needs to understand that this is not a definitive procedure and that additional future treatments are expected.

References

1. Chalmers PN, et al. Return to high-level sport after meniscal allograft transplantation. Arthroscopy. 2013;29(3):539–44.
2. Fox AJ, et al. The human meniscus: a review of anatomy, function, injury, and advances in treatment. Clin Anat. 2015;28(2):269–87.
3. Thorlund JB, et al. Changes in knee joint load indices from before to 12 months after arthroscopic partial meniscectomy: a prospective cohort study. Osteoarthr Cartil. 2016;24(7):1153–9.
4. Fairbank TJ. Knee joint changes after meniscectomy. J Bone Joint Surg Br. 1948;30B(4):664–70.
5. McNicholas MJ, et al. Total meniscectomy in adolescence. A thirty-year follow-up. J Bone Joint Surg Br. 2000;82(2):217–21.
6. Smillie I. *Injuries of the knee joint*. 4th ed. Edinburgh: Churcill Livingstone; 1972.
7. Barber-Westin SD, Noyes FR. Clinical healing rates of meniscus repairs of tears in the central-third (red-white) zone. Arthroscopy. 2014;30(1):134–46.
8. Alvarez-Diaz P, et al. Return to play after all-inside meniscal repair in competitive football players: a minimum 5-year follow-up. Knee Surg Sports Traumatol Arthrosc. 2016;24(6):1997–2001.
9. Elattar M, et al. Twenty-six years of meniscal allograft transplantation: is it still experimental? A meta-analysis of 44 trials. Knee Surg Sports Traumatol Arthrosc. 2011;19(2):147–57.
10. LaPrade RF, et al. A prospective outcomes study of meniscal allograft transplantation. Am J Sports Med. 2010;38(9):1804–12.
11. Verdonk PC, et al. Transplantation of viable meniscal allograft. Survivorship analysis and clinical outcome of one hundred cases. J Bone Joint Surg Am. 2005;87(4):715–24.
12. Milachowski KA, Weismeier K, Wirth CJ. Homologous meniscus transplantation. Experimental and clinical results. Int Orthop. 1989;13(1):1–11.
13. Mosich GM, et al. Operative treatment of isolated meniscus injuries in adolescent patients: a meta-analysis and review. Sports Health. 2018;10(4):311–6.
14. Northmore-Ball MD, Dandy DJ, Jackson RW. Arthroscopic, open partial, and total meniscectomy. A comparative study. J Bone Joint Surg Br. 1983;65(4):400–4.
15. Smith NA, Costa ML, Spalding T. Meniscal allograft transplantation: rationale for treatment. Bone Joint J. 2015;97-B(5):590–4.
16. Rosso F, et al. Meniscal allograft transplantation: a systematic review. Am J Sports Med. 2015;43(4):998–1007.
17. Trentacosta N, Graham WC, Gersoff WK. Meniscal allograft transplantation: state of the art. Sports Med Arthrosc Rev. 2016;24(2):e23–33.
18. Lubowitz JH, et al. Meniscus allograft transplantation: a current concepts review. Knee Surg Sports Traumatol Arthrosc. 2007;15(5):476–92.
19. Gelber PE, et al. Meniscal transplantation: state of the art. J ISAKOS: Jt Disord Orthopaed Sports Med. 2017;2(6):339–49.
20. Stone KR, et al. Long-term survival of concurrent meniscus allograft transplantation and repair of the articular cartilage: a prospective two- to 12-year follow-up report. J Bone Joint Surg Br. 2010;92(7):941–8.
21. Rue JP, et al. Prospective evaluation of concurrent meniscus transplantation and articular cartilage repair: minimum 2-year follow-up. Am J Sports Med. 2008;36(9):1770–8.
22. Lee BS, et al. Survivorship after meniscal allograft transplantation according to articular cartilage status. Am J Sports Med. 2017;45(5):1095–101.
23. Stone KR, et al. Meniscus transplantation in an active population with moderate to severe cartilage damage. Knee Surg Sports Traumatol Arthrosc. 2015;23(1):251–7.
24. Souza RB, et al. Cartilage MRI relaxation times after arthroscopic partial medial meniscectomy reveal localized degeneration. Knee Surg Sports Traumatol Arthrosc. 2015;23(1):188–97.
25. Andersson-Molina H, Karlsson H, Rockborn P. Arthroscopic partial and total meniscectomy: a long-term follow-up study with matched controls. Arthroscopy. 2002;18(2):183–9.
26. Getgood A, et al. International meniscus reconstruction experts forum (IMREF) 2015 consensus statement on the practice of meniscal allograft transplantation. Am J Sports Med. 2017;45(5):1195–205.
27. Yoon KH, et al. Meniscus allograft transplantation: a comparison of medial and lateral procedures. Am J Sports Med. 2014;42(1):200–7.
28. Packer JD, Rodeo SA. Meniscal allograft transplantation. Clin Sports Med. 2009;28(2):259–83. viii
29. Canham W, Stanish W. A study of the biological behavior of the meniscus as a transplant in the medial compartment of a dog's knee. Am J Sports Med. 1986;14(5):376–9.
30. Smigielski R, et al. Medial meniscus anatomy-from basic science to treatment. Knee Surg Sports Traumatol Arthrosc. 2015;23(1):8–14.

31. De Coninck T, et al. MR imaging of the anatomy of the anterior horn of the medial meniscus. Acta Radiol. 2017;58(4):464–71.
32. Kim NK, et al. Meniscal extrusion is positively correlated with the anatomical position changes of the meniscal anterior and posterior horns, following medial meniscal allograft transplantation. Knee Surg Sports Traumatol Arthrosc. 2019;27(8):2389–99.
33. Seedhom BB, Dowson D, Wright V. Proceedings: Functions of the menisci. A preliminary study. Ann Rheum Dis. 1974;33(1):111.
34. Myers P, Tudor F. Meniscal allograft transplantation: how should we be doing it? A systematic review. Arthroscopy. 2015;31(5):911–25.
35. Saltzman BM, et al. Preoperative Tibial subchondral bone marrow lesion patterns and associations with outcomes after isolated meniscus allograft transplantation. Am J Sports Med. 2018;46(5):1175–84.
36. Kazi HA, et al. Meniscal allograft with or without osteotomy: a 15-year follow-up study. Knee Surg Sports Traumatol Arthrosc. 2015;23(1):303–9.
37. Harris JD, et al. Biological knee reconstruction: a systematic review of combined meniscal allograft transplantation and cartilage repair or restoration. Arthroscopy. 2011;27(3):409–18.
38. von Lewinski G, et al. Twenty-year results of combined meniscal allograft transplantation, anterior cruciate ligament reconstruction and advancement of the medial collateral ligament. Knee Surg Sports Traumatol Arthrosc. 2007;15(9):1072–82.
39. Cotter EJ, et al. Meniscal allograft transplantation with concomitant osteochondral allograft transplantation. Arthrosc Tech. 2017;6(5):e1573–80.
40. Getgood A, et al. Combined osteochondral allograft and meniscal allograft transplantation: a survivorship analysis. Knee Surg Sports Traumatol Arthrosc. 2015;23(4):946–53.
41. Song JH, et al. Influence of Varus alignment on survivorship after lateral meniscal allograft transplantation. Am J Sports Med. 2020;48(6):1374–8.
42. Bloch B, et al. Higher survivorship following meniscal allograft transplantation in less worn knees justifies earlier referral for symptomatic patients: experience from 240 patients. Knee Surg Sports Traumatol Arthrosc. 2019;27(6):1891–9.
43. Pollard ME, Kang Q, Berg EE. Radiographic sizing for meniscal transplantation. Arthroscopy. 1995;11(6):684–7.
44. Yoon JR, et al. Is radiographic measurement of bony landmarks reliable for lateral meniscal sizing? Am J Sports Med. 2011;39(3):582–9.
45. Kim JG, et al. Arthroscopically assisted medial meniscal allograft transplantation using a modified bone plug to facilitate passage: surgical technique. J Knee Surg. 2009;22(3):259–63.
46. Moatshe G, et al. Posterior meniscal root injuries. Acta Orthop. 2016;87(5):452–8.
47. Bhatia S, et al. Meniscal root tears: significance, diagnosis, and treatment. Am J Sports Med. 2014;42(12):3016–30.
48. Lee SR, Kim JG, Nam SW. The tips and pitfalls of meniscus allograft transplantation. Knee Surg Relat Res. 2012;24(3):137–45.
49. Frank RM, Cole BJ. Meniscus transplantation. Curr Rev Musculoskelet Med. 2015;8(4):443–50.
50. Kim JM, et al. Results of meniscus allograft transplantation using bone fixation: 110 cases with objective evaluation. Am J Sports Med. 2012;40(5):1027–34.
51. Wilcox TR, Goble EM, Doucette SA. Goble technique of meniscus transplantation. Am J Knee Surg. 1996;9(1):37–42.
52. Kim JH, et al. Nonanatomic horn position increases risk of early graft failures after lateral meniscal allograft transplantation. Am J Sports Med. 2018;46(14):3407–14.
53. Rodeo SA. Meniscal allografts--where do we stand? Am J Sports Med. 2001;29(2):246–61.
54. Wang H, et al. Bone plug versus suture-only fixation of meniscal grafts: effect on joint contact mechanics during simulated gait. Am J Sports Med. 2014;42(7):1682–9.
55. De Coninck T, et al. Open versus arthroscopic meniscus allograft transplantation: magnetic resonance imaging study of meniscal radial displacement. Arthroscopy. 2013;29(3):514–21.
56. Abat F, et al. Suture-only fixation technique leads to a higher degree of extrusion than bony fixation in meniscal allograft transplantation. Am J Sports Med. 2012;40(7):1591–6.
57. Samitier G, et al. Meniscal allograft transplantation. Part 1: systematic review of graft biology, graft shrinkage, graft extrusion, graft sizing, and graft fixation. Knee Surg Sports Traumatol Arthrosc. 2015;23(1):310–22.
58. Bellisari G, Samora W, Klingele K. Meniscus tears in children. Sports Med Arthrosc Rev. 2011;19(1):50–5.
59. Shieh A, et al. Meniscus tear patterns in relation to skeletal immaturity: children versus adolescents. Am J Sports Med. 2013;41(12):2779–83.
60. Tuca M, Luderowski E, Rodeo S. Meniscal transplant in children. Curr Opin Pediatr. 2016;28(1):47–54.
61. Giuliani JR, et al. Treatment of meniscal injuries in young athletes. J Knee Surg. 2011;24(2):93–100.
62. Kocher MS, Tepolt FA, Vavken P. Meniscus transplantation in skeletally immature patients. J Pediatr Orthop B. 2016;25(4):343–8.
63. De Bruycker M, Verdonk PCM, Verdonk RC. Meniscal allograft transplantation: a meta-analysis. SICOT J. 2017;3:33.
64. Riboh JC, et al. Meniscal allograft transplantation in the adolescent population. Arthroscopy. 2016;32(6):1133–1140 e1.
65. Mahmoud A, et al. Meniscal allograft transplantation: the effect of cartilage status on survivorship and clinical outcome. Arthroscopy. 2018;34(6):1871–1876 e1.
66. Figueroa F, et al. Meniscus allograft transplantation: indications, techniques and outcomes. EFORT Open Rev. 2019;4(4):115–20.

67. Cengiz IF, et al. Orthopaedic regenerative tissue engineering en route to the holy grail: disequilibrium between the demand and the supply in the operating room. J Exp Orthop. 2018;5(1):14.
68. McCormick F, et al. Survival and reoperation rates after meniscal allograft transplantation: analysis of failures for 172 consecutive transplants at a minimum 2-year follow-up. Am J Sports Med. 2014;42(4):892–7.
69. Alhalki MM, Hull ML, Howell SM. Contact mechanics of the medial tibial plateau after implantation of a medial meniscal allograft. A human cadaveric study. Am J Sports Med. 2000;28(3):370–6.
70. Dean CS, et al. Medial meniscal allograft transplantation: the bone plug technique. Arthrosc Tech. 2016;5(2):e329–35.
71. Abat F, et al. Prospective comparative study between two different fixation techniques in meniscal allograft transplantation. Knee Surg Sports Traumatol Arthrosc. 2013;21(7):1516–22.
72. Frank RM, et al. Outcomes of osteochondral allograft transplantation with and without concomitant meniscus allograft transplantation: a comparative matched group analysis. Am J Sports Med. 2018;46(3):573–80.
73. Bendich I, et al. Evaluating meniscus allograft transplant using a cost-effectiveness threshold analysis. Knee. 2018;25(6):1171–80.
74. Zaffagnini S, et al. Is sport activity possible after arthroscopic meniscal allograft transplantation? Midterm results in active patients. Am J Sports Med. 2016;44(3):625–32.
75. Alentorn-Geli E, et al. Arthroscopic meniscal transplants in soccer players: outcomes at 2- to 5-year follow-up. Clin J Sport Med. 2010;20(5):340–3.
76. Marcacci M, et al. Arthroscopic meniscus allograft transplantation in male professional soccer players: a 36-month follow-up study. Am J Sports Med. 2014;42(2):382–8.
77. Lee DW, et al. Delayed rehabilitation after lateral meniscal allograft transplantation can reduce graft extrusion compared with standard rehabilitation. Am J Sports Med. 2018;46(10):2432–40.
78. Jiang D, et al. Comparative study on immediate versus delayed meniscus allograft transplantation: 4- to 6-year follow-up. Am J Sports Med. 2014;42(10):2329–37.
79. Hergan D, et al. Meniscal allograft transplantation. Arthroscopy. 2011;27(1):101–12.

Technique Corner: Cell-Based Cartilage Repair

29

Joshua Wright-Chisem and Andreas H. Gomoll

29.1 Introduction

Cartilage injuries of the knee represent a very common pathology, showing greater than 60% prevalence in arthroscopic procedures of the knee [1], with other studies reported focal full thickness chondral or osteochondral defects in nearly 20% of patients after an arthroscopic procedure [2]. While many of these lesions may be asymptomatic, symptomatic defects present as a difficult problem to treat. Untreated, focal chondral defects and osteochondral injuries of the knee can lead to progressive pain and osteoarthritis [3].

Operative interventions for treating cartilage defects of the knee can be grouped into palliative, reparative, restorative, and reconstructive procedures [4]. The focus of this chapter will be on restorative procedures, namely, cell-based cartilage repair. Restorative cartilage procedures have the goal of creating a hyaline-like cartilage layer [5]. Prior studies have shown that the articular surface created following these restorative procedures shows improved mechanical properties relative to prior procedures but is still inferior to native cartilage [6]. Studies have shown that MACI has similar clinical outcomes as prior generations of ACI, while MACI has been found to have a shorter tourniquet and procedural time [7].

29.2 Background

Cell-based cartilage repair was first introduced over two decades prior. Autologous chondrocyte implantation (ACI), developed in Sweden, was the first form of cell-based cartilage repair approved by the Federal Drug Administration (FDA) [8]. It was approved to treat full-thickness cartilage defects of the femoral trochlea and condyle; however, it also was frequently used to treat chondral defects in the patella and tibial plateau as an off-label use [9]. The first-generation ACI was used in conjunction with an autologous periosteal patch that secured the chondrocytes in situ. Following harvesting of the patch, it was sewn to normal, stable cartilage adjacent to the defect, and chondrocytes were introduced under the patch. This method was not without flaws, as graft hypertrophy was frequently seen postoperatively, most commonly in the patella [10]. Second generation ACI saw the development of a standardized collagen membrane rather than the first-generation periosteal flap. This development leads to improvements in both subjective and objective patient outcomes, as well as decreased rates of hypertrophy [11]. While the second-

J. Wright-Chisem · A. H. Gomoll (✉)
Division of Sports Medicine and Shoulder Surgery, Hospital for Special Surgery, New York, NY, USA
e-mail: wrightchisemj@hss.edu; gomolla@hss.edu

N. Nakamura et al. (eds.), *Advances in Knee Ligament and Knee Preservation Surgery*,
https://doi.org/10.1007/978-3-030-84748-7_29

generational development lead to improved clinical outcomes, it did not address certain technical challenges, including the potential for chondrocyte leakage or uneven distribution within the defect following implantation. This led to the development of the third and current generation ACI product, matrix-associated autologous chondrocyte implantation (MACI). Following FDA approval in 2017, MACI was becoming increasingly popular amongst surgeons as a tool to treat cartilage lesions. This was due to several factors, including decreased patient morbidity, ease of use, and fewer post-operative complications while demonstrating sustained excellent results.

MACI is a two-stage procedure that begins with a cartilage biopsy, typically during a diagnostic arthroscopic procedure. Following biopsy, the chondrocytes are isolated in the laboratory, expanded in culture and embedded unto a collagen scaffold, consisting of a porcine type I and III collagen bilayer membrane. During a second procedure, the collagen scaffold with its accompanying chondrocytes is implanted into the cartilage lesion and held into place with fibrin glue. Its ease of use is one reason why it has become marketed as a "simplified" delivery of ACI [12].

29.3 Clinical Evaluation

Patients with cartilage injuries of the knee may present acutely in the emergency department following a trauma or in the outpatient setting. These patients frequently endorse activity-related knee pain that may be exacerbated while participating in sports. Patients may present with a painful, swollen knee with an associated hemarthrosis, for example, after patellar dislocation, as studies have shown that osteochondral fractures may be responsible for up to 5% of acute post-traumatic hemarthroses [13]. It is imperative to perform a thorough history and physical exam for these patients to rule out additional pathology. The medial and lateral joint lines are assessed for potential meniscal pain, along with any other meniscal provocative tests, including but not limited to the McMurray's test. Chondral injuries of the trochlea or patella are not uncommon, and patellofemoral pain should be carefully evaluated. The integrity of the MPFL should be assessed, and patellar quadrant translation should be determined. In addition, the stability of the cruciate and collateral ligaments must be evaluated (Figs. 29.1, 29.2, and 29.3).

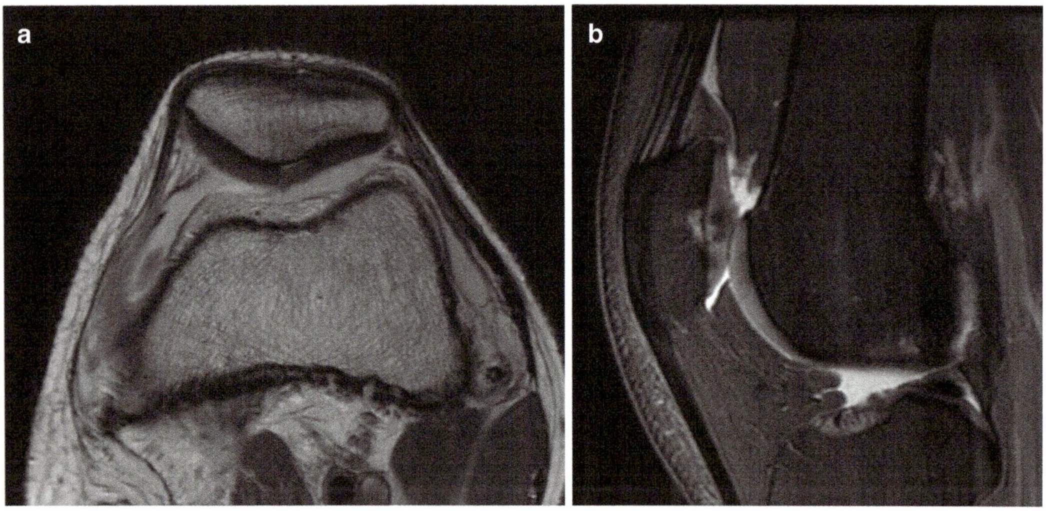

Fig. 29.1 (**a**) Axial MRI image (proton density) demonstrating delaminated cartilage flap of the median ridge. (**b**) Sagittal MRI image (Fat suppressed, inversion recovery) demonstrating subchondral edema in the central patella with signal change in the overlying articular surface

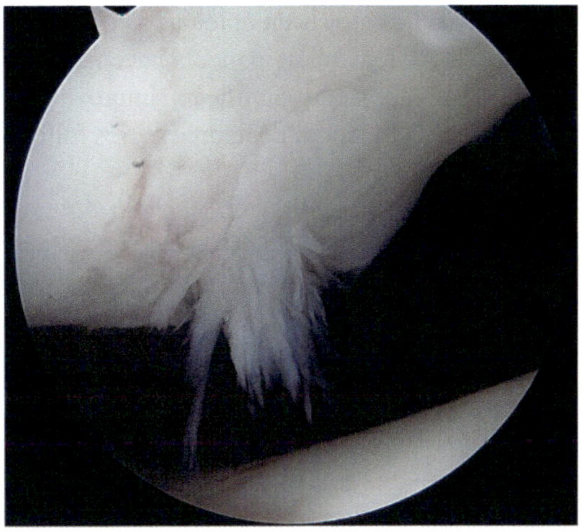

Fig. 29.2 Arthroscopic image demonstrating cartilage fraying and flap in the central patella

29.4 Imaging

Radiological studies are essential in the management of cartilage injuries of the knee. Initial evaluation begins with plain x-rays of the knee, including standing anteroposterior (AP), flexed posteroanterior (PA, Rosenberg view), lateral and merchant views of the knee. It is important to evaluate lower extremity alignment, and we recommend obtaining full length hip to ankle x-rays in patients where malalignment is in question. Cartilage injuries are frequently missed on plain radiographs; however, x-rays may demonstrate evidence of advanced osteoarthritis, which is a strict contraindication to restorative cartilage procedures [14]. Magnetic resonance imaging (MRI) is the gold standard in the diagnosis of

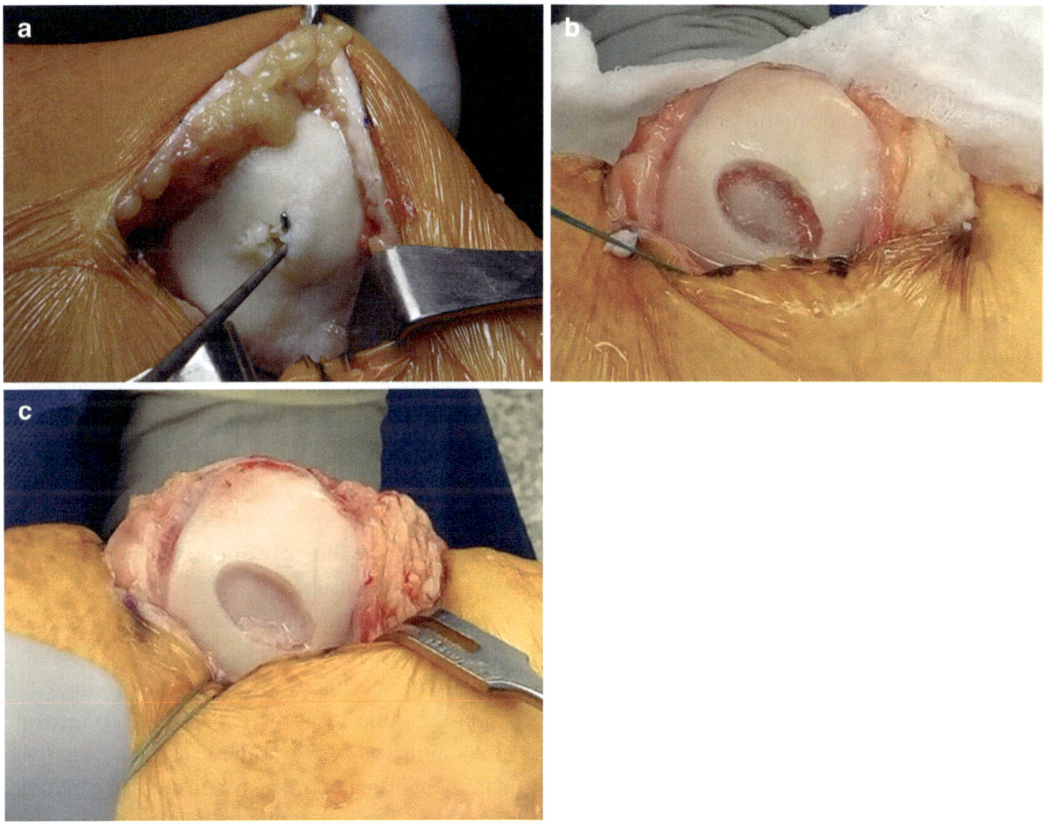

Fig. 29.3 (a) Intra-operative image showing unstable cartilage flap in the central patella. (b) Intra-operative image showing defect after debridement back to stable shoulders. (c) Intra-operative image showing defect after placement of the MACI implant and fibrin glue fixation

cartilage injuries and must be obtained in all patients during investigation of cartilage injuries. Importantly, an MRI will assist in better characterizing and identifying the cartilage lesion, while also allowing an assessment for associated soft tissue pathology, including any meniscal and ligamentous deficiency [15]. MRI studies are particularly useful in assessing the size of chondral lesions and the quality and size of subchondral bone involvement [16]. This is particularly useful as it plays an important role in determining the ideal intervention for each patient [17].

While MRI is a useful tool in the evaluation of articular cartilage defects, it may underestimate the size of the lesion. Studies have found that MRI may underestimate the size of articular cartilage lesion by as much as 70% [18]. Subsequent studies have also demonstrated that the average cartilage lesion identified on diagnostic arthroscopy was greater than 60%, larger than what was seen on MRI. For this reason, arthroscopy remains the gold standard in the evaluation of articular surfaces and should precede any procedural intervention.

29.5 Indications

A clear understanding of indications is paramount when determining whether a patient would benefit from cartilage repair with MACI. Those indications include symptomatic, full-thickness chondral and osteochondral defects, with a lesion size above 2 cm^2 [19, 20]. Anatomically, these lesions may be located anywhere in the knee and may be secondary to an acute traumatic injury, osteochondritis dissecans or focal defects from repetitive microtrauma [19]. Although there is no strict age cut off with respect to MACI, most studies have included patients between 15 and 60 years of age [20, 21].

Advanced osteoarthritis is an absolute contraindication to MACI [19]. Coronal plane malalignment, meniscal deficiency and ligamentous instability contribute to cartilage wear and degenerative changes of the knee and must be addressed prior to, or simultaneously with, any cartilage restoration procedure [19]. Elevated body mass index (BMI) has also been noted as a contraindication to MACI, as obese patients have not been shown to demonstrate significant improvement with respect to functional outcome scores following MACI [22].

29.6 Surgical Technique

MACI is a two-stage procedure, as biopsy of the articular cartilage is performed initially, and the processed chondrocytes are implanted at a later date as the second stage.

29.6.1 Biopsy Harvest

The patient is placed securely on a standard operating table, supine with care taken to pad all bony prominences. After induction of anesthesia (general or regional), a non-sterile tourniquet is placed high on the operative extremity. Next, a thorough examination under anesthesia is performed, with care taken to assess for ligamentous integrity, range of motion, and any other contributory pathology. Finally, a lateral post is placed at the level of the tourniquet.

The patient is then prepped and draped in the typical sterile fashion, and a surgical time out is performed, where surgical site, administration of preoperative antibiotics, and procedure are confirmed. A routine diagnostic arthroscopy is performed with attention paid to evaluate all articular surfaces, visualization of the suprapatellar pouch, gutters, and posterior joint recesses for any loose bodies, and the integrity of the cruciate ligaments and menisci is probed.

Using an arthroscopic shaver, the edges of the lesion are gently debrided to remove unstable flaps without uncovering the underlying subchondral bone. A probe is then used to measure the length and width of the lesion, and the measurements are recorded for insurance approval. Next, an arthroscopic gouge is introduced through the medial portal and used to perform the cartilage biopsy. The biopsy is performed on a lesser-weight bearing portion of the knee, frequently the intercondylar notch. It may also be performed on other lesser-weight bearing portions including the proximal trochlea. The biopsy is performed and then placed in a sterile container, ideally retrieving between 200 mg and 300 mg of sample cartilage [23]. The sample is then sent for processing. The portals are closed, and the patient is made weight bearing as tolerated without any restrictions with respect to range of motion.

29.6.2 Implantation

The chondrocytes from the biopsy sample are processed and stored at −80 degrees C for up to 5 years. Once insurance approval has been obtained, the implantation is scheduled, and the sample is thawed and grown for the surgical date.

The incision utilized is dependent on the area of interest; if the lesion is located along the lateral patellar facet or lateral femoral condyle, a lateral parapatellar arthrotomy is utilized. If the lesion is located along the medial patellar facet or medial femoral condyle, a medial parapatellar arthrotomy is used. A #10 blade is used for the skin incision with care taken to avoid violating the meniscus. The incision is carried down through the underlying retinaculum and joint capsule with gentle dissection performed with electrocautery. The defect is outlined with a 15 blade to include all damaged or unstable cartilage along the periphery. Alternatively, the current instrumentation provides cutters in various sizes and shapes that can be utilized to both outline the defect and prepare the graft. Utilizing curettes and rongeurs, the defect is debrided of soft tissues including the calcified layer, with care taken to preserve the subchondral plate. Once satisfied with the debridement, attention is paid to the implantation process. Historically, a piece of aluminum foil from a sterile suture pack is utilized. The foil is placed over the debrided area with the colored portion of the foil facing the subchondral bone. Utilizing fingers or a blunt instrument, the foil is pushed down gently along the defect to template the lesion. Any remaining foil that is prominent is then trimmed to ensure that the membrane of the implant is not prominent when implanted. If cutters are used, this step is not necessary.

The MACI implant is provided in *non-sterile* packaging. The container is removed from the plastic pouch by the circulator, who then opens the container. A member of the surgical team then transfers some of the culture liquid into a sterile dish on the back table with a syringe, followed by the implant itself, using non-toothed forceps. Next, the previously templated foil is placed color side up and secured with Steri-Strips to a Tegaderm. Attention is then paid toward the membrane. During processing at the facility, a portion of the implant is removed for sterility and quality testing. With the cell side facing up, the missing area is at the lower left corner, assisting in orienting the membrane appropriately during sizing and implantation. The membrane is placed on the foil template, with care taken to ensure the cell layer is facing up. The membrane is then trimmed with fine scissors according to the underlying foil template. Implant handling decreases cell viability and is therefore avoided as much as possible.

Alternatively, the same cutter utilized to outline the defect is used to trim the implant.

As an adhesive, fibrin glue is added to the base of the defect on top pf the exposed subchondral bone. The membrane is then placed with the cell layer facing down and compressed with gentle manual pressure. The implant is again trimmed to ensure that any prominent edges are removed and additional fibrin glue is added along the periphery as needed to secure the MACI. Sutures of 6–0 Vicryl can be added as necessary if there are concerns over stability. The knee is taken through full range of motion to assess stability of the implant.

The wound is then closed per routine, and the patient is placed in a hinged knee brace (HKB) post-operatively, locked in full extension. The use of a suction drain is generally discouraged.

29.7 Post-Operative Protocol

The post-operative rehabilitation will vary based on the location of the defect, while the goals of protecting the implant and maturing tissue and while restoring full range of motion and strength are the same for all patients.

A continuous passive motion (CPM) device is utilized for 6 weeks for all patients, as studies have shown that passive motion may benefit maintenance of range of motion and pain reduction and may promote tissue maturation [24]. The CPM is initiated at 40° of knee flexion with a goal of achieving 90° over the course of 2–4 weeks. Isometric quadriceps exercises and calf pumps are started immediately and maintained over the initial 2 weeks. Limited core strengthening and side raises for gluteal activation can be performed. A cold and compression therapy device is helpful for swelling and pain control. A knee brace locked in full extension is used for sleeping for the first week and for ambulation for 6 weeks. Electrical stimulation can be helpful, initially for pain control and later for muscle retraining.

Patients with a lesion on the femoral condyle are made touch down weight bearing for 4 (smaller well-contained defect) to 6 (larger defects) weeks, after which they progress to full weight bearing as tolerated. Patients with lesions located on the patella or trochlea are permitted to be fully weight bearing immediately, unless a tibial tubercle osteotomy is performed concurrently, in which case patients remain touch down weight bearing for 4–6 weeks to protect the osteotomy.

Patients are permitted to perform full range of motion at 6 weeks, and at that time, the HKB will be unlocked for ambulation and discontinued entirely when patients feel comfortable ambulating without assistive device.

Therapy is progressed at 6 weeks when patients may begin light activities with minimal resistance such as stationary biking, which can slowly be advanced. Ankle raises and gait training are started. At 12 weeks, activities are increased and rehabilitation work is focused more on core strengthening, unilateral balance, strengthening of gluteal muscles, progressive quadriceps strengthening, and hamstring isometric exercises; elliptical trainers can be incorporated into the workout, unless there were large bipolar PF grafts. At 6 months, all activities can be advanced as tolerated, based on pain and swelling, with the exception of plyometrics and impact. Patients are frequently cleared for full activities at 9–18 months, based on progression, lesion size, and location of defect. Generally, larger and multiple defects, especially when in the patellofemoral joint, will require a longer rehabilitation and delayed return to full activities than small defects in the femoral condyles.

29.8 Surgical Outcomes

The literature reports overall good functional outcomes after ACI and MACI. Kreuz et al. demonstrated that the IKDC, Lysholm, KOOS, and Noyes scores all improved significantly following MACI [25]. Furthermore, they found that MRI showed moderate to complete filling of the defect in over 70% of patients included, with significant correlation between MRI assessment and functional outcome scores. Ebert et al. evaluated which factors were predictive of improved MRI composite score and patient satisfaction [26]. They found that duration of preoperative symptoms and graft size were all significant predictors of MRI score at 5 years post-operatively. Notable, an accelerated rehab protocol with 8 rather than 12-week return to full weight bearing was associated with significantly higher patient satisfaction. Krych et al. investigated the long-term results of MACI in over 750 patients. They found that over 80% of patients had successful long-term outcomes. In this study, increasing patient age and defects greater than 4.5cm^2 were associated with higher reoperation and failure rates [27].

29.9 Conclusion

MACI is an excellent restorative option for patients with focal chondral defects of the knee. The procedure is a two-staged process with an initial diagnostic arthroscopy and cartilage biopsy, followed by implantation of the cultured chondrocytes. The post-operative rehab protocol is specific to patient and lesion characteristics and is crucial to a successful outcome. With proper patient selection and meticulous surgical technique, patients may predictably see improvements in both pain and function following MACI.

References

1. Curl WW, Krome J, Gordon ES, Rushing J, Smith BP, Poehling GG. Cartilage injuries: a review of 31,516 knee arthroscopies. Arthroscopy. 1997 Aug;13(4):456–60.
2. Hjelle K, Solheim E, Strand T, Muri R, Brittberg M. Articular cartilage defects in 1000 knee arthroscopies. Arthroscopy. 2002 Sep;18(7):730–4.
3. Perera JR, Gikas PD, Bentley G. The present state of treatments for articular cartilage defects in the knee. Ann R Coll Surg Engl. 2012;94(6):381–7. https://doi.org/10.1308/003588412X13171221592573.
4. Erickson BJ, Chalmers PN, Yanke AB, Cole BJ. Surgical management of osteochondritis dissecans of the knee. Current reviews in musculoskeletal medicine. 2013;6(2):102–14.
5. Erickson BJ, Strickland SM, Gomoll AH. Indications, Techniques, Outcomes for Matrix-Induced Autologous Chondrocyte Implantation (MACI). Oper Tech Sports Med. 2018;26(3):175–82. https://doi.org/10.1053/j.otsm.2018.06.002.
6. Nixon AJ, Rickey E, Butler TJ, Scimeca MS, Moran N, Matthews GL. A chondrocyte infiltrated collagen type I/III membrane (MACI(R) implant) improves cartilage healing in the equine patellofemoral joint model. Osteoarthritis and Cartilage. 2015;23(4):648–60.
7. Bartlett W, Skinner JA, Gooding CR, Carrington RW, Flanagan AM, Briggs TW, et al. Autologous chondrocyte implantation versus matrix-induced autologous chondrocyte implantation for osteochondral defects of the knee: a prospective, randomised study. J Bone Joint Surg Br. 2005;87(5):640–5.
8. Brittberg M, Lindahl A, Nilsson A, Ohlsson C, Isaksson O, Peterson L. Treatment of deep cartilage defects in the knee with autologous chondrocyte transplantation. N Engl J Med. 1994;331(14):889–95.
9. von Keudell A, Han R, Bryant T, Minas T. Autologous Chondrocyte Implantation to Isolated Patella Cartilage Defects. Cartilage. 2017;8(2):146–54. https://doi.org/10.1177/1947603516654944.
10. Kreuz P, Steinwachs M, Erggelet C, et al. Classification of graft hypertrophy after autologous chondrocyte implantation of full-thickness chondral defects in the knee. Osteoarthr Cartil. 2007;15(12):1339–47. https://doi.org/10.1016/j.joca.2007.04.020.
11. Niemeyer P, Salzmann G, Feucht M, et al. First-generation versus second-generation autologous chondrocyte implantation for treatment of cartilage defects of the knee: a matched-pair analysis on long-term clinical outcome. Int Orthop. 2014;38(10):2065–70. https://doi.org/10.1007/s00264-014-2368-0.
12. https://www.maci.com/healthcare-professionals/the-maci-story/product-information.html.
13. Maffulli N, Binfield PM, King JB, Good CJ. Acute haemarthrosis of the knee in athletes. A prospective study of 106 cases. J Bone Joint Surg (Br). 1993 Nov;75(6):945–9.
14. Oliver-Welsh L, Griffin JW, Meyer MA, Gitelis ME, Cole BJ. Deciding How Best to Treat Cartilage Defects. Orthopedics. 2016;39(6):343–50.
15. Rodrigues MB, Camanho GL. MRI evaluation of knee cartilage. Rev Bras Ortop. 2015;45(4):340–6. Published 2015 Nov 17. https://doi.org/10.1016/S2255-4971(15)30379-7.
16. Gorbachova T, Melenevsky Y, Cohen M, Cerniglia BW. Osteochondral Lesions of the Knee: Differentiating the Most Common Entities at MRI. Radiographics. 2018;38(5):1478–95. https://doi.org/10.1148/rg.2018180044.
17. Moyad TF. Cartilage Injuries in the Adult Knee: Evaluation and Management. Cartilage. 2011;2(3):226–36. https://doi.org/10.1177/1947603510383973.
18. Campbell AB, Knopp MV, Kolovich GP, Wei W, Jia G, Siston RA, et al. Preoperative MRI underestimates articular cartilage defect size compared with findings at arthroscopic knee surgery. Am J Sports Med. 2013;41(3):590–5.
19. Jones KJ, Cash BM. Matrix-Induced Autologous Chondrocyte Implantation with Autologous Bone Grafting for Osteochondral Lesions of the Femoral Trochlea. Arthrosc Tech. 2019;8(3):e259–66. Published 2019 Feb 11. https://doi.org/10.1016/j.eats.2018.10.022.
20. Basad E, Wissing FR, Fehrenbach P, Rickert M, Steinmeyer J, Ishaque B. Matrix-induced autologous chondrocyte implantation (MACI) in the knee: clinical outcomes and challenges. Knee Surg Sports Traumatol Arthrosc. 2015;23(12):3729–35.
21. Gille J, Behrens P, Schulz AP, Oheim R, Kienast B. Matrix-Associated Autologous Chondrocyte Implantation: A Clinical Follow-Up at 15 Years. Cartilage. 2016;7(4):309–15. https://doi.org/10.1177/1947603516638901.
22. Jaiswal PK, Bentley G, Carrington RWJ, Skinner JA, Briggs TWR. The adverse effect of elevated body mass index on outcome after autologous chondrocyte implantation. J Bone Joint Surg Br. 2012;94-B(10):1377–81. https://doi.org/10.1302/0301-620x.94b10.29388.
23. Brittberg M. Cell carriers as the next generation of cell therapy for cartilage repair: a review of the matrix-induced autologous chondrocyte implantation procedure. Am J Sports Med. 2010;38(6):1259–71.
24. Howard JS, Mattacola CG, Romine SE, Lattermann C. Continuous passive motion, early weight bearing, and active motion following knee articular cartilage repair: evidence for clinical practice. Cartilage. 2010;1(4):276–86. https://doi.org/10.1177/1947603510368055.
25. Kreuz PC, Kalkreuth RH, Niemeyer P, Uhl M, Erggelet C. Long-term clinical and MRI results of matrix-assisted autologous chondrocyte implantation for articular cartilage defects of the knee. Cartilage. 2019;10(3):305–13. https://doi.org/10.1177/1947603518756463.

26. Ebert JR, Smith A, Edwards PK, Hambly K, Wood DJ, Ackland TR. Factors predictive of outcome 5 years after matrix-induced autologous chondrocyte implantation in the tibiofemoral joint. Am J Sports Med. 2013;41(6):1245–54. https://doi.org/10.1177/0363546513484696.
27. Pareek A, Carey JL, Reardon PJ, Peterson L, Stuart MJ, Krych AJ. Long-term outcomes after autologous chondrocyte implantation: a systematic review at mean follow-up of 11.4 years. Cartilage. 2016;7(4):298–308. https://doi.org/10.1177/1947603516630786.

Technique Corner: Marrow Stimulation and Augmentation

30

Eric D. Haunschild, Ron Gilat, Theodore Wolfson, Stephanie Wong, Nolan B. Condron, Joshua T. Kaiser, and Brian J. Cole

30.1 Microfracture Technique

30.1.1 History/Physical Exam

A thorough history and physical exam should be performed on initial consultation of any candidate for surgical management of cartilage defects. Patients with focal cartilage defects commonly present with swelling, activity-related pain, and limping [1]. Any history of knee trauma, as well as symptom duration and prior surgical history in the index knee, should also be elicited. On physical exam, effusion and tenderness to palpation over a focal area of the knee are suggestive but not specific for cartilage lesions. Thorough evaluation of any ligamentous instability or limb malalignment should be examined as any deficiencies may indicate concomitant ligamentous or realignment procedures at the time of microfracture.

30.1.2 Indications and Contraindications

Evaluation for a microfracture procedure is individualized but should be considered in all isolated and symptomatic full thickness Outerbridge Grade III–IV cartilage lesions with an area less than three square centimeters [2]. The presence of large chondral lesions or diffuse degenerative osteoarthritic change in the knee is unlikely to respond to microfracture and should warrant consideration of alternative therapies. Patient age and activity level are relative indications, as active patients younger than 40 years old have demonstrated superior outcomes to older, less active patients [3, 4]. Microfracture is best suited for isolated lesions of the femoral condyles or trochlea and is relatively contraindicated in the treatment of bipolar and patellar lesions due to inferior results. In addition, lesions with significant subchondral bone involvement (ICRS Grade IV) should be evaluated on an individualized basis, as bony augmentation may be better suited to address subchondral bone insufficiency.

30.1.3 Preoperative Imaging and Evaluation

At the time of baseline evaluation, x-ray and magnetic resonance imaging (MRI) assessments should be performed in all surgical candidates. Standing AP, lateral, and Merchant view radiographs allow the identification of osteoarthritis or bony abnormalities. In addition, standing PA imaging at 45° of flexion may also be used to aid in the identification of joint space narrowing [5].

E. D. Haunschild · R. Gilat · T. Wolfson · S. Wong
N. B. Condron · J. T. Kaiser · B. J. Cole (✉)
Midwest Orthopaedics at Rush University Medical Center, Chicago, IL, USA
e-mail: brian.cole@rushortho.com

N. Nakamura et al. (eds.), *Advances in Knee Ligament and Knee Preservation Surgery*,
https://doi.org/10.1007/978-3-030-84748-7_30

MRI is critical to assess cartilage thickness and subchondral bone health and to further rule out concomitant injury. In particular, changes to cartilage including fissuring, chondral fibrillation, thinning, and focal defects can suggest the severity of chondral disease and the suitability for microfracture as a repair technique [6]. If clinically indicated, long-leg radiographs can be obtained to evaluate limb axis and any subsequent need for concomitant realignment surgery.

30.1.4 Operative Technique

30.1.4.1 Positioning and Diagnostic Arthroscopy

The preferred technique of the senior author has been described previously [7–9]. Patients are positioned supine on a standard operating table with a non-sterile thigh tourniquet placed and used at the surgeon's discretion. The leg is then prepped and draped in standard sterile fashion, and standard anteromedial and anterolateral portal incisions are made using an 11-blade. On occasion the cartilage lesion may be inaccessible using standard portals. In these cases, an accessory portal can be made after localizing the lesion site with or without the assistance of a spinal needle. A standard diagnostic arthroscopy should then be performed to evaluate the lesion of interest, the presence of any cartilage loose bodies, or other pathology.

30.1.4.2 Preparation of Defect

Once identified, the chondral lesion often requires extensive preparation and chondroplasty prior to microfracture (Fig. 30.1a). A 4.5 mm arthroscopic shaver is typically used to debride any unstable cartilage fragments. The borders of the defect should be probed and unstable flaps debrided to establish a periphery of intact and stable articular cartilage. Next, vertical walls of healthy cartilage are established circumferentially at the lesion borders, which can be achieved with a shaver or curettes. Care should be taken to not remove excess cartilage in the preparation of the defect's borders. Then, the calcified cartilage layer is removed using a curette without disrupting the underlying subchondral bone. Special care should be taken during this step as well, as surgeon reliability in removing this layer is variable, and adequate removal is needed for optimal repair, while excessive removal may lead to subchondral cyst formation [10, 11]. Once debridement is complete, all loose bodies and cartilage fragments are removed with a shaver prior to performing microfracture.

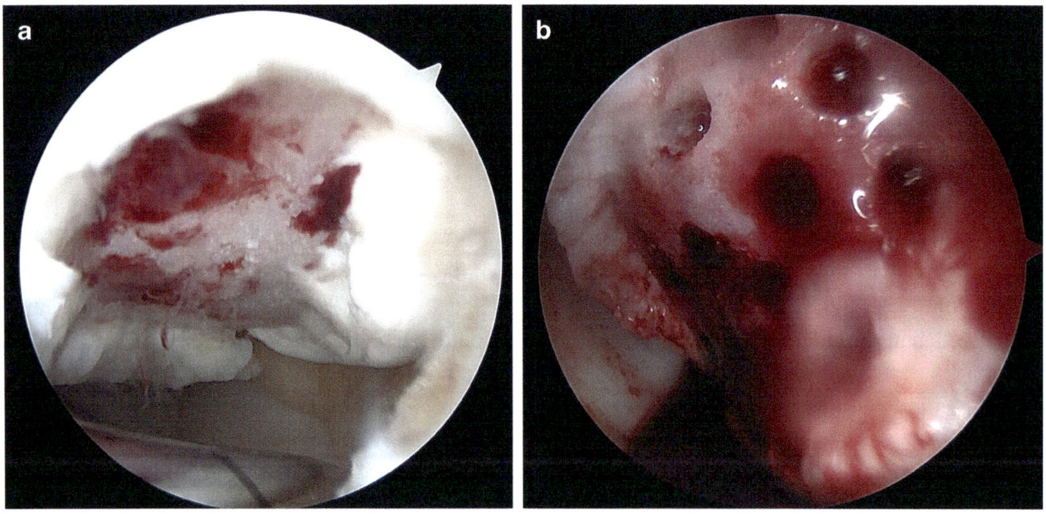

Fig. 30.1 Microfracture of a focal cartilage defect. (**a**) Prior to microfracture, focal defects often have fraying and unstable flaps requiring debridement. (**b**) Microfracture holes should be placed every 3 mm throughout the defect

30.1.4.3 Microfracture

Microfracture has been described using awls of varying angles but can also be achieved through drilling into the subchondral bone of the defect. Recently, the preferred device of the senior author has been a commercially available microdrilling marrow stimulation system (PowerPick™, Arthrex Inc., Naples, FL). This system features set angle guides of 30° and 45° to permit ideal orthogonal access to lesions, with a drill width of 1.5 mm and depth of 4.0 mm. Starting at the periphery and working toward the center of the defect, microfracture should be performed approximately every 3 mm until the entire defect has been addressed (Fig. 30.1b). Care should be taken to ensure that the microfracture holes do not communicate, and arthroscopic fluid should be turned off to allow for the accumulation of bone marrow in the defect area.

Technique Pearl.

While awls were originally used for microfracture, there is a growing body of evidence demonstrating that microdrilling provides several important benefits [12–15]. In particular, larger diameter and shallower holes as produced by awls result in increased subchondral bone trabecular compaction, cyst formation, and sclerosis when compared to the smaller diameter and deeper holes produced by microdrilling. Additionally, microdrilling has demonstrated a greater ability to create open bone marrow channels that are presumed to play an essential role in the migration of stem cells to the chondral defect for differentiation into chondrocytes. In a recent investigation, Naveen et al. found that microdrilling demonstrated superior outcomes relative to traditional microfracture awl techniques with respect to functional outcomes at 6 months and revision rates within 3 years after surgery [16]. In addition, a greater achievement rate of clinically meaningful outcome differences was evident at 6 months in the microdrilling group. Choice in technique may also impact future treatment in the event of microfracture failure. Changes to subchondral bone following microfracture have been implicated in the diminished outcomes of patients receiving subsequent ACI when compared to patients without prior microfracture [17–19]. Drilling techniques that minimize subchondral bone changes may therefore result in more favorable outcomes for those requiring subsequent ACI. For these reasons, the senior author no longer uses awls and exclusively uses microdrilling when performing microfracture.

30.2 Microfracture Augmentation Techniques

The following section provides the rationale and surgical technique for several products utilized in the augmentation of microfracture. Microfracture augmentation techniques are attractive in that they harnesses the power of emerging cell-based and cell-free restoration constructs in the first line of treatment. For many patients, this is an attractive proposition, as they wish to avoid the morbidity associated with multi-step restoration procedures. On the other hand, these techniques may not be covered by major payors given their investigational nature. Therefore, conversations with patients about the potential for out-of-pocket costs must be included in the decision-making process. Given the paucity of studies directly comparing different augmentation techniques, definitive recommendations regarding the best option cannot be made. In addition to patient preferences, surgeons should assess product availability and their comfort with the technical aspects of use when considering implementation of these techniques. The senior author has made use of each method detailed below in his practice. Owing to the novel nature and limited investigations on outcomes of each treatment modality, the relative indications and contraindications for use are generally the same as that in traditional microfracture. Each augmentation technique begins with standard microfracture, as described in the previous section, prior to product application.

30.2.1 Bone Marrow Aspirate Concentrate (BMAC)

Bone marrow aspirate concentrate (BMAC) is a biologic therapy that utilizes bone marrow as a means to obtain mesenchymal stem cells (MSCs), cytokines, and growth factors which may have anti-inflammatory and immunomodulatory effects that influence tissue regeneration. BMAC has been shown to contain high levels of interleukin-1 receptor antagonist (IL-1ra), interleukin-8 (IL-8), vascular endothelial growth factor (VEGF), and platelet-derived growth factor (PDGF) [20]. The chondrogenic potential of BMAC makes it an attractive therapy to augment procedures such as microfracture.

BMAC can be harvested from a variety of skeletal donor sites with varying morbidity and yield. For focal chondral defects in the knee, BMAC is typically harvested from the proximal tibia, which has a similar cell population and profile to the iliac crest [21]. In case of harvesting from the proximal tibia, the knee undergoes standard sterile preparation and draping. The anteromedial aspect of the proximal tibia is palpated, approximately 5–6 centimeters distal to the medial joint line of the knee, and centered on the anteromedial tibia from anterior to posterior. The sharp trochar is introduced through the skin, directly down to the bone. The trochar is directed slightly proximal and toward the fibular head, with the goal of introducing it into the medullary canal of the proximal tibia. The trochar is advanced using a mallet approximately 2–3 centimeters, during which a loss of resistance should be felt as the trochar advances past the tibial cortex into the medullary space. The central portion of the trochar is removed and a syringe preloaded with heparin is attached to the outer trochar. Bone marrow is then aspirated, while withdrawing and rotating the trochar a quarter-turn every 5–10 seconds to disrupt trabecular architecture and optimize yield [22]. The amount of bone marrow aspirated depends on the specific system used, but approximately 60 mm is generally collected. The bone marrow aspirate is then passed off the operating field for processing per the manufacturer specifications. Following processing, the BMAC is then loaded into a sterile syringe. Once the microfracture portion of the procedure is completed, the knee is drained of arthroscopic fluid, and the BMAC is injected into the knee.

Clinical studies are encouraging for the use of BMAC to augment treatment of cartilage defects in the knee. A systematic review of 11 studies using BMAC for focal chondral defects in the knee and early knee osteoarthritis (level 2–4 evidence) concluded that BMAC was safe and resulted in good to excellent overall outcomes [23]. A case series study of 23 patients with focal chondral defects treated with BMAC on a hyaluronic scaffold showed improved outcome scores (Tegner, Visual Analog Scale, and International Knee Documentation Committee score) at 6 years regardless of lesion size [24]. A prospective cohort study comparing the outcomes for osteochondral defects of the talus after microfracture versus microfracture + BMAC showed a lower revision rate (12.2% versus 28.8%, $p = 0.014$) in the BMAC-augmented group [25]. Both groups demonstrated significant improvement in pain scores, quality of life scores, and ability to participate in activities of daily living and sports. To our knowledge, there are no studies in the literature directly comparing microfracture to microfracture + BMAC for focal chondral defects in the knee. However, early studies using a scaffold augmentation with BMAC have demonstrated favorable outcomes and survivorship [26, 27] (Fig. 30.2).

30.2.2 Platelet-Rich Plasma (PRP).

In the time following its recent development, PRP has been experimented as an adjunct to a wide array of orthopedic procedures including microfracture. As with BMAC, PRP contains large quantities of key chondrogenic growth factors, such as VEGF, PDGF, and transforming growth factor beta (TGF-β) [28]. These growth factors have been identified as key modulators in the production of cartilage matrix needed for chondrocyte proliferation and cartilage growth. In addition, PRP has also been identified as stimulating the differentiation of MSCs into chondrocytes. These factors are believed to promote hyaline cartilage deposition over a focal defect rather than the fibrocartilage produced by traditional microfracture, which is postulated to cause the inferior long-term outcomes of microfracture. Despite this promise, in early investigations, PRP

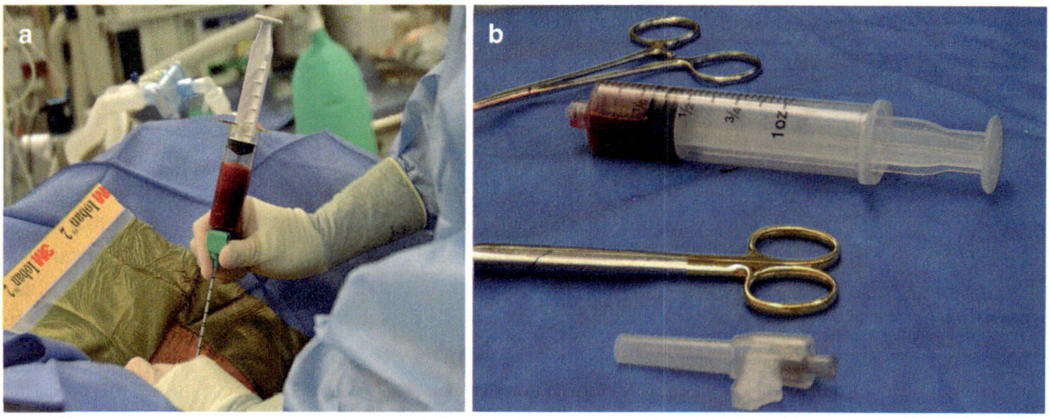

Fig. 30.2 Bone marrow aspirate concentrate (BMAC) augmentation of microfracture. (**a**) Bone marrow can be aspirated from the iliac crest or proximal tibia using a trochar. This image shows harvesting of bone marrow from the iliac crest. (**b**) Following processing per manufacturer specifications, BMAC (pictured in syringe) can then be injected over the defect area following the conclusion of microfracture

augmentation has demonstrated mixed results, with some studies [29, 30] identifying limited efficacy in the knee and others showing significant improvements when utilized in the ankle [31–33]. These mixed results can perhaps be attributed in part to the lack of standardization and widely variable PRP preparations.

The technique involved in PRP augmentation of microfracture can be performed as follows. Using any one of the several commercial PRP systems available, including the Angel® System (Arthrex Inc., Naples FL), preferred by the senior author, peripheral blood collected at the time of procedure is centrifuged into separate PRP, buffy coat, and red blood cell components. Following completion of the microfracture, all arthroscopic fluid is drained from the joint and the PRP is then injected into the area of the microfractured defect.

30.2.3 Adipose-Derived Injections.

Adipose-derived mesenchymal stem cells (ADSCs) are an alternate, attractive source of MSCs to augment microfracture. Adipose tissue is an appealing source of MSCs owing to its accessibility, limited donor site morbidity, and abundance of MSCs [34–36]. Furthermore, adipose boasts relatively high concentrations of multipotent and chondrogenic cells, is rich in anti-inflammatory cytokines and growth factors, and is minimally affected by aging [37–42]. Microfragmentation has emerged as a promising novel, nonenzymatic, technique requiring minimal manipulation to process autologous lipoaspirate through mild mechanical tissue cluster size reduction in a commercially available full-immersion, closed system (Lipogems®, Norcross, GA) [43, 44]. Microfragmentation retains critical ADSC properties and has been safely and effectively applied to a wide array of clinical applications [37, 41, 42, 45–47]. Although intra-articular injection of microfragmented adipose has shown promise for the treatment of knee chondropathy in isolation, [48–50] there has been mounting interest to harness the technology for augmentation of marrow stimulation techniques.

Augmentation of microfracture with autologous ADSC formulations remains investigational, and strict criteria for use have yet to be established. In general, ADSCs are viable substitutes for other cell-based therapies, with several unique advantages (Table 30.1). In vivo acquisition, processing, and administration of ADSCs using the microfragmentation technique have been previously described [48–54]. First, adipose tissue is emulsified and harvested with the use of tumescent liposuction from the lower abdomen

Table 30.1 Advantages of ADSCs

Property	Advantage
Availability	Accessible, abundant superficial tissue (abdomen, buttock, thigh)
Donor site morbidity	Lipoaspiration is well-tolerated with minimal complications
Cellular yield	500-fold quantity of MSCs per volume relative to BMAC [43, 44]
Cellular content	Multipotent, chondrogenic cells with anti-inflammatory paracrine exosomes [43, 44]
Cellular viability	Cellular composition and properties independent of aging [40]
Processing	Amenable to nonenzymatic microfragmentation process to minimize manipulation, maintain stromal architecture, and isolate ADSCs

or buttocks depending on patient body habitus and positioning (Fig. 30.3a). The lipoaspirate is then processed with a commercially available kit (Lipogems®) that progressively reduces the size of the adipose tissue clusters, filters out impurities, and isolates the stromal vascular fraction (SVF) and ADSCs for administration. The final processed lipoaspirate is transferred to a 10-mm syringe for subsequent injection (Fig. 30.3b). After the microfracture procedure is complete, arthroscopic fluid is evacuated, and the chondral lesion is exposed. The product is injected directly to fill the defect and can be mixed or sealed with fibrin glue to act as a structural scaffold [54]. The remaining lipoaspirate is deposited intra-articularly, and arthroscopic portals promptly closed to prevent extravasation. Post-operative restrictions and rehabilitation are unchanged from the standard microfracture protocol. An elastic compression band or abdominal binder is applied to the harvest site to limit bleeding and ecchymosis.

Several studies have demonstrated the efficacy of microfragmented adipose augmentation for microfracture of chondral defects in small animal models [45, 55, 56]. Recent studies have translated these findings to clinical application. Coughlin et al. described a technique for microfragmented ADSC transplantation after arthroscopic debridement for mild to moderate knee osteoarthritis (OA) but did not report on outcomes [53]. Cattaneo et al. investigated the results of adjuvant microfragmented adipose tissue injection after arthroscopic knee debridement for symptomatic knee OA and found progressive improvement in clinical outcome scores and patient satisfaction over 12 months [52]. Koh et al. conducted a prospective, randomized controlled trial of 80 patients with symptomatic isolated large ($\geq 3\ cm^2$) high-grade (ICRS Grade III/IV) chondral defects of the femoral condyle comparing microfracture augmented with ADSCs to microfracture alone. Clinical and radiographic follow-up at 24 months revealed improved cartilage fill and quality on MRI as well as higher Knee Injury and Osteoarthritis Outcome Score (KOOS) pain and symptom subscores in the microfracture plus ADSCs group [54]. Further research is needed to corroborate these findings and compare efficacy to other available cartilage restoration procedures.

30.2.4 Autologous Matrix-Induced Chondrogenesis (AMIC).

AMIC is another attractive augmentation technique that combines traditional microfracture with the fixation of a commercially available collagen matrix. Since its introduction by Behrens et al., AMIC has demonstrated favorable outcomes in limited comparative trials to traditional microfracture alone [57–59]. The promise of AMIC is in the ability of the matrix to serve as a stabilizing scaffold over the mesenchymal clot produced by microfracture while allowing for the differentiation of new chondrocytes over the previous defect.

To perform AMIC, the dimensions of the focal chondral defect should be measured arthroscopically using a probe after debridement of all damaged cartilage. A piece of the commercial collagen membrane (Chondro-Gide, Geistlich Pharma AG) is then cut to the specifications of the defect before completion of microfracture. Once completed, the membrane is then introduced over the microfractured defect. To secure the membrane, either commercially available or autologous fibrin glue is injected into the area between the defect and membrane. Using a probe, the membrane is then

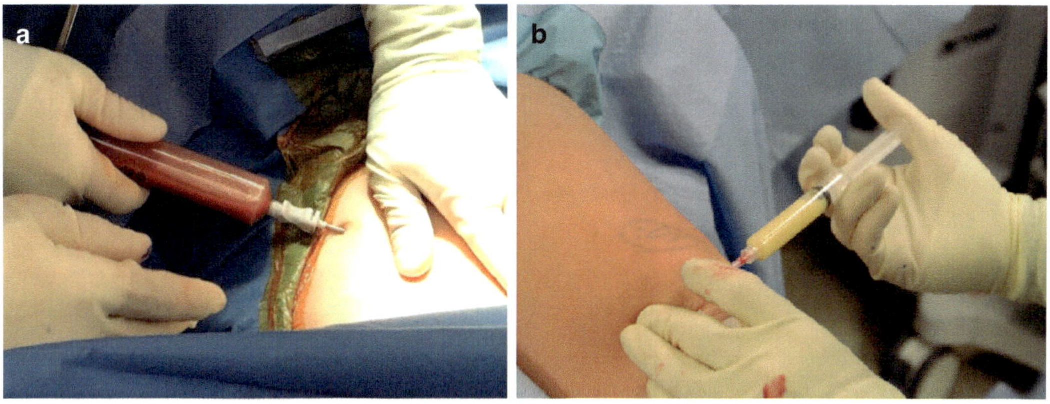

Fig. 30.3 Lipogems augmentation of microfracture. (**a**) Adipose tissue is harvested from the lower abdomen and processed into a lipoaspirate containing adipose-derived mesenchymal stem cells. (**b**) The lipoaspirate is then injected over the microfractured defect at the conclusion of the surgical case prior to closure

tapped into place over the defect, ensuring that it is partially recessed within the defect so as to avoid subsequent delamination. Finally, any excess glue at the margins of the membrane is removed with a probe, and routine closure of the surgery can proceed.

30.2.5 BioCartilage and Autologous Minced Cartilage

Another recently developed scaffold augmentation to traditional marrow stimulation techniques is the use of micronized allogeneic cartilage (Biocartilage, Arthrex Inc., Naples, FL). The senior author's preferred technique has been published previously [60, 61]. Biocartilage consists of dehydrated allogenic cartilage as well as components of native extracellular matrix including cartilage growth factors, type II collagen, and proteoglycans. As in AMIC, the promise of Biocartilage is in serving as a bioactive scaffold for the differentiation of MSCs introduced by microfracture into chondrocytes. To date clinical evaluations of Biocartilage remain limited, but the technique has demonstrated favorable outcomes when compared to traditional microfracture [62].

Application of Biocartilage requires preparation of both the chondral defect and of autologous blood products (either whole blood, BMAC, or most commonly PRP). To begin, a prepackaged 1 mm of Biocartilage is opened and placed into a designated mixing syringe. The Biocartilage is then mixed with 1 mm of PRP until a homogenized mixture of rehydrated Biocartilage, and PRP is prepared (Fig. 30.4a). After standard microfracture is performed, care should be taken to dry the defect as much as possible. Then, a Tuohy needle is introduced arthroscopically over the defect, and the Biocartilage-PRP mixture is injected into the microfractured area. Biocartilage should cover the entire defect area, which can be completed using an elevator to spread the mixture evenly. Either autologous or commercially available fibrin glue is then administered over the Biocartilage and allowed to dry for a minimum of 10 min, after which the Biocartilage implant will remain adhered over the microfractured defect (Fig. 30.4b).

In the ambition of further developing an implant conducive to chondrocyte differentiation, recent innovations to standard Biocartilage technique involve the addition of autologous minced cartilage to the Biocartilage-PRP mixture. Through the use of an autologous tissue collector (GraftNet, Arthrex Inc., Naples, FL), autologous minced cartilage can be obtained at the time of surgery during the debridement of unstable and loose cartilage. The tissue collector (Fig. 30.5a) is an add-on to arthroscopic shavers that when attached to suction collects all cartilage shavings from the debrided fragments. The collected cartilage (Fig. 30.5b) is then added in

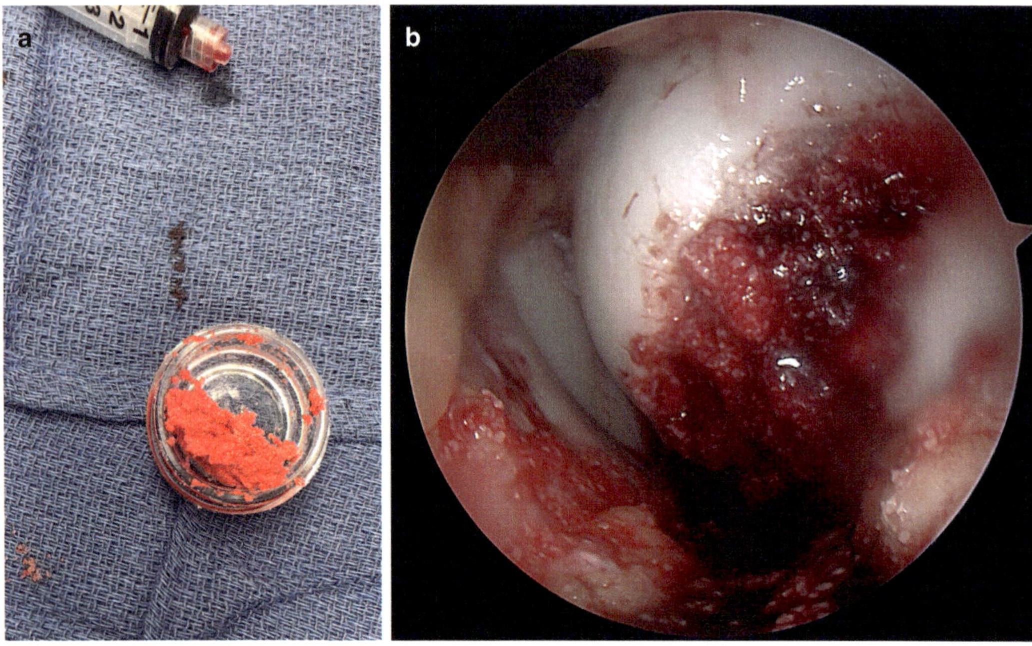

Fig. 30.4 Biocartilage preparation and augmentation of microfracture. (**a**) Dehydrated Biocartilage is rehydrated using 1 mm of PRP collected and processed at the time of surgery. (**b**) The Biocartilage-PRP mixture is injected over the defect area and secured using autologous or commercial fibrin glue

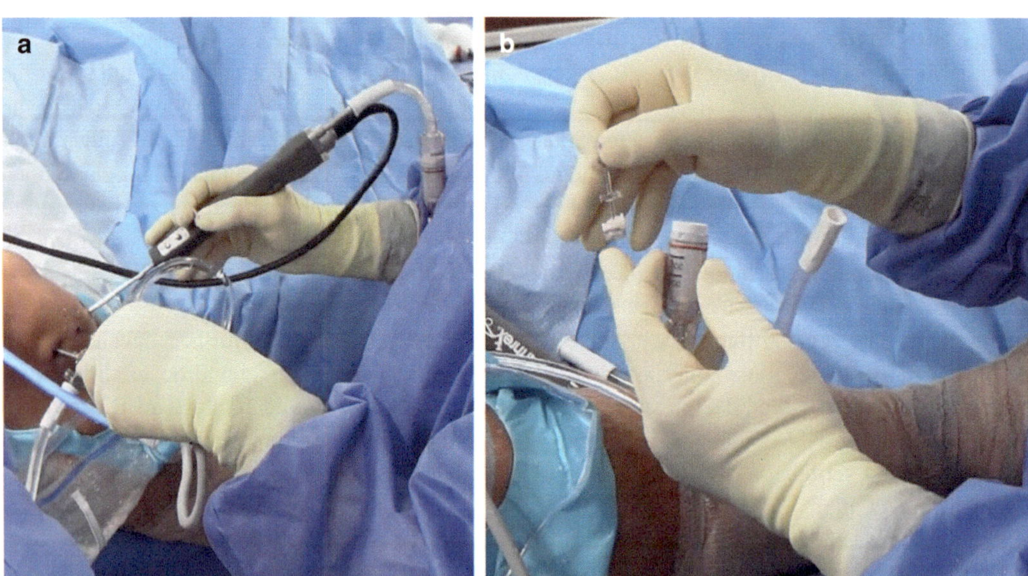

Fig. 30.5 Harvesting of minced autologous cartilage during microfracture procedures. (**a**) The GraftNet™ device is attached to a standard arthroscopic shaver with suction attached and active. (**b**) Minced autologous cartilage from debrided cartilage fragments and loose bodies is collected within the tissue collection filter, which can then be mixed into standard Biocartilage-PRP mixtures prior to injection

an equal ratio to the Biocartilage PRP mix where it can then be injected as in standard Biocartilage repair.

30.3 Post-operative Rehabilitation

Following microfracture or any augmentation technique, a standard post-operative microfracture protocol is utilized by the senior author. When the tibiofemoral joint is involved, patients are kept non-weight bearing for 6 weeks after surgery before gradually advancing to full weight bearing at 8 weeks as tolerated. For procedures addressing the patellofemoral joint, patients are allowed braced full weight bearing as tolerated immediately following surgery. When not in active rehabilitation, bracing is maintained in locked extension for the first 2 weeks before being discontinued. Beginning on the day of surgery and for 6 weeks post-operatively, continuous passive motion (CPM) is used for 6 h per day. CPM begins at 0–40° and advances 5–10° daily as tolerated. Strength and proprioception training are performed with the oversight of a physical therapist and involve advancing from initial isometric exercises to closed-chain and finally open-chain exercises as tolerated. Assuming progression of rehabilitation is tolerated without any setbacks, patients are cleared to return to sports at a minimum of 8 months after surgery.

30.4 Conclusion

While traditional microfracture remains a preferred treatment in the management of small focal cartilage defects, there is an expanding array of new augmentation modalities aimed at promoting chondrogenesis and improving long-term outcomes after microfracture. For many of these treatments, the promise of use is outpacing clinical evidence of efficacy, but preliminary studies have established a proof of concept in the growing array of marrow stimulating techniques.

References

1. Moyad TF. Cartilage injuries in the adult knee: evaluation and management. Cartilage. 2011;2(3):226–36.
2. Williams RJ 3rd, Harnly HW. Microfracture: indications, technique, and results. Instr Course Lect. 2007;56:419–28.
3. Mithoefer K, Williams RJ 3rd, Warren RF, Potter HG, Spock CR, Jones EC, et al. The microfracture technique for the treatment of articular cartilage lesions in the knee. A prospective cohort study. J Bone Joint Surg Am. 2005;87(9):1911–20.
4. Kreuz PC, Steinwachs MR, Erggelet C, Krause SJ, Konrad G, Uhl M, et al. Results after microfracture of full-thickness chondral defects in different compartments in the knee. Osteoarthr Cartil. 2006;14(11):1119–25.
5. Dervin GF, Feibel RJ, Rody K, Grabowski J. 3-foot standing AP versus 45 degrees PA radiograph for osteoarthritis of the knee. Clin J Sport Med. 2001;11(1):10–6.
6. Steadman JR, Rodkey WG, Briggs KK. Microfracture: its history and experience of the developing surgeon. Cartilage. 2010;1(2):78–86.
7. Wang KC, Frank RM, Cotter EJ, Christian DR, Cole BJ. Arthroscopic Management of Isolated Tibial Plateau Defect with Microfracture and Micronized Allogeneic Cartilage-Platelet-Rich Plasma Adjunct. Arthrosc Tech. 2017;6(5):e1613–e8.
8. Freedman KB, Nho SJ, Cole BJ. Marrow stimulating technique to augment meniscus repair. Arthroscopy. 2003;19(7):794–8.
9. Frank RMMK, Bhatia S, Cole BJ. Chapter 15: Enhanced Marrow Stimulation Techniques. In: Biologic Knee Reconstruction: a Surgeon's Guide. Thorofare: SLACK Inc.; 2015. p. 107–13.
10. Yanke AB, Lee AS, Karas V, Abrams G, Riccio ML, Verma NN, et al. Surgeon ability to appropriately address the calcified cartilage layer: an in vitro study of arthroscopic and open techniques. Am J Sports Med. 2019;47(11):2584–8.
11. Fortier LA, Cole BJ, McIlwraith CW. Science and animal models of marrow stimulation for cartilage repair. J Knee Surg. 2012;25(1):3–8.
12. Chen H, Hoemann CD, Sun J, Chevrier A, McKee MD, Shive MS, et al. Depth of subchondral perforation influences the outcome of bone marrow stimulation cartilage repair. J Orthop Res. 2011;29(8):1178–84.
13. Chen H, Sun J, Hoemann CD, Lascau-Coman V, Ouyang W, McKee MD, et al. Drilling and microfracture lead to different bone structure and necrosis during bone-marrow stimulation for cartilage repair. J Orthop Res. 2009;27(11):1432–8.
14. Eldracher M, Orth P, Cucchiarini M, Pape D, Madry H. Small subchondral drill holes improve marrow stimulation of articular cartilage defects. Am J Sports Med. 2014;42(11):2741–50.

15. Gianakos AL, Yasui Y, Fraser EJ, Ross KA, Prado MP, Fortier LA, et al. The effect of different bone marrow stimulation techniques on human Talar subchondral bone: a micro-computed tomography evaluation. Arthroscopy. 2016;32(10):2110–7.
16. Naveen N TT, Southworth T, Baker J, Beletsky A, Chahla J, Verma N, Yanke A, Cole BJ. Outcomes after Microfracture with Traditional Awl vs. Powerpick. International Cartilage Regeneration & Joint Preservation Society - 15th World Congress; 10/5/19–10/8/19; Vancouver, BC2019.
17. Gomoll AH, Madry H, Knutsen G, van Dijk N, Seil R, Brittberg M, et al. The subchondral bone in articular cartilage repair: current problems in the surgical management. Knee Surg Sports Traumatol Arthrosc. 2010;18(4):434–47.
18. Merkely G, Ogura T, Bryant T, Minas T. Severe bone marrow edema among patients who underwent prior marrow stimulation technique is a significant predictor of graft failure after autologous chondrocyte implantation. Am J Sports Med. 2019;47(8):1874–84.
19. Minas T, Gomoll AH, Rosenberger R, Royce RO, Bryant T. Increased failure rate of autologous chondrocyte implantation after previous treatment with marrow stimulation techniques. Am J Sports Med. 2009;37(5):902–8.
20. Cassano JM, Kennedy JG, Ross KA, Fraser EJ, Goodale MB, Fortier LA. Bone marrow concentrate and platelet-rich plasma differ in cell distribution and interleukin 1 receptor antagonist protein concentration. Knee Surg Sports Traumatol Arthrosc. 2018;26(1):333–42.
21. Hyer CF, Berlet GC, Bussewitz BW, Hankins T, Ziegler HL, Philbin TM. Quantitative assessment of the yield of osteoblastic connective tissue progenitors in bone marrow aspirate from the iliac crest, tibia, and calcaneus. J Bone Joint Surg Am. 2013;95(14):1312–6.
22. Piuzzi NS, Mantripragada VP, Sumski A, Selvam S, Boehm C, Muschler GF. Bone marrow-derived cellular therapies in Orthopaedics: part I: recommendations for bone marrow aspiration technique and safety. JBJS Rev. 2018;6(11):e4.
23. Chahla J, Dean CS, Moatshe G, Pascual-Garrido C, Serra Cruz R, LaPrade RF. Concentrated bone marrow aspirate for the treatment of chondral injuries and osteoarthritis of the knee: a systematic review of outcomes. Orthop J Sports Med. 2016;4(1):2325967115625481.
24. Gobbi A, Whyte GP. Long-term clinical outcomes of one-stage cartilage repair in the knee with hyaluronic acid-based scaffold embedded with mesenchymal stem cells sourced from bone marrow aspirate concentrate. Am J Sports Med. 2019;47(7):1621–8.
25. Murphy EP, McGoldrick NP, Curtin M, Kearns SR. A prospective evaluation of bone marrow aspirate concentrate and microfracture in the treatment of osteochondral lesions of the talus. Foot Ankle Surg. 2019;25(4):441–8.
26. Gigante A, Cecconi S, Calcagno S, Busilacchi A, Enea D. Arthroscopic knee cartilage repair with covered microfracture and bone marrow concentrate. Arthrosc Tech. 2012;1(2):e175–80.
27. Gobbi A, Whyte GP. One-stage cartilage repair using a hyaluronic acid-based scaffold with activated bone marrow-derived mesenchymal stem cells compared with microfracture: five-year follow-up. Am J Sports Med. 2016;44(11):2846–54.
28. Hussain N, Johal H, Bhandari M. An evidence-based evaluation on the use of platelet rich plasma in orthopedics - a review of the literature. Sicot j. 2017;3:57.
29. Mancò A, Goderecci R, Rughetti A. S DEG, Necozione S, Bernardi a, et al. microfracture versus microfracture and platelet-rich plasma: arthroscopic treatment of knee chondral lesions. A two-year follow-up study. Joints. 2016;4(3):142–7.
30. Manunta AF, Manconi A. The treatment of chondral lesions of the knee with the microfracture technique and platelet-rich plasma. Joints. 2013;1(4):167–70.
31. Gormeli G, Karakaplan M, Gormeli CA, Sarikaya B, Elmali N, Ersoy Y. Clinical effects of platelet-rich plasma and hyaluronic acid as an additional therapy for Talar osteochondral lesions treated with microfracture surgery: a prospective randomized clinical trial. Foot Ankle Int. 2015;36(8):891–900.
32. Guney A, Akar M, Karaman I, Oner M, Guney B. Clinical outcomes of platelet rich plasma (PRP) as an adjunct to microfracture surgery in osteochondral lesions of the talus. Knee Surg Sports Traumatol Arthrosc. 2015;23(8):2384–9.
33. Jazzo SF, Scribner D, Shay S, Kim K-M. Patient-reported outcomes following platelet-rich plasma injections in treating osteochondral lesions of the talus: a critically appraised topic. J Sport Rehabil. 2018;27(2):177–84.
34. Zhu Y, Liu T, Song K, Fan X, Ma X, Cui Z. Adipose-derived stem cell: a better stem cell than BMSC. Cell Biochem Funct. 2008;26(6):664–75.
35. Zuk PA, Zhu M, Ashjian P, De Ugarte DA, Huang JI, Mizuno H, et al. Human adipose tissue is a source of multipotent stem cells. Mol Biol Cell. 2002;13(12):4279–95.
36. Zuk PA, Zhu M, Mizuno H, Huang J, Futrell JW, Katz AJ, et al. Multilineage cells from human adipose tissue: implications for cell-based therapies. Tissue Eng. 2001;7(2):211–28.
37. Carelli S, Messaggio F, Canazza A, Hebda DM, Caremoli F, Latorre E, et al. Characteristics and properties of mesenchymal stem cells derived from microfragmented adipose tissue. Cell Transplant. 2015;24(7):1233–52.
38. Ceserani V, Ferri A, Berenzi A, Benetti A, Ciusani E, Pascucci L, et al. Angiogenic and anti-inflammatory properties of micro-fragmented fat tissue and its derived mesenchymal stromal cells. Vasc Cell. 2016;8:3.
39. Marfia G, Navone SE, Hadi LA, Paroni M, Berno V, Beretta M, et al. The adipose mesenchymal stem cell Secretome inhibits inflammatory responses

of microglia: evidence for an involvement of Sphingosine-1-phosphate signalling. Stem Cells Dev. 2016;25(14):1095–107.
40. Mirsaidi A, Kleinhans KN, Rimann M, Tiaden AN, Stauber M, Rudolph KL, et al. Telomere length, telomerase activity and osteogenic differentiation are maintained in adipose-derived stromal cells from senile osteoporotic SAMP6 mice. J Tissue Eng Regen Med. 2012;6(5):378–90.
41. Nava S, Sordi V, Pascucci L, Tremolada C, Ciusani E, Zeira O, et al. Long-lasting anti-inflammatory activity of human microfragmented adipose tissue. Stem Cells Int. 2019;2019:5901479.
42. Paolella F, Manferdini C, Gabusi E, Gambari L, Filardo G, Kon E, et al. Effect of microfragmented adipose tissue on osteoarthritic synovial macrophage factors. J Cell Physiol. 2019;234(4):5044–55.
43. Bianchi F, Maioli M, Leonardi E, Olivi E, Pasquinelli G, Valente S, et al. A new nonenzymatic method and device to obtain a fat tissue derivative highly enriched in pericyte-like elements by mild mechanical forces from human Lipoaspirates. Cell Transplant. 2013;22(11):2063–77.
44. Tremolada C, Colombo V, Ventura C. Adipose tissue and mesenchymal stem cells: state of the art and Lipogems(R) technology development. Curr Stem Cell Rep. 2016;2:304–12.
45. Desando G, Bartolotti I, Martini L, Giavaresi G, Nicoli Aldini N, Fini M, et al. Regenerative features of adipose tissue for osteoarthritis treatment in a rabbit model: enzymatic digestion versus mechanical disruption. Int J Mol Sci. 2019;20(11):2636.
46. Randelli P, Menon A, Ragone V, Creo P, Bergante S, Randelli F, et al. Lipogems product treatment increases the proliferation rate of human tendon stem cells without affecting their stemness and differentiation capability. Stem Cells Int. 2016;2016:4373410.
47. Vezzani B, Shaw I, Lesme H, Yong L, Khan N, Tremolada C, et al. Higher pericyte content and secretory activity of microfragmented human adipose tissue compared to enzymatically derived stromal vascular fraction. Stem Cells Transl Med. 2018;7(12):876–86.
48. Panchal J, Malanga G, Sheinkop M. Safety and efficacy of percutaneous injection of lipogems microfractured adipose tissue for osteoarthritic knees. Am J Orthop (Belle Mead NJ). 2018;47(11)
49. Russo A, Screpis D, Di Donato SL, Bonetti S, Piovan G, Zorzi C. Autologous micro-fragmented adipose tissue for the treatment of diffuse degenerative knee osteoarthritis: an update at 3 year follow-up. J Exp Orthop. 2018;5(1):52.
50. Schiavone Panni A, Vasso M, Braile A, Toro G, De Cicco A, Viggiano D, et al. Preliminary results of autologous adipose-derived stem cells in early knee osteoarthritis: identification of a subpopulation with greater response. Int Orthop. 2019;43(1):7–13.
51. Bisicchia S, Bernardi G, Pagnotta SM, Tudisco C. Micro-fragmented stromal-vascular fraction plus microfractures provides better clinical results than microfractures alone in symptomatic focal chondral lesions of the knee. Knee Surg Sports Traumatol Arthrosc. 2019;28(6):1876–84.
52. Cattaneo G, De Caro A, Napoli F, Chiapale D, Trada P, Camera A. Micro-fragmented adipose tissue injection associated with arthroscopic procedures in patients with symptomatic knee osteoarthritis. BMC Musculoskelet Disord. 2018;19(1):176.
53. Coughlin RP, Oldweiler A, Mickelson DT, Moorman CT 3rd. Adipose-derived stem cell transplant technique for degenerative joint disease. Arthrosc Tech. 2017;6(5):e1761–e6.
54. Koh Y-G, Kwon O-R, Kim Y-S, Choi Y-J, Tak D-H. Adipose-derived mesenchymal stem cells with microfracture versus microfracture alone: 2-year follow-up of a prospective randomized trial. Arthroscopy. 2016;32(1):97–109.
55. Ceylan HH, Bilsel K, Buyukpinarbasili N, Ceylan H, Erdil M, Tuncay I, et al. Can chondral healing be improved following microfracture? The effect of adipocyte tissue derived stem cell therapy. Knee. 2016;23(3):442–9.
56. Spakova T, Amrichova J, Plsikova J, Harvanova D, Hornak S, Ledecky V, et al. A preliminary study comparing microfracture and local adherent transplantation of autologous adipose-derived stem cells followed by intraarticular injection of platelet-rich plasma for the treatment of chondral defects in rabbits. Cartilage. 2018;9(4):410–6.
57. Benthien JP, Behrens P. Autologous matrix-induced Chondrogenesis (AMIC): combining microfracturing and a collagen I/III matrix for articular cartilage resurfacing. Cartilage. 2010;1(1):65–8.
58. Benthien JP, Behrens P. Autologous matrix-induced chondrogenesis (AMIC). A one-step procedure for retropatellar articular resurfacing. Acta Orthop Belg. 2010;76(2):260–3.
59. Gao L, Orth P, Cucchiarini M, Madry H. Autologous matrix-induced Chondrogenesis: a systematic review of the clinical evidence. Am J Sports Med. 2019;47(1):222–31.
60. Wang KC, Frank RM, Cotter EJ, Christian DR, Cole BJ. Arthroscopic Management of Isolated Tibial Plateau Defect With Microfracture and Micronized Allogeneic Cartilage Platelet-Rich Plasma Adjunct. Arthrosc Tech. 2017;6(5):e1613–e8.
61. Abrams G, Mall N, Fortier L, Roller B, Cole B. BioCartilage: background and operative technique. Oper Tech Sports Med. 2013;21:116–24.
62. Fortier LA, Chapman HS, Pownder SL, Roller BL, Cross JA, Cook JL, et al. BioCartilage improves cartilage repair compared with microfracture alone in an equine model of full-thickness cartilage loss. Am J Sports Med. 2016;44(9):2366–74.

Technique Corner: Particulate Cartilage

31

Theresa Diermeier and Ben Rothrauff

31.1 Particulate Cartilage

The disadvantage of autologous cartilage implantation is the necessity of a two-stage procedure especially in elderly patients. Therefore, one of the current goals in cartilage therapy is the development of a single-stage cartilage restoration technique.

Particulate cartilage is one of the current methods for single stage cartilage transplantation. Off-the-shelf allogenic particulate cartilage is implanted in a cartilage defect and secured with fibrin glue without the need for a preceding surgery during which autologous cartilage is obtained for chondrocyte isolation and expansion. In in vitro studies, juvenile particulate cartilage (<13 years) had demonstrated faster growth in monolayer culture and higher proteoglycan content compared to adult cartilage (>13 years) [1].

Implantation of particular cartilage begins with diagnostic arthroscopy in order to evaluate the cartilage defect and the status of the knee joint. Afterwards, the affected cartilage should be removed with a curette or a scalpel, either arthroscopically or in an open procedure with a mini arthrotomy, based on the location and size of the defect. It is important to create a stable cartilage shoulder around the defect (vertical wall) with normal or nearly normal cartilage around it. Thereafter, the calcified layer should be removed without damaging subchondral bone, which would induce bleeding [2]. The preparation of the defect area is similar to the preparation for a chondrocyte implantation [3]. Whenever bleeding is accidently induced, hemostasis should be achieved with epinephrine-soaked cottonoid and/or fibrin glue [2].

Then, a sterile aluminum foil is pressed into the prepared defect area to create a three-dimensional mold [2]. After the foil mold is removed from the defect, the particulate cartilage is distributed at the bottom of the mold around 1–2 mm apart in monolayer fashion. Then, fibrin glue is added on top of the cartilage pieces and allowed to cure for 5–10 min. Before the implantation of the cartilage-fibrin glue construct, a fresh layer of fibrin glue is added to the bottom of the defect area. Afterwards, the cartilage-fibrin glue construct is pressed into the cartilage defect zone, and the fibrin glue is allowed to cure for another 10 min. The cartilage-fibrin construct should be recessed to the surrounding cartilage to minimize shear forces and edge effects [2].

T. Diermeier (✉)
Department of Sportorthopedics, Klinikum rechts der Isar der Technischen Universitat Munchen, Munchen, Germany

B. Rothrauff
Department of Orthopaedic Surgery, UPMC Freddie Fu Sports Medicine Center, University of Pittsburgh Medical Center, Pittsburgh, PA, USA
e-mail: rothrauffbb3@upmc.edu

N. Nakamura et al. (eds.), *Advances in Knee Ligament and Knee Preservation Surgery*,
https://doi.org/10.1007/978-3-030-84748-7_31

In alternative option without a foil mold, first deposit a thin layer of fibrin glue on the bottom of the prepared defect area. Then, the particulate cartilage is distributed directly on the fibrin layer in the same way as described above, and the whole defect area is filled with another layer of fibrin glue [4].

31.2 Minced Cartilage

Similar to allogeneic particular cartilage, the use of minced autologous cartilage derived from non-weight-bearing articular cartilage represents an emerging technique for cartilage repair. As above, implantation of autologous minced cartilage beings with arthroscopy of the knee to identify the characteristics of the cartilage defect and to treat any coexisting pathologies. The defect size is measured before debridement, which is used to determine the size of the donor cartilage that must be obtained. On average, three cylinders are required and are harvested from the low-weight-bearing areas of the intercondylar notch.

The cartilage defect is then circumcised with blade and the defect area is further cleared with a ring curette. The whole defect area should be surrounded by healthy cartilage [5].

The calcified layer is removed with the aim to avoid bleeding of the subchondral bone. With an aluminum foil, a mold is created, and based on this a Chondro-Gide membrane is cut, slightly undersizing the defect size. Afterwards, the membrane is hydrated, with a swelling to about 115% of original area.

Simultaneously, the harvested cylinders are prepared at the back table. First, the attached bone is removed from the donor osteochondral plugs. A new No. 10 blade should be used to manually mince the cartilage into fragments with dimensions smaller than 1 × 1 × 1 mm [6, 7]. For better handling of the cartilage fragments, suspension in a drop of water is recommended. Cartilage chips that are retrieved at areas of marginally defective cartilage can be included for the preparation of the notch fragments [5]. The minced cartilage should have a paste-like consistency [5]. Before the implantation, the cartilage paste needs to be dried out completely. A thin layer of fibrin glue is placed on the bottom of the defect area, and the defect is then filled up with the minced cartilage paste. Through the use of a second fibrin glue application, the fragments are fixed within the defect [7]. The cartilage should reach the height of the surrounding cartilage or slightly beneath. Before the fibrin glue is completely dried, the Chondro-Gide membrane is placed on top and securely sutured to the surrounding healthy cartilage. Application of the membrane generates a bond between the fibrin-glue-minced-cartilage layer and the membrane. Based on the surgeon's preference, another layer of fibrin glue can be placed on top of the membrane to make it watertight.

Recently, a mechanical device has been developed to mince the cartilage automatically and change the procedure to a complete arthroscopic technique.

31.3 Rehabilitation

The rehabilitation process after cartilage repair could be divided into three phases, including graft integration, matrix production and maturation [8]. With respect to these phases, for the first 24 h after surgery, no motion is allowed. Afterwards, the protocol of the first 6 weeks included partial weight-bearing on crutches, limited range of motion depending on the localization of the repaired defect and the daily use of a continuous passive motion machine [9].

References

1. Adkisson HD, Martin JA, Amendola RL, et al. The potential of human allogeneic juvenile chondrocytes for restoration of articular cartilage. Am J Sports Med. 2010;38:1324–33.
2. Farr J, Yao JQ. Chondral defect repair with particulated juvenile cartilage allograft. Cartilage. 2011;2:346–53.
3. Steadman JR, Rodkey WG, Singleton SB, Briggs KK. Microfracture technique forfull-thickness chondral defects: technique and clinical results. Oper Tech Orthop. 1997;7:300–4.

4. Tompkins M, Hamann JC, Diduch DR, et al. Preliminary results of a novel single-stage cartilage restoration technique: particulated juvenile articular cartilage allograft for chondral defects of the patella. Arthroscopy. 2013;29:1661–70.
5. Salzmann GM, Calek AK, Preiss S. Second-generation autologous minced cartilage repair technique. Arthrosc Tech. 2017;6:e127–e31.
6. Bonasia DE, Marmotti A, Mattia S, et al. The degree of chondral fragmentation affects extracellular matrix production in cartilage autograft implantation: an in vitro study. Arthroscopy. 2015;31:2335–41.
7. Christensen BB, Foldager CB, Jensen J, Lind M. Autologous dual-tissue transplantation for osteochondral repair: early clinical and radiological results. Cartilage. 2015;6:166–73.
8. Mithoefer K, Hambly K, Logerstedt D, Ricci M, Silvers H, Della Villa S. Current concepts for rehabilitation and return to sport after knee articular cartilage repair in the athlete. J Orthop Sports Phys Ther. 2012;42:254–73.
9. Massen FK, Inauen CR, Harder LP, Runer A, Preiss S, Salzmann GM. One-step autologous minced cartilage procedure for the treatment of knee joint chondral and osteochondral lesions: a series of 27 patients with 2-year follow-up. Orthop J Sports Med. 2019;7:2325967119853773.

Osteochondral Allograft Transplantation

32

C. W. Nuelle, C. M. LaPrade, and Seth L. Sherman

32.1 Introduction

Articular cartilage injury and osteochondral damage in the knee can be debilitating conditions that lead to significant patient pain, dysfunction, and decreased activity. Restoration of the joint surface is critical to restoring overall joint mechanics and biology in order to allow patients to return to previous levels of function and prevent potential progression of osteoarthritis [1–3]. Fresh osteochondral allograft transplantation (OCA) utilizes the transfer of allograft subchondral bone and articular cartilage to a chondral or osteochondral defect. An OCA is sized matched to the patient and transfers viable chondrocytes, resulting in type II hyaline cartilage that matches the patient's native articular joint surface. With the transfer of both underlying bone and mature hyaline cartilage, OCAs offer distinct advantages over other cartilage repair techniques such as debridement, microfracture/marrow stimulation, and surface cell-based repair (i.e., autologous chondrocyte implantation (ACI)), particularly for uncontained or deep chondral or osteochondral defects [4–9]. Debridement, microfracture, and osteochondral autograft transplantation (OAT) are either impractical (OAT) or have poor long-term outcomes (debridement, microfracture) for defects >2 cm^2 [10–14]. ACI results in acceptable outcomes in larger defects but requires two separate, staged procedures and can be difficult in the setting of subchondral bone loss, failed marrow stimulation or cell-based repair, or with unshouldered defects. OCA transplantation provides a single-stage procedure for the treatment of osteochondral defects and has been shown to result in excellent mid- to long-term outcomes, with high rates of return to activity and return to sport [15–25].

Historically, issues with graft storage, chondrocyte viability, and size matching have made the availability of appropriate OCAs difficult. Novel storage methods have increased the duration of time grafts retain viable chondrocytes, and studies have shown non-orthotopic grafts (i.e., a lateral femoral graft to a native medial femoral condyle defect) to have to have excellent clinical results [26–30]. The combination of these things has greatly increased graft availability. For large defects, the procedure has historically been very technically surgical demanding, sometimes requiring multiple grafts to be stacked onto each other in a "snowman" configuration. Advances in surgical cutting guides, making

C. W. Nuelle
Department of Orthopaedic Surgery, Missouri Orthopaedic Institute, Columbia, MO, USA
e-mail: nuellec@health.missouri.edu

C. M. LaPrade · S. L. Sherman (✉)
Orthopaedic Surgery, Stanford University, Pao Alto, CA, USA
e-mail: claprade@stanford.edu; shermans@stanford.edu

N. Nakamura et al. (eds.), *Advances in Knee Ligament and Knee Preservation Surgery*,
https://doi.org/10.1007/978-3-030-84748-7_32

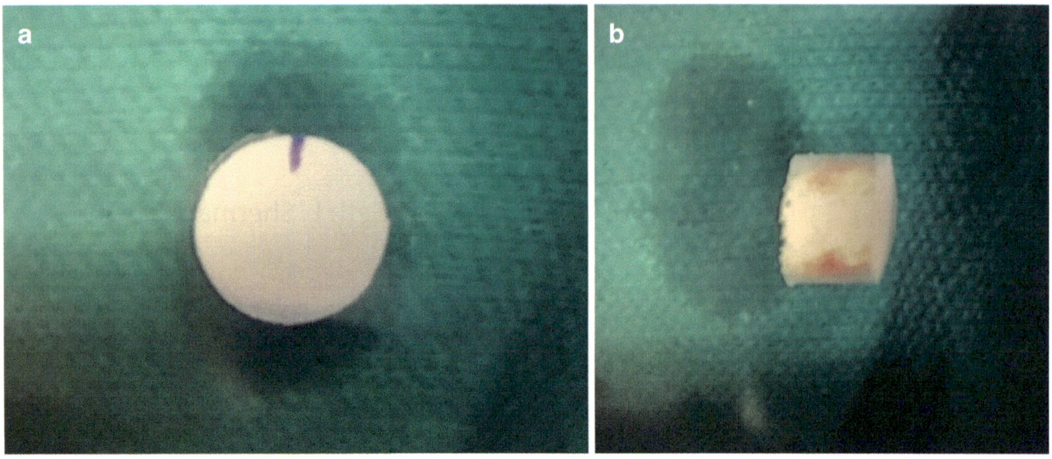

Fig. 32.1 (**a**) Intraoperative image of a fresh, precut, osteochondral allograft (OCA) core, viewing the articular cartilage with the 12 o'clock position marked on the graft. (**b**) Fresh, precut osteochondral allograft core viewed from the side, with the depth of the subchondral bone visualized

them more size specific and well contoured, have greatly decreased the technical demands.

One factor which has been cited as potential downside to OCAs is the possibility of an immune response to the graft. Articular cartilage elicits no humoral immune response, however, and studies have shown no histologic evidence of rejection, with the hyaline cartilage acting as immune privileged tissue [31–34]. The subchondral bone and, more specifically, bone marrow elements of the graft can elicit an immune response. This potential effect can be mitigated by a thorough lavage of all marrow elements prior to transplantation. This technique, combined with meticulous graft implantation, results in minimal risk of an immune response.

The combination of all the above factors previously made OCA transplantation a salvage procedure only for many surgeons, but currently it may be indicated as a first-line treatment as part of the standard joint restoration treatment algorithm.

32.2 Indications and Contraindications

The primary indications for sized matched OCA transplantation in the knee are large (>2 cm^2), symptomatic, full thickness chondral or osteochondral defects, as a salvage for previous failed cartilage restoration procedures, or in cases of significant subchondral bone loss or bony abnormality (osteonecrosis, post-traumatic). Conditions such as osteochondritis dissecans, avascular necrosis, or post-traumatic degeneration are also conditions that frequently result in large lesions that may be amenable to OCA transplantation. For smaller defects where osteochondral autograft may not be easily performed or a surgeon wishes to avoid autograft morbidity, OCA may also be performed. Fresh, pre-cut allograft cores are a viable option for isolated 10–16 mm in diameter defects [35] (Fig. 32.1). These grafts do not require size matching and thus are more readily available. In addition, they can be performed in a single stage procedure without the need for a prior staging arthroscopy.

Other indications for OCA include very large defects requiring resurfacing of a hemi-condyle or an entire condyle, as may be seen in a post-traumatic degenerative knee or a patient who has undergone tumor resection, unshouldered lesions that would not be amenable to a cell-based procedure or multifocal defects.

Primary contraindications to OCA transplantation include patients unwilling to accept allograft tissue and patients unwilling to comply with postoperative rehabilitation restrictions, inflammatory arthropathy, and diffuse degenerative arthrosis. Historically, patellofemoral defects, particularly bipolar "kissing lesions," resulted in poor outcomes with allograft transplantation [36–38]. While bipolar lesions, either within the patellofemoral or tibiofemoral compartments, still result in decreased outcomes vs. focal, solitary lesions, advanced instrumentation and fixation techniques have improved overall outcomes in these patient populations. In particular, physiologically young and active patients that require complete resurfacing of the patella or trochlea, an OCA is an excellent consideration. There are no absolute age limitations, but inferior outcomes have been reported in patients >50 years old [36, 37].

32.3 Concomitant Procedures

OCA transplantation has been shown to have excellent results when performed with concomitant procedures, such as ligament reconstruction or repair, meniscus transplantation, or limb realignment [39–43]. Thorough identification and management of each of these potential pathologies, either concurrently with OCA transplantation or in a staged fashion, is vital to the long-term success of an OCA procedure. Joint stability (stable ligaments), joint congruity and shock absorption (stable meniscus and articular cartilage surfaces), and neutral or near neutral limb alignment are important components to successful long-term outcomes. Limb malalignment is especially crucial to correct with a realignment osteotomy in order to decrease load on the graft(s). Various types of osteotomies have been described, but typically an opening wedge high tibial osteotomy is used to correct limb varus malalignment, and an opening wedge distal femoral osteotomy is used to correct limb valgus malalignment. For the patellofemoral joint, a tibial tubercle osteotomy may be necessary to decrease load from a patellofemoral graft(s) and/or correct patellar maltracking [42, 43]. A list of concomitant pathologies and subsequent procedures is listed in Table 32.1.

Table 32.1 Concomitant procedures that accompany osteochondral allograft transplantation

Procedure	Indication
Ligament reconstruction	ACL, PCL, PLC, PMC, MPFL insufficiency
Meniscus transplantation	Meniscus insufficiency
Valgus producing tibial osteotomy	Asymmetric genu varum ≥3° with medial compartment pathology
Varus producing femoral osteotomy	Asymmetric genu valgum ≥3° with lateral compartment pathology
Tibial tubercle osteotomy	Patella defect with abnormal TT-TG, abnormal Caton-Deschamps ratio
Lateral lengthening	Patella defect with fixed patella tilt with lateral retinacular tightness

ACL anterior cruciate ligament, *PCL* posterior cruciate ligament, *PLC* posterolateral corner, *PMC* posteromedial corner, *MPFL* medial patellofemoral ligament

32.4 OCA Transplantation Surgical Technique: Small Defects

The patient is placed supine on the operative table, and general anesthesia is induced after application of a regional nerve block. No tourniquet is necessary. A lateral post and foot rest or leg holder can be helpful to stabilize the leg and hold the knee in flexion when addressing condylar defects. If no prior staging arthroscopy was performed, a diagnostic arthroscopy is performed to address any concomitant pathology. For larger defects, an open standard midline skin incision is made from the superior pole of the patella to the joint line, followed by either a medial or lateral parapatellar arthrotomy to expose the affected compartment. For multifocal or multiple compartment defects, a larger skin incision and arthrotomy may be made.

For small, solitary osteochondral defects, an arthroscopic or mini-open technique may be employed to perform the OCA transplantation. Defects which are well circumscribed in easily accessible areas of the knee (mid-femoral condyle, mid-trochlea, mid-patella) are best suited for these types of approaches. After the defect size is fully assessed, recipient site preparation may begin. First, a guide pin is drilled perpendicularly in the center of the lesion. A reamer equal to the diameter of the defect is then selected, and the recipient site is reamed to a depth of 6–8 mm. The depth of the reamed socket is then measured at the 3, 6, 9, and 12 o'clock positions. An oscillating shaver is utilized to remove any remaining debris within the socket or any loose cartilage at the periphery to ensure easy graft seating during implantation. If there is sclerotic or cystic bone at the base of the defect, this is then drilled with a 2.0 mm drill bit to create multiple, small marrow stimulation tunnels.

The OCA graft is then opened on the back table. If a fresh pre-cut core is utilized, as in the case example shown in Fig. 32.2, the 12 o'clock position is marked on the graft for orientation. A ruler and marking pen are then used to mark the length of the graft at the corresponding clock positions to match previously measured depths of the recipient socket. An oscillating saw and a small rongeur are then used to precisely remove excess bone until the graft length is appropriate. The deep osseous edges of the graft may be beveled with a rasp for ease of insertion. Pulsatile lavage is then used for a minimum of 2 min to lavage any donor marrow elements out of the subchondral bone portion of the graft. Multiple small drill tunnels may be created on the backside (bony) portion of the graft to allow improved native marrow inflow and integration. The graft is then soaked in bone marrow aspirate concentrate or platelet-rich plasma on the back table.

The recipient site is debrided again to ensure smooth graft implantation. When performed arthroscopically, a hollow tube with an inner diameter equivalent to, or just slightly larger than the graft diameter, may be utilized to deliver the graft into the knee to the recipient site. Using manual pressure with a small tamp, the graft is pushed through the tube and press-fit into the recipient site. Small taps on the tamp around the edges of the graft may be utilized to ensure the graft is completely flush, with smooth transitions, but care should be taken not to exert too much force on the graft itself. If the graft does not fit flush, it may be removed, and the recipient site may be dilated with a slightly oversized tamp, or the edges of the grafts can be gently beveled with a rasp prior to re-insertion. The final position of the graft should be flush with the surrounding articular surface. It may be receded 1 mm but should not be proud relative to the surrounding cartilage.

32.5 Surgical Technique: Large Condyle Defects

The preferred surgical management for large, focal condylar defects is also a press-fit technique. For cylindrical defects, a cylindrical coring reaming system of matching size, like that described for the small defect, may be utilized to match to recipient sites ranging from 10–35 mm in diameter. Many condylar defects match the shape of the condyle, however, resulting in an oblong defect. In these cases, the Bio-Uni specialized cutting guides and preparatory system may be utilized. The steps of this cutting guide system and OCA placement are demonstrated in Fig. 32.3.

First, an appropriately sized curved (matching the condyle contour) template guide is used to cover the defect in its entirety. This guide is then placed on the graft on the back table to ensure the contour and size matches prior to any bone cuts being made. If it matches well, the graft preparation begins. A scoring guide of the exact same medial-lateral and superior-inferior lengths to the curved template guide is placed at the appropriate location on the graft. A Kirschner wire is drilled through a hole on the top over the guide (superior to the cutting portion) into the bone to hold the guide into place. A mallet is then utilized to make the oval cut into the graft osteochondral surface. Once the appropriate depth is achieved, the cutting guide block is left in place, but the handle from the cutting guide is removed and a flat saw cutting jig is assembled to it. A sagittal saw is then used to make a flat cut through the subchondral bone. This results in a smooth, flat surface of bone on the posterior bony surface of the graft. The articular cartilage joint surface still matches that of the native condyle. The oblong cut graft is then removed from the surrounding graft tissue

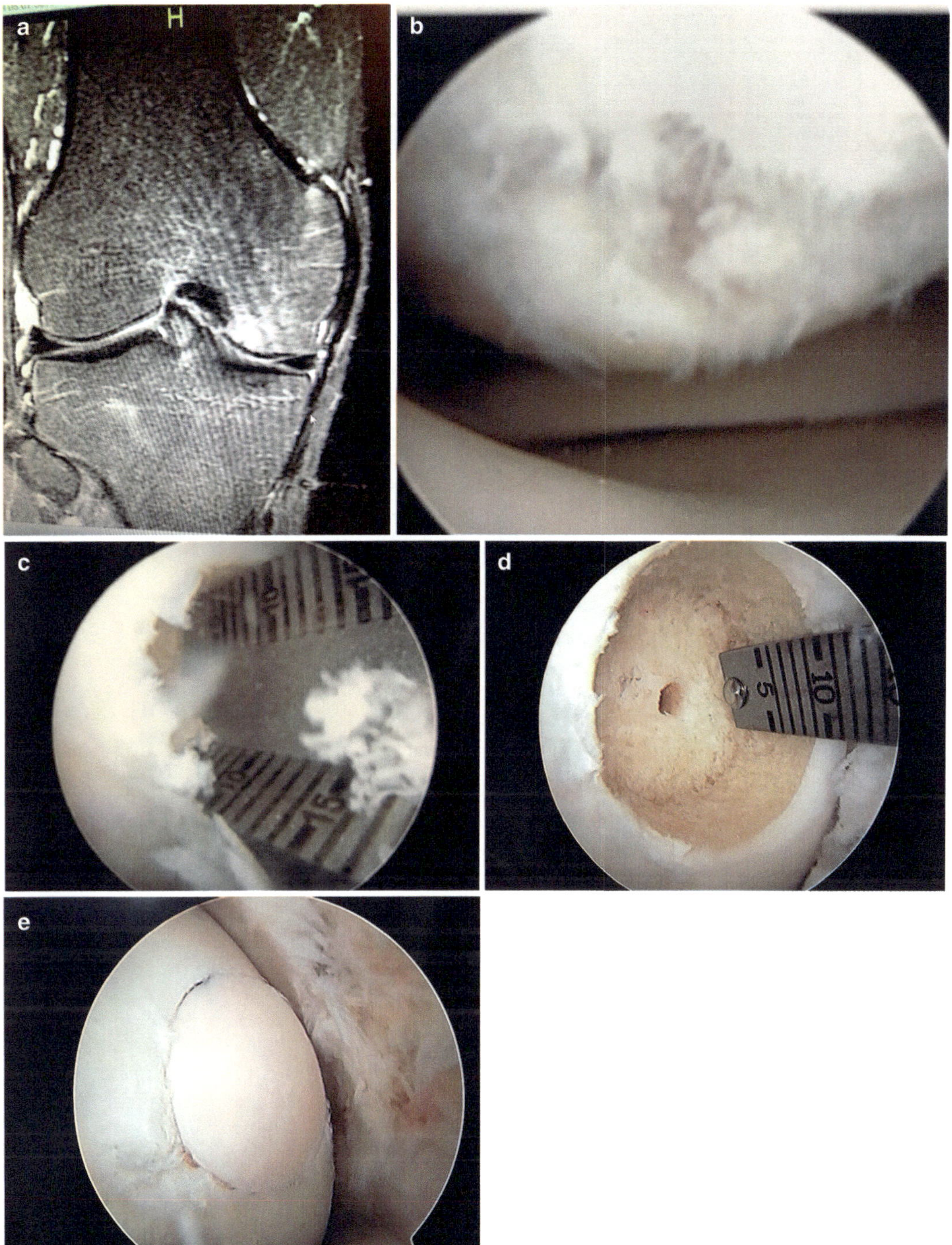

Fig. 32.2 Intraoperative images demonstrating an osteochondral allograft (OCA) transplantation single plug technique. (**a**) Coronal T2-weighted magnetic resonance image of the knee demonstrating an osteochondral defect of the medial femoral condyle. Intraoperative images of the recipient medial femoral condyle defect before (**b**) and during (**c**) reaming. The lesion is reamed to the appropriate depth of 6–10 mm and measured (**d**). The final intraoperative arthroscopic photograph (**e**) demonstrates the donor OCA after it has been press-fit into the reamed recipient defect

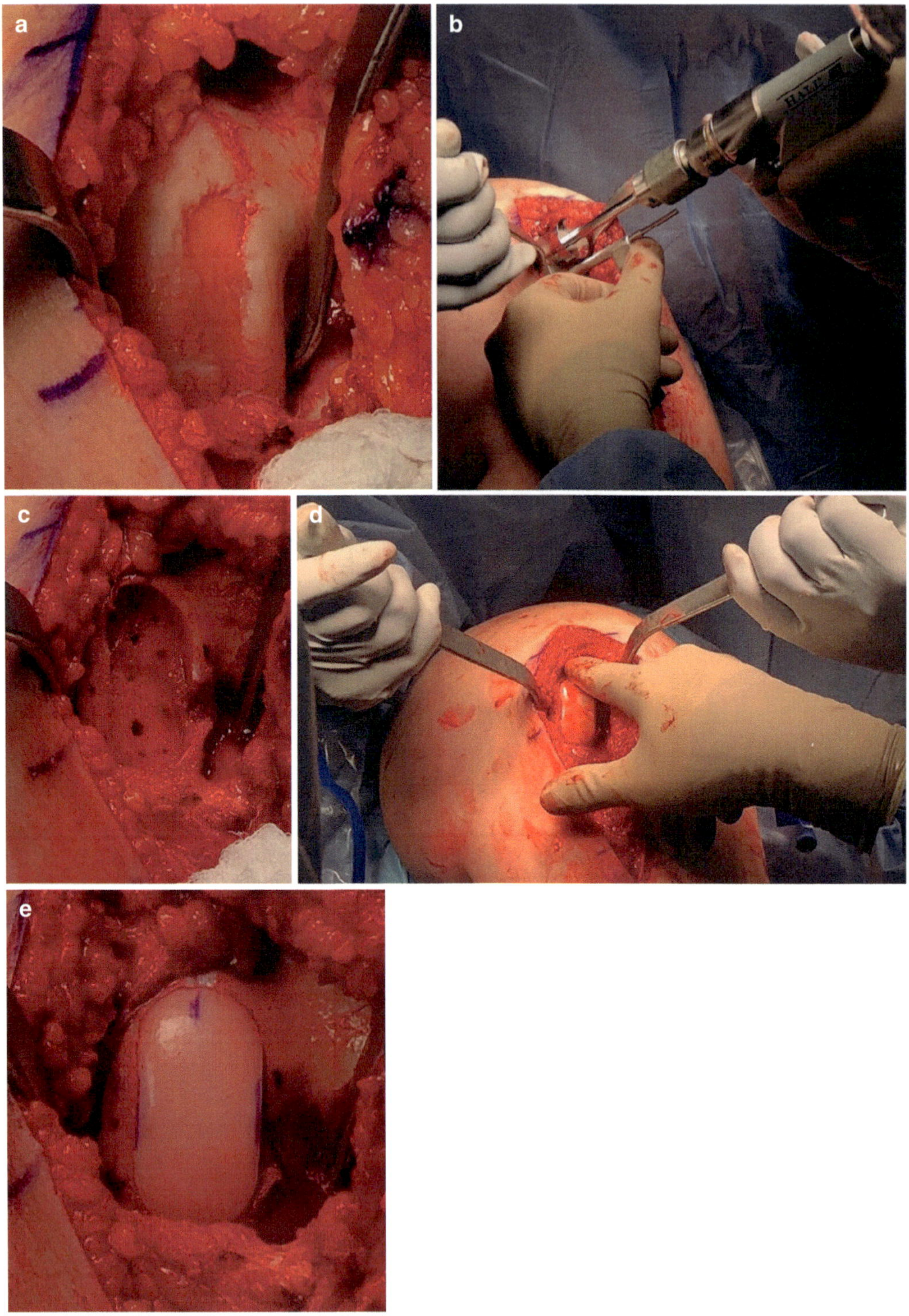

Fig. 32.3 Intraoperative images demonstrating an osteochondral allograft (OCA) large condylar defect transplantation technique in a left knee. (**a**) Native full thickness osteochondral medial femoral condyle defect. (**b**) Intraoperative image demonstrating the reamer utilized to ream the base of the recipient site. (**c**) Image demonstrating the base of the recipient after reaming and after drilling with a small drill bit to create marrow stimulation channels in the subchondral bone. (**d**) Utilizing a press-fit technique to implant the osteochondral allograft into the recipient defect site. (**e**) The final image of a large OCA in place after transplantation

and placed in a depth measuring device. If the graft is not flush around all the edges, a small rasp can be used to file down any of the proud portions. If a portion of the grafts is slightly receded, a smaller reamer is selected to ream the native recipient site. Multiple small drill tunnels are created on the backside (bony) portion of the graft to allow improved native marrow inflow and integration. The graft is then thoroughly washed with pulsatile lavage to remove blood and marrow cells to decrease the risk of a host immune response. The graft is then soaked on the back table in either bone marrow aspirate concentrate or platelet-rich plasma. The recipient defect preparation then begins.

The curved guide is placed again over the native defect, and two central guide pins are drilled perpendicular to the condyle through the guide. The curved guide is removed; a reamer depth stop guide is placed over the inferior pin. Based on the previous depth guide measurements, an appropriate reamer depth is selected. The reamer depth stop may be set at 0, or +1 or −1 mm. The reamer is then utilized to ream the superior aspect of the recipient site fully. The depth stop guide is placed over the superior guide pin, and the reamer is utilized to ream the inferior aspect of the defect recipient site in similar fashion. The depth guide is removed, and a box cutter is placed over the wires and malleted into place to remove any remaining bony debris along the edges or in the base of the defect. The recipient site is thoroughly irrigated, and the base of the defect is then drilled with a 2.0 mm drill bit to create multiple, small marrow stimulation tunnels approximately 3 mm apart. Finally, the OCA graft is brought from the back table and transplanted to the recipient site using a press-fit technique. An oblong tamp may be utilized to ensure the graft edges are flush with the native articular cartilage.

For large condyle defects that are not amenable to an oblong graft, more than one press-fit OCA graft may be required. This "snowman technique" allows coverage of a larger surface area of the condyle using a second plug. In this technique, the first graft is placed as previously described. The subchondral portion of the graft is then pinned with a K-wire in an oblique trajectory away from the articular cartilage or held in place with a small biocompression screw to prevent dislodgement during preparation and placement of the second graft. Preparation of the remaining recipient site is undertaken as before, with the reamer overlapping the previously placed graft but ensuring definitive coverage of the remaining entirety of the recipient defect site. Overlapping the grafts is preferred to leaving spaces between the grafts, as any gaps between the grafts could lead to formation of fibrocartilage or poor articular congruity. Once the second graft has been placed using the press-fit technique, stability is re-assessed. Typically, once the second graft is placed, the entire snowman construct has excellent stability, but if there is any remaining instability present, further biocompression screws may again be added to the subchondral portion of the graft to enhance stability (Fig. 32.4).

32.6 Surgical Technique: Trochlea

An OCA for the trochlea may be performed in one of two ways: using the circular reaming technique (similar to the patella or condyle) or a shell technique. For the reaming technique, the medial and lateral depths of the reamed defect will be much deeper than the proximal and distal depths. It is imperative to have enough depth proximally and distally for a press-fit graft, but not too deep to delay graft incorporation. An example of the steps of trochlear reaming is shown in Fig. 32.5.

For salvage-type procedures that result in lesions that involve an entire condyle or the entire trochlea, or that are uncontained, with very minimal shoulder of cartilage and bone, the shell technique may be employed (Fig. 32.6). For this technique an entire condyle or, in many cases, an entire distal femur is obtained. The recipient bone is cut flush with a free hand, flat cut at depth of 6–10 mm from any remaining adjacent cartilage. Creation of a basic shape at the recipient site (i.e., a trapezoid or rectangle), eases the ability to match the sizing and shape of the OCA graft. The graft is then prepped on the back table.

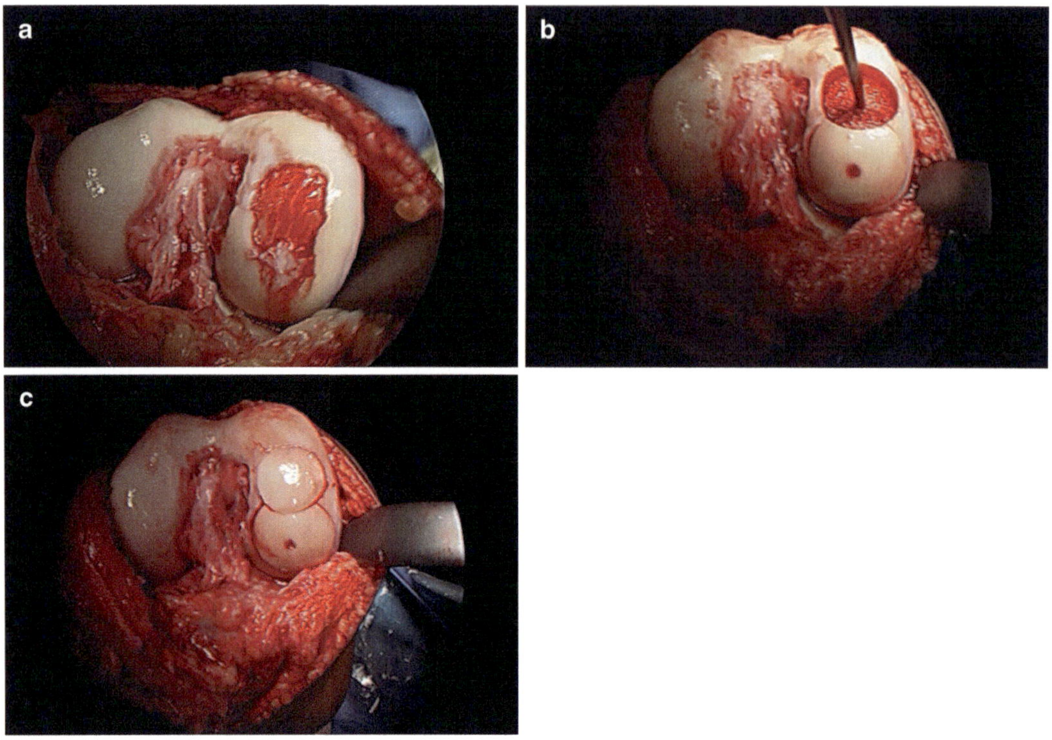

Fig. 32.4 (a) Intraoperative photograph of a medial femoral condyle defect. (b) Photograph demonstrating the reaming of the snowman technique that overlaps the previously placed inferior graft to ensure the final construct can cover the entire recipient defect site. (c) Final construct of the snowman technique using the press-fit technique to position the graft ensuring coverage of the entire medial femoral condyle defect

The graft is measured around all four edges and then stabilized in a cutting jig. The posterior aspect of the graft is then cut with a microsagittal saw to the appropriate depths so as to match recipient site exactly. When initially cutting the graft, it is best to error on oversizing the graft at first, as it can then be trimmed down to size as necessary. The donor shell graft may then be sculpted to create the best fit and then secured to the recipient site with multiple bioabsorbable or metal screws placed in oblique trajectories away from the articular cartilage.

32.7 Surgical Technique: Patella

For smaller lesions that are largely central or are well shouldered along the edges, the same cylindrical reaming and press-fit technique for the circular condyle defects is utilized for the patella as well (Fig. 32.7). For defects involving most of the patella or that result in a poor shoulder around the edges, a shell technique of the entire patellar articular surface may be employed. In this technique, a sagittal saw is utilized to make a flat cut across the entirety of the articular side of the patella. The donor allograft is the cut flush on its posterior aspect as well. It is imperative not to remove too much bone either from the native patella or the donor allograft so that each portion will be able to hold screw fixation. The depth of the native bone removed from the recipient site should be measured, and the donor graft should be cut at nearly the same depth. It is critical that the graft not be larger than the removed recipient portion, or else it will overstuff the patellofemoral joint and increase contact forces on the graft. The donor graft is then placed bone

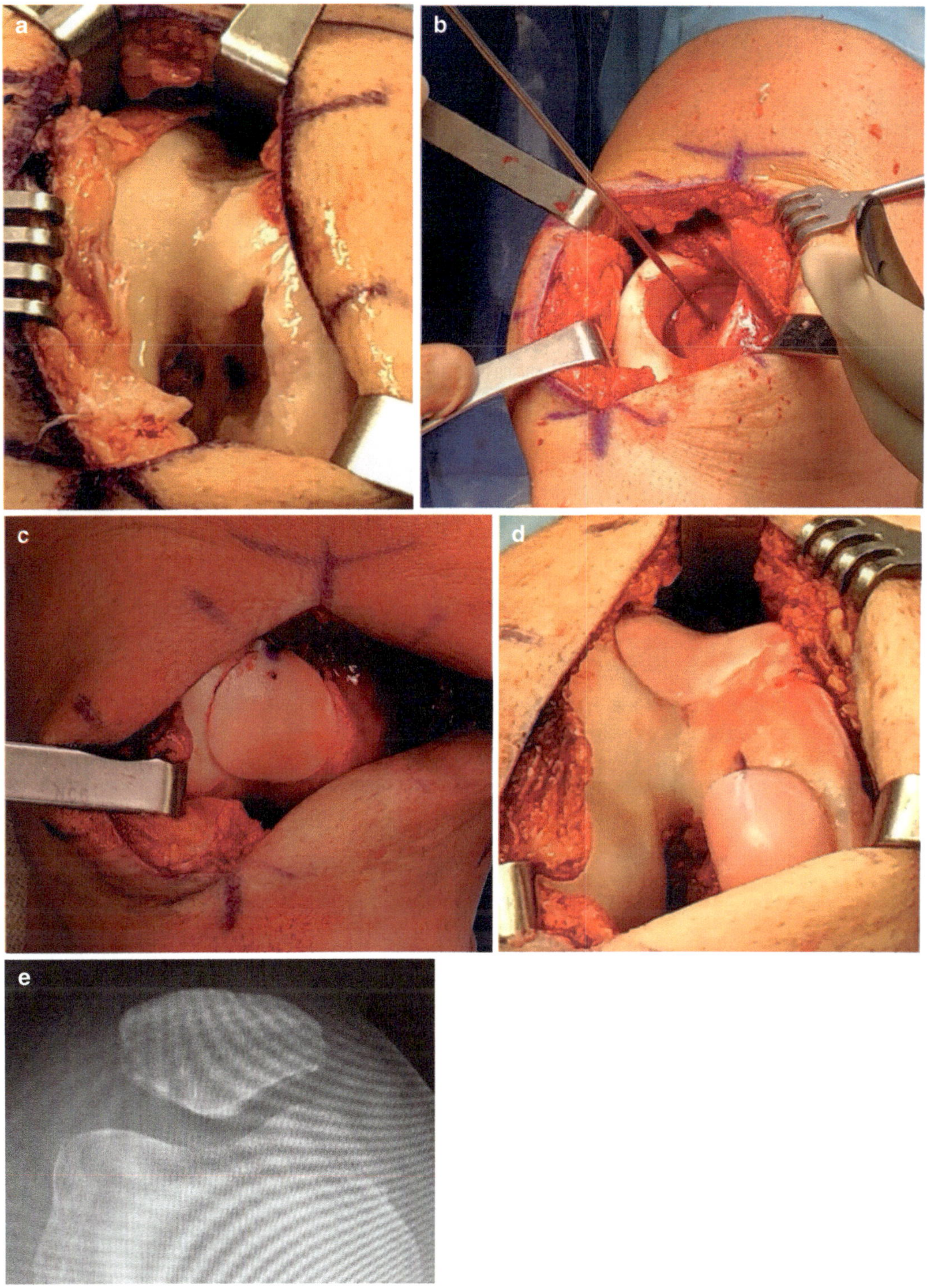

Fig. 32.5 (**a**) Intraoperative image of a right knee demonstrating multifocal but unipolar osteochondral defects of the medial femoral condyle and the trochlea. (**b**) Image demonstrating a central guide and the trochlea after circumferential reaming. (**c**) Image demonstrating the contour of the trochlear graft after placement in the knee. (**d**) Image demonstrating the final trochlea and medial femoral condyle grafts in place in the right knee. (**e**) Postoperative sunrise x-ray view of the trochlea OCA

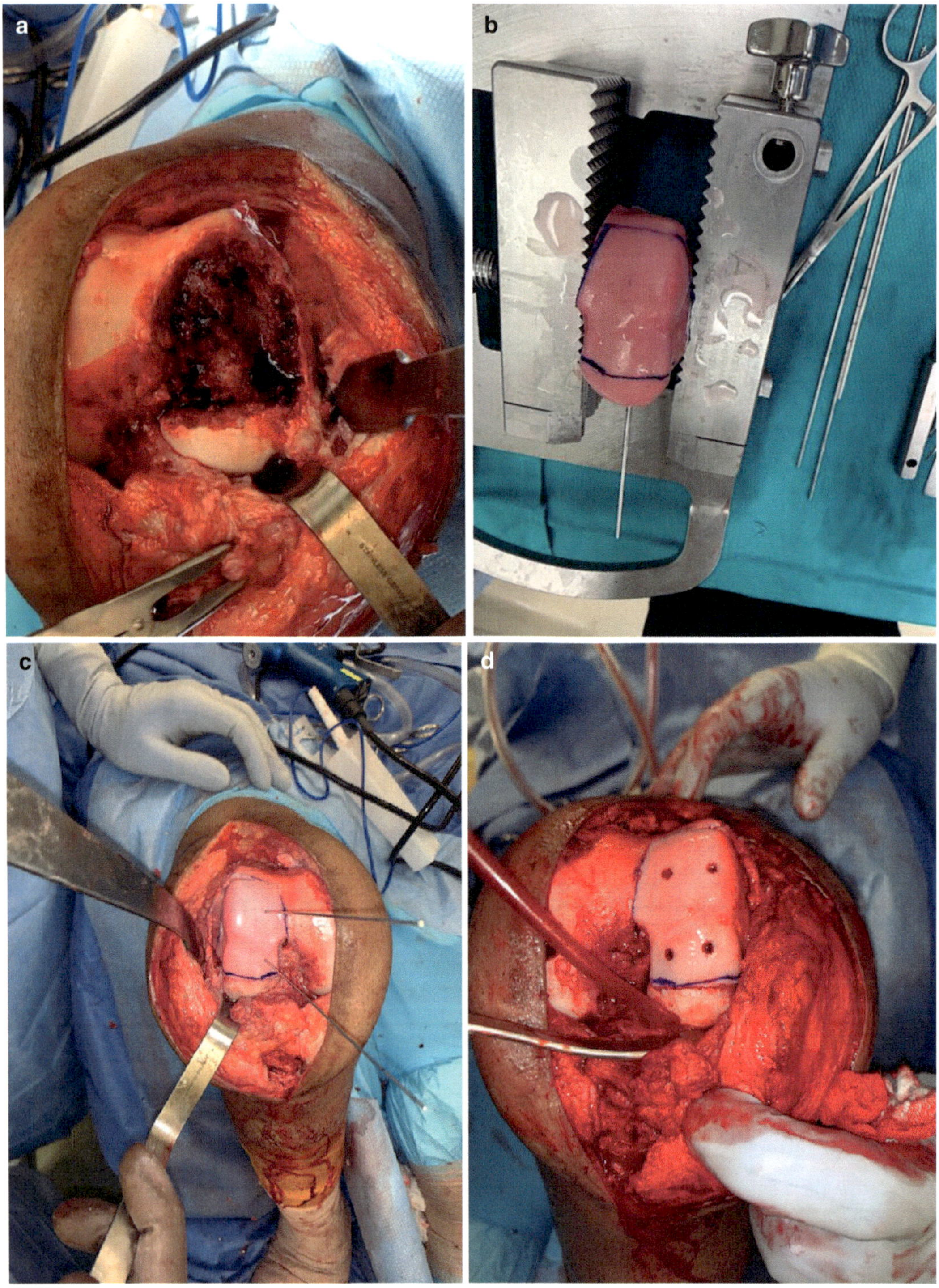

Fig. 32.6 (**a**) Intraoperative image of a trochlea and lateral femoral condyle defect after a gunshot wound to the left knee. (**b**) Image showing the preparation of a trochlea and lateral femoral condyle osteochondral allograft (OCA) transplantation shell while stabilized in a cutting jig. (**c**) Image demonstrating the OCA transplantation shell that has been sized to match the recipient site and is initially stabilized with Kirshner wires. (**d**) Final construct of the OCA shell technique fixated with headless metallic screws placed in oblique trajectories away from the articular cartilage

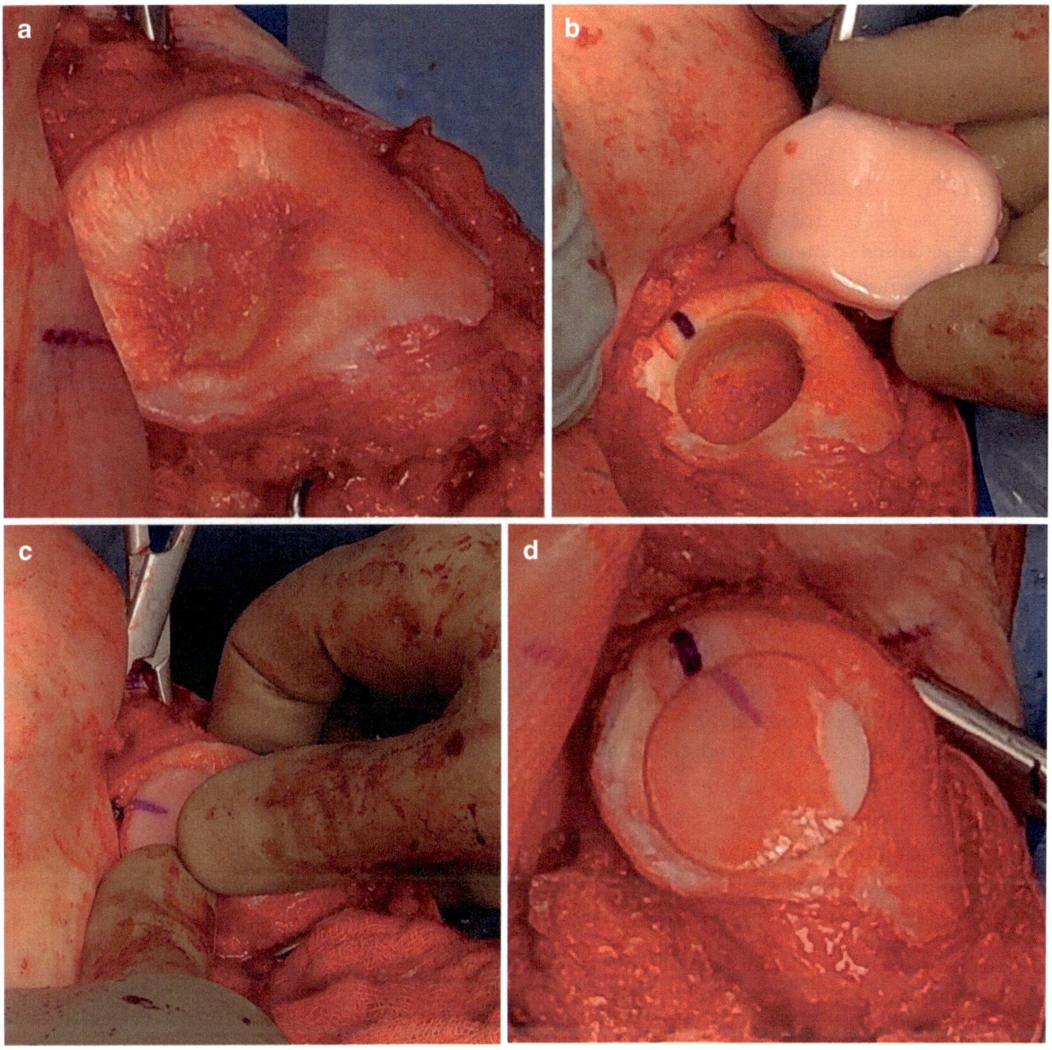

Fig. 32.7 Intraoperative images demonstrating an osteochondral allograft (OCA) transplantation to the patella technique. (**a**) Recipient site osteochondral defect. (**b**) Allograft patella demonstrated adjacent to the native patella after reaming of the base of the defect. (**c**) Image demonstrating a press-fit insertion of the patella OCA to the recipient site. (**d**) Final image after patella OCA transplantation with the lines at the 12 o'clock position of the recipient site and the graft matched up

to bone to the recipient patella and secured with either bioabsorbable or metal screws. Typically, two to three screws are utilized to ensure adequate rotational stability of the graft, and the screws are placed from anterior to posterior. Care should be taken to ensure the screws do not violate the chondral articular surface but are deep enough to have adequate fixation in the subchondral bone.

32.8 Postoperative Rehabilitation

Postoperative rehabilitation after OCA transplantation proceeds in phases, with different weight bearing restrictions for different lesions locations but with the initial phase focusing on graft protection for 0–6 weeks. The goal is to avoid excessive compressive or shear forces on the transplanted graft.

For lesions of the patellofemoral joint, weight bearing as tolerated with the knee locked in full extension in a brace is typically utilized after wound healing. Some authors recommend graduated knee flexion for the first 4–6 weeks for patellar or trochlear transplants to limit excessive pressure across the graft. For femoral condyle or tibial plateau grafts, patients are restricted to Foot Flat <10% WB until postoperative radiographs demonstrate early signs of graft incorporation. For small, well-shouldered lesions, advancement or a partial progressive weight bearing protocol may begin as soon as 4 weeks. For large lesions or poorly contained/shouldered lesions, longer weight bearing restrictions should be instituted (6 weeks or more). In general, patients may perform range of motion as tolerate for condyle lesions. Weight bearing and range of motion restrictions may also be altered based on the concomitant procedures performed (i.e., ligament reconstruction, meniscus transplantation, osteotomy). In patients with the need for prolonged period of protection, consideration can be made for the use of blood flow restriction therapy to reduce risk of muscular atrophy.

Regardless of weight bearing status, early range of motion is paramount after OCA transplantation. Early motion both supports articular cartilage viability and prevents arthrofibrosis. Use of a continuous passive motion (CPM) device can be helpful in the immediate postoperative period, particularly if weight bearing is restricted. Typical settings for CPM use would be 6 h/day, beginning at 0–40°, advanced 5–10° daily as tolerated. Gravity-assisted ROM is also encouraged.

The primary goal of the second phase of rehabilitation (6–12 weeks) is normalization of daily life activities and slow and steady strength training. Any braces utilized are discontinued with adequate quadriceps muscle control, and strength has been achieved. Some authors have advocated for use of an unloader brace to unload the affected compartment, but this has not been shown to alter long-term outcomes or graft survival rates [43, 44]. Regardless, the goal is for patients to progress to full ROM, normalized gait, and improved strength. Low-impact activities are performed in this phase (i.e., swim, bike, elliptical).

The final phase of postoperative rehabilitation (>12 weeks) is patient specific based on individual goals and expectations. In general, this phase focuses on increased strength, endurance, and a return to functional and occupational activities. In relatively sedentary patients, a transition to a home exercise program and activities of daily living may be implemented. In athletes, advanced proprioceptive and sport-specific activities may begin. Athletes should be cautioned, however, that high impact activities should be avoided for 9–12 months after surgery. Athletes should have radiographic (ideally magnetic resonance imaging) evidence of full graft incorporation, no effusion or significant pain, full knee ROM, ligamentous stability, and complete dynamic strength and endurance before return to play may be entertained. Full return to play should be evaluated on a case-by-case basis with the individual athlete and surgeon.

32.9 Potential Complications

The inherent risks of surgery (infection, arthrofibrosis) may occur and are typically prevented using standard precautions. Use of small arthrotomy (or an arthroscopic technique) and early range of motion help avoid arthrofibrosis. Allograft-related complications, such as disease transmission or immunogenic reaction, are exceedingly rare but have been documented [50, 51]. Delayed or nonunion of the graft and graft fragmentation and/or collapse may occur, especially in patients with poor bone quality. This may result from incomplete graft incorporation to the native bone due to limited revascularization. Performing marrow stimulation of the recipient site and drilling channels in the subchondral bone of the donor graft can aid in the re-vascularization process. Finally, using careful, line to line, press-fit technique helps avoid graft collapse and/or eventual fragmentation and failure. Finally, other underlying diseases processes (avascular necrosis, osteoarthritis) may result in persistent symptoms regardless of graft healing or incorporation status.

32.10 Summary

Fresh osteochondral allograft transplantation is an excellent treatment option for large, full thickness articular cartilage defects, with or without bony involvement, in the knee. It may be indicated as a first-line treatment for large defects, for defects with extensive subchondral involvement, and is an excellent salvage procedure for previously failed microfracture or other cartilage restoration procedures. OCA transplantation provides viable, mature hyaline cartilage with underlying subchondral bone to the defect area, resulting in excellent graft strength and overall joint restoration. Management of concomitant meniscus deficiency, ligament instability, and limb malalignment is vital to the success of an OCA transplant. Postoperative rehabilitation follows the general principles of cartilage restoration procedures and is modified based on concomitant pathologies and patient-specific goals. Improvements in graft storage capability, use of non-orthotopic grafts, and specialized cutting guides have greatly improved graft availability and surgical technique demands. Overall, mid- to long-term studies of OCA transplantation show good to excellent outcomes and graft survival in large series (Table 32.2). Future basic science and clinical studies continue to refine indications, graft healing and incorporation, and surgical techniques.

Table 32.2 Osteochondral allograft transplantation outcomes

Study	Lesion site	Diagnosis	Mean follow-up (years)
McCulloch et al. [8]	Multiple site	Trauma, OA, OCD, AVN	2.9
Raz et al. [15]	Femoral condyle	Trauma, OCD	22
Abrams et al. [39]	Femoral condyle	Isolated ICRS grade 3 or 4 defect of the femoral condyle	4.4
Wang et al. [45]	Femoral condyle	Previous failed cartilage repair	3.5
McCarthy et al. [19]	Femoral condyle	Idiopathic, trauma, OCD lesions >2 cm	5.9
Meric et al. [36]	Bipolar, patellofemoral	Degenerative, traumatic, OA, failed OCA, OCD, chronic	7.0
Levy et al. [9]	Femoral condyle	OCD lesions >2 cm, trauma, osteonecrosis, OA	13.5
Krych et al. [46]	Femoral condyle, trochlea, multiple locations	Trauma, nontrauma, OCD	2.5
Gracitelli et al. [37]	Patella	Idiopathic, OCD, traumatic, degenerative	9.7
Sadr et al. [47]	Femoral condyle, trochlea, multiple site	OCD	6.3
Briggs et al. [48]	Multiple sites	OCD, AVN, OA, trauma	7.6
Cameron et al. [49]	Trochlea	OCD, OA, trauma	7.0

No. of knees	Failure rate (%)	Graft survival (%)	Outcomes scores postoperative (preoperative)
25	N/A	4	Lysholm: 67 (39) IKDC total: 58 (29) SF-12: 40 (36)
58	22	91% (10 years), 84% (15 years), 69% (20 years), 59% (25 years)	Modified HSS: 87
48	46	64% (5 years), 39% (10 years)	IKDC function: 7 (3.4) IKDC pain: 4.7 (7.5) KS-F: 84 (71) Modified d'Aubigne´-Postel: 16 (12)

(continued)

Table 32.2 (continued)

32	25	N/A	Lysholm: 64 (42) IKDC: 55 (33) IKDC: 63 (43) SF-12: 47 (44)
43	9	91	SF-36: 84 (61) IKDC: 69 (46) Cincinnati: 6.5 (4.6) Marx: 6.0 (4.4) OCAM-RISS: 10.1
13	0	100	Lysholm: 64 (41) IKDC: 63 (38) Tegner: 4.5 Marx: 5.7 SF-12: 44(35) Return to sport: 77%
129	24	82% (10); 74% (15); 66% (20)	Merle d'Aubigne and Postel: 16 ± 2.2 (12.1 ± 12.1) IKDC pain: 3.8 ± 2.9 (7 ± 1.9) IKDC function: 7.2 ± 2 (3.4 ± 1.3) Knee Society function: 82.5 (65.6)
43	0	100	Limited return to sport, 88%; return to sport at preinjury level, 79% IKDC: 79.29 ± 15 (46.27 ± 14.86) KOOS ADL: 82.82 ± 14 (62 ± 15.96) Marx activity: 8.35 ± 5.9 (5.49 ± 6.35)
28	29	78 (5,10 years), 56 (15)	IKDC: 67 (37) KS-F: 81 (65) Modified d'Aubigne´-Postel: 15 (12)
149	8	95% (5), 93% (10 years)	Modified d'Aubigne´-Postel: 82(44) KS-F: 96(72)
61	18	89% (5 years), 75% (10 years)	Modified d'Aubigne´-Postel: 16.5 (12.6) IKDC: 80 (37) KS-F: 90 (67) KOOS symptoms: 85 (59)
29	21	100% (5 years) 91.7% (10 years)	Modified d'Aubigne´-Postel: 16 (13) IKDC: 72 (39) KS-F: 85 (66) UCLA: 7.9

OA osteoarthritis, *OCD* osteochondritis dissecans, *AVN* avascular necrosis

References

1. Davies-Tuck ML, Wluka AE, Wang Y, et al. The natural history of cartilage defects in people with knee osteoarthritis. Osteoarthr Cartil. 2008;16(3):337–42.
2. Tschon M, Veronesi F, Giannini S, Fini M. Fresh osteochondral allotransplants: outcomes, failures and future developments. Injury. 2017;48(7):1287–95.
3. Thomas D, Shaw KA, Waterman BR. Outcomes after fresh osteochondral allograft transplantation for medium to large chondral defects of the knee. Orthop J Sports Med. 2019;7(3):2325967119832299. Seo S-S, Kim C-W, Jung D-W. Management of focal chondral lesion in the knee joint. Knee Surg Relat Res. 2011;23(4):185–96.
4. Cole BJ, Pascual-Garrido C, Grumet RC. Surgical management of articular cartilage defects in the knee. J Bone Joint Surg Am. 2009;91(7):1778–90.
5. Bugbee W, Cavallo M, Giannini S. Osteochondral allograft transplantation in the knee. J Knee Surg. 2012;25(2):109–16.
6. Sherman S, Garrity J, Bauer K, Cook J, Stannard J, Bugbee B. Fresh osteochondaral allograft transplantation for the knee: current concepts. J Am Acad Orthop Surg. 2014;22:121–33.
7. Demange M, Gomoll AH. The use of osteochondral allografts in the management of cartilage defects. Curr Rev Musculoskelet Med. 2012;5(3):229–35.
8. McCulloch PC, Kang RW, Sobhy MH, Hayden JK, Cole BJ. Prospective evaluation of prolonged fresh osteochondral allograft transplantation of the femoral condyle: minimum 2-year follow-up. Am J Sports Med. 2007;35(3):411–20.
9. Levy YD, Görtz S, Pulido PA, McCauley JC, Bugbee WD. Do fresh osteochondral allografts successfully treat femoral condyle lesions? Clin Orthop Relat Res. 2013;471(1):231–7.
10. Ulstein S, Aroen A, Rotterud JH, Loken S, Engebretsen L, Heir S. Microfracture technique versus osteochondral autologous transplantation mosaicplasty in patients with articular chondral lesions of the knee: a prospective randomized trial with long-term follow-up. Knee Surg Sports Traumatol Arthrosc. 2014;22(6):1207–15.
11. Gudas R, Gudaite A, Pocius A, et al. Ten-year follow-up of a prospective, randomized clinical study of mosaic osteochondral autologous transplantation versus microfracture for the treatment of osteochondral defects in the knee joint of athletes. Am J Sports Med. 2012;40(11):2499–508.
12. Knutsen G, Drogset JO, Engebretsen L, et al. A randomized multicenter trial comparing autologous chondrocyte implantation with microfracture: long-term follow-up at 14 to 15 years. J Bone Joint Surg Am. 2016;98(16):1332–9.
13. Behery OA, Harris JD, Karnes JM, Siston RA, Flanigan DC. Factors influencing the outcome of autologous chondrocyte implantation: a systematic review. J Knee Surg. 2013;26(3):203–12.
14. Biant LC, Bentley G, Vijayan S, Skinner JA, Carrington RW. Long-term results of autologous chondrocyte implantation in the knee for chronic chondral and osteochondral defects. Am J Sports Med. 2014;42(9):2178–83.
15. Raz G, Safir O, Backstein D, Lee P, Gross A. Distal femoral fresh osteochondral allografts: follow-up at mean of twenty-two years. J Bone Joint Surg Am. 2014;96:1101–7.
16. Familiari F, Cinque M, Chala J, Godin J, Olesen M, Moatshe G, LaPrade R. Clinical outcomes and failure rates of osteochondral allograft transplantation in the knee: a systematic review. Am J Sports Med. 2018;46(14):3541–9.
17. Ekman E, Makela K, Kohonen I, Hiltunen A, Itala A. Favourable long-term functional and radiographical outcome after osteoautograft transplantation surgery of the knee: a minimum 10-year follow-up. Knee Surg Sports Traumatol Arthrosc. 2018;26(12):3560–5.
18. Frank RM, McCormick F, Rosas S, et al. Reoperation rates after cartilage restoration procedures in the knee: analysis of a large US commercial database. Am J Orthop (Belle Mead NJ). 2018;47(6).
19. McCarthy MA, Meyer MA, Weber AE, Levy D, Tilton A, Yanke A, Cole B. Can competitive athletes return to high-level play after osteochondral allograft transplantation of the knee? Arthroscopy. 2017;33(9):1712–7.
20. Crawford ZT, Schumaier AP, Glogovac G, Grawe BM. Return to sport and sports-specific outcomes after osteochondral allograft transplantation in the knee: a systematic review of studies with at least 2 years' mean follow-up. Arthroscopy. 2019;35(6):1880–9.
21. Balazs GC, Wang D, Burge AJ, Sinatro AL, Wong AC, Williams RJ 3rd. Return to play among elite basketball players after osteochondral allograft transplantation of full-thickness cartilage lesions. Orthop J Sports Med. 2018;6(7):2325967118786941.
22. Wang D, Kalia V, Eliasberg CD, et al. Osteochondral allograft transplantation of the knee in patients aged 40 years and older. Am J Sports Med. 2018;46(3):581–9.
23. Merkely G, Ogura T, Ackermann J, Barbieri Mestriner A, Gomoll AH. Clinical outcomes after revision of autologous chondrocyte implantation to osteochondral allograft transplantation for large chondral defects: a comparative matched-group analysis. Cartilage. 2019;1947603519833136.
24. Frank RM, Cotter EJ, Lee S, Poland S, Cole BJ. Do outcomes of osteochondral allograft transplantation differ based on age and sex? A comparative matched group analysis. Am J Sports Med. 2018;46(1):181–91.
25. Chahla J, Sweet MC, Okoroha KR, et al. Osteochondral allograft transplantation in the patellofemoral joint: a systematic review. Am J Sports Med. 2018;363546518814236.
26. Nuelle C, Nuelle J, Cook J, Stannard J. Patient factors, donor age, and graft storage duration affect osteochondral allograft outcomes in knees with or without comorbidities. J Knee Surg. 2017;30(2):179–84.

27. Schmidt K, Tirico L, McCauley J, Bugbee W. Fresh osteochondral allograft transplantation is graft storage time associated with clinical outcomes and graft survivorship? Am J Sports Med. 2017;45(10):2260–6.
28. Garrity J, Stoker A, Sims H, Cook J. Improved osteochondral allograft preservation using serum-free media at body temperature. Am J Sports Med. 2012;40(11):2542–9.
29. Stoker A, Stannard J, Kuroki K, Bozynski C, Pfeiffer F, Cook J. Validation of the Missouri osteochondral allograft preservation system for the maintenance of osteochondral allograft quality during prolonged storage. Am J Sports Med. 2017;46(1):58–65.
30. Mologne T, Cory E, Hansen B, Naso A, Chang N, Murphy M, Provencher M, Bugbee W, Sah R. Osteochondral allograft transplant to the medial femoral condyle using a medial or lateral femoral condyle allograft. Am J Sports Med. 2014;42(9):2205–14.
31. Langer F, Gross AE. Immunogenicity of allograft articular cartilage. J Bone Joint Surg Am. 1974;56(2):297–304.
32. Kandel RA, Gross AE, Ganel A, McDermott AG, Langer F, Pritzker KP. Histopathology of failed osteoarticular shell allografts. Clin Orthop Relat Res. 1985;197:103–10.
33. Stevenson S, Shaffer JW, Goldberg VM. The humoral response to vascular and nonvascular allografts of bone. Clin Orthop Relat Res. 1996;326:86–95.
34. Oakeshott RD, Farine I, Pritzker KP, Langer F, Gross AE. A clinical and histologic analysis of failed fresh osteochondral allografts. Clin Orthop Relat Res. 1988;233:283–94.
35. Jones K, Mosich G, Williams R. Fresh precut osteochondral allograft core transplantation for the treatment of femoral cartilage defects. Arthosc Tech. 2018;7(8):e791–5.
36. Meric G, Gracitelli G, Gortz S, De Young A, Bugbee W. Fresh osteochondral allograft transplantation for bipolar reciprocal osteochondral lesions of the knee. Am J Sports Med. 2015;43(3):709–15.
37. Gracitelli G, Meric G, Pulido P, Gortz S, De Young A, Bugbee W. Fresh osteochondral allograft transplantation for isolated patellar cartilage injury. Am J Sports Med. 2015;43(4):879–84.
38. Cotter EJ, Waterman BR, Kelly MP, Wang KC, Frank RM, Cole BJ. Multiple osteochondral allograft transplantation with concomitant tibial tubercle osteotomy for multifocal chondral disease of the knee. Arthrosc Tech. 2017;6(4):e1393–8.
39. Abrams GD, Hussey KE, Harris JD, Cole BJ. Clinical results of combined meniscus and femoral osteochondral allograft transplantation: minimum 2-year follow-up. Arthroscopy. 2014;30(8):964–70. e961.
40. Wang D, Eliasberg CD, Wang T, et al. Similar outcomes after osteochondral allograft transplantation in anterior cruciate ligament-intact and -reconstructed knees: a comparative matched-group analysis with minimum 2-year follow-up. Arthroscopy. 2017;33(12):2198–207.
41. Tirico LEP, McCauley JC, Pulido PA, Bugbee WD. Does anterior cruciate ligament reconstruction affect the outcome of osteochondral allograft transplantation? A matched cohort study with a mean follow-up of 6 years. Am J Sports Med. 2018;46(8):1836–43.
42. Wang D, Rebolledo BJ, Dare DM, et al. Osteochondral allograft transplantation of the knee in patients with an elevated body mass index. Cartilage. 2018;1947603518754630.
43. Cotter EJ, Hannon CP, Christian DR, et al. Clinical outcomes of multifocal osteochondral allograft transplantation of the knee: an analysis of overlapping grafts and multifocal lesions. Am J Sports Med. 2018;46(12):2884–93.
44. Görtz S, Bugbee WD. Fresh osteochondral allografts: graft processing and clinical applications. J Knee Surg. 2006;19(3):231–40.
45. Wang T, Wang D, Burge A, Pais M, Kushwaha B, Rodeo S, Williams R. Clinical and MRI outcomes of fresh osteochondral allograft transplantation after failed cartilage repair surgery in the knee. J Bone Joint Surg Am. 2018;100:1949–59.
46. Krych A, Robertson C, Williams R. Cartilage Study Group: return to athletic activity after osteochondral allograft transplantation in the knee. Am J Sports Med. 2012;40(5):1053–9.
47. Sadr KN, Pulido PA, McCauley JC, Bugbee WD. Osteochondral allograft transplantation in patients with osteochondritis dissecans of the knee. Am J Sports Med. 2016;44(11):2870–5.
48. Briggs DT, Sadr KN, Pulido PA, Bugbee WD. The use of osteochondral allograft transplantation for primary treatment of cartilage lesions in the knee. Cartilage. 2015;6(4):203–7.
49. Cameron JI, Pulido PA, McCauley JC, Bugbee WD. Osteochondral allograft transplantation of the femoral trochlea. Am J Sports Med. 2016;44(3):633–8.
50. Centers for Disease Control and Prevention (CDC). Update: allograft-associated bacterial infections-United Stated, 2002. MMWR Morb Mortal Wkly Rep. 2002;51(10):207–10.
51. American Association of Tissue Banks. Standards for tissue banking. http://www.aatb.org/.

Technique Corner: Osteochondral Autograft

33

Alexander Hundeshagen, Benedikt Brozat, and Daniel Guenther

Several surgical options for articular cartilage defects have been developed and established depending on size, location and concomitant patient-specific factors. Most of these struggle to restore composition and structure of the native hyaline cartilage and the surface topography. Osteochondral autograft transplantation (OAT) is meant to tackle this task by transferring native cartilage together with the supporting subchondral bone into the defect followed by stable bone-to-bone integration. Precursors of this technique have been applied since the mid-1950s; however, technical evolution and refined instrumentation initiated an increasing application from the 1990s on with improving results. OAT has become an effective standard procedure for cartilage defects.

As for all methods, success depends on the correct indication. OAT is usually applied for lesions between 0.5 and 3 cm^2. Smaller defects can be successfully treated with bone-marrow stimulating, reparative procedures such as micro-/nanofracturing techniques since load distribution to the adjacent cartilage is not affected [1]. For larger lesions, OAT is limited due to finite donor site availability. Defects exceeding 3 cm^2 are preferably treated with osteochondral allografts or two-staged regenerative chondrocyte transplantation methods without scaffolds or matrices (ACI/MACI) in combination with osseous defect filling using spongious bone harvested from the iliac crest or the tibial head. Comparing these methods with regard to native hyaline ratio, OAT results in superior cartilage composition [2].

A. Hundeshagen · B. Brozat · D. Guenther (✉)
Department of Trauma and Orthopaedic Surgery, Witten/Herdecke University, Cologne Merheim Medical Centre, Cologne, Germany
e-mail: HundeshagenA@kliniken-koeln.de; BrozatB@kliniken-koeln.de; guentherd@kliniken-koeln.de

33.1 Indications and Contraindications

33.1.1 Indications

- Focal full-thickness osteochondral lesions (Outerbridge/ICRS 3–4) without subchondral cyst formation
- Osteochondritis dissecans (OD)
- Lesion size 0.5–3 cm^2
- Younger, active patients (<50 years)
- Previously failed bone-marrow-stimulating techniques
- OAT is also suggested for chondral defects without bony affections. In these cases, the authors prefer scaffold supported chondrocyte procedures.

N. Nakamura et al. (eds.), *Advances in Knee Ligament and Knee Preservation Surgery*,
https://doi.org/10.1007/978-3-030-84748-7_33

33.1.2 Contraindications (Absolute and *Relative*)

- Previous joint infection (unless ruled out by culture-negative samples)
- Instabilities or malalignment of the knee (unless corrected in the same procedure)
- Degenerative osteoarthritis (Kellgren-Lawrence ≥2)
- Opposing full-thickness cartilage defects—'kissing lesions'
- Rheumatoid arthritis
- *Medical comorbidities (diabetes, immunosuppression,* etc.*)*
- *Smoking*
- *Older patients (biological age >50)*
- *Body-mass-index >40 kg/m²* [3]

33.2 Clinical Evaluation

Focal osteochondral lesions are usually seen in younger patients, either following history of trauma or in the context of osteochondrosis dissecans. Patients complain of pain, swelling, a catching or locking sensation as well as crepitations. A thorough clinical examination is imperative to rule out varus/valgus malalignment and any ligamentous instability.

33.3 Imaging

33.3.1 X-ray

- Standard a.p. and lateral views plus axial patella view are obtained to assess joint space, cystic formations, loose bodies, etc.
- The authors routinely request a weight-bearing a.p. long-leg view to rule out varus or valgus alignment.

33.3.2 MRI

- MRI is a reliable screening tool for osteochondral lesions although exact lesion size can only be estimated. Associated bone marrow oedema further confirms a decompensated loading force distribution. One should be aware that defect diameter is frequently underestimated on MRI, and appropriate alternative procedures should be readily available during surgery.
- Meniscal tears should be appreciated.
- Anterior and posterior cruciate ligament as well as collateral ligaments and patellofemoral ligament can be evaluated.

33.3.3 CT

- A CT scan is not routinely performed but is very helpful to accurately determine the extent of subchondral lesions.

Additional imaging might be useful in particular cases, e.g. if rotational malalignment of the femur or tibia is suspected or to evaluate ligamentous instability.

33.4 Surgical Technique

33.4.1 Key Principles

An autologous cylindrical osteochondral graft is harvested from a non-weight-bearing region of the ipsilateral knee and transferred to replace the affected surface area. Thus, vital hyaline chondrocytes anchored to the subchondral plate can be grafted prompting stable bone-to-bone integration.

Several preconfigured systems are commercially available to facilitate accurate and convenient graft harvesting and positioning. These consist of measuring devices, cylindric trephines or drills of different diameters and graft application devices.

It is imperative to address any ligamentous instability and bony deformity prior to OAT procedure to minimize transplant failure and overload. These procedures are described in the respective sections.

33.4.2 Positioning

The patient is placed in supine position with a footrest and a lateral thigh support so that the knee can be positioned stable in extension, 90°, and 120° of flexion. Some authors prefer a leg-holder, which, however, to our experience, either limits access to the knee or prevents flexible positioning. A thigh tourniquet is applied, and the leg is prepped and draped in the usual fashion.

33.4.3 OAT on Femoral Condyle Lesions

The procedure outlined below describes the OAT procedure for femoral lesions (see Fig. 33.1) as the most frequent indication and can be subdivided into five major steps as follows:

33.4.3.1 Arthroscopic Inspection

Two standard portals are created (anterolateral and deep anteromedial portal), and a thorough

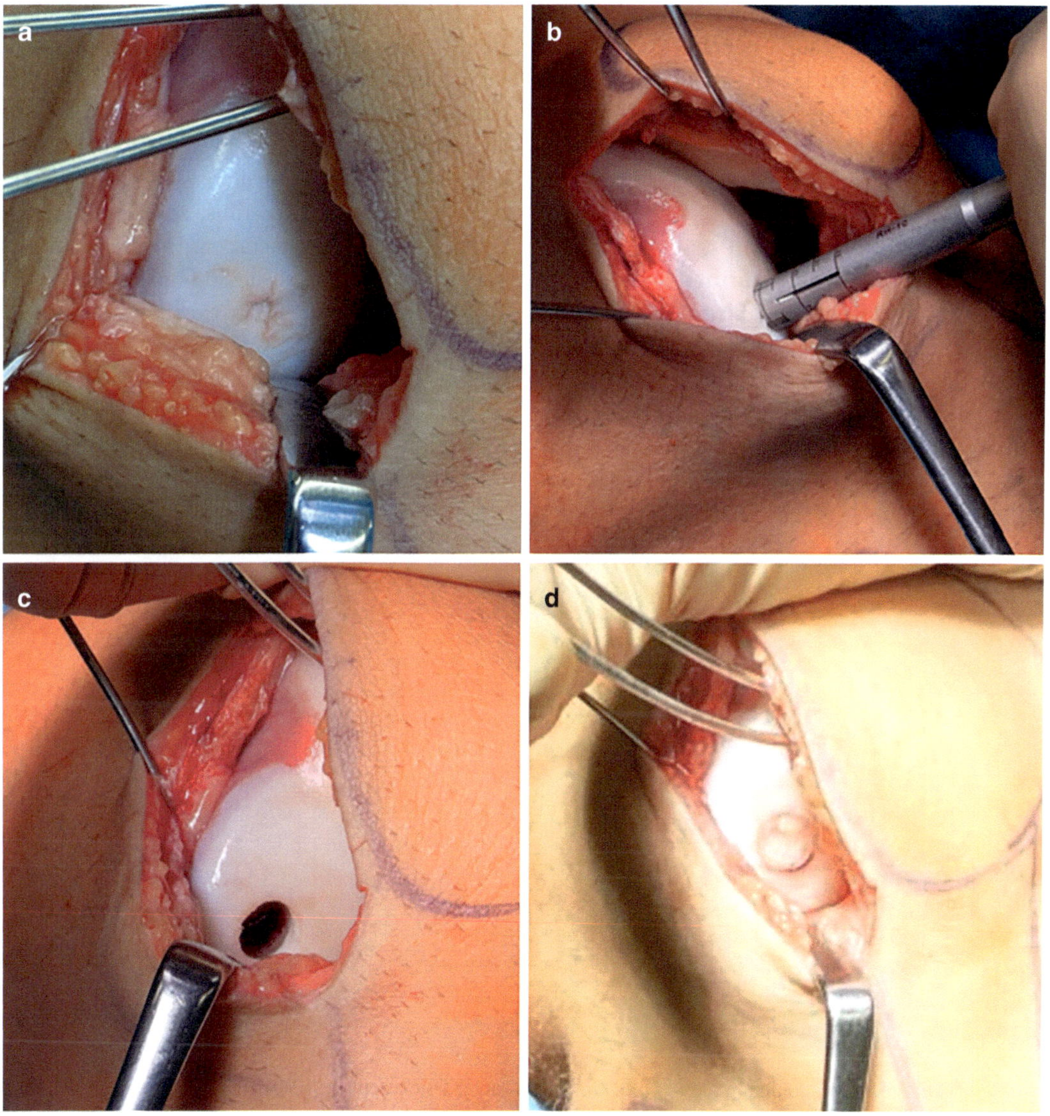

Fig. 33.1 Osteochondral autograft transplantation (open procedure). An osteochondral lesion on the lateral femoral condyle of a right knee is exposed (**a**) and excised en bloc with a trephine positioned perpendicular to the cartilage (**b**). Into the resulting socket (**c**) an autograft is implanted (**d**) to restore smooth joint surface

arthroscopic inspection is performed. Special care should be taken to rule out and address any meniscal or ligamentous instability or tear. All cartilaginous areas should be inspected and palpated with a probe to determine the degree of degenerative changes and confirm indication. Next, the osteochondral defect is localized and examined. The damaged area and any unstable adjacent cartilage have to be debrided thoroughly with a curette to not underestimate defect size. In case the lesion diameter exceeds the indication for OAT procedure, alternative techniques like allograft or ACI have to be consented and readily available. If the indication is confirmed and the aforementioned criteria are met (defect size ≤3 cm^2, stable cartilage rim), depending on the surgeon's preference, the arthroscope may then be removed and procedure be switched to mini-open approach.

33.4.3.2 Exposure of Donor and Recipient Site

To our experience, most locations can easily be accessed via a medial or lateral parapatellar mini-open arthrotomy, and the approach can be extended longitudinally if (rarely) needed.

33.4.3.3 Recipient Site Preparation

Following proper exposure, the site of the lesion is inspected. Diameter and the number of required grafts can be chosen for best coverage using sizer probes. Round lesions up to 10 mm can be covered with a single donor graft; for larger or rather longitudinally shaped lesions, several smaller plugs should be chosen either in a mosaic fashion or a trimmed figure-eight-like configuration (snowman technique) (see Fig. 33.2).

It has to be considered that single plugs fitted press-fit have proven to be more stable compared to multiple plugs. Some authors prefer a 'cobblestone' arrangement, leaving a bony socket wall for enhanced stability. The resulting gaps fill with biomechanically inferior fibrous reparation cartilage. Although complete coverage is not necessarily needed, we feel that press-fit rims, not only bony but also cartilage rims, stabilize the graft and reduce graft degeneration. Hence, when preparing the lesion-site, the authors choose a harvester trephine size slightly exceeding lesion size to get sharp cartilage edges and stable vertical rims.

Next, the chosen recipient harvester trephine is fitted to exactly cover the defect. It is of

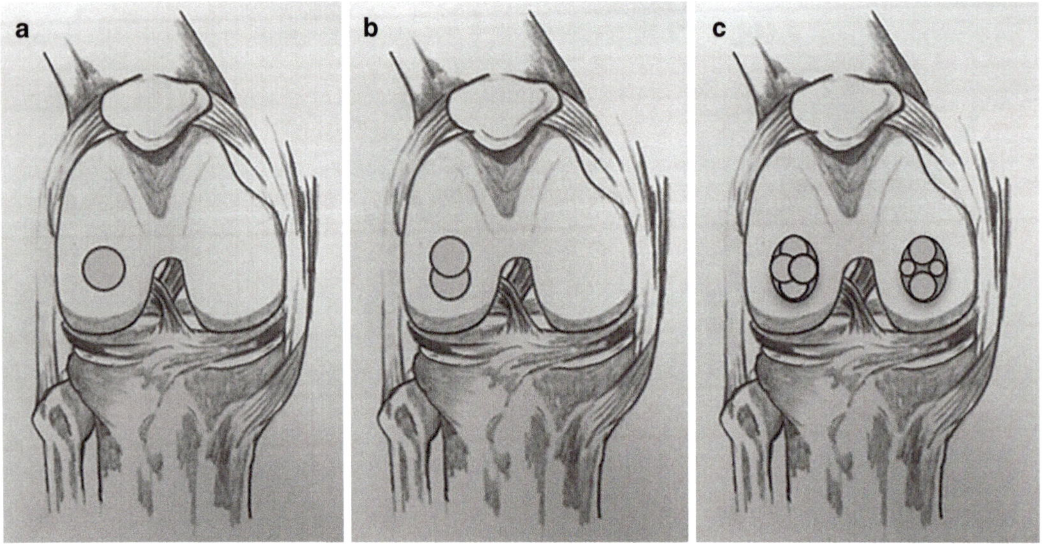

Fig. 33.2 Different graft configurations. (**a**) Single plug for smaller lesions ≤10 mm. (**b**) "Snowman" technique for longitudinal defects. (**c**) Mosaic configuration for larger lesions, either overlapping for best coverage (lateral condyle) or ‚cobblestone'arranged for superior graft stability (medial condyle)

utmost importance to place the harvester perpendicular to the surface. To ensure accurate positioning, most harvesters available show circumferential laserlines measuring penetration depth when aligned with the cartilage rim. After gently pushing through the cartilage layer to establish bony contact, the alignment of the harvester should be assessed in all planes to ensure perpendicular position. Subsequently, the harvester is impacted to a depth between 10 and 20 mm by multiple gentle mallet blows. Next, the harvester is rotated 180° clockwise and anticlockwise to free the plug from the underlying cancellous bone. The harvester is then carefully wiggled out without toggling the cavity. The plug is not discarded but kept for later. Socket depth and appropriate angle is verified with a graduated alignment rod.

33.4.3.4 Graft Harvesting

Three different harvesting sites are typically suggested based on low weight-bearing properties and surface shape matching the recipient site. For defects on the femoral condyle, the slightly convex contour is best matched by grafts from the most medial or lateral border of the trochlear flare proximal to the sulcus terminalis. Rather concave defects within the trochlea are closely mimicked by grafts from the superolateral aspect of the intercondylar notch (see Fig. 33.3) [4, 5]. For retropatellar lesions, choice of graft depends on lesion location (e.g. convex midline or medial portion versus concave lateral part). The choice of harvest site further depends on individual anatomic variances and accessibility. Some authors propose to choose the harvesting site opposing the lesion site. During follow-up, this allows to distinguish between donor and recipient site as possible origins of potential postoperative symptoms.

Similar to the aforementioned recipient site preparation, a trephine is used to harvest the graft. Graft trephine diameter should be chosen slightly bigger (≈1 mm) and graft length slightly longer (1–2 mm) than recipient site to ensure press-fit. Harvesting technique is exactly the same as described above, i.e. ensuring perpendicular access of the harvester, malleting the trephine, turning clockwise and anticlockwise and finally removing the graft.

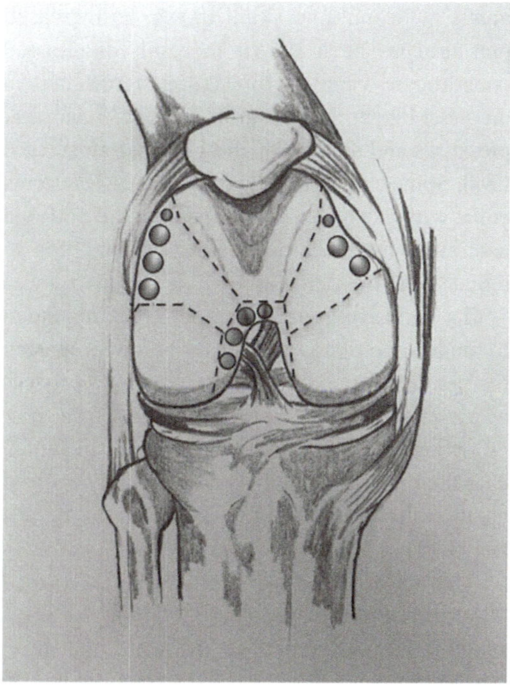

Fig. 33.3 Graft harvesting sites. Superolateral and superomedial margin of the trochlea, superolateral intercondylar notch

It is recommended to backfill the donor site to avoid excessive bleeding and increased loading forces on the adjacent cartilage. The authors prefer to utilize the recipient bone plug saved earlier and inversely graft it. Alternatively, allograft bone chips can be used, and several synthetic, biodegradable composite materials are on the market (e.g. calcium phosphate).

33.4.3.5 Graft Implantation

Before implantation, graft length should be verified to be marginally longer (1–2 mm) than recipient socket as the cancellous bone will slightly compress during impaction. The tip of the plug can be tapered with a rongeur to ease insertion. Orientation of the graft is assessed to best match the surface contour, and the graft is subsequently inserted into the recipient site. An impaction rod is used to gently mallet the graft into the socket until it is completely flush with the cartilage surface. Care must be taken to taper the graft very gently with multiple slight blows, as increased pressure has been shown to result in reduced chondrocyte viability [6]. Using a rod with a diameter bigger than the graft prevents focal peak pressures and sinking of the graft below surface level. Some commercially available OATSystems come with a delivery tube impacting the graft via a screw-home technique, rather than tapping to reduce peak pressures.

If several grafts are needed, it is recommended to finalize the first graft transfer before preparing the next socket and repeating the whole process as described. This especially holds true if a "snowman" configuration is intended, in which case the first transplant has to be partially excised again with the following socket preparation (see Fig. 33.2).

The whole procedure can also be performed all-arthroscopically (see Fig. 33.4). Via the existing portals, the femoral condyles and the intercondylar notch can be reached perpendicular by varying flexion angle of the knee. If the graft is to be harvested from the medial or lateral superior margin of the patella groove, an additional portal is needed. This should be placed after ensuring perpendicular accessibility with a 20-gauge needle. If perpendicular access cannot be accomplished, the procedure should be switched to mini-open approach.

Whenever possible, the authors perform OAT with all necessary concomitant procedures as a one-step approach to avoid repeated postoperative restrictions. With thorough planning and strict time management, this can usually be accomplished.

33.4.4 OAT on Retropatellar and Trochlear Lesions

Retropatellar lesions are challenging regardless of the technique applied due to high contact and

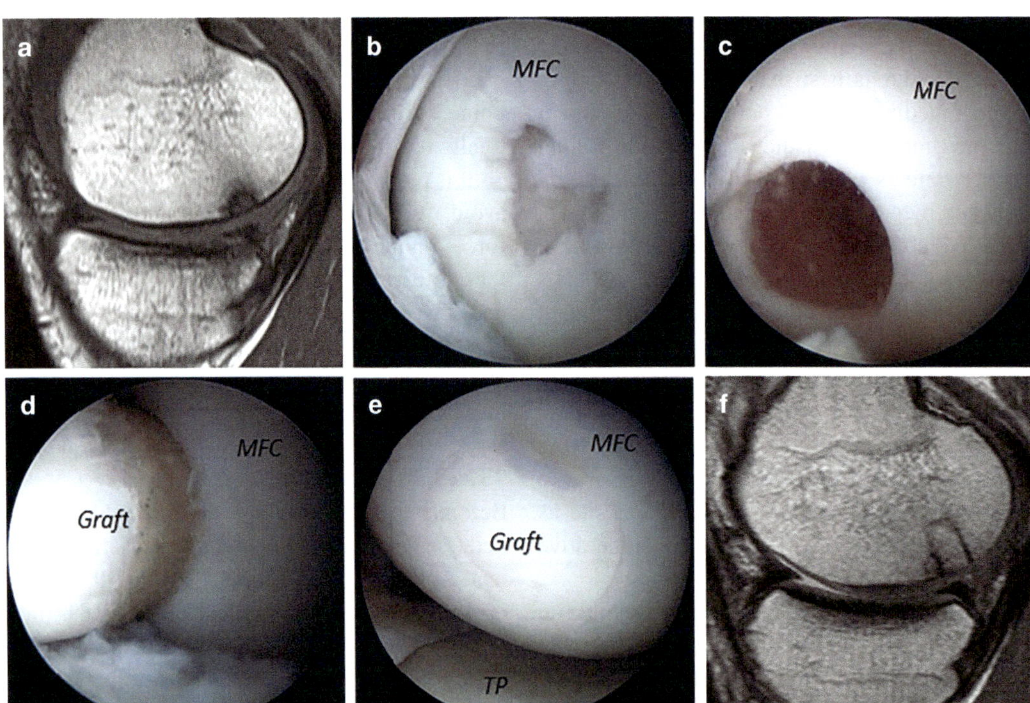

Fig. 33.4 Arthroskopic osteochondral autograft transplantation. The osteochondral lesion on the medial condyle of a left knee is inspected (**a**, **b**) and the recipient site prepared (**c**). The graft is inserted (**d**) and tapered until it sits flush (**e**). Postoperative MRI shows stable graft integration with restored cartilage layer and levelled subchondral plate (**f**). *MFC* medial femoral condyle, *TP* tibial plateau

shear forces with inconsistent outcome. The same applies to retropatellar OAT owing to a graft-recipient mismatch in cartilage thickness and hence incongruency of the subchondral plate. The technical procedure closely follows the principles outlined above, although a larger arthrotomy is needed to evert the patella. After medial parapatellar arthrotomy, the authors place two 2.0 mm Kirschner wires (K-wires) into the patella drilled horizontally from medial to lateral to evert and hold the patella retracted in a hockey-stick fashion. Care must be taken not to penetrate the retropatellar cartilage nor to exit laterally as this might scratch the femoral cartilage during the everting manoeuvre. Now, all compartments of the knee should be accessible including the retropatellar area and the whole femur. For retropatellar OAT, it is even more important to start the procedure with harvesting of the lesion-site cylinder as plug depth might be limited by patella thickness. Trephine extraction of an appropriate retropatellar cylinder can be difficult due to the dense subchondral plate. This can be tackled by reaming the socket instead. Using a slightly smaller reamer to account for toggling, the socket can be drilled carefully to maximum depth without penetrating the cortex. After measuring socket depth, an adequate graft is harvested from the intercondylar notch, prepared and impacted into the defect. Again, the cartilage surface eventually has to be flush. One has to be aware that radiographic evaluation might be misleading with the graft appearing prone due to the thinner cartilage layer.

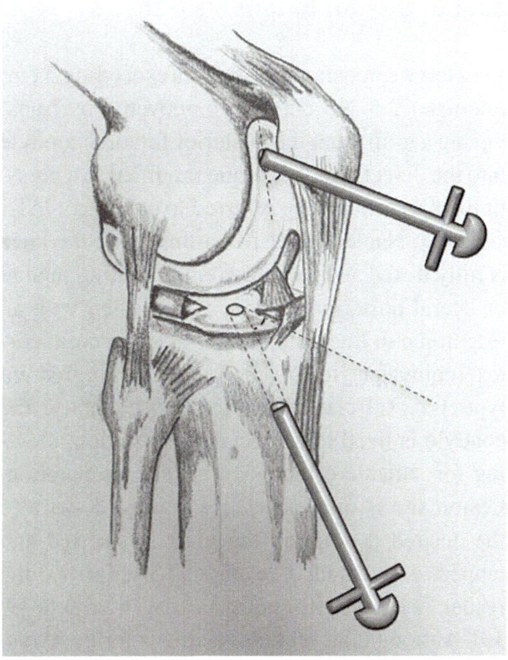

Fig. 33.5 Tibial osteochondral autograft transplantation. A steep tunnel is drilled through the tibia exiting at the defect site. The articular entry angle between the drill tunnel and the tibia plateau is determined (visually or via X-ray), and the graft is harvested with the trephine entering the harvest site at the exact same angle. This ensures flush cartilage integration after retrograde graft insertion

33.4.5 OAT on Tibial Plateau Lesions

In recent years, OAT has increasingly been applied for tibial defects with promising results [7]. The technique has been adapted with respect to the limited accessibility of the tibial plateau. Hence, the fundamental principle of perpendicular trephine instrumentation has to be abandoned, and graft implantation is achieved in a retrograde fashion (see Fig. 33.5).

Following arthroscopic debridement and measurement of the defect, a K-wire is placed obliquely, retrograde exiting at the centre of the lesion using an aimer (ideally a tip aimer) with the drill path ascending as steep as possible to avoid eroding and weakening the tibial plateau. With an appropriately sized coring reamer, a tunnel is drilled over the K-wire. The angle between the drill hole and the tibial plateau may be controlled fluoroscopically with a dilator inserted into the tunnel [8]. Next, the graft is harvested from the femoral trochlea. At this step, it is crucial to place the matching harvester obliquely, mimicking the angle of the tibial tunnel. The graft is held and rotated to correct orientation, with the graft's cartilage surface being parallel to the tibial plateau. The graft is malleted into final position until the cartilage is seated flush. Cancellous bone remains or composite fillers can be used to reline and stabilize the graft.

33.4.6 OAT for Extensive Defects

Very large osteochondral defects exceeding 5 cm^2 (diameter 2.5–3.5 cm) can be covered by transferring a graft from the posterior femoral condyle into the defect; this technique modified to a press-fit fixation often being referred to as Mega-OATS (Arthrex, Naples, Fl.) After arthrotomy, the knee is fully flexed and the non-weight-bearing medial or lateral posterior condyle is harvested with an osteotome in line with the posterior femoral cortex (equivalent to an iatrogenic Hoffa fracture type I c, CT-classification). Subsequently, the condyle is fixed to a special work station allowing for suitable osteochondral plug preparation. Lesion site is mill-cut over a centred K-wire to the desired depth, and the graft is inserted and tapped. For detailed technical description, the reader is referred to the respective literature [9, 10]. Although this technique allows for good restoration of condyle curvature, possible adverse effects due to the large donor site defect are not yet fully elucidated (damage to cartilage and posterior horn of the meniscus, inability to kneel, impeding future prosthetic implantation, etc.). Hence, the authors do not apply Mega-OATS with autograft, but rather suggest it as a salvage procedure when used with allografts [9, 11].

33.4.7 Tips and Pearls

- The key to successful osteochondral transfer is adequate visualization of both the lesion and the harvesting site to ensure perpendicular access. In doubt an arthrotomy and open procedure should be performed.
- The recipient site should be prepared first to properly assess the graft size needed and subsequently harvest the graft. This avoids graft mismatch.
- The trephines should be tapped into the bone rather than drilled to avoid heat necrosis and toggling. Meticulous handling of the graft avoids damage to the chondrocytes.
- The graft has to be flush with the surrounding cartilage making sure it does not protrude above congruency level. If the surface is sloped (due to angled graft/socket preparation), do not accept any prominence but rather countersink the graft to avoid loosening of the graft, opposing 'kissing' lesions and persistent symptoms [12–14].
- If multiple plugs are to be harvested, the single harvesting sites should be separated by at least 2 mm to avoid crossing of the drill tunnels. As these converge if each tunnel is positioned perpendicularly, the resulting graft tips might become short.
- For all-arthroscopic procedures, the authors prefer to harvest the graft from the superolateral part of the intercondylar notch. This area is easily accessible with a good surface texture and low donor site morbidity.
- For large defects, the needed mosaic plasty becomes technically demanding, and results may vary due to potential instability and incongruency of the grafts as well as increasing fibrocartilaginously filled gaps between the grafts. The authors recommend OAT procedure if the defect can be filled with one graft (≤10 mm) or a maximum of three grafts (≤8 mm). For larger lesions, other techniques like osteochondral allograft, autologous matrix induced chondroplasty or autologous chondrocyte implantation combined with cancellous bone grafting should be favoured.

33.4.8 Hazards and Pitfalls

- Perpendicular access to both donor and recipient site is crucial to avoid loosening or mismatch.
- A perfect press-fit implantation is essential to avoid graft degeneration or loosening. The following salvage strategies can be suggested in the case of graft mismatch:
  - If the graft is too short, the socket depth can be reduced with cancellous bone from the recipient plug.
  - If the graft is too slim, sitting loose in the socket, a second small graft can be harvested and placed adjacent in a figure-

eight-style to ensure press-fit fixation. Alternatively, the recipient site plug can be wedged in.
  - In case the plug is too large, the socket can be upsized. If the mismatch is noticed after impaction with the graft left proud, it should be carefully malleted further until it sits flush. If this fails, salvage is difficult and a replacement with an allograft or a second autograft has to be considered.
  - A countersunk plug can be tolerated up to 1 mm [13]. Even deeper sunk grafts are tough to correct. Sometimes a K-wire can be used in a joystick fashion to level the surface. Trephine-assisted re-extraction and graft reinsertion usually toggles the socket leaving the surgeon with a loose graft.
- When preparing the lesion site, the adequate trephine impaction depth has to be anticipated based on preoperative imaging to completely excise any avital or sclerotic bone. If the socket base is not completely cancellous, possible salvage strategies are the following:
  - Reinserting the trephine and further excavating the base.
  - Nanofracturing the cavity base with a small K-wire (e.g. 0.045 in.) under constant fluid cooling to optimize later graft integration.
- Large subchondral cysts can lead to a 'floating' graft and should be relined with cancellous bone harvested from the tibial head or iliac crest.

33.4.9 Postoperative Care

As for all cartilage treatments postoperative, the load applied is reduced to promote stable integration while prompting motion for improved nutrition and differentiation of the repaired defect and the donor site, respectively.

In this sense, the authors follow a rather conservative, restrained postoperative protocol:

- After femoral or tibial OAT procedure weight-bearing is limited to sole contact (15 kg) for 6 weeks with subsequent progressive weight-bearing as tolerated, usually achieving full load after 8 weeks. Range of motion (ROM) is not restricted and continuous passive motion (CPM) is initiated on the day of surgery complemented by physiotherapy.
- Following retropatellar or trochlear defect repair, ROM is limited to 30°, 60° and 90° for 2 weeks each by a functional rigid brace. CPM is applied within these restrictions. Weight-bearing is tolerated immediately with the orthosis fixed in full extension, progressing to normal gait after 6 weeks.
- Return-to-sports starts after 4 months given sufficient functional abilities and muscular stabilization (minimum 90% of contralateral strength).

This postoperative scheme has to be further adopted to limitations due to concomitant procedures.

More liberal regimen might be applicable and have been proposed with patients progressing to full weight-bearing after 2–4 weeks [15].

33.4.10 Outcomes

OAT is a well-established treatment option for osteochondral lesions of the knee. Results proved to be favourable with enduring significant improvement regarding clinical rating scales at both short- and long-term follow-up [16]. Return-to-play rate in athletes has been shown to be as high as 88% [17]. Histological and MRI studies further proved effectiveness regarding cartilage restoration and preservation [2, 18].

Comparison to other surgical techniques remains difficult due to differences in patient characteristics, lesion location and size as well as concomitant procedures, amongst others. Overall OAT was shown superior to microfracturing, especially on long-term outcome [16].

Best results with OAT can be achieved in young patients with small to medium defects of the femoral condyle, and success further improves when conducting necessary concomitant procedures [19].

References

1. Guettler JH, et al. Osteochondral defects in the human knee: influence of defect size on cartilage rim stress and load redistribution to surrounding cartilage. Am J Sports Med. 2004;32(6):1451–8.
2. Riboh JC, et al. Comparative efficacy of cartilage repair procedures in the knee: a network meta-analysis. Knee Surg Sports Traumatol Arthrosc. 2017;25(12):3786–99.
3. Sherman SL, Thyssen E, Nuelle CW. Osteochondral autologous transplantation. Clin Sports Med. 2017;36(3):489–500.
4. Bartz RL, et al. Topographic matching of selected donor and recipient sites for osteochondral autografting of the articular surface of the femoral condyles. Am J Sports Med. 2001;29(2):207–12.
5. Ahmad CS, et al. Biomechanical and topographic considerations for autologous osteochondral grafting in the knee. Am J Sports Med. 2001;29(2):201–6.
6. Patil S, et al. Effect of osteochondral graft insertion forces on chondrocyte viability. Am J Sports Med. 2008;36(9):1726–32.
7. Melugin HP, et al. Tibial plateau cartilage lesions: a systematic review of techniques, outcomes, and complications. Cartilage. 2019:1947603519855767.
8. Yabumoto H, et al. Surgical technique and clinical outcomes of retrograde osteochondral autograft transfer for osteochondral lesions of the tibial plateau. Arthroscopy. 2017;33(6):1241–7.
9. Agneskirchner JD, et al. Large osteochondral defects of the femoral condyle: press-fit transplantation of the posterior femoral condyle (MEGA-OATS). Knee Surg Sports Traumatol Arthrosc. 2002;10(3):160–8.
10. Minzlaff P, et al. Autologous transfer of the posterior femoral condyle for large osteochondral lesions of the knee: 5-year results of the Mega-OATS technique. Orthopade. 2010;39(6):631–6.
11. Hohmann E, Tetsworth K. Large osteochondral lesions of the femoral condyles: treatment with fresh frozen and irradiated allograft using the Mega OATS technique. Knee. 2016;23(3):436–41.
12. Koh JL, Kowalski A, Lautenschlager E. The effect of angled osteochondral grafting on contact pressure: a biomechanical study. Am J Sports Med. 2006;34(1):116–9.
13. Koh JL, et al. The effect of graft height mismatch on contact pressure following osteochondral grafting: a biomechanical study. Am J Sports Med. 2004;32(2):317–20.
14. D'Lima DL, Chen PC, Colwell JW. Osteochondral grafting: effect of graft alignment, material properties, and articular geometry. Open Orthop J. 2009;3:61–8.
15. Werner BC, et al. Accelerated return to sport after osteochondral autograft plug transfer. Orthop J Sports Med. 2017;5(4):2325967117702418.
16. Jones KJ, et al. Comparative effectiveness of cartilage repair with respect to the minimal clinically important difference. Am J Sports Med. 2019;47(13):3284–93.
17. Hurley ET, et al. Return-to-play and rehabilitation protocols following cartilage restoration procedures of the knee: a systematic review. Cartilage. 2019:1947603519894733.
18. Jungmann PM, et al. Cartilage repair surgery prevents progression of knee degeneration. Knee Surg Sports Traumatol Arthrosc. 2019;27(9):3001–13.
19. Pareek A, et al. Long-term outcomes after osteochondral autograft transfer: a systematic review at mean follow-up of 10.2 years. Arthroscopy. 2016;32(6):1174–84.

34 Technical Corner: Lateral Extra-Articular Tenodesis

Frederique Vanermen, Koen C. Lagae, Geert Declercq, and Peter Verdonk

34.1 Introduction

Anterior cruciate ligament (ACL) tears are among the most common injuries of the knee [1]. Although most authors describe good results of ACL reconstruction with modern techniques, restoring the anterolateral rotational stability completely and controlling the pivot-shift phenomenon remains a concern [2]. Recently, the anterolateral complex (ALC) has received attention as being crucial in the control of the rotational laxity [3–6]. This includes the capsule-osseous layer of the iliotibial band as well as the anterolateral ligament (ALL) [3, 7, 8]. The discovery by Claes et al. [3] of the anterolateral ligament (ALL) and the renewed interest into the Kaplan fibers have led to new insights into the important biomechanical function of the anterolateral complex for rotational control [3, 9]. Although most literature has focused predominantly on the ALL as an important stabilizing structure for rotational control, a recent cadaver study has shown that the deep fibers of the ITB make a larger contribution to the resistance to tibial internal rotation than the ALL [9, 10]. Cadaveric studies and a number of systematic reviews and meta-analyses have shown that lateral extra-articular tenodesis (LET) using an ITB strip tunnelled under the lateral collateral ligament is an excellent method to address rotational control in combination with all inside ACL reconstruction, since the rerouted ITB strip is efficiently aligned to withstand anterior movement of the lateral aspect of the tibia in comparison with an ALL graft [4, 10–13]. Historically several techniques and variations have been described to perform a LET [14–17].

In this chapter we describe our technique for a combined intra-articular ACL reconstruction together with a lateral extra-articular tenodesis we refer to as the "monoloop" technique. This technique is founded on previously described methods where an autologous iliotibial band—strip [1, 14–16] is used as a graft but is a revised and refined procedure. Although there is still no consensus on the exact indications in primary and revision ACL reconstruction, the authors perform monoloop LET (mLET) reconstruction in patients who participate in activities with high level of pivoting or cutting maneuvers, patients who present with hyperlaxity or with a high-grade of pivot shift, patients who have a high tibial slope, patients of 25 years of age or younger, and in all cases of revision ACL (Table 34.1).

F. Vanermen · K. C. Lagae · G. Declercq
P. Verdonk (✉)
Orthoca, Antwerp, Belgium

N. Nakamura et al. (eds.), *Advances in Knee Ligament and Knee Preservation Surgery*,
https://doi.org/10.1007/978-3-030-84748-7_34

Table 34.1 Indications for mLET (monoloop extra-articular tenodesis)

Indications for LET
– High-grade pivot shift on clinical examination
– Hyperlaxity
– Patient with high tibial slope
– Patients 25 years or younger
– Competing in pivoting sports or activities

34.2 Anatomy of the Anterolateral Complex

The anterolateral complex consists of the superficial and deep ITB, the capsulo-osseous layer of the ITB and the anterolateral capsule as described by Herbst et al. [7]. The superficial ITB inserts in a wide area, ranging from Gerdy's tubercle on the anterior side to the anterolateral and lateral part of the proximal tibia posteriorly. The deep part of the ITB is predominantly located on the posterior aspect of the ITB and merges with the superficial ITB distal to the lateral femoral epicondyle. Its tibial insertion is posterior to Gerdy's tubercle together with the posterior fibers of the superficial ITB. The Kaplan fibers are part of this deep layer [18]. These fibers anchor the ITB to the distal femur and, based on their anatomic orientation and biomechanical properties, make an important contribution to static restraint against tibial internal rotation in the ACL-deficient knee.

The capsulo-osseous part of the ITB can be considered a distinct layer of the ITB and is the most posterior and medial portion of the ITB. It has a triangular shape with a tibial insertion which is wider than its femoral origin. In its proximal end, it merges with the fascia of the lateral gastrocnemius muscle. On its posterior part, it is reinforced by the fascia of the biceps femoris muscle. Underneath the ITB band, the anterolateral capsule is located. The superficial layer of the anterolateral capsule encompasses the LCL, whereas the deep layer passes deep to the lateral collateral ligament (LCL). On the anterior part, both layers present as 1 continuous layer. The ALL is, as described by Claes et al. [3], a distinct ligamentous structure which has his major femoral origin on the prominence of the lateral femoral epicondyle, anterior to the origin of the LCL, and proximal and posterior to the insertion of the popliteus tendon. The tibial insertion is situated posterior to Gerdy's tubercle, with no connecting fibers to the ITB [3].

34.3 Patient Setup, Examination Pre-operatively and Intra-articular Procedure

General anesthesia is used in all patients. Patients are installed on the operating table in a supine position. Both knee stability (pivot and translation) and range of motion are tested before incision while the patients are under anesthesia. The patient is disinfected and draped in a standardized fashion. An Ioban dressing is applied to the entire surgical site. A tourniquet is used at 250 mmHg and inflated prior to the start of surgery. First, an arthroscopic intra-articular exam is performed to evaluate the ACL, and check for any additional cartilage or meniscal pathology. If present, those lesions should be addressed before preparation and placement of the ACL graft. At our institution, usually a quadrupled semitendinosus hamstring autograft in combination with a variable loop cortical system (Infinity system, Conmed Linvatec, USA, Tampa) for femoral fixation is used. On the tibial side, fixation is performed with the use of both a tibial post and a bioresorbable interference screw (Matrix, Conmed Linvatec, USA, Tampa). This tibial fixation is done after the extra-articular tenodesis is completed.

34.3.1 Step 1 and 2: Approach and Harvesting the Iliotibial Band Strip

The knee is flexed at 60°, both Gerdy's tubercle and the ITB band are palpated and marked (Fig. 34.1). An 8–10-cm curvi-linear incision is being made on the lateral aspect of the knee, starting—for a left knee—just proximally of Gerdy's tubercle, and approximately ending at the one third distal end of the ITB band. When the knee is brought to full extension, the incision

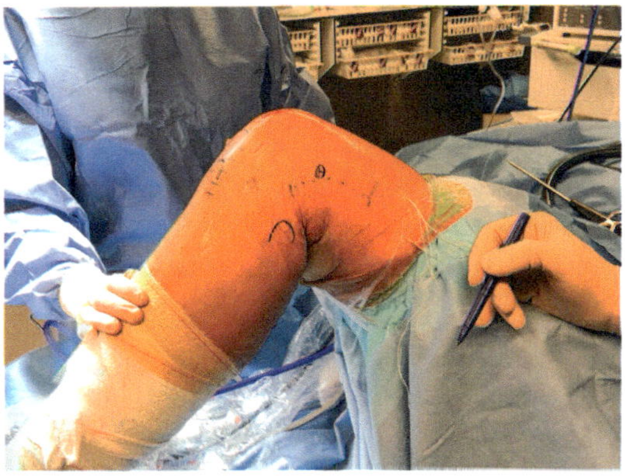

Fig. 34.1 The knee is flexed at 60°, both Gerdy's tubercle and the ITB-band are palpated and marked

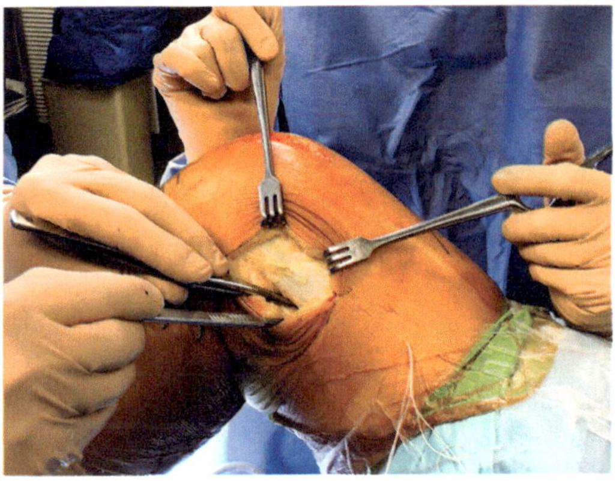

Fig. 34.2 Iliotibial band (ITB) exposed and posterior aspect identified

will become perfectly straight. Next, a subcutaneous dissection is performed, widely exposing the ITB band. Exposing the ITB proximal to the incision can be done by digital dissection.

The first landmark is the posterior edge of the ITB band on the proximal part of the incision, at the level of the underlying musculus vastus lateralis (Fig. 34.2). The posterior edge actually represents the Kaplan fiber's connection of the superficial ITB tot the posterolateral aspect of the femur. At 1 cm, anterior to the posterior edge, the ITB band is incised at the level of the underlying vastus lateralis, horizontally across the lateral epicondyle over a distance of approximately 5 cm with a no. 15 blade.

Next, a parallel, more anterior cut, creating an 8–12 mm wide band (depending on the size of the patient), is made (Fig. 34.3a). Subsequently these two cuts are lengthened proximally underneath the skin for about 5 cm with the use of dissection scissors (Fig. 34.3b, c). Therearfter, an ITB strip is created by connecting the anterior and posterior cut proximally (Fig. 34.3d). Distally, the ITB strip is incised using dissection scissors and a no. 15 blade close to its posterior insertion on Gerdy's tubercle. Care should be taken to steer the anterior and posterior cut of the ITB strip toward Gerdy's tubercle as the anterior ITB fibers tend to run toward the patella and patellar tendon and hence may result in a too wide distal part of the strip. A strip of ITB of approximately 15 cm of length can thus be created. The strip should remain attached to Gerdy's tubercle but is freed up proximally and pulled distally to provide the necessary space for the following step (Fig. 34.4).

34.3.2 Step 3: Identification of the Lateral Collateral Ligament

With the knee in 90° of flexion, the lateral femoral epicondyle can easily be identified by palpation within the window created by removal of the ITB strip. The LCL can then be identified as a strong string-like structure (Fig. 34.5) which can be put under tension with the knee in a Fig. 34.4 position. It runs from the lateral femoral epicondyle towards the fibular head. Often, the soft tissues overlying the LCL are swollen by the arthroscopy and make the dissection of the LCL more difficult. An important tip is to dissect these swollen soft tissues from the underlying LCL starting slightly distal to the LCL and progressively moving proximally. Once the LCL is identified close to the epicondyle, two 1 cm vertical incisions are made with a no. 15 blade, anterior and posterior, respectively, to the LCL in proximity of the lateral femoral epicondyle (Fig. 34.6). These two incisions form the entry and exit point for the ITB strip when tunnelled under the LCL (see step 4). Care must be taken to avoid transecting the lateral meniscus or popliteal tendon, which are both situated deep to the lateral collateral ligament.

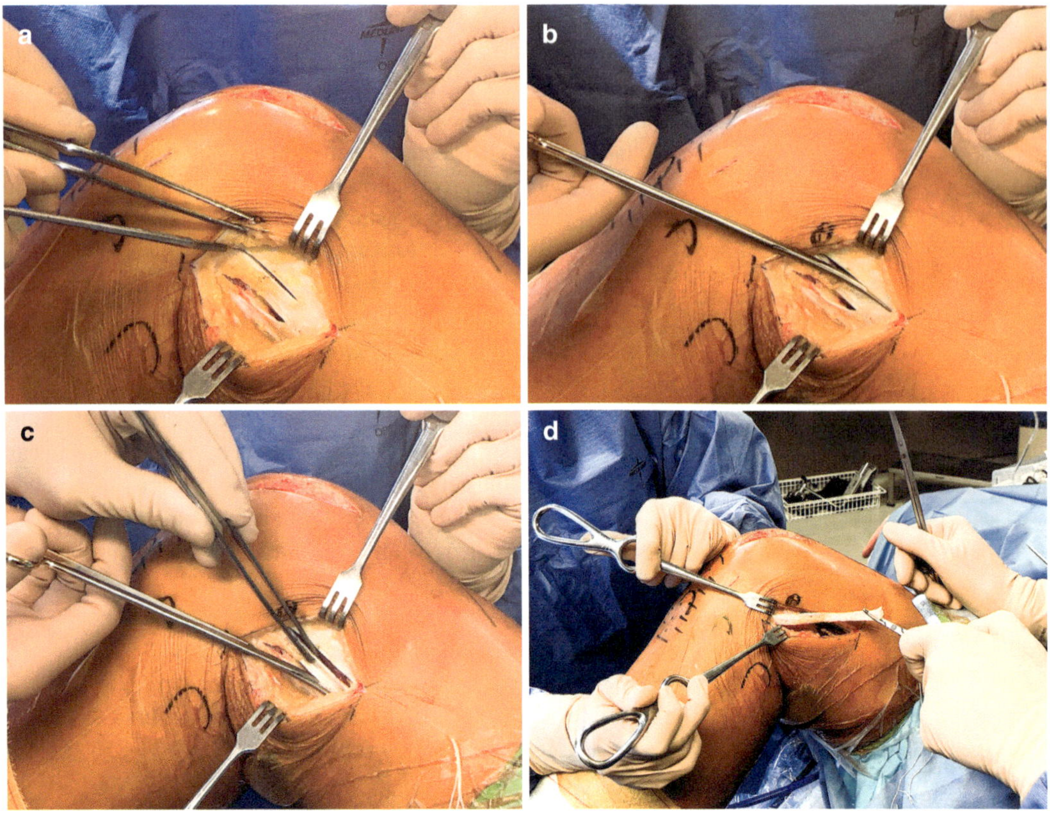

Fig. 34.3 (**a**) Two parallel cuts are made, the first one at 1 cm anterior of the posterior edge, creating an 8–12 mm wide band. (**b, c**) Both cuts are lengthened proximally underneath the skin for about 5 cm with the use of dissection scissors. (**d**) the proximal part of the ITB strip is cut loose and the free proximal end is created

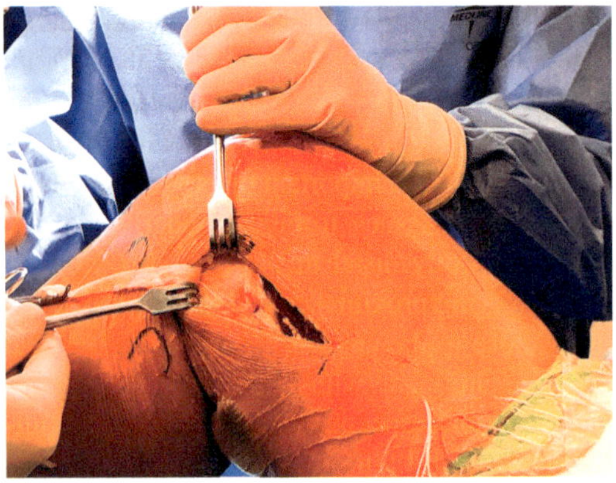

Fig. 34.4 The ITB strip can be freed up proximally and pulled distally to provide the necessary space for the following step

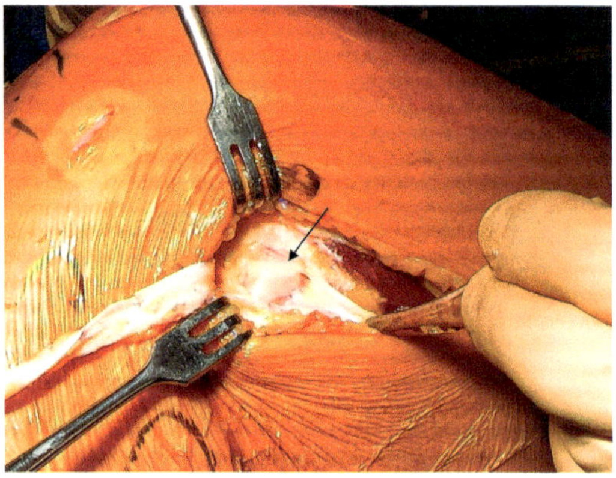

Fig. 34.5 The LCL can be identified as a strong string-like structure (arrow) in the window created by turning down the freed up ITB band

34.3.3 Step 4: Tunnelling the ITB Strip Deep to the LCL

A tunnel is created deep to the LCL using dissection scissors (Fig. 34.7a). With the use of a mosquito clamp, the ITB strip can be tunnelled deep to the LCL ligament from distal to proximal (Fig. 34.7b). The ITB strip is pushed deep to the LCL and picked up from the mosquito clamp, proximal to the LCL, using pick-ups (Fig. 34.7c). Twisting the strip while passing it deep to the LCL should be avoided.

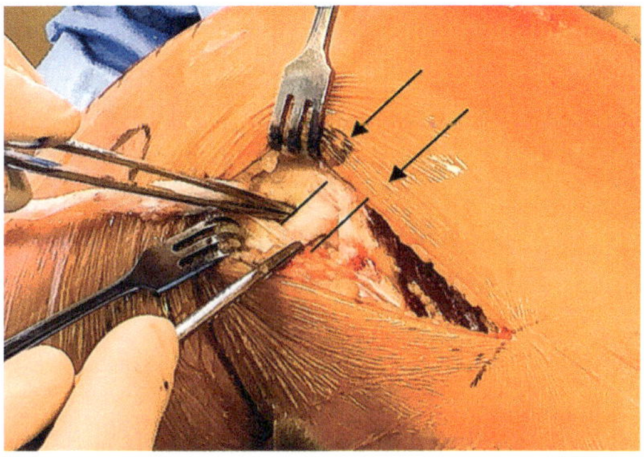

Fig. 34.6 An incision is made, both anteriorly and posteriorly of the LCL band (added arrows and markings)

34.3.4 Step 5: Proximal Dissection

Now, the distal edge of the vastus lateralis muscle can be identified and dissected. Using cautery, a small incision parallel to the distal oblique fibers of the vastus lateralis is made. Introducing a finger in this incision, allows easy access to the anterior part of the distal femur. Subsequently, a Hohmann retractor is placed underneath the vastus lateralis muscle (lateral subvastus approach) (Fig. 34.8). Now the distal femur shaft can be dissected just proximal to the lateral condyle with the use of an electric cautery. A direct lateral access to the distal femur is created with anteriorly the

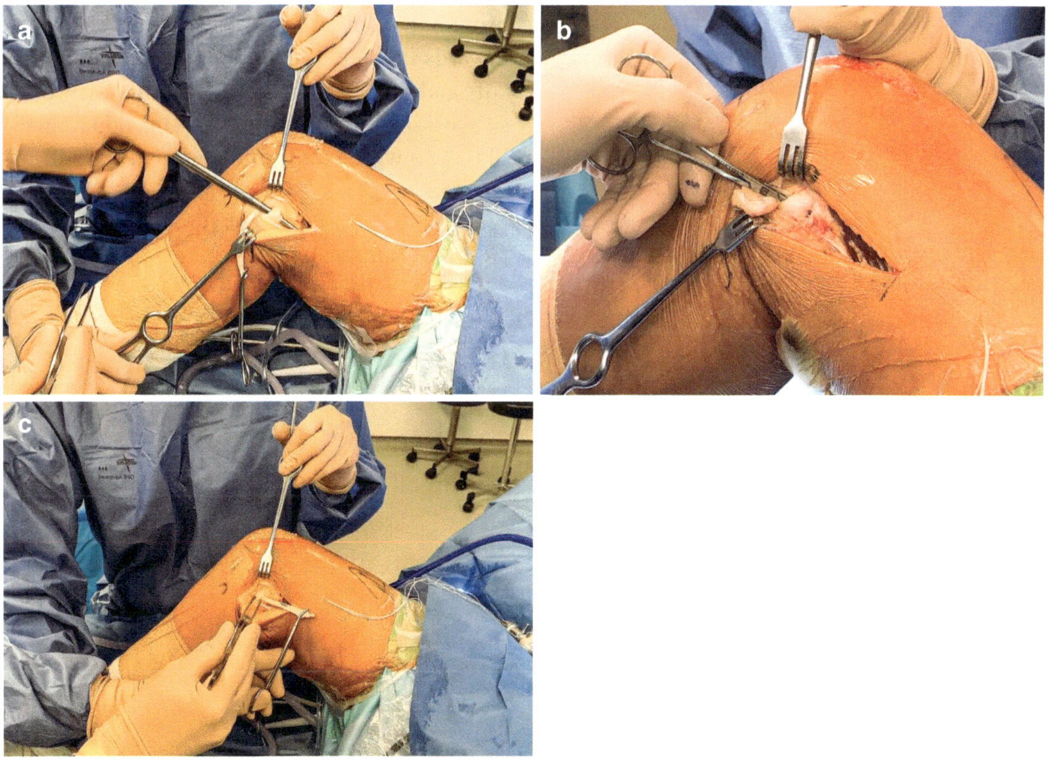

Fig. 34.7 (**a**) A tunnel can be created deep to the LCL using dissection scissors. (**b, c**) The ITB strip is pushed deep to the LCL and picked up from the mosquito clamp, proximal to the LCL, using pick-ups

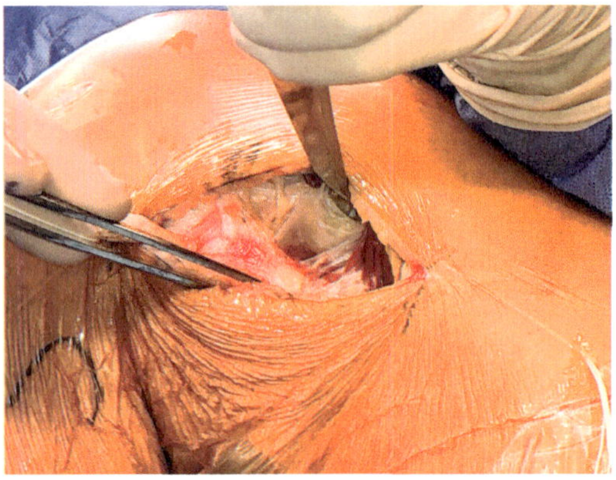

Fig. 34.8 After dissection, access to the lateral part of the femur condyle can be obtained and a Hohmann retractor is placed underneath the vastus lateralis muscle (lateral subvastus approach)

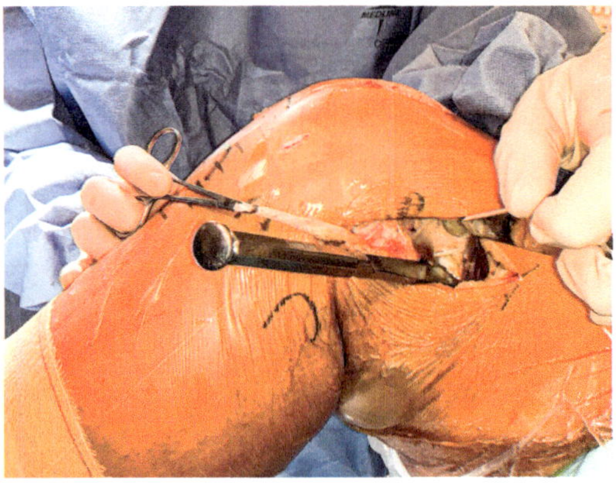

Fig. 34.9 A small horizontal cortical osteotomy can be made, just proximal to the condylar massive and just anterior to the posterior edge of the distal femur, using a 1 cm chisel

Hohmann retractor and posteriorly the Kaplan fibers connecting part of the ITB to the lateral femur. Care is taken to coagulate potential perforating vascular elements as well as periosteal vessels.

34.3.5 Step 6: Cortical Osteotomy

A small horizontal cortical osteotomy can be made, just proximal to the condylar massive and just anterior to the posterior edge of the distal femur, using a 1 cm chisel (Fig. 34.9). The created cleft is used to allow bone integration of the ITB strip. In general, the cleft is proximal to the cortical suspension button.

34.3.6 Step 7: Tunnelling the ITB Strip Underneath the Soft Tissues

With the knee in 90° of flexion, the ITB strip is now tunnelled in a straight line in a distal to proximal fashion from proximal to the LCL towards the lateral distal femur under the local soft and fatty tissues using first the dissection scissors and subsequently by introducing the ITB strip with a mosquito clamp in the soft tissue tunnel. The ITB strip is hence introduced posterior to the distal Kaplan fibers (Fig. 34.10a, b).

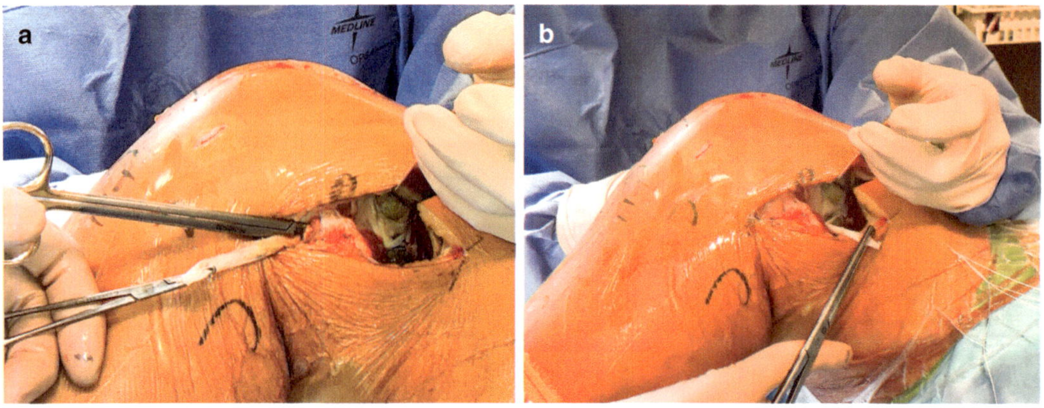

Fig. 34.10 (**a**) A tunnel underneath the soft tissues is created using dissection scissors. (**b**) The ITB strip is now tunneled under the lateral head of the gastrocnemius and underneath intermuscular septum toward the just created cortical cleft

34.3.7 Step 8: Proximal Fixation of the Free End of the ITB Strip

Now, the ITB strip is fixated using an 8 mm staple, positioned parallel to and just anterior of the posterior edge of the femur (Fig. 34.11a–d). This fixation should be done with the knee in 60° of flexion, the foot in slight external rotation and with tension on the tibial side of the ACL (which is not yet fixed) (Fig. 34.12). During fixation, minimal tension of 20 N on the ITB strip should be maintained.

34.3.8 Step 9: Final Check and Tourniquet Release

The course of the strip and the stability of the knee must be verified before closure. If the extra-articular stabilization is done in conjunc-

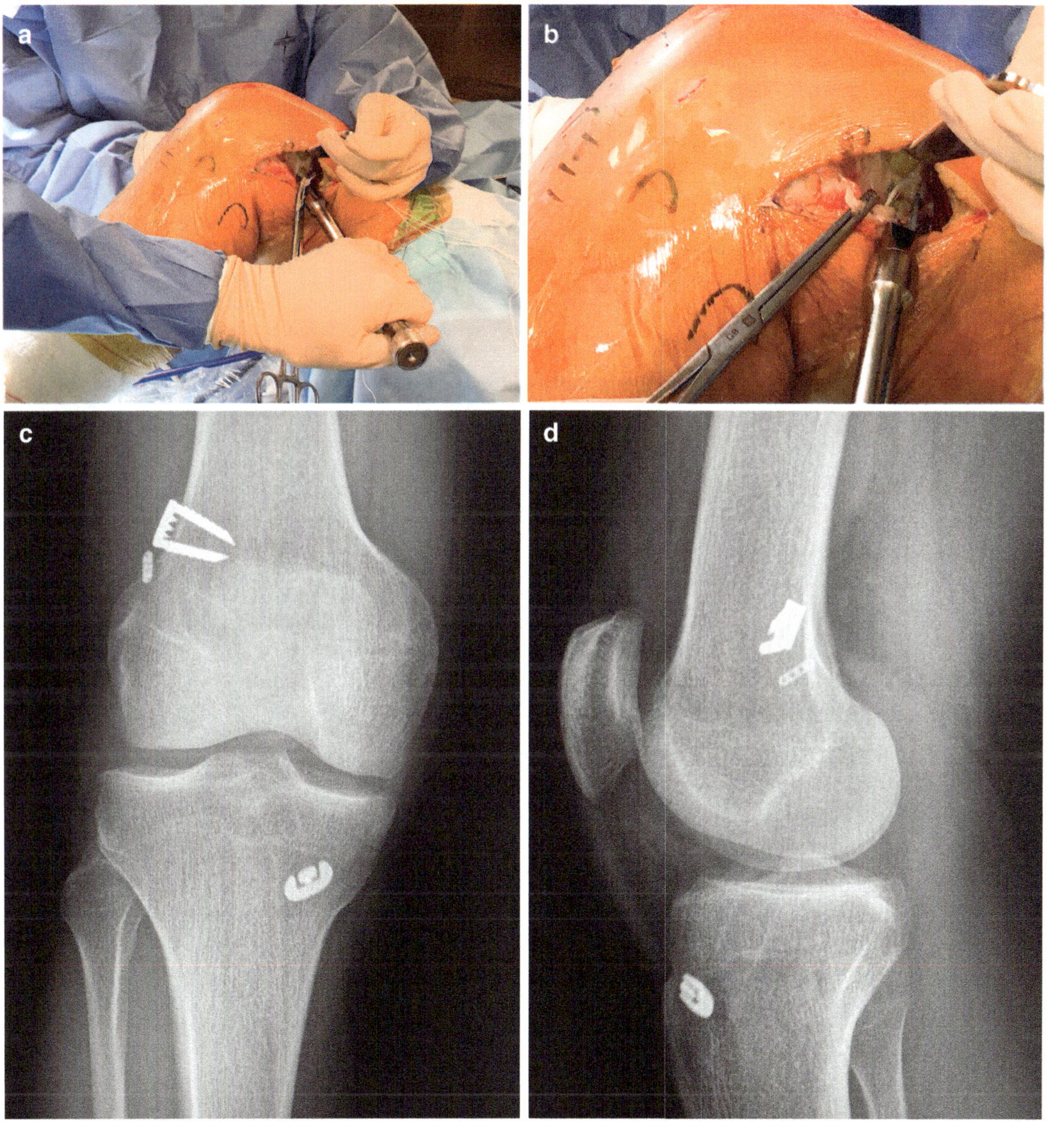

Fig. 34.11 (**a**) The ITB strip is fixated using an 8 mm staple, positioned horizontally, parallel to and just anterior of the posterior edge of the femur. (**b**) Close-up of the configuration of the staple. Maintain 20 N of tension on the ITB strip. (**c**, **d**) Post-operative imaging, AP and lateral view, showing the position of the fixation staple

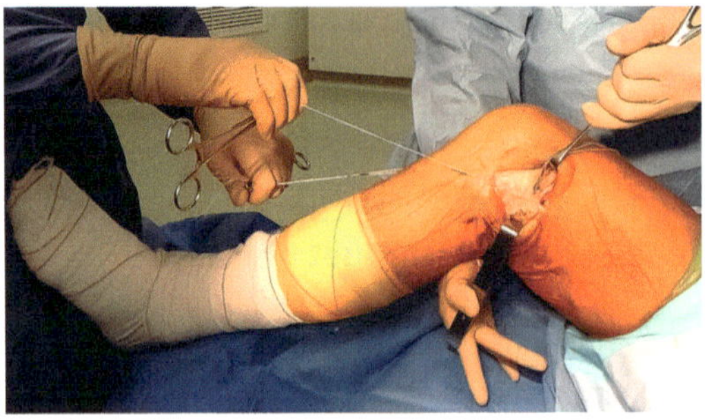

Fig. 34.12 Fixation of the ITB-strip should be done with the knee in 60° of flexion, the foot in slight external rotation and with tension on the tibial side of the ACL (which is not yet fixed)

tion with an intra-articular ACL reconstruction, the tibial fixation of the ACL graft can now be performed. At the end of the procedure, before closing, the tourniquet can be released, and a careful coagulation must be performed to avoid active bleeding and hematoma post-operatively. At our institution, a 24-h drain is left in the incision. Subsequently the proximal ITB window is closed using resorbable (Polysorb 1) sutures to the level of the lateral epicondyle to avoid muscle herniation. Distal to the lateral epicondyle, the ITB window is left open.

34.4 Conclusions

Recently, the anterolateral complex (ALC), including the capsule-osseous layer of the iliotibial band as well as the ALL, has received attention as being crucial in controlling the rotational laxity after ACL injuries. The literature supports the biomechanical benefits of providing an extra-articular restraint against internal tibial rotation and pivot shift. LET is becoming widely accepted as a useful adjunct to current ACL reconstruction techniques in specific patient populations. The above described mLET technique offers an adequate restoration of the anterolateral rotational stability as well as a low-morbidity procedure with limited complications when performed correctly.

Appendix

All figures are original.

References

1. Ferretti A, Monaco E, Fabbri M, Mazza D, De Carli A. The fascia lata anterolateral tenodesis technique. Arthrosc Tech. 2017;6(1):e81–6.
2. Tanaka M, Vyas D, Moloney G, Bedi A, Pearle AD, Musahl V. What does it take to have a high-grade pivot shift? Knee Surg Sport Traumatol Arthrosc. 2012;20(4):737–42.
3. Claes S, Vereecke E, Maes M, Victor J, Verdonk P, Bellemans J. Anatomy of the anterolateral ligament of the knee. J Anat. 2013;223(4):321–8.
4. Hewison CE, Tran MN, Kaniki N, Remtulla A, Bryant D, Getgood AM. Lateral extra-articular tenodesis reduces rotational laxity when combined with anterior cruciate ligament reconstruction: a systematic review of the literature. Arthrosc J Arthrosc Relat Surg. 2015;31(10):2022–34.
5. Getgood A, Brown C, Lording T, et al. The anterolateral complex of the knee: results from the International ALC Consensus Group Meeting. Knee Surgery, Sport Traumatol Arthrosc. 2019;27(1):166–76.
6. Musahl V, Getgood A, Neyret P, et al. Contributions of the anterolateral complex and the anterolateral ligament to rotatory knee stability in the setting of ACL Injury: a roundtable discussion. Knee Surg Sport Traumatol Arthrosc. 2017;25(4):997–1008.
7. Herbst E, Albers M, Burnham JM, et al. The anterolateral complex of the knee: a pictorial essay. Knee Surg Sport Traumatol Arthrosc. 2017;25(4):1009–14.
8. Kittl C, El-Daou H, Athwal KK, et al. The role of the anterolateral structures and the ACL in controlling laxity of the intact and ACL-deficient knee. Am J Sports Med. 2016;44(2):345–54.
9. Geeslin AG, Chahla J, Moatshe G, et al. Anterolateral knee extra-articular stabilizers: a robotic sectioning study of the anterolateral ligament and distal iliotibial band Kaplan fibers. Am J Sports Med. 2018;46(6):1352–61.
10. Lagae KC, Robberecht J, Athwal KK, Verdonk PCM, Amis AA. ACL reconstruction combined with lateral monoloop tenodesis can restore intact knee laxity. Knee Surg Sport Traumatol Arthrosc. 2020;28(4):1159–68.
11. Devitt BM, Bell SW, Ardern CL, et al. The role of lateral extra-articular tenodesis in primary anterior cruciate ligament reconstruction: a systematic review with meta-analysis and best-evidence synthesis. Orthop J Sport Med. 2017;5(10):1–12.
12. Song GY, Hong L, Zhang H, Zhang J, Li Y, Feng H. Clinical outcomes of combined lateral extra-articular tenodesis and intra-articular anterior cruci-

ate ligament reconstruction in addressing high-grade pivot-shift phenomenon. Arthrosc J Arthrosc Relat Surg. 2016;32(5):898–905.
13. Getgood AMJ, Bryant DM, Litchfield R, et al. Lateral extra-articular tenodesis reduces failure of hamstring tendon autograft anterior cruciate ligament reconstruction: 2-year outcomes from the STABILITY study randomized clinical trial. Am J Sports Med. 2020;48(2):285–97.
14. Dodds AL, Gupte CM, Neyret P, Williams AM, Amis AA. Extra-articular techniques in anterior cruciate ligament reconstruction: a literature review. J Bone Jt Surg Ser B. 2011;93(11):1440–8.
15. O'Brien SJ, Warren RF, Wickiewicz TL, et al. The iliotibial band lateral sling procedure and its effect on the results of anterior cruciate ligament reconstruction. Am J Sports Med. 1991;19(1):21–5.
16. Claes T, Declerq G, Martens M, Lefevre J. Extra-articular ligamentoplasty for chronic ACL insufficiency. Acta Orthop Belg. 1986;52(4):515–25.
17. Magnussen RA, Lustig S, Jacobi M, Elguindy A, Neyret P. The role of lateral extra-articular augmentation in revision ACL reconstruction. In: Marx RG, editor. Revision ACL reconstruction. New York: Springer; 2013. p. 151–6.
18. Kaplan EB. The iliotibial tract: clinical and morphological significance. J Bone Jt Surg Ser B. 1958;40(4):817–32.

35 Clinical Application of Scaffold-Free Tissue-Engineered Construct Derived from Synovial Stem Cells

Kazunori Shimomura, David A. Hart, Wataru Ando, and Norimasa Nakamura

35.1 Introduction

Articular cartilage injuries from joint trauma occur frequently and are quite common in clinical practice [1]. However, injured articular cartilage does not usually heal spontaneously, owing in part to its avascular and aneural environment as well as its relatively unique matrix and cell organization. Over time, such injuries can progress to osteoarthritis (OA) because of this inability of chondral lesions to heal effectively, leading to significantly reduced physical activity, chronic pain, and substantial lifestyle modifications. Therefore, a variety of approaches have been assessed to improve cartilage healing over the past few decades [2, 3].

Since the first report on autologous chondrocyte implantation (ACI) was published by Brittberg et al. in 1994 [4], chondrocyte-based therapies have been extensively studied [5–11]. However, this procedure likely has some limitations including the sacrifice of undamaged cartilage within the same joint and alterations of chondrogenic phenotype associated with the in vitro expansion of the cells. Furthermore, the availability of such cells may be limited in elderly individuals due to alterations and degenerative changes in cartilage associated with aging [12, 13].

To overcome such potential problems, stem cell therapies have become a focus to facilitate regenerative tissue repair. Mesenchymal stem cells (MSCs) have the capability to differentiate into a variety of connective tissue cells including bone, cartilage, tendon, muscle, as well as adipose tissue [14]. These cells can be isolated from various tissues such as bone marrow, skeletal muscle, synovial membrane, adipose tissue, and umbilical cord blood, as well as synovial fluid [14–20]. Pluripotent cells isolated from synovium may be well suited for cell-based therapies for cartilage because of the relative ease of harvest

K. Shimomura
Department of Orthopaedic Surgery, Osaka University Graduate School of Medicine, Osaka, Japan
e-mail: kazunori-shimomura@umin.net

D. A. Hart
McCaig Institute for Bone & Joint Health, University of Calgary, Calgary, AB, Canada
e-mail: hartd@ucalgary.ca

W. Ando
Department of Orthopaedic Medical Engineering, Osaka University Graduate School of Medicine, Osaka, Japan
e-mail: w-ando@umin.ac.jp

N. Nakamura (✉)
Institute for Medical Science in Sports, Osaka Health Science University, Osaka, Japan

Global Centre for Medical Engineering and Informatics, Osaka University, Osaka, Japan
e-mail: norimasa.nakamura@ohsu.ac.jp

N. Nakamura et al. (eds.), *Advances in Knee Ligament and Knee Preservation Surgery*,
https://doi.org/10.1007/978-3-030-84748-7_35

and their strong capacity for chondrogenic differentiation [16]. Synovium-derived stem cells are reported to exhibit the greatest chondrogenic potential among the other mesenchymal tissue-derived cells examined [17]. As other options for a cell source, allogeneic MSCs [21, 22] or induced pluripotent stem (iPS) cells [23, 24] may also be considered. However, there has not been much evidence generated to date using these cells in terms of clinical safety, and thus further studies and clinical trials would be necessary before considering clinical applications.

In addition to selection of a cell source, effective local delivery of cells to chondral lesions has been another area of concern. It is well recognized that an appropriate three-dimensional (3D) environment is important to optimize cell proliferation and chondrogenic differentiation [25]. Therefore, a 3D scaffold consisting of natural materials and/or synthetic polymers and seeded with cells is usually utilized to enhance repair of the defects [26–34]. This technique is robust and easy for surgeons to handle and was reported to significantly improve the healing of cartilage defects. On the other hand, there are still several issues associated with the long-term safety and efficacy of these materials. Thus, such scaffolds should ideally be excluded to minimize their currently unknown risks, and a scaffold-free technique could be an excellent alternative.

35.2 Scaffold-Free Techniques

Recently, several scaffold-free approaches have been assessed [35]. Regarding the fabrication methods, DuRaine et al. [36] defined these approaches into two categories: those that exhibit self-organization and those with self-assembly. Self-organization represents an approach to fabricate 3D tissue by utilizing external energy or force, such as bioprinting and cell-sheet technology. Approaches where cell aggregates are formed by applying a rotational force to the cells are categorized as one of self-organization. Self-assembly is defined as forming a 3D tissue without employing any external force. Regarding cell selection in a scaffold-free approach, chondrocytes have been mostly employed [35]. These cells readily produce their cartilage-specific ECM, especially in a 3D culture environment [37]. As an alternative, MSCs and iPS cells have also been recently tested, and an engineered tissue generated with these cells has shown feasibility for cartilage repair that is comparable to chondrocyte-based tissues [38–40].

Notably, some research has progressed to the stage of preclinical and clinical studies. Mainil-Varlet et al. developed a cartilage-like implant with chondrocytes in high density culture supported by a bioreactor and implanted the resulting neotissue onto minipig cartilage defects by press-fit fixation [41]. Histological analysis showed such an implant yielded consistent cartilage repair with a matrix predominantly composed of type II collagen. Ebihara et al. [42] used layered chondrocyte sheets prepared on a temperature-responsive culture dish and demonstrated these constructs facilitated cartilage repair in a minipig model. From the same research group, Sato et al. [43] reported the transplantation of chondrocyte sheets combined with conventional surgical treatments, including those for anterior cruciate ligament reconstruction and open-wedge high tibial osteotomy. In a small study with eight patients with knee osteoarthritis, the results of such an approach showed improved clinical outcomes up to 36 months postoperatively without serious adverse events and regeneration of cartilage confirmed by both second-look arthroscopy and MRI [43]. Also, they evaluated the tissue obtained from biopsy specimen at 12 months, revealing regeneration of hyaline cartilage as assessed by histology. Interestingly, Yamashita et al. developed a scaffold-less hyaline cartilaginous tissue (particle) from human iPS cells and demonstrated the feasibility of using these particles for hyaline cartilage regeneration based on the results of a minipig study [40].

Taken together, many promising scaffold-free approaches have been developed until now, and such technologies could become a next-generation vehicle for cartilage repair, with regard to their high level of safety. On the other hand, there are still several issues to be resolved, including complicated fabrication processes, a longer culture period, requirement of a large number of cells, and anti-adhesive properties of generated tissues, as addressed in a recent review article [35]. To overcome such potential issues, we have developed a novel scaffold-free 3D tissue-engineered construct (TEC) that is comprised of MSCs derived from synovium and an ECM synthesized by the cells themselves. Hereafter, the safety and effectiveness of the TEC methodology for cartilage repair and regeneration will be discussed.

35.3 Scaffold-Free 3D TEC

Synovial membrane harvested from either porcine or human knee joints was enzymatically digested, and synovial MSCs were isolated and subsequently expanded in growth media containing virus- and prion-free fetal bovine serum. The isolated cells showed characteristics of MSCs with regard to morphology, growth characteristics, cell surface phenotype, and multipotent differentiation capacity (to osteogenic, chondrogenic, and adipogenic lineages) [44, 45]. When synovium-derived MSCs were cultured to confluence in the basic growth medium, they did not synthesize an abundant collagenous matrix. In contrast, in the presence of >0.1 mM ascorbic acid-2 phosphate (Asc-2P), collagen synthesis significantly increased with time in culture [45]. Subsequently, the monolayer cell-matrix complex cultured in Asc-2P became a stiff sheet-like structure, which could be readily detached from the substratum by exerting mild shear stress at the cell-substratum interface using gentle pipetting. After detachment, the monolayer sheet immediately began to actively contract and evolve into a thick 3D tissue (Fig. 35.1). When the matrix folded and contracted, it was apparent that the layers were integrated into each other, leading to development of one spherical body several millimeters thick.

Histology and scanning electron microscope (SEM) assessment of this 3D tissue indicated that the cells and the corresponding ECM were three dimensionally integrated together at high cell density [45]. Immunohistochemical analysis showed that the TEC was rich in collagen I and III [44]. In contrast, there was no detectable expression of collagen II within the TEC. However, such TEC exhibited an increased synthesis of glycosaminoglycan (GAG) and collagen II synthesis when cultured in a chondrogenic media. Notably, adhesion molecules such as fibronectin and vitronectin were also abundant in these TEC, and all the molecules detected were diffusely distributed throughout the matrix. These adhesion molecules enabled TECs to readily adhere to the injured chondral surfaces without any reinforcement of fixation [44]. Thus, such adhesive properties could be significant advantage to avoid an initial fixation by suturing or fibrin glue when implanted, as articular cartilage exhibits anti-adhesive properties due to its unique matrix organization. Therefore, integration of the implanted tissue to the adjacent cartilage normal matrix has been a past issue in the treatment of chondral injuries [3, 46–48] and an issue that could be overcome with implanted TEC.

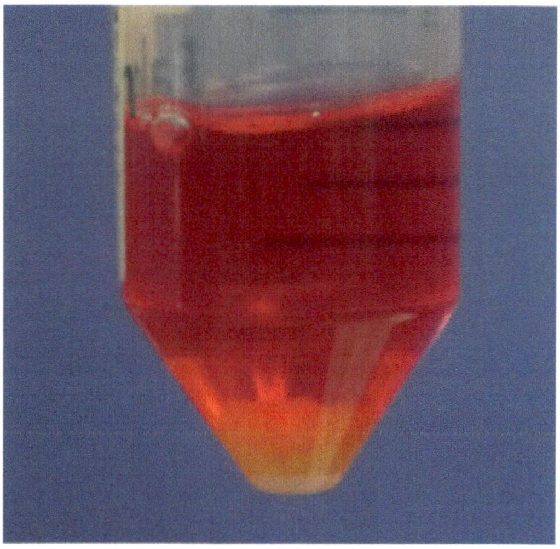

Fig. 35.1 Development of TEC from human MSCs

Biomechanical testing revealed that the tensile strength of the TEC significantly increased with time of culture in the presence of Asc-2P, and the values increased to approximately 1.3 MPa after 21 days in culture [45]. Such TECs were sufficiently robust to maintain their integrity during surgical handling.

35.4 Cartilage Repair Using TECs in Preclinical Studies

To assess the efficacy of the TEC in preclinical studies for clinical applications, a porcine model was chosen since the physiology of the pig is similar to that of humans in many respects [49], and porcine articular cartilage of the knee is sufficiently thick as to allow creation of a chondral defect without damaging the subchondral bone. In such porcine studies, both immature and mature animals were used in order to test the feasibility of using the porcine TEC approach for a wide range of recipient ages to effect repair of a chondral injury. The TECs derived from undifferentiated porcine synovial MSCs were implanted into equivalent chondral defects in the medial femoral condyle (8.5 mm in diameter, 2 mm in depth) of both immature and mature pigs, respectively. At 6 months post-implantation, regardless of maturity, untreated lesions exhibited no evidence for repair or only partial tissue coverage, while the defects treated with a TEC were totally or mostly covered with repair tissue. Histologically, the chondral lesions in the untreated control groups showed evidence of osteoarthritic changes, with loss of cartilage and destruction of subchondral bone in both skeletally immature and mature animals (Fig. 35.2). Conversely, when treated with a TEC, the defects were filled with repair tissue exhibiting good integration to the adjacent cartilage and the restoration of a smooth surface, regardless of age at the time of implantation (Fig. 35.2). Notably, higher magnification views indicated that there was good tissue integration to the adjacent cartilage obtained when the TEC were implanted in both immature and mature animals. Such repair tissue exhibited predominantly spindle-shaped fibroblast-like cells in the superficial zone of the repair tissue, while the majority of the remaining repair matrix contained round-shaped cells in lacuna. Using histological scoring, the TEC groups exhibited significantly higher scores than did the control group, regardless of the maturity of the pigs at the time of implantation. Comparing the repair tissues developing following TEC implantation in immature and mature animals, no significant differences were detected. In addition, the mechanical properties of porcine chondral defects treated with a porcine-derived TEC were assessed at 6 months post-implantation in immature and mature animals, in comparison with those of normal cartilage. These results suggested that the viscoelastic properties of the tissue in defects repaired by TEC implantation are similar to those of normal cartilage, regardless of age at the time of implantation.

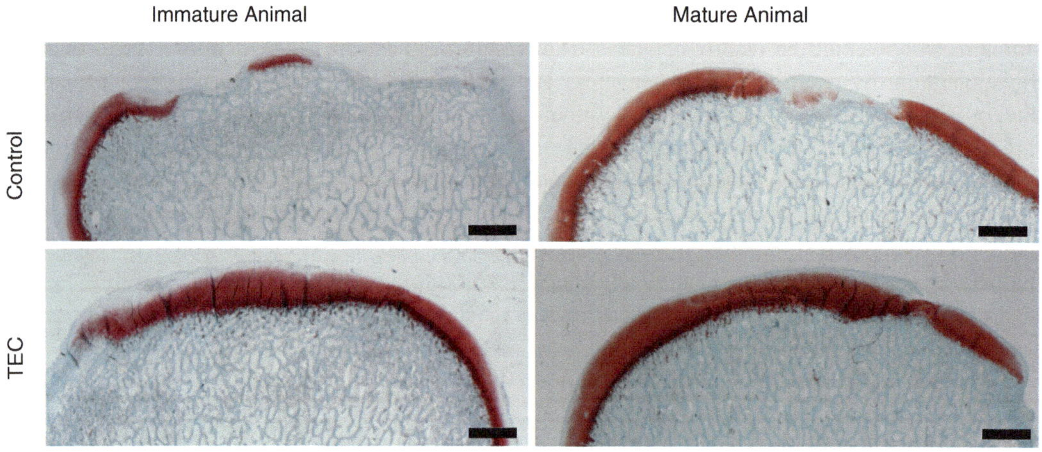

Fig. 35.2 Safranin O staining of untreated chondral lesions or lesions repaired by implantation of an autologous TEC at 6 months post-operation. Bar = 1 mm. (Quoted and modified from ref. [21] (Shimomura et al., Biomaterials 2010))

35.5 First-in-Human Clinical Trials Using a TEC for Repair of an Isolated Cartilage Defect

Based on the encouraging and promising results of the preclinical studies mentioned above, we have proceeded to pilot clinical studies under the auspices of an approved first-in-human protocol [50, 51]. A first-in-human observational study (limited to five cases) was approved as a proof of concept trial by the Japan Ministry of Health, Labor, and Welfare. Five patients aged 20–60 years with isolated full-thickness cartilage defects of the knee (<5 cm^2, International Cartilage Regeneration & Joint Preservation Society [ICRS] grade III or IV) with normal alignment were enrolled in the study.

We performed a 2-step procedure: the first for arthroscopic evaluation and synovial tissue biopsy and the second for the implantation surgery. Under general or spinal anesthesia, approximately 1 g of synovial membrane is harvested from the knee joint, which is then subjected to the isolation and culture of MSC for their separation and expansion. Following 4–6 weeks post-tissue harvest, the TECs are prepared for autologous implantation. By mini-arthrotomy or arthroscopy, the chondral lesions were debrided so as to not breach the subchondral bone. Before implantation, the TEC is washed several times with sterile phosphate-buffered saline to minimize bovine serum-related protein contamination, followed by the adjustment of the TEC size to match that of the chondral defect. Implantation was completed within 5–10 min, without any reinforcement for fixation. The knees were immobilized in a brace for 2 weeks followed by the initiation of range-of-motion exercises and muscle exercises. Full weight bearing was allowed 6–8 weeks after implantation surgery. Return to strenuous activity was allowed approximately 12 months following implantation. The duration for follow-up was 24 months, and the primary end point of this study was an analysis of safety of the procedure. The secondary end point was the assessment of the efficacy of the procedure, which consisted of subjective assessment (visual analog score [VAS] for pain, knee injury and osteoarthritis outcome score [KOOS]), and structural assessment. For the structural assessment, histologic analysis of a biopsy specimen at 12 months and magnetic resonance imaging (conventional and quantitative T2 mapping) at 1.5, 6, 12, and 24 months were performed.

All patients were followed over 24 months postoperatively and did not require additional treatment during this observational period. No serious adverse events were observed out to 24 months after TEC implantation. Joint pain, effusion, and swelling were observed in the early stages after surgery, and all symptoms were completely improved by 4 weeks. No postoperative infections were observed out to 24 months after surgery for any patient.

Arthroscopic analysis indicated cartilage repair was confirmed by the covering of implanted defects with cartilaginous tissue, with good tissue integration to adjacent cartilage in all cases at 12 months (Fig. 35.3a). In addition, no hypertrophy was observed in the repair cartilage for all cases. Histology of the biopsy specimens at 12 months showed repair with cartilaginous tissue exhibiting positive safranin O staining in all cases (Fig. 35.3b). Especially, the majority of the deeper repair matrix in all the cases showed positive staining for safranin O and contained round-shaped cells in lacuna, suggesting repair with a hyaline cartilage-like matrix. Based on MRI assessments, cartilage defects were filled with newly generated tissues over time (Fig. 35.4), and the defect filling rate reached 100% coverage without detectable hypertrophy of the repair tissues by 12 months for all patients. Some subchondral bone edema was observed around the TEC implantation sites in the early stages after surgery, but such abnormal signals disappeared by 12 months for all cases. The repair tissue exhibited good tissue integration with adjacent host cartilage. The T2 mapping showed that implanted chondral defects were becoming similar in intensity to the surrounding cartilage over time. Similar to the results of the structural assessments, the subjective assessments by VAS and KOOS were significantly improved (Fig. 35.5).

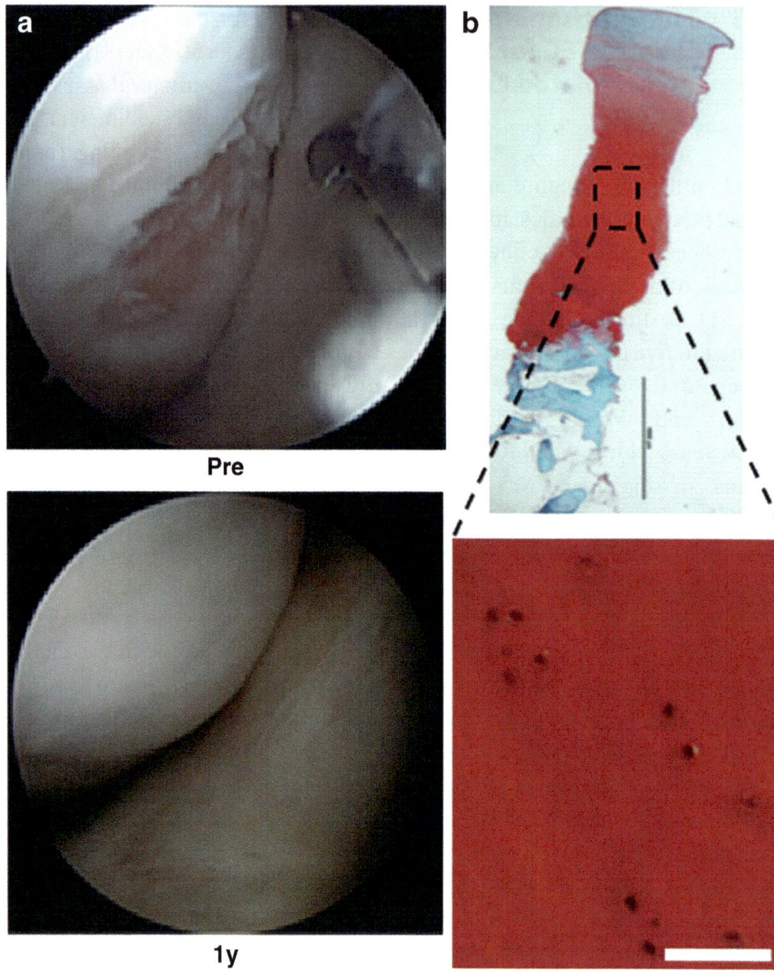

Fig. 35.3 (**a**) Arthroscopic image before implantation of tissue-engineered constructs and during a second-look arthroscopy at 1 year postoperatively. (**b**) Safranin O staining of repair cartilage from biopsy specimens obtained at 1 year post-implantation. Bar = 100 mm. (Quoted and modified from ref. [51] (Shimomura et al., Am J Sports Med 2018))

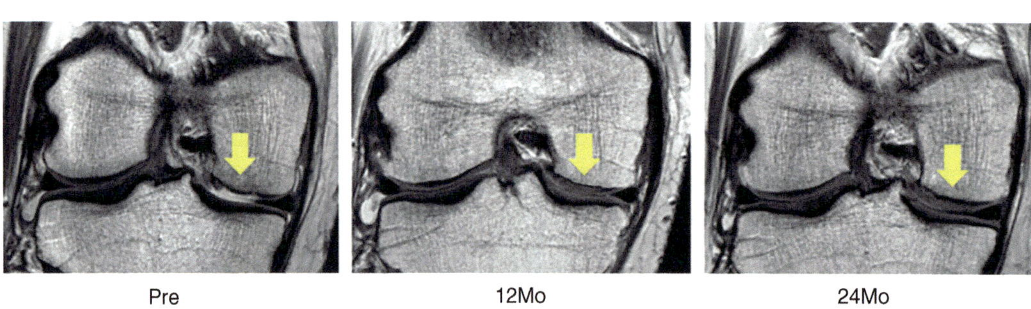

Fig. 35.4 Magnetic resonance imaging analysis preoperatively and at 1 and 2 years postoperatively. Arrows indicate cartilage defect (left) and the repair site following implantation of a TEC (center and right). (Quoted and modified from ref. [51] (Shimomura et al., Am J Sports Med 2018))

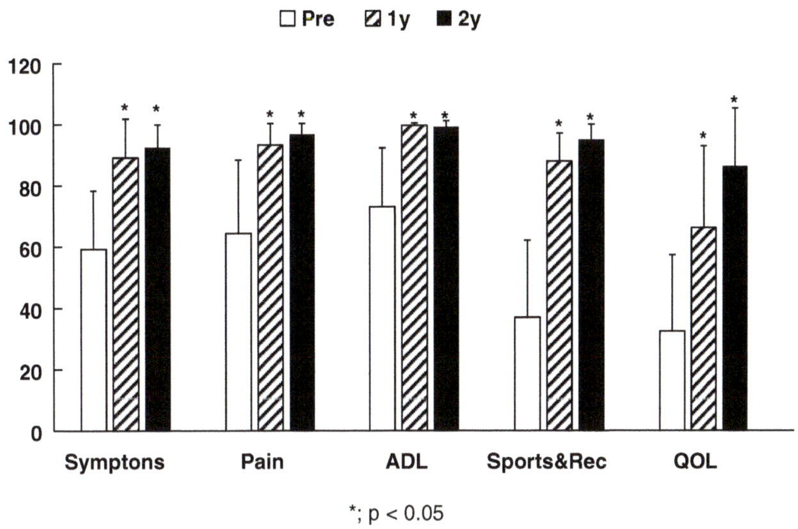

Fig. 35.5 KOOS for each patient preoperatively and at 1 and 2 years postoperatively. (Quoted and modified from ref. [51] (Shimomura et al., Am J Sports Med 2018))

35.6 Future Directions

The present chapter has demonstrated the feasibility of using a unique scaffold-free TEC generated from synovial MSCs for effective cell-based cartilage repair via suture-less and simple implantation procedure. We have addressed many of the characteristics of these scaffold-free 3D TEC to conclude they are a unique and promising implant for facilitating cartilage repair. This was demonstrated in vivo using a preclinical model with a range of ages [21, 44, 45], as well as the recent clinical trial with five patients [51]. Due to the scaffold-free nature of the in vitro generated structure, implantation of TEC could yield more long-term safety and efficacy than that derived from scaffold-based cell therapies. Furthermore, the implanted MSC in the TEC were undifferentiated and thus, appeared to differentiate in the cartilage environment.

Initially, being a collagen I rich matrix, the basic TEC construct could also potentially be suitable for augmenting repair of compromised skin or enhancing the repair of ligaments or tendons, which are also collagen I-rich environments. Since a TEC also has osteogenic or adipogenic differentiation capacity, the TEC could also be useful for potential applications to other musculoskeletal tissues, and among them, we demonstrated its feasibility for the repair of meniscus, osteochondral tissue, growth plate, and intervertebral disc in animal studies [52–57].

Moreover, TEC could be developed from MSCs derived from other tissues, such as adipose tissue that is also an abundant source of MSC and could be readily obtained without entering the injured joint. Therefore, tissue engineering using the TEC technology could potentially provide a variety of therapeutic interventions in regenerative medicine for a number of tissue applications using MSC from different sources.

Acknowledgments This work was supported by a Health and Labor Sciences Research grant from the Ministry of Health, Labor, and Welfare of Japan; a grant from the New Energy and Industrial Technology Development Organization, Japan; and a Grant-in-Aid for Scientific Research, Japan Society for the Promotion of Science.

References

1. Hjelle K, Solheim E, Strand T, Muri R, Brittberg M. Articular cartilage defects in 1,000 knee arthroscopies. Arthroscopy. 2002;18(7):730–4.
2. Buckwalter JA. Articular cartilage injuries. Clin Orthop Relat Res. 2002;402:21–37.

3. Hunziker EB. Articular cartilage repair: basic science and clinical progress. A review of the current status and prospects. Osteoarthr Cartil. 2002;10(6):432–63.
4. Brittberg M, Lindahl A, Nilsson A, Ohlsson C, Isaksson O, Peterson L. Treatment of deep cartilage defects in the knee with autologous chondrocyte transplantation. N Engl J Med. 1994;331(14):889–95.
5. Peterson L, Brittberg M, Kiviranta I, Akerlund EL, Lindahl A. Autologous chondrocyte transplantation. Biomechanics and long-term durability. Am J Sports Med. 2002;30(1):2–12.
6. Browne JE, Anderson AF, Arciero R, Mandelbaum B, Moseley JB Jr, Micheli LJ, et al. Clinical outcome of autologous chondrocyte implantation at 5 years in US subjects. Clin Orthop Relat Res. 2005;436:237–45.
7. Niemeyer P, Salzmann G, Feucht M, Pestka J, Porichis S, Ogon P, et al. First-generation versus second-generation autologous chondrocyte implantation for treatment of cartilage defects of the knee: a matched-pair analysis on long-term clinical outcome. Int Orthop. 2014;38(10):2065–70.
8. Goyal D, Goyal A, Keyhani S, Lee EH, Hui JH. Evidence-based status of second- and third-generation autologous chondrocyte implantation over first generation: a systematic review of level I and II studies. Arthroscopy. 2013;29(11):1872–8.
9. Steinwachs M, Kreuz PC. Autologous chondrocyte implantation in chondral defects of the knee with a type I/III collagen membrane: a prospective study with a 3-year follow-up. Arthroscopy. 2007;23(4):381–7.
10. Bright P, Hambly K. A systematic review of reporting of rehabilitation in articular-cartilage-repair studies of third-generation autologous chondrocyte implantation in the knee. J Sport Rehabil. 2014;23(3):182–91.
11. Gobbi A, Kon E, Berruto M, Filardo G, Delcogliano M, Boldrini L, et al. Patellofemoral full-thickness chondral defects treated with second-generation autologous chondrocyte implantation: results at 5 years' follow-up. Am J Sports Med. 2009;37(6):1083–92.
12. Iwasa J, Engebretsen L, Shima Y, Ochi M. Clinical application of scaffolds for cartilage tissue engineering. Knee Surg Sports Traumatol Arthrosc. 2009;17(6):561–77.
13. Nehrer S, Domayer S, Dorotka R, Schatz K, Bindreiter U, Kotz R. Three-year clinical outcome after chondrocyte transplantation using a hyaluronan matrix for cartilage repair. Eur J Radiol. 2006;57(1):3–8.
14. Pittenger MF, Mackay AM, Beck SC, Jaiswal RK, Douglas R, Mosca JD, et al. Multilineage potential of adult human mesenchymal stem cells. Science. 1999;284(5411):143–7.
15. Jankowski RJ, Deasy BM, Huard J. Muscle-derived stem cells. Gene Ther. 2002;9(10):642–7.
16. De Bari C, Dell'Accio F, Tylzanowski P, Luyten FP. Multipotent mesenchymal stem cells from adult human synovial membrane. Arthritis Rheum. 2001;44(8):1928–42.
17. Sakaguchi Y, Sekiya I, Yagishita K, Muneta T. Comparison of human stem cells derived from various mesenchymal tissues: superiority of synovium as a cell source. Arthritis Rheum. 2005;52(8):2521–9.
18. Wickham MQ, Erickson GR, Gimble JM, Vail TP, Guilak F. Multipotent stromal cells derived from the infrapatellar fat pad of the knee. Clin Orthop Relat Res. 2003;412:196–212.
19. Lee OK, Kuo TK, Chen WM, Lee KD, Hsieh SL, Chen TH. Isolation of multipotent mesenchymal stem cells from umbilical cord blood. Blood. 2004;103(5):1669–75.
20. Ando W, Kutcher JJ, Krawetz R, Sen A, Nakamura N, Frank CB, et al. Clonal analysis of synovial fluid stem cells to characterize and identify stable mesenchymal stromal cell/mesenchymal progenitor cell phenotypes in a porcine model: a cell source with enhanced commitment to the chondrogenic lineage. Cytotherapy. 2014;16(6):776–88.
21. Shimomura K, Ando W, Tateishi K, Nansai R, Fujie H, Hart DA, et al. The influence of skeletal maturity on allogenic synovial mesenchymal stem cell-based repair of cartilage in a large animal model. Biomaterials. 2010;31(31):8004–11.
22. Dashtdar H, Rothan HA, Tay T, Ahmad RE, Ali R, Tay LX, et al. A preliminary study comparing the use of allogenic chondrogenic pre-differentiated and undifferentiated mesenchymal stem cells for the repair of full thickness articular cartilage defects in rabbits. J Orthop Res. 2011;29(9):1336–42.
23. Takahashi K, Yamanaka S. Induction of pluripotent stem cells from mouse embryonic and adult fibroblast cultures by defined factors. Cell. 2006;126(4):663–76.
24. Tsumaki N, Okada M, Yamashita A. iPS cell technologies and cartilage regeneration. Bone. 2015;70:48–54.
25. De Bari C, Dell'Accio F, Luyten FP. Failure of in vitro-differentiated mesenchymal stem cells from the synovial membrane to form ectopic stable cartilage in vivo. Arthritis Rheum. 2004;50(1):142–50.
26. Vunjak-Novakovic G, Martin I, Obradovic B, Treppo S, Grodzinsky AJ, Langer R, et al. Bioreactor cultivation conditions modulate the composition and mechanical properties of tissue-engineered cartilage. J Orthop Res. 1999;17(1):130–8.
27. Andriano KP, Tabata Y, Ikada Y, Heller J. In vitro and in vivo comparison of bulk and surface hydrolysis in absorbable polymer scaffolds for tissue engineering. J Biomed Mater Res. 1999;48(5):602–12.
28. Guo JF, Jourdian GW, MacCallum DK. Culture and growth characteristics of chondrocytes encapsulated in alginate beads. Connect Tissue Res. 1989;19(2–4):277–97.
29. Masuda K, Takegami K, An H, Kumano F, Chiba K, Andersson GB, et al. Recombinant osteogenic protein-1 upregulates extracellular matrix metabolism by rabbit annulus fibrosus and nucleus pulposus cells cultured in alginate beads. J Orthop Res. 2003;21(5):922–30.
30. Lee CH, Singla A, Lee Y. Biomedical applications of collagen. Int J Pharm. 2001;221(1–2):1–22.
31. Homminga GN, Buma P, Koot HW, van der Kraan PM, van den Berg WB. Chondrocyte behavior in fibrin glue in vitro. Acta Orthop Scand. 1993;64(4):441–5.
32. Brun P, Cortivo R, Zavan B, Vecchiato N, Abatangelo G. In vitro reconstructed tissues on hyaluronan-

based temporary scaffolding. J Mater Sci Mater Med. 1999;10(10/11):683–8.
33. Lahiji A, Sohrabi A, Hungerford DS, Frondoza CG. Chitosan supports the expression of extracellular matrix proteins in human osteoblasts and chondrocytes. J Biomed Mater Res. 2000;51(4):586–95.
34. Shimomura K, Moriguchi Y, Murawski CD, Yoshikawa H, Nakamura N. Osteochondral tissue engineering with biphasic scaffold: current strategies and techniques. Tissue Eng Part B Rev. 2014;20(5):468–76.
35. Shimomura K, Ando W, Fujie H, Hart DA, Yoshikawa H, Nakamura N. Scaffold-free tissue engineering for injured joint surface restoration. J Exp Orthop. 2018;5(1):2.
36. DuRaine GD, Brown WE, Hu JC, Athanasiou KA. Emergence of scaffold-free approaches for tissue engineering musculoskeletal cartilages. Ann Biomed Eng. 2015;43(3):543–54.
37. Huang BJ, Hu JC, Athanasiou KA. Cell-based tissue engineering strategies used in the clinical repair of articular cartilage. Biomaterials. 2016;98:1–22.
38. Murdoch AD, Grady LM, Ablett MP, Katopodi T, Meadows RS, Hardingham TE. Chondrogenic differentiation of human bone marrow stem cells in transwell cultures: generation of scaffold-free cartilage. Stem Cells. 2007;25(11):2786–96.
39. Ishihara K, Nakayama K, Akieda S, Matsuda S, Iwamoto Y. Simultaneous regeneration of full-thickness cartilage and subchondral bone defects in vivo using a three-dimensional scaffold-free autologous construct derived from high-density bone marrow-derived mesenchymal stem cells. J Orthop Surg Res. 2014;9:98.
40. Yamashita A, Morioka M, Yahara Y, Okada M, Kobayashi T, Kuriyama S, et al. Generation of scaffoldless hyaline cartilaginous tissue from human iPSCs. Stem Cell Rep. 2015;4(3):404–18.
41. Mainil-Varlet P, Rieser F, Grogan S, Mueller W, Saager C, Jakob RP. Articular cartilage repair using a tissue-engineered cartilage-like implant: an animal study. Osteoarthritis Cartilage. 2001;9 Suppl A:S6–15.
42. Ebihara G, Sato M, Yamato M, Mitani G, Kutsuna T, Nagai T, et al. Cartilage repair in transplanted scaffold-free chondrocyte sheets using a minipig model. Biomaterials. 2012;33(15):3846–51.
43. Sato M, Yamato M, Mitani G, Takagaki T, Hamahashi K, Nakamura Y, et al. Combined surgery and chondrocyte cell-sheet transplantation improves clinical and structural outcomes in knee osteoarthritis. NPJ Regen Med. 2019;4:4.
44. Ando W, Tateishi K, Hart DA, Katakai D, Tanaka Y, Nakata K, et al. Cartilage repair using an in vitro generated scaffold-free tissue-engineered construct derived from porcine synovial mesenchymal stem cells. Biomaterials. 2007;28(36):5462–70.
45. Ando W, Tateishi K, Katakai D, Hart DA, Higuchi C, Nakata K, et al. In vitro generation of a scaffold-free tissue-engineered construct (TEC) derived from human synovial mesenchymal stem cells: biological and mechanical properties and further chondrogenic potential. Tissue Eng Part A. 2008;14(12):2041–9.
46. Lee CR, Grodzinsky AJ, Hsu HP, Spector M. Effects of a cultured autologous chondrocyte-seeded type II collagen scaffold on the healing of a chondral defect in a canine model. J Orthop Res. 2003;21(2):272–81.
47. Dorotka R, Bindreiter U, Macfelda K, Windberger U, Nehrer S. Marrow stimulation and chondrocyte transplantation using a collagen matrix for cartilage repair. Osteoarthr Cartil. 2005;13(8):655–64.
48. Benthien JP, Behrens P. The treatment of chondral and osteochondral defects of the knee with autologous matrix-induced chondrogenesis (AMIC): method description and recent developments. Knee Surg Sports Traumatol Arthrosc. 2011;19(8):1316–9.
49. Vodicka P, Smetana K Jr, Dvorankova B, Emerick T, Xu YZ, Ourednik J, et al. The miniature pig as an animal model in biomedical research. Ann N Y Acad Sci. 2005;1049:161–71.
50. Nakamura N, Hui J, Koizumi K, Yasui Y, Nishii T, Lad D, et al. Stem cell therapy in cartilage repair—culture-free and cell culture–based methods. Oper Tech Orthop. 2014;24(1):54–60.
51. Shimomura K, Yasui Y, Koizumi K, Chijimatsu R, Hart DA, Yonetani Y, et al. First-in-human pilot study of implantation of a scaffold-free tissue-engineered construct generated from autologous synovial mesenchymal stem cells for repair of knee chondral lesions. Am J Sports Med. 2018;46(10):2384–93.
52. Moriguchi Y, Tateishi K, Ando W, Shimomura K, Yonetani Y, Tanaka Y, et al. Repair of meniscal lesions using a scaffold-free tissue-engineered construct derived from allogenic synovial MSCs in a miniature swine model. Biomaterials. 2013;34(9):2185–93.
53. Shimomura K, Moriguchi Y, Ando W, Nansai R, Fujie H, Hart DA, et al. Osteochondral repair using a scaffold-free tissue-engineered construct derived from synovial mesenchymal stem cells and a hydroxyapatite-based artificial bone. Tissue Eng Part A. 2014;20(17–18):2291–304.
54. Shimomura K, Moriguchi Y, Nansai R, Fujie H, Ando W, Horibe S, et al. Comparison of 2 different formulations of artificial bone for a hybrid implant with a tissue-engineered construct derived from synovial mesenchymal stem cells. Am J Sports Med. 2017;45(3):666–75.
55. Yoshida K, Higuchi C, Nakura A, Nakamura N, Yoshikawa H. Treatment of partial growth arrest using an in vitro-generated scaffold-free tissue-engineered construct derived from rabbit synovial mesenchymal stem cells. J Pediatr Orthop. 2012;32(3):314–21.
56. Ishiguro H, Kaito T, Yarimitsu S, Hashimoto K, Okada R, Kushioka J, et al. Intervertebral disc regeneration with an adipose mesenchymal stem cell-derived tissue-engineered construct in a rat nucleotomy model. Acta Biomater. 2019;87:118–29.
57. Shimomura K, Rothrauff BB, Hart DA, Hamamoto S, Kobayashi M, Yoshikawa H, et al. Enhanced repair of meniscal hoop structure injuries using an aligned electrospun nanofibrous scaffold combined with a mesenchymal stem cell-derived tissue engineered construct. Biomaterials. 2019;192:346–54.

Osteotomies Around the Knee for Older Active Patients

36

Ryohei Takeuchi, Eiji Kondo, Takenori Akiyama, Akihiko Yonekura, Ryuichi Nakamura, and Hiroshi Nakayama

36.1 Introduction

In recent years, due to advances in regenerative medicine, an increasing number of patients want to preserve their knee joints. Therefore, osteotomies around the knee (OAK), in which the knee joints are preserved without arthroplasty, are being reconsidered. Looking back at the history of OAK, Jackson et al. [1]. in 1961 and Coventry [2] in 1965 reported that removing a wedge of bone with the base on the lateral side of the proximal tibia to correct a varus deformity to a valgus alignment in patients with osteoarthritis of the knee (knee OA) was an effective procedure for pain relief. Coventry's method is the lateral closed wedge high tibial osteotomy (CWHTO), whereas that by Jackson et al. is called distal tibial tuberosity osteotomy, which is disadvantageous for bone fusion as the contraction force of the rectus femoris does not act on the osteotomy site. In 1972, Debeyre et al. and Hernigou et al. [3, 4]. in 1987 devised a method by which osteotomy was performed from the medial part to the lateral part in the proximal tibia to open the medial osteotomy site and implant autologous bone, which is the start of the open wedge high tibila osteotomy (OWHTO) technique. In the early 2000s, TomoFix (Depuy Synthes Co.) with a locking compression plate and screw system (LCP) were developed and the fixation strength at the osteotomy site was developed, leading to the era of OWHTO [5–7]. However, some studies show that advanced knee OA is commonly accompanied by patello-femoral joint (PF joint OA) and OWHTO is unsuitable in such cases. Additionally, progression of PF joint OA after OWHTO has reportedly occurred in some cases in recent years [8, 9]; thus, the surgical procedure is being studied.

This chapter briefly explains the indications, procedures, and postoperative care of six typical types of surgeries among those currently being performed in Japan.

R. Takeuchi (✉)
Saiwai Tsurumi Hospital, Tsurumi-ku, Yokohama, Japan

E. Kondo
Centre for Sports Medicine, Hokkaido University Hospital, Kita-ku, Sapporo, Japan
e-mail: eijik@med.hokudai.ac.jp

T. Akiyama
Akiyama Clinic, Kasuyagun, Fukuoka, Japan

A. Yonekura
Nagasaki University Graduate School of Biomedical Sciences, Nagasaki, Japan
e-mail: a-yone@umin.ac.jp

R. Nakamura
Harue Hospital, Sakai, Fukui, Japan
e-mail: ryu-nakamura@msj.biglobe.ne.jp

H. Nakayama
Hyogo College of Medicine, Nishinomiya, Hyogo, Japan

N. Nakamura et al. (eds.), *Advances in Knee Ligament and Knee Preservation Surgery*,
https://doi.org/10.1007/978-3-030-84748-7_36

36.2 Preoperative Planning

Radiographs of the entire lower extremity in the standing position are used. The postoperative alignment is planned so that the Mikulicz line (a load line passing from the femoral head center to the center of the ankle joint) crosses the tibial plateau at 60–63% width from the medial edge to the lateral edge except in the case of a medial closing wedge distal femoral varus osteotomy (DFO) for valgus knee.

36.2.1 Medial Open Wedge High Tibial Osteotomy (OWHTO)

Recently, medial open wedge high tibial osteotomy (OWHTO) with a LCP has attracted a great deal of attention [5]. This procedure includes biplanar osteotomy, an incomplete fracture technique for valgus correction, and fixation with a locking plate. The most notable advantages of the OWHTO include that correction of the limb alignment is not difficult, and that the risk of peroneal nerve palsy is minimized because fibular osteotomy is not needed.

36.2.1.1 Indication

Inclusion criteria are persistent pain due to medial compartmental knee OA or spontaneous osteonecrosis of the knee in the medial femoral condyle, and active compliance with rehabilitation program. Exclusion criteria included: lateral femorotibial angle (FTA) >185°; a loss of knee extension >15°; range of knee motion <130°; history of knee infection; patello-femoral (PF) joint OA.

36.2.1.2 Surgical Technique and Rehabilitation

A 7-cm medial longitudinal incision is made in the proximal tibia. Complete release of the superficial medial collateral ligament (sMCL) is performed. Then, 2 or 3 pairs of guidewires are inserted into the tibia so that each inserted guidewire precisely reached the proximal tibiofibular joint using the parallel guide. The oblique osteotomy should be started from the upper margin of the pes anserinus and end 5 mm from the lateral cortical margin, just above the proximal tibiofibular joint. Next, a biplanar osteotomy of the tibia, which consisted of an oblique HTO and a frontal plane osteotomy behind the tibial tubercle, is performed using an oscillating saw and chisel. The oblique osteotomy site is gradually opened for the correction angle and target width using a spreader. Then, the beta-tricalcium phosphate spacers are implanted into the posterior opening space. Before implantation of the plate, the sMCL and periosteum are sutured.

Finally, the tibia is fixed with a locking plate. After surgery, full weight bearing is permitted 1–2 weeks after surgery (Fig. 36.1).

36.2.2 Open Wedge Distal Tuberosity Tibial Osteotomy (OWDTO)

A risk of onset or progression of PF joint OA have been reported to result from patella infra following OWHTO. To minimize patellar height reduction, several studies [10, 11] have recommended performing open wedge distal tuberosity tibial osteotomy (OWDTO) instead of OWHTO.

36.2.2.1 Indication

OWDTO can be performed in all patients who are indicated for OWHTO, especially patients having patella infra and those requiring a large degree of valgus correction.

36.2.2.2 Surgical Technique and Rehabilitation

Two techniques can be recommended. The biplane technique, which is most popular, was developed by Gaasbeek et al. [10]. A descending cut is made distally in the coronal plane from 1 cm behind the tuberosity directly toward the anterior distal tibial cortex. Therefore, the tuberosity remains attached to the proximal part of the tibia after osteotomy. On the other hand, Akiyama et al. [12] developed a modified technique that includes distal tuberosity arc osteotomy. In this technique, a descending cut is started parallel to the tibial shaft 10–15 mm behind the tuberosity in the coronal plane, and then, an arc cut is made around the hinge position (center of rotation).

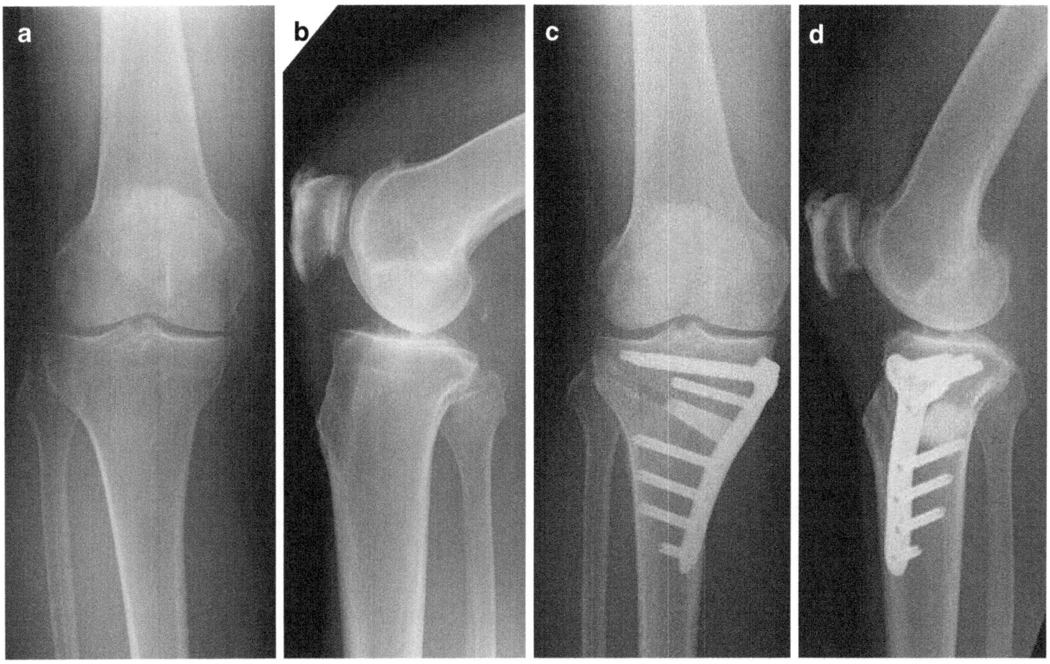

Fig. 36.1 Pre and postoperative radiographs. (**a**) Preoperative antero-posterior view. (**b**) Preoperative lateral view. (**c**) Postoperative antero-posterior view. (**d**) Preoperative lateral view

Thereafter, transverse osteotomy is initiated similar to the approach in OWHTO. After ensuring that all cuts are completed with the hinge area intact, the transverse osteotomy site is opened until the Micklicz line passes through 62.5% of the tibial plateau. Finally, fixation is accomplished in the same manner as in OWHTO. Then, bicortical screw fixation from the tuberosity to the posterior tibia is performed (Figs. 36.2 and 36.3).

The postoperative rehabilitation protocol is similar to that after OWHTO.

36.2.3 Hybrid Closed Wedge High Tibial Osteotomy (HCWHTO)

The hybrid closed wedge high tibial osteotomy (HCWHTO) surgical method has overcome the defects of the conventional CWHTO [13] (Fig. 36.4a, b). The benefits are as follows: due to positioning the hinge point on the proximal osteotomy line, a small amount of bone is removed and no leg length discrepancy is generated; walking with full weight bearing is allowed from an early stage as lateral bone cortices tightly attach to each other; and the medial soft tissue does not exfoliate, which is advantageous for bone fusion.

36.2.3.1 Indication

Indications include cases that need a relatively large degree of correction which is postoperative medial proximal tibial angle MPTA $\leq 95°$ and those that are complicated by PF joint OA or flexion contracture knee.

36.2.3.2 Surgical Technique and Rehabilitation

Arthroscopy is performed and intra-articular pathology is treated as necessary. In order to avoid peroneal nerve palsy, an appropriate length of the fibula depending on the correction is resected in the center. An approach to the proximal tibia is performed by lateral longitudinal skin incision. Transverse cut (Fig. 36.4c): K-wires (2 mm in diameter) are inserted obliquely from the lateral part (**a**) to the medial part (**c**). The target medial point (**c**) is approxi-

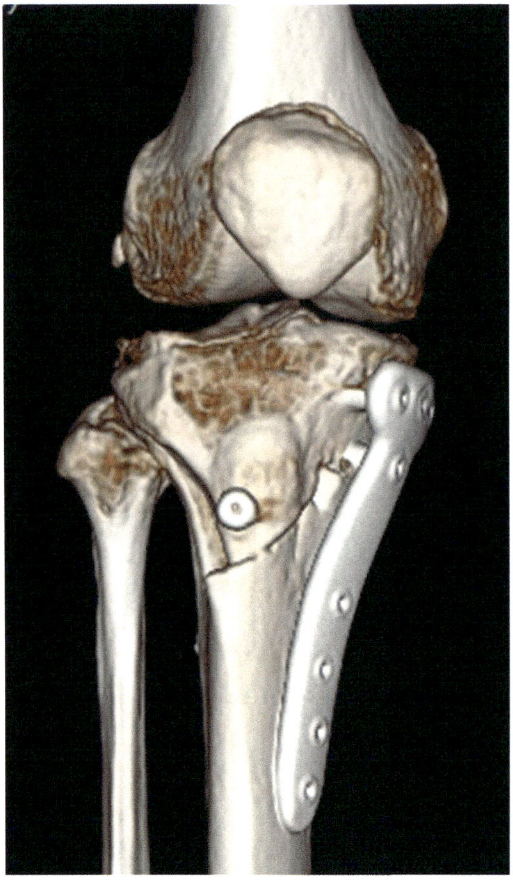

Fig. 36.2 Three-dimensional CT showing OWDTO

mately 1.5 cm distal to the tibial articular surface. A K-wire that indicates the hinge point (H) determined in the preoperative planning is percutaneously inserted perpendicular to the tibial axis. A goniometer is placed at the hinge point to determine a distal osteotomy line appropriate to the angle of the correction (α). A K-wire is inserted toward the hinge point along the distal osteotomy line (b-H). Ascending cut (Fig. 36.4d): The tibial tuberosity cut is performed from the position above the attachment site of the patella tendon to the transverse cut line (d-e-f) at an angle of approximately 110° (β). The bone in the area enclosed with the K-wires is cut using a bone saw to remove a triangular prism-shaped bone fragment. The portion inside the hinge point is completely cut off along the proximal osteotomy line (H-c). Transplantation of osteophytes (harvested from knee joint with arthroscopy) into the osteotomy site contribute to bone consolidation. Repositioning is performed so that the lateral cortical bones at osteotomy site attach to each other tightly by compressing the osteotomy site using a lag screw technique. Fixation is performed by using a long locking plate on the lateral side (Fig. 36.4e, f). In postoperative rehabilitation, patients start range of motion training and calf raise exercises in a standing position with full weight bearing on the following day, and gait training with weight bearing on day 2. Patients are allowed to walk with full weight bearing 2 weeks after surgery.

36.2.4 Tibial Condylar Valgus Osteotomy(TCVO)

Tibial condylar valgus osteotomy (TCVO) is an L-shaped intra-articular osteotomy between the medial and lateral tibial condyles [14]. By changing the shape of the tibial plateau with TCVO, not only the Micklicz line shifts laterally but also the congruency of the knee joint is improved.

36.2.4.1 Indication

1. Kellgren-Lawrence grade 3 or 4, advanced stage knee OA localized to the medial femorotibial (FT) joint.
2. Pagoda deformity [15]; significantly worn medial tibial plateau, resulting in inclination against the lateral tibial plateau.
3. Five degrees or more of joint line convergence angle (JLCA) (Fig. 36.5a).
4. Good preoperative range of motion of the knee joint; flexion as 90° or more and flexion contracture is 10° or less.
5. Active patient under 75 years old.

36.2.4.2 Surgical Technique and Rehabilitation

The differences from the open wedge high tibial osteotomy (OWHTO) are described in the following. In TCVO, an L-shaped osteotomy is

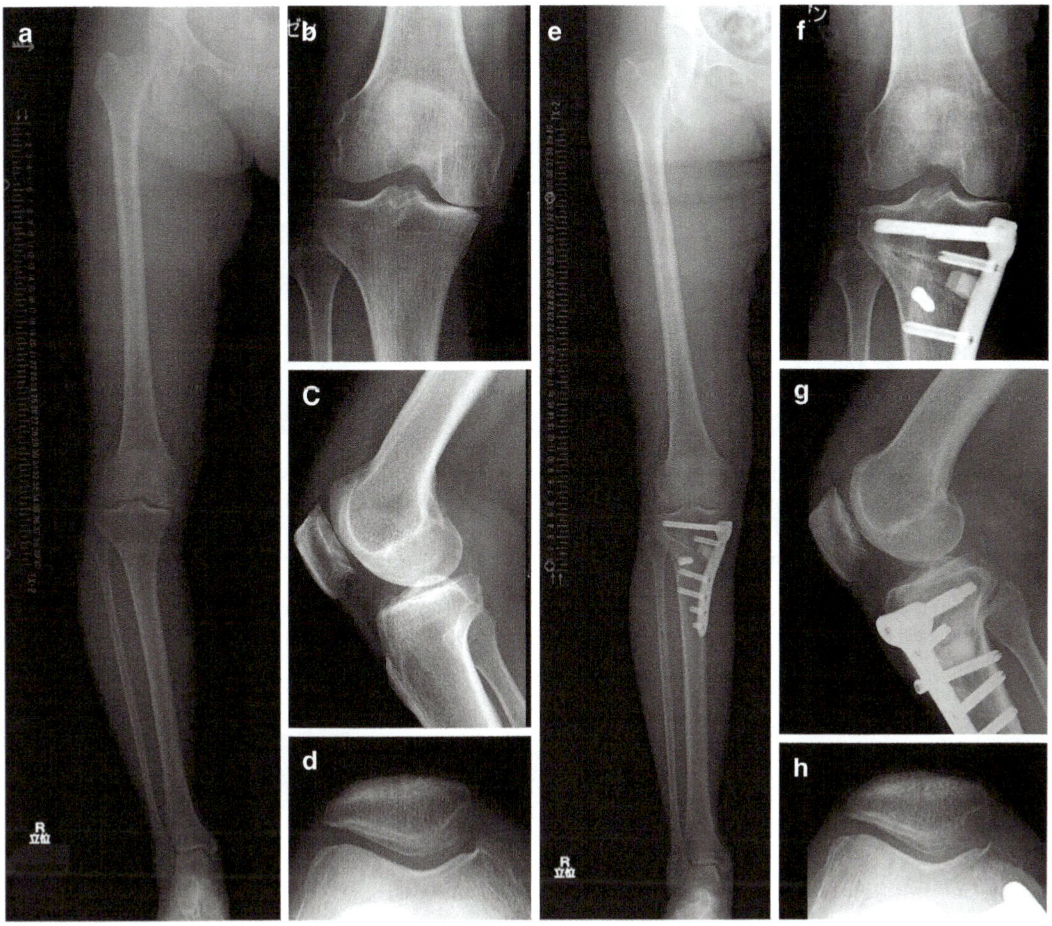

Fig. 36.3 Pre and postoperative (6 months) plane radiographs (right knee [**a–h**]). (**a**) Preoperative whole leg standing view. (**b**) Preoperative antero-posterior view. (**c**) Preoperative lateral view. (**d**) Preoperative merchant view. (**e**) Postoperative whole leg standing view. (**f**) Postoperative antero-posterior view. (**g**) Postoperative lateral view. (**h**) Postoperative merchant view

performed (Fig. 36.5a). The point A is determined at 4 cm distal from the medial tibial plateau. The point H is determined by the position of the lateral hinge point in the OWHTO. The intersection of the line AH and the medial edge of the tibial tuberosity is defined as the apex of the L-shaped osteotomy line (point B). First, the anterior bone cortex is cut from point B to the lateral intercondylar eminence (point C) with a chisel. The posterior bone cortex is cut under the lateral view of the X-ray image intensifier to prevent neurovascular injury. Next, the bone from point A to point B is cut using a bone saw. A 1.8 mm K-wire is inserted directly below the tibial plateau, the stopper devices are installed from both sides of K-wire to prevent from separating the medial and lateral tibial condyles. The osteotomy site is opened with a bone spreader forceps until the lateral FT joint become parallel in the frontal image (Fig. 36.5b). The osteotomy site is fixed by the proximal medial tibial locking plate, such as the TomoFix™, with plate bending as necessary (Fig. 36.5c). As for postoperative rehabilitation, ROM exercise and weight bearing as pain tolerated start day after surgery.

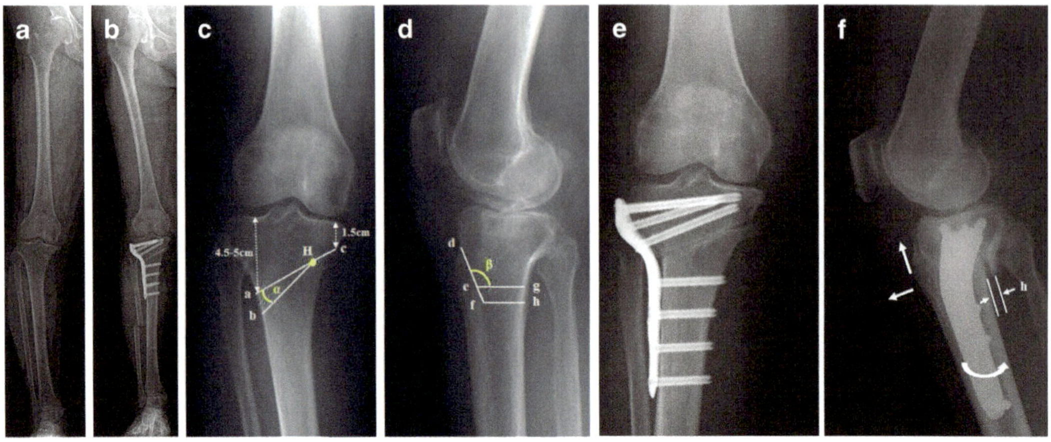

Fig. 36.4 Radiographs pre and postoperative HCWHTO. (**a**) Preoperative whole leg weight bearing radiograph. (**b**) Postoperative whole leg weight bearing radiograph. (**c**) Antero-posterior view. (*a–c*) is proximal osteotomy line and (*b–H*) is distal osteotomy line. (*H*) indicates the hinge point and (α) is correction angle. (**d**) Lateral view. (*d–f*) is the ascending cut line and (β) is the angle between ascending cut line and transverse cut line. (**e**) Antero-posterior view. Osteotomy site is fixed with TriS lateral plate (Olympus Termo Biomaterial Co.). (**f**) Lateral view. After osteotomy, the tibial tubercle moves anteriorly and proximally and the region distal to the osteotomy site rotates medially. (*h*) indicates the anterior translation distance of tibial tubercle

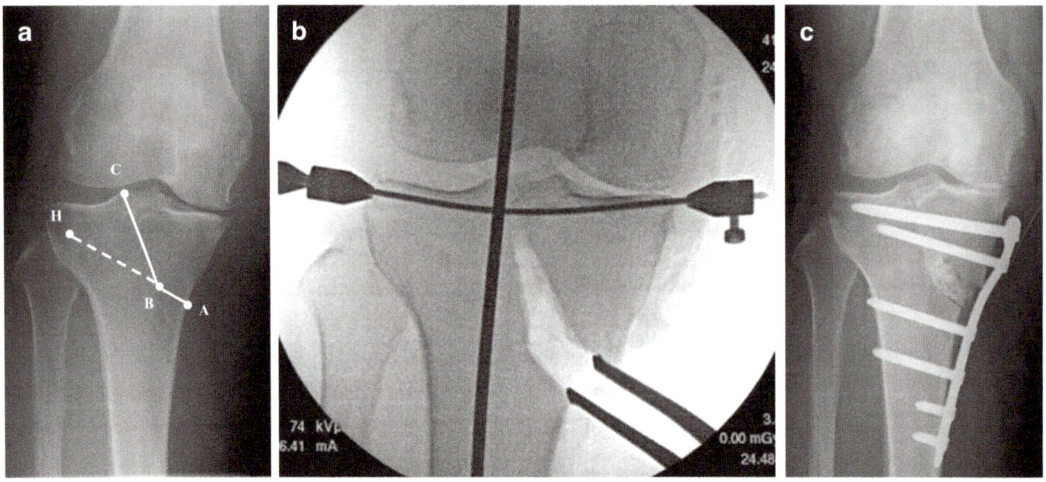

Fig. 36.5 Preoperative radiographs of TCVO. (**a**) Preplanning. (*A*) tibial cortex bone surface 4 cm distal from the medial tibial plateau, (*B*) apex of the L-shaped osteotomy line, (*C*) lateral intercondylar eminence, (*H*) lateral hinge point in the OWHTO, solid line; osteotomy line of TCVO, dotted line; virtual line when adding OWHTO. (**b**) Intraoperative radiograph. Vertical line; alignment rod after opening of osteotomy site. The L-shaped osteotomy line is opened by a bone spreader forceps until the lateral femoro-tibial joint become parallel. (**c**) Postoperative radiograph

36.2.5 Medial Closed Wedge Distal Femoral Osteotomy for Valgus Knees with Lateral Compartment Osteoarthritis (MCWDFO)

Medial closed wedge distal femoral osteotomy (MCWDFO) is the ideal correction for the majority of valgus knees because the deformity center is located mainly around the distal femur. Here, we introduce the MCWDFO with TomoFix Medial Distal Femoral Plate (TomoFix MDF; Synthes GmbH; Solothurn, Switzerland) [16].

36.2.5.1 Indication

Indications for MCWDFO are: (1) lateral compartmental knee OA with valgus alignment, (2) no medial compartmental knee OA, (3) center of deformity at the distal femur, (4) no flexion contracture, and (5) young active patient.

36.2.5.2 Surgical Technique and Rehabilitation.

The recommended mechanical femoro-tibial varus is 0–3° [16], which is equivalent to the postoperative weight bearing line between the center of the intercondylar eminence and the top of the medial eminence (Fig. 36.6). A biplanar osteotomy is recommended rather than a conventional single plane osteotomy (Fig. 36.7) [16]. After making a straight medial parapatellar incision, the vastus medialis muscle is elevated and the muscular branches from the descending genicular artery are coagulated or ligated. The first and second K-wires for the oblique osteotomies are inserted just above and below the hinge point, respectively (Fig. 36.7a). The distal oblique osteotomy is started from 4 cm proximal to the medial femoral epicondyle and the proximal oblique osteotomy is started at the planned distance from the distal oblique osteotomy. The osteotomy lines are then drawn by an electrosurgical knife (Fig. 36.7a). The ascending osteotomy and the two oblique osteotomies between the two wires are performed. After removing the bony wedge between the osteotomies, the osteotomized site is closed gradually and the plate is temporarily fixed with two K-wires (Fig. 36.7b). Two distal screws are inserted and a temporary lag screw is used to provide a compressive force to the osteotomized site. The lag screw is replaced with a bicortical locking screw and locking screws are inserted into the remaining holes. As for postoperative rehabilitation, range of motion exercise and par-

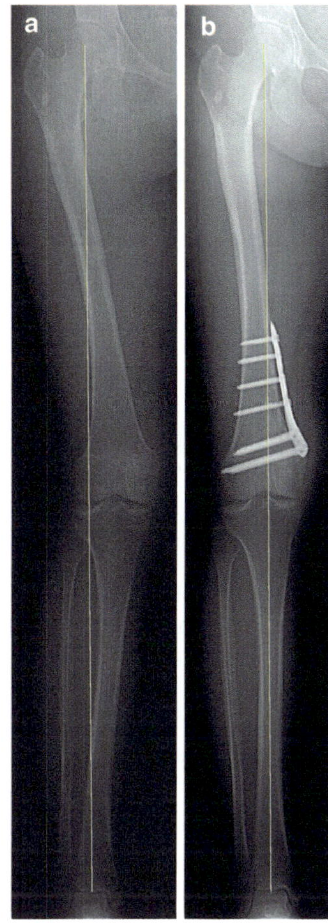

Fig. 36.6 Whole leg weight bearing radiograph of a 31-year-old female. (**a**) Preoperative radiograph. The X-ray of the right leg showed a valgus deformity caused by the femur. The hip-knee-ankle angle (HKA) was 7°, indicating a valgus angle. The weight bearing line (WBL) percentage (WBL to the medial edge /tibial plateau width × 100%) was 79%. (**b**) Three months after the osteotomy. The HKA and WBL were corrected to −1° and 45%, respectively

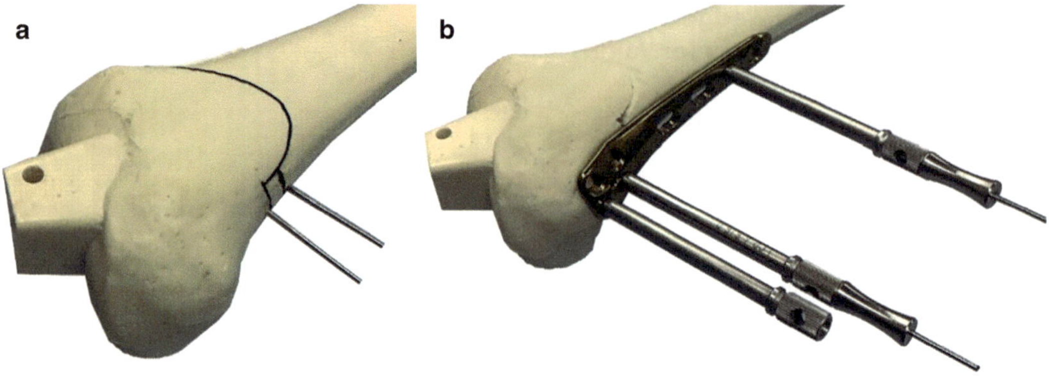

Fig. 36.7 Osteotomy and plate installation. (**a**) The positional relationship between the guide wires and the osteotomy lines. (**b**) Temporary fixation of the plate

tial weight bearing of about 1/3 of the body weight are started on the first postoperative day. Full weight bearing walk begins 6 weeks after surgery.

36.2.6 Double Level Osteotomy (DLO)

In surgical management of knees with severe varus deformity, correction by isolated OWHTO requires a large wedge opening resulting in non-physiologic lateral inclination of the joint line [17, 18]. Double level osteotomy (DLO) has been introduced with the intention of restoring physiologic joint alignment and orientation in correction of severe varus deformity [18, 19]. Recent DLO technique for varus deformity consists of biplane cut lateral closed wedge DFO (LCWDFO) and biplane cut OWHTO using the minimally invasive plate osteosynthesis (MIPO) technique with locking compression plate fixation [20, 21].

36.2.6.1 Indication

DLO is indicated for varus knee OA exhibiting combined deformities in both the femur and the tibia in active patient population who wishes to retain high activity level. If the predicted mechanical medial proximal tibial angle MPTA is 95° or greater for deformity correction with OWHTO alone, or mLDFA is greater than 90° with MPTA of smaller than 87°, DLO was considered as a surgical option [20, 21].

36.2.6.2 Surgical Technique and Rehabilitation.

DLO procedure is started with biplanar LCWDFO with a 4–5 cm longitudinal incision made at the lateral side of the femur just above the femoral epicondyle. Fixation of the osteotomy is accomplished utilizing MIPO technique. At this stage, the alignment following lateral (LCWDFO) is checked under fluoroscopy for confirmation of correspondence with preoperative planning. Afterwards, the osteotomy procedure is completed with subsequent biplanar OWHTO. Postoperatively, knee motion is started as tolerated on the following days. Weight bearing is not allowed for 3 weeks. Subsequently, partial weight bearing is started at 3 weeks with progression to full weight bearing at 4 weeks (Fig. 36.8).

36.3 Discussion

36.3.1 OWHTO

Recently, favorable short- and mid-term results after OWHTO have been reported. However, several disadvantages of OWHTO have been pointed out. First, it is technically difficult to obtain sufficient valgus correction without any

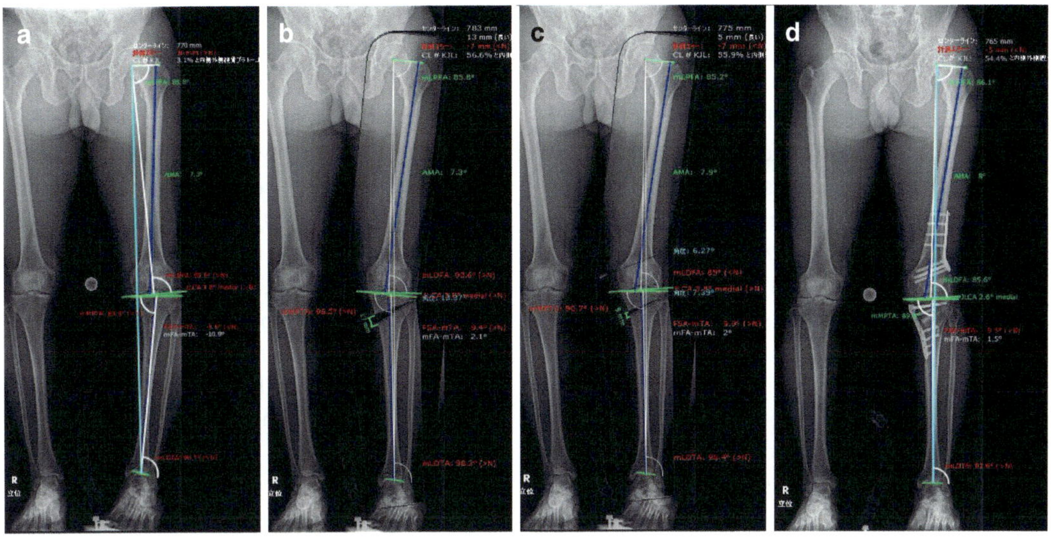

Fig. 36.8 Pre and postoperative radiographs of a 47-year-old male who underwent double level osteotomy (DLO) for the left knee exhibiting severe varus deformity. (**a**) Whole leg weight bearing radiograph. The parameters presented by digital planning software are mechanical lateral femoral distal angle (mLDFA): 90.6°, mechanical medial proximal tibial angle (MPTA): 83.6°, and mechanical tibiofemoral angle (mTFA): 10.9° varus. (**b**) Surgical simulation is started with isolated medial open wedge high tibial osteotomy (OWHTO).Osteotomy simulation aiming at mTFA of 2° valgus indicates that the required wedge size is 17 mm with resultant MPTA of 96.5° when the correction is made by OWHTO alone. Considering the amounts of wedge size (>15 mm) and MPTA value (>95°) in the simulation of isolated OWHTO, DLO is adopted as a surgical option for this case. (**c**) Surgical planning of DLO is conducted with mLDFA: 85°, MPTA: 90.7°, and mTFA: 2° valgus. Simulated DLO is composed of lateral closed wedge distal femoral osteotomy (wedge size: 5 mm, correction: 6.3°) and OWDTO [13] (wedge size: 9 mm, correction: 7.4°). (**d**) Whole leg weight bearing radiograph at 1 year after DLO. All parameters are corrected to the values corresponding to the preoperative simulation (mLDFA: 85.6°, MPTA: 89.8°, and mTFA: 1.5° valgus)

surgical problems in the knee with severe varus deformity. Second, the posterior tibial slope angle increases after surgery. Third, the patellar height decreases, and the length of the lower limb increases after surgery. Martin et al. [22] reported that the rate of major complications including the lateral hinge fracture were 25%. In addition, it is difficult for a surgeon to put the straight plate on the medial aspect of the tibia so that this plate is commonly placed on the anteromedial aspect of the tibia, resulting in potential risk of neurovascular injury and biomechanical inferiority compared with transverse insertion of the screws. Therefore, the authors have developed a newly designed locking plate (Fig. 36.1). It should be kept in mind that the leading symptom of a painful PF joint OA should be an exclusion criterion for OWHTO due to the risk of increased PF joint pressure and pain impairment postoperatively. Careful preoperative planning is needed to decrease the incidence of complications associated with OWHTO.

36.3.2 OWDTO

Patella height were evaluated in six published series [10, 11, 23–26]. Patella height did not significantly change in the OWDTO group but significantly decreased in the OWHTO group in all these papers. As for arthroscopic cartilage evaluation findings in two papers, the majority of OWDTO cases showed no progression of PF joint OA [23, 24]; however, 30–60% of OWHTO cases exhibited increased PF joint OA. According to the findings of three comparative studies [11, 23, 24] on the improvement of clinical outcomes, OWDTO can be considered superior to or equal to OWHTO. In contrast to OWHTO, OWDTO showed no change in patella height, which theoretically induced no increase in PF joint contact pressure, and moreover, the progression of PF joint OA could be consequently prevented.

Biomechanical weakness and a risk of slow bone healing at the descending osteotomy site were described in OWDTO [24].Thus, additional screw fixation in the tibial tuberosity is needed. To resolve these problems, the modified technique reported by Akiyama et al. [12] can be considered. Their technique provides anterior cortical support without a bone defect in the tibial crest having wide cancellous bone contact surfaces, and it may enhance stability and thus promote bone healing at the fixation site (Fig. 36.1).

36.3.3 HCWHTO

Postoperative outcomes, knee pain, and clinical scores after HCWHTO significantly improved as compared to before surgery [27–29]. Otsuki et al. [30] performed a comparative analysis of radiographic preoperative and postoperative images between OWHTO (24 knees) and HCWHTO (24 knees) and stated that the patella height and tibial tuberosity–trochlear groove (TT-TG) distance in the HCWHTO group significantly decreased, whereas the medial joint space of the PF joint significantly increased compared with the OWHTO group. Ishimatsu et al. [31] compared the outcomes 5 years after surgery between OWHTO (36 knees) and HCWHTO (21 knees) and reported that there was no difference in the clinical scores with respect to the postoperative clinical outcomes. In addition, the study showed that the patella height decreased in the HCWHTO group as compared to the OWHTO group, whereas the lateral joint space and congruence angle of the PF joint significantly improved. Improvement of congruity is brought by oblique osteotomy. In a recent study, more growth factors were shown to be present in osteophytes than in general cancellous bone [32, 33]. Studies have demonstrated that HCWHTO is effective and indicated in medial compartmental knee OA with relatively advanced varus deformity accompanied by PF joint OA [34].

36.3.4 TCVO

Visual analogue scale (VAS: from 73 to 13) and Western Ontario and McMaster Universities Osteoarthritis Index (WOMAC: from 52 to 14) improved in clinical results 5 years after surgery. Micklicz line improved from 1 to 60 and JLCA improved from 6 to 1 [14]. It is possible to treat the advanced stage knee OA with high activity and good range of motion at a young age by TCVO instead of unicompartmental knee arthroplasty (UKA) or TKA. However, a sufficient lateral shift of Micklicz line may not be obtained by TCVO alone in the case of excessive varus knees. The combined use of OWHTO and/or lateral closed wedge DFO is necessary in such cases.

36.3.5 MCWDFO

Although medial closed wedge high tibial osteotomy could be an option for valgus knees with lateral compartment OA, it creates medial tilting of the tibial plateau. The excessive tilting may cause medial subluxation of the femur on the tibia and the non-physiologic joint line obliquity may induce substantial shear stress [17]. Therefore, a MCWDFO is the appropriate way to correct valgus knees at the deformity center.

36.3.6 DLO

Sequential radiographic images of a representative case are shown in Fig. 36.8. Regarding the intended parameter values during the preoperative surgical simulation and planning, mechanical tibiofemoral angle (mTFA) was set to 1–2° (a slight valgus alignment), while mLDFA and MPTA were set to 85° and 90°, respectively [20, 21] (Fig. 36.8). The short-term clinical and radiological outcomes in our patient population have been reported in Reference [21]. In that study, 20 consecutive patients who underwent DLO were tracked for a minimum of 1 year. The follow-up results showed that all of the radiological parameters as well as results of patient-reported outcome measures were significantly improved after surgery.

36.4 Conclusions

A growing population of aged persons wishes to continue sports activities, and an increasing number of patients expect knee regeneration by utilizing regenerative medicine technologies. Furthermore, due to widespread internet availability, increasing number of patients search for therapies appropriate for themselves besides TKA. Before finalizing the TKA treatment, procedures such as osteotomy that aim at knee joint preservation, should also be considered.

Several types of OAK introduced here are becoming general procedures in Japan. However, for these surgical procedures to be completely established, long-term follow-ups with more cases are needed. Constructing a medical environment in which a surgical procedure can be chosen depending on the patient's age, activities, background, and needs is desirable.

References

1. Jackson JP, Waugh W. Tibial osteotomy for osteoarthritis of the knee. J Bone Joint Surg Br. 1961;43-B:746–51.
2. Coventry MB. Osteotomy of the upper portion of the tibia for degenerative arthritis of the knee. A preliminary report. J Bone Joint Surg Am. 1965;47:984–90.
3. Debeyre J, Artigou JM. Long term results of 260 tibial osteotomies for frontal deviations of the knee. Rev Chir Orthop Reparatrice Appar Mot. 1972;58(4):335–9.
4. Hernigou P, Medevielle D, Debeyre J, Goutallier D. Proximal tibial osteotomy for osteoarthritis with varus deformity. A ten to thirteen-year follow-up study. J Bone Joint Surg Am. 1987;69(3):332–54.
5. Staubli AE, De Simoni C, Babst R, Lobenhoffer P. TomoFix: a new LCP-concept for open wedge osteotomy of the medial proximal tibia—early results in 92 cases. Injury. 2003;34(Suppl 2):B55–62.
6. Floerkemeier S, Staubli AE, Schroeter S, Goldhahn S, Lobenhoffer P. Outcome after high tibial open-wedge osteotomy: a retrospective evaluation of 533 patients. Knee Surg Sports Traumatol Arthrosc. 2013;21(1):170–80.
7. Takeuchi R, Aratake M, Bito H, Saito I, Kumagai K, Sasaki Y, Akamatsu Y, Saito T. Medial opening wedge high tibial osteotomy with early full weight bearing. Arthroscopy. 2009;25:46–53.
8. Goshima K, Sawaguchi T, Shigemoto K, Iwai S, Nakanishi A, Ueoka K. Patellofemoral osteoarthritis progression and alignment changes after open-wedge high tibial osteotomy do not affect clinical outcomes at mid-term follow-up. Arthroscopy. 2017;33(10):1832–9.
9. Kim KI, Kim DK, Song SJ, Lee SH, Bae DK. Medial open-wedge high tibial osteotomy may adversely affect the patellofemoral joint. Arthroscopy. 2017;33(4):811–6.
10. Gaasbeek RDA, Sonneveld H, Van Heerwaarden RJ, Jacobs WCH, Wymenga AB. Distal tuberosity osteotomy in open wedge high tibial osteotomy can prevent patella infera: a new technique. Knee. 2004;11:457–61.
11. Longino PD, Birmingham TB, Schultz WJ, Moyer RF, Giffin JR. Combined tibial tubercle osteotomy with medial opening wedge high tibial osteotomy minimizes changes in patellar height: a prospective cohort study with historical controls. Am J Sports Med. 2013;41(12):2849–57.
12. Akiyama T, Osano K, Mizu-Uchi H, Nakamura N, Okazaki K, Nakayama H, Takeuchi R. Distal tibial tuberosity arc osteotomy in open-wedge proximal tibial osteotomy to prevent patella infra. Arthrosc Tech. 2019;8(6):e655–62.
13. Takeuchi R, Hiroyuki I, Yasuyuki M, Yohei S, Takashi K, So T. A novel closed wedge high tibial osteotomy procedure to treat osteoarthritis of the knee: hybrid technique and rehabilitation measures. Arthrosc Tech. 2014;3(4):e.431–7.
14. Chiba K, Yonekura A, Miyamoto T, Osaki M, Chiba G. Tibial condylar valgus osteotomy (TCVO) for osteoarthritis of the knee: 5-year clinical and radiological results. Arch Orthop Trauma Surg. 2017;137:303–10.
15. Lobenhoffer P, Van Heerwaarden R, Staubli AE, Jakob RP. Indications for high-tibial osteotomy, unicondylar knee arthroplasty, and total knee prosthesis. In: Osteotomies around the knee. Indications–planning-surgical techniques using plate fixators. Thieme Medical Publishers; 2008. p. 19–27.
16. van Heerwaarden R, Brinkman JM, Pronk Y. Correction of femoral valgus deformity. J Knee Surg. 2017;30(8):746–55.
17. Nakayama H, Schröter S, Yamamoto C, Iseki T, Kanto R, Kurosaka K, Kambara S, Yoshiya S, Higa M. Large correction in opening wedge high tibial osteotomy with resultant joint-line obliquity induces excessive shear stress on the articular cartilage. Knee Surg Sports Traumatol Arthrosc. 2018;26:1873–8.
18. Babis GC, An KN, Chao EY, Rand JA, Sim FH. Double level osteotomy of the knee: a method to retain joint-line obliquity. Clinical results. J Bone Joint Surg Am. 2002;84:1380–8.
19. Saragaglia D, Nemer C, Colle PE. Computer-assisted double level osteotomy for severe genu varum. Sports Med Arthrosc. 2008;16:91–6.
20. Schröter S, Nakayama H, Yoshiya S, Stöckle U, Ateschrang A, Gruhn J. Development of the double level osteotomy in severe varus osteoarthritis showed good outcome by preventing oblique joint line. Arch Orthop Trauma Surg. 2019;139:519–27.

21. Nakayama H, Iseki T, Kanto R, Kambara S, Kanto M, Yoshiya S, Schröter S. Physiologic knee joint alignment and orientation can be restored by the minimally invasive double level osteotomy for osteoarthritic knees with severe varus deformity. Knee Surg Sports Traumatol Arthrosc. 2020;28:742–50.
22. Martin R, Birmingham TB, Willits K, Litchfield R, Lebel ME, Giffin JR. Adverse event rates and classifications in medial opening wedge high tibial osteotomy. Am J Sports Med. 2014;42:1118–26.
23. Horikawa T, Kubota K, Hara S, Akasaki Y. Distal tuberosity osteotomy in open-wedge high tibial osteotomy does not exacerbate patellofemoral osteoarthritis on arthroscopic evaluation. Knee Surgery, Sport Traumatol Arthrosc. 2019;28(6):1750–6. Cited in PubMed; PMID31250057.
24. Ogawa H, Matsumoto K, Yoshioka H, Sengoku M, Akiyama H. Distal tibial tubercle osteotomy is superior to the proximal one for progression of patellofemoral osteoarthritis in medial opening wedge high tibial osteotomy. Knee Surg Sport Traumatol Arthrosc. 2020;28(10):3270–8. Cited in PubMed; PMID31875232.
25. Krause M, Drenck TC, Korthaus A, Preiss A, Frosch KH, Akoto R. Patella height is not altered by descending medial open-wedge high tibial osteotomy (HTO) compared to ascending HTO. Knee Surg Sport Traumatol Arthrosc. 2018;26:1859–66.
26. Park H, Kim HW, Kam JH, Lee DH. Open wedge high tibial osteotomy with distal tubercle osteotomy lessens change in patellar position. Biomed Res Int. 2017;2017:1–10.
27. Saito H, Kimio S, Yoichi S, Ya T, Shin Y, Takahiro S, Koji N, Kijima H, Naohisa M. Short-term results of hybrid closed-wedge high tibial osteotomy: a case series with a minimum 3-year follow-up. Knee Surg Relat Res. 2018;30(4):293–302.
28. Takahara Y, Takayuki F, Hirotaka N, Satoru I, Makoto N, Yoichiro U, Hisayoshi K, Yoshitaka T, Yuichi I, Nobuaki O. Time to bone union after hybrid closed-wedge high tibial osteotomy. Acta Med Okayama. 2019;73(6):511–6.
29. Takagawa S, Kobayashi N, Yukizawa Y, Oishi T, Tsuji M, Inaba Y. Preoperative soft tissue laxity around knee was associated with less accurate alignment correction after hybrid closed-wedge high tibial osteotomy. Knee Surg Sports Traumatol Arthrosc. 2020;28(9):3022–30. https://doi.org/10.1007/s00167-019-05762-2.
30. Otsuki S, Tomohiko M, Yoshinori O, Kosuke N, Nobuhiro O, Hitoshi W, Masashi N. Hybrid high tibial osteotomy is superior to medial opening high tibial osteotomy for the treatment of varus knee with patellofemoral osteoarthritis. Knee Surg Sports Traumatol Arthrosc. 2019;27(4):1332–8.
31. Ishimatsu T, Ryohei T, Hiroyuki I, Yuichiro Y, Akira M, Katsunari O, Woon HJ. Hybrid closed wedge high tibial osteotomy improves patellofemoral joint congruity compared with open wedge high tibial osteotomy. Knee Surg Sports Traumatol Arthrosc. 2019;27(4):1299–309.
32. Akiyama T, Okazaki K, Mawatari T, Ikemura S, Nakamura S. Autologous osteophyte grafting for open-wedge high tibial osteotomy. Arthrosc Tech. 2016;5(5):e989–95.
33. Ishihara K, Okazaki K, Akiyama T, Akasaki Y, Nakashima Y. Characterisation of osteophytes as an autologous bone graft source: an experimental study in vivo and in vitro. Bone Joint Res. 2017;6(2):73–81.
34. Kuwashima U, Takeuchi R, Ishikawa H, Shioda M, Nakashima Y, Schröter S. Comparison of torsional changes in the tibia following a lateral closed or medial open wedge high tibial osteotomy. Knee. 2019;26(2):374–81.

37 Current Introduction of the Biological Agent Derived from Adipose Tissue to the Treatment of Knee Osteoarthritis

Wataru Ando, Isabel Wolfe, Kazunori Shimomura, Stephen Lyman, Naomasa Yokota, and Norimasa Nakamura

37.1 Osteoarthritis

Osteoarthritis (OA) is the most common joint disorder worldwide [1–3], and the estimated global prevalence for knee OA is 3.8% [3, 4]. Pain and associated functional disabilities are the targets of most standard treatments, whose indications are based on disease severity. Patients with mild OA, who have a minimal disability and knee x-rays with Kellgren Lawrence (KL) grade of I (on a scale of I to IV), are treated conservatively with physical therapy to strengthen the muscles around the knee, as well as oral analgesics such as acetaminophen, aspirin, and non-steroidal anti-inflammatory drugs (NSAIDs) [3, 5–8].

By the time patients have progressed to severe OA with debilitating pain and x-rays of KL grade III or IV, surgical treatments is considered. Joint preserving surgeries including high tibial osteotomy (HTO) is the preferred treatment strategy for joint chondropathologies in young and active patients to improve pain, restore activity, and delay arthroplasty. More progressive joint degeneration was given due to the poor regenerative

W. Ando
Department of Orthopaedic Medical Engineering, Osaka University Graduate School of Medicine, Suita, Osaka, Japan
e-mail: w-ando@umin.ac.jp

I. Wolfe
Hospital for Special Surgery, New York, NY, USA
e-mail: wolfei@HSS.EDU

K. Shimomura
Department of Orthopaedic Surgery, Osaka University Graduate School of Medicine, Suita, Osaka, Japan
e-mail: kazunori-shimomura@umin.net

S. Lyman
Hospital for Special Surgery, New York, NY, USA

Department of Orthopaedic Surgery, Graduate School of Medical Sciences, Kyushu University, Fukuoka, Fukuoka, Japan

N. Yokota
Tokyo Knee Osteoarthritis Clinic Ginza, Tokyo, Japan

Tokyo Knee Osteoarthritis Clinic Shinjuku, Tokyo, Japan

N. Nakamura (✉)
Department of Orthopaedic Surgery, Osaka University Graduate School of Medicine, Suita, Osaka, Japan

Institute for Medical Science in Sports, Osaka Health Science University, Osaka, Osaka, Japan

Global Center for Medical Engineering and Informatics, Osaka University, Suita, Osaka, Japan

Department of Rehabilitation Science, Osaka Health Science University, Osaka, Osaka, Japan
e-mail: norimasa.nakamura@ohsu.ac.jp

N. Nakamura et al. (eds.), *Advances in Knee Ligament and Knee Preservation Surgery*,
https://doi.org/10.1007/978-3-030-84748-7_37

properties of cartilage, and arthroplasty is performed to provide pain relief; conversely, the range of motion in knee joints have not completely recovered, and some patients were not satisfied with the clinical outcome after TKA [9]. Baker et al. evaluated that almost 20% of patients were not satisfied with their TKA [10]. Meta-analysis showed that the 25 years survival of TKAs was 82.3% [11]. As high activity can increase the likelihood of early revision [12–15], younger active patients are often counseled to delay joint replacement for as long as can be tolerated. Furthermore, many patients may feel reluctant to undergo surgery. Most importantly, existing approaches leave a treatment gap for patients with moderate OA stage (KL grade II). One recently available therapy for moderate OA is the intra-articular injection of hyaluronic acid, which is intended to improve joint mechanical function by increasing the volume and viscosity of synovial fluid [16–18]. Unfortunately, its effectiveness has been questioned by the American Academy of Orthopaedic Surgeons based on available clinical evidence [17, 19]. The intra-articular administration of a biological agent derived from adipose tissue is one of many novel therapies currently in development to address the limitations of standard treatments and address not only the symptoms but the pathogenesis of OA.

A degenerative disorder, OA is most commonly associated with aging [1], but etiologies are varied. Mechanical factors such as deformity or previous injury as well as genetic and/or environmental factors can also increase risk [1, 2]. Pathogenesis is complex, including alterations in the subchondral bone, which may precede other changes and make overlying articular cartilage more susceptible to damage [20]; anabolic changes, which produce characteristic osteophytes at the joint periphery; as well as inflammatory and degradative processes, which change the phenotype of normally quiescent chondrocytes, leading to a breakdown in articular cartilage matrix [21]. Ultimately, hyaline cartilage on the articular surfaces completely erodes, leaving only bony sclerosis at the end-stage. These observations suggest that novel therapies, which support cartilage formation and/or limit inflammation and matrix degradation, may be effective, particularly if administered, while OA is still moderate. Intra-articular injection of the agent adipose-derived cells is one of many biologic approaches, which are being developed based on such therapeutic goals.

Not enough is yet known about the therapeutic effects or mechanisms of action of the various biologic approaches in clinical use and/or under clinical investigation to declare how they may compare with the intra-articular injection of adipose-derived cells [19, 22]. Few high-quality, randomized studies have been performed [19]. It was reported that the placebo effect has been relatively high with intra-articular injection of any agent [22]. New therapeutic approaches currently under investigation include control of inflammation by specific blockade of cytokines, such as interleukin-beta 1 and tumor necrosis factor-alpha using novel pharmacologic agents or gene therapy [19], or supporting cartilage tissue regeneration using purified growth factors (such as bone morphogenetic protein-7, and fibroblast growth factor-18) [19]. Blood and cell-based strategies are believed both to quell inflammation and bolster regeneration and have the added advantage of being available from an autologous source. Intra-articular injection of platelet-rich plasma (PRP) has become popular for a variety of musculoskeletal conditions [23, 24], and studies are under way to test efficacy in knee OA alone and/or in combination with adipose-derived cells [24–31]. Studies investigating the intra-articular injection of a variety of stem cells derived not only from adipose tissue but from a variety of sources (umbilical cord blood and bone marrow) are also under way [3, 32–36]. More researches are required before any of these approaches can be considered superior, or even efficacious, but the promise of biologic approaches remains attractive for the many patients for whom TKA or standard medical treatments are not desired.

37.2 Biological Agents Derived from Adipose Tissue

Studies by Zuk et al. [37, 38] in 2001 and 2002, reporting the isolation of multipotent cells from adipose tissue, have become the basis of their therapeutic application in orthopedic and other conditions. Compared to other types of stem cells (e.g., derived from bone marrow or umbilical cord), adipose-derived cells can be harvested from an autologous source with little morbidity. Adipose-derived cells are typically harvested from the patient's own subcutaneous tissues (often the belly or buttock) via liposuction, and while post-harvest donor site complaints are not uncommon, they are rarely serious. Approximately 500,000–2,000,000 nucleated cells are typically isolated from one gram (g) of adipose tissue or one milliliter (mL) of lipoaspirate, and 1–10% of them may be multipotent stem cells [39–42]. Cell yield and quality, as well as the proportion of different cells isolated, vary based on the method of isolation, including whether isolation was mechanical or enzymatic, as well as donor age and donor site [39, 43, 44], but the impact of these variations on therapeutic effectiveness are not known. Adipose-derived cells have been investigated in experimental models and clinical studies for treating a variety of conditions including those requiring soft or hard (bone) tissue reconstruction, immunomodulation (e.g., Crohn's disease, rheumatoid arthritis), and reversal of ischemia (e.g., myocardial infarction, stroke) [45].

Biological agents derived from adipose tissue have been applied to the knee as a component of enzymatically treated "stromal vascular fraction" (SVF) [25–31, 46–56] (Tables 37.1, 37.2, and 37.3), mechanically isolated "fractionated adipose tissue" (FAT, such as the Lipogems product) [56–65] (Tables 37.4 and 37.5), or culture-expanded adipose-derived mesenchymal stem cells (ADMSCs) [49, 66–75] (Tables 37.6 and 37.7). The SVF and FAT are so-called based on the observation that these heterogeneous distillates are enriched in pericytes or pericyte-like endothelial cell precursors [40, 44, 45, 76–78]. In contrast to FAT or SVF, ADMSCs are purified stem cells with defined cell surface markers and multi-potency [40, 45, 76]. The use of SVF, FAT, and ADMSCs in knee OA is summarized below.

37.3 Stromal Vascular Fraction (SVF) Cells

In his original studies, Zuk [37, 38] isolated plastic-adherent adipose-derived mesenchymal stem cells (ADMSCs) from the stromal vascular fraction (SVF), a collagenase digestate of adipose tissue. On this basis, in many studies, SVF is administered directly to the knee, without culture and isolation of ADSCMs. Unlike purified ADMSCs, the SVF is a heterogeneous distillate, including not only mesenchymal stem cells, but pericytes, vascular adventitial cells, fibroblasts, pre-adipocytes, monocytes, macrophages, red blood cells, fibrous tissue, and extracellular matrix [37, 38, 40, 44, 45, 76–78]. In the clinical setting, SVF cells are not usually characterized prior to administration. As such the mechanism by which SVF may mitigate knee OA progression may not be the same as adipose-derived or other mesenchymal stem cells per se. However, Kim et al. [53] have reported that mesenchymal stem cells can be cultured from aliquots of the SVF administered to patients. In addition, Koh et al. have reported improved cartilage tissue quality based on second-look arthroscopy, histology, and/or MRI, suggesting that SVFs can support tissue regeneration in the setting of knee OA [30, 52, 55]. In some of their studies, this group used cells isolated from the knee fat pad [30, 31], but in most studies using SVF (including later studies from this group), cells are harvested from the patient's subcutaneous tissues. The salient advantage of SVF over ADMSCs is that cells do not need to be cultured but can be administered immediately upon harvest.

Most studies investigating the potential benefit of intra-articular SVF injection for knee OA report modest improvements in PROs over 1–3 years (Table 37.1). Primary outcomes measured included pain and physical function, most frequently assessed using VAS and WOMAC scores. Most papers reported significant improvement in these domains in the majority of patients after an average of 12 months, with the longest follow-up being 5 years. Adverse outcomes were minor. The most common events reported were pain around the harvest site (7.7–

Table 37.1 Clinical studies investigating the intra-articular administration of adipose-derived stromal vascular fraction (SVF) for knee osteoarthritis[a]

Author (year)	*N*	Design	Associated clinical trial(s)	Kellgren Lawrence Grade (KL) of OA or other diagnosis	Treatment other than SVF	Follow-up
Berman (2019) [46]	2586	Prospective cohort	NCT10953523	KL 1–4	None	5 years
Hong (2019) [47]	16	Phase III	ChiCTR1800015125	KL 2–3	HA[b]	1 year
Michalek (2019) [48]	29	Prospective case-control		KL 2–4 in 1–4 large weight bearing joints (including hip and knee) and 0–8 other joints	None	36 months
Yokota (2019) [49]	80	Retrospective cohort		KL 2–4	None	6 months
Bansal (2017) [25]	10	Prospective cohort	NCT03089762	1 or 2 on Brandt Grading Scale	PRP[c] and collagenase digestion with flow cytometry	2 years
Nguyen (2017) [26]	30	Prospective cohort		KL 2–3	PRP and microfracture	18 months
Pintat (2017) [27]	19	Case series		Patellofemoral OA	PRP	12 months
Yokota (2017) [50]	13	Case series		KL 3–4	None	6 months
Fodor (2016) [51]	6	Case series	NCT02357485	KL 1–3	None	1 year
Koh (2016) [52]	80	Prospective cohort		KL $\leq$ 2 ICRS[b] grade 3–4 cartilage lesions	Thrombin, fibrinogen, and microfracture	26–30 months
Kim (2015) [53]	54	Retrospective cohort			Fibrin glue	24–34 months
Kim (2015) [53]	40	Retrospective cohort			Fibrin glue, PRP	Average 28.6 months
Koh (2015) [54]	30	Case series		KL 2–3		2 years
Bui (2014) [29]	21	Case series		Grade 2–3 (scale unknown)	PRP	6 months
Koh (2014) [55]	44	Prospective cohort		KL $\leq$ 3	HTO[d]	14–24 months

(continued)

Table 37.1 (continued)

Koh (2013)[e] [30]	18	Case series	KL 3 in multiple compartments	PRP	24–26 months
Koh (2012)[e] [31]	25	Retrospective cohort	KL < 4	PRP	12–18 months

[a]Studies with fewer than 5 patients or follow-up shorter than 6 months were not included
[b]*HA* hyaluronic acid
[c]*PRP* platelet-rich plasma
[d]*HTO* high tibial osteotomy
[e]Cells were harvested from the infrapatellar fat pad rather than subcutaneous tissue

Table 37.2 Clinical trials investigating the intra-articular administration of adipose-derived stromal vascular fraction (SVF) for knee osteoarthritis (OA)[a]

Number	Sponsor	Status	Study phase	Condition or treatment other than knee OA	Description of arms
NCT02726945	The GID Group	Completed	N/A	None	SVF injection (2 doses) vs. placebo
NCT04238143	Healeon Medical Inc. with collaborators	Active, not recruiting	N/A	PRP[b]	tSVF[c] in PRP vs. cSVF[d] in PRP vs. cSVF in saline
NCT03940950	Mayo Clinic	Recruiting	Phase I	None	SVF vs. placebo
NCT02967874	Russian Academy of Medical Sciences	Completed	Phase I/II	None	SVF vs. HA injection
NCT03818737	Emory University with the Marcus Foundation	Recruiting	Phase III	None	BMAC[e] vs. SVF vs. umbilical cord tissue (compared to corticosteroid)
NCT03164083	SCARM Institute	Withdrawn	Phase II	BMAC	SVF and BMAC vs. placebo
NCT02846675	Shanghai East Hospital	Completed	Phase I/II	None	SVF vs. placebo on contralateral knee
NCT03090672	Robert W. Alexander, MD with Regeneris Medical and Global Alliance for Regen Med	Recruiting	N/A	PRP	tSVF or cSVF in PRP vs. tSVF or cSVF in saline
NCT02142842	University of Science, Ho Chi Minh City	Completed	Phase I/II	PRP	SVF in PRP vs. control (not specified)

[a]Single arm trials were not included
[b]*PRP* platelet-rich plasma
[c]*tSVF* tissue stromal vascular fraction (fractionated adipose tissue product [FAT] isolated by mechanical dissociation)
[d]*cSVF* cellular stromal vascular fraction (standard SVF that is enzymatically isolated)
[e]*BMAC* bone marrow aspirate concentrate

34% of patients) and knee joint pain/effusion (3.5–37.5% of patients). These resolved over the follow-up period. Major adverse outcomes such as marked pain and swelling were rare. Increasingly, SVFs are being delivered in conjunction with other biologics, such as HA and PRP [25–31, 47] (Table 37.1) including in currently ongoing clinical trials (Table 37.2).

Several devices that facilitate the isolation of SVF are available commercially (Table 37.3). Personalized medicine's PSC-01 is a form of autologous cellular therapy resulting from the enzymatic processing of lipoaspirate, while GIDBio's GIDSVF2 is a sterile tissue-processing container designed for SVF enzymatic isolation.

Table 37.3 Commercially available products related to the administration of adipose-derived mesenchymal stem cells (ADMSCs), stromal vascular fraction (SVF), or fractionated adipose tissue (FAT)

Product	Implant type	Manufacturer/sponsor	Associated clinical trials or published study
Elixcyte	ADMSCs	UnicoCell Biomed Co., Ltd.	NCT02784964
Jointstem	ADMSCs	R-Bio Co., Ltd. Nature Cell	NCT03000712 NCT01300598 NCT03990805 NCT02674399 NCT03509025 NCT02658344
ReJoin®	ADMSCs	Cellular Biomedicine Group, Inc.	NCT02641860
GXCPC1	Allogenic ADMSCs	Gwo Xi Stem Cell Applied Technology Co., Ltd.	NCT03943576
AdipoCell™	SVF	U.S. Stem Cell, Inc.	NCT03089762
ADSC Extraction Kit	SVF	GeneWorld Co., Ltd.	Bui (2014)
Celution Centrifuge IV	SVF	Cytori Therapeutics, Inc.	Yokota (2019) [49]
GIDSVF1	SVF	The GID Group, Inc.	NCT02276833
GIDSVF2	SVF	The GID Group, Inc.	NCT02726945
PSC-01	SVF	Personalized Stem Cells, Inc.	NCT04043819
AdiPrep®	FAT	Terumo BCT, Inc.	NCT03467919
Lipogems®	FAT	Lipogems International SpA	NCT04230902 NCT03242707 NCT03922490 NCT03788265 NCT03714659 NCT03771989 NCT03527693 NCT03379168 NCT03117608 NCT02697682

37.4 Fractionated Adipose Tissue (FAT)

In some studies, mechanical, rather than enzymatic, processes have been used to isolate adipose-derived cells. Lipogems® (Lipogems International, Milano, Italy) and AdiPrep® (Terumo BCT, Japan) are technologies, which remove blood and oil while isolating a full complement of cells from fat tissue, producing "fractionated adipose tissue" (FAT) for application to the knee or other joints to treat OA [41, 79, 80] or other applications (Table 37.4). The Coleman technique, developed for fat grafting in plastic surgery, has been used in the knee in one study [60]. The FAT is sometimes also referred to as "tissue stromal vascular fraction" (tSVF) to differentiate it from the standard SVF, or cellular fraction (cSVF). Although enzymatic digestion is the original method by which the SVF cells (and from them, ADMSCs) were isolated, the US Federal Drug Administration (FDA) statement that isolation of autologous cells should "not alter the original relevant characteristics of the tissue relating to the tissue's utility for reconstruction, repair, or replacement" raise questions about the use of enzymatic digestion for isolating stem cells [41, 78, 81]. Although FAT responds to these concerns, it is clear that the compliment and quality of cells isolated via mechanical fractionation are different from that derived by enzymatic digestion [41, 78]. The question of how FAT compares to SVF or ADMSCs for knee OA or any other therapeutic application remains open.

Most papers investigating the use of FAT for knee OA reported significant improvement in the majority of patients after an average of 12 months, with the longest follow-up being 36 months [56–65] (Table 37.4). Primary outcomes measured included pain and physical function, most frequently assessed using KOOS and VAS scores. Adverse outcomes were minor and infrequent. The most common events reported were related to the fat harvest site (3.33–6.67% of patients). Major adverse outcomes were not observed. Multiple clinical studies using FAT are ongoing (Table 37.5).

Table 37.4 Clinical studies investigating the intra-articular administration of fractionated adipose tissue (FAT) for knee osteoarthritis (OA)[a]

Author (year)	*N*	Design	Associated clinical trial(s)	Kellgren Lawrence Grade (KL) of OA or other diagnosis	Treatment other than FAT	Follow-up
Barford (2019) [56]	20	Prospective cohort	NCT02697682	KL 1–4	None	12 months
Mautner (2019) [57]	76	Retrospective cohort		KL 1–4	± BMAC[b]	0.6–1.6 years
Panni (2019) [58]	52	Case series		KL 0–2	None	6–24 months
Cattaneo (2018) [59]	38	Case series	NCT03527693	KL ≤ 3; ICRS[c] grade ≥2 chondral lesion	None	12 months
Roato (2018) [60]	20	Prospective cohort		KL 1–3	None	18 months
Russo (2018) [61]	30	3-year follow-up of retrospective cohort		KL ≤ 3; ICRS grade ≥2 chondral lesion	None	36 months
Hudetz (2017) [62]	17	Prospective cohort	ISRCTN1333702	KL 2–4	None	12 months
Russo (2017) [63]	30	Retrospective cohort		KL ≤ 3; ICRS grade ≥2 chondral lesion	None	12 months
Oliver (2015) [64]	70	Prospective cohort		KL 2–4	± BMAC	180 days
Centeno (2014) [65]	681	Retrospective cohort		KL 1–4	± BMAC	Average 10.4 months

[a]Studies with fewer than 5 patients or follow-up shorter than 6 months were not included
[b]*BMAC* bone marrow aspirate concentrate
[c]*ICRS* International Cartilage Research Society

Table 37.5 Clinical trials investigating the intra-articular administration of fractionated adipose tissue (FAT) for knee osteoarthritis (OA)[a]

Number	Sponsor	Status	Study phase	Condition or treatment other than knee OA	Description of arms
NCT03467919	Stanford University	Recruiting	Phase III	None	Adiprep vs. corticosteroid
NCT04230902	American University of Beirut Medical Center	Recruiting	Phase III	None	Lipogems vs. corticosteroid
NCT03242707	University of Southern California	Recruiting	N/A	None	Lipogems vs. HA[b]
NCT03922490	Hospital for Special Surgery, NY	Not yet recruiting	Phase IV	Arthroscopic debridement	Arthroscopic debridement and Lipogems vs. arthroscopy alone
NCT03771989	Hvidovre University Hospital	Recruiting	N/A	None	Lipogems vs. placebo
NCT03379168	Dustin L. Richter, MD, University of New Mexico	Recruiting	N/A	None	Lipogems vs. corticosteroid vs. placebo
NCT03117608	Lipogems International SpA	Recruiting	Phase IV	None	Lipogems vs. PRP[c]

[a]Single arm trials were not included
[b]*HA* hyaluronic acid
[c]*PRP* platelet-rich plasma

37.5 Cultured Adipose-Derived Mesenchymal Stem Cells (ADMSCs)

Whereas the use of SVF and FAT respond to clinical and regulatory concerns, only cultured, plastic-adherent adipose-derived cells can be considered true mesenchymal stem cells. The International Federation for Adipose Therapeutics and Science (IFATS) and the International Society for Cellular Therapy (ISCT) proposed three minimal criteria for the definition of adipose-derived mesenchymal stem cells (ADMSCs): (1) plastic adherence, (2) characteristic expression of cell surface markers (CD73, CD90, and CD105-positive and CD11b, CD14, CD19, CD45, and HLA-DR-negative), and (3) differentiation potential into preadipocytes, chondrocytes, and osteoblasts [44, 82, 83]. These criteria cannot be fulfilled by SVF or FAT cells, even though the SVF cells are the source of ADMSCs. Furthermore, most pre-clinical studies showing chondrogenic potential or characterizing the paracrine and anti-inflammatory activities have been done on plastic-adherent ADMSCs and not SVFs [74, 78, 84]. Labeled ADMSCs injected intra-articularly in animal models have been shown to remain in the joint support of this mode of delivery clinically [33, 85, 86]. In *in vitro* and animal models, ADMSCs have been reported to have tumorigenic potential but have not been reported in humans [87, 88].

Other studies have reported on the potential benefits of intra-articular injection of ADMSCs for knee OA in the last several years (Table 37.6). The number of cells administered has ranged from 1.0×10^7 to 1.0×10^8 cells/mL [49, 66–75]. Primary outcomes measured included pain and physical function, most frequently assessed using WOMAC, KOOS, and VAS scores. Most papers reported significantly improved knee function and decreased pain as measured by these scales after 6–12 months, with the longest follow-up being 2 years. Adverse outcomes were

Table 37.6 Clinical studies investigating the intra-articular administration of adipose-derived mesenchymal stem cells (ADMSCs) for knee osteoarthritis (OA)[a]

Author (year)	*N*	Design	Associated clinical trial(s)	Kellgren Lawrence Grade (KL) of OA or other diagnosis	Treatment other than ADMSCs	Follow-up
Freitag (2019) [66]	30	Phase II	ACTRN12614000814673	KL 2–3	None	1 year
Jiang (2019) [67]	12	Case series			None	6 months
Lee (2019) [68]	24	Phase IIb		KL 2–4	None	6 months
Lu (2019) [69]	53	Phase II	NCT02162693	KL 1–3	None	1 year
Yokota (2019)[b] [49]	80	Retrospective cohort		KL 2–4	None	6 months
Zhao (2019) [70]	18	Phase I/IIa	NCT02641860	KL 2–3	None	48 weeks
Kyriakidis (2018) [71]	20	Case series		Grade 4 chondral lesion	None	2 years
Song (2018) [72]	18	Phase I/II	NCT01809769	KL ≥ 2	None	96 weeks
Jo (2017) [73]	18	Follow-up of a Phase I/II study	NCT01300598	KL ≥ 2	None	2 years
Pers (2016) [74]	18	Phase I	NCT0158585	KL 3–4	None	6 months
Jo (2014) [75]	18	Phase I/II		KL ≥ 2	None	6 months

[a]Studies with fewer than 5 patients or follow-up shorter than 6 months were not included

[b]Yokota et al. [49] is included in both Tables 37.1 and 37.6, as ADMSC and SVF treatments were compared in the study

Table 37.7 Clinical trials investigating the intra-articular administration of adipose-derived mesenchymal stem cells (ADMSCs) for knee osteoarthritis (OA)[a]

Number	Sponsor	Status	Study phase	Condition or treatment other than knee OA	Description of arms
NCT02784964	UnicoCell Biomed Co. Ltd. with A2 Healthcare Taiwan Corp.	Active, not recruiting	Phase 1/II	None	ADMSC (3 doses) vs. HA[b]
NCT03000712	R-Bio	Active, not recruiting	N/A	HTO[c]	ADMSC 1 week after HTO vs. HTO alone
NCT01300598	R-Bio	Completed	Phase I/II	None	ADMSC (3 doses)
NCT03869229	Medical University of Warsaw	Recruiting	Phase I/II	Hip OA, Shoulder OA	ADMSC in knee vs. hip vs. shoulder
NCT03357575	Peking University People's Hospital	Not yet recruiting	N/A	None	ADMSC vs. HA
NCT02162693	Cellular Biomedicine Group Ltd. with RenJi Hospital, Gen Hospital of Chinese Armed Police Force	Completed	Phase II	None	ADMSC vs. HA
NCT01809769	Cellular Biomedicine Group Ltd. with RenJi Hospital	Completed	Phase I/II	None	ADMSC (3 doses)
NCT02838069	University Hospital, Montpellier with collaborators	Unknown	Phase II	None	ADMSC (2 doses) vs. placebo
NCT01585857	University Hospital, Montpellier	Completed	Phase I	None	ADMSC (3 doses)
NCT03990805	R-Bio	Recruiting	Phase III	None	ADMSC vs. placebo
NCT02674399	Nature Cell Co. Ltd. with KCRN Research, LLC	Completed	Phase II	None	ADMSC vs. HA
NCT02658344	R-Bio	Completed	Phase II	None	ADMSC vs. placebo
NCT02855073	Cellular Biomedicine Group Ltd. with Shanghai Ninth People's Hospital Affiliated to Shanghai Jiao Tong University	Active, not recruiting	Phase II	Articular cartilage defect	ADMSC on day 1, 22 and HA on day 8, 15 vs. HA on days 1, 8, 15, 22
NCT03955497	Qilu Hospital of Shandong University	Recruiting	Phase I/II	None	ADMSC vs. HA
NCT02642848	Dongsik Chae	Unknown	N/A	HTO	HTO with microfracture vs. HTO with BMAC[d] vs. HTO with ADMSC
NCT04212728	Yantai Yuhuangding Hospital	Recruiting	N/A	PRP	ADMSC in PRP[e] vs. PRP at 0, 3, 6 months
NCT02641860	Cellular Biomedicine Group Ltd. with RenJi Hospital	Completed	Phase I	None	Allogenic ADMSC (3 doses)
NCT04208646	Cellular Biomedicine Group Ltd. with collaborators	Not yet recruiting	Phase II	None	Allogenic ADMSC (2 doses) vs. placebo
NCT03943576	Gwo Xi Stem Cell Applied Technology Co., Ltd.	Recruiting	Phase I/II	None	Allogenic ADMSC (2 doses) vs. HA
NCT03014401	University of Colorado, Denver with Stanford University	Recruiting	N/A	Arthroscopic debridement	Autologous fat pad with arthroscopic debridement vs. arthroscopic debridement alone

[a]Single arm trials were not included
[b]*HA* hyaluronic acid
[c]*HTO* high tibial osteotomy
[d]*BMAC* bone marrow aspirate concentrate
[e]*PRP* platelet-rich plasma

minor. The most common were joint pain and effusion. Where reported, these and other minor adverse events had an incidence ranging from 2 to 74% of patients and resolved within the follow-up period. Major adverse outcomes were not observed.

The clinical use of cultured cells presents regulatory challenges, especially in the USA [45, 78]. Elsewhere, however, clinical trials are under way (Table 37.7) with commercially available ADMSCs products (Table 37.3) such as Jointstem's K Stem Cell (Korea), CBMG's ReJoint (China), and UnicoCell Biomed's Elixcyte (Taiwan) (Table 37.3). The European consortium ADIPOA (http://adipoa2.eu) also has a current clinical trial with lab-grown ADMSCs. GXCPC1 is an allogenic adipose-derived stem cell product (Taiwan) undergoing safety trials currently.

37.6 Comparisons and Future Directions

To understand how therapies of biological agents derived from adipose tissue may compare to the standard of care or other novel pharmacologic and biologic interventions for knee OA with regard to safety, efficacy, or cost, much more research is needed. One important question to consider is the potential difference in the efficacy of plastic-adherent ADMSCs, which are better characterized, versus SVF, which are easier to isolate. We recently compared them directly and observed that patients who had received adherent cells had a more rapid and greater improvement in pain and symptoms than those treated with SVF cells [49]. The proportion of responders in the adherent cell-treated group also correlated with OA Kellgren Lawrence grade, whereas those in the SVF cell-treated group did not [49]. Differences over our 6-month follow-up period were small, but significant, supporting additional future studies comparing the two cell types concerning their relative efficacies. Other questions, some of which are already under investigation, include the comparison of adipose-derived MSCs to other types of MSCs, the effect of adipose tissue donor age and/or site, and the advantage of co-administration with other biologic treatments such as PRP or HA. In addition, longer-term studies up to 5 or 10 years will be necessary to determine whether this approach can forestall the necessity of TKA for a meaningful period in younger, active patients or older patients who cannot tolerate medications or surgery.

Intra-articular injection of biological agents derived from adipose tissue is a promising approach to the treatment of knee OA. Like other novel cell-based and pharmacologic therapies currently under investigation, intra-articular injection of SVF, FAT, or ADMSCs awaits comparative and long-term studies to establish therapeutic efficacy.

References

1. Arden N, Nevitt MC. Osteoarthritis: epidemiology. Best Pract Res Clin Rheumatol. 2006;20(1):3–25.
2. Neogi T, Zhang Y. Epidemiology of osteoarthritis. Rheum Dis Clin N Am. 2013;39(1):1–19.
3. Jevotovsky DS, Alfonso AR, Einhorn TA, Chiu ES. Osteoarthritis and stem cell therapy in humans: a systematic review. Osteoarthr Cartil. 2018;26(6):711–29.
4. Cross M, Smith E, Hoy D, Nolte S, Ackerman I, Fransen M, et al. The global burden of hip and knee osteoarthritis: estimates from the global burden of disease 2010 study. Ann Rheum Dis. 2014;73(7):1323–30.
5. Conaghan PG, Dickson J, Grant RL. Care and management of osteoarthritis in adults: summary of NICE guidance. BMJ (Clin Res Ed). 2008;336(7642):502–3.
6. McAlindon TE, Bannuru RR, Sullivan MC, Arden NK, Berenbaum F, Bierma-Zeinstra SM, et al. OARSI guidelines for the non-surgical management of knee osteoarthritis. Osteoarthritis Cartilage. 2014;22(3):363–88.
7. Nelson AE, Allen KD, Golightly YM, Goode AP, Jordan JM. A systematic review of recommendations and guidelines for the management of osteoarthritis: the chronic osteoarthritis management initiative of the U.S. bone and joint initiative. Semin Arthritis Rheum. 2014;43(6):701–12.
8. Zhang W, Nuki G, Moskowitz RW, Abramson S, Altman RD, Arden NK, et al. OARSI recommendations for the management of hip and knee osteoarthritis: part III: changes in evidence following systematic cumulative update of research published through January 2009. Osteoarthritis Cartilage. 2010;18(4):476–99.

9. Bourne RB, Chesworth B, Davis A, et al. Comparing patient outcomes after THA and TKA: is there a difference? Clin Orthop Relat Res. 2010;468:542–6.
10. Baker PN, van der Meulen JH, Lewsey J, et al. National Joint Registry for England and Wales. The role of pain and function in determining patient satisfaction after total knee replacement. Data from the National Joint Registry for England and Wales. J Bone Joint Surg (Br). 2007;89:893–900.
11. Evans JT, Walker RW, Evans JP, Blom AW, Sayers AS, Whitehouse MR. How long does a knee replacement last? A systematic review and meta-analysis of case series and national registry reports with more than 15 years of follow-up. Lancet. 2019;393(10172):647–54.
12. Hossain F, Patel S, Haddad FS. Midterm assessment of causes and results of revision total knee arthroplasty. Clin Orthop Relat Res. 2010;468(5):1221–8.
13. Odland AN, Callaghan JJ, Liu SS, Wells CW. Wear and lysis is the problem in modular TKA in the young OA patient at 10 years. Clin Orthop Relat Res. 2011;469(1):41–7.
14. Namba RS, Cafri G, Khatod M, Inacio M, et al. Risk factors for total knee arthroplasty aseptic revision. J Arthroplast. 2013;28(8):122–7.
15. Leta TH, Lygre SH, Skredderstuen A, Hallan G, Furnes O. Failure of aseptic revision total knee arthroplasties. Acta Orthop. 2015;86(1):48–57.
16. Balazs EA, Denlinger JL. Viscosupplementation: a new concept in the treatment of osteoarthritis. J Rheumatol Suppl. 1993;39:3–9.
17. Bannuru RR, Vaysbrot EE, Sullivan MC, McAlindon TE. Relative efficacy of hyaluronic acid in comparison with NSAIDs for knee osteoarthritis: a systematic review and meta-analysis. Semin Arthritis Rheum. 2014;45(5):593–9.
18. Moreland LW. Intra-articular hyaluronan (hyaluronic acid) and hylans for the treatment of osteoarthritis: mechanisms of action. Arthritis Res Ther. 2003;5(2):54–67.
19. Jones IA, Togashi R, Wilson ML, Heckmann N, Vangsness CT. Intra-articular treatment options for knee osteoarthritis. Nat Rev Rheumatol. 2019;15(2):77–90.
20. Goldring MB, Goldring SR. Articular cartilage and subchondral bone in the pathogenesis of osteoarthritis. Ann N Y Acad Sci. 2010;1192(1):230–7.
21. Sokolove J, Lepus CM. Role of inflammation in the pathogenesis of osteoarthritis: latest findings and interpretations. Therapeut Adv Musculoskel Dis. 2013;5(2):77–94.
22. Roseti L, Desando G, Cavallo C, Petretta M, Grigolo B. Articular cartilage regeneration in osteoarthritis. Cell. 2019;8(11):1305.
23. Chahla J, Cinque ME, Piuzzi NS, Mannava S, Geeslin AG, Murray IR, Dornan GJ, Muschler GF, LaPrade RF. A call for standardization in platelet-rich plasma preparation protocols and composition reporting: a systematic review of the clinical orthopaedic literature. J Bone Joint Surg. 2017;99(20):1769–79.
24. Hohmann E, Tetsworth K, Glatt V. Is platelet-rich plasma effective for the treatment of knee osteoarthritis? A systematic review and meta-analysis of level 1 and 2 randomized controlled trials. Eur J Orthop Surg Traumatol. 2020;30(6):955–67.
25. Bansal H, Comella K, Leon J, Verma P, Agrawal D, Koka P, Ichim T. Intra-articular injection in the knee of adipose derived stromal cells (stromal vascular fraction) and platelet rich plasma for osteoarthritis. J Transl Med. 2017;15(1):141.
26. Nguyen PD, Tran TD, Nguyen HT, Vu HT, Le PT, Phan NL, Vu BN, Phan NK, Pham P. Comparative clinical observation of arthroscopic microfracture in the presence and absence of a stromal vascular fraction injection for osteoarthritis. Stem Cells Transl Med. 2017;6(1):187–95.
27. Pintat J, Silvestre A, Magalon G, Gadeau AP, Pesquer L, Perozziello A, Peuchant A, Mounayer C, Dallaudiere B. Intra-articular injection of mesenchymal stem cells and platelet-rich plasma to treat patellofemoral osteoarthritis: preliminary results of a long-term pilot study. J Vasc Interv Radiol. 2017;28(1):1708–13.
28. Kim YS, Kwon OR, Choi YJ, Suh DS, Heo DB, Koh YG. Comparative matched-pair analysis of the injection versus implantation of mesenchymal stem cells for knee osteoarthritis. Am J Sports Med. 2015;43(11):2738–46.
29. Bui KH, Duong TD, Nguyen NT, Nguyen TD, Le VT, Mai VT, et al. Symptomatic knee osteoarthritis treatment using autologous adipose derived stem cells and platelet-rich plasma: a clinical study. Biomed Res Therp. 2014;1:2–8.
30. Koh YG, Jo SB, Kwon OR, Suh DS, Lee SW, Park SH, Choi YJ. Mesenchymal stem cell injections improve symptoms of knee osteoarthritis. Arthroscopy. 2013;29(4):748–55.
31. Koh YG, Choi YJ. Infrapatellar fat pad-derived mesenchymal stem cell therapy for knee osteoarthritis. Knee. 2012;19(6):902–7.
32. Freitag J, Bates D, Boyd R, Shah K, Barnard A, Huguenin L, Tenen A. Mesenchymal stem cell therapy in the treatment of osteoarthritis: reparative pathways, safety and efficacy—a review. BMC Musculoskelet Disord. 2016;17:230.
33. Pers Y, Jorgensen C. Adipose derived stem cells for regenerative therapy in osteoarticular diseases. Horm Mol Biol Clin Invest. 2016;28(3):113–20.
34. Shariatzadeh M, et al. The efficacy of different sources of mesenchymal stem cells for the treatment of knee osteoarthritis. Cell Tissue Res. 2019;378:399–410.
35. Wyles CC, Houdek M, Behfar A, Sierra RJ. Mesenchymal stem cell therapy for osteoarthritis: current perspectives. Stem Cells Cloning. 2015;8:117–24.
36. Yubo M, Yanyan L, Li L, Tao S, Bo L, Lin C. Clinical efficacy and safety of mesenchymal stem cell transplantation for osteoarthritis treatment: a meta-analysis. PLoS One. 2017;12(4):e0175449. https://doi.org/10.1371/journal.pone.0175449.

37. Zuk PA, Zhu M, Mizuno H, Huang J, Futrell JW, Katz AJ, Benhaim P, Lorenz HP, Hedrick MH. Multilineage cells from human adipose tissue: implications for cell-based therapies. Tissue Eng. 2001;7(2):211–28.
38. Zuk PA, Zhu M, Ashjian P, De Ugarte DA, Huang JI, Mizuno H, Alfonso ZC, Fraser JK, Benhaim P, Hedrick MH. Human adipose tissue is a source of multipotent stem cells. Mol Biol Cell. 2002;13(12):4279–95.
39. Baer PC, Geiger H. Adipose-derived mesenchymal stromal/stem cells: tissue localization, characterization, and heterogeneity. Stem Cells Int. 2012;2012:812693.
40. Palumbo P, et al. Methods of isolation, characterization and expansion of human adipose-derived stem cells (ASCs): an overview. Int J Mol Sci. 2018;19(7):e1897.
41. Trivisonno, et al. Intraoperative strategies for minimal manipulation of autologous adipose tissue for cell- and tissue-based therapies: concise review. Stem Cells Transl Med. 2019;8(12):1265–71.
42. Tsuji W, Rubin JP, Marra KG. Adipose-derived stem cells: implications in tissue regeneration. World J Stem Cells. 2014;6:312–21.
43. Durfane D. Impact of age on human adipose stem cells for bone tissue engineering. Cell Transplant. 2017;26(9):1496–504.
44. Pak J, Lee JH, Park KS, Park M, Kang LW, Lee SH. Current use of autologous adipose tissue- derived stromal vascular fraction cells for orthopedic applications. J Biomed Sci. 2017;24(1):9.
45. Gimble JM, Bunnell BA, Chiu ES, Guilak F. Concise review: adipose-derived stromal vascular fraction cells and stem cells: let's not get lost in translation. Stem Cells. 2011;29(5):749–54.
46. Berman M, Lander E, Grogan T, O'Brien W, Braslow J, Shawntae D, Berman S. Prospective study of autologous adipose derived stromal vascular fraction containing stem cells for the treatment of knee osteoarthritis. Int J Stem Cell Res Ther. 2019;6:064.
47. Hong Z, Chen J, Zhang S, Zhao C, Bi M, Chen X, Bi Q. Intra-articular injection of autologous adipose-derived stromal vascular fractions for knee osteoarthritis: a double-blind randomized self-controlled trial. Int Orthop. 2019;43(5):1123–34.
48. Michalek J, Vrablikova A, Darinskas A, Lukac L, Prucha J, Skopalik J, Travnik J, Cibulka M, Dudasova Z. Stromal vascular fraction cell therapy for osteoarthritis in elderly: multicenter case-control study. J Clin Orthop Trauma. 2019;10(1):76–80.
49. Yokota N, Hattori M, Ohtsuru T, Otsuji M, Lyman S, Shimomura K, Nakamura N. Comparative clinical outcomes after intra-articular injection with adipose-derived cultured stem cells or noncultured stromal vascular fraction for the treatment of knee osteoarthritis. Am J Sports Med. 2019;47(11):2577–83.
50. Yokota N, Yamakawa M, Shirata T, Kimura T, Kaneshima H. Clinical results following intra-articular injection of adipose-derived stromal vascular fraction cells in patients with osteoarthritis of the knee. Regen Ther. 2017;6:108–12.
51. Fodor PB, Paulseth SG. Adipose derived stromal cell (ADSC) injections for pain management of osteoarthritis in the human knee joint. Aesthet Surg J. 2016;36(2):229–36.
52. Koh YG, Kwon OR, Kim YS, Choi YJ, Tak DH. Adipose-derived mesenchymal stem cells with microfracture versus microfracture alone: 2-year follow-up of a prospective randomized trial. Arthroscopy. 2016;32(1):97–109.
53. Kim YS, Choi YJ, Suh DS, Heo DB, Kim Y, Ryu JS, Koh YG. Mesenchymal stem cell implantation in osteoarthritic knees: is fibrin glue effective as a scaffold? Am J Sports Med. 2015;43(1):176–85.
54. Koh YG, Choi YJ, Kwon SK, Kim YS, Yeo JE. Clinical results and second-look arthroscopic findings after treatment with adipose-derived stem cells for knee osteoarthritis. Knee Surg Sports Traumatol Arthrosc. 2015;23(5):1308–16.
55. Koh YG, Kwon OR, Kim YS, Choi YJ. Comparative outcomes of open-wedge high tibial osteotomy with platelet-rich plasma alone or in combination with mesenchymal stem cell treatment: a prospective study. Arthroscopy. 2014;30(11):1453–60.
56. Barford KW, Blond L. Treatment of osteoarthritis with autologous and microfragmented adipose tissue. Dan Med J. 2019;66(10):A5565.
57. Mautner K, Bowers R, Easley K, Fausel Z, Robinson R. Functional outcomes following microfragmented adipose tissue versus bone marrow aspirate concentrate injections for symptomatic knee osteoarthritis. Stem Cells Transl Med. 2019;14(22):1149–56.
58. Panni AS, Vasso M, Braile A, Toro G, De Cicco A, Viggiano D, Lepore F. Preliminary results of autologous adipose-derived stem cells in early knee osteoarthritis: identification of a subpopulation with greater response. Int Orthop. 2019;43:7–13.
59. Cattaneo G, De Caro A, Napoli F, Chiapale D, Trada P, Camera A. Micro-fragmented adipose tissue injection associated with arthroscopic procedures in patients with symptomatic knee osteoarthritis. BMC Musculoskelet Disord. 2018;19(1):176.
60. Roato I, Belisario DC, Compagno M, Lena A, Bistolfi A, Maccari L, et al. Concentrated adipose tissue infusion for the treatment of knee osteoarthritis: clinical and histological observations. Int Orthop. 2019;43(1):15–23.
61. Russo A, Screpis D, Di Donato SL, Bonetti S, Piovan G, Zorzi C. Autologous micro-fragmented adipose tissue for the treatment of diffuse degenerative knee osteoarthritis: an update at 3 year follow-up. J Exp Orthop. 2018;5(1):52.
62. Hudetz D, Boric I, Rod E, Jelec Z, Radic A, et al. The effect of intra-articular injection of autologous microfragmented fat tissue on proteoglycan synthesis in patients with knee osteoarthritis. Genes. 2017;8(10):E270.

63. Russo A, Screpis D, Di Donato SL, Bonetti S, Piovan G, Zorzi C. Autologous micro-fragmented adipose tissue for the treatment of diffuse degenerative knee osteoarthritis. J Exp Orthop. 2017;4(1):33.
64. Oliver KS, Bayes M, Crane D, Pathikonda C. Clinical outcome of bone marrow concentrate in knee osteoarthritis. J Prolother. 2015;7:937–46.
65. Centeno C, Pitts J, Al-Sayegh H, Freeman M. Efficacy of autologous bone marrow concentrate for knee osteoarthritis with and without adipose graft. Biomed Res Int. 2014;2014:370621.
66. Freitag J, Bates D, Wickham J, Shah K, Hugeuenin L, Tenen A, Paterson K, Boyd R. Adipose-derived mesenchymal stem cell therapy in the treatment of knee osteoarthritis: a randomized controlled trial. Regen Med. 2019;14(3):213–20.
67. Jiang Y, Iwata S, Yang C, Shirakawa K, Matsuoka T. Cartilage regeneration by autologous adipose-derived mesenchymal stem cells for the treatment of osteoarthritis. Cytotherapy. 2019;21(5):S83–4.
68. Lee WS, Kim HJ, Kim KI, Kim GB, Jin W. Intra-articular injection of autologous adipose tissue-derived mesenchymal stem cells for the treatment of knee osteoarthritis: a phase IIb, randomized, placebo-controlled clinical trial. Stem Cells Transl Med. 2019;8(6):504–11.
69. Lu L, Dai C, Zhang Z, Du H, Li S, Ye P, Fu Q, Li Z, et al. Treatment of knee osteoarthritis with intra-articular injection of autologous adipose-derived mesenchymal progenitor cells: a prospective, randomized, double-blind, active-controlled, phase IIb clinical trial. Stem Cell Res Ther. 2019;10(1):143.
70. Zhao X, Ruan J, Tang H, Li J, Shi Y, Li M, Li S, Xu C, Lu Q, Dai C. Multi-compositional MRI evaluation of repair cartilage in knee osteoarthritis with treatment of allogeneic human adipose-derived mesenchymal progenitor cells. Stem Cell Res Ther. 2019;10(1):308.
71. Kyriakidis T, Iosifidis M, Michalopoulos E, Melas I, Papadopoulos P, Stavropoulos-Giokas C. Matrix-induced adipose-derived mesenchymal stem cells implantation for knee articular cartilage repair. Two years follow-up. Acta Orthop Belg. 2018;84(4):443–51.
72. Song Y, Du H, Dai C, Zhang L, Li S, Hunter D, Lu L, Bao C. Human adipose-derived mesenchymal stem cells for osteoarthritis: a pilot study with long-term follow-up and repeated injections. Regen Med. 2018;13(3):295–307.
73. Jo CH, Chai JW, Jeong EC, Oh S, Shin JS, Shim H, Yoon KS. Intra-articular injection of mesenchymal stem cells for the treatment of osteoarthritis of the knee. A 2-year follow-up study. Am J Sports Med. 2017;45(12):2774–83.
74. Pers YM, Rackwitz L, Ferreira R, et al. Adipose mesenchymal stromal cell-based therapy for severe osteoarthritis of the knee: a phase I dose-escalation trial. Stem Cells Transl Med. 2016;5(7):847–56.
75. Jo CH, Lee YG, Shin WH, Kim H, Chai JW, Jeong EC, Kim JE, Shim H, Shin JS, Shin IS, Ra JC, Oh S, Yoon KS. Intra-articular injection of mesenchymal stem cells for the treatment of osteoarthritis of the knee: a proof-of-concept clinical trial. Stem Cells. 2014;32(5):1254–66.
76. Bateman ME, Strong AL, Gimble JM, Bunnell BA. Concise review: using fat to fight disease: a systematic review of nonhomologous adipose-derived stromal/stem cell therapies. Stem Cells. 2018;36(9):1311–28.
77. Dykstra JA, Facile T, Partrick RJ, Francis KR, Milanovich S, Weimer JM, Kota DJ. Concise review: fat and furious: harnessing the full potential of adipose derived stromal vascular fraction. Stem Cells Transl Med. 2017;6(4):1096–108.
78. Malanga GA, Bemanian S. Microfragmented adipose injections in the treatment of knee osteoarthritis. J Clin Orthop Trauma. 2019;10(1):46–8.
79. Bianchi F, Maioli M, Leonardi E, Olivi E, et al. A new nonenzymatic method and device to obtain a fat tissue derivative highly enriched in pericyte-like elements by mild mechanical forces from human lipoaspirates. Cell Transplant. 2013;22(11):2063–77.
80. Tremolada C, Colombo V, Ventura C. Adipose tissue and mesenchymal stem cells: state of the art and Lipogems® technology development. Curr Stem Cell Rep. 2016;2(3):304–12.
81. Regulatory considerations for human cells, tissues, and cellular and tissue-based products: minimal manipulation and homologous use guidance for Industry and Food and Drug Administration Staff. U.S. Department of Health and Human Services Food and Drug Administration Center for Biologics Evaluation and Research Center for Devices and Radiological Health Office of Combination Products; 2017.
82. Bourin P, Bunnell BA, Casteilla L, Dominici M, Katz AJ, March KL, Redl H, et al. Stromal cells from the adipose tissue-derived stromal vascular fraction and culture expanded adipose tissue-derived stromal/stem cells: a joint statement of the International Federation for Adipose Therapeutics and Science (IFATS) and the International Society for Cellular Therapy (ISCT). Cytotherapy. 2013;15:641–8.
83. Dominici M, Le Blanc K, Mueller I, Slaper-Cortenbach I, Marini FC, Krause DS, Deans RJ, Keating A, Prockop DJ, Horwitz EM. Minimal criteria for defining multipotent mesenchymal stromal cells. The International Society for Cellular Therapy position statement. Cytotherapy. 2006;8:315–7.
84. Kilroy GE, Foster SJ, Wu X, Ruiz J, Sherwood S, Heifetz A, Ludlow JW, Stricker DM, Potiny S, Green P, Halvorsen YDC, Cheatham B, Storms RW, Gimble JM. Cytokine profile of human adipose-derived stem cells: expression of angiogenic, hematopoietic, and pro-inflammatory factors. J Cell Physiol. 2007;212:702–9.

85. Li J, Zhu X, Shao Q, Xu F, Sun G. Allogeneic adipose-derived stem cell transplantation on knee osteoarthritis rats and its effect on MMP-13 and DDR2. Ex Ther Med. 2019;18(1):99–104.
86. Toupet K, Maumus M, Peyrafitte J-A, Bourin P, van Lent PL, Ferreira R, Orsetti B, Pirot N, Casteilla L, Jorgensen C, Noel D. Long-term detection of human adipose-derived mesenchymal stem cells after intraarticular injection in SCID mice. Arthritis Rheum. 2013;65:1786–94.
87. Prockop DJ, Brenner M, Fibbe WE, et al. Defining the risks of mesenchymal stromal cell therapy. Cytotherapy. 2010;12:576–8.
88. Rubio D, Garcia-Castro J, Martin MC, et al. Spontaneous human adult stem cell transformation. Cancer Res. 2005;65:3035–9.

GPSR Compliance

The European Union's (EU) General Product Safety Regulation (GPSR) is a set of rules that requires consumer products to be safe and our obligations to ensure this.

If you have any concerns about our products, you can contact us on ProductSafety@springernature.com

In case Publisher is established outside the EU, the EU authorized representative is:

Springer Nature Customer Service Center GmbH
Europaplatz 3
69115 Heidelberg, Germany

Batch number: 10731383

Printed by Printforce, the Netherlands